W9-BKQ-048

# Choices in Relationships

TWELFTH EDITION

# Choices in Relationships

## An Introduction to Marriage and the Family

**David Knox**

*East Carolina University*

**Caroline Schacht**

*East Carolina University*

Australia • Brazil • Mexico • Singapore • United Kingdom • United States

***Choices in Relationships: An Introduction to Marriage and the Family,* Twelfth Edition**
**David Knox and Caroline Schacht**

Product Director: Marta Lee-Perriard
Product Manager: Jennifer Harrison
Associate Content Developer: Naomi Dreyer
Product Assistant: Julia Catalano
Media Developer: John Chell
Marketing Manager: Kara Kindstrom
Content Project Manager: Cheri Palmer
Art Director: Caryl Gorska
Manufacturing Planner: Judy Inouye
Production Service, Illustrator, Compositor: MPS Limited
Photo Researcher: Nazveena Begum Syed, Lumina Datamatics Ltd.
Text Researcher: Kavitha Balasundaram, Lumina Datamatics Ltd.
Copy Editor: Susan Norton
Cover Designer: Caryl Gorska
Cover Image: Digital Vision



Library of Congress Control Number: 2014943335

ISBN: 978-1-305-09444-4

**Cengage Learning**
20 Channel Center Street
Boston, MA 02210
USA

Cengage Learning is a leading provider of customized learning solutions with office locations around the globe, including Singapore, the United Kingdom, Australia, Mexico, Brazil, and Japan. Locate your local office at **www.cengage.com/global**.

Cengage Learning products are represented in Canada by Nelson Education, Ltd.

To learn more about Cengage Learning Solutions, visit **www.cengage.com**.

Purchase any of our products at your local college store or at our preferred online store **www.cengagebrain.com**.

Printed in the United States of America
Print Number: 01 Print Year: 2014

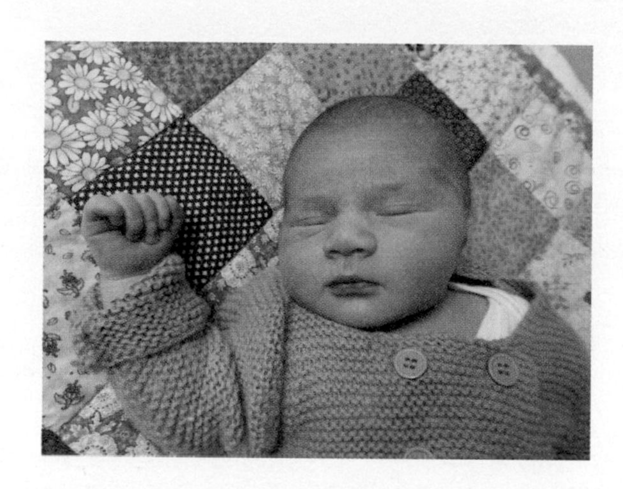

*To Sky Silverwood....and the relationship choices in his future.*

# Contents in Brief

## Chapters

# Contents

CHAPTER **8**

# Communication and Technology in Relationships 212



CHAPTER **9**

# Sexuality in Relationships 240

# Preface

Technology is a major new focus in this new edition. Sending text messages has become the new way in which relationships are initiated, maintained, and (sometimes) ended. Chapter 8 on "Communication and Technology in Relationships" explicates how relationships are constantly affected by technology.

The extent to which technology has entered relationships in American culture is illustrated in the 2013 movie *Her*, which features lonely and heartbroken Theodore Twombly's involvement in a relationship with a woman who exists only as a computer operating system. Samantha asks him how his day was, understands him....they fall in love with each other. He comes to depend on her being there for him. Alas, she is only there for him as an operating system, and in the end, technology has its limits.

While technology is the new trend in relationships, we continue our primary focus on the ***choices*** we make in reference to relationships, marriages, and families. These choices have consequences for the happiness, health, and well-being of ourselves, our partners, our marriages, our parents, and our children. By making deliberate informed choices, everyone wins. Not to take our relationship choices seriously is to limit our ability to enjoy deep, emotionally fulfilling relationships—the only game in town.

Five-hundred and eighty-three new references, new Self-Assessment scales (e.g. Toronto Alexithymia Scale), and new Applying Social Research boxes (e.g. Taking Chances in Romantic Relationships) are featured in this new edition, which continues the cutting edge platform for which *Choices in Relationships* is known. In addition, in "The Future of..." section at the end of each chapter, we predict, based on trends found in the latest research studies, what the future is likely to hold for marriage, singlehood, parenting, divorce/remarriage, and other relationship topics. Other new content added to each chapter includes the following.

## New to the Twelfth Edition: Chapter by Chapter

**Chapter 1 Choices in Relationships: An Introduction to Marriage and the Family**

*Applying Social Research: Taking Chances in Romantic Relationships*

*Self-Assessment: The Relationship Involvement Scale*

*Self-Assessment: Attitudes toward Marriage Scale*

*Is the nuclear family universal?*

*Age at marriage varies by state*

*Three-way marriages*

*Hawaiian Hanai adoptions*

*Table on various generations*

*New research on "marriage naturalists" and "marriage planners"*

**Chapter 2 Love**

*Applying Social Research: Using Relationship Surrogates to Learn the Basics of Interaction*

*Self-Assessment: Toronto Alexithymia Scale*

*Chapman's Five Languages of Love*

*Integral love-combination of autistic and empathetic love*

*Self-compassion-definition and relationship to positive relationship functioning*

*Erik Erickson's sixth stage of psychosocial development-isolation versus intimacy*

*New research on how undergraduates respond to the statement "I love you"*

*New research on effect of pornography exposure on one's love relationship*

**Chapter 3 Gender**

*Self-Assessment: Gender Role Attitudes: Family Life Index*

*Fafafini males reared as females in Samoan society*

*Applying Social Research: Different Gender Worlds: Sex, Betrayal, and Love*

*Biopsychosocial view of gender roles*

*Bioecological model*

*Gender roles in the Cuban American Family*

**Chapter 4 Singlehood, Hanging Out, Hooking Up, and Cohabitation**

*Self-Assessment: Attitudes Toward Singlehood Scale*

*Worldwide data on percentage of adults who never marry*

*New data on individuals choosing to delay marriage*

*Cross-cultural data from 27 countries comparing marriage, cohabitation, and singlehood*

*Cyclical cohabitation*

*Legal issues regarding Living Apart Together*

**Chapter 5 Selecting a Partner**

*Self-Assessment: Attitudes toward Interracial Dating Scale*

*Applied Research: The Relationship Talk*

*Cyclical relationships*

*Tinder App Use in Finding a Partner*

*Attachment anxiety may prevent relationship formation*

*Blood type as criteria for marriage among some Japanese and Koreans*

*Dark Triad Personality (females being attracted to "bad boys")*

**Chapter 6 Marriage Relationships**

*Mormon Families*

*College Marriages*

*Top reason for marriage*

*How religion is functional for marriage*

*New data from California Health Interview Survey*

*Effect of military deployment on spouses and their marriage*

**Chapter 7 GLBT Individuals, Couples, and Families**

*Chapter name change to include transgendered individuals, couples, families*

*Self-Assessment: Sexual Prejudice Scale*

*Self-Assessment: Scale for Assessing Same-Gender Couples*

*Self-Assessment: Beliefs about Children's Adjustment on Same-Sex Families Scale*

*Charter schools for GLBT youth*

*Bisexual women/men and transsexuals in heterosexual relationships*

*Fluidity as a new category of sexual orientation*

*How children react to the breakup of their GLBT parent's relationship*

*Love and sexuality*

**Chapter 8 Communication and Technology in Relationships**

*Technology Mediated Communication in Romantic Relationships*

*Texting and interpersonal communication*

*When texting/Facebook become a relationship problem*

*Principle of Least Interest*

*Nomophobia*

*Catfishing*

**Chapter 9 Sexuality in Relationships**

*Self-Assessment: The Conservative—Liberal Sexuality Scale*

*New definitions of "virginity"—a three-part approach*

*New section on "condom assertiveness"*

*Script for delaying intercourse in one's relationship*

*OraQuick home test for HIV*

*Consequences of experiencing vaginal sex by first year students*

*Seven types of friends with benefits relationships*

*Sexual problems based on sample of over 35,000*

### Chapter 10 Violence and Abuse in Relationships

*Applying Social Research: Double Victims: Sexual Coercion Via Date and Stranger*

*Case of Jaycee Dugard*

*John Gottman's definitions of violence and battering*

*Male Rape Myths*

*Seven types of relationships characterized by violence*

*Fighting back? The best strategy*

*Types of stalking*

*Effect of taking out a protective order on subsequent abuse*

*My 13 Years in an Abusive Marriage*

### Chapter 11 Planning for Children

*New data on the controversy of the psychological effects of abortion*

*China's disastrous one child policy and the reality of forced abortions*

*Effect of obesity on fertility*

*"Google Baby"*

*Consequences of delaying pregnancy to age 35 and beyond*

*Confidence in making a decision regarding abortion*

*Practice of "re-homing" adopted children*

*Contraception update*

### Chapter 12 Parenting

*Applying Social Research: Providing Financial Support for Emerging Adult Offspring*

*Boomerang children*

*Each Gender is Unique*

*Technology Use to Monitor Cell Phone/Text Messaging Use*

*Emotional competence-experience emotion, express emotion and regulate emotion*

*New data comparing cohabitant and married mothers on relationship satisfaction after birth*

*Technology use to encourage safe driving*

### Chapter 13 Money, Work, and Relationships

*Self-Assessment: Dual Employed Coping Scales (DECS)*

*New section on having a successful marriage and work*

*New section on family vacations*

*New data on preference for virtual versus office employment*

*New data on how unhappiness in one's marriage impacts job satisfaction*

*New data on how the wife's working affects the food consumed by the family*

**Chapter 14 Stress and Crisis in Relationships**

*Applied Social Research: CHEATING: Reactions to Discovery of a Partner's Cheating*

*Self-Assessment: Family Hardiness Scale*

*Discovery of a partner's affair via snooping*

*Use of cybersex*

**Chapter 15 Divorce and Remarriage**

*Applying Social Research: Saying Goodbye in Romantic Relationships*

*Effect of divorce on friendships*

*Comparison of satisfaction of sole versus joint custody*

*Positive outcomes of divorce for children*

*Psychological adjustment to breakup of marriage versus cohabitation relationship*

*Gender differences in adjusting to remarriage*

**Chapter 16 Relationships in the Later Years**

*Applying Social Research: Effect of Exercise on Depression and Mortality of the Elderly*

*Self-Assessment: Family Member Well-Being (FMWE)*

*Self-Assessment: How Will You Afford Retirement?*

*Use of technology to maintain existing relationships*

*Use of technology and partner's death*

*Divorce among middle aged and older*

*Retirement Communities— a New Niche*

# Unique Features of the Text

*Choices in Relationships* has several unique features that are included in every chapter.

## Self-Assessment Scales

Each chapter features one or more self-assessment scales that allow students to measure a particular aspect of themselves or their relationships. Scales new to this edition include "Communication Danger Signs Scale," "Satisfaction with Married Life Scale," and "Childfree Lifestyle Scale."

## Applying Social Research

To emphasize that *Choices in Relationships* is not merely a self-help trade book but a college textbook, we present a research application feature in every chapter and specify how new research may be applied to one's interpersonal relationships. Examples of new Applying Social Research features in this edition include, "Taking Chances in Romantic Relationships," "CHEATING: Reactions to Discovery of a Partner's Cheating," and "Saying Goodbye in Romantic Relationships."

## Family Policy

Congress is concerned about enacting into law "family policies that work." In each chapter we review policies relevant to marriage and the family. Examples include marriage preparation/relationship programs, Megan's Law and beyond, and whether divorce mediation should be required before litigation in divorce proceedings.

## Personal Choices

An enduring popular feature of the text is the Personal Choices—detailed discussions of personal choice dilemmas. Examples include: "Who is the Best Person for You to Marry?" "Should I Get Involved in a Long-Distance Relationship?" and "Seek a Divorce If Spouse Has an Affair?"

## What If?

To further personalize the focus of choices in relationships, we present "what ifs?" throughout the text. Examples include, "What if you fall out of love with the person you had planned to marry?" "What if you are in love with two people at the same time?" and "What if your partner is an alcoholic?"

## Diversity in the United States and Diversity in Other Countries

The "one-size-fits-all" model of relationships and marriage is nonexistent. Individuals may be described as existing on a continuum from heterosexuality to homosexuality, from rural to urban dwellers, and from being single and living alone to being married and living in communes. Emotional relationships range from being close and loving to being distant and violent. Family diversity includes two parents (other or same-sex), single-parent families, blended families, families with adopted children, multigenerational families, extended families, and families representing different racial, religious, and ethnic backgrounds. Diversity is the term that accurately describes marriage and family relationships today.

Offset in various paragraphs in each chapter, "Diversity in the United States" reveals racial, religious, same-sex, economic, and educational differences in regard to relationship phenomena. In addition, to reveal courtship, marriage, and family patterns in other societies, "Diversity in Other Countries" paragraphs are presented throughout the text. An example follows from Chapter 3 on Gender:

*Samoan society/culture provides a unique example of gender role socialization via the family. The **Fafafini** (commonly called Fafa) are males reared as females. There are about 3,000 in Samoa. The practice arose when there was a lack of women to perform domestic chores and the family had no female children. Thus, effeminate boys were identified and socialized/reared as females. Fafafini represent a third gender, neither female nor male; they are unique and valued, not stigmatized. Most Samoan families have at least one Fafafini child who takes on the role of a woman, including having sex with men (Abboud 2013).*

## National and International Data

To replace speculation and guessing with facts, we provide data from national samples as well as data from around the world. For example, while a fourth of husbands and 20% of wives in the United States report engaging in extramarital sex at least once, in China the figures are 15% percent of husbands and 5% of wives.

## Photo Essay

New photos give visual meaning to the various roles of parenting (see Chapter 12).

### Special Topics

This edition features the Special Topics section on marriage and family careers and a state of the art revision of the content on contraception.

### Chapter Summaries

Each chapter ends with a summary, formatted as questions and answers, with each question relating back to the Learning Objectives listed at the beginning of every chapter.

### Key Terms

Boldface type indicates key terms, which are defined and featured in the margin of the text as well as in the glossary at the end of the text.

### Web Links

The Internet is an enormous relationship resource. Internet addresses are provided at the end of each chapter.

### Quotes

New quotes are scattered throughout the text to give unique perspectives and to generate thought and discussion of various key topics.

## Supplements and Resources

*Choices in Relationships,* 12th Edition, is accompanied by a wide array of supplements. Some new resources have been created specifically to accompany the 12th edition, and all of the continuing supplements have been thoroughly revised and updated.

**Online Instructor's Resource Manual** This online resource provides instructors with learning objectives, a list of key concepts and terms (with page references), detailed lecture outlines, extensive student projects and classroom activities, identification of how current movies and television programs can be used in the classroom, current Internet resources, and self-assessment handouts for each chapter. The Test Bank contains 60 multiple-choice, 10 true/false, 10 short answer/discussion questions, and 5 essay questions per chapter. The Test Bank items are also available electronically through the instructor resources website, accessible through login.cengage.com

**Cengage Learning Testing Powered by Cognero** is a flexible, online system that allows you to import, edit, and manipulate test bank content from the *Choices in Relationships,* 12e, test bank or elsewhere, including your own favorite test questions; create multiple test versions in an instant; and deliver tests from your LMS, your classroom, or wherever you want.

**Online PowerPoint Lecture Slides** These ready-to-use Microsoft PowerPoint® slides enable you to assemble, edit, publish, and present custom lectures with ease. To review these activities, go to login.cengage.com to create an account and login.

**Online Activities for Courses on the Family** Created from contributions by instructors who teach Marriage and Family courses, this online resource will add new life to your lectures. Includes group exercises, lecture ideas, and homework assignments. To review these activities, go to login.cengage.com to create an account and login.

**The Wadsworth Sociology Video Library, Vol. 1–4** drives home the relevance of course topics through short, provocative clips of current and historical events. Perfect for enriching lectures and engaging students in discussion, many of the segments on this volume have been gathered from BBC Motion Gallery. Through an agreement with BBC Motion Gallery, Cengage Learning selects content from the BBC archive to enhance students' knowledge and understanding of significant events that relate to the concepts in your course. Clips are drawn from the BBC's vast library of award-winning news, science, business, humanities and social science-related programming, including biographies of notable artists, authors, scientists, and inventors from history through today. Ask your Cengage Learning representative for complete details on this amazing resource.

**Aplia** for *Choices in Relationships*, 12e, helps students learn to use their sociological imagination through compelling content and thought-provoking questions. Students complete interactive activities that encourage them to think critically in order to practice and apply course concepts. Aplia includes

- Auto-assigned, auto-graded homework that holds students accountable for the material before they come to class, increasing effort and preparation.
- Immediate, detailed explanations for every answer to enhance student comprehension of sociological theories and concepts.
- Grades automatically recorded in the instructor's Aplia Gradebook.
- Gradebook Analytics that allow instructors to monitor and address performance on a student-by-student and topic-by-topic basis.
- An electronic version of the textbook.

Go to login.cengagebrain.com to access Aplia.

# Acknowledgments

Texts are always a collaborative and collective product. Mia Dreyer, Content Developer, kept the project on track. Others who provided enormous support included John Chell, Media Developer; Caryl Gorska, Art Director; Cheri Palmer, Production Project Manager; Deanna Ettinger, Intellectual Property Analyst; Erik Fortier, Product Development Manager; and Marta Lee-Perriard, Product Director. All were superb, and we appreciate their professionalism and attention to detail. We would also like to thank Chelsea Curry, Brittany Bolen, Rachel Calisto, and E. Fred Johnson Jr. for their superb photographs.

## Reviewers for the 12th Edition

Jacki Booth, San Diego State University; Caroline Calogero, Brookdale Community College; Jerry Cook, California State University, Sacramento; Raquel Delevi, California State University, Los Angeles; Wayne Flake, Eastern Arizona College; Dale Fryxell, University of Hawaii; Terry Hatkoff, California State University, Northridge; Steve Kramer, Fort Scott Community College; Karen McCue, University of New Mexico; and Robert Nelson, Montgomery Community College

## Reviewers for Previous Editions

Sheldon Helfing, College of the Canyons; Alissa King, Iowa Central Community College; Michallene McDaniel, Gainesville State College; Jodi McKnight, Mid

Continent University; Nancy Reeves, Gloucheste County College; Linda Stone, Towson University

Grace Auyang, University of Cincinnati; Rosemary Bahr, Eastern New Mexico University; Von Bakanic, College of Charleston; Mary Beaubien, Youngstown State University; Sampson Lee Blair, Arizona State University; Mary Blair-Loy, Washington State University; David Daniel Bogumil, Wright State University; Elisabeth O. Burgess, Georgia State University; Craig Campbell, Weber State University; Michael Capece, University of South Florida; Lynn Christie, Baldwin-Wallace College; Laura Cobb, Purdue University and Illinois State University; Jean Cobbs, Virginia State University; Donna Crossman, Ohio State University; Karen Dawes, Wake Technical Community College; Susan Brown Donahue, Pearl River Community College; Doug Dowell, Heartland Community College; John Engel, University of Hawaii; Kim Farmer, Martin Community College; Mary Ann Gallagher, El Camino College; Shawn Gardner, Genesee Community College; Ted Greenstein, North Carolina State University; Heidi Goar, St. Cloud State University; Norman Goodman, State University of New York at Stony Brook; Jerry Ann Harrel-Smith, California State University, Northridge; Gerald Harris, University of Houston; Rudy Harris, Des Moines Area Community College; Terry Hatkoff, California State University, Northridge; Christina Hawkey, Arizona Western College; Sheldon Helfing, College of the Canyons; Tonya Hilligoss, Sacramento City College; Rick Jenks, Indiana University; Richard Jolliff, El Camino College; Diane Keithly, Louisiana State University; Steve Long, Northern Iowa Area Community College; Patricia B. Maxwell, University of Hawaii; Carol May, Illinois Central College; Tina Mougouris, San Jacinto College; Jane A. Nielsen, College of Charleston; Lloyd Pickering, University of Montevallo; Scott Potter, Marion Technical College; Janice Purk, Mansfield University; Cherylon Robinson, University of Texas at San Antonio; Patricia J. Sawyer, Middlesex Community College; Cynthia Schmiege, University of Idaho; Susan Schuller Friedman, California State University Los Angeles; Eileen Shiff, Paradise Valley Community College; Scott Smith, Stanly Community College; Tommy Smith, Auburn University; Beverly Stiles, Midwestern State University; Dawood H. Sultan, Louisiana State University; Elsie Takeguchi, Sacramento City College; Myrna Thompson, Southside Virginia Community College; Teresa Tsushima, Iowa State University; Janice Weber-Breaux, University of Southwestern Louisiana; Kathleen Wells, University of Arizona South; Loreen Wolfer, University of Scranton

We love the study, writing, and teaching of marriage and the family and recognize that no one has a corner on relationships. We welcome your insights, stories, and suggestions for improvement in the next edition of this text. We check our e-mail frequently and invite you to send us your thoughts and ideas. We will respond.

David Knox, *Knoxd@ecu.edu*
Caroline Schacht, *CSchacht@suddenlink.net*

## About the Authors

**David Knox,** Ph.D., is Professor of Sociology at East Carolina University, where he teaches Courtship and Marriage, Marriage and the Family, and Sociology of Human Sexuality. He is a marriage, family, and sex therapist and the author or coauthor of twelve books and over 100 professional articles. He and Caroline Schacht are married with three children and four grandchildren.

**Caroline Schacht,** M.A. in Sociology and M.A. in Family Relations, is formerly an instructor of sociology at East Carolina University where she taught Courtship and Marriage, Introduction to Sociology, and the Sociology of Food. Her clinical work includes marriage and family relationships. She is the coauthor of *Understanding Social Problems* (Cengage Learning ©2015) and a certified Romana Pilates instructor.

# Choices in Relationships: An Introduction to Marriage and the Family

*It is not in the stars to hold our destiny but in ourselves.*

WILLIAM SHAKESPEARE, *JULIUS CAESAR*

Chelsea Curry

**Learning Objectives**

- Explain a "choices" view of relationships.
- Identify the central elements and various types of marriage.
- Understand the definition and types of family.
- Describe the theoretical frameworks for studying marriage and the family.
- Review the research process and caveats.

## TRUE OR FALSE?

1. When asked to rank values, undergraduates rank "financial success" higher than "raising a family."
2. In regard to making choices, undergraduates are more likely to report that they are "in control" than "irresponsible."
3. Our society values marriage education since the societal cost of divorce is $35 billion annually.
4. Online marriage education is as worthless as no marriage education course at all—there must be face-to-face contact with an instructor for benefits to occur.
5. A societal backlash is operative in that Generation Yers are now more intent on getting married soon after college than the previous generation.

*Answers:* 1. T 2. T 3. T 4. F 5. F

J. H. Newman, cardinal and philosopher, noted that we tend to fear less that life will end but, rather, that life will never begin. His point targets the importance of love that provides an unparalleled richness, meaning, and happiness in one's life. While young adults are delaying marriage, they are not sitting alone in their rooms. They are often in search of meaningful love connections. These result in deliberate, thoughtful, considered choices in one's relationships.

Many of these intense and sustained love relationships end up in marriage and having a family—the bedrock of society. All of us were born into a family (however one defines this concept) and will end up in a family of our own.

### National **Data**

"Raising a family" remains one of the top values for undergraduates. In a nationwide study of 165,743 undergraduates in 234 colleges and universities, 73% identified raising a family as an essential objective (82% chose financial success as their top goal) (Eagan et al. 2013).

In this chapter we review the choices framework—the theme of this text—the definitions and types of marriage and family, and how researchers go about conducting research so that we can be more informed in our relationship choices.

# Choices in Relationships—View of the Text

Whatever your relationship goal, in this text we encourage a proactive approach of taking charge of your life and making wise relationship choices. Making the right choices in your relationships, including marriage and family, is critical to your health, happiness, and sense of well-being. Your times of greatest elation and sadness will be in reference to your love relationships.

The central theme of this text is choices in relationships. Although we have many choices to make in our society, among the most important are whether to marry, whom to marry, when to marry, whether to have children, whether to remain emotionally and sexually faithful to one's partner, and whether to protect oneself from sexually transmitted infections and unwanted pregnancy. Though structural and cultural influences are operative, a choices framework emphasizes that individuals have some control over their relationship destiny by making deliberate choices to initiate, nurture, or terminate intimate relationships.

### Facts about Choices in Relationships

The facts to keep in mind when making relationship choices include the following.

**Not to Decide Is to Decide** Not making a decision is a decision by default. If you are sexually active and decide not to use a condom, you have made a decision to increase your risk for contracting a sexually transmissible infection, including HIV. If you don't make a deliberate choice to end a relationship that is unfulfilling, abusive, or going nowhere, you have made a choice to continue in that relationship and have little chance of getting into a more positive and satisfying relationship. If you don't make a decision to be faithful to your partner, you have made a decision to be vulnerable to cheating. See the Personal Choices section for more examples of taking charge of your life by making deliberate choices.

*If you don't know where you are going, you may end up someplace else.*

LAURENCE PETER, *THE LAUGHTER PRESCRIPTION*

**PERSONAL CHOICES**

### *Relationship Choices—Deliberately or by Default?*

It is a myth that you can avoid making relationship decisions, because not to make a decision is to make a decision by default. Some examples follow:

- If you don't make a decision to pursue a relationship with a particular person, you have made a decision (by default) not to have a relationship with that person.
- If you do not decide to do the things that are necessary to keep or improve your current relationship, you have made a decision to let the relationship slowly disintegrate.
- If you do not make a decision to be faithful to your partner, you have made a decision to be open to situations and relationships in which you are vulnerable to being unfaithful.
- If you do not make a decision to delay having intercourse early in a new relationship, you have made a decision to have intercourse soon in a relationship (which typically has a negative outcome).
- If you are sexually active and do not make a decision to use birth control or a condom, you have made a decision to expose yourself to getting pregnant or to contracting an STI.
- If you do not make a decision to break up with an emotionally/physically abusive partner or spouse, you have made a decision to continue the relationship and incur further emotional and/or physical abuse including damage to your self-esteem.

Throughout the text, as we discuss various relationship choices, consider that you automatically make a choice by being inactive—that not to make a choice is to make one. We encourage a proactive style whereby you make deliberate relationship choices.

---

**Action Must Follow a Choice** While the private life of Woody Allen has been the subject of public dismay (e.g., he married his long-time partner's adopted daughter), he is one of the few Hollywood directors who is given complete control over all aspects of his films. His success began with a decision to become a stand-up comedian and make a name for himself, then use this influence to launch his film career. His biographer writes, "Woody is nothing if he is not deliberate. Decisions may take a long while to be made, but once his mind is made up to do something, he devotes all his effort to it" (Lax 1991, p. 156).

*If he's decided that something should happen, he's just going to make it happen.*

SAID OF STEVE JOBS IN HIS BIOGRAPHY BY WALTER ISAACSON

## APPLYING SOCIAL RESEARCH | Taking Chances in Romantic Relationships

Making choices sometimes includes taking chances—moving in together after knowing each other for a short time, changing schools to be together, and forgoing condom usage thinking "this time won't end in a pregnancy." To assess the degree to which undergraduates take chances in their relationships, 381 students completed a 64-item questionnaire posted on the Internet. The majority of respondents were female (over 80%) and white (approximately 74%). Over half of the respondents (53%) described their relationship status as emotionally involved with one person, with 4% engaged or married.

**Findings**

Of the various risk-taking behaviors identified on the questionnaire, eight were identified by 25% or more of the respondents as behaviors they had participated in. These eight are identified in Table 1.1

**Table 1.1**
**Most Frequent Risk-Taking Behaviors in a Romantic Relationship**
**N = 381**

| Risk-Taking Behavior | Percent |
|---|---|
| Unprotected sex | 70 |
| Being involved in a "friends with benefits" relationship | 63 |
| Broke up with a partner to explore alternatives | 46 |

While in his twenties, Allen performed two to three shows a night to small, 50-person audiences six nights a week ($75 to $100 a week) for two years. He was beset with the fear of "going live," exhausted at the grueling schedule, and doubted the future (should he quit?). He persevered, pushed through another six months, got his break, and moved forward. The section on Applying Social Research reveals that taking action on one's choices often means taking a chance.

*Things do not happen. Things are made to happen.*

JOHN F. KENNEDY, 35TH PRESIDENT

**Choices Involve Trade-Offs** By making one choice, you relinquish others. Every relationship choice you make will have a downside and an upside. For example, in the two years Woody Allen devoted to his stand-up career, his marriage to his wife Harlene began to unravel.

If you decide to stay in a relationship that becomes a long-distance commitment, you are continuing involvement in a relationship that is obviously important to you. However, because of the long- distance relationship, you may end up spending a lot of time alone. If you decide to marry, you will give up your freedom to pursue other emotional and/or sexual relationships, and you will also give up some control over how you spend your money. But, you may also get a wonderful companion with whom to share life for many years.

Any partner that you select will also have characteristics that must be viewed as a trade-off. One woman noted of her man, "he doesn't do text messaging or e-mail . . . he doesn't even know how to turn on a computer. But he knows how to build a house, plant a garden, and fix a car. . . . trade-offs that make him an impressive partner."

**Some Choices Require Correction** Some of our choices, although appearing correct at the time that we make them, turn out to be disasters. Once we realize that a choice has consistently negative consequences, it is important to stop defending it, make new choices, and move forward. Otherwise, we remain consistently locked into continued negative outcomes of bad choices.

| Risk-Taking Behavior | Percent |
|---|---|
| Had sex before feeling ready | 41 |
| Disconnected w/friends because of partner | 34 |
| Maintained long-distance relationship (1 year) | 32 |
| Cheated on partner | 30 |
| Lied to partner about being in love | 28 |



Almost three-fourths (72%) of the sample self-identified as being a "person willing to take chances in my love relationship." However, only slightly over one-third of the respondents indicated that they considered themselves as risk takers in general. This suggests that college students may be more likely to engage in risk-taking behavior in love relationships than in other areas of their lives. Both love and alcohol were identified as contexts for increasing one's vulnerability for taking chances in romantic relationships—60 and 66%, respectively. Both being in love and drinking alcohol (both love and alcohol may be viewed as drugs) gives one a sense of immunity from danger or allows one to deny danger.

**Source**

Adapted and abridged from L. Elliott, B. Easterling, and D. Knox. 2012. TakingChances in Romantic Relationships. Poster, Southern Sociological Society Annual Meeting, New Orleans, March.

WHAT IF?

## What If You Have Made a Commitment to Marry But Feel It Is a Mistake?

We know three individuals who reported the following: "On my wedding day, I knew it was a mistake to marry this person." Although all had their own reasons for going through with marrying, the basic reason was social pressure. All three are now divorced and clearly regret marrying. The take-home message is to "listen to your senses" and to act accordingly. Not acting on the feelings that the relationship is doomed may be to delay the inevitable. The price of ending a marriage, particularly with children involved, is much higher than ending a relationship before marriage. If you are in a relationship that does not feel right, make a choice to improve it or to end it. Otherwise you'll remain in a disastrous relationship going nowhere.

**Choices Include Selecting a Positive or a Negative View** As Thomas Edison progressed toward inventing the light bulb, he said, "I have not failed. I have found ten thousand ways that won't work." Ron Wayne, negotiating with Steve Jobs, was offered a 10% share in the computer giant Apple when it started up, but he was ambivalent since he would have to invest money that he feared he would lose. In early 2011, his 10% stake would have been worth approximately $2.6 billion. Later in life he said that he was not bitter and had made the best decision at the time (Isaacson 2011, p. 65).

*Each player must accept the cards life deals him or her. But once they are in hand, he or she alone must decide how to play the cards in order to win the game.*

VOLTAIRE, FRENCH PHILOSOPHER

In spite of an unfortunate event in your life, you can choose to see the bright side. Regardless of your circumstances, you can opt for viewing a situation in positive terms. A breakup with a partner you have loved can be viewed as the end of your happiness or an opportunity to become involved in a new, more fulfilling

*Nothing is either good or bad but thinking makes it so.*

WILLIAM SHAKESPEARE, *HAMLET*

relationship. The discovery of your partner cheating on you can be viewed as the end of the relationship or as an opportunity to examine your situation, to open up communication channels with your partner, and to develop a stronger connection. Discovering that you are infertile can be viewed as a catastrophe or as a challenge to face adversity with your partner. It is not the event but your view of it that determines its effect on you.

**Choices Involve Different Decision-Making Styles** College students tend to feel confident of their decision making. Allen et al. (2008) identified four patterns in the decision-making process of 148 college students. These patterns and the percentage using each pattern included (1) "I am in control" (45%), (2) "I am experimenting and learning" (33%), (3) "I am struggling but growing" (14%), and (4) "I have been irresponsible" (3%). Of those who reported that they were in control, about a third said that they were "taking it slow," and about 11% reported that they were "waiting it out." Men were more likely to report that they were "in control." These data reflect that the respondents knew what decision-making style they had adopted. Of note, only 3% labeled themselves as being irresponsible.

*Destiny is a name often given in retrospect to choices that had dramatic consequences.*

J.K. ROWLING, ENGLISH WRITER

*When you get to a fork in the road, take it.*

YOGI BERRA, BASEBALL PLAYER

**Most Choices Are Revocable; Some Are Not** Most choices can be changed. For example, a person who has chosen to be sexually active with multiple partners can decide to be monogamous or to abstain from sexual relations in new relationships. People who have been unfaithful in the past can elect to be emotionally and sexually committed to a new partner.

Other choices are less revocable. For example, backing out of the role of parent is very difficult. Social pressure keeps most parents engaged, but the law (e.g., forced child support) is the backup legal incentive. Hence, the decision to have a child is usually irrevocable. Choosing to have unprotected sex may also result in a lifetime of coping with sexually transmitted infections.

*Millennials are lazy, entitled narcissists who still live with their parents...but they can save us all.*

JOEL STEIN, *TIME MAGAZINE*

**Generation Y** those typically born between early 80s and early 2000s.

**Choices of Generation Y** Generations vary (see Table 1.2). Those in **Generation Y** (typically born between 1979 and 1984) are the children of the baby boomers. Numbering about 80 million, these Generation Yers (also known as the Millennial or Internet Generation) have been the focus of their parents' attention. They have been nurtured, coddled, and scheduled into day care to help them get ahead. The result is a generation of high self-esteem, self-absorbed individuals who believe they are "the best." Unlike their parents, who believe in paying one's dues, getting credentials, and sacrificing through hard work to achieve economic stability, Generation Yers focus on fun, enjoyment, and flexibility. They might choose a summer job at the beach if it buys a burger and an apartment with six friends over an internship at IBM that smacks of the corporate America sellout.

Generation Yers are also relaxed about relationship choices. Rather than forming a pair-bond, they "hang out," "hook up," and live together. They are in no hurry to find "the one," to marry, or to begin a family (indeed, many enjoy

**TABLE 1.2 Four Generations in Recent History**

Great Generation—Years of the Great Depression, World War II (veterans and civilians). Culture steeped in traditional values.

Baby Boomers—Children of WW II's Great Generation, born between 1946 and 1964. Questioning of traditional values.

Gen X (Generation X)—Children of boomers born between early 60s to early 80s. Generation of change, MTV, AIDS, diversity.

Gen Y (Generation Y)—Also known as Millennial Generation, persons born from early 80s to early 2000s. Loyalty to corporations gone, frequent job changes.

living alone) (Klinenberg 2012). Instead, they are focused on their educations and careers, and enjoying their freedom in the meantime.

**Choosing to Be in a Relationship/Knowing When You Are There** Palomer et al. (2012) asked 494 individuals in romantic relationships: "How did you know that you and your partner were in a relationship?" Five themes emerged: (1) mutual decision ("Are we boyfriend/girlfriend now?"), (2) intimacy and exclusivity (e.g., couple has sex and stops seeing other people), (3) time spent together (e.g., hang out together regularly), (4) becoming "Facebook official," and (5) deepening love or affection. (e.g., love feelings escalate). Johnson et al. (2013) also identified markers in relationship transition from "talking" to greater involvement. Analysis of the data from 362 undergraduates (ages ranged from 18 to 26) revealed that the passage of time, being invited to be exclusive, and agreeing to see only each other marked the transition to being a couple.

The Relationship Involvement Scale (see Self-Assessment) provides a way for you to assess the degree to which you are currently involved with your partner. While you may not be in a relationship, you may use the scale to assess your level of involvement in a future relationship.

*Do you think we enjoy hearing about your brand-new million-dollar home when we can barely afford to eat Kraft Dinner sandwiches in our own grimy little shoe boxes and we're pushing thirty?*

**DOUGLAS COUPLAND,**
***GENERATION X: TALES FOR AN ACCELERATED CULTURE***

## SELF-ASSESSMENT | The Relationship Involvement Scale

This scale is designed to assess the level of your involvement in a current relationship. Please read each statement carefully, and write the number next to the statement that reflects your level of disagreement or agreement, using the following scale.

| 1 | 2 | 3 | 4 | 5 | 6 | 7 |
|---|---|---|---|---|---|---|
| Strongly Disagre | | | | | | Strongly Agree |

_____ 1. I have told my friends that I love my partner.
_____ 2. My partner and I have discussed our future together.
_____ 3. I have told my partner that I want to marry him/her.
_____ 4. I feel happier when I am with my partner.
_____ 5. Being together is very important to me.
_____ 6. I cannot imagine a future with anyone other than my partner.
_____ 7. I feel that no one else can meet my needs as well as my partner.
_____ 8. When talking about my partner and me, I tend to use the words "us," "we," and "our."
_____ 9. I depend on my partner to help me with many things in life.
_____ 10. I want to stay in this relationship no matter how hard times become in the future.

### Scoring

Add the numbers you assigned to each item. A 1 reflects the least involvement and a 7 reflects the most involvement. The lower your total score (10 is the lowest possible score), the lower your level of involvement; the higher your total score (70 is the highest possible score), the greater your level of involvement. A score of 40 places you at the midpoint between a very uninvolved and a very involved relationship.

### Other Students Who Completed the Scale

Students from two universities completed the scale—31 male and 86 female undergraduate psychology students at Valdosta State University and 60 male and 129 female undergraduate students from the Courtship and Marriage courses at East Carolina University.

### Scores of Participants

When students' scores from both universities were combined, the men's average score was 50.06 (SD = 14.07) and the women's average score was 52.93 (SD = 15.53), reflecting moderate involvement for both sexes. There was no significant difference between men and women in level of involvement. However, there was a significant difference ($p < .05$) between whites and non-whites, with whites reporting greater relationship involvement (M = 53.37; SD = 14.97) than non-whites (M = 48.33; SD = 15.14).

In addition, there was a significant difference between the level of relationship involvement of seniors compared with juniors ($p < .05$) and freshmen ($p < .01$). Seniors reported more relationship involvement (M = 57.74; SD = 12.70) than did juniors (M = 51.57; SD 5 = 15.37) or freshmen (M = 50.31; SD = 15.33).

### Source

"The Relationship Involvement Scale" 2004 by Mark Whatley, Ph.D., Department of Psychology, Valdosta State University, Valdosta, GA 31698-0100. Used by permission. Other uses of this scale by written permission of Dr. Whatley only (mwhatley@valdosta.edu). Information on the reliability and validity of this scale is available from Dr. Whatley.

**TABLE 1.3 "Best" and "Worst" Choices Identified by University Students**

| Best Choice | Worst Choice |
|---|---|
| Waiting to have sex until I was older and involved. | Cheating on my partner. |
| Ending a relationship with someone I did not love. | Getting involved with someone on the rebound. |
| Insisting on using a condom with a new partner. | Making decisions about sex when drunk. |
| Ending a relationship with an abusive partner. | Staying in an abusive relationship. |
| Forgiving my partner and getting over cheating. | Changing schools to be near my partner. |
| Getting out of a relationship with an alcoholic. | Not going after someone I really wanted. |



*All the world's a stage, and all the men and women merely players: they have their exits and their entrances; and one man in his time plays many parts.*

WILLIAM SHAKESPEARE, *AS YOU LIKE IT*

*In any moment of decision, the best thing you can do is the right thing, the next best thing is the wrong thing, and the worst thing you can do is nothing.*

THEODORE ROOSEVELT, 26TH PRESIDENT

*When you have collected all the facts and fears and made your decision, turn off all your fears and go ahead!*

GEORGE S. PATTEN, WW II GENERAL

**Choices Are Influenced by the Stage in the Family Life Cycle** The choices a person makes tend to be individualistic or familistic, depending on the stage of the family life cycle that the person is in. Before marriage, individualism characterizes the thinking and choices of most individuals. Individuals are concerned only with their own needs. Once married, and particularly after having children, the person's familistic values ensue as the needs of a spouse and children begin to influence behavioral choices. For example, evidence of familistic values and choices is reflected in the fact that spouses with children are less likely to divorce than spouses without children.

All of us are proud of some of the choices we have made. We also regret other choices. We asked our students to identify their "best" and "worst" relationship choices (see Table 1.3).

## Global, Structural/Cultural, and Media Influences on Choices

Choices in relationships are influenced by global, structural/cultural, and media factors. This section reviews the ways in which globalization, social structure, and culture impact choices in relationships. Although a major theme of this book is the importance of taking active control of your life in making relationship choices, it is important to be aware that the social world in which you live restricts and channels such choices. For example, social disapproval for marrying someone of another race is part of the reason that over 95% of adults in the United States have married someone of the same race (*Statistical Abstract of the United States, 2012–2013,* Table 60).

**Globalization** Families exist in the context of globalization. Indeed, "globalization is the critical driving force that is fundamentally restricting the social order around the world, and families are at the center of this change" (Trask 2010, p. v). Economic, political, and religious happenings throughout the world affect what happens in your marriage and family in the United States. When the price of oil per barrel increases in the Middle East, gasoline costs more, leaving fewer dollars to spend on other items. When the stock market in Hong Kong drops 500 points, Wall Street reacts, and U.S. stocks drop.

The country in which you live also affects your happiness and well-being. For example, in a study, citizens of 13 countries were asked to indicate their level of life satisfaction on a scale from 1 (dissatisfied) to 10 (satisfied): Citizens in Switzerland averaged 8.3, those in Zimbabwe averaged 3.3, and those in the United States averaged 7.4 (Veenhoven 2007). The Internet, CNN, and mass communications provide global awareness so that families are no longer isolated units.

**Social Structure** The social structure of a society consists of institutions, social groups, statuses, and roles.

**1. *Institutions.*** The largest elements of society are social **institutions**, which may be defined as established and enduring patterns of social relationships. The institution of the family in the United States is held as a strong value, as reflected by tax deductions for parents, family-friendly work policies, and government benefits for young mothers and their children (e.g., the WIC—Women, Infants and Children—program).

**Institutions** established and enduring patterns of social relationships.

In addition to the family, major institutions of society include the economy, education, and religion. Institutions affect individual decision making. For example, you live in a capitalist society where economic security is paramount—it is the number one value held by college students (Eagan et al. 2013). In effect, the more time you spend focused on obtaining money, the less time you have for relationships. You are now involved in the educational institution that will impact your choice of a mate (e.g., college-educated people tend to select and marry one another). Religion also affects relationship choices (e.g., devout members select each other as a mate). Spouses who "believe in the institution of the family" are less likely to divorce.

**2. *Social groups.*** Institutions are made up of social groups, defined as two or more people who share a common identity, interact, and form a social relationship. Most individuals spend their days going between social groups. You may awaken in the context (social group) of a roommate, partner, or spouse. From there you go to class with other students, lunch with friends, work with other employees, and text with your parents. Within 24 hours you have participated in at least five social groups. These social groups have various influences on your choices. Your roommate influences what other people you can have in your room for how long, your friends may want to eat at a particular place, your fellow workers will ignore you or interact with you, and your parents may want you to come home for the weekend.

*Students sometimes argue that they—as individuals—make choices. In reality, the choices they make are only the ones the social context permits. For example, a Mormon woman married to a Mormon man in the Mormon Church has almost no choice to be "childfree." Although the Mormon woman will assert that she is "choosing" to have seven children, her illusion of choice is context controlled. Change her context (so that she is no longer a member of the Mormon Church and is married to a man who wants to be childfree) and she is now free to be childfree (but only because her context has changed) (Zusman 2014).*

**Child marriage** young females (ages 8 to 12) are required to marry an older man selected by their parents.

Your interpersonal choices are influenced mostly by your partner and peers (e.g., your sexual values, use of contraception and the amount of alcohol you consume). Thus, selecting a partner and peers is important. As Falstaff, one of Shakespeare's characters, said, "Company, villainous company, hath been the spoil of me" (*Henry IV, Pt. 1*, 1.3).

**DIVERSITY IN OTHER COUNTRIES**

In countries such a Nepal and Afghanistan **child marriage** occurs, whereby young females (ages 8 to 12) are required to marry an older man selected by their parents. Suicide is the only alternative "choice" for these children.

As a first-year female student, you will be in demand by freshmen, sophomore, junior, and senior men. But because of the **mating gradient** which gives social approval to men who seek out younger, less-educated, less financially secure women, as a senior female you are no longer "younger" and only senior men (or graduate students) are typically interested in you. Hence, your pool of eligible men shrinks each year.

**Mating gradient** gives social approval to men who seek out younger, less-educated, financially secure women.

As a first-year male student, just the opposite occurs. During your first year, only other freshmen women are potential partners (and of course, women from your high school). But as you become a senior, all women on campus become options. In effect, your pool of eligibles increases every year. Social structure (your place in student class rank) affects the individuals you can hang out with, not your "personality."

**Primary group** social group that tends to involve small numbers of individuals in a close emotional relationship.

**Secondary group** social group that may be small or large, characterized by interaction that is impersonal and formal.

**Status** a position a person occupies within a social group.

Social groups may be categorized as primary or secondary. **Primary groups**, which tend to involve small numbers of individuals, are characterized by interaction that is intimate and informal. A family is an example of a primary group. Persons in our primary groups are those who love us and have lifetime relationships with us. In contrast to primary groups, **secondary groups**, which may be small or large, are characterized by interaction that is impersonal and formal. Your classmates, teachers, and coworkers are examples of individuals in your secondary groups. Unlike your parents, siblings, and spouse, members of your secondary groups do not have an enduring emotional connection with you and are more transient.

**3. *Statuses.*** Just as institutions consist of social groups, social groups consist of statuses. A **status** is a position a person occupies within a social group. The statuses we occupy largely define our social identity. The statuses in a family may consist of mother, father, child, sibling, and stepparent. In discussing family issues, we refer to statuses such as teenager, partner, and spouse. Statuses are relevant to choices in that many choices can significantly change one's status. Making decisions that change one's status from single person to spouse to divorced person can influence how people feel about themselves and how others treat them.

**4. *Roles.*** Every status is associated with many roles, or sets of rights, obligations, and expectations. Our social statuses identify who we are; our roles identify what we are expected to do. Roles guide our behavior and allow us to predict the behavior of others. Spouses adopt a set of obligations and expectations associated with their status. By doing so, they are better able to influence and predict each other's behavior.

Because individuals occupy a number of statuses and roles simultaneously, they may experience role conflict. For example, the role of the parent may conflict with the role of the spouse, employee, or student. If your child needs to be driven to the math tutor, your spouse needs to be picked up at the airport, your employer wants you to work late, and you have a final exam all at the same time, you are experiencing role conflict.

David Knox

Cultural values are translated into family values that parents teach their children.

**Culture** Just as social structure refers to the parts of society, culture refers to the meanings and ways of living that characterize people in a society. Two central elements of culture are beliefs and values.

**1. *Beliefs.*** Beliefs refer to definitions and explanations about what is true. The beliefs of an individual or couple influence the choices they make. Couples who believe that young children flourish best with a full-time parent in the home make different job and child-care decisions than do couples who believe that day care offers opportunities for enrichment. Hall (2010) noted that individuals have different beliefs about marriage (e.g., that successful marriages are destined to succeed or that successful marriages grow naturally) and these respective views impact choices.

**2. *Values.*** Values are standards regarding what is good and bad, right and wrong, desirable and undesirable. J. D. Salinger, author of *The Catcher in the Rye*, noted that he felt powerless *not* to buy into society's value for fame. "I'm sick of not having the courage to be an absolute nobody. I'm sick of myself and everybody else that wants to make some kind of a splash" (Lacayo 2010, p. 66).

Values influence choices. Valuing **individualism** leads to making decisions that serve the individual's interests rather than the family's interests (**familism**). "What makes me happy" is the focus of the individualist, not "what makes my family happy." Kacey Musgraves reflects an individualistic theme in her song "Follow Your Arrow":

*You're damned if you do*
*And you're damned if you don't*
*So, you might as well just do*
*Whatever you want*

**Individualism** making decisions that serve the individual's interests rather than the family's interest.

**Familism** making decisions that serve the family's interests rather than the individual's interests.

European Americans are characteristically individualistic, whereas Hispanic Americans are characteristically familistic. Cherlin (2009) noted a paradox in the values of Americans—they value marriage and abhor divorce (familism), but they are also individualistic, which ensures a high divorce rate. Forty-two percent of 4,581 undergraduates reported that they would divorce their spouse if they fell out of love (Hall and Knox 2013). Allowing one's personal love feelings to dictate the stability of a marriage is a highly individualistic value. Aside from individualism is **collectivism**, which emphasizes doing what is best for the group (not specific to the family group); collectivism is characteristic of traditional Chinese families.

**Collectivism** philosophy whereby the interests of the larger group take precedence over the interests of the individual.

Those who remain single, who live together, who seek a childfree lifestyle, and who divorce are more likely to be operating from an individualistic perspective than those who marry, do not live together before marriage, rear children, and stay married (a familistic value). Collectivistic values are illustrated by an Asian child on a swim team who would work for the good of the team, not for personal acclaim.

These elements of social structure and culture play a central role in making interpersonal choices and decisions. One of the goals of this text is to emphasize the influence of social structure and culture on your interpersonal decisions. Sociologists refer to this awareness as the **sociological imagination** (or sociological mindfulness). For example, though most people in the United States assume that they are free to select their own sex partner, this choice (or lack of it) is heavily influenced by structural and cultural factors. Most people hang out with, date, have sex with, and marry a person of the same racial background. Structural forces influencing race relations include segregation in housing, religion, and education. The fact that African Americans and European Americans live in different neighborhoods, worship in different churches, and often attend different schools makes meeting a person of a different race unlikely. When such encounters occur, prejudices and bias may influence these interactions to the point that individuals are hardly "free" to act as they choose. Hence, cultural values (transmitted by parents and peers) generally do not support or promote mixed racial interaction, relationship formation, and marriage. Consider the last three relationships in which you were involved, the racial similarity, and the structural and cultural influences on your choices.

**Sociological imagination** the influence of social structure and culture on your interpersonal decisions.

**Media** A Kaiser Family Foundation (2010) study of 8 to 18 year olds revealed that daily consumption of television averages 270 minutes, listening to music averages 151 minutes, and playing video games averages 73 minutes. Media in all of its forms (television, music, video games, print) influences how we view gender roles and make relationship choices. Chaney and Fairfax (2013) emphasized the profound influence media coverage of Barack Obama's marriage and family life could have on black marriage rates. The authors suggested that the affectionate, hand-holding Obamas in the context of their obvious devotion to their children serve as positive role models for marriage and family life (particularly among blacks).

## Other Influences on Relationship Choices

Aside from structural and cultural influences on relationship choices, other influences include family of origin (the family in which you were reared), unconscious motivations, habit patterns, individual personality, and previous experiences.

**Family of Origin** Your family of origin is a major influence on your subsequent choices and relationships. Roger Ebert (2011), the late film critic, revealed the impact of his mother's approval on his choice of a wife:

> *...Ingrid and I were close and loving companions, but I had no desire to face the wrath of my mother by declaring any serious plans. This is shameful and I cringe to write it, but I have to face the truth: I couldn't deal with the tears, denunciations and scenes from my mother....I guessed that I would never marry before my mother died (p. 566).*

*She's Lady Macbeth! My marriages failed because she fucked me up!*

JOHNNY CARSON, OF HIS MOTHER

Valle and Tillman (2014) found that being reared in a "nontraditional" family (e.g., stepfamily or single-parent) was associated with the development of romantic relationships during adolescence. Cui and Fincham (2010) found that individuals with divorced parents reported lower marital satisfaction in that they were less committed to their partner. In addition, marital conflict of one's parents was associated with conflict in one's own romantic relationships. Finally, Fish et al. (2011) examined data on 353 respondents and concluded: "Individuals who reported that they were aware of their parents' affair were significantly more likely to report having been unfaithful themselves: physically, emotionally, and composite. This makes sense considering the tendency of individuals to use family patterns through generations." Finally, Nesteruk and Gramescu (2012) emphasized the influence immigrant parents have on the endogamous mate selection of their children, who were expected to marry someone of their own cultural background.

**Personality** One's personality (e.g., introverted, extroverted; passive, assertive) also influences choices. For example, people who are assertive are more likely than those who are passive to initiate conversations with someone they are attracted to at a party. People who are very quiet and withdrawn may never choose to initiate a conversation even though they are attracted to someone. People with a bipolar disorder who are manic one part of the semester (or relationship) and depressed the other part are likely to make different choices when each phase is operative.

# Marriage

*The better part of one's life consists of friendships.*

ABRAHAM LINCOLN

**Marriage** a legal relationship that binds a couple together for the reproduction, physical care, and socialization of children.

While young adults think of marriage as "love" and "commitment" (Muraco and Curran 2012), the federal government regards **marriage** as a legal relationship that binds a couple together for the reproduction, physical care, and socialization of children. Each society works out its own details of what marriage is. In the United States, marriage is a legal contract between a heterosexual couple (although an increasing number of states are now recognize same-sex marriage) and the state in which they reside, that specifies the economic relationship between the couple (they become joint owners of their income and debt) and encourages sexual fidelity. See the self-assessment to assess the degree to which you have a positive view of marriage.

Various elements implicit in the marriage relationship in the United States are discussed in the following.

## SELF-ASSESSMENT | Attitudes toward Marriage Scale

The purpose of this survey is to assess the degree to which you view marriage positively. Read each item carefully and consider what you believe. There are no right or wrong answers. After reading each statement, select the number that best reflects your level of agreement, using the following scale:

| 1 | 2 | 3 | 4 | 5 | 6 | 7 |
|---|---|---|---|---|---|---|
| Strongly Disagree | | | | | | Strongly Agree |

_____ 1. I am married or plan to get married.
_____ 2. Being single and free is not as good as people think it is.
_____ 3. Marriage is NOT another word for being trapped.
_____ 4. Single people are more lonely than married people.
_____ 5. Married people are happier than single people.
_____ 6. Most of the married people I know are happy.
_____ 7. Most of the single people I know think marriage is less attractive than singlehood.
_____ 8. The statement that singles are more lonely and less happy than marrieds is mostly true.
_____ 9. It is better to be married than to be single.
_____ 10. Married people enjoy their lifestyle more than single people.
_____ 11. Marrieds have more close intimate relationships than singles.
_____ 12. Marrieds have a greater sense of joy than singles.
_____ 13. Being married is a more satisfying lifestyle than being single.
_____ 14. People who think that married people are happier than single people are correct.
_____ 15. Single people struggle with avoiding loneliness.
_____ 16. Married people are not as lonely as single people.
_____ 17. The companionship of marriage is a major advantage of the lifestyle.
_____ 18. Married people have better sex than singles.
_____ 19. The idea that singlehood is a happier lifestyle than being married is nonsense.
_____ 20. Singlehood as a lifestyle is overrated.

### Scoring

After assigning a number from 1 (strongly disagree) to 7 (strongly agree), add the numbers. The higher your score (140 is the highest possible score), the more positive your view of marriage. The lower your score (20 is the lowest possible score), the more negatively you view marriage. The midpoint is 60 (scores lower than 60 suggest more a negative view of marriage; scores higher than 60 suggest a more positive view of marriage).

### Norms

The norming sample of this self-assessment was based on 32 males and 174 females at East Carolina University. The average score of the males was 92 and the average score of the females was 95, suggesting a predominantly positive view of marriage (with females more positive than males).

### Source

"Attitudes toward Marriage Scale" was developed for this text by David Knox. It is to be used for general assessment and is not designed to be a clinical diagnostic tool or as a research instrument.

### National **Data**

Of adult women and men in the United States over the age of 65, 96% have married at least once (*Statistical Abstract of the United States* 2012–2013, Table 34).

*The first bond of society is marriage.*

MARCUS TULLIUS CICERO, ROMAN PHILOSOPHER

## Elements of Marriage

No one definition of marriage can adequately capture its meaning. Rather, marriage might best be understood in terms of its various elements. Some of these include the following:

**Legal Contract** Marriage in our society is a legal contract into which two people of different sexes and legal age may enter when they are not already married to someone else. The age required to marry varies by state and is usually from 16 to 18 (most states set 17 or 18 as the requirement). In some states (e.g., Alabama) individuals can marry at age 14 with parental or judicial consent. In California, individuals can marry at any age with parental consent.

The marriage license certifies that a legally empowered representative of the state perform the ceremony, often with two witnesses present. The marriage contract actually gives more power to the government and its control over the couple (Aulette 2010). The government not only dictates who may marry (e.g., persons of certain age, not currently married) but also stipulates the conditions of divorce (e.g., division of property, custody of children, and child support).

Under the laws of the state, the license means that spouses will jointly own all future property acquired and that each will share in the estate of the other. In most states, whatever the deceased spouse owns is legally transferred to the surviving spouse at the time of death. In the event of divorce and unless the couple has a prenuptial agreement, the property is usually divided equally regardless of the contribution of each partner. The license also implies the expectation of sexual fidelity in the marriage. Though less frequent because of no-fault divorce, infidelity is a legal ground for both divorce and alimony in some states.

**Common-law marriage** a marriage not based upon a license, ceremony, or any other legal formality but upon the couple's agreement to have a marital relationship. 14 states recognize common-law marriage

The marriage license is also an economic authorization that entitles a spouse to receive payment from a health insurance company for medical bills if the partner is insured, to collect Social Security benefits at the death of one's spouse, and to inherit from the estate of the deceased. Spouses are also responsible for each other's debts.

Brittany Bolen

Love and companionship are the primary reasons individuals in the United States marry.

Though the courts are reconsidering the definition of what constitutes a "family," the law is currently designed to protect spouses, not lovers or cohabitants. An exception is **common-law marriage**, in which a heterosexual couple cohabits and presents themselves as married; they will be regarded as legally married in those states that recognize such marriages. Common-law marriages exist in fourteen states (Alabama, Colorado, Georgia, Idaho, Iowa, Kansas, Montana, New Hampshire, Ohio, Oklahoma, Pennsylvania, Rhode Island, South Carolina, and Texas) and the District of Columbia. Even in these states, not all persons can marry by common-law—they must be of sound mind, be unmarried, and must have lived together for a certain period of time (e.g., three years). Persons married by common law who move to a non-common law state are recognized as being married in the state to which they move.

**Emotional Relationship** Ninety-three percent of married adults in the United States point to love as their top reason for getting married. Other reasons include making a lifelong commitment (87%), having companionship (81%), and having children (59%) (Cohn 2013). American emphasis on love as a reason to marry is not shared throughout the world. Individuals in other cultures (for example, India and Iran) do not require feelings of love to marry—love is expected to follow, not precede, marriage. In these countries, parental approval and similarity of religion, culture, education, and family background are considered more important criteria for marriage than love.

**Sexual Monogamy** Marital partners expect sexual fidelity. Over two-thirds (68%) of 4,539 undergraduates agreed with the statement, "I would divorce a spouse who had an affair" (Hall and Knox 2013).

*We want things not because we have reasons for them, we have reasons for them because we want them.*

ARTHUR SCHOPENHAUER, PHILOSOPHER

**Legal Responsibility for Children** Although individuals marry for love and companionship, one of the most important reasons for the existence of marriage from the viewpoint of society is to legally bind a male and a female for the nurture and support of any children they may have. In our society, child rearing is the primary responsibility of the family, not the state.

Marriage is a relatively stable relationship that helps to ensure that children will have adequate care and protection, will be socialized for productive roles in society, and will not become the burden of those who did not conceive them. Even at divorce, the legal obligation of the noncustodial parent to the child is maintained through child-support payments.

**Announcement** The legal binding of a couple in a public ceremony is often preceded by an engagement announcement. Following the ceremony there is a wedding announcement in the newspaper. Public knowledge of the event helps to solidify the commitment of the partners to each other and helps to marshal social and economic support to launch the couple into married life.

*Marriage. It's like a cultural hand-rail. It links folks to the past and guides them to the future.*

DIANE FROLOV AND ANDREW SCHNEIDER, *NORTHERN EXPOSURE, OUR WEDDING*, 1992

## Benefits of Marriage

Most (over 90%) people in our society get married, which has enormous benefits. Researchers Holt-Lunstad et al. (2010) emphasized the value of marriage as a social context conducive to one's mental as well as physical health and mortality. Their conclusion is based on a review of 148 studies involving 308,849 participants, which showed a 50% increase in the likelihood of survival for both women and men who had high social connections with family, friends, neighbors, and colleagues. Low social interaction/involvement had negative consequences and could be compared to smoking 15 cigarettes a day, being an alcoholic, not exercising, and being obese.

When married people are compared with singles, the differences are strikingly in favor of the married (see Table 1.4). The advantages of marriage over singlehood have been referred to as the **marriage benefit** and are true for first as well as subsequent marriages. Explanations for the marriage benefit include economic resources (e.g., higher income/wealth/can afford health care), social control (e.g., spouses ensure partner moderates alcohol/drug use, does not ride a motorcycle), and social/emotional/psychosocial support (e g., in-resident counselor, loving and caring partner) (Rauer 2013; Carr and Springer 2010). However, just being married is not beneficial to all individuals. Being married is associated with obesity and spouses often do not sleep as well as singles since a spouse may snore or bed hog (Rauer 2013). In addition, being in a stressful marriage is associated with elevated risk for cardiovascular disease, diabetes, stroke, and early death (Whisman et al. 2010).

**Marriage benefit** the advantages of marriage over singlehood.

**DIVERSITY IN OTHER COUNTRIES**

Australian youth plan to delay marriage but the overwhelming majority do not want to remain single forever. Skrbis et al. (2012) surveyed 6,937 adolescents with less than 3.24% reporting that they never planned to marry.

## Types of Marriage

Although we think of marriage in the United States as involving one man and one woman, other societies view marriage differently. **Polygamy** is a generic term for marriage involving more than two spouses. Polygamy occurs "throughout the

**Polygamy** marriage involving more than two spouses.

**TABLE 1.4 Benefits of Marriage and the Liabilities of Singlehood**

| | Benefits of Marriage | Liabilities of Singlehood |
|---|---|---|
| **Health** | Spouses have fewer hospital admissions, see a physician more regularly, and are sick less often. They recover from illness/surgery more quickly. | Single people are hospitalized more often, have fewer medical checkups, and are sick more often. |
| **Longevity** | Spouses live longer than single people. | Single people die sooner than married people. |
| **Happiness** | Spouses report being happier than single people. | Single people report less happiness than married people. |
| **Sexual satisfaction** | Spouses report being more satisfied with their sex lives, both physically and emotionally. | Single people report being less satisfied with their sex lives, both physically and emotionally. |
| **Money** | Spouses have more economic resources than single people. | Single people have fewer economic resources than married people. |
| **Lower expenses** | Two can live more cheaply together than separately. | Cost is greater for two singles than one couple. |
| **Drug use** | Spouses have lower rates of drug use and abuse. | Single people have higher rates of drug use and abuse. |
| **Connected** | Spouses are connected to more individuals who provide a support system—partner, in-laws, etc. | Single people have fewer individuals upon whom they can rely for help. |
| **Children** | Rates of high school dropouts, teen pregnancies, and poverty are lower among children reared in two-parent homes. | Rates of high school dropouts, teen pregnancies, and poverty are higher among children reared by single parents. |
| **History** | Spouses develop a shared history across time with significant others. | Single people may lack continuity and commitment across time with significant others. |
| **Crime** | Spouses are less likely to be involved in crime. | Single people are more likely to be involved in crime. |
| **Loneliness** | Spouses are less likely to report loneliness. | Single people are more likely to report being lonely. |



world . . . and is found on all continents and among adherents of all world religions" (Zeitzen 2008). There are three forms of polygamy: polygyny, polyandry, and pantagamy.

**Polygyny** marriage involving one husband and two or more wives.

### Polygyny in the United States

**Polygyny** involves one husband and two or more wives and is practiced illegally in the United States by some religious fundamentalist groups. These groups are primarily in Arizona, New Mexico, and Utah (as well as Canada), and have splintered off from the Church of Jesus Christ of Latter-day Saints (commonly known as the Mormon Church). To be clear, the Mormon Church does not practice or condone polygyny (the church outlawed it in 1890). Those that split off from the Mormon Church represent only about 5% of Mormons in Utah. The largest offshoot is called the Fundamentalist Church of Jesus Christ of the Latter-day Saints (FLDS). Members of the group feel that the practice of polygyny is God's will. Joe Jessop, age 88, is an elder of the FLDS. He has 5 wives, 46 children, and 239 grandchildren. Although the practice is illegal, polygynous individuals are rarely prosecuted because a husband will have only one legal wife while the others will be married in a civil ceremony. Women are socialized to bear as many children as possible to build up the "celestial family" that will remain together for eternity (Anderson 2010).

## National **Data**

There are an estimated 38,000 breakaway Mormon fundamentalists who practice plural marriage in North America. The largest group, FLDS, was founded in Hildale and Colorado City astride the Utah-Arizona border. There are about 10,000 FLDS members who live in the Western United States and Canada (Anderson 2010).

It is often assumed that polygyny in FLDS marriages exists to satisfy the sexual desires of the man, that the women are treated like slaves, and that jealousy among the wives is common. In most polygynous societies, however, polygyny has a political and economic rather than a sexual function. Polygyny, for members of the FLDS, is a means of having many children to produce a celestial family. In other societies, a man with many wives can produce a greater number of children for domestic or farm labor. Wives are not treated like slaves (although women have less status than men in general); all household work is evenly distributed among the wives; and each wife is given her own house or private sleeping quarters. In FLDs households, jealousy is minimal because the female is socialized to accept that her husband is not hers alone but is to be shared with other wives "according to God's plan." The spouses work out a rotational system for conjugal visits, which ensures that each wife has equal access to sexual encounters, while the other wives take care of the children.

**Three-way marriage** marriage where one male is "married" to two females.

Independent of polygynous marriage, some couples want a **three-way marriage**. Examples have existed in Brazil and the Netherlands whereby one male was "married" to two females. While these are not legal marriages, they reflect the diversity of lifestyle preferences and patterns. Theoretically, the arrangement could be of any sex, gender, and sexual orientation. The example in the Netherlands was of a heterosexual man "married" to two bisexual women.

**DIVERSITY IN OTHER COUNTRIES**

Jacob Zuma, the president of South Africa, has 5 wives, 19 children, and a fiancée. In the West African country of Mali, 43% of women live in polygamous marriages. Though the majority of the women disapprove of polygyny unions, divorce is not an option (Tabi et al. 2010).

**Polyandry** Tibetan Buddhists foster yet another brand of polygamy, referred to as **polyandry**, in which one wife has two or more (up to five) husbands. These husbands, who may be brothers, pool their resources to support one wife. Polyandry is a much less common form of polygamy than polygyny. The major reason for polyandry is economic. A family that cannot afford wives or marriages for each of its sons may find a wife for the eldest son only. Polyandry allows the younger brothers to also have sexual access to the one wife that the family is able to afford.

**Polyandry** one wife with two or more (up to five) husbands.

**Pantagamy** **Pantagamy** describes a group marriage in which each member of the group is "married" to the others. Pantagamy is a formal arrangement that was practiced in communes (for example, Oneida) in the 19th and 20th centuries. Pantagamy is, of course, illegal in the United States. Some polyamorous individuals see themselves in a group marriage.

**Pantagamy** a group marriage in which each member of the group is "married" to the others.

Our culture emphasizes monogamous marriage and values stable marriages. One cultural expression of this value is the existence of family policies in the form of laws, policies, and services (Rose and Humble 2013). (See the Family Policy section for marriage education which is designed to improve mate selection and strengthen marriage relationships.)

FAMILY POLICY | Marriage Preparation/Relationship Programs

While the terms (e. g., marriage preparation, premarital counseling, marriage education, relationship education) and content (e.g., focus on communication skills) (Childs and Duncan 2012) differ, the federal government has a vested interest in premarriage, marriage, and relationship education programs. One motivation is economic. The estimated cost for divorce to U.S. society is almost $35 billion (Wilmont et al. 2010) because divorce often plunges individuals into poverty, making them dependent on public resources. The philosophy behind marriage preparation education is that building a fence at the top of a cliff is preferable to putting an ambulance at the bottom. Hence, to the degree that people select a mate wisely and stay married, there is greater economic stability for the family and less drain on social services in the United States for single-parent mothers and the needs of their children (Schramm et al. 2013a).

Over 2,000 public schools nationwide offer a marriage preparation course. In Florida, all public high school seniors are required to take a marriage and relationship skills course. Toews and Yazedjian (2013) emphasized that these programs provide the tools necessary for building and maintaining healthy relationships. They noted that the most positive effects occur when both partners attend and when the programs are integrated into the existing educational curriculum. Ma et al. (2013) found in a study of 1,604 students that those exposed to relationship material had fewer faulty relationship views (e.g., one true love, love conquers all) and more realistic views about cohabitation (e.g., cohabitation does not ensure marital success).

Schramm et al. (2013b) emphasized that child welfare offices might also provide an excellent entry into relationship education. Carroll et al. (2013) reviewed a marriage education program specific for spouses/individuals in military families, the content of which included managing finances, garnering social support, and accessing community resources. Regardless of the source—school, child welfare, communication, or clergy—marriage education supplies positive outcomes at any time in a couple's relationship, and in either first or subsequent marriages (Lucier-Greer et al. 2012).

There is opposition to marriage preparation education in the public school system. Opponents question using school time for relationship courses. Teachers are viewed as overworked, and an additional course on marriage seems to press the system to the breaking point. In addition, some teachers lack the training to provide relationship courses. Although training teachers would stretch already-thin budgets, many schools at present have programs in family and consumer sciences, and teachers in these programs are trained in teaching about marriage and the family. A related concern with teaching about marriage and the family in high school is the fear on the part of some

# Family

Most people who marry choose to have children and become a family. However, the definition of what constitutes a family is sometimes unclear. This section examines how families are defined, their numerous types, and how marriages and families have changed in the past fifty years.

*Call it a clan, call it a network, call it a tribe, call it a family. Whatever you call it, whoever you are, you need one.*

JANE HOWARD, JOURNALIST

## Definitions of Family

The U.S. Census Bureau defines **family** as a group of two or more people related by blood, marriage, or adoption. This definition has been challenged because it does not include foster families or long-term couples who live together. Marshall (2013) surveyed 105 faculty members from 19 Ph.D. marriage and family therapy programs and found no universal agreement on the definition of the family. Same-gender couples, children of same gender couples, and children with nonresidential parents were sometimes excluded from the definition of the family.

**Family** a group of two or more people related by blood, marriage, or adoption.

The answer to the question "Who is family?" is important because access to resources such as health care, Social Security, and retirement benefits is parents that the course content may be too liberal. Some parents who oppose teaching sex education in the public schools fear that such courses lead to increased sexual activity.

The Marriage and Responsible Fatherhood Act provides $150 million a year through 2014 (with the potential for renewal) for relationship programs primarily targeted toward low income/disadvantaged populations. Attendance has been low and results mixed. Some outcome studies have shown no difference in treatment and control groups while others have shown a reduced divorce rate, lower domestic violence, and higher child well-being (Ooms and Hawkins, 2012).

**Your Opinion?**

1. To what degree do you believe marriage education belongs in the public school system?
2. What evidence reveals that marriage education is effective?
3. Should marriage be encouraged by the federal government?

### Sources


Carroll, E. B., D. K. Orthner, A. Behnke, C. M. Smith, S. Day, and M. Raburn. 2013. Integrating life skills into relationship and marriage education: The Essential Life Skills for Military Families Program. *Family Relations* 62: 559–570.

Childs, G. R. and S. F. Duncan. 2012. Marriage preparation education programs: An assessment of their components. *Marriage & Family Review* 48: 59–81.

Lucier-Greer, M., F. Adler-Baeder, S. A. Ketring, K. T. Harcourt, and T. Smith. 2012. Comparing the experiences of couples in first marriages and remarriages in couple and relationship education. *Journal of Divorce and Remarriage* 53: 55–75.

Ma, Y., J. Pittman, J. Kerpelman, and F. Adler-Baeder. 2013. Relationship education and classroom climate impact on faulty relationship views. National Council on Family Relations annual meeting. San Antonio, November 5–9.

Ooms, T. and A. Hawkins. 2012. Marriage and relationship education: A promising strategy for strengthening low-income vulnerable families. *The State of Our Unions.* Charlottesville, VA.: National Marriage Project and Institute for American Values.

Schramm, D. G., S. M. Harris, J. B. Whiting, A. J. Hawkins, M. Brown, and R. Porter. 2013a. Economic Costs and policy implications associated with divorce: Texas as a case study. *Journal of Divorce and Remarriage* 54: 1–24.

Schramm, D. G., T. G. Futris, A. M. Galovan, and K. Allen. 2013b. Is relationship and marriage education relevant and appropriate to child welfare? *Children and Youth Services Review* 35: 429–438.

Toews, M. L. and A. Yazedjian. 2013. An evaluation of a relationship education program for adolescent parents. Poster, National Council on Family Relations Annual Meeting. San Antonio, November 5–9.


involved. Unless cohabitants are recognized by the state in which they reside as in a "domestic partnership," cohabitants are typically not viewed as "family" and are not accorded health benefits, Social Security, and retirement benefits of the partner. Indeed, the "live-in partner" may not be allowed to see the beloved in the hospital, which limits visitation to "family only."

The definition of who counts as family is being challenged. In some cases, families are being defined by function rather than by structure—what is the level of emotional and financial commitment and interdependence between the partners? How long have they lived together? Do the partners view themselves as a family?

## National **Data**

Eighty-six percent of U.S adults say a single parent and child constitute a family; nearly as many (80%) say an unmarried couple living together with a child is a family. Sixty-three percent say a gay or lesbian couple raising a child constitutes a family (Pew Research Center 2010).

Friends sometimes become family. Due to mobility, spouses may live several states away from their respective families. Although they may visit their families

## DIVERSITY IN OTHER COUNTRIES

While Hawaii is one of the United States, their cultural tradition of "**hanai adoptions**" is unlike any system of adoption on the mainland. Hawaiian culture allows a child to be "hanai'd out," which means the child may be adopted by someone in the extended family or by a childless couple. Typically no papers are signed but the new adoptive parents rear, educate, and love the child as though he or she were a biological child (the relationship between the child and birth parents is permitted and encouraged).

**Hanai adoption** Hawaiian culture allows a child to be adopted by someone in the extended family or by a childless couple.

**Civil union** pair-bonded relationship given legal significance in terms of rights and privileges.

**Domestic partnership** relationship in which cohabitating individuals are given some kind of official recognition by a city or corporation so as to receive partner benefits.

for holidays, they often develop close friendships with others on whom they rely locally for emotional and physical support on a daily basis. Persons in the military who are separated from their parents and siblings often form close "family" relationships with other military individuals, couples, and families.

Sociologically, a family is defined as a kinship system of all relatives living together or recognized as a social unit, including adopted people. The family is regarded as the basic social institution because of its important functions of procreation and socialization, and because it is found in some form in all societies.

Same-sex couples (e.g., Ellen DeGeneres and her partner Portia de Rossi) certainly define themselves as family. An increasing number of states recognize marriages between same-sex individuals. With the Supreme Court recognition of same-sex marriage, more states will grant marriage licenses to same-sex couples. Short of marriage, other states (the number keeps changing) may recognize committed gay relationships as **civil unions** (pair-bonded relationship given legal significance in terms of rights and privileges). In France, civil unions are becoming increasingly popular—there are now two civil unions for every three marriages. In France a civil union can be dissolved with a registered letter (Sayare & De La Baume 2010).

Over twenty-four cities, 11 states, and numerous countries (including Canada) recognize some form of domestic partnership. **Domestic partnerships** are relationships in which cohabitating individuals are given some kind of official recognition by a city or corporation so as to receive partner benefits (for example, health insurance). The Walt Disney Company recognizes domestic partnerships; Walmart offers benefits to same sex partners. Domestic partnerships do not confer any federal recognition or benefits.

David Knox

This family dog thinks he is the family baby . . . and is carried just like one.

About 60% of Americans own a pet. Some view their pets as part of their family. In a Gallo Family Vineyard survey of 691 pet owners, 93% agreed that their pet was a part of the family (Payne and Bravo 2013). Examples of treating pets like children include living only where there is a fenced-in backyard, feeding the pet a special diet, hanging a stocking and/or buying presents for the pet at Christmas, buying "clothes" for the pet, and leaving money in one's will for the care of the pet. Mooney et al. (2010) studied attachment as the benchmark of the human-companion animal bond and found that, in a sample of 250 adults who owned an animal, women, those not married, those over 65, and dog-only owners reported the greatest attachment. Some pet owners buy accident insurance—Progressive© insurance covers pets. Gregory (2010) noted that pets are now the legal subject of divorce—the divorcing parties are granted custody and visitation rights to the animals of the couple.

### Types of Families

There are various types of families.

**Family of Origin** Also referred to as the **family of orientation**, this is the family into which you were born or the family in which you were reared. It involves you, your parents, and your siblings. When you go to your parents' home for the holidays, you return to your *family of origin*. Your experiences in your family of origin have an impact on your own relationships. Chiu and Busby (2010) found that individuals whose parents had a high-quality relationship are likely to have a similar high-quality marriage and to have a familistic orientation. Similarly, Rechcck et al. (2010) documented how the relationship with one's parents affects the quality of one's own marriage.

Siblings in one's family of origin provide a profound influence on one another's behavior and emotional development and adjustment (McHale et al. 2012). Meinhold et al. (2006) noted that the relationship with one's siblings, particularly the sister-sister relationship, represents the most enduring relationship in a person's lifetime. Sisters who lived near one another and who did not have children reported the greatest amount of intimacy and contact. Myers (2011) studied 124 adults who provided an explanation for why they maintained their relationship with one of their siblings. "We are family" was one of the major categories of reasons.

Lisa Emmott

The relationship these sisters have with each other will likely be the most enduring relationship each has in their respective lifetimes.

**Family of Procreation** The **family of procreation** represents the family that you will begin should you marry and have children. Of U.S. citizens living in the United States 65 years old and over, 96% have married and established their own family of procreation (*Statistical Abstract of the United States* 2012–2013, Table 34). Across the life cycle, individuals move from the family of orientation to the family of procreation.

**Family of orientation** the family into which you were born or the family in which you were reared.

**Family of procreation** the family that you will begin should you marry and have children.

**Nuclear family** either a family or origin or a family of procreation.

**Nuclear Family** The **nuclear family** refers to either a family of origin or a family of procreation. In practice, this means that your nuclear family consists either of you, your parents, and your siblings or of you, your spouse, and your children. Generally, one-parent households are not referred to as nuclear families. They are binuclear families if both parents are involved in the child's life, or single-parent families if only one parent is involved in the child's life.

**Is the Nuclear Family Universal?** Sociologist George Peter Murdock's classic study (1949) emphasized that the nuclear family is a "universal social grouping" found in all of the 250 societies he studied. The nuclear family channels sexual energy between two adult partners who reproduce and also cooperate in the care of offspring and their socialization to be productive members of society. "This universal social structure, produced through cultural evolution in every human society, as presumably the only feasible adjustment to a series of basic needs, forms a crucial part of the environment in which every individual grows to maturity" (p. 11).

The universality of the nuclear family has been questioned. In *Sex at Dawn*, Ryan and Jetha (2010) reviewed cross-cultural data and emphasized that the terms marriage and family do not have universal meanings. In some groups, adults have sexual relationships with various partners throughout their life and view themselves as mothers and fathers to all of the children in the community. Children in these villages view all adults as their mother and father.

*If we don't shape our kids, they will be shaped by outside forces that don't care what shape our kids are in.*

LOUISE HART, PARENT EDUCATOR

Dr. Robert Bunger (2014) is a premier anthropologist. His reaction to the thesis of *Sex at Dawn* follows:

> *In my opinion the idea that everyone had sex with whomever and that all adults were parents of everyone's children is utter nonsense. Louis Henry Morgan in Ancient Society (published in the 19th century) suggested that early humans lived in a state of "primitive promiscuity" and the idea was taken up by Marx and Engels. I do not know of any society that actually lives that way. The Muria Ghond of India have a system whereby the young people, between puberty and marriage, live in a group marriage where everyone is allowed to have sex with everyone else of the other sex. At some point they drop out and settle into monogamous marriage. I do not think that there is any traditional society where group marriage for adults is the norm. I think that some communal movements like the Amana society tried group marriage but later gave it up.*

**Traditional family** two-parent nuclear family, with the husband as breadwinner and the wife as homemaker.

**Modern family** dual career family in which both spouses work outside the home.

**Postmodern family** lesbian or gay couples and mothers who are single by choice.

**Binuclear family** family in which the members live in two separate households.

**Blended family** parents remarry and bring additional children into the respective units.

**Extended family** includes not only the nuclear family but other relatives as well.

**Traditional, Modern, and Postmodern Family** Sociologists have identified three central concepts of the family (Silverstein and Auerbach 2005). The **traditional family** is the two-parent nuclear family, with the husband as breadwinner and the wife as homemaker. The **modern family** is the dual-earner family, in which both spouses work outside the home. **Postmodern families** represent a departure from these models and include lesbian or gay couples and mothers who are single by choice.

**Binuclear Family** A **binuclear family** is a family in which the members live in two separate households. This family type is created when the parents of the children divorce and live separately, setting up two separate units, with the children remaining a part of each unit. Each of these units may also change again when the parents remarry and bring additional children into the respective units (**blended family**). Hence, the children may go from a nuclear family with both parents, to a binuclear unit with parents living in separate homes, to a blended family when parents remarry and bring additional children into the respective units.

**DIVERSITY IN OTHER COUNTRIES**

Asians are more likely than European Americans to live with their extended families. Among Asians, the status of the elderly in the extended family derives from religion. Confucian philosophy, for example, prescribes that all relationships are of the superordinate–subordinate type—that is, husband-wife, parent-child, and teacher-pupil. For traditional Asians to abandon their elderly rather than include them in larger family units would be unthinkable. However, commitment to the elderly may be changing as a result of the westernization of Asian countries such as China, Japan, and Korea.

**Extended Family** The **extended family** includes not only the nuclear family (or parts of it) but other relatives as well. These relatives include grandparents, aunts, uncles, and cousins. An example of an extended family living together would be a husband and wife, their children, and the husband's parents (the children's grandparents). The extended family is particularly important for African-American married couples.

## When Families Are Destroyed—The Australian Aboriginal Example

In Australia, between 1885 and 1969, between 50,000 and 100,000 "half-caste" (one white parent) Aboriginal children were taken by force from their parents by the Australian police. The white society wanted to convert these children to Christianity and to destroy their Aboriginal culture, which was viewed as primitive and without value. The children walked, were taken by camel (depending on the territory in Australia where they lived), or put on a train and taken hundreds of miles away from their parents to church missions. Australian

Bob Randall, an Australian Aboriginal, was taken from his parents (whom he never saw again) by the government at age 7 and brought up in missionary camps. He and his partner Barbara Schacht work together to fight for the rights of the Aborigines.

David Knox

government destruction of aboriginal families is the theme of *Rabbit-Proof Fence* (available on DVD). Randall wrote of his experience:

> *Instead of the wide open spaces of my desert home, we were housed in corrugated iron dormitories with rows and rows of bunk beds. After dinner we were bathed by the older women, put in clothing they called pajamas, and then tucked into one of the iron beds between the sheets. This was a horrible experience for me. I couldn't stand the feel of the cloth touching my skin (Randall 2008, p. 35).*

The Australian government subsequently apologized for the laws and policies of successive parliaments and governments that inflicted profound grief, suffering, and loss on the Aborigines. However, Randall notes that the Aborigines continue to be marginalized and that nothing has been done to compensate them for the horror of taking children from their families.

## Differences between Marriage and Family

Marriage can be thought of as a social relationship that leads to the establishment of a family. Indeed, every society or culture has mechanisms (from "free" dating to arranged marriages) for guiding their youth into permanent emotionally, legally, or socially bonded relationships (marriage) that are designed to result in procreation and care of offspring. Although the concepts of marriage and the family are sometimes used synonymously, they are distinct. The late sociologist Dr. Lee Axelson identified some of these differences in Table 1.5.

### Changes in Marriage and the Family in the Last 60 Years

Enormous changes have occurred in marriage and the family since the 1950s. Table 1.6 reflects some of these changes.

**TABLE 1.5 Differences between Marriage and the Family in the United States**

| Marriage | Family |
|---|---|
| Usually initiated by a formal ceremony | Formal ceremony not essential |
| Involves two people | Usually involves more than two people |
| Ages of the individuals tend to be similar | Individuals represent more than one generation |
| Individuals usually choose each other | Members are born or adopted into the family |
| Ends when spouse dies or is divorced | Continues beyond the life of the individual |
| Sex between spouses is expected and approved | Sex between near kin is neither expected nor approved |
| Requires a license | No license needed to become a parent |
| Procreation expected | Consequence of procreation |
| Spouses are focused on each other | Focus changes with addition of children |
| Spouses can voluntarily withdraw from marriage | Parents cannot divorce themselves from obligations via divorce to children |
| Money in unit is spent on the couple | Money is used for the needs of children |
| Recreation revolves around adults | Recreation revolves around children |

Reprinted by permission of the estate of Dr. Lee Axelson.

**TABLE 1.6 Changes in Marriages and Families—1950 and 2015**

| | 1950 | 2015 |
|---|---|---|
| **Family Relationship Values** | Strong values for marriage and the family. Individuals who wanted to remain single or childfree are considered deviant, even pathological. Husband and wife should not be separated by jobs or careers. | Individuals who remain single or childfree experience social understanding and sometimes encouragement. Single and childfree people are no longer considered deviant or pathological but are seen as self-actuating individuals with strong job or career commitments. Husbands and wives can be separated for reasons of job or career and live in a commuter marriage. Married women in large numbers have left the role of full-time mother and housewife to join the labor market. |
| **Gender Roles** | Rigid gender roles, with men dominant and earning income while wives stay home, taking care of children. | Egalitarian gender roles with both spouses earning income. Greater involvement of men in fatherhood. |
| **Sexual Values** | Marriage was regarded as the only appropriate context for intercourse in middle-class America. Living together is unacceptable, and a child born out of wedlock was stigmatized. Virginity is sometimes exchanged for marital commitment. | For many, concerns about safer sex have taken precedence over the marital context for sex. Virginity is rarely exchanged for anything. Living together is regarded as not only acceptable but sometimes preferable to marriage. For some, unmarried single parenthood is regarded as a lifestyle option. Hooking up is new courtship norm. |
| **Homogamous Mating** | Strong social pressure exists to date and marry within one's own racial, ethnic, religious, and social class group. Emotional and legal attachments are heavily influenced by obligation to parents and kin. | Dating and mating have become more heterogamous, with more freedom to select a partner outside one's own racial, ethnic, religious, and social class group. Attachments are more often by choice. |

**TABLE 1.6 Continued**

| | 1950 | 2015 |
|---|---|---|
| **Cultural Silence on Intimate Relationships** | Intimate relationships are not an appropriate subject for the media. | Individuals on talk shows, interviews, and magazine surveys are open about sexuality and relationships behind closed doors. |
| **Divorce** | Society strongly disapproves of divorce. Familistic values encouraged spouses to stay married for the children. Strong legal constraints keep couples together. Marriage is forever. | Divorce has replaced death as the endpoint of a majority of marriages. Less stigma is associated with divorce. Individualistic values lead spouses to seek personal happiness. No-fault divorce allows for easy severance. Marriage is tenuous. Increasing numbers of children are being reared in single-parent households apart from other relatives. |
| **Familism versus Individualism** | Families are focused on the needs of children. Mothers stay home to ensure that the needs of their children are met. Adult concerns are less important. | Adult agenda of work and recreation has taken on increased importance, with less attention being given to children. Children are viewed as more sophisticated and capable of thinking as adults, which frees adults to pursue their own interests. Day care is used regularly. |
| **Homosexuality** | Same-sex emotional and sexual relationships are a culturally hidden phenomenon. Gay relationships are not socially recognized. | Supreme Court legalized same-sex marriage. Gay relationships are, increasingly, a culturally open phenomenon (e.g., television sitcoms, gay athletes, etc.). |
| **Scientific Scrutiny** | Aside from Kinsey's, few studies are conducted on intimate relationships. | Acceptance of scientific study of marriage and intimate relationships. |
| **Family Housing** | Husbands and wives live in same house. | Husbands and wives may "live apart together" (LAT), which means that, although they are emotionally and economically connected, they (by choice) maintain two households, houses, condos, or apartments. |
| **Technology** | Nonexistent except phone. | Use of iPhones, texting, sexting, Facebook. |



# Theoretical Frameworks for Viewing Marriage and the Family

Although we emphasize choices in relationships as the framework for viewing marriage and the family, other conceptual theoretical frameworks are helpful in understanding the context of relationship decisions. All **theoretical frameworks** are the same in that they provide a set of interrelated principles designed to explain a particular phenomenon and provide a point of view. In essence, theories are explanations.

**Theoretical framework** provides a set of interrelated principles designed to explain a particular phenomenon and provide a point of view.

## Social Exchange Framework

The **social exchange framework** is one of the most commonly used theoretical perspectives in marriage and the family. The framework views interaction and choices in terms of cost and profit.

**Social exchange framework** theoretical perspective that views interaction and choices in terms of cost and profit.

The social exchange framework also operates from a premise of **utilitarianism**—the theory that individuals rationally weigh the rewards and costs associated with behavioral choices. Vespa (2013) studied cohabitants age 50 and older and found that unhealthy but wealthy males were more likely to marry—they trade their wealth/agreement to marry for care giving by a female who needs economic support/wants to be married.

**Utilitarianism** theory that individuals rationally weigh the rewards and costs associated with behavioral choices.

A social exchange view of marital roles emphasizes that spouses negotiate the division of labor on the basis of exchange. For example, a man participates in child care in exchange for his wife earning an income, which relieves him of the total financial responsibility. Social exchange theorists also emphasize that power in relationships is the ability to influence, and avoid being influenced by, the partner.

Albert Einstein's second marriage to Elsa Einstein provides another example of exchange. "She was an efficient and lively woman, who was eager to serve and protect him.....He was pleased to be looked after...which allowed him to spend hours in a rather dreamy state, focusing more on the cosmos than on the world around him" (Isaacson 2007, p. 247).

## Family Life Course Development Framework

**Family life course development** framework that emphasizes the important role transitions of individuals that occur in different periods of life and in different social contexts.

The **family life course development** framework emphasizes the important role transitions of individuals that occur in different periods of life and in different social contexts. For example, young unmarried lovers may become cohabitants, then parents, grandparents, retirees, and widows. The family life cycle is a basic set of stages through which not all individuals pass (e.g., the child-free) and in which there is great diversity, particularly in regard to race and education (e.g., African Americans are less likely to marry; the highly educated are less likely to divorce) (Cherlin 2010).

**Family life cycle** based in psychology, which emphasizes the various developmental tasks family members face across time.

The family life course developmental framework has its basis in sociology (e.g., role transitions), whereas the **family life cycle** has its basis in psychology, which emphasizes the various developmental tasks family members face across time (e.g., marriage, childbearing, preschool, school-age children, teenagers, and so on). If developmental tasks at one stage are not accomplished, functioning in subsequent stages will be impaired. For example, one of the developmental tasks of early American marriage is to emotionally and financially separate from one's family of origin. If such separation from parents does not take place, independence as individuals and as a couple may be impaired.

## Structure-Function Framework

**Structure-function framework** emphasizes how marriage and family contribute to society.

The **structure-function framework** emphasizes how marriage and family contribute to society. Just as the human body is made up of different parts that work together for the good of the individual, society is made up of different institutions (e.g., family, religion, education, economics) that work together for the good of society. Functionalists view the family as an institution with values, norms, and activities meant to provide stability for the larger society. Such stability depends on families performing various functions for society.

First, families serve to replenish society with socialized members. Because our society cannot continue to exist without new members, we must have some way of ensuring a continuing supply. However, just having new members is not enough. We need socialized members—those who can speak our language and know the norms and roles of our society. So-called **feral** (meaning wild, not domesticated) **children** are those who are thought to have been reared by animals. Newton (2002) detailed nine such children, the most famous of whom was Peter the Wild Boy found in the Germanic woods at the age of 12 and brought to London in 1726. He could not speak; growling and howling were his modes of expression. He lived until the age of 70 and never learned to talk. Feral children emphasize that social interaction and family context make us human. Girgis et al. (2011) emphasized that "societies rely on families to produce upright people who make for conscientious, law-abiding citizens lessening the demand for governmental policing and social services" (p. 245).

**Feral children** those who are thought to have been reared by animals.

Genie is a young girl who was discovered in the 1970s; she had been kept in isolation in one room in her California home for 12 years by her abusive father (James 2008). She could barely walk and could not talk. Although provided intensive therapy at UCLA and the object of thousands of dollars of funded research, Genie progressed only slightly. Today, she is in her late 50s, institutionalized, and speechless. Her story illustrates the need for socialization; the legal bond of marriage and the obligation to nurture and socialize offspring help to ensure that this socialization will occur.

Second, marriage and the family promote the emotional stability of the respective spouses. Society cannot provide enough counselors to help us whenever we have emotional issues/problems. Marriage ideally provides in-residence counselors who are loving and caring partners with whom people share (and receive help for) their most difficult experiences.

Children also need people to love them and to give them a sense of belonging. This need can be fulfilled in a variety of family contexts (two-parent families, single-parent families, extended families). The affective function of the family is one of its major offerings. No other institution focuses so completely on meeting the emotional needs of its members as marriage and the family.

Third, families provide economic support for their members. Although modern families are no longer self-sufficient economic units, they provide food, shelter, and clothing for their members. One need only consider the homeless in our society to be reminded of this important function of the family.

In addition to the primary functions of replacement, emotional stability, and economic support, other functions of the family include the following:

- ***Physical care***—Families provide the primary care for their infants, children, and aging parents. Other agencies (neonatal units, day care centers, assisted-living residences, shelters) may help, but the family remains the primary and recurring caretaker. Spouses also show concern about the physical health of each other by encouraging each other to take medications and to see a doctor.
- ***Regulation of sexual behavior***—Spouses are expected to confine their sexual behavior to each other, which reduces the risk of having children who do not have socially and legally bonded parents, and of contracting or spreading sexually transmitted infections.
- ***Status placement***—Being born into a family provides social placement of the individual in society. One's family of origin largely determines one's social class, religious affiliation, and future occupation. Baby Prince George Alexander Louis, son of Kate Middleton and Prince William of the royal family of Great Britain, was born into the upper class and is destined to be in politics by virtue of being born into a political family.
- ***Social control***—Spouses in high-quality, durable marriages provide social control for each other that results in less criminal behavior. Parole boards often note that the best guarantee against recidivism is a spouse who expects the partner to get a job and avoid criminal behavior and who reinforces these behaviors.

## Conflict Framework

**Conflict framework** views individuals in relationships as competing for valuable resources (time, money, power). Conflict theorists recognize that family members have different goals and values that produce conflict. Adolescents want freedom (e.g., stay out all night with new love interest) while parents want their child to get a good night's sleep, not get pregnant, and stay on track in school.

**Conflict framework** views individuals in relationships as competing for valuable resources (time, money, power).

Conflict theorists also view conflict not as good or bad but as a natural and normal part of relationships. They regard conflict as necessary for the change and growth of individuals, marriages, and families. Cohabitation relationships, marriages, and families all have the potential for conflict. Cohabitants are in conflict about commitment to marry, spouses are in conflict about the division of labor, and parents are in conflict with their children over rules such as curfew, chores, and homework. These three units may also be in conflict with other systems. For example, cohabitants are in conflict with the economic institution for health benefits for their partners. Similarly, employed parents are in conflict with their employers for flexible work hours, maternity or paternity benefits, and day care facilities.

*Nonviolence is the answer to the crucial political and moral questions of our time.*

MARTIN LUTHER KING, JR.

Conflict theory is also helpful in understanding choices in relationships with regard to mate selection and jealousy. Unmarried individuals in search of a partner are in competition with other unmarried individuals for the scarce resources of a desirable mate. Such conflict is particularly evident in the case of older women in competition for men. At age 85 and older, there are twice as many women (3.7 million) as there are men (1.8 million) (*Statistical Abstract of the United States* 2012–2013, Table 7). Jealousy is also sometimes about scarce resources. People fear that their "one and only" will be stolen by someone else who has no partner. Thus wives are aware of how much time their husbands spend talking to the attractive newly divorced female at a social gathering.

## Symbolic Interaction Framework

**Symbolic interaction framework** views marriages and families as symbolic worlds in which the various members give meaning to one another's behavior.

The **symbolic interaction framework** views marriages and families as symbolic worlds in which the various members give meaning to one another's behavior. Human behavior can be understood only by the meaning attributed to behavior. Curran et al. (2010) assessed the meaning of marriage for 31 African Americans of different ages and found that the two most common meanings were commitment and love. Herbert Blumer (1969) used the term *symbolic interaction* to refer to the process of interpersonal interaction. Concepts inherent in this framework include the definition of the situation, the looking-glass self, and the self-fulfilling prophecy.

**Definition of the Situation** Two people who have just spotted each other at a party are constantly defining the situation and responding to those definitions. Is the glance from the other person (1) an invitation to approach, (2) an approach, or (3) a misinterpretation—was he or she looking at someone else? The definition used will affect subsequent interaction.

Getting married also has different definitions/meanings. For "marriage naturalists" it is an event that is a natural progression of a relationship (often begun in high school) and is expected of oneself, one's partner, and both of their families. Persons in rural areas more often have this view. In contrast, "marriage planners" are more metropolitan and view marriage as an event one "gets ready for" by completing one's college or graduate school education, establishing oneself in a job/career, and maturing emotionally and psychologically. These individuals may cohabit and have children before they decide to marry (Kefalas et al. 2011).

**Looking-Glass Self** The image people have of themselves is a reflection of what other people tell them about themselves (Cooley 1964). People develop an idea of who they are by the way others act toward them. If no one looks at or speaks to them, they will begin to feel unsettled, according to Charles Cooley. Similarly, family members constantly hold up social mirrors for one another into which the respective members look for definitions of self.

G. H. Mead (1934), a classic symbolic interactionist, believed that people are not passive sponges but that they evaluate the perceived appraisals of others, accepting some opinions and not others. Although some parents teach their children that they are worthless, these children may reject the definition by believing in more positive social mirrors from friends, teachers, and lovers.

**Self-Fulfilling Prophecy** Once people define situations and the behaviors in which they are expected to engage, they are able to behave toward one another in predictable ways. Such predictability of behavior affects subsequent behavior. If you feel that your partner expects you to be faithful, your behavior is likely to conform to these expectations. The expectations thus create a self-fulfilling prophecy.

## Family Systems Framework

The **family systems framework** views each member of the family as part of a system and the family as a unit that develops norms of interacting, which may be explicit (e.g., parents specify when their children must stop texting for the evening and complete homework) or implicit (e.g., spouses expect fidelity from each other). These rules serve various functions, such as the allocation of keeping the education of offspring on track and solidifying the emotional bond of the spouses.

*Language, after all, is only the use of symbols.*

GEORGE HENRY LEWES, ENGLISH PHILOSOPHER

**Family systems framework** views each member of the family as part of a system and the family as a unit that develops norms of interaction.

Rules are most efficient if they are flexible (e.g., they should be adjusted over time in response to a child's growing competence). A rule about not leaving the yard when playing may be appropriate for a 4 year old but inappropriate for a 16 year old.

Family members also develop boundaries that define the individual and the group and separate one system or subsystem from another. A boundary is a "border between the system and its environment that affects the flow of information and energy between the environment and the system" (White and Klein 2002, p. 124). A boundary may be physical, such as a closed bedroom door, or social, such as expectations that family problems will not be aired in public. Boundaries may also be emotional, such as communication, which maintains closeness or distance in a relationship. Some family systems are cold and abusive; others are warm and nurturing.

In addition to rules and boundaries, family systems have roles (leader, follower, scapegoat) for the respective family members. These roles may be shared by more than one person or may shift from person to person during an interaction or across time. In healthy families, individuals are allowed to alternate roles rather than being locked into one role. In problem families, one family member is often allocated the role of scapegoat, or the cause of all the family's problems (e.g., an alcoholic spouse).

Family systems may be open, in that they are receptive to information and interaction with the outside world, or closed, in that they feel threatened by such contact. The Amish have a closed family system and minimize contact with the outside world. Some communes also encourage minimal outside exposure. Twin Oaks Intentional Community of Louisa, Virginia, does not permit any of its almost 100 members to own a television or keep one in their room. Exposure to the negative drumbeat of the evening news is seen as harmful.

Holmes et al. (2013) used a family systems perspective to explain the transition of spouses and their marriage to parenthood. The researchers noted that it is the context that must be considered to understand changes. For example, having a daughter is associated with more conflict for fathers across time and this impacts the interaction of the wife with her husband.

## Feminist Framework

Although a **feminist framework** views marriage and family as contexts of inequality and oppression for women, there are eleven feminist perspectives, including lesbian feminism (emphasizing oppressive heterosexuality), psychoanalytic feminism (focusing on cultural domination of men's phallic-oriented ideas and repressed emotions), and standpoint feminism (stressing the neglect of women's perspective and experiences in the production of knowledge) (Lorber 1998). Regardless of which feminist framework is being discussed, all feminist frameworks have the themes of inequality and oppression. According to feminist theory, gender structures our experiences (e.g., women and men will experience

**Feminist framework** views marriage and family as contexts of inequality and oppression for women.

*Feminism is the radical notion that women are people.*

ANI DIFRANCO, AMERICAN SINGER

life differently because there are different expectations for the respective genders) (White and Klein 2002). Feminists seek equality in their relationships with their partners.

The major theoretical frameworks for viewing marriage and the family are summarized in Table 1.7.

**TABLE 1.7 Theoretical Frameworks for Marriage and the Family**

| Theory | Description | Concepts | Level of Analysis | Strengths | Weaknesses |
|---|---|---|---|---|---|
| Social Exchange | In their relationships, individuals seek to maximize their benefits and minimize their costs. | Benefits<br>Costs<br>Profit<br>Loss | Individual<br>Couple<br>Family | Provides explanations of human behavior based on outcome. | Assumes that people always act rationally and all behavior is calculated. |
| Family Life Course Development | All families have a life course that is composed of all the stages and events that have occurred within the family. | Stages<br>Transitions<br>Timing | Institution<br>Individual<br>Couple Family | Families are seen as dynamic rather than static. Useful in working with families who are facing transitions in their life courses. | Difficult to adequately test the theory through research. |
| Structure Function | The family has several important functions within society; within the family, individual members have certain functions. | Structure<br>Function | Institution | Emphasizes the relation of family to society, noting how families affect and are affected by the larger society. | Families with nontraditional structures (single-parent, same-sex couples) are seen as dysfunctional. |
| Conflict | Conflict in relationships is inevitable, due to competition over resources and power. | Conflict<br>Resources<br>Power | Institution | Views conflict as a normal part of relationships and as necessary for change and growth. | Sees all relationships as conflictual, and does not acknowledge cooperation. |
| Symbolic Interaction | People communicate through symbols and interpret the words and actions of others. | Definition of the situation<br>Looking-glass self; Self-fulfilling prophecy | Couple | Emphasizes the perceptions of individuals, not just objective reality or the viewpoint of outsiders. | Ignores the larger social interaction context and minimizes the influence of external forces. |
| Family Systems | The family is a system of interrelated parts that function together to maintain the unit. | Subsystem<br>Roles Rules<br>Boundaries<br>Open system<br>Closed system | Couple<br>Family | Very useful in working with families who are having serious problems (violence, alcoholism). Describes the effect family members have on each other. | Based on work with systems, troubled families, and may not apply to nonproblem families. |
| Feminism | Women's experience is central and different from man's experience of social reality. | Inequality<br>Power<br>Oppression | Institution<br>Individual<br>Couple Family | Exposes inequality and oppression as explanations for frustrations women experience. | Multiple branches of feminism may inhibit central accomplishment of increased equality. |

# Research Process and Caveats

Research is valuable since it helps to provide evidence for or against a hypothesis. For example, there is a stigma associated with persons who have tattoos and it is often assumed that students who have tattoos make lower grades than those who do not have tattoos. But Martin and Dula (2010) compared the GPA of persons who had tattoos and those who did not and found no significant differences.

Researchers follow a standard sequence when conducting a research project and there are certain caveats to be aware of when reading any research finding.

## Steps in the Research Process

Several steps are used in conducting research.

*It is a capital mistake to theorize before one has data.*

ARTHUR CONAN DOYLE (VIA SHERLOCK HOLMES).

**1.** ***Identify the topic or focus of research.*** Select a subject about which you are passionate. For example, are you interested in studying cohabitation of college students? Give your project a title in the form of a question— "Do People Who Cohabit Before Marriage Have Happier Marriages Than Those Who Do Not Cohabit?"

**2.** ***Review the literature.*** Go online to the various databases of your college or university and read research that has already been published on cohabitation. Not only will this prevent you from "reinventing the wheel" (you might find that a research study has already been conducted on exactly what you want to study), but it will also give you ideas for your study.

**3.** ***Develop hypotheses.*** A **hypothesis** is a suggested explanation for a phenomenon. For example, you might hypothesize that cohabitation results in greater marital happiness and less divorce because the partners have a chance to "test-drive" each other and their relationship.

**Hypothesis** a suggested explanation for a phenomenon.

**Cross-sectional research** studying the whole population at once as compared to studying a segment of the population across time.

**4.** ***Decide on type of study and method of data collection.*** The type of study may be **cross-sectional**, which means studying the whole population at one time—in this case, finding out from persons now living together about their experience—

State of the art research is presented at professional conferences.

**Longitudinal** studying the same group across time.

or **longitudinal,** which means studying the same group across time—in this case, collecting data from the same couple each of their four years of living together during college. The method of data collection could be archival (secondary sources such as journals), survey (questionnaire), interview (one or both partners), or case study (focus on one couple). A basic differentiation in method is quantitative (surveys, archival) or qualitative (interviews/case study).

**5.** ***Get IRB approval.*** To ensure the protection of people who agree to be interviewed or who complete questionnaires, researchers must obtain IRB approval by submitting a summary of their proposed research to the Institutional Review Board (IRB) of their institution. The IRB reviews the research plan to ensure that the project is consistent with research ethics and poses no undue harm to participants. When collecting data from individuals, it is important that they are told that their participation is completely voluntary, that the study maintains their anonymity, and that the results are confidential. Respondents under age 18 need the consent of their parents. Community members were aghast when some parents of students at Memorial Middle School in Massachusetts reported that they never received consent forms for their seventh graders to take a survey on youth risk behavior that included items on oral sex and drug use (Starnes 2011).

**6.** ***Collect and analyze data.*** Various statistical packages are available to analyze data to discover if your hypotheses are true or false.

**7.** ***Write up and publish results.*** Writing up and submitting your findings for publication are important so that your study becomes part of the academic literature.

*Research is what I'm doing when I don't know what I'm doing.*

WERNHER VON BRAUN, GERMAN SCIENTIST

## Caveats to Consider in Evaluating Research Quality

"New Research Study" is a frequent headline in popular magazines such as *Cosmopolitan*, *Glamour*, and *Redbook* promising accurate information about "hooking up," "what women want," "what men want," or other relationship, marriage, and family issues. As you read such articles, as well as the research in texts such as these, be alert to their potential flaws. The various issues to keep in mind when evaluating research are identified in Table 1.8.

**TABLE 1.8 Potential Research Problems in Marriage and Family**

| Weakness | Consequences | Example |
|---|---|---|
| Sample not random | Cannot generalize findings | Opinions of college students do not reflect opinions of other adults. |
| No control group | Inaccurate conclusions | Study on the effect of divorce on children needs control group of children whose parents are still together. |
| Age differences between groups of respondents | Inaccurate conclusions | Effect may be due to passage of time or to cohort differences. |
| Unclear terminology | Inability to measure what is not clearly defined | What is definition of cohabitation, marital happiness, sexual fulfillment, good communication, quality time? |
| Researcher bias | Slanted conclusions | A researcher studying the value of a product (e.g., The Atkins Diet) should not be funded by the organization being studied. |
| Time lag | Outdated conclusions | Often-quoted Kinsey sex research is over 65 years old. |
| Distortion | Invalid conclusions | Research subjects exaggerate, omit information, and/or recall facts or events inaccurately. Respondents may remember what they wish had happened. |
| Deception | Public misled | Dr. Anil Potti (Duke University) changed data on research reports and provided fraudulent results. (Darnton 2012) |

## Future of Marriage

While there is a decline of marriage among middle Americans (defined as the nearly 60 percent of Americans aged 25 to 60 who have a high school but not a four-year college degree), marriage remains the dominant choice for most Americans, particularly for college-educated individuals with a good income (National Marriage Project 2012). Though these individuals will increasingly delay getting married until their late twenties/early thirties (to complete their educations, launch their careers, and/or become economically independent), there is no evidence that marriage will cease to be a life goal. Indeed six in ten never-married adults say they want to get married.

*Every saint has a past and every sinner has a future.*

OSCAR WILDE, IRISH WRITER

With the Supreme Court legalizing same sex marriage and federal tax codes permitting the filing of joint returns by same sex couples, an increasing number of states will legalize same sex marriage. Marriage education programs traditionally designed for heterosexual couples will then be modified to include same sex partners (Whitton and Buzzellla 2012).

*The future seems to me no unified dream but a mince pie, long in the baking, never quite done.*

E. B. WHITE, AMERICAN WRITER

## Summary

***What is the view/theme of this text?***

A central theme of this text is to encourage you to be proactive—to make conscious, deliberate relationship choices to enhance your own well-being and the well-being of those in your intimate groups. The most important choices are whether to marry, whom to marry, when to marry, whether to have children, whether to remain emotionally and sexually faithful to one's partner, and whether to use a condom. Important issues to keep in mind about a choices framework for viewing marriage and the family are that (1) not to decide is to decide, (2) some choices require correcting, (3) all choices involve trade-offs, (4) choices include selecting a positive or negative view, (5) making choices produce ambivalence, and (6) some choices are not revocable. For today's youth, they are in no hurry to find "the one," to marry, and to begin a family.

***What is marriage?***

Marriage is a system of binding a man and a woman together for the reproduction, care (physical and emotional), and socialization of offspring. The federal government regards marriage as a legal contract between a couple and the state in which they reside that regulates their economic and sexual relationship. Other elements of marriage involve emotion, sexual monogamy, and a formal ceremony. Types of marriage include monogamy and polygamy. Various forms of polygamy are polygyny, polyandry, and pantagamy.

***What is family?***

The U.S. Census Bureau defines family as a group of two or more people related by blood, marriage, or adoption. In recognition of the diversity of families, the definition of family is increasingly becoming two adult partners whose interdependent relationship is long term and characterized by an emotional and financial commitment. The family of origin is the family into which you were born or the family in which you were reared. The family of procreation represents the family that you will begin should you marry and have children. Central concepts of the family are traditional, modern, and post modern. Types of family include nuclear, binuclear, extended, and blended.

Changes in the family in the last 60-plus years include: divorce has replaced death as the endpoint for the majority of marriages; marriage and intimate relations have emerged as legitimate objects of scientific study; feminism and changes in gender roles in marriage have risen; the age at marriage has increased; some spouses may now live apart (LAT); and the acceptance of singlehood, cohabitation, and childfree marriages has increased.

***What are differences between marriage and the family?***

Differences between marriage and the family include that marriage involves a ceremony (the family does not have a ceremony which signals its beginning); the ages of the spouses are similar in marriage (in the family, the ages of parents and children are different); spouses choose each other in marriage (in the family, children do not choose their parents); marriage ends by death or divorce (the family continues even though a parent dies or the parents' divorce); sex between spouses is expected in marriage (in the family, parents and children are not expected to have sex); and marriage involves a license (in the family, there is no license to have children).

***What are the theoretical frameworks for viewing marriage and the family?***

Theoretical frameworks used to study the family include the (1) social exchange framework (spouses exchange resources, and decisions are made on the basis of perceived profit and loss), (2) structural-functional framework (how the family functions to serve society), (3) conflict framework (family members are in conflict over scarce resources of time and money), (4) symbolic interaction framework (symbolic worlds in which the various family members give meaning to each other's behavior), (5) family systems framework (each member of the family is part of a system and the family as a unit develops norms of interaction), (6) family life course development framework (the stages and process of how families change over time), (7) feminist framework (inequality and oppression). The framework used in most empirical family studies is the social exchange framework.

***What are steps in the research process and what caveats should be kept in mind?***

Steps in the research process include identifying a topic, reviewing the literature, deciding on methods and data collection procedures, ensuring protection of subjects via getting IRB (Institutional Review Board) approval, analyzing the data, and submitting the results to a journal for publication.

Caveats that are factors to be used in evaluating research include a random sample (the respondents providing the data reflect those who were not in the sample), a control group (the group not subjected to the experimental design for a basis of comparison), terminology (the phenomenon being studied should be objectively defined), researcher bias (present in all studies), time lag (takes two years from study to print), and distortion or deception (although rare, some researchers distort their data). Few studies avoid all research problems.

***What is the future of marriage?***

Marriage will continue to be the lifestyle of choice for the majority (85–90%) of U.S. adults. Though individuals will increasingly delay getting married until their late twenties to early thirties (to complete their educations, launch their careers, and/or become economically independent) and there will be an increase in those who never marry, there is no evidence that marriage will cease to be a life goal for most.

# Key Terms

Binuclear family
Blended family
Child marriage
Civil union
Collectivism
Common-law marriage
Conflict framework
Cross-sectional research
Domestic partnerships
Extended family
Familism
Family
Family life course development
Family life cycle
Family of orientation
Family of procreation
Family systems framework
Feminist framework
Feral children
Generation Y
Hanai adoption
Hypothesis
Individualism
Institution
Longitudinal
Marriage
Marriage benefit
Mating gradient
Modern family
Nuclear family
Pantagamy
Polyandry
Polygamy
Polygyny
Postmodern family
Primary groups
Secondary groups
Social exchange framework
Sociological imagination
Status
Structure-function framework
Symbolic interaction framework
Theoretical frameworks
Three-way marriage
Traditional Family
Utilitarianism

# Web Links

Brown Skin Baby (stolen children of Australia)
http://www.youtube.com/watch?v=v3ytJioxKzI

Family Process
http://www.trinity.edu/~mkearl/family.html

Gilder Lehrman Institute of American History—History of the Family
http://www.digitalhistory.uh.edu/historyonline/familyhistory.cfm

National Council on Family Relations
http://www.ncfr.org/

National Healthy Marriage Resource Center
http://twoofus.org/index.aspx

National Marriage Project University of Virginia
http://www.virginia.edu/marriageproject/
http://www.stateofourunions.org

U.S. Census Bureau
http://www.census.gov/

# CHAPTER 2 Love

Eric Ilg

> *I was about half in love with her by the time we sat down. That's the thing about girls. Every time they do something pretty... you fall half in love with them, and then you never know where the hell you are.*
>
> J. D. SALINGER, THE CATCHER IN THE RYE

## Learning Objectives

- Know three ways of viewing the concept of love.
- Review how the person you love is influenced by social factors.
- Identify and explain six theories of love.
- Understand the conditions under which a person falls in love.
- Describe how love is a context for problems.
- Learn the sources and consequences of jealousy.
- Review gender differences in jealousy.
- Discuss compersion and polyamory.

## TRUE OR FALSE?

1. Solo pornography use is associated with love relationship breakups.
2. Ludic love represents the most frequent type of love style among undergraduates.
3. When a new partner says "I love you" the most frequent response is "Me too."
4. Relationships in which the persons fell in love at first sight are happier than those in which the partners fell in love over time.
5. Love is like alcohol in that it creates a context in which the individual takes chances.

*Answers:* 1. T 2. F 3. F 4. F 5. T

---

*Leaving* is a riveting movie in which Suzanne (played by Kristin Scott Thomas) says to her physician husband "It hit me." She is speaking of her new obsessive love for the carpenter/ex-convict who works for her husband. Being together becomes her goal at all costs, and the costs are enormous . . . the carpenter/convict returns to prison because he is caught selling the ex-husband's paintings and Suzanne ends up shooting her husband since he demands her return to him as the price for her new lover's freedom from prison. The movie ends with the lovers on a hilltop sharing one last embrace with police sirens in the distance signaling an end to their life together: both will go to prison, she faces the death penalty, and they will be separated forever. The viewer is struck with how love influences one's choices. Indeed, the lovers are out of control.

While being captured by the spell of love had disastrous consequences for this couple, love is also associated with positive outcomes (e.g., greater individual and relationship happiness, fewer mental health problems) [Demir 2008; Graham 2011; Braithwaithe et al. 2010]. This chapter reviews not only the consequences of love but also its conceptions, styles, and theories. Along the way we ask the basic question whether you can choose the person you fall in love with. . . . or do social factors dictate your love choice? We begin by reviewing basic ways of conceptualizing love.

### National **Data**

Ninety-three percent of 1,306 spouses and 83% of 1,385 unmarried individuals said that "love" was an important reason for getting married (Cohn 2013).

# Ways of Conceptualizing Love

Love is very much a part of an individual's life. Eighty-five percent of 5,500 single individuals (no gender difference) reported that they had been in love (Walsh 2013). A common class exercise among professors who teach marriage and the family is to randomly ask class members to identify one word they most closely associate with love. Invariably, students identify different words (e.g., commitment, feeling, trust, altruism), suggesting great variability in the way we think about love.

The word love is polysemous—the meanings are numerous (Berscheid 2010). Love is elusive and incapable of being defined by those caught in its spell. Hegi and Bergner (2010) provide a standard definition—*"Investment in the well-being of the other for his or her own sake."* Hatfield et al. (2012) noted that passionate love (similar to what most tend to think of as romantic love) is "*A state of intense longing for the union with another.*" Passionate love is not unique to Western society but is a universal emotion/motivation (p. 154).

Love is often confused with lust and infatuation. Love (and attachment) is about deep, abiding feelings with a focus on the long term (Langeslag et al. 2013; Foster 2010); **lust** is about sexual desire and the present. The word **infatuation** comes from the same root word as *fatuous*, meaning "silly" or "foolish," and refers to a state of passion or attraction that is not based on reason. Infatuation is characterized by euphoria (Langeslag et al. 2013) and by the tendency to idealize the love partner. People who are infatuated magnify their lovers' positive qualities ("My partner is always happy") and overlook or minimize their negative qualities ("My partner doesn't have a problem with alcohol; he just likes to have a good time").

**Lust** sexual desire and the present.

**Infatuation** a state of passion or attraction that is not based on reason.

In the following section, we look at the various ways of conceptualizing love.

## Romantic versus Realistic Love

*There are two sorts of romantics: those who love, and those who love the adventure of loving.*

LESLEY BLANCH, ENGLISH WRITER

Love may also be described as being on a continuum from romanticism to realism (see self-assessment scale on Love Attitudes). For some people, love is romantic; for others, it is realistic. **Romantic love**, said to have appeared in all human groups at all times in human history (Berscheid 2010), is characterized in modern America by such beliefs as "love at first sight," "one true love," and "love conquers all."

**Romantic love** characterized in modern America by such beliefs as "love at first sight," "one true love," and "love conquers all".

Regarding love at first sight, 36% of 1,094 undergraduate males and 25% of 3,448 undergraduate females reported that they had experienced love at first sight (Hall and Knox 2013). One explanation for men falling in love more quickly than women is that (from a biological/evolutionary perspective) men must be visually attracted to young, healthy females to inseminate them. This biologically based reproductive attraction is interpreted as a love attraction so that the male feels immediately drawn to the female, but what he may actually see is an egg needing fertilization. Dotson-Blake et al. (2008) found further evidence that males are more romantic than females in that men were significantly more likely (85% versus 73%) than women to believe that they could solve any relationship problem as long as they were in love. However, Chatizow (2011) studied love attitudes of intimate partners and found that women were more romantic and men more rational.

*Love looks not with the eyes, but with the mind;*
*And therefore is wing'd cupid painted blind.*

SHAKESPEARE, *A MIDSUMMER NIGHT'S DREAM*

In regard to love at first sight, Barelds and Barelds-Dijkstra (2007) studied the relationships of 137 married couples or cohabitants (together for an average of twenty-five years) and found that the relationship quality of those who fell in love at first sight was similar to that of those who came to know each other more gradually. Huston et al. (2001) found that, after two years of marriage, the couples who had fallen in love more slowly were just as happy as couples who fell in love at first sight.

*Each moment of the happy lover's hour is worth an age of dull and common life.*

APHRA BEHN, ENGLISH WRITER

An openness to falling in love and developing an intimate relationship is Erik Erikson's sixth stage of psychosocial development. He noted that between the ages of 19 and 40, most individuals move from "isolation to intimacy" wherein they establish committed loving relationships. Failure to do so is to leave one vulnerable to loneliness and depression.

Berscheid (2010) also noted that the intensity of romantic love decreases across time in a relationship. It is not unusual for lovers to vary in the intensity of their love for each other. The interesting question is whether being the person who loves more in a relationship is better than being the person who loves less. The person who loves more may suffer more anguish. Some regard the anguish as worth the price as captured in the quote, "Tis better to have loved and lost, than never to have loved at all" (Alfred Lord Tennyson).

### DIVERSITY IN OTHER COUNTRIES

The theme of American culture is individualism, which translates into personal fulfillment, emotional intimacy, and love as the reason for marriage. In Asian cultures (e.g., China) the theme is collectivism, which focuses on "family, comradeship, obligations to others, and altruism" with love as secondary (Riela et al. 2010). While arranged marriages are and have been the norm in Eastern societies, love marriages are becoming more frequent (Allendorf 2013).

Bob Bradley

This couple, married for over 25 years, enjoy the conjugal love of sharing experiences together (in this photo they are in Antarctica).

In contrast to romantic love is realistic love. Realistic love is also known as conjugal love. **Conjugal love** is the love between married people characterized by companionship, calmness, comfort, and security. Conjugal love is in contrast to romantic love, which is characterized by excitement and passion. Stanik et al. (2013) interviewed 146 African-American couples who had been married from three to 25 years and confirmed a decrease in the intensity of love feelings across time. However, couples with a traditional division of labor (compared to egalitarian couples) showed a greater decline. The Love Attitudes Scale provides a way for you to assess the degree to which you tend to be romantic or realistic (conjugal) in your view of love. When you determine your score from the Love Attitudes Scale, be aware that your tendency to be a romantic or a realist is neither good nor bad. Both romantics and realists can be happy individuals and successful relationship partners. Love also conveys enormous benefits, including positive mental health. Plant et al. (2010) emphasized that these benefits are so pronounced for the individual that he or she will protect these benefits by diverting themselves when they feel attracted to an alternative.

**Conjugal love** love between married people characterized by companionship, calmness, comfort, and security.

*A lady of forty-seven who had been married twenty-seven years and has six children knows what love really is and once described it for me like this: "Love is what you've been through with somebody."*

JAMES THURBER, WRITER

## SELF-ASSESSMENT | The Love Attitudes Scale

This scale is designed to assess the degree to which you are romantic or realistic in your attitudes toward love. There are no right or wrong answers.

| 1 | 2 | 3 | 4 | 5 |
|---|---|---|---|---|
| Strongly agree | Mildly agree | Undecided | Mildly disagree | Strongly disagree |

**Directions**

After reading each sentence carefully, write the number that best represents the degree to which you agree or disagree with the sentence.

_____ 1. Love doesn't make sense. It just is.

_____ 2. When you fall "head over heels" in love, it's sure to be the real thing.

_____ 3. To be in love with someone you would like to marry but can't is a tragedy.

*(continued)*

## SELF-ASSESSMENT (continued)

_____ 4. When love hits, you know it.
_____ 5. Common interests are really unimportant; as long as each of you is truly in love, you will adjust.
_____ 6. It doesn't matter if you marry after you have known your partner for only a short time as long as you know you are in love.
_____ 7. If you are going to love a person, you will "know" after a short time.
_____ 8. As long as two people love each other, the educational differences they have really do not matter.
_____ 9. You can love someone even though you do not like any of that person's friends.
_____ 10. When you are in love, you are usually in a daze.
_____ 11. Love "at first sight" is often the deepest and most enduring type of love.
_____ 12. When you are in love, it really does not matter what your partner does because you will love him or her anyway.
_____ 13. As long as you really love a person, you will be able to solve the problems you have with the person.
_____ 14. Usually you can really love and be happy with only one or two people in the world.
_____ 15. Regardless of other factors, if you truly love another person, that is a good enough reason to marry that person.
_____ 16. It is necessary to be in love with the one you marry to be happy.
_____ 17. Love is more of a feeling than a relationship.
_____ 18. People should not get married unless they are in love.
_____ 19. Most people truly love only once during their lives.
_____ 20. Somewhere there is an ideal mate for most people.
_____ 21. In most cases, you will "know it" when you meet the right partner.
_____ 22. Jealousy usually varies directly with love; that is, the more you are in love, the greater your tendency to become jealous will be.
_____ 23. When you are in love, you are motivated by what you feel rather than by what you think.
_____ 24. Love is best described as an exciting rather than a calm thing.
_____ 25. Most divorces probably result from falling out of love rather than failing to adjust.
_____ 26. When you are in love, your judgment is usually not too clear.
_____ 27. Love comes only once in a lifetime.
_____ 28. Love is often a violent and uncontrollable emotion.
_____ 29. When selecting a marriage partner, differences in social class and religion are of small importance compared with love.
_____ 30. No matter what anyone says, love cannot be understood.

### Scoring

Add the numbers you wrote to the left of each item. 1 (strongly agree) is the most romantic response and 5 (strongly disagree) is the most realistic response. The lower your total score (30 is the lowest possible score), the more romantic your attitudes toward love. The higher your total score (150 is the highest possible score), the more realistic your attitudes toward love. A score of 90 places you at the midpoint between being an extreme romantic and an extreme realist. Of 45 undergraduate males, 85.3 was the average score; of 193 undergraduate females, 85.5 was the average score. These scores reflect a slight lean toward romanticism with no gender difference.

### Source

Knox, D. "Conceptions of Love at Three Developmental Levels." Dissertation, Florida State University. Permission to use the scale for research available from David Knox at davidknox2@yahoo.com or by contacting Dr. Knox, Department of Sociology, East Carolina University, Greenville, NC 27858.

In addition to the romantic/realistic continuum of viewing love, Sleeth (2013) conceptualized another continuum "involving a dialectical sequence of development—**autistic love** (all about me), **empathetic love** (all about you), and **integral love** (all about us)—the last combining the prior two" (p. 5).

**Autistic love** all about me.

**Emphatic love** all about you.

**Integral love** all about us.

## Love Styles

Theorist John Lee (1973, 1988) identified a number of styles of love that describe the way lovers relate to each other. Keep in mind that individuals may view love in different ways at different times. These love styles are also independent of one's sexual orientation—no one love style is characteristic of heterosexuals or homosexuals.

**Ludic love style** views love as a game in which the player has no intention of getting seriously involved.

**1.** ***Ludic.*** The **ludic love style** views love as a game in which the player has no intention of getting seriously involved. The ludic lover refuses to become dependent on any one person and does not encourage another's intimacy. Two essential skills of the ludic lover are to juggle several partners at the same time and to manage each relationship so that no one partner is seen too often.

These strategies help to ensure that the relationship does not deepen into an all-consuming love. Don Juan represented the classic ludic lover, embodying the motto of "Love 'em and leave 'em." Tzeng et al. (2003) found that whereas men were more likely than women to be ludic lovers, ludic love characterized the love style of college students the least.

*Love is just a system for getting someone to call you darling after sex.*

JULIAN BARNES, WRITER

While ludic lovers may sometimes be characterized as manipulative and uncaring (Jonason and Kavanagh 2010), they may also be compassionate and very protective of another's feelings. For example, some uninvolved, soon-to-graduate seniors avoid involvement with anyone new and become ludic lovers so as not to encourage anyone whom they would soon leave to fall in love with them.

**2. *Pragma.*** The **pragma love style** is the love of the pragmatic—that which is logical and rational. Pragma lovers assess their partners on the basis of assets and liabilities. One undergraduate female hung out with a guy because he had a car and could drive her home on weekends to see her boyfriend. An undergraduate male dated his partner because she would write his term papers and do his laundry.

**Pragma love style** the love of the pragmatic—that which is logical and rational.

Pragma lovers do not become involved in interracial, long-distance, or age-discrepant relationships because logic argues against doing so. See the Personal Choices about making decisions with one's heart or head.

*One ought to hold on to one's heart; for if one lets it go, one soon loses control of the head too.*

FRIEDRICH NIETZSCHE, PHILOSOPHER

## PERSONAL CHOICES

### *Do You Make Relationship Choices with Your Heart or Head?*

Lovers are frequently confronted with the need to make decisions about their relationships, but they are divided on whether to let their heart or their head rule in such decisions. Some evidence suggests that the heart rules. Almost 60% (57%) of 4,561 undergraduates agreed with the statement, "I make relationship decisions more with my heart than my head" (Hall and Knox 2013), suggesting that the heart tends to rule in relationship matters. We asked students in our classes to fill in the details about deciding with their heart or their head. Some of their answers follow:

### *Heart*

Those who relied on their hearts for making decisions (women more than men) felt that emotions were more important than logic and that listening to their heart made them happier. One woman said:

> *In deciding on a mate, my heart would rule because my heart has reasons to cry and my head doesn't. My heart knows what I want, what would make me most happy. My head tells me what is best for me. But I would rather have something that makes me happy than something that is good for me.*

Some men also agreed that the heart should rule. One said:

> *I went with my heart in a situation, and I'm glad I did. I had been hanging out with a girl for two years when I decided she was not the one I wanted and that my present girlfriend was. My heart was saying to go for the one I loved, but my head was telling me not to because if I broke up with the first girl, it would hurt her, her parents, and my parents. But I decided I had to make myself happy and went with the feelings in my heart and started dating the girl who is now my fiancée.*

Relying on one's emotions does not always have a positive outcome, as the following experience illustrates:

*Last semester, I was dating a guy I felt more for than he did for me. Despite that, I wanted to spend any opportunity I could with him when he asked me to go somewhere with him. One day he had no classes, and he asked me to go to the park by the river for a picnic. I had four classes that day and exams in two of them. I let my heart rule and went with him. He ended up breaking up with me on the picnic.*

## Head

Some undergraduates make relationship choices based on their head as some of the following comments show:

*In deciding on a mate, I feel my head should rule because you have to choose someone that you can get along with after the new wears off. If you follow your heart solely, you may not look deep enough into a person to see what it is that you really like. Is it just a pretty face or a nice body? Or is it deeper than that, such as common interests and values? The "heart" sometimes can fog up this picture of the true person and distort reality into a fairy tale.*

Another student said:

*Love is blind and can play tricks on you. Two years ago, I fell in love with a man whom I later found out was married. Although my heart had learned to love this man, my mind knew the consequences and told me to stop seeing him. My heart said, "Maybe he'll leave her for me," but my mind said, "If he cheated on her, he'll cheat on you." I broke up with him and am glad that I listened to my head.*

Some individuals feel that both the head and the heart should rule when making relationship decisions.

*When you really love someone, your heart rules in most of the situations. But if you don't keep your head in some matters, then you risk losing the love that you feel in your heart. I think that we should find a way to let our heads and hearts work together.*

There is an adage, "Don't wait until you find the person you can live with; wait and find the person that you can't live without!" In your own decisions you might consider the relative merits of listening to your heart or head and moving forward recognizing there is not one "right" answer for all individuals on all issues.

**Eros love style** love imbued with passion and sexual desire.

*Meeting you was fate, becoming your friend was a choice, but falling in love with you I had no control over.*

ANONYMOUS

**Mania love style** out-of-control love where the person "must have" the love object.

**Storge love style** calm, soothing, nonsexual love devoid of intense passion.

**3. *Eros.*** Just the opposite of the pragmatic love style, the **eros love style** (also known as romantic love) is imbued with passion and sexual desire. Eros is the most common love style of college women and men (Tzeng et al. 2003). The idea of a soul mate is part of American culture. In a study of over 1,000 adults, 66% reported "belief in the idea of a soul mate"; 34% disagreed (Carey and Gellers 2010).

**4. *Mania.*** The **mania love style** is the out-of-control love whereby the person "must have" the love object. Jealousy, possessiveness, dependency, and controlling are symptoms of manic love. Zamora et al. (2013) identified the love styles of 72 gay men who revealed that a mania love style was the most frequent. These gay men were anxious for reassurance (mania) from their partner.

**5. *Storge.*** The **storge love style**, also known as companionate love, is a calm, soothing, nonsexual love devoid of intense passion. Respect, friendship, commitment, and familiarity are characteristics that help to define the storge love relationship. The partners care deeply about each other but not in a romantic or lustful sense. Their love is also more likely to endure than fleeting romance. One's grandparents who have been married fifty years are likely to have a storge type of love. Neto (2012) compared love perceptions by age group and found that the older the individual, the more important love became and the less important sex became.

**6.** ***Agape.*** **Agape love style** (also known as compassionate love) is characterized by a focus on the well-being of the person who is loved, with little regard for reciprocation. Key qualities of agape love are not responding to a partner's negativity and not expecting an exchange for positives but believing that the other means well and will respond kindly in time. Berscheid (2010) suggested that compassionate love may be the most enduring of all loves since it does not depend on immediate reciprocation. The love parents have for their children is often described as agape love.

*Love moderately; long love doth so*
*Too swift arrives as tardy as too slow.*

SHAKESPEARE, *ROMEO AND JULIET*, II, IV, 14

**Agape love style** characterized by a focus on the well-being of the person who is loved, with little regard for reciprocation.

## Triangular View of Love

Sternberg (1986) developed the "triangular" view of love, which consists of three basic elements: intimacy, passion, and commitment. The presence or absence of these three elements creates various types of love experienced between individuals, regardless of their sexual orientation. These various types include the following:

**1.** ***Nonlove***—the absence of intimacy, passion, and commitment. Two strangers looking at each other from afar are experiencing nonlove.

**2.** ***Liking***—intimacy without passion or commitment. A new friendship may be described in these terms of the partners liking each other.

**3.** ***Infatuation***—passion without intimacy or commitment. Two people flirting with each other in a bar may be infatuated with each other.

**4.** ***Romantic love***—intimacy and passion without commitment. Love at first sight reflects this type of love.

**5.** ***Conjugal love*** (also known as married love)—intimacy and commitment without passion. A couple married for fifty years are said to illustrate conjugal love.

**6.** ***Fatuous love***—passion and commitment without intimacy. Couples who are passionately wild about each other and talk of the future but do not have an intimate connection with each other have a fatuous love.

**7.** ***Empty love***—commitment without passion or intimacy. A couple who stay together for social (e.g., children) and legal reasons but who have no spark or emotional sharing between them have an empty love.

**8.** ***Consummate love***—combination of intimacy, passion, and commitment; Sternberg's view of the ultimate, all-consuming love.

Individuals bring different combinations of the elements of intimacy, passion, and commitment (the triangle) to the table of love. One lover may bring a predominance of passion, with some intimacy but no commitment (romantic love), whereas the other person brings commitment but no passion or intimacy (empty love). The triangular theory of love allows lovers to see the degree to which they are matched in terms of passion, intimacy, and commitment in their relationship (see Figure 2.1)

*The rules for marriage are the same as those for lifeboat passengers: stay in your place, no sudden moves, and keep all disastrous thoughts to yourself.*

GARRISON KEILLOR, AUTHOR

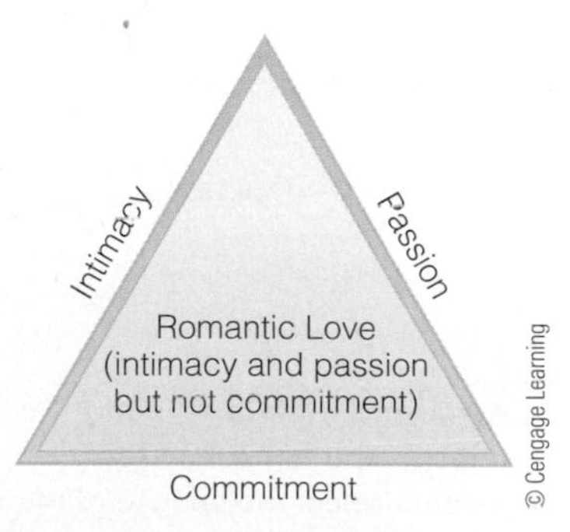



**Figure 2.1** Romantic love is characterized by eros—intimacy and passion without commitment—and is the type of love most prevalent among college students.

## Love Languages

Gary Chapman's (2010) **five love languages** have become part of American love culture. These five languages are gifts, quality time, words of affirmation, acts of service, and physical touch. Chapman encourages individuals to use the language of love most desired by the partner rather than the one preferred by the individual providing the love. Chapman has published nine books (e.g., for couples, singles, men, military) to illustrate the application

## FAMILY POLICY | Love in the Workplace

With women being almost half (47%) of the American workforce, individuals delaying marriage until they are older (late 20s), and workers being around each other for eight or more hours a day, the workplace has become a predictable context for romantic relationships to develop. More future spouses may meet at work than in academic, social, or religious settings. Barack and Michelle Obama, Bill and Melinda Gates, and Brad Pitt and Angelina Jolie met "on the job." Fifty-seven percent of a national survey of workers revealed that they had become involved in a romance with someone at work (Vault Careers 2013).

Although such relationships are most often between peers, sometimes a love relationship develops between individuals occupying different status positions. And it can get ugly. Such was the case of Harry Stonecipher (a 68-year-old married man and head of Boeing) and a 48-year-old divorced employee, which resulted in Stonecipher being fired. His dismissal was not because of the affair (there were no company rules) but because of the negative publicity he brought to Boeing when his steamy e-mails became public. These types of love relationships are sometimes problematic in the workplace. Four-star general David Petraeus, presidential candidate John Edwards, and talk show host David Letterman also had affairs with women they met at work. Only Letterman survived and maintained his professional position.

### Advantages of an Office Romance

The energy that both fuels and results from intense love feelings can also fuel productivity at work. If the coworkers eventually marry or enter a nonmarital but committed, long-term relationship, they may be more satisfied with and committed to their jobs than spouses whose partners work elsewhere. Working at the same location enables married couples to commute together, go to company-sponsored events together, and talk shop together. Workplaces such as academia often try to hire both spouses since they are likely to become more permanent employees.

Recognizing the potential benefits of increased job satisfaction, morale, productivity, creativity, and commitment, some companies encourage love relationships among employees. Aware that their single employees are interested in relationships, in Tokyo, Japan, Hitachi Insurance Service provides a dating service for its 400,000 employees (many of whom are unmarried) called Tie the Knot. Those interested in finding a partner complete an application and a meeting or lunch is arranged with a suitable candidate through the Wedding Commander. In America, some companies hire two employees who are married, reflecting a focus on the value of each employee to the firm rather than on their love relationship outside work.

### Disadvantages of an Office Romance

However, workplace romances can also be problematic for the individuals involved as well as for their employers. When a workplace romance involves a supervisor/subordinate relationship, other employees might make claims of favoritism or differential treatment. In a typical differential-treatment allegation,

of "five love languages" in different contexts. No empirical studies have been conducted on the five languages.

# Social Control of Love

**Arranged marriage** mate selection pattern whereby parents select the spouse of their offspring.

The ultimate social control of love is **arranged marriage**—mate selection pattern whereby parents select the spouse of their offspring. Parents arrange 80% of marriages in China, India, and Indonesia (three countries representing 40% of the world's population). In most Eastern countries marriage is regarded as the linking of two families; the love feelings of the respective partners are irrelevant. Love is expected to follow marriage, not precede it. Arranged marriages not only help to guarantee that cultural traditions will be carried on and passed to the new generation, but they also link two family systems together for mutual support of the couple.

an employee (usually a woman) claims that the company denied her a job benefit because her supervisor favored another female coworker—who happens to be the supervisor's girlfriend.

If a workplace relationship breaks up, it may be difficult to continue to work in the same environment (and others at work may experience the fallout). A breakup that is less than amicable may result in efforts by partners to sabotage each other's work relationships and performance, incidents of workplace violence, harassment, and/or allegations of sexual harassment. In a survey of 774 respondents who had experience in the workplace, over a third (36.3%) recommended avoiding involvement in an office romance (Merrill and Knox 2010).

### Company Policies on Office Romances

Some companies such as Walt Disney, Universal, and Columbia have "anti fraternization" clauses that impose rules on workers talking about private issues or sending personal e-mails. Some British firms have "love contracts" that require workers to tell their managers if they are involved with anyone in the office.

Most companies (Walmart is an example) do not prohibit romantic relationships among employees. However, the company may have a policy prohibiting open displays of affection between employees in the workplace and romantic relationships between a supervisor and a subordinate. Most companies have no policy regarding love relationships at work and generally regard romances between coworkers as "none of their business." There are some exceptions to the general permissive policy regarding workplace romances. Many companies have written policies prohibiting intimate relationships when one member of the couple is in a direct supervisory position over the other. These policies may be enforced by transferring or dismissing employees who are discovered in romantic relationships.

### Your Opinion?

1. To what degree do you believe corporations should develop policies in regard to workplace romances?
2. What are the advantages and disadvantages of a workplace romance for a business?
3. What are the advantages and disadvantages for individuals involved in a workplace romance?
4. How might an office romance of peers affect coworkers?

### Sources

Merrill, J., and D. Knox. 2010. *Finding Love from 9 to 5: Trade Secrets of Office Romance.* Santa Barbara, CA: Praeger.

Vault Careers, 2013. Vault.Com Office Romance Survey 2013. http://www.vault.com/blog/workplace-issues/the-results-are-in-2013-office-romance-survey/ Retrieved January 18, 2014.

Parents may know a family who has a son or daughter whom they would regard as a suitable partner for their offspring. If not, they may put an advertisement in newspaper identifying the qualities they are seeking. The prospective mate is then interviewed by the family to confirm his or her suitability. Or a third person—a matchmaker—may be hired to do the screening and introducing.

Selecting a spouse for a daughter may begin early in the child's life. In some countries (e.g., Nepal and Afghanistan), child marriage occurs whereby young females (ages 8 to 12) are required to marry an older man selected by their parents. Suicide is the only alternative "choice" for these children.

#### DIVERSITY IN OTHER COUNTRIES

Zaman (2014) identified another form of arranged marriage in the rural community of Kabirwala in South Punjab in Pakistan:

> *The exchange may include two or more families, which are often blood related. One family gives a daughter to be married to the son of the second family, and in response to this, the second family will give their daughter to be married to the son of the first family. If a family is unable to reciprocate a woman immediately to the other family for the son's marriage, the family promises to return a woman to the wife-giver family in the next generation (p. 69).*

While parents in Eastern societies exercise direct control by selecting the partner for their son or daughter, American parents influence mate choice by moving to certain neighborhoods, joining certain churches, and enrolling their children in certain schools. Doing so increases the chance that their offspring will "hang out" with, fall in love with, and marry people who are similar in race, religion, education, and social class. Parents want their offspring to meet someone who will "fit in" and with whom they will feel comfortable.

Peers also exert an influence on homogenous mating by approving or disapproving of certain partners. Their motive is similar to that of parents—they want to feel comfortable around the people their peers bring with them to social encounters. Both parents and peers are influential, as most offspring and friends end up falling in love with and marrying people of the same race, education, and social class. Social approval of one's partner is normally important for a love relationship to proceed. If your parents and peers disapprove of the person you are involved with, it is difficult for you to continue the relationship.

Diamond (2003) emphasized that individuals are biologically wired and capable of falling in love and establishing intense emotional bonds with members of their own or opposite sex. Discovering that one's offspring is in love with and wants to marry someone of the same sex is a challenge for many parents.

*How delicious is the winning of a kiss at love's beginning.*

THOMAS CAMPBELL, ENGLISH POLITICIAN

The social control of love may also occur in the workplace (see the Family Policy section).

## Theories on the Origins of Love

*To the world you may be one person, but to one person you may be the world.*

HEATHER CORTEZ, POET

Various theories have been suggested with regard to the origins of love.

### Evolutionary Theory

Gillath et al. (2008) provided evidence that sexual interest and arousal are associated with motives to form and maintain a close relationship, to fall in love. They suggested these motives are hardwired to ensure a stable relationship for producing offspring. Although these motives are subject to distraction of new sexual opportunities, they nevertheless suggest a broader relationship motivation to have sex with a love object. In effect, love has an evolutionary purpose by providing a bonding mechanism between the parents during the time their offspring are dependent infants.

David Knox

Two Siamang gibbons sitting on a log; these apes mate for life. While not typical among lower animals, there is evolutionary precedence for stable mating relationships.

Love's strongest bonding lasts about four years after birth, the time when children are most dependent and when two parents are most beneficial to the developing infant. "If a woman was carrying the equivalent of a twelve-pound bowling ball in one arm and a pile of sticks in the other, it was ecologically critical to pair up with a mate to rear the young," observed anthropologist Helen Fisher (Toufexis 1993). The "four-year itch" is Fisher's term for the time at which parents with one child are most likely to divorce—the time when the woman can more easily survive without parenting help from the male. If the couple has a second child, doing so resets the clock, and "the seven-year itch" is the next most vulnerable time.

Love may also have a physiological basis. Dopamine is a chemical in the brain that is associated with the development of love. The prairie vole is a monogamous rodent. If a drug that reduces the effect of dopamine is injected into the brain of a monogamous vole, it loses the preference to "commit" to one mate. If the animal is then placed with a new mate and given a drug that increases dopamine levels, it develops the equivalent of romantic feelings for the new partner (Fisher et al. 2006).

## Learning Theory

Unlike evolutionary theory, which views the experience of love as innate, learning theory emphasizes that love feelings develop in response to certain behaviors engaged in by the partner. Individuals in a new relationship who look at each other, smile at each other, compliment each other, touch each other endearingly, do things for each other, and do enjoyable things together are engaging in behaviors that encourage the development of love feelings. In effect, love can be viewed as a feeling that results from a high frequency of positive behavior and a low frequency of negative behavior. One high-frequency behavior is positive labeling whereby the partners flood each other with positive statements. We asked one of our students who reported that she was deliriously in love to identify the positive statements her partner had said to her. She kept a list of these statements for a week, which included:

*Angel, Sweetie, Cinderella, Sleeping Beauty.*
*You understand me so well.*
*Precious jewel, Sugar bear, Snow White, Honey.*
*You are my best friend.*
*I never thought you were out there.*
*You know me better than anyone in my whole life.*
*You're always safe in my arms.*
*I would sell my guitar for you if you needed money.*

People who "fall out of love" may note the high frequency of negatives on the part of their partner and the low frequency of positives. People who say, "this is not the person I married," are saying the ratio of positives to negatives has changed dramatically.

## Sociological Theory

Sixty years ago, Ira Reiss (1960) suggested the wheel model as an explanation for how love develops. Basically, the wheel has four stages—rapport, self-revelation, mutual dependency, and fulfillment of personality needs. In the rapport stage, each partner has the feeling of having known the partner before, feels comfortable with the partner, and wants to deepen the relationship.

Such desire leads to self-revelation or self-disclosure, whereby each reveals intimate thoughts to the other about oneself, the partner, and the relationship. Such revelations deepen the relationship because it is assumed that the confidences are shared only with special people, and each partner feels special when listening to the revelations of the other. Indeed, Shelon et al. (2010) confirmed that intimacy develops when individuals disclose personal information about themselves and perceive that the listener understands/validates and cares about the disclosure. As the level of self-disclosure becomes more intimate, a feeling of mutual dependency develops. Each partner is happiest in the presence of the other and begins to depend on the other for creating the context of these euphoric feelings. "I am happiest when I am with you" is the theme of this stage.

The feeling of mutual dependency involves the fulfillment of personality needs. The desires to love and be loved, to trust and to be trusted, and to support and be supported are met in the developing love relationship.

## Psychosexual Theory

According to psychosexual theory, love results from blocked biological sexual desires. In the sexually repressive mood of his time, Sigmund Freud (1905–1938) referred to love as "aim-inhibited sex." Love was viewed as a function of the sexual desire a person was not allowed to express because of social restraints. In Freud's era, people would meet, fall in love, get married, and have sex. Freud felt that the socially required delay from first meeting to having sex resulted in the development of "love feelings." By extrapolation, Freud's theory of love suggests that love dies with marriage (access to one's sexual partner).

## Biochemical Theory

"Love is deeply biological" wrote Carter and Porges (2013), who reviewed the biochemistry involved in the development and maintenance of love. They noted that oxytocin and vasopressin are hormones involved in the development and maintenance of social bonding. The hormones are active in forging emotional connections between adults and infants and between adults. They are also necessary for our social and physiological survival.

**Oxytocin** released from the pituitary gland during the expulsive stage of labor.

**Oxytocin** is released from the pituitary gland during the expulsive stage of labor that has been associated with the onset of maternal behavior in lower animals (but oxytocin may be manufactured in both women and men when an infant or another person is present—hence it is not dependent on the birth process) (Carter and Porges 2013). Oxytocin has been referred to as the "cuddle chemical" because of its significance in bonding. Later in life, oxytocin seems operative in the development of love feelings between lovers during sexual arousal. Oxytocin may be responsible for the fact that more women than men prefer to continue cuddling after intercourse.

Phenylethylamine (PEA) is a natural, amphetamine-like substance that makes lovers feel euphoric and energized. The high that they report feeling just by being with each other is from PEA that the brain releases into their bloodstream. The natural chemical high associated with love may explain why the intensity of passionate love decreases over time. As with any amphetamine, the body builds up a tolerance to PEA, and it takes more and more to produce the special kick. Hence, lovers develop a tolerance for each other. "Love junkies" are those who go from one love affair to the next to maintain the high. Alternatively, some lovers break up and get back together frequently as a way of making the relationship new again and keeping the high going.

Zeki (2007) emphasized the neurobiology of love in that both romantic love and maternal love are linked to the perpetuation of the species. Romantic love bonds the male and female together to reproduce, take care of, and socialize new societal members, whereas maternal love ensures that the mother will prioritize the care of her baby over other needs. Because of the social functions of these love states, neurobiologists have learned via brain imaging techniques that both types of attachment activate regions of the brain that access the brain's reward system (areas rich in oxytocin and vasopressin receptors). At the same time, negative cognitions or emotions about these relationships are shut down to allow the positive to predominate. No wonder both lovers and mothers seem very happy and focused. They are on a biological mission, and the reward center of their brain keeps them on track.

Meyer (2007) noted that taking selective serotonin reuptake inhibitor (SSRI) medications commonly used for depression and anxiety can negatively affect relationship satisfaction (for example, blunt emotions and decrease sexual interest); Meyer also emphasized the importance of checking with one's physician. In some cases, other medications can be used to block these negative side effects.

## Attachment Theory

The attachment theory of love emphasizes that a primary motivation in life is to be emotionally connected with other people. Children abandoned by their parents and placed in foster care (400,000 are in foster care in the United States) are vulnerable to having their early emotional attachment to their parents disrupted and developing "reactive attachment disorder" (Stinehart et al. 2012). This disorder involves a child who is anxious and insecure since he or she does not feel to be in a safe and protected environment. Such children find it difficult to connect emotionally, ever. Conversely, a secure emotional attachment with loving adults as a child is associated with the capacity for later involvement in satisfying, loving, communicative adult relationships. Mohr et al. (2010) used attachment theory to note that undergraduates who measured high on anxiety (presumably from lack of secure attachments in their family) predicted challenges in long-term romantic relationships. Men and singles were particularly more likely to predict difficulties over time with a romantic partner. Similarly, Shoemann et al. (2012) found that attachment anxiety was related to regret in relationship choices.

Attachment theory has its basis in the work of Rene Spitz and Harry Harlow. The former emphasized the importance of infants being held and nurtured for their physical and emotional development. Dr. Harlow studied infant rhesus monkeys and found that they preferred soft mother-like dummies that offered no food over dummies that provided a food source but were made of wire and were less pleasant to the touch.

Each of the theories of love presented in this section has critics (see Table 2.1).

**TABLE 2.1 Love Theories and Criticisms**

| Theory | Criticism |
|---|---|
| **Evolutionary**—love is the social glue that bonds parents with dependent children and spouses with each other to care for offspring. | The assumption that women and children need men for economic/emotional survival is not true today. Women can have and rear children without male partners. |
| **Learning**—positive experiences create love feelings. | The theory does not account for (1) why some people share positive experiences but do not fall in love, and (2) why some people stay in love despite negative behavior by their partner. |
| **Psychosexual**—love results from blocked biological drive. | The theory does not account for love couples who report intense love feelings and have sex regularly. |
| **Sociological**—the wheel theory whereby love develops from rapport, self-revelation, mutual dependency, and personality need fulfillment. | Not all people are capable of rapport, revealing oneself, and so on. |
| **Biochemical**—love is chemical. Oxytocin is an amphetamine-like chemical that bonds mother to child and produces a giddy high in young lovers. | The theory does not specify how much of what chemicals result in the feeling of love. Chemicals alone cannot create the state of love; cognitions are also important. |
| **Attachment**—primary motivation in life is to be connected to others. Children bond with parents and spouses to each other. | Not all people feel the need to be emotionally attached to others. Some prefer to be detached. |

## APPLYING SOCIAL RESEARCH | Using Surrogates to Learn Relationship Basics

**Client and Problem**

Not everyone has the basic skills of being able to talk with, touch, and kiss a potential love partner. One example is a 29-year-old virgin male who sought therapy to enable him to learn the basic skills of talking with, touching, and kissing a woman. To assist the client in achieving his goal, the therapist enlisted the aid of three female relationship surrogates. A surrogate is person who works with clients in therapy as part of a three-way therapeutic team, consisting of the therapist, the surrogate partner, and the client.

**Relationship Surrogates 1 & 2—Talking and Touching**

The first stage of therapy involved the use of two female surrogates who met with the client in a social context (e.g., Starbucks) and taught him the basic communication skills of how to talk with a woman (e.g., use of open-ended questions, reflective statements, eye contact, etc.). He met with the first surrogate on five occasions and the second surrogate on four occasions to develop skills in talking with different women. These appointments were scheduled through the therapist, who paid the surrogates ($25 a session—money provided by the client) after they gave feedback about their meeting with the client. No independent contact was ever made between the client and the surrogates. The surrogates were part of the therapeutic team, not potential girlfriends. After six weeks the client was able to talk to and touch (e.g., hold hands with) each of the surrogates. His self-esteem soared.

**Relationship Surrogate 3—Kissing**

Having developed comfort (anxiety 3 or below) in talking with and touching the two female surrogates, the client

# Falling in Love

*Love comes unseen; we only see it go.*

AUSTIN DOBSON, POET

Various social, physical, psychological, physiological, and cognitive conditions affect the development of love relationships.

## Social Conditions for Love

Love is a social label given to an internal feeling. Our society promotes love through popular music, movies, and novels. These media convey the message that love is an experience to pursue, enjoy, and maintain. People who fall out of love are encouraged to try again: "Love is lovelier the second time you fall." Unlike people reared in Eastern cultures, Americans grow up in a cultural context which encourages them to turn on their radar for love.

## Physical Conditions for Love

The probability of being involved in a love relationship is influenced by approximating the cultural ideal of physical appearance. Halpern et al. (2005) analyzed data on a nationally representative sample of 5,487 African-American, Caucasian, and Hispanic adolescent females and found that, for each one-point increase in body mass index (BMI), the probability of involvement in a romantic relationship dropped by 6%. Hence, to the degree that a woman approximates the cultural ideal of being trim and "not being fat," she increases the chance of attracting a partner and becoming involved in a romantic love relationship. In another study of weight and marital status, Sobal and Hanson (2011) found that never-married women were more likely to be obese than married women. The researchers suggested that weight for a woman was associated with not being selected as a marriage partner. Married men were heavier than men who were separated or divorced.

**DIVERSITY IN OTHER COUNTRIES**

The preference for thin and trim is not universal. Two researchers compared body-mass preferences among 300 cultures throughout the world and found that 81% of cultures preferred a female body size that would be described as "plump" (Brown and Sweeney 2009).

wanted to overcome his fear of kissing. To assist the client in achieving his goal, a third relationship surrogate (a 42-year-old woman) was identified and socialized as a member of the therapeutic team. She agreed (with the prior permission of her husband) to assist the client in overcoming his anxiety about kissing a woman. The client and the third surrogate met for a total of four sessions over a period of two months. The early sessions involved only talking and touching (hand holding) with kissing being gradually introduced to keep the client's anxiety low. He would move toward the woman while constantly assessing his anxiety level. If he felt anxiety at a level of three or above, he would move back until his anxiety dropped. By the fourth session with the third surrogate, the client was able to kiss her with low anxiety.

**Outcome of Therapy**

As with relationship surrogates one and two, the third relationship surrogate reported complete success with her client. The client was also proud of his accomplishment and his self-esteem soared. Subsequently, he was able to make dates with non-surrogate women in his social network, talk with them, touch them, and kiss them (and more, he reported). This case history illustrates that learning the basics with surrogates is sometimes helpful to be able to feel comfortable with non surrogates.

**Source**

Adapted from Zentner, M. and D. Knox. 2013. Surrogates in relationship therapy: A case study in learning how to talk, touch, and kiss. *Psychology Journal* 10: 63–68.

## Psychological Conditions for Love

Five psychological conditions associated with falling in love are perception of reciprocal liking, personality, high self-esteem, self compassion, self-disclosure, etc.

**Perception of Reciprocal Liking** Riela et al. (2010) conducted two studies on falling in love using both American and Chinese samples. The researchers found that one of the most important psychological factors associated with falling in love was the perception of reciprocal liking. When one perceives that he or she is desired by someone else, this perception has the effect of increasing the attraction toward that person. Such an increase is particularly strong if the person is very physically attractive (Greitemeyer 2010).

**Personality Qualities** The personality of the love object has an important effect on falling in love (Riela et al. 2010). Viewing the partner as intelligent or having a sense of humor is an example of a quality that makes the lover want to be with the beloved.

**Self-Esteem** High self-esteem is important for falling in love because it enables individuals to feel worthy of being loved. Feeling good about yourself allows you to believe that others are capable of loving you. Individuals with low self-esteem doubt that someone else can love and accept them.

*He's lucky to be going out with me.*

KATE MIDDLETON, TO A FRIEND, OF HER DATING WILLIAM, PRINCE OF WALES

People who have never felt loved and wanted may require constant affirmation from a partner as to their worth, and may cling desperately to that person out of fear of being abandoned. Such dependence (the modern term is *codependency*) may also encourage staying in unhealthy relationships (e.g., abusive relationships) because the person may feel "this is all I deserve."

Although having positive feelings about oneself when entering into a love relationship is helpful, sometimes these feelings develop after one becomes involved in the relationship. "I've always felt like an ugly duckling," said one woman. "But once I fell in love with him and him with me, I felt very different. I felt very

*From what I can see of the people like me, we get better but we never get well.*

PAUL SIMON, MUSICIAN

good about myself because I knew I was loved." High self-esteem, then, is not necessarily a prerequisite for falling in love. People who have low self-esteem may fall in love with someone else as a result of feeling deficient. The love they perceive the other person has for them may compensate for the perceived deficiency and improve their self-esteem. This phenomenon can happen with two individuals with low self-esteem—love can elevate the self-concepts of both of them.

**Self-compassion** being kind, caring and understanding towards oneself when feelings of suffering are present.

**Self-Compassion** Related to self-esteem is **self-compassion**, which refers to being "kind, caring and understanding towards oneself when feelings of suffering are present" (Neff and Beretvas 2013). The self-compassionate person views one as part of the total human condition and is accepting of one's failures (rather than being self-critical and blaming oneself). Persons with high levels of self-compassion also report higher levels of relationship satisfaction. Not only are they kind to themselves but also to their partner and to their relationship.

**Self-Disclosure** Disclosing oneself is necessary if one is to fall in love—to feel invested in another. Ross (2006) identified eight dimensions of self-disclosure: (1) background and history, (2) feelings toward the partner, (3) feelings toward self, (4) feelings about one's body, (5) attitudes toward social issues, (6) tastes and interests, (7) money and work, and (8) feelings about friends. Disclosed feelings about the partner included "how much I like the partner," "my feelings about our sexual relationship," "how much I trust my partner," "things I dislike about my partner," and "my thoughts about the future of our relationship"—all of which are associated with relationship satisfaction. Of interest in Ross's findings is that disclosing one's tastes and interests was negatively associated with relationship satisfaction. By telling a partner too much detail about what he or she liked, partners discovered something that turned them off and lowered their relationship satisfaction.

**Alexithymia** personality trait which describes a person with little affect.

It is not easy for some people to let others know who they are, what they feel, or what they think. **Alexithymia** is a personality trait which describes a person with little affect. The term means "lack of words for emotions," which suggests that the person

David Knox

These spouses are very open with each other about what they are thinking and feeling.

## SELF-ASSESSMENT | Toronto Alexithymia Scale

### Introduction

Alexithymia is an inability to understand, describe, or process emotions. Its name comes from the Greek words *lexis thumos* which translates as "without words for emotions."

The TAS was created in 1994 by R. Michael Bagby, a leading alexithymia researcher, and colleagues. It yields scores on three scales as well as an overall score.

### Procedure

The scale has twenty items that you must rate on a scale of 1 to 5, 1 meaning you completely disagree and 5 being you completely agree. It should not take most people more than three minutes.

| | Disagree | | Neutral | Agree | |
|---|---|---|---|---|---|
| I am often confused about what emotion I am feeling. | 1 | 2 | 3 | 4 | 5 |
| It is difficult for me to find the right words for my feelings. | 1 | 2 | 3 | 4 | 5 |
| I have physical sensations that even doctors don't understand. | 1 | 2 | 3 | 4 | 5 |
| I am able to describe my feelings easily. | 1 | 2 | 3 | 4 | 5 |
| I prefer to analyze problems rather than just describe them. | 1 | 2 | 3 | 4 | 5 |
| When I am upset, I don't know if I am sad, frightened, or angry. | 1 | 2 | 3 | 4 | 5 |
| I find it hard to describe how I feel about people. | 1 | 2 | 3 | 4 | 5 |
| I prefer to just let things happen rather than to understand why they turned out that way. | 1 | 2 | 3 | 4 | 5 |
| I have feelings that I can't quite identify. | 1 | 2 | 3 | 4 | 5 |
| Being in touch with emotions is essential. | 1 | 2 | 3 | 4 | 5 |
| I am often puzzled by sensations in my body. | 1 | 2 | 3 | 4 | 5 |
| People tell me to describe my feelings more. | 1 | 2 | 3 | 4 | 5 |
| I don't know what's going on inside me. | 1 | 2 | 3 | 4 | 5 |
| I often don't know why I am angry. | 1 | 2 | 3 | 4 | 5 |
| I prefer talking to people about their daily activities rather than their feelings. | 1 | 2 | 3 | 4 | 5 |
| I prefer to watch "light" entertainment shows rather than psychological dramas. | 1 | 2 | 3 | 4 | 5 |
| It is difficult for me to reveal my innermost feelings, even to close friends. | 1 | 2 | 3 | 4 | 5 |
| I can feel close to someone, even in moments of silence. | 1 | 2 | 3 | 4 | 5 |
| I find examination of my feelings useful in solving personal problems. | 1 | 2 | 3 | 4 | 5 |
| Looking for hidden meanings in movies or plays distracts from their enjoyment. | 1 | 2 | 3 | 4 | 5 |

### Scoring

Add the numbers you circled and write your score here ____.

The following scoring has been suggested (Ciarrochi and Bilich 2006):

| Score | Result |
|---|---|
| 20–51 | Non-alexithymia |
| 52–60 | Possible alexithymia |
| 61–100 | Alexithymia |

Below is a graph of how others who have taken this test here have scored.

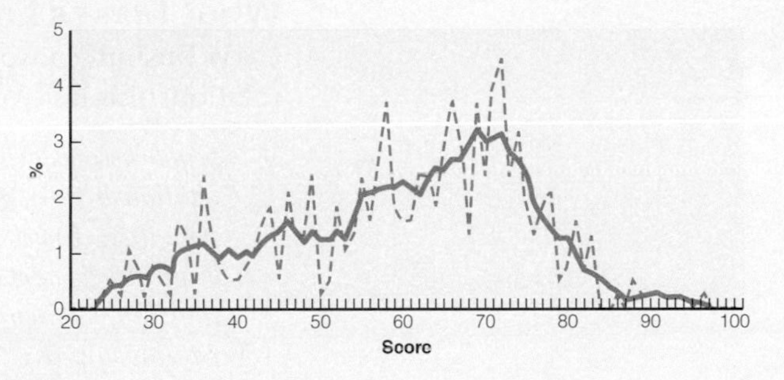


### Sources


Ciarrochi, J.; L. Bilich. (2006). "Process measures of potential relevance to ACT." http://www.uow.edu.au/content/groups/public/@web/@health/documents/doc/uow039223.pdf

Bagby, R., J. Parker, G. Taylor. (1994). "The twenty-item Toronto Alexithymia Scale-I. Item selection and cross-validation of the factor structure." *Journal of Psychosomatic Research*, 38, 23–32.

does not experience or convey emotion. Persons with alexithymia are not capable of experiencing psychological intimacy. Frye-Cox (2012) studied 155 couples who had been married an average of 18.6 years and found that an alexithymic spouse reported lower marital quality, as did his or her spouse. Alexithymia also tends to repel individuals in mate selection in that persons who seek an emotional relationship are not reinforced by alexithymics. The self-assessment on page 63 provides a way to measure the degree to which one is alexithymic.

Trust is the condition under which people are most willing to disclose themselves. When people trust someone, they tend to feel that whatever feelings or information they share will not be judged and will be kept safe with that person. If trust is betrayed, people may become bitterly resentful and vow never to disclose themselves again. One woman said, "After I told my partner that I had had an abortion, he told me that I was a murderer and he never wanted to see me again. I was devastated and felt I had made a mistake telling him about my past. You can bet I'll be careful before I disclose myself to someone else."

## Physiological and Cognitive Conditions for Love

Arousal (strong physiological reactions when in the presence of the other) is associated with falling in love (Riela et al. 2010). This physical chemistry is powerful. Partners who feel strong chemistry toward each other escalate their relationship; those who do not have less motivation to become or to stay involved.

Another physiological change is feeling excited or anxious and interpreting this stirred-up state as love (Walster & Walster 1978). But cognitive function must be operative or love feelings cannot develop. Individuals with brain cancer who have had their frontal lobe removed are incapable of love (Ackerman 1994).

The social, physical, psychological, physiological, and cognitive conditions are not the only factors important for the development of love feelings. Timing must also be right. There are only certain times in life (e.g., when educational and career goals are met or within sight) when some people seek a love relationship. When those times occur, a person is likely to fall in love with another person who is there and who is also seeking a love relationship. Hence, many love pairings exist because each of the individuals is available to the other at the right time—not because they are particularly suited for each other.

## What Makes Love Last

Caryl Rusbult's investment model of commitment has been used to identify why relationships last. While love is important, there are other factors involved....

> *people become dependent on their relationships because they (a) are satisfied with the relationship—it gratifies important needs, including companionship, intimacy, and sexuality; (b) believe their alternatives are poor—their most important needs could not be gratified independent of the particular relationship (e.g., in an alternative relationship, by friends and kin); and (c) have invested many important resources in the relationship (e.g., time, effort, shared friendship network, and material possessions). The model thereby includes not only internal factors (i.e., satisfaction) to explain why partners stick with each other but also external, structural factors to capture individuals in their interpersonal context* (Finkenauer 2010, p. 162).

Hence, people stay in a relationship because their needs are being met, they have no place to go, and they have made considerable investment in getting where they are and don't want to give it all up. Lambert et al. (2012) also identified a behavior associated with weakening one's love/commitment to a partner—pornography. In a study of 240 undergraduates, higher use of pornography was associated with more flirting and infidelity in the form of hooking up. "Our research suggests that there is a relationship cost associated with pornography" (p. 432).

# Love as a Context for Problems

"Love" answers Hester when asked by her husband Bill, "What has happened to you?" She has had an affair with Freddie, fallen hopelessly in love, with divorce and attempted suicide in the wake. *The Deep Blue Sea*, a 2012 movie, reveals up close and personal, how love can become a context for problems. In this section we review six such problems including the following:

*The course of true love never did run smooth.*

WILLIAM SHAKESPEARE, *A MIDSUMMER NIGHT'S DREAM*

## Unrequited/Unreciprocated Love

**Unrequited love** is a one-sided love where one's love is not returned. In the 16th century it was called "lovesickness"; today it is known as erotomania or erotic melancholy (Kem 2010). An example is from the short story *Winter Dreams* by F. Scott Fitzgerald. Dexter Green is in love with Judy Jones. "He loved her, and he would love her until the day he was too old for loving—but he could not have her."

Blomquist and Giuliano (2012) assessed the reactions of a sample of adults and college students to a partner telling them "I love you." The predominant response by both men and women was "I'm just not there yet." Both genders acknowledged that while this response was honest, it would hurt the individual who was in love.

*The sweetest joy, the wildest woe is love.*

PEARL BAILEY, SINGER

**Unrequited love** one-sided love where one's love is not returned.

## Making Risky, Dangerous Choices

Plato said that "love is a grave mental illness," and some research suggests that individuals in love make risky, dangerous, or questionable decisions. Nonsmokers who become romantically involved with a smoker are more likely to begin smoking (Kennedy et al. 2011). Similarly, couples in love and in a stable relationship are less likely to use a condom (doing so isn't very romantic) (Warren et al. 2012). Of 381 undergraduates, 70% reported that they had engaged in unprotected sex with their romantic partner (Elliott et al. 2012).

*Love is a fire. But whether it is going to warm your heart or burn down your house, you can never tell.*

JOAN CRAWFORD, ACTRESS

## Ending the Relationship with One's Parents

Some parents disapprove of the partner their son or daughter is involved with (e.g., race, religion, age) to the point that their offspring will end the relationship with their parents. "They told me I couldn't come home if I kept dating this guy, so I stopped going home," said one of our students who was involved with a partner of a different race. Choosing to end a relationship with one's parents is a definite downside of love.

## Simultaneous Loves

Sometimes an individual is in love with two or more people at the same time. While this is acceptable in polyamorous relationships where the partners agree on multiple relationships (discussed later in the chapter), simultaneous loves becomes a serious problem.

*Love is always trouble.*

BETTE DAVIS, ACTRESS

## Abusive/Stalking Relationships

Twenty-one percent of 1,101 undergraduate males and 37% of 3,469 undergraduate females reported that they had been involved in an emotionally abusive relationship with a partner. As for physical abuse, 3% of the males and 11% of the females reported such previous involvement (Hall and Knox 2013).

Rejected lovers (more often men) may stalk (repeated pursuit of a target victim that threatens the victim's safety) a partner because of anger and jealousy and try to win the partner back. Eighteen percent of the males and 28% of the females in the above sample reported that they had been stalked. We discuss both abusive relationships and stalking in detail in Chapter 13.

WHAT IF?

## What If You Are in Love with Two People at the Same Time?

One answer to the dilemma of simultaneous love is to let the clock run. Most love relationships do not have a steady course. Time has a way of changing them. If you maintain both relationships, one is likely to emerge as more powerful, and you will have your answer. Alternatively, if you feel guilty for having two loves, you may make the conscious choice to spend your time and attention with one partner and let the other relationship go in terms of actual time spent with the partner. Although you can have emotions for two people at the same time, you cannot psychically be with more than one person at a time. The person with whom you choose to spend New Year's Eve is likely to be the person you love "a little bit more" and with whom your love feelings are likely to increase.

### Profound Sadness/Depression when a Love Relationship Ends

*If love be good, from whennes comth my wo?*

CHAUCER, TROILUS'S SONG

Fisher et al. (2010) noted that "romantic rejection causes a profound sense of loss and negative affect. It can induce clinical depression and in extreme cases lead to suicide and/or homicide." The researchers studied brain changes via magnetic resonance imaging of 10 women and 5 men who had recently been rejected by a partner but reported they were still intensely "in love." Participants alternately viewed a photograph of their rejecting beloved and a photograph of a familiar individual interspersed with a distraction-attention task. Their responses while looking at the photo of the person who rejected them included feelings of love, despair, good and bad memories, and wondering about why this happened. Brain image reactions to being rejected by a lover are similar to withdrawal from cocaine.

There is also physical pain. "The pain one experiences in response to an unwanted breakup is identical to the pain one experiences when physically hurt" noted Dr. Steven Richeimer (2011) of the University of Southern California Pain Center. He was speaking in reference to a University of Michigan study where researchers asked people who recently had an unwanted romantic breakup to look at a picture of their ex or hold a hot cup of coffee. The pain reaction in the brain was exactly the same.

*Romantic love is mental illness. But it's a pleasurable one. It's a drug. It distorts reality, and that's the point of it. It would be impossible to fall in love with someone that you really saw.*

FRAN LEBOWITZ, AMERICAN WRITER

While there is no recognized definition or diagnostic criteria for "love addiction," some similarities to substance dependence include: euphoria and unrestrained desire in the presence of the love object or associated stimuli (drug intoxication); negative mood and sleep disturbance when separated from the love object (drug withdrawal); intrusive thoughts about the love object; and problems associated with love which may lead to clinically significant impairment or distress (Reynaud et al. 2011).

Problems associated with love begin early. Starr et al. (2012) conducted a study of 83 seventh and eighth grade (mean age 13.8) girls and found an association between those who reported involvement in romantic activities (e.g., flirting, dating, kissing) and depression, anxiety, and eating disorders. The researchers pointed out that it was unclear if romance led to these outcomes or if the girls who were depressed, anxious, etc., sought romance to help cope with their depressive symptoms.

WHAT IF?

## What If You Fall out of Love with the Person You Had Planned to Marry?

In a sample of 1,103 undergraduate males, 42% agreed that "I would divorce my spouse if I no longer loved him or her"; of 3,478 undergraduate females, 43% agreed with the statement (Hall and Knox 2013). Although Asians more typically view love as a feeling that is to follow marriage, Americans have been socialized to expect being in love with their spouse-to-be, and to feel embarrassed about and to hide it if they are not. Although falling in love with someone after marriage is more than possible, someone socialized in America with individualistic values might be very cautious in proceeding with a wedding (because such an enormous value is placed on love in U.S. society). To do so is to run the risk of being in a "loveless" marriage that may make one vulnerable to "falling in love" outside the marriage.

# Jealousy in Relationships

**Jealousy** can be defined as an emotional response to a perceived or real threat to an important or valued relationship. People experiencing jealousy fear being abandoned and feel anger toward the partner or the perceived competition. People become jealous when they fear replacement.

**Jealousy** an emotional response to a perceived or real threat to an important or valued relationship.

Jealousy is a learned emotion and varies with social context. Polygynous Mormons (see Chapter 1) reflect the context where wives are socialized to "share" the husband. Not only is the principle of multiple wives "God's will," but also the wives embrace the presence of other wives who help out with childcare and housework. While her research findings have been questioned, Margaret Mead noted that the Samoans were absent feelings of jealousy (Freeman 1999). Some individuals in polyamorous/open relationships report zero feelings of jealousy for their partner's emotional/sexual relationship with others.

*The wounds invisible*
*That love's keen arrows make.*

SHAKESPEARE, *AS YOU LIKE IT*, III, V, 30

Data on the jealously of college students reflects that, of 4,582 undergraduates, 52% agreed with the statement, "I am a jealous person" (women were slightly more jealous than men) (Hall and Knox 2013). In another study, 185 students gave information about their experience with jealousy (Knox et al. 1999). On a continuum of 0 ("no jealousy") to 10 ("extreme jealousy"), with 5 representing "average jealousy," these students reported feeling jealous at a mean level of 5.3 in their current or last relationship. Students who had been dating a partner for a year or less were significantly more likely to report higher levels of jealousy (mean = 4.7) than those who had dated 13 months or more (mean = 3.3). Hence, jealously is more likely to occur early in a couple's relationship. Consistent with this theme, Gatzeva and Paik (2011) compared noncohabiting, cohabiting, and married couples in regard to satisfaction as related to jealous conflict. They found that married couples were less likely to report jealous conflict and that satisfaction was higher in the absence of such conflict.

*To demand of love that it be without jealousy is to ask of light that it cast no shadows.*

OSCAR HAMMLING, WRITER

## Types of Jealousy

Barelds-Dijkstra and Barelds (2007) identified three types of jealousy as reactive jealousy, anxious jealousy, and possessive jealousy. **Reactive jealousy** consists of

**Reactive jealousy** feelings that are a reaction to something the partner is doing.

**Anxious jealousy** obsessive ruminations about the partner's alleged infidelity.

**Possessive jealousy** an attack on the partner or the alleged person to whom the partner is showing attention.

feelings that are a reaction to something the partner is doing (e.g., texting a former lover). **Anxious jealousy** is obsessive ruminations about the partner's alleged infidelity that make one's life a miserable emotional torment. **Possessive jealousy** involves an attack on the partner or the alleged person to whom the partner is showing attention. Jealousy is a frequent motive when one romantic partner kills another. Less dramatic, note M. Fisher and Cox (2011) is "competitor derogation," whereby the person talks negatively about the alleged competing suitor.

## Causes of Jealousy

Jealousy can be triggered by external or internal factors.

**External Causes** External factors refer to behaviors a partner engages in that are interpreted as (1) an emotional and/or sexual interest in someone (or something) else, or (2) a lack of emotional and/or sexual interest in the primary partner. In the study of 185 students previously referred to, the respondents identified "actually talking to a previous partner" (34%) and "talking about a previous partner" (19%) as the most common sources of their jealousy. Also, men were more likely than women to report feeling jealous when their partner talked to a previous partner, whereas women were more likely than men to report feeling jealous when their partner danced with someone else. Buunk et al. (2010) found that what makes a woman jealous is the physical attractiveness of a rival; what makes a man jealous is the rival's physical dominance.

**Internal Causes** Internal causes of jealousy refer to characteristics of individuals that predispose them to jealous feelings, independent of their partner's behavior. Examples include being mistrustful, having low self-esteem, being highly involved in and dependent on the partner, and having no perceived alternative partners available (Pines 1992). The following are explanations of these internal causes of jealousy:

1. ***Mistrust.*** If an individual has been cheated on in a previous relationship, that individual may have learned to be mistrustful in subsequent relationships. Such mistrust may manifest itself in jealousy. Mistrust and jealousy may be intertwined. Indeed, one must be careful if cheated on in a previous relationship not to transfer those feelings to a new partner. Disregarding the past is not easy but constantly reminding oneself of reality ("My new partner has given me *zero* reason to be distrustful") may help.
2. ***Low self-esteem.*** Individuals who have low self-esteem tend to be jealous because they lack a sense of self-worth and hence find it difficult to believe anyone can value and love them. Feelings of worthlessness may contribute to suspicions that someone else is valued more. It is devastating to a person with low self-esteem to discover that a partner has, indeed, selected someone else.
3. ***Lack of perceived alternatives.*** Individuals who have no alternative person or who feel inadequate in attracting others may be particularly vulnerable to jealousy. They feel that if they do not keep the person they have they will be alone.
4. ***Insecurity.*** Individuals who feel insecure (e.g., no commitment from the partner) in a relationship may experience higher levels of jealousy. They feel at any moment their partner could find someone more attractive/desirable and end the relationship.

*O! Beware, my lord, of jealousy*
*It is the green-eyed monster which doth mock*
*The meat it feeds on.*

SHAKESPEARE, *OTHELLO*

## Consequences of Jealousy

Jealousy can have both desirable and undesirable consequences.

**Desirable Outcomes** Barelds-Dijkstra and Barelds (2007) studied 961 couples and found that reactive jealousy is associated with a positive effect on the relationship. Not only may reactive jealousy signify that the partner is cared for (the implied message is "I love you and don't want to lose you to someone else"), but also the partner may learn that the development of other romantic and sexual relationships is unacceptable.

One wife said:

*When I started spending extra time with this guy at the office, my husband got jealous and told me he thought I was getting in over my head and asked me to cut back on the relationship because it was "tearing him up." I felt he really loved me when he told me this, and I chose to stop having lunch with the guy at work.*

The researchers noted that making the partner jealous may also have the positive function of assessing the partner's commitment and of alerting the partner that one could leave for greener mating pastures. Hence, one partner may deliberately evoke jealousy to solidify commitment and ward off being taken for granted. In addition, sexual passion may be reignited if one partner perceives that another would take the love object away. That people want what others want is an adage that may underlie the evocation of jealousy.

**Undesirable Outcomes** Shakespeare referred to jealousy as the "green-eyed monster," suggesting that it sometimes leads to undesirable outcomes for relationships. Anxious jealousy with its obsessive ruminations about the partner's alleged infidelity can make individuals miserable. They are constantly thinking about the partner being with the new person, which they interpret as confirmation of their own inadequacy. And, if the anxious jealousy results in repeated unwarranted accusations, a partner can tire of such attacks and seek to end the relationship with the accusing partner.

In its extreme form, jealousy may have fatal consequences. Possessive jealousy involves an attack on a partner or an alleged person to whom the partner is showing attention. In the name of love, people have stalked or killed the beloved and then killed themselves in reaction to rejected love. An example is the 2010 murder of University of Virginia student Yeardley Love by George Hugely. His jealousy over her having broken up with him and her seeing someone else were contributing motives to the brutal murder—he beat her to death.

*Love is strong as death; jealousy is cruel as the grave.*

SONG OF SOLOMON 8:6

### Gender Differences in Coping with Jealousy

Zengel et al. (2013) studied a national sample of women and men and found that women reported higher levels of jealousy than men. The researchers also noted that men were more jealous when their partner engaged in sexual intercourse with another man than when their partner was emotionally involved with someone else. Evolutionary theorists point out that men are wired to care about the paternity of their offspring ("Am I the biological father of this child?"), which is the basis of their focus on physical fidelity.

**Strategies Used to Cope with Jealousy** Women tend to turn to food and men to alcohol as strategies to cope with feelings of jealously. Both might consider exercise as a way of relieving their stress.

*And as for living a life with one woman, that's an intriguing illusion, because in a long marriage, you live with many; and they must suffer many of you.*

RON CARLSON, *LONGEVITY*

## Compersion and Polyamory

**Compersion**, sometimes thought of as the opposite of jealousy, is the approval (indeed embracing) of a partner's emotional and sexual involvement with another person. Polyamory means multiple loves (poly = many; amor = love). **Polyamory** is a lifestyle in which two lovers embrace the idea of having multiple lovers. By agreement, each partner may have numerous emotional and sexual relationships. About half of the 100 members of Twin Oaks Intentional Community in Louisa, Virginia, are polyamorous in that each partner may have

**Compersion** the approval of a partner's emotional and sexual involvement with another person.

**Polyamory** a lifestyle in which two lovers embrace the idea of having multiple lovers.

These individuals are in a polyamorous relationship. The two males have an emotional and sexual relationship with the female (and she with them). The three individuals also have emotional and sexual relationships with others not in the photo.

David Knox

several emotional or physical relationships with others at the same time. Although not legally married, these adults view themselves as emotionally bonded to each other and may even rear children together.

**Swinging** a married or pair-bonded couple agree that each will have recreational sex with others while maintaining an emotional allegiance to each other.

Polyamory is not **swinging** (a married or pair-bonded couple agree that each will have recreational sex with others while maintaining an emotional allegiance to each other), as polyamorous lovers are concerned about enduring, intimate relationships that include sex. People in polyamorous relationships seek to rid themselves of jealous feelings and to increase their level of compersion. To feel happy for a partner who delights in the attention and affection of—and sexual involvement with—another person is the goal of polyamorous couples. Morrison et al. (2013) compared polyamorous and monoamorous individuals and found that the former had higher intimacy needs and interests than the latter.

*That I love my husband, that I am free to love other men, and that I have no designs on my husband's own freedom, all this strikes me as only good—there is no time for frivolous anger.*

HANNAH PINE (PSEUDONYM)

### Advantages and Disadvantages of Polyamory

Embracing polyamory has both advantages and disadvantages (Fennell 2014). Advantages of polyamory include greater variety in one's emotional and sexual life; the avoidance of hidden affairs and the attendant feelings of deception, mistrust, betrayal or guilt; and the opportunity to have different needs met by different people. The disadvantages of polyamory involve having to manage one's feelings of jealousy and limited time with each partner. Of the latter, one polyamorous partner said, "With three relationships and a full-time job, I just don't have much time to spend with each partner so I'm frustrated about who I'll be with next. And managing the feelings of the other partners who want to spend time with me is a challenge."

## The Future of Love Relationships

Love will continue to be one of the most treasured experiences in life. Love will be sought, treasured, and when lost or ended, will be met with despair and sadness. After a period of recovery, a new search will begin. As our society becomes more diverse, the range of potential love partners will widen to include those with demographic characteristics different from oneself. Romantic love will continue and love will maintain its innocence as those getting remarried love just as deeply and invest in the power of love all over again.

## Summary

***What are some ways of conceptualizing love?***

Love remains an elusive and variable phenomenon. Researchers have conceptualized love as a continuum from romanticism (e.g., from belief in love at first sight, one true love, and love conquers all) to realism, as a style (e.g., ludic, eros, storge, mania), and as a triangle consisting of three basic elements (intimacy, passion, and commitment).

***How is love under social control?***

All parents attempt to influence and control the person their children fall in love with. Love may be blind, but offspring are socialized to know what color a person's skin is (about 90% of Americans fall in love with/marry someone of their same racial background). Because romantic love is such a powerful emotion and marriage such an important relationship, mate selection is not left to chance when connecting an outsider with an existing family and peer network. Unlike Eastern parents who arrange the marriage of their children (in 40% of the world's population, marriages are arranged by the parents), American parents move to certain neighborhoods, join certain churches, and enroll their children in certain schools. Doing so increases the chance that their offspring will "hang out" with, fall in love with, and marry people who are similar in race, education, and social class.

***What are the various theories of love?***

Theories of love include evolutionary (love provides the social glue needed to bond parents with their dependent children and spouses with each other to care for their dependent offspring), learning (positive experiences create love feelings), sociological (Reiss's "wheel" theory), psychosexual (love results from a blocked biological drive), and biochemical (love involves feelings produced by biochemical events). For example, the neurobiology of love emphasizes that, because romantic love and maternal love are linked to the perpetuation of the species, biological wiring ensures the bonding of the male and female to rear offspring and of the mother to the infant. Finally, attachment theory focuses on the fact that a primary motivation in life is to be connected with other people.

***What is the process of "falling in love"?***

Love occurs under certain conditions. Social conditions include a society that promotes the pursuit of love, peers who enjoy it, and a set of norms that link love and marriage. Body type is related to falling in love in that the closer one's body type (physical condition) matches the cultural ideal, the more likely the person is to fall in love. Psychological conditions involve high self-esteem, a willingness to disclose oneself to others, perception that the other person has a reciprocal interest, and gratitude. Physiological and cognitive conditions imply that the individual experiences a stirred-up state and labels it "love." People stay in a relationship because it meets important emotional needs (e.g., satisfaction), they have few alternatives (e.g., no place to go), and they have already invested resources (e.g., time, money, friendship networks).

***How is love a context for problems?***

For all of its joy, love is associated with problems, which include unrequited/non-reciprocated love, making dangerous/destructive choices, ending the relationship with one's parents, simultaneous loves, involvement in an abusive relationship, and profound sadness/depression when a love relationship ends.

***What is jealousy—the various types and consequences (positive and negative)?***

Jealousy is an emotional response to a perceived or real threat to a valued relationship. Types of jealousy are reactive (partner shows interest in another), anxious (ruminations about partner's unfaithfulness), and possessive (striking back at a partner or another). Jealous feelings may have both internal and external causes and may have both positive and negative consequences for a couple's relationship.

***What is compersion and polyamory?***

Compersion is the opposite of jealousy and involves feeling positive about a partner's emotional and physical relationship with another person. Polyamory ("many loves") is an arrangement whereby lovers agree to have numerous emotional relationships (which may include sex) with others at the same time.

***What is the future of love relationships?***

Love will continue to be one of the most treasured experiences in life. As our society becomes more diverse, the range of potential love partners will widen to include those with demographic characteristics different from oneself. Romantics will continue and love will maintain its innocence as those getting remarried will love just as deeply as those in first marriages.

## Key Terms

Agape love style
Alexithymia
Anxious jealousy
Arranged marriage
Autistic love
Compersion
Conjugal love
Emphatic love
Eros love style
Infatuation
Integral love
Jealousy
Ludic love style
Lust
Mania love style
Oxytocin
Polyamory
Possessive jealousy
Pragma love style
Reactive jealousy
Romantic love
Self-compassion
Storge love style
Swinging
Unrequited love

## Web Links

The Polyamory Society
http://www.polyamorysociety.org/

Third Age
http://www.thirdage.com/romance/

# CHAPTER 3 Gender

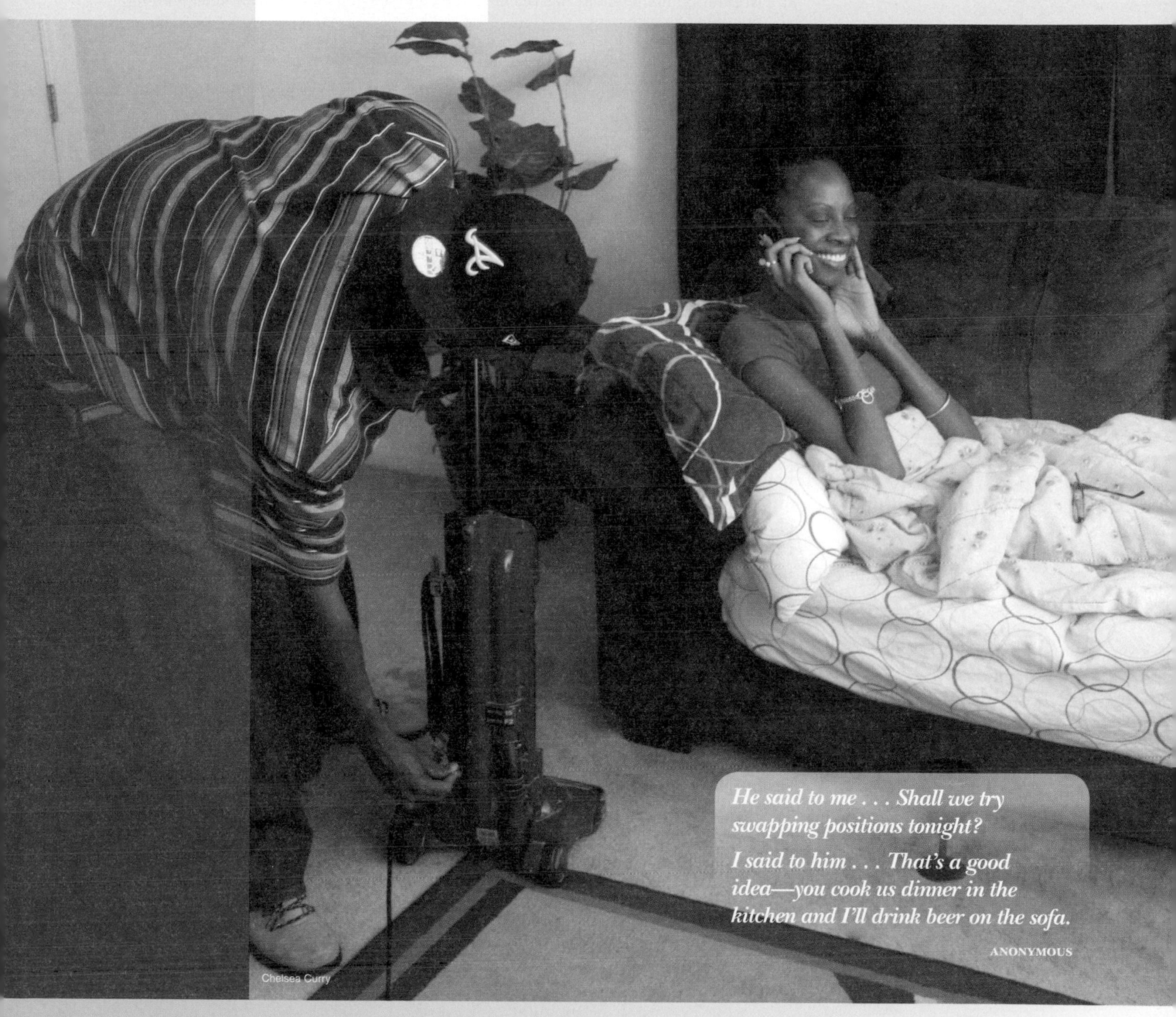


Chelsea Curry

## Learning Objectives

- List and explain the five primary gender role terms.
- Identify five theories of gender role development.
- Review seven agents which influence gender role learning.
- Describe in detail gender roles in four other societies.
- Discuss the consequences of gender role socialization for women and men.
- Know how gender roles are changing and the future of gender roles.

## TRUE OR FALSE?

1. A higher percentage of young women (18–34) than young men say that achieving success in a high-paying career or profession is important in their lives.
2. Caribbean families are known for the centrality of the father in the family.
3. Greater gender equality has resulted in women having equal power in negotiating condom use with men.
4. Men continue to earn more undergraduate college degrees than women.
5. Women who have the same education as men are typically paid three-fourths of what men are paid.

*Answers:* 1. T 2. F 3. F 4. F 5. T

---

Chaz Bono enhanced the visibility of gender/transgender issues when he appeared on "Dancing with the Stars." The daughter of Sonny and Cher, Chastity, now a female-to-male transsexual, noted on "The Jay Leno Show" that he had always thought of himself as a male. His story reminds us that gender roles and sexuality are on a continuum.

Sociologists note that one of the defining moments in an individual's life is when his or her sex as a fetus (in the case of an ultrasound) or an infant (in the case of a birth) is announced. The phrase "It's a boy" or "It's a girl" immediately summons an onslaught of cultural programming affecting the color of the nursery (e.g., blue for a boy and pink for a girl), the name of the baby (there are few gender-free names such as Chris), and occupational choices (Hillary Clinton and Nancy Pelosi are two of the few upper-echelon women in politics). In this chapter, we examine variations in gender roles and the choices individuals make in how they express their gender behavior. We begin by looking at the terms used to discuss gender issues.

# Terminology of Gender Roles

In common usage, the terms *sex* and *gender* are often used interchangeably, but social scientists do not regard these terms as synonymous. After clarifying the distinction between *sex* and *gender,* we clarify other relevant terminology, including *gender identity, gender role,* and *gender role ideology.*

## Sex

**Sex** the biological distinction between females and males.

*Women. They are a complete mystery.*

STEPHEN HAWKING, BRITISH PHYSICIST

**Sex** refers to the biological distinction between females and males. Hence, to be assigned as a female or male, several factors are used to determine the biological sex of an individual:

- *Chromosomes:* XX for females; XY for males
- *Gonads:* Ovaries for females; testes for males
- *Hormones:* Greater proportion of estrogen and progesterone than testosterone in females; greater proportion of testosterone than estrogen and progesterone in males
- *Internal sex organs:* Fallopian tubes, uterus, and vagina for females; epididymis, vas deferens, and seminal vesicles for males
- *External genitals:* Vulva for females; penis and scrotum for males

Even though we commonly think of biological sex as consisting of two dichotomous categories (female and male), biological sex exists on a continuum. Sometimes not all of the five bulleted items just listed are found neatly in one person (who would be labeled as a female or a male). Rather, items typically associated with females or males might be found together in one person, resulting in mixed or ambiguous genitals; such persons are called **intersexed individuals**. Indeed, the genitals in these intersexed (or middlesexed) individuals (about 2% of all births) are not clearly male or female. Intersex development refers to congenital variations in the reproductive system, sometimes resulting in ambiguous genitals.

*In the last couple of weeks I have seen the ads for the Wonder Bra. Is that really a problem in this country? Men not paying enough attention to women's breasts?*

HUGH GRANT, ACTOR

## Gender

The term **gender** is a social construct and refers to the social and psychological characteristics associated with being female or male. Women are often thought of as soft, passive, and cooperative; men as rough, aggressive, and forceful. Robnett and Susskind (2010) found that 3rd and 4th grade children valued their own sex's personality traits, but that boys with many same-sex friends tended to denigrate female traits. Boys who acted like girls were criticized.

**Intersexed individuals** items typically associated with females or males might be found together in one person, resulting in mixed or ambiguous genitals.

**Gender** a social construct referring to the social and psychological characteristics associated with being female or male.

**Socialization** the process through which we learn attitudes, values, beliefs, and behaviors appropriate to the social positions we occupy.

In popular usage, gender is dichotomized as an either-or concept (feminine or masculine). Each gender has some characteristics of the other. However, gender may also be viewed as existing along a continuum of femininity and masculinity.

Gender differences are a consequence of biological (e.g., chromosomes and hormones) and social factors (e.g., male/female models such as parents, siblings, peers). The biological provides a profound foundation for gender role development. Evidence for this biological influence is the experience of the late John Money, psychologist and former director of the now-defunct Gender Identity Clinic at Johns Hopkins University School of Medicine, who encouraged the parents of a boy (Bruce Reimer) to rear him as a girl (Brenda) because of a botched circumcision that left the infant without a penis. Money argued that social mirrors dictate one's gender identity, and thus, if the parents treated the child as a girl (e.g., name, clothing, toys), the child would adopt the role of a girl and later that of a woman. The child was castrated and sex reassignment began.

*We have a secret in our culture, and it's not that birth is painful. It's that women are strong.*

LAURA STAVOE HARM, FREELANCE WRITER

However, the experiment failed miserably; the child as an adult (now calling himself David) reported that he never felt comfortable in the role of a girl and had always viewed himself as a boy. He later married and adopted his wife's three children.

In the past, David's situation was used as a textbook example of how nurture is the more important influence in gender identity, if a reassignment is done early enough. Today, his case makes the point that one's biological wiring dictates gender outcome. Indeed, David Reimer noted in a television interview, "I was scammed," referring to the absurdity of trying to rear him as a girl. Distraught with the ordeal of his upbringing and beset with financial difficulties, he committed suicide in May 2004 via a gunshot to the head.

The story of David Reimer emphasizes the power of biology in determining gender identity. Other research supports the critical role of biology. Nevertheless, **socialization** (the process through which we learn attitudes, values, beliefs, and behaviors appropriate to the social positions we occupy) does impact gender role behaviors, and social scientists tend to emphasize the role of social influences in gender differences.

**DIVERSITY IN OTHER COUNTRIES**

Although her research is controversial, Margaret Mead (1935) focused on the role of social learning in the development of gender roles in her study of three cultures. She visited three New Guinea tribes in the early 1930s, and observed that the Arapesh socialized both men and women to be feminine, by Western standards. The Arapesh people were taught to be cooperative and responsive to the needs of others. In contrast, the Tchambuli were known for dominant women and submissive men—just the opposite of our society. Both of these societies were unlike the Mundugumor, which socialized only ruthless, aggressive, "masculine" personalities. The inescapable conclusion of this cross-cultural study is that human beings are products of their social and cultural environments and that gender roles are learned.

Ayden is a female to male transsexual.

## Gender Identity

**Gender identity** is the psychological state of viewing oneself as a girl or a boy, and later as a woman or a man. Such identity is largely learned and is a reflection of society's conceptions of femininity and masculinity. Some individuals experience gender dysphoria, a condition in which one's gender identity does not match one's biological sex. An example of gender dysphoria is transgender or transsexualism.

The word **transgender** is a generic term for a person of one biological sex who displays characteristics of the other sex. **Transsexuals** are individuals with the biological and anatomical sex of one gender (e.g., female) but the self-concept of the other sex (e.g., male). "I am a female trapped in a man's body" reflects the feelings of the male-to-female transsexual (MtF), who may take hormones to develop breasts and reduce facial hair and may have surgery to artificially construct a vagina. Such a person lives full-time as a woman.

The female-to-male transsexual (FtM) is a biological and anatomical female who feels "I am a man trapped in a female's body." This person may take male hormones to grow facial hair and deepen her voice and may have surgery to create an artificial penis. This person lives full-time as a man. Individuals need not take hormones or have surgery to be regarded as transsexuals. The distinguishing variable is living full-time in the role of the gender opposite one's biological sex. A man or woman who presents full-time as the opposite gender is a transsexual by definition.

**Gender identity** the psychological state of viewing oneself as a girl or a boy, and later as a woman or a man.

*As long as she thinks of a man, nobody objects to a woman thinking.*

VIRGINIA WOOLF, *ORLANDO*

**Transgender** a person of one biological sex who displays characteristics of the other sex.

**Transsexual** individual with the biological and anatomical sex of one gender but the self-concept of the other sex.

*The woman most in need of liberation is the woman in every man and the man in every woman.*

WILLIAM SLOAN COFFIN, *CLERGYMAN AND POLITICAL ACTIVIST*

**Gender roles** social norms which specify the socially appropriate behavior for females and males in a society.

## Gender Roles

**Gender roles** are social norms which specify the socially appropriate behavior for females and males in a society. All societies have expectations of how boys and girls, men and women "should" behave. Gender roles influence women and men in virtually every sphere of life, including dating. Of 2,467 adults between the ages of 18–59, 69% of the men and 55% of the women reported that the man should pay for the date (Jayson 2014). Regarding family roles, even in those situations where women and men have equal pay and education, women still do more housework (Lachance-Grzela and Bouchard 2010).

And, even with transgender male-to-female individuals, women end up doing more domestic work. Pfeffer (2010) interviewed 50 women partners of transgender and transsexual men to assess the division of labor. Women often spoke of a nonegalitarian division of labor, but rationalized the reasons for this division. Ani stated: "I do the dishes; but I'm so neurotic about having a clean house and he is not.... I definitely do more than he does but, again, I'm the one that happens to be a neat freak."

Another prevalent norm in family life is that women end up devoting more time to child rearing and child care. Barclay (2013) noted that even in countries which promote equality in division of labor, women end up taking care of children more often.

The term *sex roles* is often confused with and used interchangeably with the term *gender roles*. However, whereas gender roles are socially defined and

## SELF-ASSESSMENT | Gender Role Attitudes: Family Life Index

Read each item and select the number that reflects your belief/attitude.

1 = strongly agree

2 = agree

3 = disagree

4 = strongly disagree

_____ 1. The husband should be the head of the family.
_____ 2. Babies and young children need to have their mothers around most of the day.
_____ 3. It is much better for everyone involved if the man is the achiever outside the home and the woman takes care of the home and family.
_____ 4. A woman's most important task in life is being a mother.
_____ 5. By nature, women are better than men at making a home and caring for children.
_____ 6. A preschool child is likely to suffer if his or her mother works outside the home.
_____ 7. A husband should earn a larger salary than his wife.
_____ 8. A woman should not be employed if her husband can support her.
_____ 9. All in all, family life suffers when the wife has a full-time job.

### Scoring

Add the numbers you selected for each item. 1 (strongly agree) is the most traditional response and 4 (strongly disagree) is the most egalitarian response. The lower your total score (9 is the lowest possible score), the more traditional your gender role attitudes. The higher your total score (36 is the highest possible score), the more egalitarian your gender role attitudes. A score of 22.5 places you at the midpoint between being very traditional and very egalitarian.

### Norms

The norming sample of this self-assessment was based on 37 males and 172 females. The average score of the males was 23 and the average score of the females was 25, suggesting a lean toward being egalitarian by both males and females with females being more egalitarian.

### Source

Adapted from Erarslan, A. B. and B. Rankin. 2013. Gender role attitudes of female students in single-sex and coeducational high schools in Istanbul. *Sex Roles* 69: 455–469. Used by permission of A. B. Erarslan.

can be enacted by either women or men, **sex roles** are defined by biological constraints and can be enacted by members of one biological sex only—for example, wet nurse, sperm donor, child bearer. You might want to complete the Self-Assessment on Gender Role Attitudes to find out the degree to which you are traditional or egalitarian. Erarslan and Rankin (2013) found that high school students in Turkey attending single sex schools were more likely to have egalitarian gender role attitudes than students attending coeducational classes.

**Sex roles** defined by biological constraints and can be enacted by members of one biological sex only.

*Despite my thirty years of research into the feminine soul, I have not yet been able to answer the great question that has never been answered: What does a woman want?*

SIGMUND FREUD, PSYCHOANALYST

## Gender Role Ideology

**Gender role ideology** refers to beliefs about the proper role relationships between women and men in a society. Traditionally, men initiated relationships, called women for dates, and were expected to be the ones who proposed. New norms include that women may be first to initiate an interaction, text men for time together, and ask men when they can marry. However, traditional gender experiences have not disappeared (see the section on Different Gender Worlds in the Applying Social Research section).

Gender role ideology is also operative in cross-sex friendships. Women are more likely to regard cross-sex friendships as devoid of romantic attraction and sexual possibility, particularly if they are involved in a relationship. Men, on the other hand, are more likely to experience attraction in cross-sex friendships whether or not they are involved in a relationship. Hence when men and women friends are hanging out, he may be thinking

**Gender role ideology** beliefs about the proper role relationships between women and men in a society.

## APPLYING SOCIAL RESEARCH | Different Gender Worlds: Sex, Betrayal and Love*

In spite of egalitarian changes in our society, women and men report significantly different experiences. The table below reflects some of these differences in a large non-random sample of 4,570 undergraduates.

**Percent Agreement On Sex, Betrayal And Love**

| Item | Female | Male | Sig. |
|---|---|---|---|
| | N = 3468 | N = 1102 | |
| **Sexual Experiences** | | | |
| I have masturbated. | 65.2% | 96.5% | .000 |
| I could hook up and have sex with someone I liked. | 23.4% | 59.7% | .000 |
| I have hooked up/had sex with a person I just met. | 22.4% | 33.5% | .000 |
| I regret my choice for sexual intercourse the first time. | 27.2% | 15.4% | .000 |
| **Betrayed** | | | |
| I have been involved with someone who cheated on me. | 51.5% | 38.8% | .000 |
| **Love at First Sight** | | | |
| I have experienced love at first sight. | 25.4% | 35.5% | .000 |

**Discussion**

Differences in female masturbation reflect anatomy, biology, and socialization. Unlike the penis, which is an external appendage that lends itself to being rubbed, touched, and handled, the clitoris is hidden and embedded in the woman's vaginal lips. In addition to lower anatomical availability, lower testosterone levels in females may result in a lower biological drive to reduce the sex drive. In addition, the socialization of women includes more caveats about sex. Women are taught negatives about "down there" and to regard menstruation as something to deal with. Also,

*The Mr. Musgroves had their game to guard, and to destroy, their horses, dogs, and newspapers to engage them; and the females were fully occupied in all the other common subjects of housekeeping, neighbors, dress, dancing, and music.*

JANE AUSTEN, *PERSUASION*

of a romantic/sex potential while she typically does not share his view (Bleske-Rechek et al. 2012).

Traditional American gender role ideology has perpetuated and reflected patriarchal male dominance and male bias in almost every sphere of life. Even our language reflects this male bias. For example, the words *man* and *mankind* have traditionally been used to refer to all humans. There has been a growing trend away from using male-biased language. Dictionaries have begun to replace *chairman* with *chairperson* and *mankind* with *humankind*.

# Theories of Gender Role Development

Various theories attempt to explain why women and men exhibit different characteristics and behaviors.

## Biosocial/Biopsychosocial

**Biosocial theory** social behaviors are biologically based and have an evolutionary survival function.

In the discussion of gender at the beginning of the chapter, we noted the profound influence of biology on one's gender. **Biosocial theory** emphasizes that masturbation is less often a topic shared by female peers than by male peers. Finally, women are encouraged to experience sex in a relationship context, unlike men who are socialized to experience sexuality independent of relationship factors. The result of these influences not only reflect an overall lifetime lower frequency of masturbation for females but also an annual lower frequency. Based on a survey of 5,865 respondents in the United States of adults ages 20–29, 68% of women compared to 84% of men reported having masturbated alone (rather than with a partner) in the past 12 months (Reece et al. 2012).

The lower frequency of females hooking up with someone they just met reflects female socialization to have sex in the context of a relationship. Later in the chapter we review sociobiological theory which suggests that females are wired to be more selective about their sexual behavior due to the potential consequence of becoming pregnant and being socialized for the role of caretaker of the infant/child. Men can drop their sperm and move on. Women, in order to protect themselves and their offspring, must be more selective. The higher regret at first intercourse among women is often over the lack of insistence on a meaningful relationship as the context.

Greater experience in being betrayed is reflected in other data which point out that men are more likely to cheat. About a quarter of husbands and 20 percent of wives report having had sex with someone outside the marriage at some time during the marriage (Russell et al. 2013). Having a partner who is unfaithful often results in the woman learning to develop feelings of mistrust in their new relationship.

Finally, women reporting a lower frequency of experiencing love at first sight than men speaks to the fact that men may be more attentive to visual cues in partner selection. Gervais et al. (2013) confirmed that compared to women, men showed an increased tendency to exhibit "the objectifying gaze" toward women who represented the cultural ideal (i.e., hourglass-shaped women with large breasts and small waist-to-hip ratios).

### References


Gervais, S. J., A. M. Holland, and M. D. Dodd. 2013. My eyes are up here: The nature of the objectifying gaze toward women. *Sex Roles* 69: 557–570.

Reece, M., D. Herbenick, J. D. Fortenberry et al. 2012. The National Survey of Sexual Health and Behavior. http://www.nationalsexstudy.indiana.edu /graph.html Retrieved January 20, 2014.

Russell, V. M, L. R. Baker, and J. K. McNulty. 2013. Attachment insecurity and infidelity in marriage: Do studies of dating relationships really inform us about marriage? *Journal of Family Psychology* 27: 241–251.


* Based on original data from Hall, S. and D. Knox. 2013. Relationship and sexual behaviors of a sample of 4,590 university students. Department of Family and Consumer Sciences, Ball State University and Department of Sociology, East Carolina University.

social behaviors (e.g., gender roles) are biologically based and have an evolutionary survival function. For example, women tend to select and mate with men whom they deem will provide the maximum parental investment in their offspring. The term **parental investment** refers to any venture by a parent that increases the offspring's chance of surviving and thus increases reproductive success of the adult. Parental investments require time and energy. Women have a great deal of parental investment in their offspring (including nine months of gestation), and they tend to mate with men who have high status, economic resources, and a willingness to share those economic resources. As we will see in the parenting chapter, economic resources are not inconsequential as the average cost today of rearing a child from birth to age 18 is almost $250,000.

The biosocial explanation (also referred to as **sociobiology**) for mate selection is extremely controversial. Critics argue that women may show concern for the earning capacity of a potential mate because they have been systematically denied access to similar economic resources, and selecting a mate with these resources is one of their remaining options. In addition, it is argued that both women and men, when selecting a mate, think more about their partners as companions to have fun with than as future parents of their offspring. Related to the biosocial view of gender roles is the biopsychosocial view which includes psychological aspects such as level of self-control, emotions, and

*The ladies here probably exchanged looks which meant, "Men never know when things are dirty or not;" and the gentlemen perhaps thought each to himself, "Women will have their little nonsense and needless cares."*

JANE AUSTEN, *EMMA*

**Parental investment** any venture by a parent that increases the offspring's chance of surviving and thus increases reproductive success of the adult.

**Sociobiology** biological explanations for social behavior.

thinking as they impact gender role expression. How the individual perceives the social aspects of socioeconomic status, culture, poverty, technology, and religion will impact gender role expression. For example, a transgender person brings a personality to the culture that reflects various levels of acceptance or disapproval.

## Bioecological Model

The bioecological model, proposed by Urie Bronfenbrenner, emphasizes the importance of understanding bidirectional influences between an individual's development and his or her surrounding environmental contexts. The focus is on the combined interactive influences so that the predispositions of the individual interact with the environment/culture/society, resulting in various gender expressions. For example, the individual will read what gender role behavior his or her society will tolerate and adapt accordingly.

## Social Learning

Derived from the school of behavioral psychology, the social learning theory emphasizes the roles of reward and punishment in explaining how a child learns gender role behavior. This is in contrast to the biological explanation for gender roles. For example, consider the real life example of two young brothers who enjoyed playing "lady"; each of them put on a dress, wore high-heeled shoes, and carried a pocketbook. Their father came home early one day and angrily demanded, "Take those clothes off and never put them on again. Those things are for women." The boys were punished for "playing lady" but rewarded with their father's approval for boxing and playing football (both of which involved hurting others).

Reward and punishment alone are not sufficient to account for the way in which children learn gender roles. They also learn gender roles when parents or peers offer direct instruction (e.g., "girls wear dresses" or "a man stands up and shakes hands"). In addition, many of society's gender rules are learned through modeling. In modeling, children observe and imitate another's behavior. Gender role models include parents, peers, siblings, and characters portrayed in the media.

The impact of modeling on the development of gender role behavior is controversial. For example, a modeling perspective implies that children will tend to imitate the parent of the same sex, but children in all cultures are usually reared mainly by women. Yet this persistent female model does not seem to interfere with the male's development of the behavior that is considered appropriate for his gender. One explanation suggests that boys learn early that our society generally grants boys and men more status and privileges than it does girls and women. Therefore, boys devalue the feminine and emphasize the masculine aspects of themselves.

Regardless of the source, Witt and Wood (2010) found that expectations for following gender role standards are so powerful that one's well-being is related to the degree that one's behavior reflects these standards. Persons who accept the standards but who do not live up to them feel an impaired sense of well-being.

## Identification

Although researchers do not agree on the merits of Freud's theories (and students question its relevance), Freud was one of the first theorists to study gender role acquisition. He suggested that children acquire the characteristics and behaviors of their same-sex parent through a process of identification. Boys identify with their fathers, and girls identify with their mothers.

*All women become like their mothers. That is their tragedy. No man does. That's his.*

OSCAR WILDE, NOVELIST

## Cognitive-Developmental

The cognitive-developmental theory of gender role development reflects a blend of the biological and social learning views. According to this theory, the biological readiness of the child, in terms of cognitive development, influences how the child responds to gender cues in the environment (Kohlberg 1966). For example, gender discrimination (the ability to identify social and psychological characteristics associated with being female or male) begins at about age 30 months. However, at this age, children do not view gender as a permanent characteristic. Thus, even though young children may define people who wear long hair as girls and those who never wear dresses as boys, they also believe they can change their gender by altering their hair or changing clothes.

Not until age six or seven do children view gender as permanent (Kohlberg 1966; 1969). In Kohlberg's view, this cognitive understanding involves the development of a specific mental ability to grasp the idea that certain basic characteristics of people do not change. Once children learn the concept of gender permanence, they seek to become competent and proper members of their gender group. For example, a child standing on the edge of a school playground may observe one group of children jumping rope while another group is playing football. That child's gender identity as either a girl or a boy connects with the observed gender-typed behavior, and the child joins one of the two groups. Once in the group, the child seeks to develop behaviors that are socially defined as gender-appropriate.

# Agents of Socialization

Three of the four theories discussed in the preceding section emphasize that gender roles are learned through interaction with the environment. Indeed, though biology may provide a basis for one's gender identity, cultural influences in the form of various socialization agents (parents, peers, religion, and the media) shape the individual toward various gender roles. These powerful influences, in large part, dictate what people think, feel, and do in their roles as men or women. In the next section, we look at the different sources influencing gender socialization.

*In every conceivable manner, the family is the link to our past and the bridge to our future.*

ALEX HALEY, HISTORIAN

## Family

The family is an institution with female and male roles highly structured by gender. The names parents assign to their children, the clothes they dress them in, and the toys they buy them all reflect gender. Parents (particularly African-American mothers) may also be stricter on female children—determining the age they are allowed to leave the house at night, the time of curfew, and using directives such as "text me when you get to the party." Female children are also assigned more chores (Mandara et al. 2010).

**Fafafini** in Samoan society, biological males who are reared as females.

Siblings also influence gender role learning. As noted in Chapter 1, the relationship with one's sibling (particularly in sister-sister relationships) is likely to be the most enduring of all relationships (Meinhold et al. 2006). Also, growing up in a family of all sisters or all brothers intensifies social learning experiences toward femininity or masculinity. A male reared with five sisters and a single-parent mother is likely to reflect more feminine characteristics than a male reared in a home with six brothers and a stay-at-home dad.

**DIVERSITY IN OTHER COUNTRIES**

Samoan society/culture provides a unique example of gender role socialization via the family. The **Fafafini** (commonly called Fafa) are males reared as females. There are about 3,000 Fafafini in Samoa. The practice arose when there was a lack of women to perform domestic chores and the family had no female children. Thus, effeminate boys were identified and socialized/reared as females. Fafafini represent a third gender, neither female nor male; they are unique and valued, not stigmatized. Most Samoan families have at least one Fafafini child who takes on the role of a woman, including having sex with men (Abboud 2013).

## Peers

Though parents are usually the first socializing agents that influence a child's gender role development,

*But if the while I think on thee, dear friend,*
*All losses are restored and sorrows end.*
SHAKESPEARE, *SONNET 30*

peers become increasingly important during the school years. In regard to the degree to which peers influence the use of alcohol, 371 adolescents ages 11 to 13 years old participated in a study (Trucco et al. 2011). The researchers found that having peers who used alcohol and who approved of alcohol use by the participant predicted initiation of alcohol use. In another study, having one close friend who engaged in a high frequency of deviant behavior (e.g., smoking, drinking, driving recklessly) resulted in a greater influence than having several friends, only one of whom engaged in deviant behavior (Rees and Pogarsky 2011).

## Religion

*We are punished by our sins, not for them.*
ELBERT HUBBARD, AMERICAN WRITER

An example of how religion impacts gender roles involves the Roman Catholic Church, which does not have female clergy. Men dominate the 19 top positions in the U.S. dioceses. In addition, the Mormon Church is dominated by men and does not provide positions of leadership for women. Maltby et al. (2010) also observed that the stronger the religiosity for men, the more traditional and sexist their view of women. This association was not found for women.

### National **Data**

Of Americans aged 18 and older, 56% of adults in the United States say that religion is "very important" in their lives (Pew Research 2010).

*If you think education is expensive, try ignorance.*
ANDY MCINTYRE AND DEREK BOK

### DIVERSITY IN OTHER COUNTRIES

Boehnke (2011) examined the gender role attitudes among 20- to 55-year-old women and men across 24 highly developed countries and found that those with a higher level of educational attainment and those who were themselves offspring of a working mother were more approving of egalitarian gender roles.

## Education

The educational institution is another socialization agent for gender role ideology. However, such an effect must be considered in the context of the society or culture in which the "school" exists and of the school itself. Schools are basic cultures of transmission in that they make deliberate efforts to reproduce the culture from one generation to the next.

## Economy

**Occupational sex segregation** concentration of men and women in different occupations which has grown out of traditional gender roles.

The economy of the society influences the roles of the individuals in the society. The economic institution is a very gendered institution. **Occupational sex segregation** is the concentration of men and women in different occupations which has grown out of traditional gender roles. Men dominate as airline pilots, architects, and auto mechanics; women dominate as elementary school teachers, florists, and hair stylists (Weisgram et al. 2010). Only recently have women become NASCAR drivers (e.g., Danica Patrick).

Female-dominated occupations tend to require less education and have lower status. England (2010) noted that gains of women have been uneven because women have had a strong incentive to enter traditionally male jobs, but men have had little incentive to take on female jobs or activities. The salaries in these occupations (e.g., elementary education) have remained relatively low. Women are aware that there are gender inequities in pay and they are resigned to being paid less than men often at both the beginning and the peak of their careers (Hogue et al. 2010).

Increasingly, occupations are becoming less segregated on the basis of gender, and social acceptance of nontraditional career choices has increased. In 1960, 98% of persons entering veterinary medicine were male; today only 49% are male (Lincoln 2010).

Danica Patrick is one of the few females to break into Nascar racing.

## Mass Media

Mass media, such as movies, television, magazines, newspapers, books, music, computer games, and music television videos, both reflect and shape gender roles. Media images of women and men typically conform to traditional gender stereotypes, and media portrayals depicting the exploitation, victimization, and sexual objectification of women are common. As for music, Ter Gogt et al. (2010) studied 410 thirteen- to sixteen-year-old students and found that a preference for hip-hop music was associated with gender stereotypes (e.g., men are sex driven and tough, women are sex objects). In regard to music television videos, Wallis (2011) conducted a content analysis of 34 music videos and found that significant gender displays reinforced stereotypical notions of women as sexual objects, females as subordinate, and males as aggressive.

Pompper (2010) noted that media may project gender images that become unsettling for some men. The researcher conducted a series of interviews and noted how the media may threaten traditional conceptions of masculinity for some males. No longer is "the man" the tough, rugged cowboy. Real men can also be clean shaven, metrosexual "pretty boys." The cumulative effects of family, peers, religion, education, the economic institution, and mass media perpetuate gender stereotypes. Each agent of socialization reinforces gender roles that are learned from other agents of socialization, thereby creating a gender role system that is deeply embedded in our culture. All of these influences effect relationship choices (see Table 3.1).

**TABLE 3.1 Effects of Gender Role Socialization on Relationship Choices**

**Women**

1. A woman who is not socialized to pursue advanced education (which often translates into less income) may feel pressure to stay in an unhappy relationship with someone on whom she is economically dependent.
2. Women who are socialized to play a passive role and not initiate relationships are limiting interactions that might develop into valued relationships.
3. Women who are socialized to accept that they are less valuable and less important than men are less likely to seek, achieve, or require egalitarian relationships with men.
4. Women who internalize society's standards of beauty and view their worth in terms of their age and appearance are likely to feel bad about themselves as they age. Their negative self-concept, more than their age or appearance, may interfere with their relationships.
5. Women who are socialized to accept that they are solely responsible for taking care of their parents, children, and husband are likely to experience role overload. In this regard, some women may feel angry and resentful, which may have a negative impact on their relationships.
6. Women who are socialized to emphasize the importance of relationships in their lives will continue to seek relationships that are emotionally satisfying.

**Men**

1. Men who are socialized to define themselves in terms of their occupational success and income may discover their self-esteem and masculinity vulnerable if they become unemployed, retired, or work in a low-income job.
2. Men who are socialized to restrict their experience and expression of emotions are denied the opportunity to discover the rewards of emotional interpersonal involvement.
3. Men who are socialized to believe it is not their role to participate in domestic activities (child rearing, food preparation, house cleaning) will not develop competencies in these life skills. Potential partners often view domestic skills as desirable.
4. Heterosexual men who focus on cultural definitions of female beauty overlook potential partners who might not fit the cultural beauty ideal but who would be wonderful life companions.
5. Men who are socialized to have a negative view of women who initiate relationships will be restricted in their relationship opportunities.
6. Men who are socialized to be in control of relationship encounters may alienate their partners, who may desire equality.



# Gender Roles in Other Societies

Because culture largely influences gender roles, individuals reared in different societies typically display the gender role patterns of those societies. The following subsections discuss how gender roles differ in various societies.

## Gender Roles in Mexican, Latino/Hispanic Families

While all trace their roots to the Spanish and Mexican settlers in the Southwest before the arrival of the Pilgrims in New England, Hispanic families are both native born and those who immigrated from elsewhere. Hispanics today represent about 50 million, or 15 percent, of the U.S. population. Mexican Americans are the fastest growing and largest ethnic group in the United States (Becerra 2012).

Although the traditional family model in Spain calls for men as providers and women as homemakers and mothers, Hispanic families have been influenced by social, economic, and political change. Mobility and urbanization have also made their impact. Not only have the marital roles moved away from domination by the male but also there has been a loss of extended family in terms of number and influence. Today, much of the culture survives but change continues. Intermarriage of Hispanics to non-Hispanics (the highest of all intermarriage combinations) is another influence eroding traditional Hispanic patterns and values.

### Gender Roles in the Cuban American Family

As diplomatic dialogue between the United States and Cuba continues, there is interest in the Cuban American family. About two million Cuban Americans live in the United States. Traditional family patterns have undergone change from the original immigrant families strongly rooted in Cuban culture to the second generation with half of their influence coming from American norms to the third generation with most of their influence coming from American culture (Suarez and Perez 2012).

Some Cuban families have insulated themselves inside the "enclave," which reinforces their Cuban identity and a rich source of social capital: jobs. But as assimilation into American culture has continued, cultural traditions have been challenged. Female labor force participation has increased the power of the wife and challenged male authority in the home. Divorce rates have increased and singlehood has become more acceptable. A gradual transition from familism to individualism has occurred. "Although Cuban Americans can still be said to have strong family bonds, family patterns and values sustaining them seem to have weakened" (p. 120).

### Gender Roles in Caribbean Families

For spring break, college students sometimes go to the Bahamas, Jamaica, or other English-speaking islands in the Caribbean (e.g., Barbados, Trinidad, Guyana, or the Dominican Republic) and may wonder about the family patterns and role relationships of the people they encounter. The natives of the Caribbean number more than 30 million, with a majority being of African ancestry. Their family patterns are diverse but are often characterized by centrality of women and their children as the primary family unit—the fathers of these children rarely live in the home. For example, in Dominican society, the woman's importance is expressed in the phrase "Madre solo una y padre cualquiera" ("Mother is only one, and father anyone") (Hernandez, 2012). Hence, men may have children with different women and be psychologically and physically absent from their children's' lives. When a man does live with a woman, traditional division of labor prevails, with the woman taking care of domestic and child-care tasks.

### Gender Roles in East and South African Families

Africa is a diverse continent with more than 50 nations. The cultures range from Islamic and Arab cultures of northern Africa to industrial and European influences in South Africa. In some parts of East Africa (e.g., Kenya), gender roles are in flux.

Meredith Kennedy (2013) has lived in East Africa and makes the following observations of gender roles:

> *The roles of men and women in most African societies tend to be very separate and proscribed, with most authority and power in the men's domain. For instance, Maasai wives of East Africa do not travel much, since when a husband comes home he expects to find his wife (or wives) waiting for him with a gourd of sour milk. If she is not, he has the right to beat her when she shows up. Many African women who believe in and desire better lives will not call themselves "feminists" for fear of social censure. Change for people whose lives are based on tradition and "fitting in" can be very traumatic.*

In South Africa, where more than 75% of the racial population is African and around 10% is white, the African family is also known for its traditional role relationships and patriarchy. African men were socialized by the Dutch with firm patriarchal norms and adopted this style in their own families. In addition, women were subordinate to men "within a wider kinship system, with the chief as the controlling male."

# Consequences of Traditional Gender Role Socialization

This section discusses positive and negative consequences of traditional female and male socialization in the United States.

## Consequences of Traditional Female Role Socialization

In this section we summarize some of the negative and positive consequences of being socialized as a woman in U.S. society. Each consequence may or may not be true for a specific woman. For example, although women in general have less education and income, a particular woman may have more education and a higher income than a particular man.

**Negative Consequences of Traditional Female Role Socialization** There are several negative consequences of being socialized as a woman in our society.

**1. *Less Income.*** Although women earn more college degrees than men (Wells et al. 2011) and earn 47% of Ph.D.s (National Science Foundation 2012), they have lower academic rank and earn less money. The lower academic rank is because women give priority to the care of their children and family over their advancement in the workplace. Women still earn about three-fourths of what men earn, even when their level of educational achievement is identical (see Table 3.2). Their visibility in the ranks of high corporate America is also still low. Less than 5% of CEOs at Fortune 500 companies are women (Sandberg 2013). An exception is Virginia Rometty, CEO of IBM—the first female head of the company in 100 years.

However, the value women place on a high-income career is changing. The Pew Research Center (2012) reports a reversal of traditional gender roles with young women surpassing young men in saying that achieving success in a high-paying career or profession is important in their lives. Two-thirds (66%) of young women ages 18 to 34 rate career success high on their list of life priorities, compared with 59% of young men. In 1997, 56% of young women and 58% of young men felt the same way. Today's young men are also supportive of women having careers equal to their own. Fifty-six percent of 1,095 undergraduate men agreed that it was important to them that their wives had a career of equal status to their own (Hall and Knox 2013).

*Who she marries is the most important career decision a woman makes.*

SHERRYL SANDBERG, CHIEF OPERATING OFFICER AT FACEBOOK

**TABLE 3.2 Women's and Men's Median Income with Similar Education**

| | Bachelor's | Master's | Doctoral Degree |
|---|---|---|---|
| Men | $54,091 | $69,825 | $89,845 |
| Women | $35,972 | $50,576 | $65,587 |

Source: *Statistical Abstract of the United States*, 2012–2013. 131th ed. Washington, DC: U.S. Bureau of the Census, Table 702.

**2. *Feminization of Poverty.*** Another reason many women are relegated to a lower income status is the **feminization of poverty**. This term refers to the disproportionate percentage of poverty experienced by women living alone or with their children. Single mothers are particularly associated with poverty.

**Feminization of poverty** the disproportionate percentage of poverty experienced by women living alone or with their children.

When head-of-household women are compared with married-couple households, the median income is $32,597 versus $71,830 (*Statistical Abstract of the United States, 2012–2013,* Table 692). The process is cyclical—poverty contributes to teenage pregnancy because teens have limited supervision and few alternatives to parenthood (the median income for head-of-household men is $48,084). Such early childbearing interferes with educational advancement and restricts women's earning capacity, which keeps them in poverty. Their offspring are born into poverty, and the cycle begins anew.

Low pay for women is also related to the fact that they tend to work in occupations that pay relatively low incomes. Indeed, women's lack of economic power stems from the relative dispensability (it is easy to replace) of women's labor and how work is organized (men occupy and control positions of power). Women also live longer than men, and poverty is associated with being elderly.

When women move into certain occupations, such as teaching, there is a tendency in the marketplace to segregate these occupations from men's, and the result is a concentration of women in lower-paid occupations. The salaries of women in these occupational roles increase at slower rates. For example, salaries in the elementary and secondary teaching profession, which is predominately female, have not kept pace with inflation.

Conflict theorists assert that men are in more powerful roles than women and use this power to dictate incomes and salaries of women and "female professions." Functionalists also note that keeping salaries low for women keeps women dependent and in child-care roles so as to keep equilibrium in the family. Hence, for both conflict and structural reasons, poverty is primarily a feminine issue. One of the consequences of being a woman is to have an increased probability of feeling economic strain throughout life.

**3. *Higher Risk for Sexually Transmitted Infections.*** Due to the female anatomy, women are more vulnerable to sexually transmitted infections and HIV (they receive more bodily fluids from men). Some women also feel limited in their ability to influence their partners to wear condoms (East et al. 2011).

**4. *Negative Body Image.*** Just as young girls tend to have less positive self-concepts than boys (Yu and Zi 2010), they also feel more negatively about their bodies due to the cultural emphasis on being thin and trim (Grogan 2010). Darlow and Lobel (2010) noted that overweight women who endorse cultural values of thinness have lower self-esteem. There are more than 3,800 beauty pageants annually in the United States. The effect for many women who do not match the cultural ideal is to have a negative body image. Hollander (2010) reported on a study whereby teenage girls who viewed themselves as overweight had an increased risk of having their first sexual experience by age 13.

**DIVERSITY IN OTHER COUNTRIES**

While women in the United States have a more negative body image than do men, Mellor et al. (2010) studied a sample of 513 Malay, Indian, and Chinese adolescent boys and girls in Malaysia and found no differences between the genders in regard to body satisfaction. The researchers noted that most Asian females tend to have trim bodies which approximate the Western cultural ideal. However, the researchers observed that boys reported greater interest in strategies to increase muscles. Again, Asian males tend to be smaller and may be more motivated to work out to achieve the male Western cultural ideal of a muscular body.

American women also live in a society that devalues them in a larger sense. Their lives and experiences are not taken as seriously as men's. **Sexism** is an attitude, action, or institutional structure that subordinates or discriminates against individuals or groups because of their sex. Sexism against women reflects the tradition of male dominance and presumed male superiority in American society.

**Sexism** an attitude, action, or institutional structure that subordinates or discriminates against individuals or groups because of their sex.

## FAMILY POLICY | Female Genital Alteration

**Female genital alteration,** more commonly known as FGC (female genital cutting), female genital mutilation, or female circumcision, involves cutting off the clitoris or excising (partially or totally) the labia minora. "World-wide about 130 million women have undergone FGC. In the USA, more than 168,000 females have had or are at risk for this procedure and the number may be increasing as the admission ceiling for African refugees is raised. Federal law criminalizes the performance of FGC on females under 18 in the United States; however, the procedure is not unknown in this country. More commonly, young women are sent back to their country of origin for the procedure. Over 90% of women from Djibouti, Egypt, Eritrea, Ethiopia, Mali, Sierra Leone, and Northern Sudan have had the procedure" (Nicoletti 2007) as well as 98% of the women in Somalia (Simister, 2010). The American Academy of Pediatrics condemns all types of female genital cutting (Policy Statement 2010).

The practice of FGC is not confined to a particular religion. The reasons for the practice include:

- **a.** Sociological/cultural—parents believe that female circumcision makes their daughters lose their desire for sex, which helps them maintain their virginity and helps to ensure their marriageability and fidelity to their husbands. Hence, the "circumcised" female is seen as one whom males will desire as a wife. FGC is seen as a "rite of passage" that initiates a girl into womanhood and increases her bonding and social cohesion with other females.
- **b.** Hygiene/aesthetics—female genitalia are considered dirty and unsightly so their removal promotes hygiene and provides aesthetic appeal.
- **c.** Religion—some Muslim communities practice FGC in the belief that the Islamic faith demands it, but it is not mentioned in the Qur'an.
- **d.** Myths: FGC is thought to enhance fertility and promote child survival (Nicoletti 2007).

Elnashar and Abdelhady (2007) compared the sexuality of married women who had been circumcised with those who had not. The researchers found statistically significant differences such that the former were more likely to report pain during intercourse, loss of libido, and failure to orgasm. The wives who had been circumcised also reported more physical complaints, anxiety, and phobias.

Changing a country's deeply held beliefs and values concerning this practice cannot be achieved by legislation alone. Efforts in Senegal have been less than effective where there has been resistance and, in some cases, the practice has been driven underground (Bettina et al. 2013). More effective approaches to discourage the practice include:

1. Respect the beliefs and values of countries that practice female genital operations. Calling the practice "genital mutilation" and "barbaric" and referring to it as a form of "child abuse" and "torture" convey disregard for the beliefs and values of the cultures where it is practiced. In essence, we might adopt a culturally relativistic point of view (without moral acceptance of the practice).
2. Remember that genital operations are arranged and paid for by loving parents who deeply believe that the surgeries are for their daughters' welfare.

**Benevolent sexism** reflects the belief that women are innocent creatures who should be protected and supported.

**Benevolent sexism** (reviewed by Maltby et al. 2010) is a related term and reflects the belief that women are innocent creatures who should be protected and supported. While such a view has positive aspects, it assumes that women are best suited for domestic roles and need to be taken care of by a man since they are not capable of taking care of themselves.

**5.** ***Less Personal/Marital Satisfaction.*** Demaris et al. (2012) analyzed data on 707 marriages and found that couples characterized by more traditional attitudes toward gender roles were significantly less satisfied than others. Wives in traditional marriages are particularly likely to report lower marital satisfaction (Bulanda 2011). Such lower marital satisfaction of wives is attributed to power differentials in the marriage. Traditional husbands expect to be dominant and expect their wives to take care of the house and children. Women have been socialized to believe that they can have both a family and a career—that they can have it all. In reality, they have discovered that they "can have two things

3. It is important to be culturally sensitive to the meaning of being a woman. Indeed, genital cutting is mixed up with how a woman sees herself.

Simister (2010) studied national samples of genital alteration in Kenya and found that the higher the education of the mother, the lower the incidence of genital alteration in her daughters. Hence, increasing the educational level of women in a community is a structural way to reduce genital alteration in females.

Braun (2009) discussed the practice of female genital cosmetic surgery (FGCS) where Western women voluntarily have their genitals altered for aesthetic reasons. Such surgeries include those designed to reduce size of the labia majora (labiaplasty) and those designed to tighten the vagina (so-called vaginal rejuvenation or vaginoplasty). These operations are sometimes called "designer vaginas" and have been critiqued by the American College of OBGYN. Researcher Braun finds it interesting that surgery on the genitals can be both a cause for public outcry and an issue of personal cosmetics.

### Your Opinion?

1. To what degree do you feel the United States should become involved in the practice of female genital alterations of U.S. citizens?
2. To what degree can you regard the practice from the view of traditional parents and daughters?
3. How could not having the operation be a liability and a benefit for the woman whose culture supports the practice?
4. How does education influence the probability of genital alteration?
5. How might women who have female genital cutting view women who elect to have vaginal surgeries?

### Sources


Bettina, S., K. Wander, Y. Hernlund, and A. Moreau. 2013. Legislating change? Responses to criminalizing female genital cutting in Senegal. *Law & Society* 47: 803–835.

Braun, V. 2009. "The women are doing it for themselves": The rhetoric of choice and agency around female genital "cosmetic surgery." *Australian Feminist Studies* 24: 233–249.

Elnashar, A., and R. Abdelhady. 2007. The impact of female genital cutting on health of newly married women. *International Journal of Gynecology & Obstetrics* 97: 238–44.

Nicoletti, A. 2007. Female genital cutting. *Journal of Pediatric & Adolescent Gynecology* 20: 261–62.

Policy Statement — Ritual Genital Cutting of Female Minors. 2010. *Pediatrics* 125: 1088–1093.

Simister, J. G. 2010. Domestic violence and female genital mutilation in Kenya: Effects of ethnicity and education. *Journal of Family Violence* 25: 247–257.


half way" and that men are not much help (Haag 2011, p. 47). In large nonclinical samples comparing wives and husbands on marital satisfaction (which include egalitarian marriages), wives do not report lower marital satisfaction.

Poorer mental health (Read and Grundy 2011) and higher levels of anger (Simon and Lively 2010) among women are also associated with traditional gender roles. One interpretation of these associations is related to the sense of powerlessness that women feel in America due to inequitable division of labor, their likelihood of holding lower status/lower wage jobs, less power in relationships, etc.

Before ending this section on negative consequences of being socialized as a woman, look at the Family Policy feature on female genital alteration. This is more of an issue for females born in some African, Middle Eastern, and Asian countries than for women in the United States. However, the practice continues even here.

**Positive Consequences of Traditional Female Role Socialization** We have discussed the negative consequences of being born and socialized as a woman. However, there are also decided benefits.

1. ***Longer Life Expectancy.*** Women have a longer life expectancy than men. It is not clear if their greater longevity is related to biological or to social factors.

## National **Data**

Females born in the year 2020 are expected to live to the age of 81.9, in contrast to men, who are expected to live to the age of 77.1 (*Statistical Abstract of the United States, 2012–2013,* Table 104).

2. ***Stronger Relationship Focus.*** Women prioritize family over work and do more child care than men (Craig & Mullan 2010). Mothers provide more "emotion work," helping children with whatever they are struggling with (Minnottea et al. 2010).

3. ***Keeps Relationships on Track.*** Women are more likely to initiate "the relationship talk" to ensure that the relationship is moving forward (Nelms et al. 2012). In addition, when there is a problem in the relationship, it is the woman who moves the couple toward help (Eubanks Fleming and Córdova 2012).

*For the woman, the man is a means: the end is always the child.*

FRIEDRICH NIETZSCHE, PHILOSOPHER

4. ***Bonding with Children.*** Another advantage of being socialized as a woman is the potential to have a closer bond with children. In general, women tend to be more emotionally bonded with their children than men. Although the new cultural image of the father is of one who engages emotionally with his children, many fathers continue to be content for their wives to take care of their children, with the result that mothers, not fathers, become more emotionally bonded with their children. Table 3.3 summarizes the consequences of traditional female role socialization.

## Consequences of Traditional Male Role Socialization

Male socialization in American society is associated with its own set of consequences. As with women, each consequence may or may not be true for a specific man.

**Negative Consequences of Traditional Male Role Socialization** There are several negative consequences associated with being socialized as a man in U.S. society.

1. ***Identity Synonymous with Occupation.*** Ask men who they are, and many will tell you what they do. Society tends to equate a man's identity with his occupational role. Male socialization toward greater involvement in the labor force is evident in governmental statistics.

That men work more and play less may translate into fewer friendships and relationships. While friendships are important to both men and women

**TABLE 3.3 Consequences of Traditional Female Role Socialization**

| Negative Consequences | Positive Consequences |
|---|---|
| Less income (more dependent) | Longer life |
| Feminization of poverty | Stronger relationship focus |
| Higher STD/HIV infection risk | Keeps relationships on track |
| Negative body image | Bonding with children |
| Less personal/marital satisfaction | Identity not tied to job |

Fishing is a pastime in which males bond.

## National Data

Seventy-one percent of men, compared with 59% of women, were estimated to be in the civilian workforce in 2008 (*Statistical Abstract of the United States,* 2012–2013, Table 587).

(Meliksah et al. 2013), men have fewer sustained friendships than women across time. This is due not only to their work focus but also to the cultural values of independence which support men being the loner (Way 2013). Grief (2006) suggested other reasons men have fewer friends—homophobia, lack of role models, fear of being vulnerable, and competition between men.

**2.** ***Limited Expression of Emotions.*** Most men not only cry less (Barnes et al. 2012) but are also pressured to disavow any expression that could be interpreted as feminine (e.g., emotional). Lease et al. (2010) confirmed that the socialization of men (particularly white men) involves men "being more emotionally isolated and competitive in relationships and less competent at providing support to others, disclosing their own feelings, or managing conflict effectively."

**3.** ***Fear of Intimacy.*** Garfield (2010) reviewed men's difficulty with emotional intimacy and noted that their emotional detachment stems from the provider role which requires them to stay in control. Being emotional is seen as weakness. Men's groups where men learn to access their feelings and express them have been helpful in increasing men's emotionality. Murphy et al. (2010) emphasized that men profit from becoming involved in the emotional labor of maintaining a relationship—their satisfaction increases.

*The majority of husbands remind me of an orangutan trying to play the violin.*

HONORE DE BALZAC, FRENCH NOVELIST AND PLAYWRIGHT

**4.** ***Custody Disadvantages.*** Courts are sometimes biased against divorced men who want custody of their children. Because divorced fathers are typically regarded as career focused and uninvolved in child care, some are relegated to seeing their children on a limited basis, such as every other weekend or four evenings a month.

**5.** ***Shorter Life Expectancy.*** Men typically die five years sooner (at age 77) than women. One explanation is that the traditional male role emphasizes achievement, competition, and suppression of feelings, all of which may produce stress. Not only is stress itself harmful to physical health, but it may lead to

**TABLE 3.4 Consequences of Traditional Male Role Socialization**

| Negative Consequences | Positive Consequences |
|---|---|
| Identity tied to work role | Higher income and occupational status |
| Limited emotionality | More positive self-concept |
| Fear of intimacy; more lonely | Less job discrimination |
| Disadvantaged in getting custody | Freedom of movement; more partners to select from; more normative to initiate relationships |
| Shorter life | Happier marriage |



*The true man wants two things: danger and play. For that reason he wants woman, as the most dangerous plaything.*

FRIEDRICH NIETZSCHE, PHILOSOPHER

compensatory behaviors such as smoking, alcohol, or other drug abuse, and dangerous risk-taking behavior (e.g., driving fast, binge drinking).

**Benefits of Traditional Male Socialization** As a result of higher status and power in society, men tend to have a more positive self-concept and greater confidence in themselves. In a sample of 1,104 undergraduate men, 29.7% "strongly agreed" with the statement, "I have a very positive self-concept." In contrast, 19% of 3,473 undergraduate women strongly agreed with the statement (Hall and Knox 2013). Men also enjoy higher incomes and an easier climb up the good-old-boy corporate ladder; they are stalked/followed/harassed less often than women (18.2% versus 27.8%). Other benefits are the following:

1. ***Freedom of Movement.*** Men typically have no fear of going anywhere, anytime. Their freedom of movement is unlimited. Unlike women (who are taught to fear rape, be aware of their surroundings, walk in well-lit places, and not walk alone after dark) men are oblivious to these fears and perceptions. They can go anywhere alone and are fearless about someone harming them.

2. ***Greater Available Pool of Potential Partners.*** Because of the mating gradient (men marry "down" in age and education whereas women marry "up"), men tend to marry younger women so that a 35-year-old man may view women from 20 to 40 years of age as possible mates. However, a woman of age 35 is more likely to view men her same age or older as potential mates. As she ages, fewer men are available.

3. ***Norm of Initiating a Relationship.*** Men are advantaged because traditional norms allow them to be aggressive in initiating relationships with women. In addition, men tend to initiate marriage proposals and the bride more often takes the last name of her husband rather than keeping her last name (Robert and Leaper 2013). Table 3.4 summarizes the consequences of being socialized as a male.

# Changing Gender Roles

Androgyny, gender role transcendence, and gender postmodernism emphasize that gender roles are changing.

**Androgyny** a blend of traits that are stereotypically associated with masculinity and femininity.

## Androgyny

**Androgyny** is a blend of traits that are stereotypically associated with masculinity and femininity. Androgynous celebrities include Lady Gaga,

David Bowie, Boy George, Patti Smith, and Annie Lennox. Two forms of androgyny are:

**1.** Physiological androgyny refers to intersexed individuals, discussed earlier in the chapter. The genitals are neither clearly male nor female, and there is a mixing of "female" and "male" chromosomes and hormones.

**2.** Behavioral androgyny refers to the blending or reversal of traditional male and female behavior, so that a biological male may be very passive, gentle, and nurturing and a biological female may be very assertive, rough, and selfish.

Androgyny may also imply flexibility of traits; for example, an androgynous individual may be emotional in one situation, logical in another, and assertive in another. Gender role identity (androgyny, masculinity, femininity) was assessed in a sample of Korean and American college students with androgyny emerging as the largest proportion in the American sample and femininity in the Korean sample (Shin et al. 2010).

**Gender role transcendence** abandoning gender frameworks and looking at phenomena independent of traditional gender categories.

## Gender Role Transcendence

Beyond the concept of androgyny is that of gender role transcendence. We associate many aspects of our world, including colors, foods, social or occupational roles, and personality traits, with either masculinity or femininity. The concept of **gender role transcendence** means abandoning gender frameworks and looking at phenomena independent of traditional gender categories.

Transcendence is not equal for women and men. Although females are becoming more masculine, in part because our society values whatever is masculine, men are not becoming more feminine. Indeed, adolescent boys may be described as very gender entrenched. Beyond gender role transcendence is gender postmodernism.

David Knox

Fishing is thought of as a "man's sport." But gender roles are in flux and women are free to enjoy fishing.

## Gender Postmodernism

Gender postmodernism abandons the notion of gender as natural and emphasizes that gender is socially constructed. Fifteen years ago, Monro (2000) noted that people in the postmodern society would no longer be categorized as male or female but be recognized as capable of many identities—"a third sex" (p. 37). A new conceptualization of "trans" people calls for new social structures, "based on the principles of equality, diversity and the right to self determination" (p. 42). No longer would our society telegraph transphobia but would instead embrace pluralization "as an indication of social evolution, allowing greater choice and means of self-expression concerning gender" (p. 42).

## PERSONAL CHOICES

### *Choosing gender behavior that fits*

Being aware that gender role behavior is socially constructed gives one the freedom to engage in whatever occupational and relationship gender role behavior that seems a natural fit for one's personality. Occupational choices traditionally reserved for women (e.g., elementary school teacher) or men (e.g., athletic coach) need no longer be off the table for the other sex. Similarly, dating roles whereby the woman initiates and the man is passive or marital roles whereby the woman is the primary breadwinner and the man is the child-focused homemaker become options.

## The Future of Gender Roles

Imagine a society in which women and men each develop characteristics, lifestyles, and values that are independent of gender role stereotypes. Characteristics such as strength, independence, logical thinking, and aggressiveness are no longer associated with maleness, just as passivity, dependence, emotions, intuitiveness, and nurturance are no longer associated with femaleness. Both sexes are considered equal, and women and men may pursue the same occupational, political, and domestic roles. These changes are occurring...slowly. Lucier-Greer and Adler-Baeder (2010) provided data which showed that, compared to 2000, gender role attitudes of today are becoming more egalitarian. Fisher (2010) emphasized that peer marriage or marriage between equals is the most profound change in marriage in recent years. McGeorge et al. (2012) noted that couple therapists are being trained to be sensitive to and encourage relationship equality in terms of power and division of labor in relationships.

Another change in gender roles is the independence and ascendency of women. Women will less often require marriage for fulfillment, will increasingly take care of themselves economically, and will opt for having children via adoption or donor sperm rather than foregoing motherhood. That women are slowly outstretching men in terms of education will provide the impetus for these changes.

## Summary

***What are the important terms related to gender?***

*Sex* refers to the biological distinction between females and males. One's biological sex is identified on the basis of one's chromosomes, gonads, hormones, internal sex organs, and external genitals, and exists on a continuum rather than being a dichotomy. *Gender* is a social construct and refers to the social and psychological characteristics associated with being female or male. Other terms related to gender include *gender identity* (one's self-concept as a girl or boy), *gender role* (social norms of what a girl or boy "should" do), *gender role ideology* (how women and men "should" interact), and *transgender* (expressing characteristics different from one's biological sex).

***What theories explain gender role development?***

Biosocial theory emphasizes that social behaviors (e.g., gender roles) are biologically based and have an evolutionary survival function. Biopsychosocial theory includes the psychological/personality dimension of gender role expression. Ecobiological emphasizes the interaction of the individual with the environment. Social learning theory emphasizes the roles of reward and punishment in explaining how children learn gender role behavior. Identification theory says that children acquire the characteristics and behaviors of their same-sex parent through a process of identification. Cognitive-developmental theory emphasizes biological readiness, in terms of cognitive development, of the child's responses to gender cues in the environment. Once children learn the concept of gender permanence, they seek to become competent and proper members of their gender group.

***What are the various agents of socialization?***

Various socialization influences include parents and siblings (representing different races and ethnicities), peers, religion, the economy, education, and mass media. These shape individuals toward various gender roles and influence what people think, feel, and do in their role as woman or man.

***How are gender roles expressed in other societies?***

In Mexican, Latino, and Hispanic families, not only have the marital roles moved away from domination by the male but also there has been a loss of extended family in number and influence. Today, much of the culture survives but change continues. Intermarriage of Hispanics to non-Hispanics (the highest of all intermarriage combinations) is another influence eroding traditional Hispanic patterns and values.

For the Cuban American family, as assimilation into American culture has continued, cultural traditions have been challenged. Female labor force participation has increased the power of the wife and challenged male authority in the home. Divorce rates have increased and singlehood has become more acceptable. A gradual transition from familism to individualism has occurred. In the Caribbean, family patterns are diverse but are often characterized by women and their children as the primary family unit, with men often not living in the home.

***What are the consequences of traditional gender role socialization?***

Traditional female role socialization may result in negative outcomes such as less income, negative body image, and lower marital satisfaction but positive outcomes such as a longer life, a stronger relationship focus, keeping relationships on track, and a closer emotional bond with children. Traditional male role socialization may result in the fusion of self and occupation, a more limited expression of emotion, disadvantages in child custody disputes, and a shorter life but higher income, greater freedom of movement, a greater available pool of potential partners, and greater acceptance in initiating relationships.

***How are gender roles changing?***

Androgyny refers to a blend of traits that are stereotypically associated with both masculinity and femininity. It may also imply flexibility of traits; for example, an androgynous individual may be emotional in one situation, logical in another, assertive in another, and so forth. The concept of gender role transcendence involves abandoning gender schema so that personality traits, social and occupational roles, and other aspects of our lives become divorced from gender categories. However, such transcendence is not equal for women and men. Although females are becoming more masculine partly because our society values whatever is masculine, men are not becoming more feminine.

***What is the future of gender roles?***

Imagine a society in which women and men each develop characteristics, lifestyles, and values that are independent of gender role stereotypes. Characteristics such as strength, independence, logical thinking, and aggressiveness are no longer associated with maleness, just as passivity, dependence, emotions, intuitiveness, and nurturance are no longer associated with femaleness. Both sexes are considered equal, and women and men may pursue the same occupational, political, and domestic roles. These changes are occurring, albeit slowly. Women are also becoming more ascendant—they are earning more college degrees than men; they increasingly can take care of themselves economically; and they less often require marriage for their personal fulfillment.

## Key Terms

Androgyny
Benevolent sexism
Biosocial theory
Fafafini
Female genital alteration
Feminization of poverty
Gender
Gender identity
Gender role ideology
Gender role transcendence
Gender roles
Intersexed individuals
Occupational sex segregation
Parental investment
Sex
Sex roles
Sexism
Socialization
Sociobiology
Transgender
Transsexual

## Web Links

American Men's Studies Association
http://www.mensstudies.org/

Equal Employment Opportunity Commission
http://www.eeoc.gov/

Intersex Society of North America
http://www.isna.org/

National Organization for Women (NOW)
http://www.now.org/

Transgender Forum
http://www.tgforum.com/

Transsexuality
http://www.transsexual.org/

# CHAPTER 4 Singlehood, Hanging Out, Hooking Up, and Cohabitation

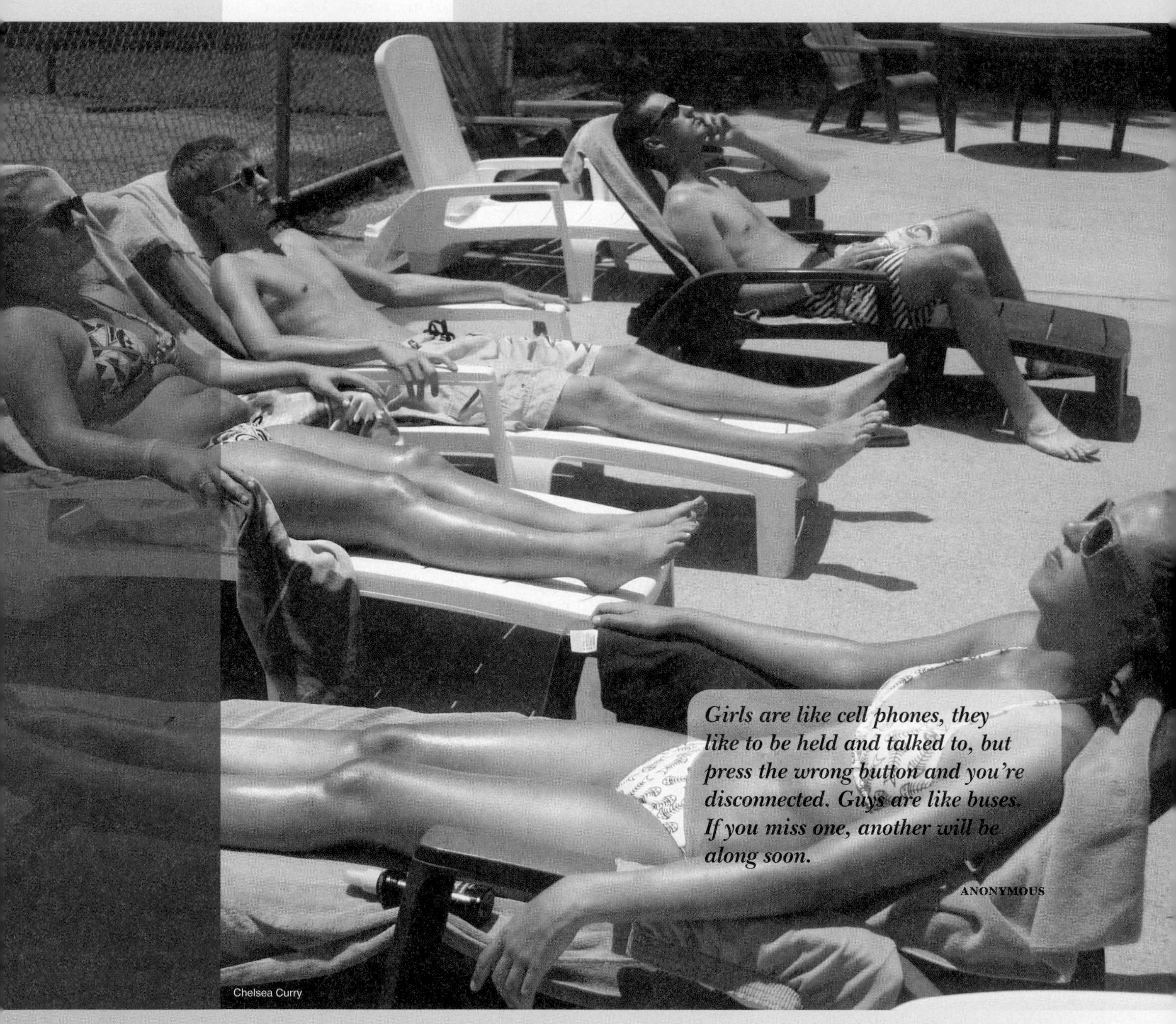

*Girls are like cell phones, they like to be held and talked to, but press the wrong button and you're disconnected. Guys are like buses. If you miss one, another will be along soon.*

ANONYMOUS

Chelsea Curry

## Learning Objectives

- Explain the types of singles and the reasons individuals give for delaying marriage.
- Identify the changes in dating in the last 60 years.
- Understand the goals and outcomes of "hooking up."
- Describe the types of cohabitant relationships and the relationship consequences.
- Review the positives and negatives of living apart together.
- Discuss the future of singlehood.

## TRUE OR FALSE?

1. Over half of women and men today report that they never want to get married.
2. Engaged cohabitants are happier than cohabitants who are just testing their relationship.
3. Individuals in France, Germany, and Italy are very focused on getting married early.
4. Cohabitants in the United States live together a much shorter period of time than cohabitants in Canada.
5. When romantic partners disagree about whether to cohabit, the preferences of the one who wants to cohabit wins out and the couple end up living together.

*Answers:* 1. F  2. T  3. F  4. T  5. F

A major change in values regarding relationships in America is that being single has become more acceptable. Rather than the never married being stigmatized and pitied because of their presumed loneliness and unhappiness, married couples are often viewed as trapped with no freedom to enjoy their lives. "My friends are all getting married and their life is over" is one way some singles view marriage.

Though six in ten U.S. single adults report wanting to get married (Cohn 2013), youth today are in no hurry. They enjoy the freedom of singlehood and most put off marriage until their late twenties. Young American adults reflect the same pattern of youth in other countries: individuals in France, Germany, and Italy are engaging in a similar pattern of delaying marriage. In the meantime, the process of courtship has evolved, with various labels and patterns, including hanging out (undergraduates less often use the term "dating" but say they are "seeing someone"), hooking up (the new term for "one-night stand"), and "being in a relationship" for seeing each other exclusively (and changing one's Facebook status), which may include cohabitation as a prelude to marriage. We begin with examining singlehood versus marriage.

# Singlehood

In this section, we discuss the various categories of singles, why individuals are delaying marriage, the choice to be permanently unmarried, and the legal blurring of the married and unmarried.

## Categories of Singles

The term singlehood is most often associated with young unmarried individuals. However, there are three categories of single people: the never-married, the divorced, and the widowed.

*When I'm single, I don't focus. I focus on a guy if he's a boyfriend, but I don't focus on finding a boyfriend. They're never around when you want them.*

SCARLETT JOHANSSON, ACTRESS

**Never-Married Singles** It is rare for an adult in the United States to remain unmarried his or her entire life (see the National Data insert). Eventually, almost all adults marry (Bill Maher and Oprah Winfrey are examples of well-known never marrieds). Indeed, even feminist Gloria Steinem married at age 66. Nevertheless, more individuals are remaining single longer. They are completing their educations and building their careers while enjoying the freedom of singlehood and its attendant companionship, sex, cohabitation, and procreation without being married. Some ask, "Is marriage obsolete?" (Eck 2013).

Chelsea Curry

Chelsea Curry

These never-married heterosexuals enjoy their life/lifestyle—they report that a major benefit is the freedom to live their life as they choose.

## International **Data**

In a review of worldwide singlehood, of over 150 countries reporting, all but 16 of the countries revealed that 20% had never married by age 49 (United Nations 2011).

What are the characteristics of those who never marry? Single men are likely to be less educated (Murray 2012) and to have lower incomes (Ashwin and Isupova 2014) than married men. Research on never-married women women reveals that, as a group, they tend to be poor, to have mental/physical health issues, to use drugs, and to have children with multiple partners (Manning et al. 2010). Those who are never married (both women and men) also tend to be obese (Sobal and Hanse, 2011) and to have been reared in single-parent families (Valle and Tillma, 2014). The unmarried disproportionately live in large cities—New York, DC, Los Angeles, Atlanta, and Boston. These individuals are often young adults seeking adventure, careers, and relationships.

Are singles less happy than spouses? Yes. While over three-quarters of 5,500 single individuals reported that they were happy with their personal life (Walsh 2013), they were less happy when compared to those who were married. Wienke and Hill (2009) compared single people with married people and cohabitants (both heterosexual and homosexual) and found that single people were less happy regardless of sexual orientation. Rauer (2013) also compared the unmarried (she included divorced and widowed) with the married and found that the latter were happier and healthier and that these benefits were particularly pronounced in the elderly....that "the negative effects of being unmarried are disproportionately felt by older adults."

*Being single entitles you to not ask someone's opinion before you do something to mess up your own room, your credit report or even your life.*

SAYALI PATIL, INDIAN BUSINESSWOMAN

Nevertheless, increasingly, individuals are living alone: 28% of all households were single households in 2011, compared to 9% in 1950. And "why not?" asks sociologist Klinenberg (2012): "living alone helps to pursue sacred modern values such as individual freedom, personal control and self-realization." His book *Going Solo* reflects interviews with 300 adults who live alone and, for the most part, enjoy doing so. Eck (2013) interviewed single individuals and found most content in being unmarried. Like all lifestyles, never being married is a complex issue, as revealed by a single woman in Table 4.1.

## TABLE 4.1 A Never-Married Single Woman's View of Singlehood

A never-married female, 40 years of age, spoke to our marriage and family class about her experience as a single woman. The following is from the outline she developed and the points she made about each topic.

**Stereotypes about Never-Married Women**

Various assumptions are made about the never-married woman and why she is single. These include the following:
*Unattractive*—She's either overweight or homely, or else she would have a man.
*Lesbian*—She has no real interest in men and marriage because she is homosexual.
*Workaholic*—She's career driven and doesn't make time for relationships.
*Poor interpersonal skills*—She has no social skills, and she embarrasses men.
*History of abuse*—She has been turned off to men by the sexual abuse of, for example, her father, a relative, or a date.
*Negative previous relationships*—She's been rejected again and again and can't hold a man.
*Man-hater*—Deep down, she hates men.
*Frigid*—She hates sex and avoids men and intimacy.
*Promiscuous*—She is indiscriminate in her sexuality so that no man respects or wants her.
*Too picky*—She always finds something wrong with each partner and is never satisfied.
*Too weird*—She would win the Miss Weird contest, and no man wants her.

**Positive Aspects of Being Single**

1. Freedom to define self in reference to own accomplishments, not in terms of a man.
2. Freedom to pursue own personal and career goals without the time restrictions posed by a spouse and children.
3. Freedom to come and go as you please and to do what you want, when you want.
4. Freedom to establish relationships with members of both sexes at the desired level of intensity and intimacy.
5. Freedom to travel and explore new cultures, ideas, values.

**Negative Aspects of Being Single**

1. Increased extended-family responsibilities. The unmarried sibling is assumed to have the time to care for elderly parents.
2. Increased job expectations. The single employee does not have marital or family obligations and is often expected to work at night, on weekends, and on holidays.
3. Isolation. Too much time alone does not allow others to give feedback such as "Are you drinking too much?", "Have you had a checkup lately?", or "Are you working too much?"
4. Decreased privacy. Others assume the single person is always at home and always available. They may call late at night or drop in whenever they feel like it.
5. Less safe. A single woman living alone is more vulnerable than a married woman with a man in the house.
6. Feeling different. Many work-related events are for couples, husbands, and wives. A single woman stands out.
7. Lower income. Single women have much lower incomes than married couples.
8. Less psychological intimacy. The single woman often does not have an emotionally intimate partner.
9. Negotiation skills lie dormant. Because single people do not negotiate issues with someone on a regular basis, they may become deficient in compromise and negotiation skills.
10. Patterns become entrenched. Because no other person is around to express preferences, the single person may establish a very repetitive lifestyle.
11. Since single women typically have a higher number of sexual partners than a married woman, they are at greater risk for an STI and an HIV infection.

**Maximizing One's Life as a Single Person**

1. Frank discussion. Talk with parents about your commitment to and enjoyment of the single lifestyle and request that they drop marriage references. Talk with siblings about joint responsibility for aging parents and your willingness to do your part. Talk with employers about spreading workload among all workers, not just those who are unmarried and childfree.
2. Relationships. Develop and nurture close relationships with parents, siblings, extended family, and friends to have a strong and continuing support system.
3. Participate in social activities. Go to social events with or without a friend. Avoid becoming a social isolate.
4. Be cautious. Be selective in sharing personal information such as your name, address, and phone number. This caveat is particularly important if the Internet is used to find dating partners.
5. Money. Pursue education to maximize income; set up a retirement plan. Plan for the later years.
6. Health. Exercise, have regular checkups, and eat healthy food. Take care of yourself.

*For I shall never give up the state of living alone, which has manifested itself as an indescribable blessing.*

EINSTEIN

**Divorced Singles** Another category of those who are single is the divorced. While some divorced singles may have ended the marriage, other unions were terminated by their spouse. Hence, some are voluntarily single again while others are forced into being single again.

**National Data**

There were 13.7 million divorced females and 9.9 million divorced males in the United States in 2009 (*Statistical Abstract of the United States, 2012–2013,* Table 5.7).

The divorced have a higher suicide risk. Denney (2010) examined the living arrangements of over 800,000 adults and found that being married or living with children decreases one's risk of suicide. Spouses are more likely to be connected to intimates; this connection seems to insulate a person from suicide. Of course, intimate connections can occur outside of marriage but marriage tends to ensure these connections over time.

Married people also look out for the health of their mate. Spouses often prod each other to "go to the doctor," "have that rash on your skin looked at," and "remember to take your medication." Single people often have no one in their life to nudge them toward regular health maintenance.

**Widowed Singles** Although divorced people have often chosen to leave their spouses and be single again, the widowed are forced into singlehood. The stereotype of the widow and widower is utter loneliness, even though there are compensations (e.g., escape from an unhappy marriage, Social Security). Kamiya et al. (2013) found that widowhood for men was associated with depressive symptoms. But this association was mitigated by income.

**National Data**

There are 11.4 million widowed females and 2.9 widowed males in the United States (*Statistical Abstract of the United States, 2012–2013,* Table 57).

## Individuals Are Delaying Marriage Longer

Having identified the three categories of singles as never married, divorced, and widowed, we now focus on the never married, the largest group. A primary question is whether or not more people are choosing to never marry. We do not know the answer. What we do know is that individuals are *delaying marriage.* As evidence, in 1960, two-thirds (68%) of all twenty-somethings were married. More recently, just over a fourth (26%) in this age category was married (Pew Research Center 2010). Will those who are delaying marriage eventually marry? We suspect most will but don't know. We need to wait till the current cohort of youth reach age 75 and beyond to see if as high a percentage (96%) eventually marry as is true of the current percent of those age 75 and older (*Statistical Abstract of the United States, 2012–1013, Table 57*). The self-assessment to follow allows you to evaluate your attitude toward singlehood.

## Reasons for Delaying Marriage

Muraco and Curran (2012) identified 13 reasons individuals delay marriage. These include the need for financial stability, inability to pay for a wedding,

## SELF-ASSESSMENT | Attitudes Toward Singlehood Scale

The purpose of this survey is to assess the degree to which students view remaining single (never getting married) positively. Read each item carefully and consider what you believe. There are no right or wrong answers, so please give your honest reaction and opinion. After reading each statement, select the number that best reflects your level of agreement, using the following scale:

| 1 | 2 | 3 | 4 | 5 | 6 | 7 |
|---|---|---|---|---|---|---|
| Strongly Disagree | | | | | | Strongly Agree |

____ 1. I plan to remain single.
____ 2. Getting married is not as advantageous as it used to be.
____ 3. Marriage is another word for being trapped.
____ 4. Singlehood is another word for being free.
____ 5. Single people are happier than married people.
____ 6. Most of the married people I know are unhappy.
____ 7. Most of the single people I know enjoy their lifestyle.
____ 8. The statement that singles are lonely and unhappy is not true.
____ 9. It is better to be single than to be married.
____ 10. Of the two lifestyle choices, single or married, single is better.
____ 11. Singles have more friendships/social connections than spouses.
____ 12. Singles have a greater sense of independence than spouses.
____ 13. Being single is a more satisfying lifestyle than being married.
____ 14. People who think you must be married to be happy are wrong.
____ 15. You can be alone and not be lonely.
____ 16. The freedom of singlehood outweighs any advantage of marriage.
____ 17. The companionship of marriage is overrated.
____ 18. Singles have better sex than spouses.
____ 19. The idea that only spouses are fulfilled is nonsense.
____ 20. Marriage as a lifestyle is overrated.
____ 21. Most people who are married envy for those who are single.
____ 22. Spouses lose control of their lives while singles maintain control.
____ 23. There is entirely too much social pressure to get married.
____ 24. People are finding out that being single is a better deal than being married.
____ 25. The singles I know are happier than the spouses I know.

### Scoring

After assigning a number from 1 (strongly disagree) to 7 (strongly agree), add the numbers. The higher your score (175 is the highest possible score), the more positively you view remaining single. The lower your score (25 is the lowest possible score), the more negatively you view singlehood. A score of 75 is the midpoint. Scores below 75 tend to reflect a negative view of singlehood while scores above 75 reflect a positive view of singlehood.

### Norms

The norming sample of this self-assessment was based on the responses of 187 undergraduates (24 men and 163 women) at a large southeastern university. The mean score for the total sample was 86. The mean for the men was 95; for the women, 85. Hence, the total sample tended to regard singlehood positively with men having more positive attitudes toward singlehood than women.

### Source

"Attitudes Toward Singlehood Scale," 2014, by Mark Whatley, Ph.D., Department of Psychology, Valdosta State University, Valdosta, GA 31698-0100. Used by permission. Other uses of this scale only by written permission of Dr. Whatley (mwhatley@valdosta.edu). This scale is intended to provide basic feedback about one's view of singlehood. It is not designed to be a sophisticated research instrument.

doubts about self as a potential spouse, doubts about partner as spouse, quality of relationship, doubts about self as parent, doubts about partner as parent, capability of being economic provider, partner capability of being economic provider, fear of divorce, infidelity, in-laws, and bringing children from own and partner's previous relationships together.

Table 4.2 identifies the benefits of singlehood and the limitations of marriage. The primary advantage of remaining single is freedom and control over one's life. Once a decision has been made to involve another in one's life, one's choices become vulnerable to the influence of that other person. The person who chooses to remain single may view the needs and influence of another person as something to avoid.

**TABLE 4.2 Benefits of Singlehood and Limitations of Marriage**

| Benefits of Singlehood | Limitations of Marriage |
|---|---|
| Freedom to do as one wishes | Restricted by spouse or children |
| Variety of lovers | One sexual partner |
| Spontaneous lifestyle | Routine, predictable lifestyle |
| Close friends of both sexes | Pressure to avoid close other-sex friendships |
| Responsible for one person only | Responsible for spouse and children |
| Spend money as one wishes | Expenditures influenced by needs of spouse and children |
| Freedom to move as career dictates | Restrictions on career mobility |
| Avoid being controlled by spouse | Potential to be controlled by spouse |
| Avoid emotional and financial stress of divorce | Possibility of divorce |



Never-married educated black women report a particularly difficult time finding eligible (similar education and income) men from whom to choose. Hence, black women are more likely to be single than white women (Chambers and Kravitz 2011). There are also more never-married black men than never-married white men.

*The man who goes alone can start today; but he who travels with another must wait till that other is ready.*

HENRY DAVID THOREAU, AMERICAN AUTHOR AND POET

## National **Data**

Among adults 18 years and older, about 41% of black women and 45% of black men have never married, in contrast to about 21% of white women and 28% of white men (*Statistical Abstract of the United States, 2012–2013,* Table 56).

### DIVERSITY IN OTHER COUNTRIES

In France, singlehood is gaining prominence as there are now two civil unions (pair-bonded relationships given legal significance in terms of rights and privileges) for every three marriages. Reasons include disenchantment with marriage, fear of divorce, and the fact that civil unions provide similar tax advantages and other benefits of marriage. France is the only country which allows civil unions between heterosexual couples (begun in 1999). A civil union can be dissolved with a registered letter (Sayare and la De La Baume 2010).

But men in general, not just black men, may be scarce or become less valuable as spouses. Rosin (2010) in her article on "The End of Men" noted that women now hold the majority of the nation's jobs. The result is that men are being marginalized and bring less to the table economically. Rosin also states that modern industrial society no longer values men's size and strength, that "social intelligence, open communication, the ability to sit still and focus are, at a minimum, not predominantly male"—yet these are precisely the qualities now demanded in the global economy.

Some people do not set out to be single but drift into singlehood longer than they anticipated—and discover that they like it. Others delay getting married because they are afraid.

*In many ways, single women are under constant social surveillance. They are constantly being questioned: So what's new? Are you seeing anyone? What are you waiting for?! They are constantly being warned that they are liable to miss their train or die alone.*

KINNERET LAHAD, SOCIOLOGIST

Unmarried Equality is an organization that supports individuals who value being single. According to the mission statement identified on Unmarried Equality's website (http://www.unmarried.org/aboutus.php), the emphasis of the organization is to advocate

> *...for equality and fairness for unmarried people, including people who are single, choose not to marry, cannot marry, or live together before marriage. We provide support and information for this fast-growing constituency, fight discrimination on the basis of marital status, and educate the public and policymakers about relevant social and economic issues. We believe that marriage is only one of many acceptable family forms, and that society should recognize and support healthy relationships in all their diversity.*

PERSONAL CHOICES

### Is Singlehood for You?

Singlehood is not a one-dimensional concept. Whereas some are committed to singlehood as a permanent lifestyle, others enjoy it for now but intend to marry eventually. An essential difference between traditional marriage and singlehood is the personal, legal, and social freedom to do as you wish. Of 1,101 undergraduate males, 1.2% reported that maintaining their freedom and never marrying was their top life goal; of 3,471 undergraduate females, 9% identified singlehood as their top life goal (Hall and Knox 2013). Those who prefer to never marry may experience various challenges including loneliness, less money, and establishing an identity independent of being in the role of spouse.

**1. *Loneliness?*** "Not having someone to share my life with" was identified as the most challenging part of being single by a sample of 5,500 single individuals (Walsh 2013).

**Loneliness** is defined as the subjective evaluation that the number of relationships a person has is smaller than the person desires or that the intimacy the person wants has not been realized. Fokkema et al. (2012) noted that loneliness is associated with being single. But for some singles, being alone is a desirable, enjoyable, and sought-after experience. For them, being alone is not to feel alone. Henry David Thoreau, who never married, spent two years alone on 14 acres bordering Walden Pond in Massachusetts. He said of his experience, "I love to be alone. I have never found the companion that was as companionable as solitude." Musick and Bumpass (2012) compared spouses, cohabitants and non-pair-bonded singles and found few differences in terms of social ties. Indeed, singles often reported having more connections.

**2. *Less money.*** Unmarried individuals who live alone typically have less income than married people. The median income of the single female householder with no husband is $32,597 (for the male single householder with no wife, $48,084) compared to a married couple with an income of $71,830 (*Statistical Abstract of the United States, 2012–2013,* Table 692). Married couples typically have two incomes.

**3. *Social Identity.*** Single people must establish a social identity—a role—that defines who they are independent of their role of spouse. A career provides structure, relationships with others, and a strong sense of identity. "I am a veterinarian—I love my work," said one single female.

**4. *Children.*** Some individuals want to have a child but not a spouse. We will examine this issue in Chapter 11 in a section on Single Mothers by Choice. There are no data on single men seeking the role of parent. Although some custodial divorced men are single fathers, this is not the same as never having married and having a child.

In evaluating the single lifestyle, to what degree, if any, do you feel that loneliness is or would be a problem for you? What is your economic situation? Your social identity and your work role satisfaction? What are your emotional and structural needs for marriage? For children? The old idea that you can't be happy unless you are married is no longer credible. Whereas marriage will be the first option for some, it will be the last option for others. As one 76-year-old single-by-choice person said, "A spouse would have to be very special to be better than no spouse at all."

#### Sources

Fokkema, T., J. D. Gierveld, and P. A. Dykstra. 2012. Cross-national differences in older adult loneliness. *The Journal of Psychology* 146: 201–228.

Hall, S. and D. Knox. 2013. Relationship and sexual behaviors of a sample of 4,567 university students. Unpublished data collected for this text. Department of Family and Consumer Sciences, Ball State University and Department of Sociology, East Carolina University.

Musick, K. and L. Bumpass. 2012. Reexamining the case for marriage: Union formation and changes in well-being. *Journal of Marriage and Family* 74: 1–18.

*Statistical Abstract of the United States, 2012–2013.* 131th ed. Washington, DC: U.S. Census Bureau.

*Singleness would be recognized as a vital stage of the journey to maturation, a time to learn about who we are, to learn responsibility and self-sufficiency, to identify our true desires, and to confront our inner strengths and demons.*

HARVILLE HENDRIX, AUTHOR OF *GETTING THE LOVE YOU WANT*

# Functions and Changes in Dating

*Then we sat on the edge of the earth, with our feet dangling over the edge, and marveled that we had found each other.*

ERICK DILLARD, WRITER

Dating in the traditional sense (the male calling the female several days in advance and asking her to a specific event) has given way to hanging out and meeting someone. Sometimes one individual will send a message to another on Facebook, such as "Hey—wanna hang out Thursday night at the...?" The person will text back "sure" and typically show up. Jayson (2014) noted that there is ambiguity in whether this meeting would be a date. Indeed, 68% of a sample of 2,467 adults between the ages of 18–59 said they would be confused. However, 80% said the event would be a date if the individuals "planned a one-on-one hangout."

After the partners hang out on several occasions they will get more specific about when they will see each other again. In effect, they slide into the structure of seeing each other at predictable times rather than impose the structure initially. Whether it is called "hanging out" or "seeing someone" (the current term for dating), the couple typically spend increasing amounts of time together and become involved. Such involvement has various functions.

## Functions of Involvement with a Partner

Meeting and becoming involved with someone has at least six functions: (1) confirmation of a social self; (2) recreation; (3) companionship, intimacy, and sex; (4) anticipatory socialization; (5) status achievement; and (6) mate selection.

**1.** ***Confirmation of a social self.*** In Chapter 1, we noted that symbolic interactionists emphasize the development of the self. Parents are usually the first social mirrors in which we see ourselves and receive feedback about who we are; new partners continue the process. When we are hanging out with a person, we are continually trying to assess how that person sees us: What does the person think of me? Does the person like me? Will the person want to be with me again? When the person gives us positive feedback through speech and gesture, we feel good about ourselves and tend to view ourselves in positive terms.

**2.** ***Recreation.*** The focus of hanging out and pairing off is fun. Being fun is often a reason partners select each other. Individuals who just met make only small talk and learn very little about each other—what seems important is that they have fun together.

**3.** ***Companionship, intimacy, and sex.*** Beyond fun, other qualities for becoming involved are companionship, intimacy, and sex. The impersonal environment of a large university makes a secure relationship very appealing. "My last two years have been the happiest ever," remarked a senior in interior design. "But it's because of the involvement with my partner. During my freshman and sophomore years, I felt alone. Now I feel loved, needed, and secure."

**4.** ***Anticipatory socialization.*** Before puberty, boys and girls interact primarily with same-sex peers. A fifth grader may be laughed at if he or she shows an interest in someone of the other sex. Even when boy-girl interaction becomes the norm at puberty, neither sex may know what is expected of the other. "Seeing someone" provides a context for individuals to learn how to interact with a partner in a relationship. Though the manifest function of seeing someone is to teach partners how to negotiate differences (e.g., how much sex and how soon), the latent function is to help them learn the skills necessary to maintain long-term relationships (e.g., empathy, communication, and negotiation). In effect, pairing off involves a form of socialization that anticipates a more permanent union in one's life. Individuals may also try out different role patterns, like dominance or passivity, and try to assess the feel and comfort level of each.

**5.** ***Status achievement.*** Being involved with someone is associated with more status than being unattached and alone. In a couples' world, sometimes there is embarrassment with "I'm not seeing anyone." Some may seek such involvement because of the associated higher status. Others may become involved out of conformity to gender roles, not for emotional reasons.

**6.** ***Mate selection.*** Finally, seeing someone may eventually lead to marriage, which remains a major goal in our society. Selecting a mate is big business. The bookstore chain Barnes & Noble lists more than 276,000 books on "relationships" and more than 9,000 books on "finding a partner" on their website.

## Changes in Dating in the Past 60 Years

There have been numerous changes to dating and marriage patterns since 1950. The changes include an increase in the age at marriage. Marrying at age 29 rather than 22 provides more time and opportunity to hang out with more people.

The dating pool today also includes an increasing number of individuals in their thirties who have been married before. Usually divorced, these individuals often have children, which changes the context of being together from hanging out alone to watching a DVD with the kids.

As we will note later in this chapter, cohabitation has become more normative. For some couples, the sequence of dating, falling in love, and getting married has been replaced by dating, falling in love, and living together. Such a sequence results in the marriage of couples that are more relationship-savvy than those who dated and married just after high school.

Not only do individuals now "see" more partners and live together more often, but also gender role relationships have become more egalitarian. Though the double standard still exists, women today are more likely to ask men out than women in the 1950s (see Applying Social Research), to have sex without being in love or requiring a commitment, and to postpone marriage until meeting their own educational and career goals. Women no longer feel desperate to marry but consider marriage one of many goals they have for themselves.

*There is a path from me to you that I am constantly looking for.*

RUMI, 13TH CENTURY PERSIAN POET

Unlike during the 1950s, both sexes today are more aware of and cautious of becoming HIV infected. The 1950s fear of asking a druggist for a condom has been replaced by the commonplace buying of condoms along with one's groceries.

## Dating after Divorce

Becoming involved in a new relationship is not just for youth. Some individuals have been married and divorced and are back on the market. Dating for these older individuals differs from that of people becoming involved for the first time.

**1.** ***Internet for new partner.*** Aside from traditional ways to meet new partners (through friends, at work, at health or exercise clubs, or religious services) the Internet has become a valuable tool for divorced people in finding someone new. With full-time jobs and single parenthood, they may have little time for traditional dating. The Internet provides a quick way to sift through hundreds of people after work/at home and start text messaging or e-mailing someone new.

**2.** ***Older population.*** Divorced individuals are, on the average, ten years older than people in the marriage market who have never been married before. Hence, divorced people tend to be in their mid-to-late thirties. Widows and widowers are usually 40 and 30 years older, respectively (around ages 65 and 55), when they begin to date the second time around. Most divorced people date and marry others who are divorced. Since they are older they may also be learning the new norms of hanging out and seeing each other; some will miss the structure and comfort of traditional dating.

**3.** ***Fewer potential partners for the divorced.*** Most men and women who are dating the second time around find fewer partners from whom to choose than

## APPLYING SOCIAL RESEARCH | Women Who Initiate Relationships with Men

Mae West, film actress of the 1930s and 1940s, is remembered for being very forward with men. Two classic phrases of hers are "Is that a pistol in your pocket or are you glad to see me?" and "Why don't you come up and see me sometime? I'm home every night." As a woman who went after what she wanted, West was not alone—then or now. There have always been women not bound by traditional gender role restrictions. This study is concerned with women who initiate relationships with men—women who, like Mae West, have ventured beyond the traditional gender role expectations of the passive female.

### Sample and Methodology

Data for this study was provided by 692 undergraduate women who answered "yes" or "no" to the question, "I have asked a new guy to go out with me"—a nontraditional gender role behavior.

### Findings and Discussion

Of the 692 women surveyed, 39.1% reported that they had asked a new guy out on a date; 60.9% had not done so. Analysis of the data revealed seven statistically significant findings in regard to the characteristics of those women who had initiated a relationship with a man and those who had not done so.

1. ***Nonbeliever in "one true love."*** Of the women who asked men out, 42.3% did not believe in one true love in contrast to 31.7% who believed in one true love—a statistically significant difference ($p < .01$). These assertive women felt they had a menu of men from which to choose.
2. ***Experienced "love at first sight."*** Of the women who had asked a man out, 45.7% had experienced falling in love at first sight in contrast to 28.2% who had not had this experience. Hence, women who let a man know they were interested in him were likely to have already had a "sighting" of the man with whom they fell in love.
3. ***Sought partner on the Internet.*** Only a small number (59 of 692; 8.5%) of women reported that they had searched for a partner using the Internet. However, 54.2% of those who had done so (in contrast to only 37.5% who had not used the Internet to search for a partner) reported that they had asked a man to go out ($p < .02$). Because both seeking a partner on the Internet and asking a partner to go out verbally are reflective of nontraditional gender role behavior, these women were clearly in charge of their lives and moved the relationship forward rather than waiting for the man to make the first move.
4. ***Nonreligious.*** Respondents who were not religious were more likely to ask a guy out than those who were religious (74.6% versus 64.2%; $p < .001$). These women may have felt inclined to go after their man rather than have him sent from Heaven.
5. ***Nontraditional sexual values.*** Consistent with the idea that women who had asked a guy out were also nonreligious (a nontraditional value) is the finding that these

when they were dating before their first marriage. The large pools of never-married people (27% of the population) and currently married people (56% of the population) are usually not considered an option (*Statistical Abstract of the United States, 2012–2013,* Table 56). Most divorced people (10% of the population) date and marry others who have also been married before.

**4. *Increased HIV risk.*** As might be assumed, the older unmarried people are, the greater the likelihood that they have had multiple sexual partners, which is associated with increased risk of contracting HIV and other STIs. Therefore, individuals entering the dating market for the second time are advised to *always* use a condom and to assume that one's partner has been sexually active and is at risk.

**5. *Children.*** More than half of the divorced people who are dating again have children from a previous marriage. How these children feel about their parents dating, how the partners feel about each other's children, and how the partners' children feel about each other are challenging issues. Deciding whether to have intercourse when one's children are in the house, when a new partner should be introduced to the children, and what the children should call the new partner are other issues familiar to parents dating for the second time.

same women tended to have nontraditional sexual values. Of those women who had initiated a relationship with a guy, 44.4% reported having a hedonistic sexual value ("If it feels good, do it") compared to 25.7% who regarded themselves as having absolutist sexual values ("Wait until marriage to have intercourse"). Hence, women with nontraditional sexual values were much more likely to be aggressive in initiating a new relationship with a man.

6. ***Open to cohabitation.*** Of the women who reported that they had asked a man to go out, 44.6% reported that they would cohabit with a man compared to 26.0% who would not cohabit. This finding is consistent with the liberal female who goes after her man and who does not require marriage.
7. ***White.*** Over 40% (41.4%) of white women, compared to 28.2% of black women in the sample, reported that they had asked a guy out. This finding may be more reflection of context than race. Since most of the men available to these women were white, black women would be crossing racial lines to ask a white male to go out. If this study were conducted at a predominately black university, we anticipate that there would be a higher percentage of black females who would ask out a black male.

**Implications**

Analysis of these data revealed that 39.1% of the undergraduate women at a large southeastern university had asked a guy to go out (a nontraditional gender role behavior). There are implications of this finding for both women and men. Women who feel uncomfortable asking a man out, who fear rejection for doing so, or who lack the social skills to do so ("Want to hang out some time?") may be less likely to find the man who will be whisked away by women who have such comfort and who make known their interest in a partner.

The implication of this study for men is not to be surprised when a woman makes a direct request to hang out—relationship norms are changing. For some men, this comes as a welcome trend in that they feel burdened that they must always be the first one to indicate interest in a partner and to move the relationship forward. Men might also reevaluate their negative stereotypes of women who initiate relationships ("they are loose") and be reminded that the women in this study who had asked men out were *more* likely to have been faithful in previous relationships than those who had not.

**Source**

*Abridged from C. Ross, D. Knox, and M. Zusman. 2008. "Hey Big Boy": Characteristics of university women who initiate relationships with men. Poster, Southern Sociological Society Annual Meeting, April, Richmond, VA.

**6. *Ex-spouse issues.*** Ex-spouses may be uncomfortable with their former partner's involvement in a new relationship. They may not only create anxiety on the part of their former spouse but may also directly attack the new partner. Dealing with an ex-spouse may be challenging. Ties to an ex-spouse, in the form of child support or alimony, and phone calls may also have an influence on the new relationship. Some individuals remain psychologically and sexually involved with their exes. In other cases, if the divorce was bitter, partners may be preoccupied or frustrated in their attempts to cope with a harassing ex-spouse (and feel emotionally distant).

**7. *Brief courtship.*** Divorced people who are dating again tend to have a shorter courtship period than people married for the first time. In a study of 248 individuals who remarried, the median length of courtship was nine months, as opposed to 17 months the first time around (O'Flaherty and Eells 1988). A shorter courtship may mean that sexual decisions are confronted more quickly—timing of first intercourse, discussing the use of condoms and contraceptives, and clarifying whether the relationship is to be monogamous or if it has a future.

People who have been previously divorced or widowed derive a new sense of well-being from becoming involved with a new partner—"I'm getting used to being loved again" is the lyric and title to an old country western song by Gene Watson. Caution may be prudent because the old rules still apply. Knowing a partner at least two years before marrying that person is predictive of a happier and more durable relationship than marriage after a short courtship. Divorced people pursuing a new relationship might consider slowing down their relationship.

# Hanging Out and Hooking Up

**DIVERSITY IN OTHER COUNTRIES**

Americans are not alone in looking for a partner. There are 180 million single individuals in China looking for a partner. While there are more males than females in China, men view women who reach age 28 as "old" (Macleod 2010).

Whether an individual has the goal of remaining unmarried or eventually getting married, most have the goal of finding someone to have fun with/share their lives with. One of the unique qualities of the college or university environment is that it provides a context in which to meet hundreds of potential partners of similar age, education, and social class. This context will likely never recur following graduation. Although people often meet through friends (including through Facebook) or on their own in school, work, or recreation contexts, an increasing number are open to a range of alternatives including hanging out and meeting online.

## Hanging Out

**Hanging** out going out in groups where the agenda is to meet others and have fun.

*I met her in Monterey in old Mexico. Stars and steel guitars and luscious lips as red as wine. Broke somebody's heart and I'm afraid it was mine.*

MABEL WAYNE AND BILLY ROSE, *IT HAPPENED IN MONTEREY*

**Hanging out**, also referred to as getting together, means going out in groups where the agenda is to meet others and have fun. The individuals may watch television, rent a DVD, go to a club or party, and/or eat out. Of 4,562 undergraduates, 91% reported that "hanging out for me is basically about meeting people and having fun" (Hall and Knox 2013). Hanging out may occur in group settings such as at a bar, a sorority or fraternity party, or a small gathering of friends that keeps expanding. Friends may introduce individuals, or they may meet someone "cold," as in initiating a conversation. There is usually no agenda beyond meeting and having fun. Of the respondents noted above, only 6% said that hanging out was about beginning a relationship that may lead to marriage (Hall and Knox 2013).

## Hooking Up

**Hooking up** sexual encounter that occurs between individuals who have no relationship commitment.

**Hooking up** is a sexual encounter that occurs between individuals who have no relationship commitment. Lewis et al. (2013) defined hooking up in their survey as: "event where you were physically intimate (any of the following: kissing, touching, oral sex, vaginal sex, anal sex) with someone whom you were not dating or in a romantic relationship with at the time and in which you understood there was no mutual expectation of a romantic commitment." Their sample of 1,468 individuals revealed that, while definitions vary, most define hooking up as involving some type of sex (vaginal, oral, anal), not just kissing.

For those who hook up there is generally no expectation of seeing each other again, and alcohol is often involved. Chang et al. (2012) identified the unspoken rules of hooking up—hooking up is not dating, hooking up is not a romantic relationship, hooking up is physical, hooking up is secret, one who hooks up is to expect no subsequent phone calls from their hooking up partner, and condom/protection should always occur (though only 57% of their sample reported condom use on hookups). Aubrey and Smith (2013) also noted that there is a set of cultural beliefs about hooking up. These beliefs include that hooking up is shameless/fun, will enhance one's status in one's peer group, and reflects one's sexual freedom/control over one's sexuality.

Data on the frequency of college students having experienced a hookup varies. Barriger and Velez-Blasini (2013) in their review of literature found that between 77% and 85% of undergraduates reported having hooked up within the previous year. Reporting on more than 17,000 students at 20 colleges and universities (from Stanford University's Social Life Survey), sociologist Paula England revealed that 72% of both women and men reported having hooked up (40% intercourse; 35% kiss, touch; 12% hand genital; 12% oral sex). Men reported having ten hookups, women seven (Jayson, 2011).

LaBrie et al. (2014) noted the effect of drinking alcohol on hooking up. Of 828 college students, over half (55%) reported hooking up in the last year. Of those who hooked up 31% of the females and 28% of the males reported that they would not have hooked up had they not been drinking. A similar percent reported they would not have gone as far physically had they not been drinking. Females who had been drinking and hooked up were more likely to feel discontent with their hookup decisions.

Fieldera and Careya (2010) surveyed 118 first semester female college students on their hookup experiences (60% had done so). Most hookups involved friends (47%) or acquaintances (23%) rather than strangers. Olmstead et al. (2012) surveyed 158 first semester men to assess those factors predictive of hooking up their first semester on campus. Those who had hooked up before coming to college (77% had done so), those who had a pattern of binge drinking, and those who had casual sexual attitudes were more likely to hook up their first semester on campus.

Chang et al. (2012) surveyed 369 undergraduates (69% reported having hooked up) and found that those with pro-feminist attitudes were less likely to hook up. However, women who hooked up were more likely than men to agree with the unspoken rules of hooking up—no commitment, no emotional intimacy, and no future obligation to each other. They were clear that hooking up was about sex with no future. These data reflect that hooking up is becoming the norm on college and university campuses. Not only do female students outnumber men students (60 to 40) (which means women are less able to bargain sex for commitment since men have a lot of options), individuals want to remain free for summer internships, study abroad, and marriage later (Uecker and Regnerus 2011). Kalish and Kimmel (2011) suggested that hooking up is a way of confirming one's heterosexuality.

The hooking up experience is also variable. Bradshaw et al. (2010) compared the experiences of women and men who hooked up. Men benefited more since they were able to have casual sex with a willing partner and no commitment. Women were more at risk for feeling regret/guilt, becoming depressed, and defining the experience negatively. Women in hookup contexts were less likely to experience cunnilingus but were often expected to provide fellatio (Blackstrom et al. 2012). Both women and men experienced the hook up in a context of deception (neither being open about their relationship goals) and may have exposed themselves to an STI. While few long-term relationships begin with a hook up, there is a belief that it is not unusual. When 4,567 undergraduates were presented with the statement "People who 'hook up' and have sex the first night don't end up in a stable relationship," 50% agreed (Hall and Knox 2013).

Maggie Carter

This couple met at a Disney summer work program. He lived in Kansas and she, in North Carolina. Their long-distance relationship survived and resulted in his moving to North Carolina.

PERSONAL CHOICES

### Should I Get Involved in a Long-Distance Relationship?

One outcome of online dating is that you may meet someone who does not live close to you so you end up in a long-distance relationship. Alternatively, you may already be in a relationship and the separation occurs. Long-distance relationships must be considered with great caution. In a sample of 5,500 never-married undergraduates, about half (51% of the men and 47% of the women) reported that a long-distance relationship was "out of the question" (Walsh 2013).

**Long-distance relationship** being separated from a romantic partner by 500 or more miles.

*How like the winter hath my absence been From thee?*

SHAKESPEARE, *SONNET*, 97.1

*Love knows not its own depth until the hour of separation.*

KABLIL GIBRAN, PHILOSOPHER

*In true love the smallest distance is too great, and the greatest distance can be bridged.*

HANS NOUWENS, WRITER

The primary advantages of **long-distance relationships** (defined here as being separated from a romantic partner by 500 or more miles, which precludes regular weekly face-to-face contact) include: positive labeling ("even though we are separated, we care about each other enough to maintain our relationship"), keeping the relationship "high" since constant togetherness may result in the partners being less attentive to each other, having time to devote to school or a career, and having a lot of one's own personal time and space. Pistole (2010) in a review of the literature on long-distance relationships noted that they are as satisfying and as stable as geographically close relationships.

People suited for such relationships have developed their own autonomy or independence for the times they are apart, have a focus for their time such as school or a job, have developed open communication with their partner to talk about the difficulty of being separated, and have learned to trust each other because they spend a lot of time away from each other. Another advantage is that the partner may actually look better from afar than up close. One respondent noted that he and his partner could not wait to live together after they had been separated—but "when we did, I found out I liked her better when she wasn't there."

The primary disadvantages of long-distance relationships include being frustrated over not being able to be with the partner and loneliness. Pistole et al. (2010) noted that being attached to someone is the strongest of social behaviors and that there is a "separation protest" when the two are separated. This longing for each other may be stronger than either anticipated. When the long- distance lovers are reunited, they spend higher order quality time together.

Other disadvantages of involvement in a long-distance relationship are missing out on other activities and relationships, less physical intimacy, spending a lot of money on phone calls/ travel, and not discussing important relationship topics. Regarding the latter, Stafford (2010) compared the communication topics of individuals in long-distance dating relationships (LDDR) with those in geographically close-distance relationships (GCDR) and found that the former avoided sensitive topics such as what household roles men and women should fulfill, the importance of marriage, if and how many children were desired, the importance of religion, etc. "Accentuating intimacy and positive affect in their talk, and avoiding discussion of potentially problematic or taboo topics could allow geographically separated couples to maintain a positive outlook on their relationship" (p. 292).

For couples who have the goal of maintaining their relationship and reducing the chance that the distance will result in their breaking up, some specific behaviors to engage in include:

1. ***Maintain daily contact via text messaging.*** Texting allows individuals to stay in touch throughout the day. A husband we interviewed who must travel four to five days a week said, "Texting allows us stay connected with what is going on in each other's lives throughout the day so when I get home on weekends we've been 'together' as much as possible during the week."

2. ***Enjoy or use the time when apart.*** While separated, it is important to remain busy with study, friends, work, sports, and personal projects. Doing so will make the time pass faster.

3. ***Avoid arguing during phone conversations.*** Talking on the phone should involve the typical sharing of events. When the need to discuss a difficult topic arises (e.g., trust), the phone is not the best place for such a discussion. Rather, it may be wiser to wait and have the discussion face to face. If you decide to settle a disagreement over the phone, stick to it until you have a solution acceptable to both of you.

4. ***Stay monogamous.*** Agreeing not to be open to other relationships is crucial to maintaining a long-distance relationship. Individuals who say, "Let's date others to see if we are really meant to be together," often discover that they are capable of being attracted to and becoming involved with numerous "others." Such other involvements usually predict the end of an LDR. Lydon et al. (1997) studied 69 undergraduates who were involved in LDRs and found that "moral commitment" predicted the survival of the relationships. Individuals committed to maintaining their relationships are not open to becoming involved with others.

*In short, I will part with anything for you, except you.*

LADY MARY WORTLEY MONTAGU

5. ***Skype.*** Skyping allows the partners to see and hear each other. Frequent Skype encounters allow the partners to "date" even though they cannot touch each other. Some partners also become sexual by stripping for each other, which is a way of being intimate with each other while physically separated.

6. ***Be Creative.*** Some partners in long-distance relationships watch Netflix movies together—they each pull up the movie on their computer and talk on the phone while they watch it. Others send video links, photos, etc., throughout the day. One coed says "I wear his shirts" and "he has my pillow." Some find that keeping a journal of their relationship/feelings during the LDDR adventure is helpful. Be creative.

### Sources


Lydon, J., T. Pierce, and S. O'Regan. 1997. Coping with moral commitment to long-distance dating relationships. *Journal of Personality and Social Psychology* 73: 104–13.

Pistole, M. C. 2010. Long-distance romantic couples: An attachment theoretical perspective. *Journal of Marital and Family Therapy* 36: 115–125.

Pistole, M. C., A. Roberts, and M. L. Chapman. 2010. Attachment, relationship maintenance, and stress in long-distance and geographically close romantic relationships. *Journal of Social and Personal Relationships* 27: 535–552.

Stafford, L. 2010. Geographic distance and communication during courtship. *Communication Research* 37: 275–297.

Walsh, S. 2013. Match's 2012 singles in America survey. http://www.hookingupsmart.com/2013/02/07/hookinguprealities/matchs-2012-singles-in-america-survey/


# Cohabitation

**Cohabitation**, also known as living together, involves two adults, unrelated by blood or by law, involved in an emotional and sexual relationship, who sleep in the same residence at least four nights a week for three months. While not enforced, cohabitation is illegal in some states (e.g., Michigan and New York). Not all cohabitants are college students. Indeed, only 18% of all cohabitants are under the age of 25. The largest percentage (36%) are between the ages of 25 and 34 (Jayson 2012). Most cohabitants are other sex (1% same sex) and white (87%) (*Statistical Abstract of the United States, 2012–2013*).

**Cohabitation** two adults, unrelated by blood or law, involved in an emotional and sexual relationship, who sleep in the same residence at least four nights a week for three months.

Reasons for the increase in cohabitation include career or educational commitments; increased tolerance of society, parents, and peers; improved birth control technology; desire for a stable emotional and sexual relationship without legal ties; avoiding loneliness (Kasearu 2010); and greater disregard for convention. Two-thirds of 122 cohabiters reported concerns about divorce (Miller et al. 2011).

## DIVERSITY IN OTHER COUNTRIES

Virtually all marriages in Sweden are preceded by the couple living in a cohabitation relationship (Thomson and Bernardt 2010); however, only 12% of first marriages in Italy are preceded by cohabitation (Kiernan 2000). Italy is primarily Catholic, which helps to account for the low cohabitation rate. Religious affiliation is also associated with lower rates of cohabitation in the United States (Gault-Sherman and Draper 2012).

In a study of 1,624 mothers cohabiting at the time of a nonmarital birth, two-thirds ended their relationship within five years (young, black, multipartnered women with several children) (Kamp Dush 2011). The median duration that a never-married woman cohabits with a partner in the United States is 14 months. In contrast, in France it is 51 months; Canada, 40 months; Germany, 27 months. In European countries never-married women often have children during cohabitation, which is less common in the United States (Cherlin 2010). Urquia et al. (2013) found that cohabitation in a national sample of Canadians was associated with a greater likelihood of intimate partner violence.

## Nine Types of Cohabitation Relationships

There are various types of cohabitation:

**1.** ***Here and now.*** These new partners have a fun relationship and are focused on the here and now, not the future of the relationship. They want to be together more often and living together is one way to do so.

**2.** ***Testers.*** These couples are involved in a relationship and want to assess whether they have a future together. Sassler and Miller (2011) indentified such a couple:

> *We're trying to see how it is going to work. So if our relationship's going to continue to grow and prosper, this would be one way for us to kind of gauge, you know, maybe we will get married. And if we can't live together, we're not going to get married. So this will help us make future decisions and stuff like that.*

Women and men cohabitants often value the relationship differently. Rhoades et al. (2012) studied 120 cohabiting couples and found women more committed than men to the relationship in about half the couples (46%). Such discrepancy was associated with a lower-quality relationship.

**3.** ***Engaged.*** These cohabiting couples are in love and are planning to marry. Among those who report having lived together, about two-thirds (64%) say they thought of their living arrangement as a step toward marriage (Pew Research Center, 2010). Engaged cohabitant couples who have an agreed upon future report the highest level of satisfaction, the lowest level of conflict, and, in general, have a higher quality relationship than other types of cohabitants (Willoughby et al. 2012). After three years, 40% of first premarital cohabitants end up getting married, 32% are still cohabiting, and 27% have broken up.

**4.** ***Money savers.*** These couples live together primarily out of economic convenience. They are open to the possibility of a future together but regard such a possibility as unlikely. Sassler and Miller (2011) noted that working class individuals tend to transition more quickly than middle class individuals to cohabitation out of economic necessity or to meet a housing need. Dew (2011) noted that financial disagreements were an important factor in predicting the breakup of cohabitation couples.

**5.** ***Pension partners.*** These cohabitation partners are older, have been married before, still derive benefits from their previous relationships, and are living with someone new. Getting married would mean giving up their pension benefits from the previous marriage. An example is a widow from the war in Afghanistan who was given military benefits due to a spouse's death. If remarried, the widow would forfeit both health and pension benefits, but now lives with a new partner and continues to get benefits from the previous marriage.

**6.** ***Alimony maintenance.*** Related to widows who cohabit are the divorced who are collecting alimony which they would forfeit should they remarry. To maintain the benefits, they live with a new partner instead of marrying. An example is a divorced woman receiving a hefty alimony check from her ex, a successful attorney. She was involved in a new relationship with a partner who wanted to marry her. She did not marry him in order to keep the alimony flowing. In effect, her ex-husband was paying the bills for his former wife and her new lover, who had moved in.

**7.** ***Security blanket cohabiters.*** Some of the individuals in these cohabitation relationships are drawn to each other out of a need for security rather than mutual attraction. Being alone is not an option. They want somebody, anybody, in the house.

**8.** ***Rebellious cohabiters.*** Some couples use cohabitation as a way of making a statement to their parents that they are independent and can make their own choices. Their cohabitation is more about rebelling from parents than being drawn to each other.

**9.** ***Marriage never (cohabitants forever).*** Ten percent of cohabitants view their living together not as a prelude to marriage but as a way of life (Sommers et al. 2013). Ortyl et al. (2013) interviewed 48 long-term heterosexual cohabiters to identify their motives for cohabiting as a permanent lifestyle. Six themes emerged:

**1.** Marriage free—the largest percent—38%–believed that marriage is unnecessary to their happiness (they used the term "marriage free" much like one would use the word "child free"). Such couples may feel a moral commitment to each other and to their relationship yet have no interest in marriage (Pope and Cashwell 2013).

**2.** Risk aversion—the cohabitants had parents or siblings in disastrous marriages and wanted to avoid the same fate.

**3.** Marriage boycott—the cohabitants rejected the government defining marriage as heterosexual, thus supporting gays who are denied same sex marriage in all states.

**4.** Sexism dissent—cohabitants rejected the patriarchal history of marriage which controlled women.

**5.** American dreamer—some cohabitants saw the day when their school debts would be paid off and their career established as the day it would be OK to get married.

**6.** Economic disincentives—cohabitants knew that marriage involves being responsible for the debts of the spouse and wanted to avoid (by living together) being economically burdened by a partner with an unstable job and money history.

Chelsea Curry

These individuals are in love with and committed to each other. They live together and are having a child together. They have no interest in marriage.

Some couples who view their living together as "permanent" seek to have it defined as a **domestic partnership,** a relationship involving two adults who have chosen to share each other's lives in an intimate and committed relationship of mutual caring (see the following Family Policy feature).

**Domestic partnership** a relationship involving two adults who have chosen to share each other's lives in an intimate and committed relationship of mutual caring.

## Consequences of Cohabitation

Although living together before marriage does not ensure a happy, stable marriage, it has some potential advantages.

**Advantages of Cohabitation** Many unmarried couples who live together report that it is an enjoyable, maturing experience. Other potential benefits of living together include the following:

**1.** ***Sense of well-being.*** Compared to uninvolved individuals or those involved but not living together, cohabitants are likely to report a sense of well-being (particularly if the partners see a future together). They are in love, the relationship is new, and the disenchantment that frequently occurs in long-term relationships has not had time to surface. One student reported, "We have had to make some adjustments in terms of moving all our stuff into one place, but we very much enjoy our life together." Although young cohabitants report high levels of enjoyment compared to single people, cohabitants in midlife who have

## FAMILY POLICY | Domestic Partnership

Domestic partners, both heterosexual and homosexual, want their employers, whether governmental or corporate, to afford them the same rights as spouses. Specifically, employed people who pay for health insurance would like their domestic partner to be covered in the same way that one's spouse would be. Gonzales and Blewett (2014) confirmed that employer-sponsored insurance (ESI) coverage access is improved in those states with domestic partnerships. Employers are reluctant to provide such coverage since to do so is an added expense.

Aside from the economic issue, fundamentalist religious groups have criticized domestic partner benefits as eroding family values by giving nonmarital couples the same rights as married couples. Over half the states recognize domestic partnerships in some cities with the law providing rights and responsibilities in areas as varied as child custody, legal claims, housing protections, bereavement leave, and state government benefits.

To receive benefits, domestic partners must register, which involves signing an affidavit of domestic partnership verifying that they are a nonmarried, cohabitating couple 18 years of age or older, unrelated by blood, jointly responsible for debts to third parties, and they intend to remain in the intimate committed relationship indefinitely. Should they terminate their domestic

never married, when compared to married spouses, report lower levels of relational and subjective well-being.

**2.** ***Delayed marriage.*** Another advantage of living together is remaining unmarried—the longer one waits to marry, the better. Being older at the time of marriage is predictive of marital happiness and stability, just as being young (particularly 18 years and younger) is associated with marital unhappiness and divorce. Hence, if a young couple who have known each other for a short time is faced with the choice of living together or getting married, their delaying marriage while they live together seems to be the better choice. Also, if they break up, the split will not go on their record as would a divorce.

About 60% of whites who cohabit and 40% of blacks who cohabit end up getting married. For whites, cohabitation is more often viewed as a transition to marriage whereas for blacks, it is more often viewed as a permanent lifestyle (Rinelli and Brown 2010).

**3.** ***Knowledge about self and partner.*** While living together before marriage does not make a couple immune to the increase in interpersonal conflict (Hall and Adams 2011), it does provide couples with an opportunity for learning more about themselves and their partner. For example, a person's behavior with family (calling parents daily) or friends (having a beer on weekends with a buddy), habits ("neat freak"), and relationship expectations (how emotionally close or distant) are sometimes more fully revealed when living together than in a traditional dating context.

**4.** ***Safety.*** Particularly for heterosexual females, living with a partner provides a higher level of safety not enjoyed by single females who live alone—presumably the male would deter someone who broke into the apartment of the female. Of course, living with a roommate or group of friends would provide a similar function of safety.

#### DIVERSITY IN OTHER COUNTRIES

Iceland is a homogeneous country of 250,000 descendants of the Vikings. Their sexual norms include early protected intercourse, living together before marriage, and non-marital parenthood (Halligan et al. 2014). Indeed, a wedding photo often includes not only the couple but also the children they have already had. One American woman who was involved with an Icelander noted, "My parents were upset with me because Ollie and I were thinking about living together, but his parents were upset that we were not already living together" (personal communication).

**Disadvantages of Cohabitation** There is a downside for individuals and couples who live together.

**1.** ***More problems than spouses.*** There are a number of differences in the frequency of reported relationship problems between cohabiting and married individuals. Cohabiting individuals tend to argue more, find their partnership, they are required to file notice of such termination.

Domestic partnerships offer a middle ground between those states that want to give full legal recognition to same-sex marriages and those that deny any legitimacy to same-sex unions. Such relationships also include long-term, committed heterosexuals who are not married.

### Your Opinion?

1. To what degree do you believe benefits should be given to domestic partners?
2. What criteria should be required for a couple to be regarded as domestic partners?
3. How can abuses of those claiming to be domestic partners be eliminated?

### Source

Gonzales, G. and L. A. Blewett. 2014. National and state-specific health insurance disparities for adults in same sex-relationships. *American Journal of Public Health* 104: 95–104.

relationships more unstable or insecure, and have more differences over future goals and values (Hsueh et al. 2009). Partners also report more emotional and physical abuse after they begin living together than before.

**2.** ***Feeling used or tricked.*** When expectations differ, the more invested partner may feel used or tricked if the relationship does not progress toward marriage. One partner said, "I always felt we would be getting married, but it turns out that he never saw a future for us."

**3.** ***Parental problems.*** Parents are an important influence on whether a couple cohabits—when parents disapprove, the couple is less likely to cohabit (Sommers et al. 2013). Some cohabiting couples report that they must contend with parents who disapprove of or do not fully accept their living arrangement. For example, cohabitants sometimes report that, when visiting their parents' homes, they are required to sleep in separate beds in separate rooms. Some cohabitants who have parents with traditional values respect these values, and sleeping in separate rooms is not a problem. Other cohabitants feel resentful of parents who require them to sleep separately. Some parents express their disapproval of their child's cohabiting by cutting off communication, as well as economic support, from their child. Other parents display lack of acceptance of cohabitation in more subtle ways. One woman who had lived with her partner for two years said that her partner's parents would not include her in the family's annual photo. Emotionally, she felt very much a part of her partner's family and was deeply hurt that she was not included in the family portrait. Still other parents are completely supportive of their children's cohabiting and support their doing so. "I'd rather my kid live together than get married and, besides, it is safer for her and she's happier," said one father.

**4.** ***Economic disadvantages.*** Some economic liabilities exist for those who live together instead of getting married. In the Family Policy section on domestic partnerships, we noted that cohabitants typically do not benefit from their partner's health insurance, Social Security, or retirement benefits. In most cases, only spouses qualify for such payoffs.

Given that most relationships in which people live together are not long term and that breaking up is not uncommon, cohabitants might develop a written and signed legal agreement should they purchase a house, car, or other costly items together. The written agreement should include a description of the item, to whom it belongs, how it will be paid for, and what will happen to the item if the relationship terminates. Purchasing real estate together may require a separate agreement, which should include how the mortgage, property taxes, and repairs will be shared. The agreement should also specify who gets the house if the partners break up and how the value of the departing partner's share will be determined.

If the couple has children, another agreement may be helpful in defining custody, visitation, and support issues in the event the couple terminates the relationship. Such an arrangement may take some of the romance out of the cohabitation relationship, but it can save a great deal of frustration should the partners decide to go their separate ways.

In addition, couples who live together instead of marrying can protect themselves from some of the economic disadvantages of living together by specifying their wishes in wills; otherwise, their belongings will go to next of kin or to the state. They should also own property through joint tenancy with rights of survivorship. This stipulation means that ownership of the entire property will revert to one partner if the other partner dies. Finally, the couple should save for retirement, because live-in companions may not access Social Security benefits, and some company pension plans bar employees from naming anyone other than a spouse as the beneficiary.

**5. *Effects on children.*** About 40% of children will spend some time in a home where the adults are cohabitating. In addition to being disadvantaged in terms of parental income and education, they are likely to experience more disruptions in family structure. In a study of the health differences of children born to cohabiting versus married parents, the former had worse health (Schmeer 2011). Should the cohabitation relationship break, the parents report more depressive symptoms (Kamp Dush 2013), which may impact children.

**Cyclical cohabitation** a couple live together off and on.

Nepomnyaschy and Teitler (2013) examined the concept of **cyclical cohabitation**—whereby the couple live together off and on—over a nine-year period and found that 25% of cohabitants with children fit this category (10% of all cohabitants). Children of cyclically cohabiting parents fared no worse than children of stably cohabiting biological parents.

WHAT IF?

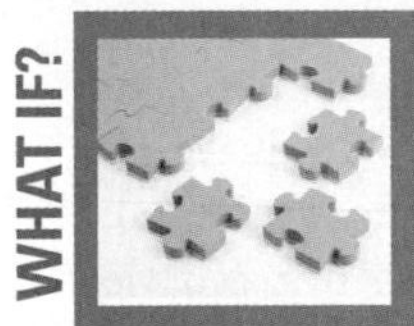

## What If Your Partner Wants To Live Together And You Do Not?

It is not unusual that partners in love have different values and goals for the relationship. Wanting to live together as well as not wanting to do so are equally valid positions.

However, the costs to the person who is asked to live together who feels this is wrong or unwise is greater than the cost to the person for not having the person live with him or her. The consequence of going against one's values is to continually feel uncomfortable while cohabiting and to risk blaming the other person if something goes wrong. Win-lose decisions in relationships are never a good idea. Manning et al. (2011) provided data on this question and found that when romantic partners disagreed about whether to cohabit, the feelings of the partner against cohabitation trumped the desires of the other—so the couple did not live together.

## Having Children while Cohabitating?

As noted above, most U.S. undergraduates believe they should be married (not just cohabitating) before they begin a family (Sommers et al. 2013). Twenty percent of first-time premarital cohabitants experience a pregnancy (Copen et al. 2013), which is associated with the couple getting married (Guzzo and Hayford 2014).

**DIVERSITY IN OTHER COUNTRIES**

In an comparison of 36,889 individuals who were married, cohabitating, or single in 27 countries, higher levels of happiness were reported by spouses, followed by cohabitants, followed by the never married (Lee et al. 2012).

## PERSONAL CHOICES

### *Will Living Together Ensure a Happy, Durable Marriage?*

Couples who live together before marrying assume that doing so will increase their chances of having a happy and durable marriage relationship. But will it? The answer is, "It depends." For individuals (particularly women—Manning and Cohen 2012) who have only one cohabitation experience with the person they eventually marry, there is no increased risk of divorce (Jose et al. 2010). The period of time while these engaged couples are cohabiting is superior to the time spent by couples who are not committed to the future (Willoughby et al. 2012).

Because people commonly have more than one cohabitation experience, the term **cohabitation effect** applies. This means that those who have multiple cohabitation experiences prior to marriage are more likely to end up in marriages characterized by violence, lower levels of happiness, lower levels of positive communication, and higher levels of depression (Booth et al. 2008). Liat and Havusha-Morgenstern (2011) compared the marital adjustment among women who cohabited with their spouses before marriage versus those who did not. They found that cohabiting women reported lower levels of adjustment of spousal cohesion and display of affection than did women who did not cohabit.

**Cohabitation effect** those who have multiple cohabitation experiences prior to marriage are more likely to end up in marriages characterized by violence, lower levels of happiness, lower levels of positive communication, and higher levels of depression.

What is it about serial cohabitation relationships that predict negatively for future marital happiness and durability? One explanation is that cohabitants tend to be people who are willing to violate social norms by living together before marriage. Once they marry, they may be more willing to break another social norm and divorce if they are unhappy than are unhappily married people who tend to conform to social norms and have no history of unconventional behavior.

Kuperberg (2014) provided data to confirm that previous research linking cohabitation with divorce did not account for the age at which coresidence began. She suggested that it is the age at which individuals begin their lives together (coresidence) which impacts divorce, not cohabitation per se. She suggested that individuals delay their marriage into their mid twenties "when they are older and more established in the lives, goals and careers, whether married or not at the time of coresidence rather than avoiding premarital cohabitation altogether" (p. 368).

Another reason is the restraints inherent in cohabitation. Rhoades et al. (2012) studied 120 cohabiting couples and found that restraints often keep a couple together. These include signing a lease, having a joint bank account, and having a pet. In some cases couples may move forward toward marriage for reasons of constraint rather than emotional desire. Whatever the reason, cohabitants should not assume that cohabitation will make them happier spouses or insulate them from divorce.

Not all researchers have found negative effects of cohabitation on relationships. Reinhold (2010) found that among more recent cohabitant cohorts, the negative association between living together and marital instability is weakening. Additional research is needed to confirm the effect of cohabitation on subsequent marriage.

### Sources

Booth, A., E. Rustenbach, and S. McHale. 2008. Early family transitions and depressive symptom changes from adolescence to early adulthood. *Journal of Marriage and Family* 70: 3–14.

Jose, A., K. D. O'Leary, and A. Moyer. 2010. Does premarital cohabitation predict subsequent marital stability and marital quality? A meta-analysis. *Journal of Marriage and Family* 72: 105–116.

Kuperberg, A. 2014. Age at coresidence, premarital cohabitation, and marital dissolution 1985–2009. *Journal of Marriage and the Family* 76: 352–369.

Liat, K. and H. Havusha-Morgenstern. 2011. Does cohabitation matter? Differences in initial marital adjustment among women who cohabited and those who did not. *Families in Society* 92: 120–127.

Manning, W. D. and J. A. Cohen. 2012. Premarital cohabitation and marital dissolution: An examination of recent marriages. *Journal of Marriage and the Family* 74: 377–387.

Reinhold, S. 2010. Reassessing the link between premarital cohabitation and marital instability. *Demography* 47: 719–733.

Rhoades, G. K., S. M. Stanley, and H. J. Markman. 2012. A longitudinal investigation of commitment dynamics in cohabiting relationships. *Journal of Family Issues* 33: 369–390.

Willoughby, B. J., J. S. Carroll and D. M. Busby. 2012. The different effects of "living together": Determining and comparing types of cohabiting couples. *Journal of Social and Personal Relationships* 29: 397–419.

## Legal Aspects of Living Together

In recent years, the courts and legal system have become increasingly involved in relationships in which couples live together. Some of the legal issues concerning cohabiting partners include common-law marriage, palimony, child support, and child inheritance. Lesbian and gay couples also confront legal issues when they live together (as opposed to getting married).

Technically, cohabitation is against the law in some states. For example, in North Carolina, cohabitation is a misdemeanor punishable by a fine not to exceed $500, imprisonment for not more than six months, or both. Most law enforcement officials view cohabitation as a victimless crime and feel that the general public can be better served by concentrating upon the crimes that do real damage to citizens and their property.

**Common-Law Marriage** The concept of common-law marriage dates to a time when couples who wanted to be married did not have easy or convenient access to legal authorities (who could formally sanction their relationship so that they would have the benefits of legal marriage). Thus, if the couple lived together, defined themselves as husband and wife, and wanted other people to view them as a married couple, they would be considered married in the eyes of the law.

Despite the assumption by some that heterosexual couples who live together a long time have a common-law marriage, only 14 states recognize such marriages (see Figure 4.1). In these states a heterosexual couple may be considered married if they are legally competent to marry, if the partners agree that they are married, and if they present themselves to the public as a married couple. A ceremony or compliance with legal formalities is not required.

In common-law states, individuals who live together and who prove that they were married "by common law" may inherit from each other or receive alimony and property in the case of relationship termination. They may also receive health and Social Security benefits, as would other spouses who have a marriage license. In states not recognizing common-law marriages, the individuals who live together are not entitled to benefits traditionally afforded married individuals. More than three-quarters of the states have passed laws prohibiting the recognition of common-law marriages within their borders.

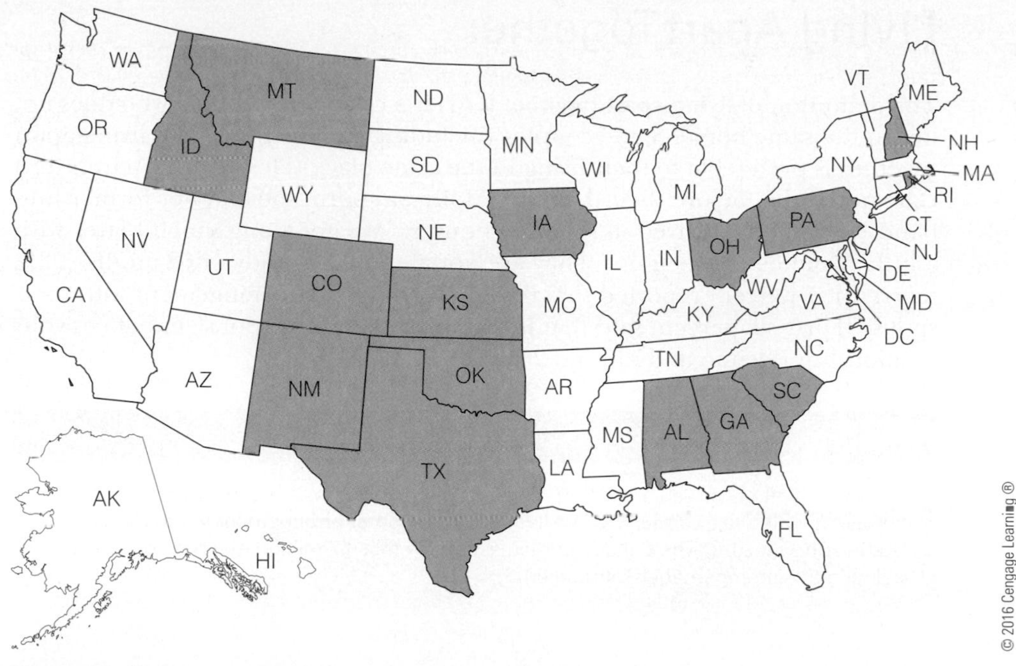


**Figure 4.1** States recognizing common law marriage.

A takeoff on the word alimony, **palimony** refers to the amount of money one "pal" who lives with another "pal" may have to pay if the partners end their relationship. For example, comedian Bill Maher has been the target of a $9 million palimony suit by ex-girlfriend Coco Johnsen. Similarly, Norman Greenbaum (known for his song "Spirit in the Sky") was the target of a palimony suit in 2010 by his ex-girlfriend and fan club leader Tracy E. Outlaw of South Carolina. She accused Greenbaum of fraud for failing to live up to the terms of an agreement in which the former Clemson University employee would quit her job and move across country in exchange for one-half interest in Greenbaum's Saddleback Court home and a third of his estate. South Carolina is one of the states which recognize common law marriages–palimony suits.

**Palimony** refers to the amount of money one "pal" who lives with another "pal" may have to pay if the partners end their relationship.

**Child Support** Heterosexual individuals who conceive children are responsible for those children whether they are living together or married. In most cases, the custody of young children will be given to the mother, and the father will be required to pay child support. In effect, living together is irrelevant with regard to parental obligations. However, a woman who agrees to have a child with her lesbian partner cannot be forced to pay child support if the couple breaks up. The Massachusetts Supreme Judicial Court ruled that their informal agreement to have a child together did not constitute an enforceable contract.

Couples who live together or who have children together should be aware that laws traditionally applying only to married couples are now being applied to many unwed relationships. Palimony, distribution of property, and child support payments are all possibilities once two people cohabit or parent a child.

**Child Inheritance/Access to Parents after Breakup** Children born to unmarried cohabitants are able to inherit from their respective parents. However, if there is a breakup, the biological parent (re: gay couple) has more power (e.g., will be given custody).

# Living Apart Together

**Living Apart Together (LAT)** a committed couple who does not live in the same home.

The definition of **living apart together (LAT)** is a committed couple who does not live in the same home. Some couples (including spouses) find that living apart together is preferable to their living in the same place (Hess 2012). Actress Teri Garr noted that before their daughter Molly was born, she and her former husband (John O'Neil) lived in separate houses: "We got along much better with a little distance between us." They are not alone. In a study of 68 adults (93% married), 7 percent reported that they preferred a LAT arrangement with their spouse. Forty-six percent said that living apart from your spouse enhances your relationship (Jacinto and Ahrend 2012).

## National **Data**

Seven percent of women and 6% of men age 23 and older who are in a romantic relationship with their partner are in a living apart together arrangement in the United States (Cherlin 2010). The Census Bureau estimates that 1.7 million married couples are living in this arrangement (Gottman 2013).

Three criteria must be met for a couple to be defined as a "living apart together" couple: (1) they must define themselves as a committed couple; (2) others must define the partners as a couple; and (3) they must live in separate

Chelsea Curry

These individuals are committed to each other but enjoy their privacy by living apart.

domiciles. The lifestyle of living apart together involves partners in loving and committed relationships (married or unmarried) identifying their independent needs in terms of the degree to which they want time and space away from each other. People living apart together exist on a continuum from partners who have separate bedrooms and baths in the same house to those who live in a separate place (apartment, condo, house) in the same or different cities. LAT couples are not those couples who are forced by their career or military assignment to live separately. Rather, LAT partners choose to live in separate domiciles and feel that their relationship benefits from the LAT structure. You might take the Self-Assessment on LAT to assess the degree to which this lifestyle is compatible with your needs.

**DIVERSITY IN OTHER COUNTRIES**

Another version of living apart together is the "walking marriage." High in the Tibetan Himalayas, "walking marriages" occur in the Mosuo culture (Kingdom of Women) in China, which does not have traditional marriage (no husbands or wives). Rather, in this matrilineal society, women live with other women (and raise the children) and men "walk by at night and visit and leave the next morning." The women and men can have as many "walking marriages" as they want with the men never living with the woman, only visiting her (Mosuo 2010).

The living apart together lifestyle or family form is not unique to couples in the United States (e.g., the phenomenon is more prevalent in European countries such as France, Sweden, and Norway). Couples choose this pattern for a

## SELF-ASSESSMENT | Living Apart Together (LAT) Scale

This scale will help you assess the degree to which you might benefit from living in a separate residence from your spouse or partner with whom you have a lifetime commitment. There are no right or wrong answers.

### Directions

After reading each sentence carefully, circle the number that best represents the degree to which you agree or disagree with the sentence.

| 1 | 2 | 3 | 4 | 5 |
|---|---|---|---|---|
| Strongly agree | Mildly agree | Undecided | Mildly disagree | Strongly disagree |

| | SA | MA | U | MD | SD |
|---|---|---|---|---|---|
| 1. I prefer to have my own place (apart from my partner) to live. | 1 | 2 | 3 | 4 | 5 |
| 2. Living apart from my partner feels "right" to me. | 1 | 2 | 3 | 4 | 5 |
| 3. Too much togetherness can kill a relationship. | 1 | 2 | 3 | 4 | 5 |
| 4. Living apart can enhance your relationship. | 1 | 2 | 3 | 4 | 5 |
| 5. By living apart you can love your partner more. | 1 | 2 | 3 | 4 | 5 |
| 6. Living apart protects your relationship from staleness. | 1 | 2 | 3 | 4 | 5 |
| 7. Couples who live apart are just as happy as those who don't. | 1 | 2 | 3 | 4 | 5 |
| 8. Couples who LAT are just as much in love as those who live together in the same place. | 1 | 2 | 3 | 4 | 5 |
| 9. People who LAT probably have less relationship stress than couples who live together in the same place. | 1 | 2 | 3 | 4 | 5 |
| 10. LAT couples are just as committed as couples who live together in the same residence. | 1 | 2 | 3 | 4 | 5 |

### Scoring

Add the numbers you circled. The lower your total score (10 is the lowest possible score), the more suited you are to the Living Apart Together lifestyle. The higher your total score (50 is the highest possible score), the least suited you are to the Living Apart Together lifestyle. A score of 25 places you at the midpoint between being the extremes. One-hundred and thirty undergraduates completed the LAT scale with an average score of 28.92, which suggests that both sexes view themselves as less rather than more suited (30 is the midpoint between the lowest score of 10 and the highest score of 50) for a LAT arrangement with females registering greater disinterest than males.

### Source

number of reasons, including the desire to maintain some level of independence, to enjoy their time alone, and to keep their relationship exciting.

## Advantages of LAT

The benefits of LAT relationships include the following:

**1.** ***Space and privacy.*** Having two places enables each partner to have a separate space to read, watch TV, talk on the phone, or whatever. This not only provides a measure of privacy for the individuals, but also for the couple. When the couple has overnight guests, the guests can stay in one place while the partners stay in the other place. This arrangement gives guests ample space and the couple private space and time apart from the guests.

**2.** ***Career or work space.*** Some individuals work at home and need a controlled quiet space to work on projects, talk on the phone, and focus on their work without the presence of someone else. The LAT arrangement is particularly appealing to musicians for practicing, artists to spread out their materials, and authors for quiet (Hemingway built a separate building where he wrote in Key West).

**3.** ***Variable sleep needs.*** Although some partners enjoy going to bed at the same time and sleeping in the same bed, others like to go to bed at radically different times and to sleep in separate beds or rooms. A frequent comment from LAT partners is, "My partner thrashes throughout the night and kicks me, not to mention the wheezing and teeth grinding, so to get a good night's sleep, I need to sleep somewhere else."

**4.** ***Allergies.*** Individuals who have cat or dog allergies may need to live in a separate antiseptic environment from their partner who loves animals and won't live without them. "He likes his dog on the bed," said one woman.

**5.** ***Variable social needs.*** Partners differ in terms of their need for social contact with friends, siblings, and parents. The LAT arrangement allows for the partner who enjoys frequent time with others to satisfy that need without subjecting the other to the presence of a lot of people in one's life space. One wife from a family of seven children enjoyed frequent contact with both her siblings and parents. The LAT arrangement allowed her to continue to enjoy her family at no expense to her husband who was upstairs in a separate condo.

**6.** ***Blended family needs.*** LAT works particularly well with a blended family in which remarried spouses live in separate places with their children from previous relationships. An example is a remarried couple who bought a duplex with each spouse living with his/her own children on the respective sides of the duplex. The parents could maintain a private living space with their children, the spouses could be next door to each other, and the stepsiblings were not required to share living quarters—a structural answer to major stepfamily blending problems.

**7.** ***Keeping the relationship exciting.*** Zen Buddhists remind us of the necessity to be in touch with polarities, to have a perspective where we can see and appreciate the larger picture—without the darkness, we cannot fully appreciate the light. The two are inextricably part of a whole. This is the same with relationships; time apart from our beloved can make time together more precious. **Satiation** is a well-established psychological principle. The term means that a stimulus loses its value with repeated exposure. Just as we tire of eating the same food, listening to the same music, or watching the same movie, we can become satiated in a relationship. Indeed, couples who are in a long-distance dating relationship know the joy of missing each other and the excitement of being with each other again. Similarly, individuals in a LAT relationship help to ensure that they will not become satiated with each other but will maintain some of the excitement in seeing or being with each other.

**8.** ***Self-expression and comfort.*** Partners often have very different tastes in furniture, home décor, music, and temperature. With two separate places, each can arrange and furnish their respective homes according to their individual preferences. The respective partners can also set the heat or air conditioning according to their own preference, and play whatever music they like.

**9.** ***Cleanliness or orderliness.*** Separate residences allow each partner to maintain the desired level of cleanliness and orderliness without arguing about it. Some individuals like their living space to be as ordered as a cockpit. Others simply don't care.

**10.** ***Elder care.*** One partner may be taking care of an elderly parent in his or her own house. The other partner may prefer not to live with an elderly person. A LAT relationship allows for the partner taking care of the elderly parent to do so while providing a place for the couple to be alone.

**11.** ***Maintaining one's lifetime residence.*** Some retirees, widows, and widowers meet, fall in love, and want to enjoy each other's companionship. However, they don't want to move out of their own home. The LAT arrangement does not require that the partners move.

**12.** ***Leaving inheritances to children from previous marriages.*** Having separate residences allows the respective partners to leave their family home or residential property to their biological children from an earlier relationship without displacing their surviving spouse.

## Disadvantages of LAT

There are also disadvantages to the LAT lifestyle.

**1.** ***Stigma or disapproval.*** Because the norm that married couples move in together is firmly entrenched, couples who do not do so are suspect. "People who love each other should want to live in their own house together... those who live apart aren't really committed" is the traditional perception of people in a committed relationship.

**2.** ***Cost.*** The cost of two separate living places can be more expensive than two people living in one domicile. But there are ways LAT couples afford their lifestyle. Some live in two condominiums that are cheaper than owning one larger house. Others buy housing outside of high-priced real estate areas. One partner said, "We bought a duplex 10 miles out of town where the price of housing is 50% cheaper than in town. We have our LAT context and it didn't cost us a fortune."

**3.** ***Inconvenience.*** Unless the partners live in a duplex or two units in the same condominium, going between the two places to share meals or be together can be inconvenient.

**4.** ***Lack of shared history.*** Because the adults are living in separate quarters, a lot of what goes on in each house does not become a part of the life history of the couple. For example, children in one place don't benefit as much from the other adult who lives in another domicile most of the time.

**5.** ***No legal protection.*** The legal nature of the LAT relationship is ambiguous. Currently this relationship does not have legal protection in the United States. Lyssens-Danneboom et al. (2013) noted that LAT partners in Belgium "believe they should be granted the same family-based benefits as those enjoyed by their cohabiting or married counterparts."

# The Future of Singlehood, Long-Term Relationships, and Cohabitation

Singlehood will (in the cultural spirit of diversity) lose some of its stigma; more young adults will choose this option; and those who remain single will, increasingly, find satisfaction in this lifestyle.

Individuals will continue to be in no hurry to get married. Completing their education, becoming established in their career, and enjoying hanging out and hooking up will continue to delay serious consideration of marriage. The median age for women getting married is 26; for men, 28. This trend will continue as individuals keep their options open in America's individualistic society.

Cohabitation will increase not just in the percent of those living together before marriage (now about two-thirds) but also in the prevalence of serial cohabitation (Vespa 2014). The link between cohabitation and negative marital outcome will dissolve as more individuals elect to cohabit before marriage. Previously, only risk takers and people willing to abandon traditional norms lived together before marriage. In the future, mainstream individuals will cohabit.

# Summary

***What is the status of singlehood today?***

Singles consist of the never married, the divorced, and the widowed. The never married are delaying marriage in favor of completing their education, establishing themselves in a career, and enjoying hanging out and hooking up. The primary attraction of singlehood is the freedom to do as one chooses. Others fear getting into a bad marriage or having to go through a divorce.

***What are the functions of and changes in dating?***

The various functions of dating include: (1) confirmation of a social self; (2) recreation; (3) companionship, intimacy, and sex; (4) anticipatory socialization; (5) status achievement; (6) mate selection; and (7) health enhancement. The changes in dating in the last 60 years include: an increase in the age at marriage, more formerly divorced in the dating pool, cohabitation becoming more normative, and women becoming more assertive in initiating relationships.

***How have hanging out and hooking up replaced traditional dating?***

Couples today often meet each other while hanging out and hooking up. While the latter may be a hoped-for beginning of a continuing relationship, it rarely is.

***What is cohabitation like among today's youth?***

Cohabitation, also known as living together, is becoming an expected stage in courtship with about two-thirds of college students reporting having lived together. Of 100 weddings in the United States almost 60% of American brides report that they had cohabited before marriage. Reasons for an increase in the number of couples living together include a delay of marriage for educational or career commitments, fear of marriage, increased tolerance of society for living together, and a desire to avoid the legal entanglements of marriage. Types of relationships in which couples live together include the here-and-now, testers (testing the relationship), engaged couples (planning to marry), and cohabitants forever (never planning to marry). Most people who live together eventually marry but not necessarily each other.

Domestic partners are two adults who have chosen to share each other's lives in an intimate and committed relationship. Such cohabitants, both heterosexual and homosexual, want their employers, whether governmental or corporate, to afford them the same rights as spouses.

Although living together before marriage does not ensure a happy, stable marriage, it has some potential advantages. These include a sense of well-being, delayed marriage while learning about yourself and your partner, and being able to end the relationship with minimal legal hassle. Disadvantages include feeling exploited, feeling guilty about lying to parents, and not having the same economic benefits as those who are married. Social Security and retirement benefits are paid to spouses, not live-in partners.

***What are the pros and cons of "living apart together"?***

LAT means that monogamous committed partners—whether married or not—carve out varying degrees of physical space between themselves. People living apart together exist on a continuum from partners who have separate bedrooms and baths in the same house to those who live in separate apartments, condos, or houses in the same or different cities. Couples choose this pattern for a number of reasons, including the desire to maintain some level of independence, to enjoy their time alone, and to keep their relationship fresh and exciting. Disadvantages include being confronted with stigma or disapproval, cost, and inconvenience.

***What is the future of singlehood?***

Singlehood will lose some of its stigma; more young adults will choose this option; and those who remain single will, increasingly, find satisfaction in this lifestyle. While marriage will remain the lifestyle choice, individuals will continue to be in no hurry to get married.

## Key Terms

Cohabitation
Cohabitation effect
Cyclical cohabitation
Domestic partnership
Hanging out
Hooking up
Living apart together (LAT)
Long-distance relationship
Palimony
Satiation

## Web Links

Unmarried Equality
http://www.unmarried.org

8minuteDating
http://www.8minutedating.com/

Independent Women's Forum
http://www.happinessonline.org/BeFaithfulToYourSexualPartner/p17.htm

Selective Search
http://www.selectivesearch.com/

# CHAPTER 5 Selecting a Partner

Rachel Calisto

*All those love songs on the radio, all they talk about is that first chemical flush of love. I don't know why they emphasize that part so much. What's more important to me is what comes after those first few months of jangly nerves and starry eyes. Does the guy do the dishes, does he respect my work, is he considerate?*

AUDREY SCHULMAN, WRITER

## Learning Objectives

- Identify the cultural, sociological, and psychological factors operative in selecting a partner.
- Describe sociobiology, discuss how it explains mate selection, and review the criticisms.
- Review premarital counseling, premarital education, and prenuptial agreements.
- Explain the various ways of using technology to find a partner and know the pros and cons.
- Discuss the conditions under which you should consider calling off the wedding.
- Predict what will occur in reference to mate selection in the future.

## TRUE OR FALSE?

1. Parents are more disapproving of their daughters' than of their sons' interracial involvement.
2. Women value intelligence more in a potential mate than men.
3. Being unhappy in one's personal life is predictive of being unhappy in a relationship.
4. Half of the partners who are asked by their partner about the future of the relationship confirm that they see a future for the relationship.
5. Men are more deceptive in Internet dating than women.

*Answers:* 1. T 2. T 3. T 4. T 5. F

---

Kate and Prince William and their son continue to be culturally visible in the media. But their premarital relationship included a split instigated by William who said that the relationship was "confining" and "claustrophobic." In response, Kate quickly went into action to win her Prince back—she dropped a dress size, put on a mini dress/knee-high boots, and was photographed "dancing and flirting for hours" with high-status men who were all pleased to be with her. Within six weeks the Prince was back to reclaim her—clearly she was playing a game...and clearly she had won (Andersen 2011).

"I don't want to play games" is often heard by individuals finding their way in a new relationship. When a game is defined as making behavioral choices to increase the likelihood of achieving one's relationship goals, we all play games. This chapter is about playing the game of choosing and winning one's partner.

No other choice in life is as important as your choice of a marriage partner. Your day-to-day happiness, health, and economic well-being will be significantly influenced by the partner with whom you choose to share your life. Most have high hopes and some believe in the perfect mate.

Because the heart of this text is about making wise choices in relationships, this chapter begins with the examination of the cultural, sociological, and psychological factors involved in mate selection. We then discuss how the Internet is being used to assist individuals in finding their partner. Once a partner is identified, we identify specific questions to ask a potential partner to increase your knowledge of the partner and the potential for a loving and durable relationship. We end the chapter with wrong reasons to get married and the future of mate selection.

### National **Data**

Almost 85% of 5,500 single individuals reported that they were open for involvement in a relationship and actively looking for a partner (Walsh 2013).

*I found Miss Right.... I just didn't know her first name was Always.*

ANONYMOUS

# Cultural Aspects of Mate Selection

Individuals are not free to become involved with/marry whomever they want. Rather, their culture and society radically restrict and influence their choice. The best example of mate choice being culturally and socially controlled is the fact that *less than* 1% of the over 63 million marriages in the United States consist of a black spouse and a white spouse (*Statistical Abstract of*

## DIVERSITY IN OTHER COUNTRIES

In most cultures, mate selection is not left to chance and parents are involved in the selection of their offspring's partner. However, who picks the best mate: parents or the individual? India is made up of 29 states, reflecting a range of languages, dialects, and religions. Because Indian culture is largely familistic (focusing on family unity and loyalty) rather than individualistic (focusing on personal interests and freedom), most Indian youth see parents as better able to select a lifetime marital partner than themselves and defer to their parents' judgment.

*the United States 2012–2013,* Table 60). Indeed, up until the late sixties, such marriages in most states were a felony and mixed spouses were put in jail. In 1967, the Supreme Court ruled that mixed marriages (e.g., between whites and blacks) were legal. Only recently have homosexuals received federal recognition for same sex marriage.

Endogamy and exogamy are also two forms of cultural pressure operative in mate selection.

### Endogamy

**Endogamy** cultural expectation to select a marriage partner within one's own social group.

**Endogamy** is the cultural expectation to select a marriage partner within one's own social group, such as in the same race, religion, and social class. Endogamous pressures involve social approval from parents (Nesteruk and Gramescu 2012) for selecting a partner within one's own group and disapproval for selecting someone outside one's own group. The pressure toward an endogamous mate choice is especially strong when race is involved. Love may be blind, but it knows the color of one's partner as the overwhelming majority of individuals end up selecting someone of the same race to marry.

### Exogamy

**Exogamy** the cultural pressure to marry outside the family group.

**Exogamy** is the cultural pressure to marry outside the family group (e.g., you cannot marry your siblings). Woody Allen experienced enormous social disapproval because he fell in love with and married in 1997 his long-time partner's adopted daughter, Soon-Yi Previn (they remain married).

## DIVERSITY IN THE UNITED STATES

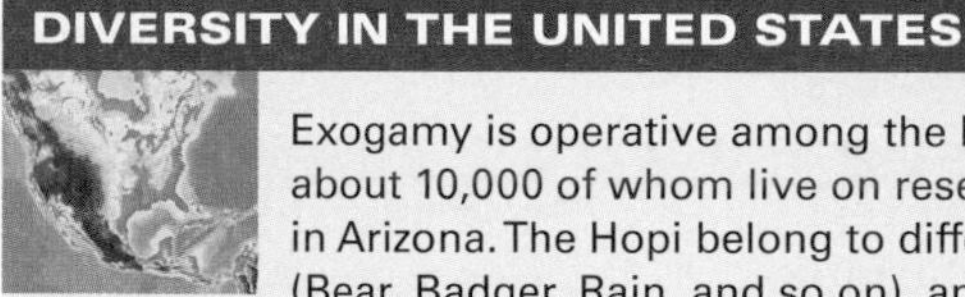

Exogamy is operative among the Hopi Indians, about 10,000 of whom live on reservations in Arizona. The Hopi belong to different clans (Bear, Badger, Rain, and so on), and the youth are expected to marry someone of a different clan. By doing so, they bring resources from another clan into their own.

Incest taboos are universal; in addition, children are not permitted to marry the parent of the other sex in any society. In the United States, siblings and (in some states) first cousins are also prohibited from marrying each other. The reason for such restrictions is fear of genetic defects in children whose parents are too closely related.

**Pool of eligibles** the population from which a person may select a mate.

Once cultural factors have determined the general **pool of eligibles** (the population from which a person may select a mate), individual mate choice becomes more operative. However, even when individuals feel that they are making their own choices, social influences are still operative (e.g., approval from one's peers).

# Sociological Factors Operative in Mate Selection

Numerous sociological factors are at work in bringing two people together who eventually marry.

### Homogamy

**Homogamy** the tendency for the individual to seek a mate with similar characteristics.

Whereas endogamy refers to cultural pressure to select within one's own group, **homogamy** ("like selects like") refers to the tendency for the individual to seek a mate with similar characteristics (e.g., age, race, education, etc.). In general, the more couples have in common, the higher the reported relationship satisfaction and the more durable the relationship (Amato et al. 2007). Mcintosh et al. (2011) studied older adults who posted online ads and found that they were concerned about homogamous factors in terms of age, race, religion, etc., and they were willing to travel great distances to find such a partner. A small contingent of

Japanese and Koreans believe that similarity of blood type is an important factor in selecting one's partner. For them, "what is your blood type?" is an important and appropriate question to ask a person they consider as a potential mate.

**Race** Race refers to physical characteristics that are given social significance. Yoo et al. (2010) noted the difference between blatant and subtle racism. Blatant racism involves outright name calling and discrimination. Tiger Woods was once denied the right to play golf on a Georgia golf course because he is black. Subtle racism involves omissions, inactions, or failure to help, rather than a conscious desire to hurt. Not sitting at a table where persons of another race are sitting or failing to stop to help a person because of his or her race is subtle racism. The roots or racism run deep, the horror of which was depicted in the Academy Award-winning movie *12 Years a Slave*.

*Racial superiority is a mere pigment of the imagination.*

AUTHOR UNKNOWN

## National **Data**

About 13% of American married couples involve someone whose partner is of a different race/ethnicity (Burton et al. 2010).

Blacks, younger individuals, and the politically liberal are more willing to cross racial lines for dating and for marriage than are whites, older individuals, and conservatives (Tsunokai & McGrath 2011). These findings emerged from a study of 1,335 respondents on Match.com. Being from the South and identifying with religion (Catholic, Protestant, Jewish) was associated with less openness toward interracial relationships. The greatest resistance to interracial relationships of whites was to African Americans, then Hispanics, then Asians. Gays also tended to select partners of the same racial and ethnic background. When an African-American male marries a white female, he more often has higher educational credentials than she does (Hou and Myles 2013).

The Self-Assessment feature allows you to assess your openness to involvement in an interracial relationship.

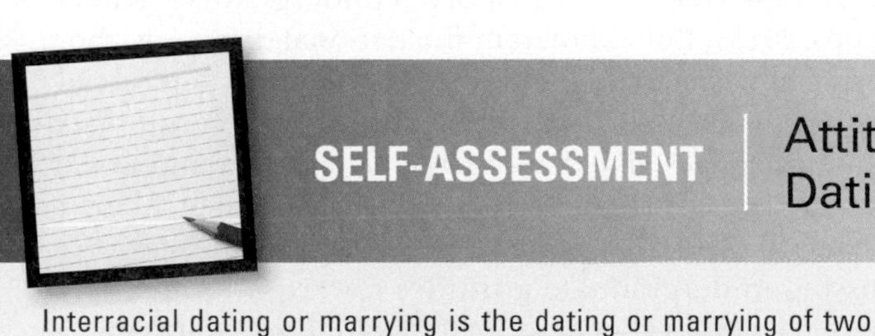

## SELF-ASSESSMENT | Attitudes toward Interracial Dating Scale

Interracial dating or marrying is the dating or marrying of two people from different races. The purpose of this survey is to gain a better understanding of what people think and feel about interracial relationships. Please read each item carefully, write a number from 1 to 7 in the space to the left of each item, and add the items to get a total score.

| 1 | 2 | 3 | 4 | 5 | 6 | 7 |
|---|---|---|---|---|---|---|
| Strongly Disagree | | | | | | Strongly Agree |

There are no right or wrong answers to any of these statements.

_____ 1. I believe that interracial couples date outside their race to get attention.

_____ 2. I feel that interracial couples have little in common.

_____ 3. When I see an interracial couple, I find myself evaluating them negatively.

_____ 4. People date outside their own race because they feel inferior.

_____ 5 Dating interracially shows a lack of respect for one's own race.

_____ 6. I would be upset with a family member who dated outside our race.

_____ 7. I would be upset with a close friend who dated outside our race.

_____ 8. I feel uneasy around an interracial couple.

_____ 9 People of different races should associate only in non-dating settings.

_____ 10. I am offended when I see an interracial couple.

*(continued)*

## SELF-ASSESSMENT (*continued*)

_____ 11. Interracial couples are more likely to have low self-esteem.
_____ 12. Interracial dating interferes with my fundamental beliefs.
_____ 13. People should date only within their race.
_____ 14. I dislike seeing interracial couples together.
_____ 15. I would not pursue a relationship with someone of a different race, regardless of my feelings for that person.
_____ 16. Interracial dating interferes with my concept of cultural identity.
_____ 17. I support dating between people with the same skin color, but not with a different skin color.
_____ 18. I can imagine myself in a long-term relationship with someone of another race.
_____ 19. As long as the people involved love each other, I do not have a problem with interracial dating.
_____ 20. I think interracial dating is a good thing.

### Scoring

First, reverse the scores for items 18, 19, and 20 by switching them to the opposite side of the spectrum. For example, if you selected 7 for item 18, replace it with a 1; if you selected 3, replace it with a 5, and so on. Next, add your scores and divide by 20. Possible final scores range from 1 to 7, with 1 representing the most positive attitudes toward interracial dating and 7 representing the most negative attitudes toward interracial dating.

### Norms

The norming sample was based upon 113 male and 200 female students attending Valdosta State University. The participants completing the Attitudes toward Interracial Dating Scale (IRDS) received no compensation for their participation. All participants were U.S. citizens. The average age was 23.02 years (standard deviation = 5.09), and participants ranged in age from 18 to 50 years old. The average score on the IRDS was 2.88 (SD = 1.48), and scores ranged from 1.00 to 6.60, suggesting very positive views of interracial dating. Men scored an average of 2.97 (SD = 1.58), and women, 2.84 (SD = 1.42). There were no significant differences between the responses of women and men.

### Source

"Attitudes Toward Interracial Dating Scale," 2004 by Mark Whatley, Ph.D., Department of Psychology, Valdosta State University, Valdosta, GA 31698-0100. Used by permission. Other uses of this scale only by written permission of Dr. Whatley (mwhatley@valdosta.edu). Information on the reliability and validity of this scale is available from Dr. Whatley.

In general, as one moves from dating to living together to marriage, the willingness to marry interracially decreases. Of 1,095 undergraduate males, 33% reported that they had dated interracially; 39% of 3,435 undergraduate females had done so (Hall and Knox 2013). But as noted in the national data, only about 15% actually marry interracially. Some individuals prefer and seek interracial/interethnic relationships (Yodanis et al. 2012). But these are often not black-white mergers. Black/white marriages, like that of Grace Hightower and Robert De Niro (as well as the marriages of Chaz Bono and the late Roger Ebert to black women), are less than 1% of all marriages.

We have been discussing undergraduate attitudes toward interracial dating. But what about their parents? Ozay et al. (2012) analyzed data on 251 parents of undergraduates and found that that over a third (35%) disapproved of their child's involvement in an interracial relationship. Parental disapproval (which did not vary by gender of the parent) increased with the seriousness of relationship involvement (e.g., dating or marriage) and parents were more disapproving of their daughters' than of their sons' interracial involvement.

In regard to parental approval of their son's or daughter's interracial involvement, black mothers and white fathers have different roles in their respective black and white communities, in terms of setting the norms of interracial relationships. Hence, the black mother who approves of her son's or daughter's interracial relationship may be less likely to be overruled by the father than the white mother. The white husband may be more disapproving and have more power over his white wife than the black father/husband over his partner.

Chelsea Curry

Interracial partners are highly varied, including Hispanic/Filipino.

## National **Data**

Of the over 63 million married couples in the United States, less than 1% (0.009%) consists of a black spouse and white spouse (*Statistical Abstract of the United States 2012–2013,* Table 60).

**Age** Most individuals select a partner who is relatively close in age. Men tend to select women three to five years younger than themselves. The result is the **marriage squeeze**, which is the imbalance of the ratio of marriageable-aged men to marriageable-aged women. In effect, women have fewer partners to select from because men choose from not only their same age group but also from those younger than themselves. One 40-year-old recently divorced woman said, "What chance do I have with all these guys looking at all these younger women?"

**Marriage squeeze** the imbalance of the ratio of marriageable-aged men to marriageable-aged women.

**Intelligence** Dijkstra et al. (2012) studied the mate selection preferences of the intellectually gifted and found that intelligence was one of the primary qualities sought in a potential partner. While both genders valued an intelligent partner, women gave intelligence a higher priority.

*If you live to be a hundred, I want to live to be a hundred minus one day so I never have to live without you.*

A.A. MILNE, ENGLISH AUTHOR

**Education** Educational homogamy also operates in mate selection. Not only does college provide an opportunity to meet, hang out with, and cohabit with potential partners, it also increases one's chance that only a college-educated partner becomes acceptable as a potential cohabitant or spouse. Education becomes an important criterion to find in a partner.

**Open minded** receptive to understanding alternative points of view, values, and behaviors.

**Open Mindedness** People vary in the degree to which they are **open minded** (receptive to understanding alternative points of view, values, and behaviors). Homogamous pairings in regard to open mindedness reflect that partners seek like-minded individuals. Such open mindedness translates into tolerance of various religions, political philosophies, and lifestyles.

*It was comfortable to date someone who drank from wineglasses instead of from jelly jars, who didn't case fast-food restaurants to relieve them of their surplus napkins, ketchup packets, and plastic forks.*

ZZ PACKER, WRITER

**Social Class** Brown (2010) interviewed undergraduate students to assess their awareness of social class differences. Having a large number of designer clothes and owning one's own new car were viewed as characteristics of middle- or upper-class students in contrast to students in the working or lower class. Also, the degree to which an item was sought because it was extravagant or functional was an indicator of class. Upper-class students have expensive, fancy computers; middle-class students have computers that are functional—they work, but are not the latest model. Having discretionary income to frequently replace items such as cars, iPhones, or laptops was also seen as an indicator of one's social class.

Social class reflects your parents' occupations, incomes, and educations as well as your residence, language, and values. If you were brought up by parents who were physicians, you probably lived in a large house in a nice residential area—summer vacations and a college education were givens. Alternatively, if your parents dropped out of high school and worked blue-collar jobs, your home was likely smaller and in a less expensive part of town, and your opportunities (such as education) were more limited. Social class affects one's comfort in interacting with others. We tend to select as friends and mates those with whom we feel most comfortable. One undergraduate from an upper-middle-class home said, "When he pulled out coupons at Subway for our first date, I knew this was going nowhere."

*When you go to work, if your name is on the building, you're rich. If your name is on the desk, you're middle class. If your name is on your shirt, you're poor.*

RICH HALL, WRITER

The **mating gradient** refers to the tendency for husbands to be more advanced than their wives with regard to age, education, and occupational success. Indeed, husbands are typically older than their wives, have more advanced education, and earn higher incomes (*Statistical Abstract of the United States 2012–2013*, Table 702).

**Mating gradient** the tendency for husbands to be more advanced than their wives with regard to age, education, and occupational success.

**Physical Appearance** Homogamy is operative in regard to physical appearance in that people tend to become involved with those who are similar in degree of physical attractiveness. However, a partner's attractiveness may be a more important consideration for men than for women. Meltzer and McNulty (2014) emphasized that "body valuation by a committed male partner is positively associated with women's relationship satisfaction when that partner also values them for their nonphysical qualities, but negatively associated with women's relationship satisfaction when that partner is not committed or does not value them for their nonphysical qualities" (p. 68).

*Beauty itself doth of itself persuade*
*The eyes of men without an orator.*

SHAKESPEARE, *THE RAPE OF LUCRECE*

**Career** Individuals tend to meet and select others who are in the same career. In 2012, country western singers Blake Shelton and Miranda Lambert won the Male and Female Vocalist of the Year awards, respectively (she won it again in 2014). They are married. Michelle and Barack Obama are both attorneys. Angelina Jolie and Brad Pitt are both movie celebrities. Danica Patrick and Ricky Stenhouse Jr. are both NASCAR drivers.

**Marital Status** Never-married people tend to select other never-married people as marriage partners; divorced people tend to select other divorced people; and widowed people tend to select other widowed people.

Rachel Calisto

These individuals mirror each other in terms of physical appearance...both value piercings and tattoos.

Melissa Zentner

These individuals (now engaged) sought each other out on Facebook as they were both varsity athletes—she in volleyball, he in tennis.

**Religion/Spirituality/Politics** Religion may be defined as a set of beliefs in reference to a supreme being which involves practices or rituals (e.g., communion) generally agreed upon by a group of people. Of course, some individuals view themselves as "not religious" but "spiritual," with spirituality defined as belief in the spirit as the seat of the moral or religious nature that guides one's decisions and behavior. Religious/spiritual homogamy is operative in that people of a similar religion or spiritual philosophy tend to seek out each other. Over a third (35%) of 1,098 undergraduate males and 44% of 3,453 undergraduate

This couple share the value of prayer, which increases their emotional bond with each other.

Chelsea Curry

females agreed that "It is important that I marry someone of my same religion." Ellison et al. (2010) observed that couples who have shared religious beliefs and who practice in-home family devotional activities report higher relationship quality than couples without common religious values/no ritual sharing.

Alford et al. (2011) emphasized that homogamy is operative in regard to politics. An analysis of national data on thousands of spousal pairs in the United States revealed that homogamous political attitudes were the strongest of all social, physical, and personality traits. Further, this similarity derived from initial mate choice "rather than persuasion and accommodation over the life of the relationship."

**Personality** Gonzaga et al. (2010) asked both partners of 417 eHarmony couples who married to identify the degree to which various personality trait terms reflected who they, as individuals, were. Examples of the terms included warm, clever, dominant, outgoing, quarrelsome, stable, energetic, affectionate, intelligent, witty, content, generous, wise, bossy, kind, calm, outspoken, shy, and trusting. Results revealed that those who were more similar to each other in personality characteristics were also more satisfied with their relationship.

**Circadian preference** an individual's preference for morningness-eveningness in regard to intellectual and physical activities.

**Circadian Preference** **Circadian preference** refers to an individual's preference for morningness-eveningness in regard to intellectual and physical activities. In effect, some prefer the morning while others prefer late afternoon or evening hours. In a study of 84 couples, Randler and Kretz (2011) found that partners in a romantic relationship tended to have similar circadian preferences. The couples also tended to have similar preferences as to when they went to bed in the evening and when they rose in the morning.

**Traditional Roles** Partners who have similar views of what their marital roles will be are also attracted to each other. Abowitz et al. (2011) found that about a third of the 692 undergraduate women wanted to marry a "traditional" man—one who viewed his primary role as that of provider and who would be supportive of his wife staying home to rear the children. In a related study of 1,027 undergraduate men, 31% reported that they wanted a traditional wife—one who viewed her primary role as wife and mother, staying home and rearing the children (not having a career) (Knox and Zusman 2007).

**Geographic Background** Haandrikman (2011) studied Dutch cohabitants and found that the romantic partners tended to have grown up within six kilometers of each other (**spatial homogamy**). While the university context draws people from different regions of the country, the demographics of state universities tend to reflect a preponderance of those from within the same state.

**Spatial homogamy** romantic partners who grow up within six kilometers of each other.

**Economic Values, Money Management, and Debt** Individuals vary in the degree to which they have money, spend money, and save money. Some are deeply in debt and carry significant educational debt. The average debt for those with a bachelor's degree is $23,000. Almost half (48%) of those who borrowed money for their undergraduate education report that they are burdened by this debt (Pew Research Center 2011). Of 1,094 undergraduate males, 4.3% and 3.5% of 3,446 undergraduate females owed more than a thousand dollars on a credit card (Hall and Knox 2013). Money becomes an issue in mate selection in that different economic values predict conflict. One undergraduate male noted, "There is no way I would get involved with this woman—she jokes about maxing out her credit cards. She is going to be someone else's nightmare."

WHAT IF?

## What If the Partner You Love Has a Dealbreaker Flaw?

A dealbreaker flaw is a behavior that you find unacceptable. Examples include "does drugs," "is abusive," "requires that we live no more than two hours from his parents," or "previously married with kids." Although Ann Landers's phrase that "everyone comes with a catch" is a fact of mate selection, some factors should, indeed, eliminate a partner from consideration. Only you can decide the degree to which the behavior warrants ending the relationship. Of course, beliefs that the "partner will change because of their love for me" or "will change after marriage" are nonsense.

# Psychological Factors Operative in Mate Selection

*Our similarities bring us to a common ground; our differences allow us to be fascinated by each other.*

TOM ROBBINS, AMERICAN NOVELIST

Psychologists have focused on complementary needs, exchanges, parental characteristics, and personality types with regard to mate selection.

## Complementary-Needs Theory

**Complementary-needs theory** (also known as "opposites attract") states that we tend to select mates whose needs are opposite yet complementary to our own. For example, some partners may be drawn to each other on the basis of nurturance versus receptivity. These complementary needs suggest that one person likes to give and take care of another, whereas the other likes to be the benefactor of such care. Other examples of complementary needs are responsibility versus irresponsibility, peacemaker versus troublemaker, and disorder versus order. Former *Tonight* Show host Jay Leno revealed in his autobiography that he and his wife of over 30 years are very different:

**Complementary-needs theory** we tend to select mates whose needs are opposite yet complementary to our own.

*He felt now that he was not simply close to her, but that he did not know where he ended and she began.*

LEO TOLSTOY

> *We were, and are, opposites in most every way. Which I love. There's a better balance...I don't consider myself to have much of a spiritual side, but Mavis has*

*almost a sixth sense about people and situations. She has deep focus, and I fly off in 20 directions at once. She reads 15 books a week, mostly classic literature. I collect classic car and motorcycle books. She loves European travel; I don't want to go anywhere people won't understand my jokes...(Leno 1996, 214–215).*

## Exchange Theory

**Exchange theory** theory that emphasizes that relations are formed and maintained based on who offers the greatest rewards at the lowest costs.

*The meeting of two personalities is like the contact of two chemical substances. If there is any reaction, both are transformed.*

CARL JUNG, SWISS PSYCHIATRIST

*Blessed are those who give without remembering and take without forgetting.*

ELIZABETH BIBESCO, ENGLISH WRITER

**Exchange theory** in mate selection is focused on finding the partner who offers the greatest rewards at the lowest cost. The following five concepts help to explain the exchange process in mate selection:

**1. *Rewards.*** Rewards are the behaviors (your partner looking at you with the eyes of love), words (saying "I love you"), resources (being beautiful or handsome, having a car, condo at the beach, and money), and services (cooking for you, typing for you) your partner provides that you value and that influence you to continue the relationship. Similarly, you provide behaviors, words, resources, and services for your partner that he or she values. Relationships in which the exchange is equal are happiest. Increasingly, men are interested in women who offer "financial independence" and women are interested in men who "cook and do the dishes."

**2. *Costs.*** Costs are the unpleasant aspects of a relationship. A woman identified the costs associated with being involved with her partner: "He abuses drugs, doesn't have a job, and lives nine hours away." The costs her partner associated with being involved with this woman included "she nags me," "she doesn't like sex," and "she insists that we live in the same town they live in if we marry."

**3. *Profit.*** Profit occurs when the rewards exceed the costs. Unless the couple described in the preceding paragraph derive a profit from staying together, they are likely to end their relationship and seek someone else with whom there is a higher profit margin. Biographer Thomas Maier (2009) said of Virginia Johnson (of the famous Masters and Johnson team) that she was motivated by money to marry Dr. William Masters. "I never wanted to be with him, but when you are making $200,000 a year, you don't walk" said Johnson (p. 235).

Profit may also refer to nonmonetary phenomenon. Moore et al. (2013) found that being altruistic (giving selflessly) was a characteristic of a potential mate valued by both women and men. In effect, such a person provides a higher profit margin for any relationship.

**4. *Loss.*** Loss occurs when the costs exceed the rewards. Partners who feel that they are taking a loss in their relationship are vulnerable to looking for another partner who offers a higher profit.

**5. *Alternative.*** Is another person currently available who offers a higher profit? Individuals on the marriage market have an understanding of what they are worth and whom they can attract (Bredow et al. 2011). Oberbeek et al. (2013) found that facially unattractive individuals with high body mass indexes were less selective in a speed dating context. In effect, they had less to offer so adjusted their expectations in terms of what they felt they could get. You will stay in a relationship where you have a high profit at a low cost. You will leave a relationship where your costs are high and you have an alternative partner who offers you a relationship with high rewards.

## Role Theory

**Role theory of mate selection** a son or daughter models after the parent of the same sex by selecting a partner similar to the one the parent selected as a mate.

Freud suggested that the choice of a love object in adulthood represents a shift in libidinal energy from the first love objects, the parents. **Role theory of mate selection** emphasizes that a son or daughter models after the parent of the same sex by selecting a partner similar to the one the parent selected as a mate. This means that a man looks for a wife who has similar characteristics to those of his mother and that a woman looks for a husband who is very similar to her father.

## Attachment Theory

The **attachment theory of mate selection** emphasizes the drive toward intimacy and a social and emotional connection (Sassler 2010). One's earliest experience as a child is to be emotionally bonded to one's parents (usually the mother) in the family context. The emotional need to connect remains as an adult and expresses itself in relationships with others, most notably the romantic love relationship. Children diagnosed with oppositional-defiant disorder (ODD) or post-traumatic stress disorder (PTSD) have had disruptions in their early bonding and frequently display attachment problems, possibly due to early abuse, neglect, or trauma. Children reared in Russian orphanages in the fifties where one caretaker was assigned to multiple children learned "no one cares about me" and "the world is not safe." Reversing early negative or absent bonding is difficult.

McClure and Lydon (2014) noted that anxiety attachment may express itself in meeting a new partner in the form of verbal disfluences and social awkwardness. The result is that would-be partners may be put off by this anxiety, disengage, and prevent a relationship from developing, the very context which could assist the anxious person in becoming more comfortable.

**Attachment theory of mate selection** emphasizes the drive toward intimacy and a social and emotional connection.

*The magic of first love is our ignorance that it can ever end.*

BENJAMIN DISRAELI, ENGLISH NOVELIST

## Undesirable Personality Characteristics of a Potential Mate

Researchers have identified several personality factors predictive of relationships which end in divorce or are unhappy (Burr et al. 2011; Foster 2008). Potential partners who are observed to consistently display these characteristics might be avoided.

**1. *Controlling.*** The behavior that 60% of a national sample of adult single women reported as the most serious fault of a man was his being "too controlling" (Edwards 2000).

**2. *Narcissistic.*** Individuals who are narcissistic view relationships in terms of what they get out of them. When satisfactions wane and alternatives are present, narcissists are the first to go. Because all relationships have difficult times, a narcissist is a high risk for a durable marriage relationship.

**3. *Poor impulse control.*** Lack of impulse control is problematic in relationships because such individuals are less likely to consider the consequences of their actions. Having an affair is an example of failure to control one's behavior to insure one's fidelity. Such people do as they please and worry about the consequences later.

*Never talk about yourself.*

BALTASAR GRACIAN, SPANISH WRITER IN 1600S

**4. *Hypersensitive.*** Hypersensitivity to perceived criticism involves getting hurt easily. Any negative statement or criticism is received with a greater impact than a partner intended. The disadvantage of such hypersensitivity is that a partner may learn not to give feedback for fear of hurting the hypersensitive partner. Such lack of feedback to the hypersensitive partner blocks information about what the person does that upsets the other and what could be done to make things better. Hence, the hypersensitive one has no way of learning that something is wrong, and the partner has no way of alerting the hypersensitive partner. The result is a relationship in which the partners can't talk about what is wrong, so the potential for change is limited.

**5. *Inflated ego.*** An exaggerated sense of oneself is another way of saying a person has a big ego and always wants things to be his or her way. A person with an inflated sense of self may be less likely to consider the other person's opinion in negotiating a conflict and prefer to dictate an outcome. Such disrespect for the partner can be damaging to the relationship.

**6. *Perfectionistic.*** Individuals who are perfectionists may require perfection of themselves and others. They are rarely satisfied and always find something wrong with their partner or relationship. Living with a perfectionist will be a challenge since whatever one does will not be good enough.

**7. *Insecure.*** Feelings of insecurity also make relationships difficult. The insecure person has low self-esteem, constantly feels that something is wrong, and feels disapproved of by the partner. The partner must constantly reassure him or her that all is well—a taxing expectation over time.

**8. *Controlled.*** Individuals who are controlled by their parents, former partner, child, or anyone else compromise the marriage relationship because their allegiance is external to the couple's relationship. Unless the person is able to break free of such control, the ability to make independent decisions will be thwarted, which will both frustrate the spouse and challenge the marriage. An example is a wife whose father dictated that she spend all holidays with him. The husband ultimately divorced her since he felt she was more of a daughter than a wife and that she had never emotionally left home. The late film critic Roger Ebert did not marry until he was fifty—after his mother was dead—since he said she approved of none of the women he wanted to marry and he could not break free of her disapproval.

Individuals may also be controlled by culture. The example is a Muslim man who fell in love with a woman and she with him. They dated for four years and developed an intense emotional and sexual relationship. But he was not capable of breaking from his socialization to marry a woman of another faith.

**9. *Substance abuser.*** Heavy drug use does not predict well for relationship quality or stability. Blair (2010) studied a national sample of adolescent females and noted a negative relationship between substance abuse (alcohol and marijuana) and lower likelihood of marriage. The researcher reasoned that being labeled as a "drunk" or a "stoner" as a female alienated her from her peers and reduced her chance of being selected as a romantic partner. Wiersma et al. (2011) also noted negative relationship outcomes for the female cohabitant who drinks significantly more than her partner.

**10. *Unhappy.*** Personal happiness is associated with relationship happiness. Stanley et al. (2012) noted that being unhappy in one's personal life is predictive of having an unhappy relationship once married. Conversely, having high life satisfaction before marriage is predictive of high relationship satisfaction once married. Hence, selecting an upbeat, optimistic, happy individual predicts well for a future marriage relationship with this person.

In addition to undesirable personality characteristics, Table 5.1 reflects some particularly troublesome personality disorders and how they may impact a relationship.

**TABLE 5.1 Personality Disorders Problematic in a Potential Partner**

| Disorder | Characteristics | Impact on Partner |
|---|---|---|
| Paranoid | Suspicious, distrustful, thin-skinned, defensive | Partners may be accused of everything. |
| Schizoid | Cold, aloof, solitary, reclusive | Partners may feel that they can never connect and that the person is not capable of returning love. |
| Borderline | Moody, unstable, volatile, unreliable, suicidal, impulsive | Partners will never know what their Jekyll-and-Hyde partner will be like, which could be dangerous. |
| Antisocial | Deceptive, untrustworthy, conscienceless, remorseless | Such a partner could cheat on, lie to, or steal from a partner and not feel guilty. |
| Narcissistic | Egotistical, demanding, greedy, selfish | Such a person views partners only in terms of their value. Don't expect such a partner to see anything from your point of view; expect such a person to bail in tough times. |
| Dependent | Helpless, weak, clingy, insecure | Such a person will demand a partner's full time and attention, and other interests will incite jealousy. |
| Obsessive compulsive | Rigid, inflexible | Such a person has rigid ideas about how a partner should think and behave and may try to impose them on the partner. |
| Neurotic | Worries, obsesses about negative outcomes | This individual will impose negative scenarios on the partners and couple. |

## Female Attraction to "Bad Boys"

Carter et al. (2014) reviewed the **dark triad personality** in some men and confirmed that some women are attracted to these "bad boys." The dark triad is a term for inter-correlated traits of narcissism (sense of entitlement and grandiose self-view), Machiavellianism (deceptive and insincere), and psychopathy (callous and no empathy). These men are socially skilled and manipulative, have a high number of sex partners, and engage in mate poaching. The researchers analyzed data from 128 British undergraduate females and found that they were attracted to these bad boys. Explanations for such attraction was the self-confidence of the bad boys as well as their skill in manipulating the females. And there is the challenge. "I always knew I wasn't the only one, but I wanted to be the girl that changed him," said one undergraduate who dated a bad boy.

**Dark triad personality** term for inter-correlated traits of narcissism (sense of entitlement and grandiose self- view), Machiavellianism (deceptive and insincere), and psychopathy (callous and no empathy).

Chelsea Curry

Some women find themselves attracted to bad boys. The ending is often sad.

**PERSONAL CHOICES**

### *Who Is the Best Person for You to Marry?*

Although no perfect mate exists, some individuals are more suited for you as a marriage partner than others. As we have seen in this chapter, people who have a big ego, poor impulse control, and an oversensitivity to criticism should be considered with great caution.

Just as important as avoiding certain partners, it is important to seek partners with certain characteristics. Someone with whom you have a great deal in common is predictive of your future marital happiness and stability. "Marry yourself" is a worthy guideline in selecting a mate. Homogamous matings with regard to race, education, age, values, religion, social class, and marital status (e.g., never-married people marry never-married people) are more likely to result in more durable, satisfying relationships. "Marry your best friend" is another worthy guideline for selecting the person you marry.

Finally, marrying someone with whom you have a relationship of equality and respect is associated with marital happiness. Relationships in which one partner is exploited or intimidated engender negative feelings of resentment and distance. One man said, "I want a co-chair, not a committee member, for a mate." He was saying that he wanted a partner to whom he related as an equal.

# Sociobiological Factors Operative in Mate Selection

In contrast to cultural, sociological, and psychological aspects of mate selection, which reflect a social learning assumption, the sociobiological perspective suggests that biological or genetic factors may be operative in mate selection.

## Sociobiology

Based on Charles Darwin's theory of natural selection, which states that the strongest of the species survive, **sociobiology** holds that men and women select each other as mates on the basis of their innate concern for producing offspring who are most capable of surviving.

**Sociobiology** theory which emphasizes the biological basis for all social behavior, including mate selection.

According to sociobiologists, men look for a young, healthy, attractive, sexually conservative woman who will produce healthy children and invest in taking care

Figure 5.1 Cultural, sociological, and psychological factors involved in mate selection

**Cultural Factors**

For two people to consider marriage to each other,

Endogamous factors (same race, age) and Exogamous factors (not blood-related)
↓
must be met.
↓

After the cultural prerequisites have been satisfied, sociological and psychological factors become operative.

**Sociological Factors**

Spatial homogamy = the tendency to marry someone from the same geographic area.

Homogamy = the tendency to select a mate similar to oneself with regard to the following:

| | |
|---|---|
| Race | Physical appearance |
| Education | Circadian preference compatibility |
| Social class | Religion |
| Age | Marital status |
| Intelligence | Interpersonal values |

**Psychological Factors**

Complementary needs
Reward-cost ratio for profit
Parental characteristics
Desired personality characteristics



of them. Women, in contrast, look for an ambitious man with stable economic resources who will invest his resources in her children. That men emphasize physical attractiveness and women emphasize social/economic status is true not only in the United States but also in other countries (e.g., China) (Li et al. 2011).

### Criticisms of the Sociobiological Perspective

The sociobiological explanation for mate selection is controversial. Critics argue that women may show concern for the earning capacity of men because women have been systematically denied access to similar economic resources, and selecting a mate with these resources is one of their remaining options. In addition, it is argued that both women and men, when selecting a mate, think about their partners more as companions than as future parents of their offspring.

Figure 5.1 summarizes the various cultural, sociological, and psychological filters involved in selecting a partner.

## Engagement

*Love is an irresistible desire to be irresistibly desired.*

ROBERT FROST, AMERICAN POET

Being identified as a couple occurs after certain events have happened. Chaney and Marsh (2009) interviewed 62 married and 60 cohabiting couples to find out when they first identified themselves as a couple. There were four markers: relationship events, affection/sex, having or rearing children, and time and money.

**1.** Relationship events included a specific event such as visiting the parents of one's partner, becoming engaged, or moving in together.

**2.** Affection/sexual events such as the first time the couple had sex. Losing one's virginity was a salient event.

**3.** Children—becoming pregnant, having a child together, or the first time the partner assumed a parenting role.

**4.** Time/money—spending a lot of time together, sharing funds, or exchanging financial support.

Some engagements happen very fast. Paul Jobs (father of Steve Jobs) and his wife, Clara, were engaged 10 days after they met (the marriage lasted over 40 years and ended by death). Mariah Carey and Nick Cannon married within two months of their first date.

Engagement moves the relationship of a couple from a private, love-focused experience to a public, parent-involved experience. Unlike casual dating, **engagement** is a time in which the romantic partners are sexually monogamous, committed to marry, and focused on wedding preparations. The engagement period molds the intimate relationship of the couple by means of the social support and expectations of family and friends. It is the last opportunity before marriage to systematically examine the relationship—to become confident in one's decision to marry a particular person. Johnson and Anderson (2011) studied 610 newly married couples and found that those spouses who were more confident in their decision to marry ended up investing more in their relationship with greater marital satisfaction.

**Engagement** a time in which the romantic partners are sexually monogamous, committed to marry, and focused on wedding preparations.

## Premarital Counseling

Jackson (2011) developed and implemented a unique evidence-informed treatment protocol for premarital counseling including six private sessions and two postmarital booster sessions. Some clergy require one or more sessions of premarital counseling before they agree to marry a couple.

Premarital counseling is a process of discovery. Because partners might have hesitated to reveal information they feel may have met with disapproval during casual dating, the sessions provide the context to be open with the other about their thoughts, feelings, values, goals, and expectations. The Involved Couple's Inventory (see Self-Assessment feature) is designed to help individuals in committed relationships learn more about each other by asking specific questions.

There is no shortage of advice for those about to marry. Allgood and Gordon (2010) studied 56 couples who were given advice from 412 individuals. The most frequent advice givers were friends (32%), followed by parents (30%) and religious leaders (20%). Results revealed that the closer the individual was to

Mary: *What happened when you showed the girls in the office your new engagement ring?*

Sara: *Four of them recognized it!*

***FROM HUMOR IS TREMENDOUS,* CHARLIE JONES AND BOB PHILLIPS**

Chelsea Curry

Some clergy require several premarital sessions with a couple before they agree to marry them.

*Keep your eyes wide open before marriage...and half shut afterwards.*

BEN FRANKLIN, STATESMAN

the advice giver, the more likely the person was to use the advice and to regard it as useful. Both men and women regarded their close friends (who happened to be of the same sex) as sources of advice they listened to and found most helpful.

## SELF-ASSESSMENT | Involved Couple's Inventory

The following questions are designed to increase your knowledge of how you and your partner think and feel about a variety of issues. Assume that you and your partner are considering marriage. Each partner should ask the other the following questions:

### Partner Feelings and Issues

1. If you could change one thing about me, what would it be?
2. On a scale of 0 to 10, how well do you feel I respond to criticism or suggestions for improvement?
3. What would you like me to say or not say that would make you happier?
4. What do you think of yourself? Describe yourself with three adjectives.
5. What do you think of me? Describe me with three adjectives.
6. What do you like best about me?
7. On a scale of 0 to 10, how jealous do you think I am? How do you feel about my level of jealousy?
8. How do you feel about me?
9. To what degree do you feel we each need to develop and maintain outside relationships so as not to focus all of our interpersonal expectations on each other? Does this include other-sex individuals?
10. Do you have any history of abuse or violence, either as an abused child or adult or as the abuser in an adult relationship?
11. If we could not get along, would you be willing to see a marriage counselor? Would you see a sex therapist if we were having sexual problems?
12. What is your feeling about prenuptial agreements?
13. Suppose I insisted on your signing a prenuptial agreement?
14. To what degree do you enjoy getting and giving a massage?
15. How important is it to you that we massage each other regularly?
16. On a scale of 0 to 10, how emotionally close do you want us to be?
17. How many intense love relationships have you had, and to what degree are these individuals still a part of your life in terms of seeing them or sending text messages?
18. Have you lived with anyone before? Are you open to our living together? What would be your understanding of the meaning of our living together—would we be finding out more about each other or would we be committed to marriage?
19. What do you want for the future of our relationship? Do you want us to marry? When?
20. On a ten-point scale (0 = very unhappy and 10 = very happy), how happy are you in general? How happy are you about us?
21. How depressed have you been? What made you feel depressed?
22. What behaviors that I engage in upset you and you want me to stop them?
23. What new behaviors do you want me to develop or begin to make you happier?
24. What quality for a future partner would be a requirement for you?
25. What quality for a future partner would be a deal breaker—you would not marry this person?
26. Why did your last relationship end?
27. What would your last partner say was your worst characteristic?
28. How many past sexual partners is too many for a person you would be interested in?
29. What are your texting expectations—how often do you expect a text message from me? How soon after you text me do you expect a response?

### Feelings about Parents and Family

1. How do you feel about your mother? Your father? Your siblings?
2. On a 10-point scale, how close are you to your mom, dad, and each of your siblings?
3. How close were your family members to one another? On a 10-point scale, what value do you place on the opinions or values of your parents?
4. How often do you have contact with your father or mother? How often do you want to visit your parents and/or siblings? How often would you want them to visit us? Do you want to spend holidays alone or with your parents or mine?
5. What do you like and dislike most about each of your parents?
6. What do you like and dislike about my parents?
7. What is your feeling about living near our parents? How would you feel about my parents living with us? How do you feel about our parents living with us when they are old and cannot take care of themselves?
8. How do your parents get along? Rate their marriage on a scale of 0 to 10 (0 = unhappy, 10 = happy).
9. To what degree do your parents take vacations together? What are your expectations of our taking vacations alone or with others?
10. To what degree did members of your family consult one another on their decisions? To what degree do you expect me to consult you on the decisions that I make?
11. Who was the dominant person in your family? Who had more power? Who do you regard as the dominant partner in our relationship? How do you feel about this power distribution?

12. What problems has your family experienced? Is there any history of mental illness, alcoholism, drug abuse, suicide, or other such problems?
13. What did your mother and father do to earn an income? How were their role responsibilities divided in terms of having income, taking care of the children, and managing the household? To what degree do you want a job and role similar to that of the same-sex parent?

### Social Issues, Religion, and Children

1. How do you feel about Obama as President? How do you feel about America being in Afghanistan?
2. What are your feelings about women's rights, racial equality, and homosexuality?
3. To what degree do you regard yourself as a religious or spiritual person? What do you think about religion, a Supreme Being, prayer, and life after death?
4. Do you go to religious services? Where? How often? Do you pray? How often? How important is prayer to you? How important is it to you that we pray together? What do you pray about? When we are married, how often would you want to go to religious services? In what religion would you want our children to be reared? What responsibility would you take to ensure that our children had the religious training you wanted them to have?
5. How do you feel about abortion? Under what conditions, if any, do you feel abortion is justified?
6. How do you feel about children? How many do you want? When do you want the first child? At what intervals would you want to have additional children? What do you see as your responsibility in caring for the children—changing diapers, feeding, bathing, playing with them, and taking them to lessons and activities? To what degree do you regard these responsibilities as mine?
7. Suppose I did not want to have children or couldn't have them. How would you feel? How do you feel about artificial insemination, surrogate motherhood, in vitro fertilization, and adoption?
8. To your knowledge, can you have children? Are there any genetic problems in your family history that would prevent us from having normal children? How healthy (physically) are you? What health problems do you have? What health problems have you had? What operations have you had? How often have you seen a physician in the last three years? What medications have you taken or do you currently take? What are these medications for? Have you seen a therapist, psychologist, or psychiatrist? What for?
9. How should children be disciplined? Do you want our children to go to public or private schools?
10. How often do you think we should go out alone without our children? If we had to decide between the two of us going on a cruise to the Bahamas alone or taking the children camping for a week, what would you choose?
11. What are your expectations of me regarding religious participation with you and our children?

### Sex

1. How much sexual intimacy do you feel is appropriate in casual dating, involved dating, and engagement?
2. Does "having sex" mean having sexual intercourse? If a couple has experienced oral sex only, have they had sex?
3. What sexual behaviors do you most and least enjoy? How often do you want to have intercourse? How do you want me to turn you down when I don't want to have sex? How do you want me to approach you for sex? How do you feel about just being physical together—hugging, massaging, holding, but not having intercourse?
4. By what method of stimulation do you experience an orgasm most easily?
5. What do you think about masturbation, oral sex, homosexuality, sadism and masochism (S & M), and anal sex?
6. What type of contraception do you suggest? Why? If that method does not prove satisfactory, what method would you suggest next?
7. What are your values regarding extramarital sex? If I had an affair, would you want me to tell you? Why? If I told you about the affair, what would you do? Why?
8. How often do you view pornographic videos or pornography on the Internet? How do you feel about my viewing porno? How much is too much? How do you feel about our watching porn together?
9. How important is our using a condom to you?
10. Do you want me to be tested for human immunodeficiency virus (HIV)? Are you willing to be tested?
11. What sexually transmitted infections (STIs) have you had?
12. How much do you want to know about my sexual behavior with previous partners?
13. How many friends with benefits relationships have you been in? What is your interest in our having such a relationship?
14. How much do you trust me in terms of my being faithful or monogamous with you?
15. How open do you want our relationship to be in terms of having emotional or sexual involvement with others, while keeping our relationship primary?
16. What things have you done that you are ashamed of?
17. What emotional, psychological, or physical health problems do you have? What issues do you struggle with?
18. What are your feelings about your sexual adequacy? What sexual problems do you or have you had?
19. Give me an example of your favorite sexual fantasy.
20. In what ways have you cheated/been unfaithful in a relationship?
21. In what ways has a previous partner cheated/been unfaithful to you?
22. To what degree can you guarantee me that you will not cheat/will be faithful to me?

### Careers and Money

1. What kind of job or career will you have? What are your feelings about working in the evening versus being home with the family? Where will your work require that we live? How often do you feel we will be moving? How much travel will your job require?

*(continued)*

## SELF-ASSESSMENT (continued)

2. To what degree did your parents agree on how to deal with money? Who was in charge of spending, and who was in charge of saving? Did working, or earning the bigger portion of the income, connect to control over money?
3. What are your feelings about a joint versus a separate checking account? Which of us do you want to pay the bills? How much money do you think we will have left over each month? How much of this do you think we should save?
4. When we disagree over whether to buy something, how do you suggest we resolve our conflict?
5. What jobs or work experience have you had? If we end up having careers in different cities, how do you feel about being involved in a commuter marriage?
6. What is your preference for where we live? Do you want to live in an apartment or a house? What are your needs for a car, television, cable service, phone plan, entertainment devices, and so on? What are your feelings about us living in two separate places, the living apart together idea whereby we can have a better relationship if we give each other some space and have plenty of room?
7. How do you feel about my having a career? Do you expect me to earn an income? If so, how much annually? To what degree do you feel it is your responsibility to cook, clean, and take care of the children? How do you feel about putting young children or infants in day-care centers? When the children are sick and one of us has to stay home, who will that be?
8. To what degree do you want me to account to you for the money I spend? How much money, if any, do you feel each of us should have to spend each week as we wish without first checking with the other partner? What percentage of income, if any, do you think we should give to charity each year?
9. What assets or debts will you bring into the marriage? How do you feel about debt? How rich do you want to be?
10. If you have been married before, how much child support or alimony do you get or pay each month? Tell me about your divorce.
11. In your will, what percentage of your assets, holdings, and retirement will you leave to me versus anybody else (siblings, children of a previous relationship, and so on)?

### Recreation and Leisure

1. What is your idea of the kinds of parties or social gatherings you would like for us to go to together?
2. What is your preference in terms of us hanging out with others in a group versus being alone?
3. What is your favorite recreational interest? How much time do you spend enjoying this interest? How important is it for you that I share this recreational interest with you?
4. What do you like to watch on television? How often do you watch television and for what periods of time?
5. What are the amount and frequency of your current use of alcohol and other drugs (for example, beer and/or wine, hard liquor, marijuana, cocaine, crack, meth, etc.)? What, if any, have been your previous alcohol and other drug behaviors and frequencies? What are your expectations of me regarding the use of alcohol and other drugs?
6. Where did you vacation with your parents? Where will you want us to go? How will we travel? How much money do you feel we should spend on vacations each year?
7. What pets do you own and what pets do you want to live with us? To what degree is it a requirement that we have one or more pets? To what degree can you adapt to my pets so that they live with us?

### Relationships with Friends and Coworkers

1. How do you feel about my three closest same-sex friends?
2. How do you feel about my spending time with my friends or coworkers, such as one evening a week?
3. How do you feel about my spending time with friends of the opposite sex?
4. What do you regard as appropriate and inappropriate affection behaviors with opposite-sex friends?

### Remarriage Questions

1. How and why did your first marriage end? What are your feelings about your former spouse now? What are the feelings of your former spouse toward you? How much "trouble" do you feel your former spouse will want to cause us? What relationship do you want with your former spouse? May I read your divorce settlement agreement?
2. Do you want your children from a previous marriage to live with us? What are your emotional and financial expectations of me in regard to your children? What are your feelings about my children living with us? Do you want us to have additional children? How many? When?
3. When your children are with us, who will be responsible for their food preparation, care, discipline, and driving them to activities?
4. Suppose your children do not like me and vice versa. How will you handle this? Suppose they are against our getting married?
5. Suppose our respective children do not like one another. How will you handle this?

It would be unusual if you agreed with each other on all of your answers to the previous questions. You might view the differences as challenges and then find out the degree to which the differences are important for your relationship. You might need to explore ways of minimizing the negative impact of those differences on your relationship. It is not possible to have a relationship with someone in which there is total agreement. Disagreement is inevitable; the issue becomes how you and your partner manage the differences.

### Note

This self-assessment is intended to be thought-provoking and fun. It is not intended to be used as a clinical or diagnostic instrument.

## Visiting Your Partner's Parents

Fisher and Salmon (2013) identified the reasons individuals take their potential partners home to meet their parents—to seek parental approval and feedback and to confirm to their partner that they are serious about the relationship. They also want to meet their partner's parents—to see how their potential mate will look when older, their future health, and potential familial resources that will be available.

Seize the opportunity to discover the family environment in which your partner was reared and consider the implications for your subsequent marriage. When visiting your partner's parents, observe their standard of living and the way they interact (e.g., level of affection, verbal and nonverbal behavior, marital roles) with one another. How does their standard of living compare with that of your own family? How does the emotional closeness (or distance) of your partner's family compare with that of your family? Such comparisons are significant because both you and your partner will reflect your respective family or origins. "This is the way we did it in my family" is a phrase you will hear your partner say from time to time.

If you want to know how your partner is likely to treat you in the future, observe the way your partner's parent of the same sex treats and interacts with his or her spouse. If you want to know what your partner may be like in the future, look at your partner's parent of the same sex. There is a tendency for a man to become like his father and a woman to become like her mother. A partner's parent of the same sex and the parents' marital relationship are the models of a spouse and a marriage relationship that the person is likely to duplicate.

After (and sometimes before) sizing each other up, individuals like to know where the relationship is headed. The Applying Social Research to follow looks at partner's having "the relationship talk."

## Prenuptial Agreement

Senator John McCain has a prenuptial agreement with his wife Cindy, heiress to a beer fortune estimated to be $100 million. Britney Spears and Kevin Federline had a prenuptial agreement whereby he was awarded only $300,000 of her over-$100 million in assets.

Mel Gibson did not have a prenuptial agreement with Robyn Denise Moore Gibson, to whom he was married for 28 years. When they divorced, California law required that assets be split evenly—in their case, Gibson's worth was $900 million. Catherine Zeta-Jones, who married Michael Douglas in 2000, has a "straying fee" of $5 million in their prenuptial agreement if he ever cheats on her.

Some couples, particularly those with considerable assets or those in subsequent marriages, might consider a prenuptial agreement. To reduce the chance that the agreement will later be challenged, each partner should hire their own attorney (months before the wedding) to develop and/or review the agreement.

The primary purpose of a **prenuptial agreement** (also referred to as a premarital agreement, marriage contract, or ante-nuptial contract) is to specify how property will be divided if the marriage ends in divorce or when it ends by the death of one partner. In effect, the value of what you take into the marriage is the amount you are allowed to take out of the marriage. For example, if you bring $250,000 into the marriage and buy the marital home with this amount, your ex-spouse is not automatically entitled to half the value of the house at divorce. Some agreements may also contain clauses of no spousal support (alimony) if the marriage ends in divorce (but some states prohibit waiving alimony). See Appendix C for an example of a prenuptial agreement developed

*Women are saying: "I have a place in the world. I won't just wait around and expect you to be kind and generous. Let's nail this down."*

ARKANSAS ONLINE

*A prenup says "I hope I'll be able to trust you, but I'm not sure yet."*

ERICA JONG, AUTHOR

**Prenuptial agreement** specifies how property will be divided if the marriage ends in divorce or when it ends by the death of one partner.

## APPLYING SOCIAL RESEARCH | The "Relationship Talk": Assessing Partner Commitment

For dating couples "The Talk" is culturally understood to mean a discussion whereby both parties reveal their feelings about each other and their commitment to the future together. Typically, one partner feels a greater need to clarify the future and instigates the talk. The goal of the talk is to confirm that the partner is interested in and committed to a future with the other. This study examined the talk in terms of how long partners are involved before they have the talk, specific words/strategies used in having the talk, and the context (e.g., during sex? after sex? while watching TV? eating dinner?) of the talk. Other research questions included how the partner responded and the effect of the talk on the couple's relationship.

### Sample

Data for the study was based on a sample of 211 undergraduate students who completed a 15-item questionnaire. A majority of respondents, 77.7%, were female (22.3% were male) , of which 78% were white (11.4% were African American).

### Context of the Talk

Over thirty percent (30.5%) of the respondents reported that they had the talk during a meal. Other contexts were "after sexual intimacy" (16.4%), "before sexual intimacy" (11.3%), and "while we were on a trip" (10%). Alcohol was usually not involved; only 3% of those initiating the talk said that were drinking alcohol; .6% reported that their partner was drinking when they initiated the talk. In regard to being anxious about initiating the talk, on a 10- point scale with 1 being minimal anxiety and 10 indicating extreme anxiety, the mean level of anxiety was 4.6. Hence the respondents were moderately anxious when they initiated the talk.

### Strategies involved in the Talk

Various strategies were used to initiate the talk and assess the level of commitment of the partner. The top ten are presented below.

Direct question about future (30%) "What do you see as far as the future of our relationship?"

Questioning of motives (15%) "What do you want out of this relationship?"

Direct question about marriage (8%) "Do you see a future for us—do you ever think about us getting married?"

Assessment of level of interest in relationship (7%) "How serious are you about this relationship?"

Assumption of marriage (7%) "We would just randomly be talking and say things like, 'when we get married.'"

Modeling (6%) "I mentioned to my partner what my feelings were toward him and looked to see how he

*This is going to be a great year, isn't it?....or it is going to be our last?*

FEMALE EIGHT YEARS INTO A RELATIONSHIP (THE MALE PROPOSED AT THE END OF THAT YEAR)

by a husband and wife who had both been married before and had assets and children.

Reasons for a prenuptial agreement include the following.

**1. *Protecting assets for children from a prior relationship.*** Some children encourage their remarrying parent to draw up a prenuptial agreement with the new partner so that their (the offspring's) inheritance, house, or whatever will not automatically go to the new spouse upon the death of their parent.

**2. *Protecting business associates.*** A spouse's business associate may want a member of a firm or partnership to draw up a prenuptial agreement with a soon-to-be-spouse to protect the firm from intrusion by the spouse if the marriage does not work out.

Prenuptial agreements are not very romantic ("I love you, but sign here and see what you get if you don't please me") and may serve as a self-fulfilling prophecy ("We were already thinking about divorce"). Indeed, 23% of 1,087 undergraduate males (and 25% of 3,440 undergraduate females) agreed with the statement "I would not marry someone who required me to sign a prenuptial agreement" and 27% of the males (23% of the females) feel that couples who have a prenuptial agreement are more likely to get divorced (Hall and Knox 2013). Prenuptial contracts are almost nonexistent in first marriages and are still rare in second marriages.

responded to what I said. He told me he actually felt the same way."

Question and evaluate (6%) "Asked if there was a future for us and we talked about the pros and cons."

Assessment of view of partner (5%) "Am I your soul mate?"

Soft cotton approach (4%) "I mentioned something about the future without putting any pressure on him. I tried to avoid creating an awkward situation and to help him feel comfortable."

Ultimatum (4%) "I just asked what he wanted out of this relationship and if it wasn't the same thing I wanted, I would then end the relationship."

### Reactions of the Partner to the Talk

Reactions of the partner who was asked about the future of the relationship fell into four categories.

Commitment to the future (50.5%) The most frequent response to the talk was a clear statement that the respondent wanted a future with the partner. Over half were quick to confirm that they wanted a future with the partner. While the word *marriage* was often not spoken, the assumption on the part of both partners was that the partners would eventually get married.

Uncertain (32.3%) The second most frequent response to the talk was uncertainty. The partner simply said that he or she did not know about the future and kept the partner and the relationship in limbo.

No future (9%) The third most frequent response was a "subtle revelation that there was no future." The partner was not brutal but implied that the relationship would not go anywhere.

Other responses (4%) Aside from "yes, "not sure," and "no," other responses included "made a joke out of it" or (1%) was "brutal in making clear that there was no future."

### Relationship Effects of Having the Talk

The effects of having the talk on the relationship were variable. Over a third (34.7%) of the respondents reported that the effect was "to move us closer and forward." Fourteen percent ended the relationship and 11.7% said that their relationship became strained since they had different feelings about the future. Another 10% said they were still talking about the issue.

### Source

Abridged and adapted from Nelms, B. J., D. Knox, and B. Easterling. 2012. The Relationship Talk: Assessing Partner Commitment. *College Student Journal* 46: 178–182.

The laws regulating marriage and divorce vary by state, and only attorneys in those states can help ensure that the prenup document drawn up will be honored. Individuals may not waive child support or dictate child custody. Full disclosure of assets is also important. If one partner hides assets, the prenuptial can be thrown out of court. One husband recommended that the issue of the premarital agreement should be brought up and that it be signed a minimum of six months before the wedding: "This gives the issue time to settle rather than being an explosive emotional issue if it is brought up a few weeks before the wedding." Indeed, as noted previously, if a prenuptial agreement is signed within two weeks of the wedding, that is grounds enough for the agreement to be thrown out of court because it is assumed that the document was executed under pressure.

Although individuals are deciding whether to have a prenuptial agreement, states are deciding whether to increase marriage license requirements, this chapter's family policy issue.

*Burning the prenup has become the commitment "I do" was for our grandparents.*

ERICA JONG, AUTHOR

*Life will not break your heart. It will crush it.*

HENRY ROLLINS, ROCK SINGER

**Covenant marriage** spouses sign a Declaration of Intent specifying that they have had premarital education/counseling, that they will seek marriage counseling before seeking a divorce, and that if they have children they will have a "cooling off" period of two years before they seek divorce.

## Covenant Marriage: A Stronger Commitment?

**Covenant marriage** is an agreement between persons getting married that reflects a serious regard for their marriage and their future (Cade 2010). In a covenant marriage, the spouses sign a Declaration of Intent specifying that they have had premarital education/counseling, that they will seek marriage

## FAMILY POLICY | Increasing Requirements for a Marriage License

Should marriage licenses be obtained so easily? Should couples be required, or at least encouraged, to participate in premarital education before saying "I do"? Given the high rate of divorce today, policy makers and family scholars are considering this issue. Data confirm the value of marriage education exposure (Childs and Duncan 2012; Kalinka et al. 2012; and Lucier-Greer et al. 2012).

Several states have proposed legislation requiring premarital education. For example, an Oklahoma statute provides that parties who complete a premarital education program may pay a reduced fee for their marriage license. Also, in Lenawee County, Michigan, local civil servants and clergy have made a pact: they will not marry a couple unless that couple has attended marriage education classes. Other states that are considering policies to require or encourage premarital education include Arizona, Illinois, Iowa, Maryland, Minnesota, Mississippi, Missouri, Oregon, and Washington. Proposed policies include not only mandating premarital education and lowering marriage license fees for those who attend courses but also imposing delays on issuing marriage licenses for those who refuse premarital education.

Advocates of mandatory premarital education emphasize that such courses reduce marital discord. However, questions remain about who will offer what courses and whether couples will take the content of such courses with intent. Indeed, people contemplating

counseling before seeking a divorce, and that if they have children they will have a "cooling off" period of two years before they seek divorce. Furthermore, they agree that divorce is not acceptable just because they are unhappy; rather, adultery, abandonment, or abuse are the only reasons regarded as legitimate.

Louisiana, Arizona, and Arkansas are the only three states to recognize covenant marriage. Couples opting for covenant marriage are rare; only 2% of individuals getting married in Louisiana choose this option.

DeMaris et al. (2012) studied 707 Louisiana marriages and compared the marital satisfaction of covenant and standard marriage during the first seven years. There was no difference in the decline of marital satisfaction over time. In another study of 600 Louisiana marriages, a unique finding between the two groups was that covenant wives reported an increase in marital quality following the birth of a child compared to wives in standard marriages who reported decreased marital quality (including increased thoughts about divorce) (Nock et al. 2008). Since many of those who chose covenant marriage were religious and traditional, it is not known if the positive effects were due to the covenant marriage itself or to the fact that the participants were religious or conventional.

# Internet and Other Ways of Finding a Partner

Increasingly, individuals are using the Internet (and attendant technology) to find partners for fun, companionship, and marriage. "In the past 15 years, the rise of the Internet has partly displaced not only family and school, but also neighborhood, friends, and the workplace as venues for meeting partners. The Internet increasingly allows individuals to meet and form relationships with perfect strangers" (Rosenfeld and Thomas 2012).

marriage are often narcotized with love feelings and may not take any such instruction seriously. Love myths such as "divorce is something that happens to other people" and "our love will overcome any obstacles" work against the serious consideration of such courses.

**Your Opinion?**

1. To what degree do you believe premarital education should be required before the state issues a marriage license?
2. How effective do you feel such programs are for people in a hurry to marry?
3. How receptive do you feel individuals in love are to marriage education?

**Sources**


Childs, G. R. and S. F. Duncan. 2012. Marriage preparation education programs: An assessment of their components. *Marriage & Family Review* 48: 59–81.

Kalinka, C. J., F. Fincham, and A. H. Hirsch. 2012. A randomized clinical trial of online-biblio relationship education for expectant couples. *Journal of Family Psychology* 26: 159–164.

Lucier-Greer, M., F. Adler-Baeder, S. A. Ketring, K. T. Harcourt, and T. Smith. 2012. Comparing the experiences of couples in first marriages and remarriages in couple and relationship education. *Journal of Divorce and Remarriage* 53: 55–7.5.


Kreager et al. (2014) analyzed data on individuals who sought a partner online from a sample of 8,259 men and 6,274 women and found that being aggressive in contacting someone seemed to pay off for men as male initiators connected with more desirable partners than men who waited to be contacted. Female initiators connected with equally desirable partners as women who waited to be contacted. However, female-initiated contacts were more than twice as likely as male-initiated contacts to result in a connection.

*I am part of everyone I ever dated on OK Cupid.*

SLASH COLEMAN, THE BOHEMIAN LOVE DIARIES: A MEMOIR

Sociologist Michael Rosenfeld of Stanford University followed 926 unmarried couples over a three year period—those who met online were twice as likely to marry as those who met offline, 13% to 6%. Twenty percent of all new relationships begin online (Jayson 2013).

### National **Data**

Based on a sample of 2,252 adults, 38% of Americans who are currently single and actively looking for a partner have used online dating. Of these almost two-thirds (66%) have been on a date with someone they met through a dating site or app. Almost a quarter have met their spouse or a long-term partner through these sites (Smith and Duggan 2013).

There are both advantages and disadvantages for meeting someone online.

## Meeting Online—Advantages

Online dating services have become clear in their mission—to provide a place where people go to "shop" for potential romantic partners and to "sell" themselves in hopes of creating a successful romantic relationship. In interviews with 34 persons who had used online dating services, the respondents revealed that they used economic metaphors to describe the experience—"supermarket,"

This couple found each other on a site for artists. They report their meeting was "magic."

Megan Collins

"catalog," etc. (Heino et al. 2010). Indeed, there were five themes: assessing the market worth of the various people online, determining one's own market worth, shopping for perfect parts, maximizing inventory, and calibrating selectivity. The latter referred to assessing one's own market worth by the number of e-mails received and changing their presentation of self to elicit more interest, e.g., changing photos or putting up more photos (Heino et. al. 2010). Such calibration also involved comparing what one had to offer with what one could ask for. One online dater said that since she had put on weight she had to be willing to accept older partners who might be divorced with kids. Saltes (2013) studied Internet dating as used by the disabled and found that while some viewed the Internet as a godsend since it connected them with persons aware of yet unconcerned about their disability, others felt that disclosing their disability was a turnoff and that afterward, they might never hear from some people again.

**DIVERSITY IN OTHER COUNTRIES**

Chih-Chien and Ya-Ting (2010) interviewed 36 Internet users and surveyed 248 students in Taiwan to discover their motives for using the Internet to create a new relationship (cyber relationship motives). Factors included the desire to meet new people, the ease of finding someone new with a click, and the psychological comfort of doing so (less anxiety over meeting someone online than in person). Respondents also mentioned that they wanted to escape from reality into the virtual world, were curious about the experience of making an Internet friend, and were interested in finding emotional support for a specific interest (e.g., a person who likes computer games might find others who share their interest). Finding a romantic or sexual partner were also motives.

Online meetings will continue to increase as people delay getting married and move beyond contexts where hundreds of potential partners (e.g., college) are available. Busy people who feel they don't have time for traditional dating, or who find that people in their age bracket choose not to hang out in predictable meeting places (e.g., bars or clubs) may also be attracted to finding a partner online.

As noted previously, a primary attraction of meeting someone online is efficiency. It takes time and effort to meet someone at a coffee shop for an hour, only to discover that the person has habits (e.g., does or does not smoke) or values (e.g., religious or agnostic) that would eliminate them as a potential partner. On the Internet, one can spend a short period of time and scan literally hundreds of profiles of potential partners. For noncollege people who are busy in their job or career, the Internet offers the chance to meet someone outside their immediate social circle. "There are only six guys in my office," noted one female Internet user. "Four are married and the other two are alcoholics. I don't go to church and don't like bars so the Internet has become my guy store."

Another advantage of looking for a partner online is that it removes emotion/chemistry/first meeting magic from the mating equation so that individuals can focus on finding someone with common interests, background, values, and goals. In real life, you can "fall in love at first sight" and have zero in common (Heino et al. 2010). Some websites exist to target persons with specific interests such as black singles (BlackPlanet.com), Jewish singles (Jdate.com), and gay singles (Gay.com).

## Meeting Online: The Downside

Lying occurs in Internet dating (as it does in non-Internet dating). Ellison and Hancock (2013) noted that persons using the Internet to find a partner fear that others are lying in their profiles, that "fudging" (or small deceptions) is common though "big lies" are relatively rare. However, serious misrepresentation was reported by half of the Pew Research Center respondents who had experience with online dating (Smith and Duggan 2013). Hall et al. (2010) analyzed data from 5,020 individuals who posted profiles on the Internet in search of a date and who revealed seven categories of misrepresentation. These included personal assets ("I own a house at the beach"), relationship goals ("I want to get married"), personal interests ("I love to exercise"), personal attributes ("I am religious"), past relationships ("I have only been married once"), weight, and age. Men were most likely to misrepresent personal assets, relationship goals, and personal interests whereas women were most likely to misrepresent weight. Heino et al. (2010) interviewed 34 online dating users and found that there is the assumption of exaggeration and a compensation for such exaggeration. The female respondents noted that men exaggerate how tall they are, so the women downplay their height. If a man said he was 5'11" the woman would assume he was 5'9". Lo et al. (2013) noted that deception is motivated by the level of attractiveness of the target person...higher deception if the target person is particularly attractive. In addition, women in Heino's research were more deceptive than men—lying about weight and height, not concerned about marriage but looking for a good time, etc.

*Dating is a give and take. If you only see it as "taking," you are not getting it.*

HENRY CLOUD, CLINICAL PSYCHOLOGIST

Some online users also lie about being single. They are married, older, and divorced more times than they reveal. But to suggest that the Internet is the only place where deceivers lurk is to turn a blind eye to those people met through traditional channels. Be suspicious of everyone until you know otherwise.

It is important to be cautious of meeting someone online. Although the Internet is a good place to meet new people, it also allows someone you rejected or an old lover to monitor your online behavior. Most sites note when you have been online most recently, so if you reject someone online by saying, "I'm really not ready for a relationship," that same person can log on and see that you are still looking. Some individuals become obsessed with a person they meet online and turn into a cyber stalker when rejected. A quarter of the respondents in the Pew Research Center study said they were harassed or made to feel uncomfortable by someone they had met online (Smith and Duggan 2013). Some people also use the Internet to try on new identities. For example a person who feels he or she is attracted to same sex individuals may present a gay identity online.

Other disadvantages of online meeting include the potential to fall in love too quickly as a result of intense mutual disclosure; not being able to assess "chemistry" or how a person interacts with your friends or family; the tendency to move too quickly (from texting to phone to meeting to first date) toward marriage, without spending much time to get to know each other; and not being able to observe nonverbal behavior. In regard to the nonverbal issue, Kotlyar and Ariely (2013) emphasized the importance of using Skype (which allows one to see the partner/assess nonverbal cues) as soon as possible and as a prelude to meeting in person to provide more information about the person behind the profile.

*Internet dating has made people more disposable.*

MARK BROOKS, CONSULTANT

Another disadvantage of using the Internet to find a partner is that having an unlimited number of options sometimes results in not looking carefully at the options one has. Wu and Chiou (2009) studied undergraduates looking for romantic partners on the Internet who had 30, 60, and 90 people to review and found that the more options the person had, the less time the undergraduate spent carefully considering each profile. The researchers concluded that it was better to examine a small number of potential online partners carefully than to be distracted by a large pool of applicants, which does not permit the time for close scrutiny.

It is also important to use Internet dating sites safely, including not giving out home or business phone numbers or addresses, always meeting the person in one's own town with a friend, and not posting photos that are too revealing, as these can be copied and posted elsewhere. Take it slow—after connecting in an e-mail through the dating site, move to instant messages, texting, phone calls, Skyping, then meet in a public place with friends nearby. Also, be clear about what you want (e.g., "If you are looking for a hookup, keep moving. If you are looking for a lifetime partner, 'I'm your gal'").

The Internet may also be used to find out information about a partner. Beenverified.com provides a way to confirm that the person is who he or she says via public records. Argali.com can be used to find out where the Internet mystery person lives, Zabasearch.com for how long the person has lived there, and Zoominfo.com for where the person works. The person's birth date can be found at Birthdatabase.com.

LuLu, specifically for females only, is a website (www.onlulu.com) that rates males. Women download the app anonymously, linked through one's Facebook friends, to compare notes on different aspects of a guy's "datability." LuLu first gives statistics about the male and a picture, asks the female's relationship to the male, asks for an overall rating (multiple choice list), asks which positives and negatives apply, and provides an overall score out of 100%.

Sample Review:

> John Doe, Age: 33, College: Coast Guard, Last Seen: Los Angeles
>
> Average Score: 8.4, # Reviewers: 1
>
> Best qualities: #Panty Dropper, #Loves His Family, #Great Listener
>
> Worst qualities: #Questionable Search History, #Self Absorbed, #Sketchy Call Log, #Wandering Eye
>
> Appearance: 9.0, Humor: 10, Manners: 5.0, Sex: 8.0, First Kiss: 9.0, Ambition: 9.0,
>
> Commitment: 7.0

## Finding a Partner Online: It's Work

*Finding a partner on the Internet is work; waiting for a partner to magically appear can be forever.*

ANONYMOUS

One of our former students revealed the reality of the work involved in finding a partner online. She likened doing so as a "job"—setting up a profile, culling through page after page of postings, "winking" at guys she was interested in, texting/e-mailing/Skyping, and meeting them for coffee or dinner.

> *I signed up for a month on Match.com for $31.00. During that time I messaged/e-mailed/talked to 21 guys and had five dates. I was exhausted and ready to quit. Reluctantly, I signed up for one more month and met a guy worth all the effort the second month.*

There are also some suggestions about how to use the Internet. An example is a woman who found her husband who wrote (author's files):

> *The first strategy I employed was to be more efficient in my search for a new person to hang out with. I had to stop trying to "bump into" a guy I wanted to date at a bar or class or grocery store. I no longer had the time to hang out in those places. And looking across a sea of men in a bar or classroom or mall would not tell me the important things I wanted to know: which ones were married? which ones were happy in their current relationships? which ones had kids? which ones were smokers? Getting onto Match.com helped me be MUCH more efficient in my search. I could bypass the guys that were smokers, didn't have kids or meet whatever other criteria I had established.*
>
> *The second strategy I employed was to increase the number of men I met. While my original search criteria was very narrow, I found that broadening the salary range and age range got me more prospects. I also stopped having dinner as a first date. I found that I could usually tell in the first 15 minutes if I was going to enjoy the guy's company or not/ want to spend more time with him. Dinner dates often dragged on for hours and inevitably led my date to hoping we would spend hours after dinner together. Better to just meet for coffee. This was a much better use of my time and I found that I could meet at least 4 guys in a weekend instead of just one. Some I chose to never meet again. Some I chose to have a brief physical relationship with. A few I got emotionally involved with.*
>
> *And, of course, one was Brad, who after 2 years of dating, proposed marriage. We have now been married for 5 years. We just had a great Thanksgiving weekend with our three boys (his two plus my one).*

## Apps

### National **Data**

Seven percent of cell phone app users say they have used a dating app on their cell phone (Smith and Duggan, 2013).

Online dating is moving from the Web to mobile devices—apps. POF (Plentyoffish), Skout.com, and Zoosk.com allow one to identify and connect with someone sitting in the same Starbucks; all have over a million subscribers. No longer must one be on a clunky desktop or laptop—they can simply hold their iPhone in their hand and see who they are talking to.

Tinder.com is another app which allows one to instantly select (on the basis of a photo) and be selected by someone in the same area. If both select each other, there is a connect. Users refer to Tinder as "the new hookup app."

*Online dating is just as murky and full of lemons as finding a used car in the classifieds. Once you learn the lingo, it's easier to spot the models with high mileage and no warranty.*

LAURIE PERRY, CRAZYAUNTPURL.COM

## Speed-Dating

Dating innovations that involve the concept of speed include the eight-minute date. The website http://www.8minutedating.com/ identifies these "Eight-Minute Dating Events" throughout the country, where a person has eight one-on-one "dates" at a bar that last eight minutes each. If both parties are interested in seeing each other again, the organizer provides contact information so that the individuals can set up another date. Speed-dating saves time because it allows daters to meet face to face without burning up a whole evening.

## International Dating

Go to Google.com and type in "international brides," and you will see an array of sites dedicated to finding foreign women for American men. American males often seek women from Asian countries as they are thought to be more traditional. Women from other countries seek American males as a conduit for entry into U.S. citizenship.

## Factors Which Suggest You Might Delay or Call Off the Wedding

While the Internet is a valuable tool for helping to identify partners, one must then evaluate the partner and the relationship in terms of a life together. "No matter how far you have gone on the wrong road, turn back" is a Turkish proverb. Behavioral psychologist B. F. Skinner noted that one should not defend a course of action that does not feel right, but stop and reverse directions. If your engagement is characterized by the following factors, consider delaying your wedding at least until the most distressing issues have been resolved. Alternatively, break the engagement (which happens in 30% of formal engagements), which will have fewer negative consequences and involve less stigma than ending a marriage.

### Age 18 or Younger

The strongest predictor of getting divorced is getting married during the teen years. Individuals who marry at age 18 or younger have three times the risk of divorce than those who delay marriage into their late twenties or early thirties. Teenagers may be more at risk for marrying to escape an unhappy home and may be more likely to engage in impulsive decision making and behavior. Early marriage is also associated with an end to one's education, social isolation from close friends, early pregnancy or parenting, and locking oneself into a low income. Increasingly, individuals are delaying when they marry. As noted earlier, the median age at first marriage in the United States is almost 29 for men and 27 for women.

*She bid me take love easy as the leaves grow on the tree;*
*But I, being young and foolish, with her would not agree.*

W. B. YEATS, ENGLISH POET

### Known Partner Less Than Two Years

Thirty percent of 1,093 undergraduate males and 32% of 3,450 undergraduate females agreed, "If I were really in love, I would marry someone I had known for only a short time" (Hall and Knox 2013). Impulsive marriages in which the partners have known each other for less than a month are associated with a higher-than-average divorce rate. Indeed, partners who date each other for at least two years (25 months to be exact) before getting married report the highest level of marital satisfaction and are less likely to divorce (Huston et al. 2001). A short courtship does not allow partners enough time to learn about each other's background, values, and goals and does not permit opportunity to observe and scrutinize each other's behavior in a variety of settings (e.g., with one's close friends and family).

*Marry'd in haste, we may repent at leisure.*

WILLIAM CONGREVE, PLAYWRIGHT

To increase the knowledge you and your partner have about each other, find out each other's answers to the questions in the Involved Couple's Inventory, take a five-day "primitive" camping trip, take a 15-mile hike together, wallpaper a small room together, or spend several days together when one partner is sick. If the couple plans to have children, they may want to offer to babysit a six month old of their friend's for a weekend.

### Abusive Relationship

Abusive lovers become abusive spouses, with predictable negative outcomes. Though extricating oneself from an abusive relationship is difficult before the wedding, it becomes even more difficult after marriage, particularly when children are involved. Abuse is a serious red flag of impending relationship doom that should not be overlooked and one should seek the exit ramp as soon as possible (see Chapter 10 on the details of leaving an abusive relationship).

### High Frequency of Negative Comments/Low Frequency of Positive Comments

Markman et al. (2010) studied couples across the first five years of marriage and found that more negative and less positive communication before marriage tended to be associated with subsequent divorce. In addition, the researchers emphasized that "negatives tend to erode positives over time." Individuals who criticize each other end up damaging their relationship in a way which does not make it easy for positives to erase.

### Numerous Significant Differences

Relentless conflict often arises from numerous significant differences. Though all spouses are different from each other in some ways, those who have numerous differences in key areas such as race, religion, social class, education, values, and goals are less likely to report being happy and to divorce. Amato et al. (2007) also found that the less couples had in common, the more their marital distress.

*Before you run in double harness, look well to the other horse.*

**OVID, POET**

### On-and-Off Relationship

A roller-coaster premarital relationship is predictive of a marital relationship that will follow the same pattern. Partners in **cyclical relationships** (break up and get back together several times) have developed a pattern in which the dissatisfactions in the relationship become so frustrating that separation becomes the antidote for relief. Dailey et al. (2013) identified five types of on and off relationships (the percent who represent each are from Dailey et al. 2013):

**Cyclical relationships** partners who break up and get back together several times.

**1.** controlling partner (26%)—one partner more persistent to keep the relationship together

**2.** capitalized on transitions (22%)—partners improve the relationship after a breakup

**3.** mismatched (19%)—differences in personalities and desires or geographic distance

**4.** habitual (14%)—partners break up but fall back together with little or no negotiation since it was convenient to resume the relationship

**5.** gradual separators (11%)—partners recognized relationship was over and ended it (7% could not be categorized)

Vennum (2011) studied individuals in cyclical relationships and found that they reported lower-quality relationships compared to those in relationships which were uninterrupted. Partners in cyclical relationships tended to be African Americans in a long-distance dating relationship who expressed uncertainty about the future of the relationship and had less constructive communication.

### Dramatic Parental Disapproval

Parents usually have an opinion of their son's or daughter's mate choice. Mothers disapprove of the mate choice of their daughters if they predict that he will be a lousy father, and fathers disapprove if they predict the suitor will be a poor provider (Dubbs & Buunk 2010). When student and parent ratings of mate selection traits were compared, parents ranked religion higher than did their offspring, whereas offspring ranked physical attractiveness higher than did parents (Perilloux et al. 2011). Parents were also more focused on earning capacity and education (e.g., college graduate) in their daughter's mate selection than in their son's.

## Low Sexual Satisfaction

Sexual satisfaction is linked to relationship satisfaction, love, and commitment. Sprecher (2002) followed 101 dating couples across time and found that low sexual satisfaction for both women and men was related to reporting low relationship quality, less love, lower commitment, and breaking up. Hence, couples who are dissatisfied with their sexual relationship might explore ways of improving it (alone or through counseling) or consider the impact of such dissatisfaction on the future of their relationship.

## Limited Relationship Knowledge

Individuals and couples are most likely to have a positive future together if they have relationship knowledge. Bradford et al. (2012) validated The Relationship Knowledge Questionnaire as a way of assessing relationship knowledge. Such knowledge included knowing how to listen effectively, settle disagreements/solve problems/reach compromise, deepen a loving relationship, develop a strong friendship, and spend time together.

## Marrying for the Wrong Reason

Some reasons for getting married are more questionable than others. These reasons include the following.

*Oh, what a tangled web we weave when first we practice to conceive.*

**DON HEROLD, HUMORIST**

**1.** Rebound. A rebound marriage results when you marry someone immediately after another person has ended a relationship with you. It is a frantic attempt on your part to reestablish your desirability in your own eyes and in the eyes of the partner who dropped you. To marry on the rebound is usually a bad decision because the marriage is made in reference to the previous partner and not to the current partner. In effect, you are using the person you intend to marry to establish yourself as the "winner" in the previous relationship.

Barber and Cooper (2014) used a longitudinal, online diary method to examine trajectories of psychological recovery and sexual experience following a romantic relationship breakup among 170 undergraduate students. Consistent with stereotypes about individuals on the rebound, those respondents who had been "dumped" used sex to cope with feelings of distress, anger, and diminished self-esteem. And those who had sex for these reasons were more likely (not initially, but over time) to continue having sex with different new partners. Caution about becoming involved with someone on the rebound may be warranted. One answer to the question, "How fast should you run from a person on the rebound?" may be "as fast as you can." Waiting until the partner has 12 to 18 months distance from the previous relationships provides for a more stable context for the new relationship.

**2.** Escape. A person might marry to escape an unhappy home situation in which the parents are oppressive, overbearing, conflictual, alcoholic, and/or abusive. Marriage for escape is a bad idea. It is far better to continue the relationship with the partner until mutual love and respect become the dominant forces propelling you toward marriage, rather than the desire to escape an unhappy situation. In this way you can evaluate the marital relationship in terms of its own potential and not as an alternative to unhappiness.

**3.** Unplanned pregnancy. Getting married because a partner becomes pregnant should be considered carefully. Indeed, the decision of whether to marry should be kept separate from decisions about a pregnancy. Adoption, abortion, single parenthood, and unmarried parenthood (the couple can remain together as an unmarried couple and have the baby) are all alternatives to simply deciding to marry if a partner becomes pregnant. Avoiding feelings of being trapped or later feeling that the marriage might not have happened without the pregnancy are two reasons for not rushing into marriage because of pregnancy. Couples who marry when the woman becomes pregnant have an increased chance of divorce.

**4.** Psychological blackmail. Some individuals get married because their partner takes the position that "I will commit suicide if you leave me." Because the person fears that the partner may commit suicide, he or she may agree to the wedding. The problem with such a marriage is that one partner has been reinforced for threatening the other to get what he or she wants. Use of such power often creates resentment in the other partner, who feels trapped in the marriage. Escaping from the marriage becomes even more difficult. One way of coping with a psychological blackmail situation is to encourage the person to go with you to a therapist to discuss the relationship. Once inside the therapist's office, you can tell the counselor that you feel pressured to get married because of the suicide threat. Counselors are trained to respond to such a situation. Alternatively, another response to a partner who threatens suicide is to call the police and say, "Name, address, and phone number has made a serious threat on his or her own life." The police will dispatch a car to have the person picked up and evaluated.

**5.** Insurance benefits. In a poll conducted by the Kaiser Family Foundation, 7% of adults said someone in their household had married in the past year to gain access to insurance. In effect, marital decisions had been made to gain access to health benefits. "For today's couples, 'in sickness and in health' may seem less a lover's troth than an actuarial contract. They marry for better or worse, for richer or poorer, for co-pays and deductibles" (Sack 2008). While selecting a partner who has resources (which may include health insurance) is not unusual, to select a partner solely because he or she has health benefits is dubious. Both parties might be cautious if the alliance is more about "benefits" than the relationship.

**6.** Pity. Some partners marry because they feel guilty about terminating a relationship with someone whom they pity. The fiancée of an Afghanistan soldier reported that "when he came back with his legs blown off I just changed inside and didn't want to stay in the relationship. I felt guilty for breaking up with him. ..." Regardless of the reason, if one partner becomes brain-damaged or fails in the pursuit of a major goal, it is important to keep the issue of pity separate from the advisability of the marriage. The decision to marry should be based on factors other than pity for the partner.

**7.** Filling a void. A former student in the authors' classes noted that her father died of cancer. She acknowledged that his death created a vacuum, which she felt driven to fill immediately by getting married so that she would have a man in her life. Because she was focused on filling the void, she had paid little attention to the personality characteristics of the man who had asked to marry her. She discovered on her wedding night that her new husband had several other girlfriends whom he had no intention of giving up. The marriage was annulled.

In deciding whether to continue or terminate a relationship, listen to what your senses tell you ("Does it feel right?"), listen to your heart ("Do you love this person or do you question whether you love this person?"), and evaluate your similarities ("Are we similar in terms of core values, goals, view of life?"). Also, be realistic. Indeed, most people exhibit some negative and some positive indicators before they marry.

# Selecting a Lifetime Partner

Selecting a lifetime partner will involve the same cultural constraints and sociological and psychological factors identified earlier in the chapter. Individuals are not "free" to select their partner but do so from the menu presented by their culture. Once at the relationship buffet, factors of homogamy and exchange come into play. These variables will continue to be operative.

Selecting a partner with characteristics similar to one's own will continue to be associated with happy and durable relationships. Internet profiles are scanned to identify persons with similar interests, values, and goals. Partners for fun, companionship, and marriage will increasingly be sought through the use of technology. Indeed, technology permeates the lives of today's youth. Just as iPhones and text messaging are commonplace, the use of Match.com and other such sites and apps will become normative . The Internet use will lose its stigma and will no longer be hidden by Internet couples who are asked "how did you meet?"

## Summary

***What are the cultural factors that affect your selection of a mate?***

Two types of cultural influences in mate selection are endogamy (to marry someone inside one's own social group such as race, religion, social class) and exogamy (to marry someone outside one's own family).

***What are the sociological factors that influence mate selection?***

Sociological aspects of mate selection involve homogamy—"like attracts like"—or the tendency to be attracted to people similar to oneself. Variables include race, age, religion, education, social class, personal appearance, attachment, personality, and open-mindedness.

***What are the psychological factors operative in mate selection?***

Psychological aspects of mate selection include complementary needs, exchange theory, and parental characteristics. Complementary-needs theory ("opposites attract") suggests that people select others who have characteristics opposite to their own.

Exchange theory suggests that one individual selects another on the basis of rewards and costs. As long as an individual derives more profit from a relationship with one partner than with another, the relationship will continue with the "higher profit" person. Exchange concepts influence who dates whom, the conditions of the dating relationship, and the decision to marry. Parental characteristics theory suggests that individuals select a partner similar to the opposite-sex parent.

Undesirable personality characteristics of a potential mate include being too controlling, being narcissistic, having poor impulse control, being hypersensitive to criticism, having an inflated ego, etc. Paranoid, schizoid, and borderline personality disorders may also require one to be cautious.

***What are the sociobiological factors operative in mate selection?***

The sociobiological view of mate selection suggests that men and women select each other on the basis of their biological capacity to produce and support healthy offspring. Men seek young women with healthy bodies, and women seek ambitious men who will provide economic support for their offspring. There is disagreement about the validity of this theory. Critics argue that women may show concern for the earning capacity of men because women have been systematically denied access to similar economic resources, and selecting a mate with these resources is one of their remaining options.

***What factors should be considered when becoming engaged?***

The engagement period is the time to ask specific questions about the partner's values, goals, and marital agenda, to visit each other's parents to assess parental models, and to consider involvement in a premarital counseling program.

Some couples (particularly those with children from previous marriages) develop a prenuptial agreement to specify who gets what and the extent of spousal support in the event of a divorce. To be valid, the document should be developed by an attorney in accordance with the laws of the state in which the partners reside. Last-minute prenuptial agreements put enormous emotional strain on the couple and are often considered invalid by the courts. Discussing a prenuptial agreement six months in advance of a planned wedding is recommended.

***What are the advantages and disadvantages of meeting online?***

Internet dating, app dating, and speed-dating are increasingly being used to find a partner. The Internet is efficient, allows one to screen multiple partners quickly, and to disappear at will. Downsides include that one cannot assess "chemistry" through a computer screen; Internet relationships allow no observation of the potential partner with his/her friends and family; and there is considerable deception by both parties.

***What factors suggest you might consider calling off the wedding?***

Factors suggesting that a couple may not be ready for marriage include being in their teens, having known each other less than two years, and having a relationship characterized by significant differences and/or dramatic parental disapproval. Some research suggests that partners with the greatest number of similarities in values, goals, and common interests are most likely to have happy and durable marriages. Negative reasons for getting married include being on the rebound, escaping from an unhappy home life, psychological blackmail, and pity.

***What is the future of selecting a partner?***

Cultural, sociological, and psychological factors will continue to be involved in the selection of a marital partner. Once a person is directed outside the family to look for a partner and experiences various endogamous pressures, homogamy and exchange variables come into play. Indeed, Internet profiles are scanned to identify persons with similar interests, values, and goals.

The future of mate selection will involve the increased use of dating sites to find a marriage partner. Use of technology permeates the lives of today's youth, so use of Match.com and other such sites will become normative and lose their stigma.

## Key Terms

Attachment theory of mate selection
Circadian preference
Complementary-needs theory
Covenant marriage
Cyclical relationships
Endogamy
Engagement
Exchange theory
Exogamy
Homogamy
Marriage squeeze
Mating gradient
Open minded
Pool of eligibles
Prenuptial agreement
Role theory of mate selection
Dark triad personality
Sociobiology
Spatial homogamy

## Web Links

PAIR Project
http://www.utexas.edu/research/pair

RELATE Institute
https://www.relate-institute.org/

# CHAPTER 6 Marriage Relationships

*What makes for a good marriage isn't necessarily what makes for a good romantic relationship. Marriage isn't a passion-fest; it's a partnership formed to run a very small, mundane and often boring non-profit business. And I mean this in a good way.*

LORI GOTTLIEB, *MARRY HIM: THE CASE FOR SETTLING FOR MR. GOOD ENOUGH*

Chelsea Curry

## Learning Objectives

- Discuss the individual motivations for and societal functions of marriage.
- Review how marriage is a commitment to one's partner, family, and state.
- Identify how weddings and honeymoons are rites of passage.
- Summarize the various changes from being a lover to being a spouse.
- Review Hispanic, Mormon, age-discrepant, and military marriages.
- List the characteristics of successful marriages.
- Identify how couples vary in marital happiness across time.
- Predict the future of marriage relationships.

## TRUE OR FALSE?

1. The more money a couple spend on their wedding, the happier their marriage six years later.
2. Levels of distress and psychological well-being are essentially the same for spouses and for those who are not married.
3. Being unhappy and bored was the top reason identified by 478 undergraduates as to why their most recent romantic relationship ended.
4. Quitting college seems to have a positive effect on a married couple's relationship satisfaction.
5. The happier a person as an individual, the more likely that person will have a happy marriage.

*Answers:* 1. F 2. F 3. T 4. F 5. T

If you are unmarried, you may be one phone call away from meeting your spouse. In her book *Ali in Wonderland,* Ali Wentworth related her story of calling up George Stephanopoulos on a whim to ask if he wanted to go out—she had no expectation of a future beyond the Starbuck's coffee. "I didn't even shower or shave my legs," she said (Wentworth 2012). The last thing she was interested in was getting involved with "one of these Washington political guys." But their date turned into a marriage (within six months); they now have two daughters, and are still going strong after 15 years.

All marriages are different. *Diversity* is the term that best describes relationships, marriages, and families today. No longer is there a one-size-fits-all cultural norm of what a relationship, marriage, or family should be. Rather, individuals, couples, and families select their own path. In this chapter, we review the diversity of relationships. We begin with looking at some of the different reasons people marry.

# Motivations for and Functions of Marriage

In this section, we discuss both why people marry and the functions that getting married serve for society.

## Individual Motivations for Marriage

We have defined marriage in the United States as a legal contract between two adults that regulates their economic and sexual interaction. However, individuals in the United States tend to think of marriage in personal more than legal terms. The following are some of the reasons people give for getting married.

*If there were not two of us, the question of why would never occur.*

**B F. SKINNER, PSYCHOLOGIST**

### National **Data**

Ninety-three percent of 1,306 spouses and 83% of 1,385 unmarried individuals (in the United States) said that love was an important reason for getting married (Cohn 2013).

**Love** Many couples view marriage as the ultimate expression of their love for each other—the desire to spend their lives together in a secure, legal, committed relationship. In U.S. society, love is expected to precede marriage—thus,

*Love: a temporary insanity curable by marriage.*

AMBROSE BIERCE, AMERICAN JOURNALIST-WRITER

only couples in love consider marriage. Those not in love would be ashamed to admit it.

**Personal Fulfillment** We also marry because we anticipate a sense of personal fulfillment by doing so. We were born into a family (family of origin) and want to create a family of our own (family of procreation). We remain optimistic that our marriage will be a good one. Even if our parents divorced or we have friends who have done so, we feel that our relationship will be different.

**Companionship** Talk show host Oprah Winfrey once said that lots of people want to ride in her limo, but what she wants is someone who will take the bus when the limo breaks down. One of the motivations for marriage is to enter a structured relationship with a committed companion, a person who will stick with us when the going gets rough.

Although marriage does not ensure it, companionship is the greatest expected benefit of marriage in the United States. Coontz (2000) noted that it has become "the legitimate goal of marriage" (p. 11). Johnson et al. (2010) noted that people often marry when they "find the right person."

**Parenthood** Most people want to have children. In response to the statement, "Someday, I want to have children," 90% of 1,089 undergraduate males and 90% of 3,447 undergraduate females answered "yes" (Hall and Knox 2013).

Although some people are willing to have children outside marriage (e.g., in a cohabiting relationship or in no relationship at all), most Americans prefer to have them in a marital context. Previously, a strong norm existed in our society (particularly among whites) that individuals should be married before they have children. This norm is becoming more relaxed, with more individuals willing to have children without being married.

**Economic Security** Married people report higher household incomes than do unmarried people. Indeed, almost 80% of wives work outside the home so that two incomes are available to the couple. One of the disadvantages of remaining single is that the lifestyle is associated with lower income (see Table 6.1).

**Psychological Well-Being** Regardless of sexual orientation, being married is associated with lower levels of distress and higher levels of psychological wellbeing than being single. This conclusion is based on analysis of data from the California Health Interview Survey of 2009 adults ages 18 to 70 (Wright et al. 2013).

## Societal Functions of Marriage

As noted in Chapter 1, important societal functions of marriage are to bind a male and a female together who will reproduce, provide physical care for their dependent young, and socialize them to be productive members of society who will replace those who die (Murdock 1949). Marriage helps protect children by giving the state legal leverage to force parents to be responsible to their

**TABLE 6.1 Median Income of Married Couple and Single Households**

| | |
|---|---|
| Married couple (two incomes) | $71,830 |
| Single female household | $25,269 |
| Single male household | $36,611 |

Source: *Statistical Abstract of the United States, 2012–2013.* 131th ed. Washington, DC: U.S. Bureau of the Census. Table 692.

**TABLE 6.2 Traditional versus Egalitarian Marriages**

| Traditional Marriage | Egalitarian Marriage |
|---|---|
| Limited expectation of husband to meet emotional needs of wife and children. | Husband is expected to meet emotional needs of his wife and children. |
| Wife is not expected to earn income. | Wife is expected to earn income. |
| Emphasis is on ritual and roles. | Emphasis is on companionship. |
| Couples do not live together before marriage. | Couples often live together before marriage. |
| Wife takes husband's last name. | Wife may keep her maiden name. In some cases, he will take her last name. |
| Husband is dominant; wife is submissive. | Neither spouse is dominant. |
| Roles for husband and wife are rigid. | Roles for spouses are flexible. |
| Husband initiates sex; wife complies. | Either spouse initiates sex. |
| Wife takes care of children. | Fathers more involved in child rearing. |
| Education is important for husband, not for wife. | Education is important for both spouses. |
| Husband's career decides family residence. | Career of either spouse may determine family residence. |



offspring whether or not they stay married. If couples did not have children, the state would have no interest in regulating marriage.

Additional functions of marriage include regulating sexual behavior (spouses are expected to be faithful, which results in less exposure to sexually transmitted infections than being single) and stabilizing adult personalities by providing a companion and "in-house" counselor. In the past, marriage and family served protective, educational, economic, and religious functions for its members. These functions have been taken over by the legal, educational, economic and religious institutions of our society. Only the companionship-intimacy function of marriage/family has remained virtually unchanged.

In today's social world, which consists mainly of impersonal, secondary relationships, living in a context of mutual emotional support is particularly important. Indeed, the companionship and intimacy needs of contemporary U.S. marriage have become so strong that many couples consider divorce when they no longer feel "in love" with their partner. Of 1,092 undergraduate males, 68% (and 67% of 3,447 undergraduate females) reported that they would divorce their spouse if they no longer loved him or her (Hall and Knox 2013).

*God created sex. Priests created marriage.*

VOLTAIRE, FRENCH ENLIGHTENMENT WRITER

The very nature of the marriage relationship has also changed from being very traditional or male-dominated to being very modern or egalitarian. A summary of these differences is presented in Table 6.2. Keep in mind that these are stereotypical marriages and that only a small percentage of today's modern marriages have all the traditional or egalitarian characteristics that are listed.

# Marriage As a Commitment

Marriage represents a multilevel commitment—person-to-person, family-to-family, and couple-to-state.

*Men marry because they are tired; women because they are curious. Both are disappointed.*

OSCAR WILDE, ENGLISH PLAYWRIGHT

**National Data**

Eighty-four percent of 5,500 single individuals reported that they have been in a committed relationship in the past (Walsh 2013).

## Person-to-Person Commitment

**Commitment** the intent to maintain a relationship.

**Commitment** is the intent to maintain a relationship. Persons express commitment by telling one another ("I love you and want to spend my life with you"), telling friends ("We have been going together for years and have discussed a June wedding"), doing things for each other ("I'll get the oil changed in your car"), and providing economic resources ("I'll pay off your student loans"). One way, particularly for males, to increase their commitment to the relationship is to make sacrifices for their partners. Totenhagen et al. (2012) found that men who engaged in behavior (made sacrifices) consistent with their partners' wishes (e.g., taking her to a movie she was particularly interested in) reported an increased commitment effect. Goodman et al. (2013) emphasized that the function of religion was to encourage couples to take their commitment to each other seriously. Mutual commitment is associated with higher relationship quality (Weigel 2010). One partner also tends to mirror the commitment of the other. In a study of 112 couples in marriage counseling, a commitment change in one partner was associated with commitment change in the other (Bartle-Haring 2010).

WHAT IF?

### What if My Partner is Not as Interested in the Relationship as I Am?

Both partners in a relationship rarely have the same level of interest, involvement, and commitment. These differences are due to experiences in the relationship, previous relationships, and personality. In regard to personality, individuals who have been betrayed in a former relationship are slow to trust and become involved again. Unless the different levels of interest are dramatic, such that one partner is unsure whether to remain in the relationship at all, one option is to continue the relationship to give time for that partner to catch up while the other partner slows down so that the partners are closer in their walk together.

## Family-to-Family Commitment

Whereas love is private, marriage is public. Marriage is the second of three times that one's name can be expected to appear in the local newspaper (the other two are birth and death). When individuals marry, the parents and extended kin also become involved. In many societies (e.g., Kenya), the families arrange for the marriage of their offspring, and the groom is expected to pay for his new bride. How much is a bride worth? In some parts of rural Kenya, premarital negotiations include the determination of **bride wealth**—also known as bride price or bride payment, the amount of money or goods paid by the groom or his family to the wife's family for giving her up. Such a payment is not seen as buying the woman but, rather, as compensating the parents for the loss of labor their daughter provides. Forms of payment include livestock ("I am worth many cows," said one Kenyan woman), food, and/or money. The man who pays a high price for his bride not only demonstrates that he is ready to care for a wife and children but also that he has the resources to do so.

**Bride wealth** the amount of money or goods paid by the groom or his family to the wife's family for giving her up.

Marriage also involves commitments by each of the marriage partners to the family members of the spouse. Married couples are often expected to divide their holiday visits between both sets of parents.

### Couple-to-State Commitment

In addition to making person-to-person and family-to-family commitments, spouses become legally committed to each other according to the laws of the state in which they reside. This means they cannot arbitrarily decide to terminate their own marital agreement.

Just as the state says who can marry (not close relatives or the insane) and when (usually at age 18 or older), legal procedures must also be instituted in order for the spouses to divorce. The state's interest is that a married couple with children stays married and takes care of the children. Should they divorce, the state will dictate how the parenting is to be divided (e.g., joint custody) and paid for (e.g., child support).

Social policies designed to strengthen marriage through divorce law reform reflect the value the state places on stable, committed relationships.

*What counts in making a happy marriage is not so much how compatible you are, but how you deal with incompatibility.*

LEO TOLSTOY, RUSSIAN WRITER

*Is not marriage an open question, when it is alleged, from the beginning of the world, that such as are in the institution wish to get out, and such as are out wish to get in?*

RALPH WALDO EMERSON, POET

## Marriage As a Rite of Passage

A **rite of passage** is an event that marks the transition from one social status to another. Starting school, getting a driver's license, and graduating from high school or college are events that mark major transitions in status (to student, to driver, and to graduate). The wedding itself is another rite of passage that marks the transition from fiancée and fiancé to spouse. Preceding the wedding is the traditional bachelor party for the soon-to-be groom. Somewhat new on the cultural landscape is the bachelorette party (sometimes wilder than the bachelor party), which conveys the message that both soon-to-be spouses can have their version of a pre-marital celebration—their last hurrah!

**Rite of passage** event that marks the transition from one social status to another.

*Advice father gave son on his wedding day—"It's your fault."*

*"If you leave a glass on a table too close to the edge and your wife knocks it off, tell your wife, 'That was my fault, honey.'"*

*"If your wife leaves the glass too close to the edge and you knock it off, tell your wife, 'That was my fault, honey.'"*

REX FIELDS, CAR SALESMAN

### Weddings

To obtain a marriage license, some states require the partners to have blood tests to certify that neither has an STI. The document is then taken to the county courthouse, where the couple applies for a marriage license. Two-thirds of states require a waiting period between the issuance of the license and the wedding. A member of the clergy marries 80% of couples; the other 20% (primarily remarriages) go to a justice of the peace, judge, or magistrate.

Chelsee Curry

The wedding is a time of celebration and joy.

## FAMILY POLICY | Strengthening Marriage through Divorce Law Reform

Some family scholars and policymakers advocate strengthening marriage by reforming divorce laws to make divorce harder to obtain. Since California became the first state to implement no-fault divorce laws in 1969, every state has passed similar laws allowing couples to divorce without proving in court that one spouse was at fault for the marital breakup. The intent of no-fault divorce legislation was to minimize the acrimony and legal costs involved in divorce, making it easier for unhappy spouses to get out of a marriage. Under the system of no-fault divorce, a partner who wanted a divorce could get one, usually by citing irreconcilable differences, even if their spouse did not want a divorce.

Implementation of the no-fault system brought with it the fear that divorce rates would increase—if spouses could get out of a marriage easily, they would. But data on divorce law reform and divorce rates suggest

### DIVERSITY IN OTHER COUNTRIES

Wedding ceremonies still reflect traditional cultural definitions of women as property. For example, the father of the bride usually walks the bride down the aisle and "gives her away" to her husband. In some cultures, the bride is not even present at the time of the actual marriage. For example, in the upper-middle-class Muslim Egyptian wedding, the actual marriage contract signing occurs when the bride is in another room with her mother and sisters. The father of the bride and the new husband sign the marriage contract (identifying who is marrying whom, the families they come from, and the names of the two witnesses). The father will then place his hand on the hand of the groom, and the maa'zun, the presiding official, will declare that the marriage has transpired.

*What greater thing is there for two human souls, than to feel that they are joined for life—to strengthen each other in all labor, to rest on each other in all sorrow, to minister to each other in all pain, to be one with each other in silent unspeakable memories at the moment of the last parting?*

MARY ANNE EVANS, BETTER KNOWN BY HER PEN NAME GEORGE ELIOT, ENGLISH NOVELIST

**Artifact** concrete symbol that reflects the existence of a cultural belief or activity.

The wedding is a rite of passage that is both religious and civil. To the Catholic Church, marriage is a sacrament that implies that the union is both sacred and indissoluble. According to Jewish and most Protestant faiths, marriage is a special bond between the husband and wife sanctified by God, but divorce and remarriage are permitted.

That marriage is a public event is emphasized by weddings in which the couple invites their family and friends to participate. The wedding is a time for the respective families to learn how to cooperate with each other for the benefit of the couple. Conflicts over the number of bridesmaids and ushers, the number of guests to invite, and the place of the wedding are not uncommon. Campbell et al. (2011) surveyed 610 spouses and found that those who had an elaborate wedding reported less present day satisfaction and commitment, suggesting that individuals who idealize their relationship may enact elaborate weddings, and when their high relational expectations go unmet, satisfaction and commitment decline.

Klos and Sobal (2013) discussed weight management in anticipation of one's wedding. While women typically engage in losing weight to approximate the desired cultural image of the bride, men also try to lose weight. Of 163 engaged men, 39% reported they were actively trying to lose weight (about nine pounds).

Brides often wear traditional **artifacts** (concrete symbols that reflect the existence of a cultural belief or activity): something old, new, borrowed, and blue. The old wedding artifact represents the durability of the impending marriage (e.g., an heirloom gold locket). The new wedding artifact, perhaps in the form of new undergarments, emphasizes the new life to begin. The borrowed wedding artifact is something that has already been worn by a currently happy bride (a wedding veil). The blue wedding artifact represents fidelity; "those dressed in blue or in blue ribbons have lovers true" (Henderson 1866). When the bride throws her floral bouquet, the single woman who catches it will be the next to be married; the rice thrown by the guests at the newly married couple signifies fertility.

Couples now commonly have weddings that are neither religious nor traditional. In the exchange of vows, the couple's relationship may be spelled out by otherwise. Coelho and Garoupa (2014) studied this issue in Portugal and found that divorce rates did not increase with a no-fault system. Indeed, divorce rates were already increasing when the no-fault system was put into place in response to the burden of the existing fault system on the already overworked courts. Under the fault system, the investigation of every accusation jammed court dockets and took forever. Under the no-fault system, couples would file and be on their way.

**Your Opinion?**

1. To what degree do you believe the government can legislate successful marriage relationships?
2. Do you think no-fault divorce encourages divorce?
3. Will there be a return to laws which make it more difficult to divorce?

the partners rather than by tradition, and neither partner may promise to obey the other. Vows often include the couple's feelings about equality, individualism, humanism, and openness to change.

In 2014, the average cost of a wedding for a couple getting married for the first time was estimated to be around $30,000 (http://www.theknot.com). Ways in which couples lower the cost of their wedding include marrying any day but Saturday, or marrying off-season (not June) or off-locale (in Mexico or on a Caribbean Island where fewer guests will attend). They may also broadcast their wedding over the Internet (http://www.webcastmywedding.net/). Streaming capability means that the couple can get married in Hawaii and have their ceremony beamed back to the mainland where well-wishers can see the wedding without leaving home.

### Honeymoons

Traditionally, another rite of passage follows immediately after the wedding—the **honeymoon** (the time following the wedding whereby the couple isolates themselves to recover from the wedding and to solidify their new status change from lovers to spouses). The functions of the honeymoon are both personal and social.

**Honeymoon** time following the wedding whereby the couple isolates themselves to recover from the wedding and to solidify their new status change from lovers to spouses.

The personal function is to provide a period of recuperation from the usually exhausting demands of preparing for and being involved in a wedding ceremony and reception. The social function is to provide a time for the couple to be alone to solidify their new identity from that of an unmarried to a married couple. Now that they are married, their sexual expression and childbearing with each other achieves full social approval and legitimacy.

Not all couples take a honeymoon. Thirty-nine percent of Campbell's et al. (2011) 610 spouses did not take a honeymoon. The most common reasons identified by spouses included financial (59%) and lack of time (39%). Only 6% said they and their partners (5%) were not interested in taking a honeymoon.

*Don't expect that what you have now will be what you have down the road. You won't even recognize why you first fell in love.*

NELL CASEY, WRITER

## Changes after Marriage

After the wedding and honeymoon, the new spouses begin to experience changes in their legal, personal, and marital relationship.

## APPLYING SOCIAL RESEARCH | The Wedding Night

There is considerable interest about a couple's wedding night. Stereotypes are that they are either wonderful or disastrous, but what do the data tell us?

### Sample

This sample consisted of 95 spouses, 75% women and 25% men. Over three-fourths were in their first marriage, with 20% in their second marriage. The respondents were asked to report on their "most recent wedding night." Over 80% (81.1%) were virgins on their wedding night.

### Findings

The questionnaire was designed to assess the quality of the wedding night experience, the best and worst experiences, and some recommendations.

1. ***Rating.*** When asked "On a scale of one to ten with zero being awful and ten being wonderful, what number would you select to describe your wedding night experience?" the average was 6.81, looking at responses from both men and women. Grooms reported more positive experiences than brides, 7.19 to 6.68, respectively. When second marriages were the focus, the average was 8.4.
2. ***Summaries.*** When asked to "summarize your wedding night experience" some of the comments were:
   a. "We only went away for one night as the next day was Christmas. I was 17, it was lots of fun, we ate out, and had a great night."
   b. "It was a fulfilling, happy occasion, knowing that I was starting a new chapter in life."
   c. "It was a disaster. He was thinking about his old girlfriend and wishing he had married her."
   d. "We had an amazing night. It was so hard to keep our hands off each other."
   e. "It was so wonderful but we were tired. It could have been better."

*Marriage is like a phone call in the night; first the ring, and then you wake up.*

EVELYN HENDRICKSON, ACTRESS

## Legal Changes

Unless the partners have signed a prenuptial agreement specifying that their earnings and property will remain separate, the wedding ceremony makes each spouse part owner of what the other earns in income and accumulates in property. Although the laws on domestic relations differ from state to state, courts typically award to each spouse half of the assets accumulated during the marriage (even though one of the partners may have contributed a smaller proportion).

For example, if a couple buys a house together, even though one spouse invested more money in the initial purchase, the other will likely be awarded half of the value of the house if they divorce. Having children complicates the distribution of assets because the house is often awarded to the custodial parent. In the case of death of the spouse, the remaining spouse is legally entitled to inherit between one-third and one-half of the partner's estate, unless a will specifies otherwise.

## Personal/Health Changes

New spouses experience an array of personal changes in their lives. One initial consequence of getting married may be an enhanced self-concept. Parents and close friends usually arrange their schedules to participate in your wedding and give gifts to express their approval of you and your marriage. In addition, the strong evidence that your spouse approves of you and is willing to spend a lifetime with you also tells you that you are a desirable person.

Married people also begin adopting new values and behaviors consistent with the married role. Although new spouses often vow that "marriage won't

3. ***Best part.*** When asked to identify "the best part of the wedding night," 29.7% listed "just being with my new partner"; 17.8% listed "sex"; 10.5% said "nothing"; and 9.4% listed the "reception."
4. ***Worst part.*** When asked to identify the "worst part of the wedding night," 23.1% listed "accommodations and the partner's demeanor"; 21% said "being so tired"; 9.4% said the "end of celebration"; 8.4% said "nothing"; 6.3% listed "sex"; and 3.1% listed "pain."
5. ***Change.*** To the question, "If you could replay your wedding night, what would you change?" 34.7% of respondents answered, "nothing"; 18.9% responded "not be tired;" 14.7% listed "different time/place;" and 9.5% listed "different person."

The data suggested that the wedding night for these respondents was predominately a positive experience. A recommendation was to plan a wedding early in the day so that the reception is over early. Also, the couple might plan to spend the first night a short drive from the reception. "Avoid leaving a reception at 11:00 P.M. and driving for hours." Another suggestion was "Avoid an early-morning flight the day after your wedding."

Shaley et al. (2013) reported the experience of 12 Modern Orthodox couples on their wedding night. They had had no previous sexual experience and the transition from "this is forbidden" to "everything is OK" was difficult and challenging. Two sources of help were sought—close friends and the Internet.

### Sources

This research is based on unpublished data collected for this text. Appreciation is expressed to Kelly Woody for distribution of the questionnaires and to Emily Richey for tabulating the data.

Shaley, O., N. Baum and H. Itzhaki. 2013. "There's a man in my bed": The first sex experience among modern-orthodox newlyweds in Israel. *Journal of Sex & Marital Therapy* 39: 40–55.

change me," it does. For example, rather than stay out all night at a party, which is not uncommon for single people who may be looking for a partner, spouses (who are already paired off) tend to go home early. Their roles of spouse, employee, and parent result in their adopting more regular, alcohol- and drug-free hours. Averett et al. (2013) confirmed that being married is not only associated with lower alcohol use and improved mental health but with a higher BMI. In effect, spouses put on weight after they say "I do." Weight gain is particularly true for married men when compared to single men (Berge et al. 2014).

## Friendship Changes

Marriage affects relationships with friends of the same and the other sex. Although time with same-sex friends will continue (Hall & Adams 2011), it will decrease because of the new role demands of the spouses. More time will be spent with other married couples who will become powerful influences on the new couple's relationship. Indeed, couples who have the same friends report increased marital satisfaction.

What spouses give up in friendships, they gain in developing an intimate relationship with each other. However, abandoning one's friends after marriage may be problematic because one's spouse cannot be expected to satisfy all of one's social needs. Because many marriages end in divorce, friendships that have been maintained throughout the marriage can become a vital source of support for a person adjusting to a divorce. "Don't forget your friends on your way up, you'll need them on your way down" reflects the sentiment of maintaining one's friends after getting married.

## PERSONAL CHOICES

### *Is "Partner's Night Out" a Good Idea?*

Although spouses enjoy spending time together, they also want to spend time with their friends. Some spouses have a flexible policy based on trust with each other. Other spouses are very suspicious of each other. One husband said, "I didn't want her going out to bars with her girlfriends after we were married. You never know what someone will do when they get three drinks in them." For partner's night out to have a positive impact on the couple's relationship, it is important that the partners maintain emotional and sexual fidelity to each other, that each partner have a night out, and that the partners spend some nights alone with each other.

Friendships can enhance a marriage relationship by making the individual partners happier, but friendships cannot replace the marriage relationship. Spouses must spend time alone together to nurture their relationship. Khalil Gabran said, "Let there be spaces in your togetherness."

*Men are April when they woo, December when they wed. Maids are May when they are maids, but the sky changes when they are wives.*

SHAKESPEARE, *AS YOU LIKE IT*, IV, I, 140

## Relationship Changes

Totenhagen et al. (2011) studied the variability in the relationships of 328 individuals on seven variables (satisfaction, commitment, closeness, maintenance, love, conflict, and ambivalence) and found that there was greater variability for newer couples than for longer-term couples. But variability is to be expected. A couple happily married for 45 years spoke to our class and began their presentation with, "Marriage is one of life's biggest disappointments." They spoke of the difference between all the hype and the cultural ideal of what marriage is supposed to be ... and the reality.

**Disenchantment** the transition from a state of newness and high expectation to a state of mundaneness tempered by reality.

One effect of getting married is **disenchantment**—the transition from a state of newness and high expectation to a state of mundaneness tempered by reality. It may not happen in the first few weeks or months of marriage, but it is almost inevitable. Whereas courtship is the anticipation of a life together, marriage is the day-to-day reality of that life together—and reality does not always fit the dream. "Moonlight and roses become daylight and dishes" is an old adage reflecting the realities of marriage. Musick and Bumpass (2012) compared spouses, cohabitants, and singles and noted that the advantages of being married over not being married tended to dissipate over time.

*She would make his life a living hell.*

SAID OF MIRIAM NOEL, SECOND WIFE OF FRANK LLOYD WRIGHT FROM WRIGHT'S BIOGRAPHY BY ADA HUXTABLE

Disenchantment after marriage is also related to the partners shifting their focus away from each other to work or children; each partner usually gives and gets less attention in marriage than in courtship. College students are not oblivious to the change after marriage. Twenty-seven percent of 1,090 undergraduate males and 23% of 3,442 undergraduate females agreed that "most couples become disenchanted with marriage within five years" (Hall and Knox 2013).

In addition to disenchantment, a couple will experience numerous changes once they marry:

**1.** ***Loss of freedom.*** Single people do as they please. They make up their own rules and answer to no one. Marriage changes that as the expectations of the spouse impact the freedom of the individual. In a study of 1,001 married adults, 26% reported that what they missed most about being single was not being able to live by their own rules (Cadden and Merrill 2007).

*Their lives were ruined, he thought; ruined by the fundamental error of their matrimonial union: that of having based a permanent contract on a temporary feeling.*

THOMAS HARDY, ENGLISH NOVELIST AND POET

**2.** ***More responsibility.*** Single people are responsible for themselves only. Spouses are responsible for the needs of each other and sometimes resent it. In the study of 1,001 married adults, 25% reported that having less responsibility was what they missed most about being single.

**3.** ***Less alone time.*** Aside from the few spouses who live apart, most live together. They wake up together, eat their evening meals together, and go to bed together. Each may feel too much togetherness. "This altogether, togetherness thing is something I don't like," said one spouse. In the study of 1,001 spouses, 24% reported that "having time alone for myself" was what they missed most about being single (Cadden and Merrill 2007). One wife said, "My best time of the day is at night when everybody else is asleep."

**4.** ***Change in how money is spent.*** Entertainment expenses in courtship become allocated to living expenses and setting up a household together. In the study of over a thousand spouses, 17% reported that they missed spending money the way they wanted to (Cadden and Merrill 2007).

**5.** ***Sexual changes.*** The frequency with which spouses have sex with each other decreases after marriage (Hall and Adams 2011). But marital sex is still the most satisfying of all sexual contexts. Of married people in a national sample, 85% reported that they experienced extreme physical pleasure and extreme emotional satisfaction with their spouses. In contrast, 54% of individuals who were not married or not living with anyone said that they experienced extreme physical pleasure with their partners, and 30% said that they were extremely emotionally satisfied (Michael et al. 1994).

**6.** ***Power changes.*** The power dynamics of the relationship change after marriage with men being less patriarchal/collaborating more with their wives while women change from deferring to their husbands' authority to challenging their authority (Huyck and Gutmann 1992). In effect, with marriage, men tend to lose power and women gain power. However, such power changes may not always occur. In abusive relationships, abusive partners have more power and attempt to keep their spouses under control.

**7.** ***Discovering that one's mate is different from one's date.*** Courtship is a context of deception. Marriage is one of reality. Spouses sometimes say, "He (she) is not the person I married." Jay Leno once quipped, "It doesn't matter who you marry since you will discover that you have married someone else." One's date is always attentive, focused on your needs, willing to please. One's mate has other interests, not fearful of focusing on their own needs, and sees you in the context of other relationships/interests.

*Marriage is like a violin. After the music is over you still have the strings.*

UNKNOWN

*To catch a husband is an art; to hold him is a job.*

SIMONE DE BEAUVOIR, FRENCH WRITER

*Don't you think it's sad that kissing, the simplest form of affection, is one of the first things to go in a marriage? Couples on dates spend hours making out, but after they're married, most get by with a quick peck on their way out the door. Finding time for intimacy becomes a chore, an obligation.*

ANONYMOUS

## Parents and In-Law Changes

Marriage affects relationships with parents. Time spent with parents and extended kin radically increases when a couple has children. Indeed, a major difference between couples with and without children is the amount of time they spend with relatives. Parents and kin rally to help with the newborn and are typically there for birthdays and family celebrations. In spite of all the in-law jokes, only a minority of spouses (3 to 4%) report that they do not get along with their in-laws (Amato et al. 2007).

Emotional separation from one's parents is an important developmental task in building a successful marriage. When choices must be made between one's parents and one's spouse, more long-term positive consequences for the married couple are associated with choosing the spouse over the parents. However, such choices become more complicated and difficult when one's parents are elderly, ill, or widowed.

## PERSONAL CHOICES

### *Parents Living with the Married Couple?*

This question is more often asked by individualized Westernized couples who live in isolated nuclear units. Asian couples reared in extended-family contexts expect to take care of their parents and consider it an honor to do so. American spouses with aging parents must make a decision about whether to have the parents live with them. Usually the parent in need of care is the mother of either spouse because the father is more likely to die first. One wife said, "We didn't have a choice. His mother is 82 and has Alzheimer's disease. We couldn't afford to put her in a nursing home at $5,200 a month, and she couldn't stay by herself. So we took her in. It's been a real strain on our marriage, since I end up taking care of her all day. I can't even leave her alone to go to the grocery store. But I love her and it breaks my heart to watch her die." The role of taking care of an aging parent (either one's own or the spouse's) most often falls to the married daughter (Barnett 2013).

Some elderly people have resources for nursing home care, or their married children can afford such care. However, even in these circumstances, some spouses decide to have their parents live with them. "I couldn't live with myself if I knew my mother was propped up in a wheelchair eating Cheerios when I could be taking care of her," said one daughter.

When spouses disagree about parents in the home, the result can be challenging. According to one wife, "I told my husband that Mother was going to live with us. He told me she wasn't and that he would leave if she did. She moved in, and he moved out (we divorced). Five months later, my mother died." While this example reflects the end of a marriage over the issue of parental care, more often the result is strain/distress on the part of both spouses who are challenged by the care of the aging parent (Strauss 2013).

### Financial Changes

An old joke about money in marriage is that "two can live as cheaply as one as long as one doesn't eat." The reality behind the joke is that marriage involves the need for spouses to discuss and negotiate how they are going to get and spend money in their relationship. Some spouses bring considerable debt into the marriage or amass great debt during the marriage. Such debt affects marital interaction. In one study, the researcher observed that spouses who were in debt reported spending less leisure time together and argued more about money (Dew 2008).

Marriage is also associated with the male becoming more committed to earning money. Ashwin and Isupova (2014) noted that husbands "implicitly commit themselves to a 'responsible' version of masculine identity" (rather than hard drinking, for example) and that wives monitor their behavior to ensure a productive outcome. One example is a wife who made it clear that she did not want her husband to quit his job regardless of how unhappy he was. Her focus on keeping the family income coming in and her desire to avoid the "my husband is unemployed" stigma, kept her husband on the job.

## Diversity in Marriage

Any study of marriage relationships emphasizes the need to understand the diversity of marriage/family life. Researchers (Ballard and Taylor 2011; Wright et al. 2012) have emphasized the various racial, ethnic, structural, geographic location, and contextual differences in marriage and family relationships. In this section, we review Hispanic, Mormon, Muslim American, and military families. We also look at other examples of family diversity: interracial, interreligious, cross-national, and age discrepant.

### Hispanic Families

The panethnic term *Hispanic* refers to both immigrants and U.S. natives with an ancestry to one of twenty Spanish-speaking countries in Latin America and the Caribbean. There are about 53 million Hispanics in the United States who represent 17% of the population (United States Census Bureau 2014). Hispanic families vary not only by where they are from but by whether they were born in the United States. About 40% of U.S. Hispanics are foreign born and immigrated here, 32% have parents who were born in the United States, and 28% were born here of parents who were foreign born.

Great variability exists among Hispanic families. Although it is sometimes assumed that immigrant Hispanic families come from rural impoverished Mexico where family patterns are traditional and unchanging, immigrants may also come from economically developed urbanized areas in Latin America (Argentina, Uruguay, and Chile), where family patterns include later family formation, low fertility, and nuclear family forms.

Hispanics tend to have higher rates of marriage, early marriage, higher fertility, nonmarital child rearing, and prevalence of female householder. They also have two micro family factors: male power and strong familistic values.

1. ***Male power.*** The husband and father is the head of the family in most Hispanic families. Rodriguez et al. (2010) also found that Mexican husbands reported more marital satisfaction than their wives. The children and wife respect the husband as the source of authority in the family. The wife assumes the complementary role where her focus is taking care of the home and children. Sayer and Fine (2011) found that Hispanic women do more cooking and cleaning compared with white and black women. Rodriguez et al. (2010) also found that husbands had more prestigious jobs, worked more hours, and earned more money (by $9,000) than their wives.

2. ***Strong familistic values.*** The family is the most valued social unit in the society—not only the parents and children but also the extended family. Hispanic families have a moral responsibility to help family members with money, health, or transportation needs. Children are taught to respect their parents as well as the elderly. Indeed, elderly parents may live with the Hispanic family where children may address their grandparents in a formal way. Spanish remains the language spoken in the home as a way of preserving family bonds.

### National **Data**

What about marital happiness or divorce risk of Hispanic families compared to white and black families? Using data from the National Survey of Families and Households (N = 6,231), Mexican Americans and whites have similar levels of marital quality, whereas blacks report lower marital quality than these two groups (Bulanda and Brown 2007). Strong familistic values and the fact that Hispanic wives are less of an economic threat to their husband's provider role are influential in this finding.

## Mormon Families

In 2012 presidential nominee Mitt Romney raised the visibility of the Mormon faith. Questions about the six million Mormons in the United States (and the 14 million worldwide) included interest in Mormon families. Also known as the Church of Jesus Christ of Latter-day Saints, the Mormons have been associated with polygyny. But this practice was disavowed in 1890 by mainstream Mormons. Sects of the Mormon Church continuing the practice are not included in the following discussion ( Dollahite and Marks 2012).

Unique characteristics of Mormon beliefs/families include:

1. ***Eternal marriage.*** Mormon doctrine holds that Mormon spouses are married not only until death but throughout eternity. Mormon spouses also believe that their children become a permanent part of their family both in this life and in the afterlife. Hence, the death of a spouse or child is viewed as a family

member who has gone to Heaven only to be reunited with other family members at their earthly death.

**2.** ***Family rituals.*** Mormons are expected to pray both as spouses and as a family, to study the scripture (*Book of Mormon*), and to observe family home evening every Monday night. The latter involves the family praying, singing, having a lesson taught by a parent or older child, experiencing a fun activity (e.g., board game/charades), and enjoying refreshments (e.g., homemade cookies).

**3.** ***Frequent Prayer.*** While most religions encourage prayer, the Mormon faith encourages the family to pray together three times a day—morning, at mealtimes, and at bedtime. The ritual provides an emotional bond of family members to each other.

**4.** ***Substance prohibitions.*** Mormons avoid alcohol, tobacco, coffee, and some teas (e.g., caffeinated). The health benefits include lower cancer rates and increased longevity (8 to 11 years).

**5.** ***Extended family/intergeneration support.*** Family reunions, family web pages, and ties with parents and grandparents are core to Mormon family norms. The result is a close family system of children, parents, and grandparents.

**6.** ***Intramarriage.*** Selecting another Mormon to marry is encouraged. Only devout members of the church may be married in the temple, which seals the couple for this life and for eternity.

**7.** ***Early marriage/large families.*** Mormons typically marry younger and have a higher number of children than the national average. One result of larger families is that adults stay in the role of parents for a longer period of their lives.

**8.** ***Lower divorce rate.*** It comes as no surprise that the Mormon emphasis on family rituals and values results in strong family ties and a lower divorce rate. While 40 to 50% of marriages in general in the United States end in divorce, only about 10% of Mormon marriages do so.

The Mormon faith permeates the lives of its believers. Individuals are socialized to view life as having meaning/purpose, which is to be married "for time and all eternity." Close family bonds provide support/direction to children and adolescents as they move toward adulthood.

## Muslim American Families

Although Islam (the religious foundation for Muslim families in 60 nations) is the second largest religion (next to Christianity) in North America, 9/11 increased the awareness that Muslim families are part of American demographics. These Muslim American families hardly represent the extremists responsible for terrorism, but more than six million adults in the United States and 1.3 billion worldwide self-identify with the Islamic religion (there are now more Muslims than Christians in the world). The three largest American Muslim groups in the United States are African Americans, Arabs, and South Asians (e.g., from Pakistan, Bangladesh, Afghanistan, and India).

Islamic tradition emphasizes close family ties with the nuclear and extended family, social activities with family members, and respect for the authority of the elderly and parents. Religion and family are strong sources of a Muslim's personal identity. Muslim families provide a strong sense of emotional and social support. Breaking from one's religion and family comes at a great cost because alternatives are perceived as limited. Parents of Muslim children who are reared in America struggle to maintain traditional values while allowing their children (particularly sons) to pursue higher education and professional training.

One of the striking features of Muslim American families is the strong influence parents have over the behavior of their children. Because the families control the property and economic resources and generally provide total financial support to the children, and because the offspring may not be able to find adequate work outside the family system, the children generally acquiesce to their parents' wishes. Such acquiescence does not imply the nonexistence of genuine love and affection children may have for their parents, however.

## Military Families

Although the war in Iraq is over and U.S. troops are being withdrawn from Afghanistan, approximately 1.5 million U.S. citizens are active-duty military personnel. Over half (56%) are married (726,000) (Lacks et al. 2013). Another 819,000 are in the military reserve and National Guard (*Statistical Abstract of the United States, 2012–2013,* Table 508).

There are three main types of military marriages. In one type, an individual falls in love, gets married, and subsequently joins the military. A second type of military marriage is one in which one or both of the partners is already a member of the military before getting married. The final and least common type is known as a **military contract marriage**, in which a military person will marry a civilian to get more money and benefits from the government. For example, a soldier might decide to marry a platonic friend and split the money for the additional housing allowance (which is sometimes a relatively small amount of money and varies depending on geographical location and rank). Other times, the military member keeps the extra money and the civilian will take the benefit of health insurance. Often, in these types of military marriages, the couple does not reside together. There is no emotional connection because the marriage is mercenary. Contract military marriages are not common but they do exist.

**Military contract marriage** a military person marries a civilian to get more money and benefits from the government.

Some ways in which military families are unique include:

**1.** ***Traditional sex roles.*** Although both men and women are members of the military service, the military has considerably more men than women (85% versus 15%). In the typical military family, the husband is deployed (sent away to serve) and the wife is expected to understand his military obligations and to take care of the family in his absence. Her duties include paying the bills, keeping up the family home, and taking care of the children; a military wife must often play the role of both spouses due to the demands of her husband's military career and obligations. The wife often has to sacrifice her career to follow (or stay behind in the case of deployment) and support her husband in his fulfillment of military duties.

In the case of wives or mothers who are deployed, the rare husband is able to switch roles and become Mr. Mom. One military career wife said of her husband, whom she left behind when she was deployed, "What a joke. He found out what taking care of kids and running a family was really like and he was awful. He fed the kids SpaghettiOs for the entire time I was deployed."

There are also circumstances in which both parents are military members, and this can blur traditional sex roles because the woman has already deviated from a traditional "woman's job." Military families in which both spouses are military personnel are rare.

**2.** ***Loss of control—deployment.*** Military families have little control over their lives as the chance of deployment is ever-present. Where one of the spouses will be next week and for how long are beyond the control of the spouses and parents. Easterling and Knox (2010) surveyed 259 military wives (whose husbands had been deployed) who reported feelings of loneliness, fear, and sadness (other researchers have identified stress) (Lacks et al. 2013). Some women had gone for extended periods of time without communicating with their husbands and in constant worry over their well-being. Talking with other military wives who understood was the primary mechanism for coping with the husband's deployment. Getting a job, participating in military-sponsored events, and living with a family were also helpful. On the positive side, wives of deployed husbands reported feelings of independence and strength. They were the sole family member available to take care of the house and children, and they rose to the challenge.

*None of us guys have seen our families in almost a year. I'm really looking forward to getting back home, being able to just relax with the family, and not having to worry on a day-to-day basis, "Am I to lose my legs or die today?" That's probably one of the biggest things I miss about being a civilian, just the sense of security you have.*

LANCE CORPORAL BRIAN SHEARER

Saying goodbye to a spouse who is going on deployment is the heart-wrenching experience of military marriage.

John Lambert

Adjusting to the return of the deployed spouse has its own challenges. Some deployed spouses who were exposed to combat have had their brain chemistry permanently altered and are never the same again. "These spouses rarely recover completely—they need to accept that their symptoms (e.g., depression, anxiety) can be managed but not cured" noted Theron Covin, who specializes in treating PTSD among the combat deployed (Covin 2013). Spouses have a particular challenge of adjusting to their altered spouse. "You have to learn to dance all over again" said one wife (Aducci et al. 2012). A team of researchers observed an increased incidence of spousal violence related to PTSD as a result of having been deployed (Teten et al. 2010). Foran et al. (2013) examined military marriages after the deployment of a combat exposed spouse; the greater the combat exposure, the greater the intent to divorce.

**3. *Infidelity.*** Although most spouses are faithful to each other, being separated for months (sometimes years) increases the vulnerability of both spouses to infidelity. The double standard may also be operative, whereby "men are expected to have other women when they are away" and "women are expected to remain faithful and be understanding." Separated spouses try to bridge the time they are apart with text messages, e-mails, Skype, and phone calls, but sometimes the loneliness becomes more difficult than anticipated. One enlisted husband said that he returned home after a year-and-a-half deployment to be confronted with the fact that his wife had become involved with someone else. "I absolutely couldn't believe it," he noted. "In retrospect, I think the separation was more difficult for her than it was for me."

**4. *Frequent moves and separation from extended family or close friends.*** Because military couples are often required to move to a new town, parents no longer have doting grandparents available to help them rear their children. And although other military families become a community of support for each other, the consistency of such support may be lacking. "We moved seven states away from my parents to a town in North Dakota," said one wife. "It was very difficult for me to take care of our three young children with my husband deployed."

Similar to being separated from parents and siblings is the separation from one's lifelong friends. Although new friendships and supportive relationships develop within the military community to which the family moves, the relationships are sometimes tenuous and temporary as the new families move on. The result is the absence of a stable, predictable social structure of support, which may result in a feeling of alienation and not belonging in either the military or the civilian community. The more frequent the moves, the more difficult the transition and the more likely the alienation of new military spouses.

**5. *Lower marital satisfaction and higher divorce rates among military families.*** Solomon et al. (2011) compared 264 veterans who experienced combat stress reaction (CSR) with 209 veterans who did not experience such stress. Results show that traumatized veterans reported lower levels of marital adjustment and more problems in parental functioning. Wick and Nelson Goff (2014) reaffirmed the challenge miltary marriages experience when a deployed spouse returns with PTSD. The divorce rate is also higher in military than civilian marriages (Lundquist 2007).

**6. *Employment of spouses.*** Military spouses (primarily wives) are at a disadvantage when it comes to finding and maintaining careers or even finding a job they can enjoy. Employers in military communities are often hesitant to hire military spouses because they know that the demands that are placed on them in the absence of the deployed military member can be enormous. They are also aware of frequent moves that military families make and may be reluctant to hire employees for what may be a relatively short amount of time. The result is a disadvantaged wife who has no job and must put her career on hold. Military spouses, when they do find employment, are often underemployed, which can lead to low levels of job satisfaction. They also make less, on average, than their civilian counterparts with similar skills. All of these factors can contribute to distress among military spouses (Easterling 2005).

**7. *Resilient military families.*** In spite of these challenges, there are also enormous benefits to being involved in the military, such as having a stable job (one may get demoted but it is much less frequent that one is fired) and having one's medical bills paid for. In addition, most military families are amazingly resilient. Not only do they anticipate and expect mobilization and deployment as part of their military obligation, they respond with pride. Indeed, some reenlist eagerly and volunteer to return to military life even when retired. One military captain stationed at Fort Bragg, in Fayetteville, North Carolina, noted, "It is part of being an American to defend your country. Somebody's got to do it and I've always been willing to do my part." He and his wife made a presentation to our classes. She said, "I'm proud that he cares for our country and I support his deployment. And most military wives that I know feel the same way."

*Miles mean nothing when you are in love, and love means everything when you are miles apart.*

MILITARY WIFE

## Interracial Marriages

### National **Data**

About 15% of all marriages in the United States are mixed racially, with Hispanic–non-Hispanic being the most frequent. Nine percent of whites, 16% of blacks, and 26% of Hispanics marry someone whose race or ethnicity is different from their own (Passel et al. 2010). The least likely intermarriage (1.3%) is between black women and white men (Qian & Lichter 2011).

Multiracial parents today tend to identify their children as multiracial rather than being one race.

Chelsea Curry

In discussing interracial marriages, a complicating factor is that one's racial identity may be mixed. Tiger Woods refers to his race as "Cablinasian," which combines Caucasian, black, Native American (Indian), and Asian origins—he is one-quarter Chinese, one-quarter Thai, one-quarter black, one-eighth Native American, and one-eighth Dutch. Some individuals seek partners with a different racial/ethnic heritage (Yodanis et al. 2012).

Black-white marriages are the most infrequent. Fewer than 1% of the over 63 million marriages in the United States are between a black person and a white person (*Statistical Abstract of the United States,* 2012–2013, Table 60). Segregation in religion (the races worship in separate churches), housing (white and black neighborhoods), and education (white and black colleges), not to speak of parental and peer endogamous pressure to marry within one's own race, are factors that help to explain the low percentage of interracial black and white marriages. Perry (2013) found that interracial friendships are associated with positive intermarriage attitudes. Living in a neighborhood or attending church with other racial groups is associated with racial intermarriage tolerance primarily when friendships develop in these contexts.

Field et al. (2013) examined interracial attitudes among 1,173 college students at five universities. Attitudes at historically black universities were less positive than at predominantly white universities with black students disapproving more of black/white relationships than whites (no gender differences). White students perceived their parents as being more disapproving of black/white relationships than black students did. The spouses in black and white couples are more likely to have been married before, to be age discrepant, to live far away from their families of orientation, to have been reared in racially tolerant homes, and to have educations beyond high school. Some may also belong to religions that encourage interracial unions. The Baha'i religion, which has more than 6 million members worldwide and 84,000 in the United States, teaches that God is particularly pleased with interracial unions. Finally, interracial spouses may tend to seek contexts of diversity. "I have been reared in a military family, been everywhere, and met people of different races and nationalities throughout my life. I seek diversity," noted one student.

Kennedy (2003) notes, "The argument that intermarriage is destructive of racial solidarity has been the principal basis of black opposition" (p. 115). There is also the concern for the biracial identity of offspring of mixed-race parents. Although most mixed-race parents identify their child as having minority race status, there is a trend toward identifying their child as multiracial. This trend may be increasing because of Barack Obama's racial heritage—he had a black father and a white mother.

Interracial partners sometimes experience negative reactions to their relationship. Black people partnered with white people have their blackness and racial identity challenged by other black people. White people partnered with black people may lose their white status and have their awareness of whiteness heightened more than ever before. At the same time, one partner is not given full status as a member of the other partner's race (Hill & Thomas 2000). Other researchers have noted that the pairing of a black male and a white female is regarded as "less appropriate" than that of a white male and a black female (Gaines & Leaver 2002). In the former, the black male "often is perceived as attaining higher social status (i.e., the white woman is viewed as the black man's 'prize,' stolen from the more deserving white man)" (p. 68). In the latter, when a white male pairs with a black female, "no fundamental change in power within the American social structure is perceived as taking place" (p. 68).

Hohmann-Marriott and Amato (2008) found that individuals (both men and women) in interethnic (includes interracial relationships such as Hispanic-white, black-Hispanic, and black-white) marriages and cohabitation relationships have lower-quality relationships than those in same-ethnic relationships. Lower relationship quality was defined in terms of reporting less satisfaction, more problems, higher conflict, and lower commitment to the relationship. The divorce rate is also higher in interracial relationships (Bratter and King 2008).

Black-white interracial marriages are likely to increase—slowly. Not only has white prejudice against African Americans in general declined, but also segregation in school, at work, and in housing has decreased, permitting greater contact between the races. One-third of 1,095 undergraduate males (39% of 3,435 undergraduate females) reported that they have dated someone of another race (Hall and Knox 2013). Most Americans say they approve of racial or ethnic intermarriage—not just in the abstract, but in their own families. More than six in ten say it "would be fine" with them if a family member told them they were going to marry someone from any of three major race/ethnic groups other than their own (Passel et al. 2010).

## Interreligious Marriages

### National **Data**

Of married couples in the United States, 37% have an interreligious marriage (Pew Research 2008).

Pang Kwang

This couple met on campus. She is Hmong and he is Mexican. He is dressed here in traditional Hmong attire.

Although religion may be a central focus of some individuals and their marriage, Americans in general have become more secular, and as a result religion has become less influential as a criterion for selecting a partner. In a survey of 1,098 undergraduate males and 3,453 undergraduate females, 36% and 44%, respectively, reported that marrying someone of the same religion was important for them (Hall and Knox 2013).

Are people in interreligious marriages less satisfied with their marriages than those who marry someone of the same faith? The answer depends on a number of factors. First, people in marriages in which one or both spouses profess "no religion" tend to report lower levels of marital satisfaction than those in which at least one spouse has a religious tie. People with no religion are often more liberal and less bound by traditional societal norms and values; they feel less constrained to stay married for reasons of social propriety.

The impact of a mixed religious marriage may also depend more on the devoutness of the partners than on the fact that the partners are of different religions. If both spouses are devout in their respective religious beliefs, they may expect some problems in the relationship. Less problematic is the relationship in which one spouse is devout but the partner is not. If neither spouse in an interfaith marriage is devout, problems regarding religious differences may be minimal or nonexistent. In their marriage vows, one interfaith couple who married (he was Christian, she was Jewish) said that they viewed their different religions as an opportunity to strengthen their connections to their respective faiths and to each other. "Our marriage ceremony seeks to celebrate both the Jewish and Christian traditions, just as we plan to in our life together."

## International Marriages

With increased globalization, international match-making Internet opportunities, and travel abroad programs, there is greater opportunity to meet/marry someone from another country. Levchenko and Solheim (2013) studied international marriages between Eastern European-born women and U.S.-born men. They found that these pairings reflected homogamy in race and having been previously married. However, complementarity was evident in terms of age—with women being about nine years younger. Eastern European women were also more willing to be traditional in their role relationships—a trait American men seek in their wives.

Over three-fourths of 1,095 undergraduate males (78%) and over two-thirds of 3,447 undergraduate females (67%) reported that they would be willing to marry someone from another country (Hall and Knox 2013). The opportunity to meet that someone is increasing, as upwards of 800,000 foreign students are studying at American colleges and universities.

Some people from foreign countries marry an American citizen to gain citizenship in the United States, but immigration laws now require the marriage to last two years before citizenship is granted. If the marriage ends before two years, the foreigner must prove good faith (that the marriage was not just to gain entry into the country) or he or she will be asked to leave the country.

When the international student is male, more likely than not his cultural mores will prevail and will clash strongly with his American bride's expectations, especially if the couple should return to his country. One female American student described her experience of marriage to a Pakistani, who violated his parents' wishes by not marrying the bride they had chosen for him in childhood. The marriage produced two children before the family returned to Pakistan.

The woman felt that her in-laws did not accept her and were hostile toward her. The in-laws also imposed their religious beliefs on her children and took control of their upbringing. When this situation became intolerable, the wife wanted to return to the United States. Because the children were viewed as being "owned" by their father, she was not allowed to take them with her and was banned from even seeing them. Like many international students, the husband was from a wealthy, high-status family, and his wife was powerless to fight the family. The woman has not seen her children in several years.

Cultural differences may be less extreme but, nevertheless, trying. An American woman fell in love with and married a man from Fiji. She was unaware of a norm from his culture that anyone from his large kinship system could visit at any time and stay as long as they liked—indeed it would be impolite to ask them to leave. On one occasion two "cousins" showed up unannounced at the home of the couple. The wife was in the last week of exams for her medical degree and could not tolerate visitors. The husband told her, "I cannot be impolite." But with the marriage threatened, the husband told his cousins they had to leave.

## Age-Discrepant Relationships and Marriages

Although people in most pairings are of similar age, sometimes the partners are considerably different in age. In marriage, these are referred to as ADMs (age-dissimilar marriages) and are in contrast to ASMs (age-similar marriages). ADMs are also known as **May-December marriages.** Typically, the woman is in the spring of her youth (May) whereas the man is in the later years of his life (December). There have been a number of May-December celebrity marriages, including that of Celine Dion, who is 26 younger than René Angelil (in 2015, they were aged 47 and 73). Michael Douglas is 25 years older than his wife, Catherine Zeta-Jones, and Ellen DeGeneres is 15 years older than her spouse Portia de Rossi.

**May-December marriage** the woman is in the spring of her youth whereas the man is in the later years of his life.

Sociobiology suggests that men select younger women since doing so results in healthier offspring. Women benefit from obtaining sperm from older men who have more resources for their offspring. Burrows (2013) studied age discrepant patterns in both heterosexual and homosexuals and found the same pattern regardless of sexual orientation. Burrows noted a strong cultural norm in the gay community whereby younger gay males seek to pair bond with older gay males for economic security.

In a study of 433 spouses where the husband was older, marital effects were less time spent together and more marital problems. Having less in common was the presumed reason for the negative effects (Wheeler et al. 2012).

Perhaps the greatest example of a May-December marriage that worked is of Oona and Charles Chaplin. She married him when she was 18 (he was 54). Their alliance was expected to last the requisite six months, but they remained together for 34 years (until his death at 88) and raised eight children, the last of whom was born when Chaplin was 73.

There are definite benefits for the older man (but not the younger woman) in the age-discrepant relationship. Drefahl (2010) studied the relationship between age gap and longevity and found that having a younger spouse is beneficial for men but detrimental for women. One explanation is health selection—healthier males are able to attract younger partners. Hence, an older male married to a younger woman may have lower mortality since he is healthier. He may also benefit since the younger woman provides health care and social support. Women tend not to benefit from having younger partners since social norms are less open for women marrying younger men and women have more social relationships/contacts/supports than men (so they have less need for men).

This 21 year old guy and 41 year old woman are on an evening date in her home. She is referred to as a cougar, a woman who enjoys the company of younger men.

David Knox

**Cougar** woman, usually in her 30s and 40s, who is financially stable and mentally independent and looking for a younger man to have fun with.

Although less common, some age-discrepant relationships are those in which the woman is older than her partner. Mariah Carey is 11 years older than her husband, Nick Cannon. Valerie Gibson (2002), the author of *Cougar: A Guide for Older Women Dating Younger Men,* notes that the current use of the term **cougars** refers to "women, usually in their 30s and 40s, who are financially stable and mentally independent and looking for a younger man to have fun with." Gibson noted that one-third of women between the ages of 40 and 60 are dating younger men. Financially independent women need not select a man in reference to his breadwinning capabilities. Instead, these women are looking for men not to marry but to enjoy. The downside of such relationships comes if the man gets serious and wants to have children, which may spell the end of the relationship.

## College Marriages

Cottle et al. (2013) analyzed data from 429 currently and formerly married college students. The ages ranged from 18 to 62. The newlyweds reported significantly greater life satisfaction, marital satisfaction, relationships with in-laws, communication about sex, working out problems, etc., than the older college-married students. A major finding was that students who quit college were less likely to report satisfaction in these areas.

# Marriage Quality

A successful marriage is the goal of most couples. But what is a successful marriage and what are its characteristics?

## Definition and Characteristics of Successful Marriages

**Marital success** refers to the quality of the marriage relationship measured in terms of stability and happiness. Stability refers to how long the spouses have been married and how permanent they view their relationship to be, whereas happiness refers to more subjective/emotional aspects of the relationship. In describing marital success, researchers have used the terms *satisfaction, quality, adjustment, lack of distress,* and *integration.* Marital success is often measured by asking spouses how happy they are, how often they spend their free time together, how often they agree about various issues, how easily they resolve conflict, how sexually satisfied they are, how equitable they feel their relationship is, and how often they have considered separation or divorce.

**Marital success** the quality of the marriage relationship measured in terms of stability and happiness.

*The reason I've been married this long and will be married for the whole deal is—yes, in fact partly because my wife Elaine is utterly resourceful and large-hearted—but the part of the credit I will take is simply this: I never imagined anything else.*

RON CARLSON, *LONGEVITY*

Are wives or husbands happier? Jackson et al. (2014) reported data on 226 independent samples comprising 101,110 participants and found statistically significant yet very small gender differences in marital satisfaction between wives and husbands, with wives slightly less satisfied than husbands. However, the researchers noted that this difference was due to the inclusion of clinical samples, with wives in marital therapy 51% less likely to be satisfied. When nonclinical community-based samples were used, no significant gender differences were observed in marital satisfaction among couples in the general population.

Researchers have also identified characteristics associated with enduring happy marriages (Choi and Marks 2013; Stanley et al. 2012; DeMaris 2010; Amato et al. 2007). Their findings and those of other researchers include the following:

David Knox

This couple is lighting the candle of a cake on their 60th wedding anniversary.

**1.** ***Intimacy/partner attachment.*** Patrick et al. (2010) found that intimacy (feeling close to spouse, showing physical affection, sharing ideas/events, sharing hobbies/leisure activities) was related to marital satisfaction. They studied both spouses in 124 marriages of employees at two major state universities in the Midwest. Most were in the mid-to-late forties, married an average of 20 years, and 80% in their first marriages. Wilson and Huston (2011) confirmed that partner similarity of depth of love feelings in courtship was predictive of remaining married. Finally, Kilmann et al. (2013) found that securely emotionally attached spouses reported higher levels of relationship satisfaction.

**2.** ***Communication/humor.*** Gottman and Carrere (2000) studied the communication patterns of couples over an 11-year period and emphasized that those spouses who stayed together were five times more likely to lace their arguments with positives ("I'm sorry I hurt your feelings") and to consciously choose to say things to each other that nurture the relationship. Successful spouses also feel comfortable telling each other what they want and not being defensive at feedback from the partner. Duba et al. (2012) assessed the marital satisfaction of 30 couples married at least 40 years and found the area of affective communication was particularly problematic. Hence, keeping the emotional connection on one's relationship is important. Successful spouses also have a sense of humor. Indeed, a sense of humor is associated with marital satisfaction across cultures—in the United States, China, Russia, and elsewhere (Weisfeld et al. 2011).

**3.** ***Common interests/Positive self-concepts.*** Spouses who have similar interests, values, and goals, as well as positive self-concepts, report higher marital success (Arnold et al. 2011).

*He wanted not association with glittering things and glittering people—he wanted the glittering things themselves.*

F. SCOTT FITZGERALD, ***WINTER DREAMS***

**4.** ***Not materialistic.*** Being nonmaterialistic is characteristic of happily married couples (Carroll et al. 2011). Although a couple may live in a nice house and have expensive toys (e.g., a boat and an RV), they are not tied to material comforts. "You can have my things, but don't take away my people," is a phrase from one husband reflecting his feelings about his family.

**5.** ***Role models.*** Successfully married couples speak of having positive role models in their parents. Good marriages beget good marriages—good marriages run in families. It is said that the best gift you can give your children is a good marriage.

**6.** ***Religiosity.*** A strong religious orientation and practicing one's religion is associated with being committed to one's marriage (Jorgensen et al. 2011). Religion provides spouses with a strong common value. In addition, religion provides social, spiritual, and emotional support from church members and with moral guidance in working out problems.

**7.** ***Trust.*** Trust in the partner provided a stable floor of security for the respective partners and their relationship. Neither partner feared that the other partner would leave or become involved in another relationship. "She can't take him anywhere he doesn't want to go" is a phrase from a country-and-western song that reflects the trust that one's partner will be faithful.

**8.** ***Personal and emotional commitment to stay married.*** Divorce was not considered an option. The spouses were committed to each other for personal reasons rather than societal pressure.

**9.** ***Sexual satisfaction.*** Barzoki et al. 2012 studied wives and found that marital dissatisfaction leads to sexual dissatisfaction but that this connection can become reciprocal—sexual dissatisfaction can lead to marital dissatisfaction.

**10.** ***Equitable relationships.*** Amato et al. (2007) observed that the decline in traditional gender attitudes and the increase in egalitarian decision making were related to increased happiness in today's couples. DeMaris (2010) also found that spouses who regarded their relationships as those with equal contribution reported higher quality marriages.

**11.** ***Marriage/connection rituals.*** **Marriage rituals** are deliberate repeated social interactions that reflect emotional meaning to the couple. **Connection rituals** are those which occur daily in which the couple share time and attention. Campbell et al. (2011) studied 129 unmarried individuals (involved with a partner) who identified 13 different types of rituals (average of 6). The most frequent was enjoyable activities (23%) such as having meals together and watching TV together. Intimacy expressions (19%) included "taking a shower together every morning and washing each other's hair." Communication ritual (14%) examples were sending frequent text messages and having pet names for each other.

**Marriage rituals** deliberate repeated social interactions that reflect emotional meaning to the couple.

**Connection rituals** those which occur daily in which the couple share time and attention.

**12.** ***Absence of negative statements and attributions.*** Not making negative remarks to the spouse is associated with higher marital quality (Woszidlo and Segrin 2013). In addition, spouses who do not attribute negative motives to their partner's behavior reported higher levels of marital satisfaction than spouses who ruminated about negative motives. Dowd et al. (2005) studied 127 husbands and 132 wives and found that the absence of negative attributions was associated with higher marital quality.

**13.** ***Forgiveness.*** At some time in all marriages, each spouse engages in behavior that hurts or disappoints the partner. Forgiveness rather than harboring and nurturing resentment allows spouses to move forward. Spouses who do not "drop the lowest test score" (an academic metaphor) of their partner find that they inadvertently create an unhappy marital context which they endure. However, Woldarsky and Greenberg (2014) noted that it is not forgiveness but the perception that the partner experiences shame for the transgression that is restorative to the couple. Perhaps the two work in combination?

*Lovers do not finally meet somewhere. They are in each other all along.*

RUMI

**14.** ***Economic security.*** Although money does not buy happiness, having a stable, secure economic floor is associated with marital quality (Amato et al. 2007) and marital happiness (Mitchell 2010). Indeed, higher incomes are associated with marital happiness (Choi and Marks 2013).

**15.** ***Physical Health.*** Increasingly, research emphasizes that the quality of family relationships affects family member health and that the health of family members influences the quality of family relationships and family functioning (Choi and Marks 2013; Proulx and Snyder 2009). Indeed, such an association begins early as the stress of a dysfunctional family environment can activate the physiological responses to stress, change the brain structurally, and leave children more vulnerable to negative health outcomes. High conflict spouses also experience chronic stress, high blood pressure, and depression.

The family is also the primary socialization unit for physical health in reference to eating nutritious food, avoiding smoking and getting regular exercise. The first author of this text was reared in a home where a high fat diet was routine and his father was a chronic smoker who never exercised. He died of a coronary at age 46, illustrating the need for attending to one's health.

**16.** ***Psychological Health.*** One's personal psychological health is also related to marital success. Having a positive self-concept is associated with viewing one's marriage as a success (Arnold et al. 2011). Similarly, being happy as an individual is associated with being happy in one's marital relationship. Stanley et al. (2012) found that an important predictor of later marital satisfaction is being happy in one's own life.

**17.** ***Flexibility.*** Nicoleau et al. (2012) studied 21 heterosexual couples (married 10 or more years) who represented a range of cultures/ethnicities (African American, Asians, Causacians, Africans, Guianese, and Hispanic) and found that flexibility was the most important quality for long-term stability. Flexible couples make mutual decisions, accommodate as necessary to each other's schedule, and ensure that they spend time together. Hence, whatever the issue of contention, their relationship is more important.

## SELF-ASSESSMENT | Satisfaction with Married Life Scale

The Satisfaction with Married Life Scale (SMLS) consists of five items which measure marital satisfaction.

Five statements with which you may agree or disagree follow. Using the 1–7 scale, indicate your agreement with each item by writing the appropriate number on the line in front of that item. Please be open and honest in responding to each item; there are no right or wrong answers.

| | |
|---|---|
| Strongly disagree = 1 | Slightly agree = 5 |
| Disagree = 2 | Agree = 6 |
| Slightly disagree = 3 | Strongly agree = 7 |
| Neither agree nor disagree = 4 | |

_____ 1. In most ways my married life is close to ideal.
_____ 2. The conditions of my married life are excellent.
_____ 3. I am satisfied with my married life.
_____ 4. So far I have gotten the important things I want in my married life.
_____ 5. If I could live my married life over, I would change almost nothing.

### Scoring

Add the numbers you wrote. The marital satisfaction score will range from 5 to 35. The midpoint is 20. A couple's combined marital satisfaction score is a result of summing both partners' scores with a possible score range of 10 to 70 (40 is the midpoint). Higher scores reflect greater marital satisfaction for the couple. The internal consistency of the SWML has been reported with a Cronbach's alpha of .92 along with some evidence of construct validity (Johnson et al. 2006).

### Source

Johnson, H. A., R. B. Zabriskie, and B. Hill. 2006. The contribution of couple leisure involvement, leisure time, and leisure satisfaction to marital satisfaction. *Marriage and Family Review* 40: 69–91.

## Theoretical Views of Marital Happiness and Success

Interactionists, developmentalists, exchange theorists, and functionalists view marital happiness and success differently. Symbolic interactionists emphasize the subjective nature of marital happiness and point out that the definition of the situation is critical. Indeed, a happy marriage exists only when spouses define the verbal and nonverbal behavior of their partner as positive, and only when they label themselves as being in love and happy. Marital happiness is not defined by the existence of specific criteria (time together) but by the subjective definitions of the respective partners. Indeed, spouses who work together may spend all of their time together but define their doing so as a negative. Shakespeare's phrase, "Nothing is either good or bad but thinking makes it so" reflects the importance of perception.

Family developmental theorists emphasize developmental tasks that must be accomplished to enable a couple to have a happy marriage. Wallerstein and Blakeslee (1995) identified several of these tasks, including separating emotionally from one's parents, building a sense of "we-ness," establishing an imaginative and pleasurable sex life, and making the relationship safe for expressing differences.

Exchange theorists focus on the exchange of behavior of a kind and at a rate that is mutually satisfactory to both spouses. When spouses exchange positive behaviors at a high rate (with no negatives), they are more likely to feel marital happiness than when the exchange is characterized by high-frequency negative behavior (and no positives).

Structural functionalists regard marital happiness as contributing to marital stability, which is functional for society. When two parents are in love and happy, the likelihood that they will stay together to provide for the physical care and emotional nurturing of their offspring is increased. Furthermore, when spouses take care of their own children, society is not burdened with having to pay for their care through welfare payments, paying foster parents, or paying for institutional management (group homes).

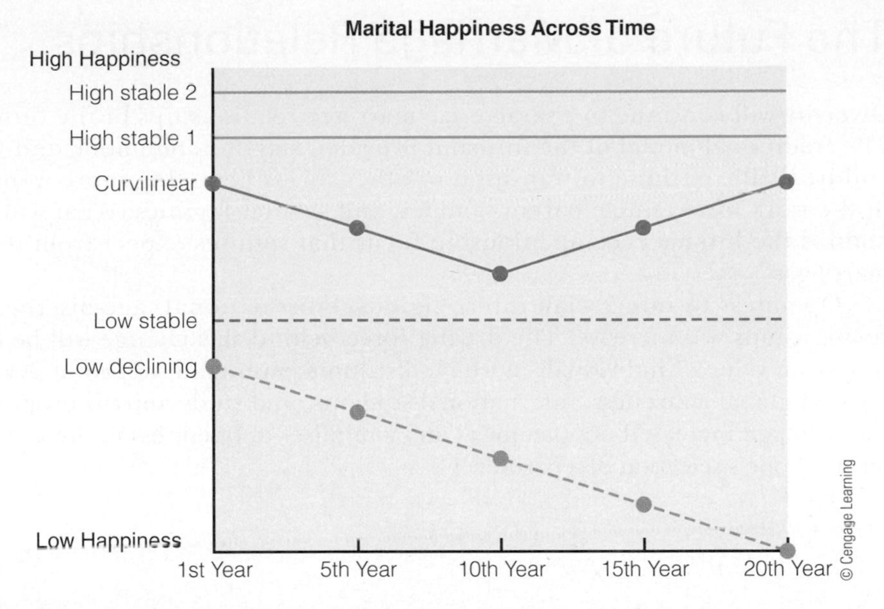


**Figure 6.1** Trajectories of Marital Happiness.

## Marital Happiness across Time

*It is not a lack of love, but a lack of friendship that makes unhappy marriages.*

FRIEDRICH NIETZSCHE, PHILOSOPHER

Anderson et al. (2010) analyzed longitudinal data of 706 individuals over a 20-year period. Over 90% were in their first marriage; most had two children and 14 years of education. Reported marital happiness, marriage problems, time spent together, and economic hardship were assessed. Five patterns emerged:

**1.** High stable 2 (started out happy and remained so across time) = 21.5%

**2.** High stable 1 (started out slightly less happy and remained so across time) = 46.1%

**3.** Curvilinear (started out happy, slowly declined, followed by recovery) = 10.6%

**4.** Low stable (started out not too happy and remained so across time) = 18.3%

**5.** Low declining (started out not too happy and declined across time) = 3.6%

The researchers found that, for couples who start out with a high level of happiness, they are capable of rebounding if there is a decline. But for those who start out at a low level, the capacity to improve is more limited.

Spencer and Amato (2011) emphasized the importance of using a number of variables such as interaction and conflict rather than just marital happiness in the examination of marital relationships over time. Based on data from couples married over 20 years they found that marital happiness shows a U-shape distribution but interaction declined across time. The researchers hypothesized that couples do not interact less because they are unhappy but because they have other interests.

Plagnol and Easterlin (2008) conceptualized happiness as the ratio of aspirations and attainments in the areas of family life and material goods. They studied 47,000 women and men and found that up to age 48, women are happier than men in both domains. After age 48, a shift causes women to become less satisfied with family life (their children are gone) and material goods (some are divorced and have fewer economic resources). In contrast, men become more satisfied with family life (the empty nest is more a time of joy) and finances (men typically have more economic resources than women, whether married or divorced). Individuals who are black and/or with lower education report less happiness.

## The Future of Marriage Relationships

Diversity will continue to characterize marriage relationships of the future. The traditional model of the husband provider, stay-at-home mom, and two children will continue to transition to other forms including more women in the work force, single parent families, and smaller families. What will remain is the intimacy/companionship focus that spouses expect from their marriages.

Openness to interracial, interreligious, cross-national, age-discrepant relationships will increase. The driving force behind this change will be the American value of individualism which discounts parental disapproval. An increased global awareness, international students, and study abroad programs will facilitate increased opportunities and a mindset of openness to diversity in terms of one's selection of a partner.

## Summary

***What are the individual motivations for and societal functions of marriage?***

Individual motives for marriage include personal fulfillment, companionship, legitimacy of parenthood, and emotional and financial security. Societal functions include continuing to provide society with socialized members, regulating sexual behavior, and stabilizing adult personalities.

***What are three levels of commitment in marriage?***

Marriage involves a commitment—person-to-person, family-to-family, and couple-to-state.

***What are two rites of passage associated with marriage?***

While weddings are often a source of stress, they are a rite of passage that is both religious and civil. To the Catholic Church, marriage is a sacrament that implies that the union is both sacred and indissoluble. According to Jewish and most Protestant faiths, marriage is a special bond between the husband and wife sanctified by God, but divorce and remarriage are permitted. Wedding ceremonies still reflect traditional cultural definitions of women as property.

The wedding is a rite of passage signifying the change from the role of fiancé/fiancée to the role of spouse. Women, more than men, are more invested in preparation for the wedding, the wedding is more for the bride's family, and women prefer a traditional wedding. Most spouses report a positive wedding night experience with exhaustion from the wedding or reception often being a problem. The honeymoon is a time of personal recuperation and making the transition to the new role of spouse.

***What changes might a person anticipate after marriage?***

Changes after the wedding are legal (each becomes part owner of all income and property accumulated during the marriage), personal (enhanced self-concept), social (less time with friends), economic (money spent on entertainment in courtship is diverted to living expenses and setting up a household), sexual (less frequency), and parental (improved relationship with parents). The greatest change is disenchantment—moving from a state of exhilaration to a state of mundaneness. New spouses also report experiencing a loss of freedom, increased responsibility, and changes in how money is spent.

***What are examples of diversity in marriage relationships?***

Hispanic families tend to marry earlier and have higher rates of marriage and higher fertility. Male power and strong familistic values also characterize Hispanic families. In Muslim American families, norms involve no premarital sex, close monitoring of children, and intense nurturing of the parental-child bond. Military families cope with deployment and the double standard.

Mixed marriages include interracial, interreligious, and age-discrepant. About 15% of all marriages in the United States are interracial with Hispanic–non-Hispanic the most frequent pairing. Interracial marriages are also more likely to dissolve than same-race marriages. The success of interreligious marriage is related to the degree of devoutness of the partners in the respective religions. If both are very devout, the conflict is more likely to surface. In regard to age-discrepant relationships, when age-discrepant and age-similar marriages are compared, there are no differences in regard to marital happiness. Men who marry younger women seem to benefit in terms of living longer.

***What are the characteristics associated with successful marriages?***

Marital success is defined in terms of both quality and durability. Characteristics associated with marital success include intimacy, commitment, common interests, communication, religiosity, trust, and nonmaterialism, and having positive role models, low stress levels, and sexual desire.

Married couples report the most enjoyment with their relationship in the beginning, followed by less enjoyment during the child-bearing stages, and a return to feeling more satisfied after the children leave home.

***What is the future of marriage relationships?***

Marriages and families will continue to be characterized by diversity. The traditional model of the family will continue to transition to other forms. What will remain is the intimacy/companionship focus that spouses expect from their marriages. Openness to interracial, interreligious, cross-national, age-discrepant relationships will increase. An increased global awareness, international travel, and international students will facilitate a mindset of openness to diversity.

## Key Terms

Artifact
Bride wealth
Commitment
Connection rituals
Cougar
Disenchantment
Honeymoon
Marital success
Marriage rituals
May-December marriage
Military contract marriage
Rite of passage

## Web Links

Bridal Registry
http://www.theknot.com

Brides and Grooms
http://www.bridesandgrooms.com/

Facts about Marriage
http://www.cdc.gov/nchs/fastats/divorce.htm

Groves Conference in Marriage and the Family
http://www.grovesconference.org/

National Healthy Marriage Resource Center
http://www.healthymarriageinfo.org/

Marriage Success Training
http://www.gottmaninstitute.com/

Military Marriages—The National Military Family Association
http://www.nmfa.org/

National Council on Family Relations
http://www.ncfr.org/

The Science Behind a Happy Relationship
http://www.happify.com/hd/the-science-behind-a-happy-relationship/

Smart Marriages
http://www.smartmarriages.com/

Wedding Webcasts
http://www.webcastmywedding.net/

CHAPTER 7

# GLBT Individuals, Couples, and Families

Chelsea Curry

*If you were expecting Prince Charming, I'm sorry. He's with his boyfriend.*

SHAYLA BLACK, WICKED TIES

## Learning Objectives

- Know the definition of homosexuality and the difficulties of identifying and classifying the term.
- Explain the various origins of homosexuality.
- Define homonegativity, homophobia, biphobia, and transphobia.
- Review GLBT and mixed-orientation relationships.
- Identify the category of person a gay person is likely to "come out" to.
- Understand the pros and cons of same-sex marriage.
- Know the various GLBT parenting issues.
- Predict the future for same-sex relationships in the United States.

## TRUE OR FALSE?

1. When a gay person "comes out," the first person who is told is likely to be a friend.
2. Federal law in all states prohibits discrimination against a person because of sexual orientation.
3. In a study of 1,229 transgender individuals, most were alone/not in a relationship.
4. Gay, lesbian, and bisexual individuals who come out to their parents report more positive relationships with them, have fewer alcohol/substance abuse problems, and have healthier psychological adjustment.
5. The American Psychological Association has developed a new treatment program for homosexuals who want to be heterosexual ("My Choice to Change").

*Answers:* 1. T 2. F 3. F 4. T 5. F

U.S. culture has turned the corner and become more accepting of gay individuals and their lifestyle. The Supreme Court has legalized same-sex marriage, mainstream television programs feature gay characters, the Boy Scouts of America has lifted their ban on admitting gays into their ranks, and *Time* magazine featured a cover story on "How Gay Marriage Won" (Von Drehle 2013).

In this chapter, we discuss lesbian, gay, bisexual, and transgender individuals, couples, and families. In many ways they are similar to heterosexuals. A major difference is that LGBT individuals, couples, and families are subjected to prejudice and discrimination.

**Sexual orientation** (also known as **sexual identity**) is a classification of individuals as heterosexual, bisexual, or gay, based on their emotional, cognitive, and sexual attractions and self-identity. **Heterosexuality** refers to the predominance of cognitive, emotional, and sexual attraction to individuals of the other sex. **Homosexuality** refers to the predominance of cognitive, emotional, and sexual attraction to individuals of the same sex, and **bisexuality** is cognitive, emotional, and sexual attraction to members of both sexes. The term **lesbian** refers to women who prefer same-sex partners; **gay** can refer to either women or men who prefer same-sex partners.

**Sexual orientation** classification of individuals as heterosexual, bisexual, or gay, based on their emotional, cognitive, and sexual attractions and self-identity.

**Sexual identity** term used synonymously with sexual orientation.

**Heterosexuality** the predominance of cognitive, emotional, and sexual attraction to individuals of the other sex.

**Homosexuality** predominance of cognitive, emotional, sexual attraction to individuals of the same sex.

**Bisexuality** cognitive, emotional, and sexual attraction to members of both sexes.

**Lesbian** women who prefer same-sex partners.

**Gay** either men or women who prefer same-sex partners.

### National **Data**

Of U.S. adults, about 1% of females self-identify as lesbian, 2% of males self-identify as gay, and 1.5% of adults self-identify as bisexual. Hence, about 3.5% or about ten million individuals in the United States are LGB (Mock and Eibach, 2012).

Increasingly universities are now providing GLBT Centers to emphasize diversity of acceptance.

**DIVERSITY IN OTHER COUNTRIES**

A national sample to assess the sexual orientation of adults in England revealed that 5.3% of men and 5.6% of women reported they were not entirely heterosexual (Hayes et al. 2012).

**Transgender** generic term for a person of one biological sex who displays characteristics of the other sex.

The word **transgender** is a generic term for a person of one biological sex who displays characteristics of the other sex. Kuper et al. (2012) identified 292 transgendered individuals online. Most self-identified as gender queer (their gender identity was neither male nor female) and pansexual/queer (they were attracted to men, women, bisexuals) as their sexual orientation.

## National **Data**

An estimated 0.3% adults or about 700,000 individuals in the United States self-identify as transgender (Gates 2011).

**Transsexual** individual with the biological and anatomical sex of one gender but the self-concept of the other sex.

*I think that there is a high level of survival instinct in the trans culture in general. As transgender people, we have to be resilient. We have to be strong. Because when we say, "I am going ahead and making this transition," well, we know we could lose everything—our family, our children, our friends, our employment, our places of worship, our standing in the community. And in some cases, we could even lose our lives.*

CHRISTINE (PSEUDONYM), A TRANSGENDER WOMAN

**Transsexuals** are individuals with the biological and anatomical sex of one gender (e.g., male) but the self-concept of the other sex (e.g., female). In the Kuper et al. (2012) sample of 292 transgendereds, most did not desire to (or were unsure of their desire to) take hormones or undergo sexual reassignment surgery.

"I am a woman trapped in a man's body" reflects the feelings of the male-to-female transsexual (MtF), who may take hormones to develop breasts and reduce facial hair and may have surgery to artificially construct a vagina. Such a person lives full time as a woman. The female-to-male transsexual (FtM) is one who is a biological and anatomical female but feels "I am a man trapped in a woman's body." This person may take male hormones to grow facial hair and deepen her voice and may have surgery to create an artificial penis. This person lives full time as a man. Thomas Beatie, born a biological woman, viewed himself as a man and transitioned from living as a woman to living as a man. His wife, Nancy, had two adult children before having a hysterectomy (which rendered her unable to bear more children) so her husband agreed to be artificially inseminated (because he had ovaries), became pregnant, and delivered their baby. Because Thomas Beatie was legally a male, the media referred to him as "The Pregnant Man"; they asked him if he gave birth to the child, could he be the father? Technically, Oregon law defines birth as an expulsion or extraction from the mother, so Tom is the technical mother. However, the new parents could petition the courts and have Tom declared as the father and his wife as the mother (Heller 2008).

Individuals need not take hormones or have surgery to be regarded as transsexuals. The distinguishing variable is living full time in the role of the other biological sex. A man or woman who presents full time as the other gender is a transsexual by definition.

Johnson et al. (2014) noted the complexity of gender, sexuality, and sexual orientation issues as experienced by transgender, queer, and questioning individuals. One of the participants in their study who identified themselves as TQQ, explained:

> *I would consider myself to be bi-gendered or gender fluid. Which is probably like the most complicated thing or decision that I have ever made . . . because there aren't very many people that understand it. Being bi-gendered makes a lot more sense for me just because like my sexuality in general is just really fluid and it's really hard to identify myself in one particular box for very long at all 'cause it's always changing.*

**Cross-dresser** a broad term for individuals who may dress or present themselves in the gender of the other sex.

**Cross-dresser** is a broad term for individuals who may dress or present themselves in the gender of the other sex. Some cross-dressers are heterosexual adult

Chelsea Curry

"I was born a biological female but always knew I was a male" says this FtM transsexual. He has taken hormones and has a girlfriend.

males who enjoy dressing and presenting themselves as women. Cross-dressers may also be women who dress as men and present themselves as men. Cross-dressers may be heterosexual, homosexual, or bisexual.

The term **queer** is typically used by a male (but it could be used by a female) as a self- identifier to indicate that the person has a sexual orientation other than heterosexual (**genderqueer** means that the person does not identify as either male or female since they do not feel sufficiently like one or the other). Traditionally, the term *queer* was used to denote a gay person, and the connotation was negative. More recently, individuals have begun using the term queer with pride, much the same way African Americans called themselves black during the 1960s Civil Rights era as part of building ethnic pride and identity. Hence, LGBT people took the term queer, which was used to demean them, and started to use it with pride. The term also has shock value, which some people who identify strongly with being queer seem to savor when they introduce themselves as being queer.

**Queer** self-identifier to indicate that the person has a sexual orientation other than heterosexual.

**Genderqueer** the person does not identify as either male or female since they do not feel sufficiently like one or the other.

Queer theory refers to a movement or theory dating from the early 1990s. Queer theorists advocate for less labeling of sexual orientation and a stronger "anyone can be anything he or she wants" attitude. The theory is that society should support a more fluid range of sexual orientations and that individuals can move through the range as they become more self-aware. Queer theory may be seen as a very specific subset of gender and human sexuality studies.

**GLTB** refers collectively to lesbians, gays, bisexuals, and transgenered individuals.

The terms LGBT or **GLBT** are often used to refer collectively to gays, lesbians, transgendered individuals, and bisexuals. A more inclusive term today is LGBITQ (lesbian, gay, bisexual, intersexed, transgender, queer) or GLBITQ (gay, lesbian, bisexual, intersexed, transgender, queer). Some researchers regard the Q as meaning "questioning."

## Problems with Identifying and Classifying Sexual Orientation

The classification of individuals into sexual orientation categories (e.g., heterosexual, homosexual, bisexual) is problematic for several reasons. First, because of the social stigma associated with nonheterosexual identities, many individuals conceal or falsely portray their sexual-orientation identities to avoid prejudice and discrimination. There is the belief that "a gay person is obvious." But what are the data?

Lyons et al. (2014) investigated the accuracy of one's "gaydar" or the ability to identify sexual orientation by looking at a person. In a study of heterosexual (N = 80) and homosexual (N = 71) women who rated the faces of heterosexual/homosexual men and women, detection accuracy was better than chance but male targets were more likely to be falsely labeled as homosexual than were female targets.

*I'm not even kind of a lesbian.*

OPRAH WINFREY

Second, not all people who are sexually attracted to or have had sexual relations with individuals of the same-sex view themselves as homosexual or bisexual. For example, 11% of U.S. adults (26 million) report experiencing same-sex attraction, a percent five times greater than the number of self-identified gay/lesbian individuals (Gates 2011). Priebe and Svedin (2013) surveyed 3,432 Swedish high school seniors who completed an anonymous survey about sexuality. Prevalence rates of sexual minority orientation varied between 4.3% for sexual behavior (males 2.9%, females 5.6%) and 29.4% for emotional or sexual attraction (males 17.7%, females 39.5%). Hence, five times as many reported same-sex emotional/sexual attraction as actually engaged in same-sex sexual behavior.

A fourth difficulty in labeling a person's sexual orientation is that an individual's sexual attractions, behavior, and identity may change across time. One's sexual orientation is, indeed, fluid. For example, in a longitudinal study of 156 lesbian, gay, and bisexual youth, 57% consistently identified as gay or lesbian and 15% consistently identified as bisexual over a one-year period, but 18% transitioned from bisexual to lesbian or gay (Rosario et al. 2006). Maulsby et al. (2013) identified men who have sex with both men and women.

*It would encourage clearer thinking on these matters if persons were not characterized as heterosexual or homosexual, but as individuals who have had certain amounts of heterosexual experience and certain amounts of homosexual experience.*

ALFRED KINSEY, SEX RESEARCHER

In a study of 243 undergraduates, Vrangalova and Savin-Williams (2010) found that 84% of the heterosexual women and 51% of the heterosexual men reported the presence of at least one same-sex quality—sexual attraction, fantasy, or behavior. Kinsey suggested that heterosexuality and homosexuality represent two ends of a sexual-orientation continuum and that most individuals are neither entirely homosexual nor entirely heterosexual, but fall somewhere along this continuum (Drucker 2010). Ross et al. (2013) studied a sample of 652 men and 1,250 women and suggested that sexual orientation be conceptualized as heterosexual, homosexual, bisexual, and fluid. Six percent of men in the Ross et al. sample reported fluidity in terms of having sex with both women and men; 15% of the women reported having sex with both women and men. In regard to sexual fantasies, the fluidity percentages were much higher—15% of the men and 49% of the women.

The Heterosexual-Homosexual Rating Scale that Kinsey et al. (1953) developed allows individuals to identify their sexual orientation on a continuum.

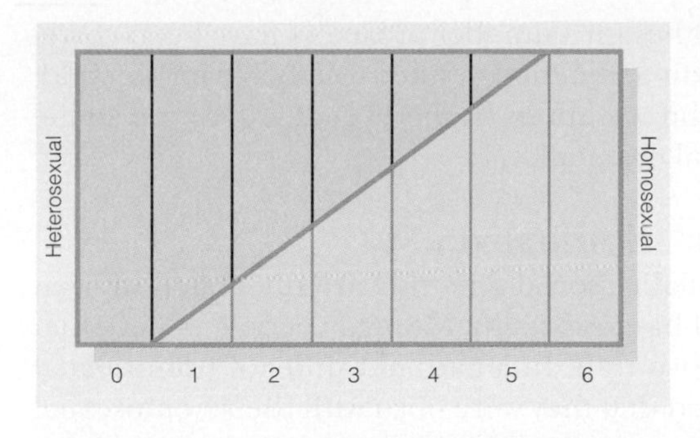


Based on both psychologic reactions and overt experience, individuals rate as follows:

0. Exclusively heterosexual with no homosexual
1. Predominantly heterosexual, only incidentallly homosexual
2. Predominantly heterosexual, but more than incidentally homosexual
3. Equally heterosexual and homosexual
4. Predominantly homosexual, but more than incidentally heterosexual
5. Predominantly homosexual, but incidentally heterosexual
6. Exclusively homosexual

**Figure 7.1** The Heterosexual-Homosexual Rating Scale.

Individuals with ratings of 0 or 1 are entirely or largely heterosexual; 2, 3, or 4 are more bisexual; and 5 or 6 are largely or entirely homosexual (see self-assessment). Very few individuals are exclusively a 0 or a 6, prompting Kinsey to believe that most individuals are bisexual.

Finally, sexual-orientation classification is also complicated by the fact that sexual behavior, attraction, love, desire, and sexual-orientation identity do not always match. For example, "research conducted across different cultures and historical periods (including present-day Western culture) has found that many individuals develop passionate infatuations with same-gender partners in the absence of same-gender sexual desires . . . whereas others experience same-gender sexual desires that never manifest themselves in romantic passion or attachment" (Diamond 2003, p. 173).

## National **Data**

There are 800,000 same-sex couples in the United States (Natale and Miller-Cribbs 2012). About 600,000 of these live together in a same-sex couple household (*Statistical Abstract of the United States*, 2012–2013, Table 63).

Why are data on the numbers of LGBT individuals and couples in the United States relevant? The primary reason is that census numbers on the prevalence of LGBT individuals and couples can influence laws and policies that affect gay individuals and their families. "The more we are counted, the more we count" is the slogan that points out the value of gay individuals being visible (emphasized in the movie *Milk* starring Sean Penn as Harvey Milk, a politically active gay man in San Francisco). Visibility and acceptance are increasing. In 2010, 107 openly gay candidates were elected to office . . . up a third from 2008. According to a Gallup Poll, almost 70% (67%) say they would vote for a president who is gay (Page 2011).

**DIVERSITY IN OTHER COUNTRIES**

Lyons et al. (2013) reported on 840 Australian gay men over the age of 60. They were more likely to be living alone than men in their 40s and 50s. Having good physical health, a satisfying sex life, feeling supported, and fewer experiences of discrimination were key factors in self-esteem and subjective well-being of these men.

# Origins of Sexual-Orientation Diversity

*Question: How do you know you are homosexual?*

*Answer: When you can't think straight.*

ANONYMOUS

Same-sex behavior has existed throughout human history. Much of the biomedical and psychological research on sexual orientation attempts to identify one or more "causes" of sexual-orientation diversity. The driving question behind this research is this: "Is sexual orientation inborn or is it learned or acquired from environmental influences?" Although a number of factors have been correlated with sexual orientation, including genetics, gender role behavior in childhood, fraternal birth order, and child sex abuse (Roberts et al. 2013), no single theory can explain diversity in sexual orientation.

## Beliefs about What "Causes" Homosexuality

*No, I've never thought that I was gay. And that's not something you think. It's something you know.*

ROBERT PLANT, ENGLISH MUSICIAN

Aside from what "causes" homosexuality, social scientists are interested in what people believe about the "causes" of homosexuality. Most gay people believe that homosexuality is an inherited, inborn trait. In a national study of homosexual men, 90% reported that they believed that they were born with their homosexual orientation; only 4% believed that environmental factors were the sole cause (Lever 1994). Overby (2014) analyzed Internet data of over 20,000 respondents (primarily heterosexual) and found that roughly half (52%) thought of homosexuality as being based primarily in "biological make-up" compared to 32% who saw sexual orientation as more of a lifestyle choice.

Individuals who believe that homosexuality is genetically determined tend to be more accepting of homosexuality and are more likely to be in favor of equal rights for lesbians and gays (Tyagart 2002). In contrast, "those who believe homosexuals choose their sexual orientation are far less tolerant of gays and lesbians and more likely to feel that homosexuality should be illegal than those who think sexual orientation is not a matter of personal choice" (Rosin and Morin 1999, p. 8).

Although the terms *sexual preference* and *sexual orientation* are often used interchangeably, the term *sexual orientation* avoids the implication that homosexuality, heterosexuality, and bisexuality are determined. Hence, those who believe that sexual orientation is inborn more often use the term *sexual orientation*, and those who think that individuals choose their sexual orientation use *sexual preference* more often. The term sexual identity is, increasingly, being used since it connotes more about the whole person than does sexuality.

## Can Homosexuals Change Their Sexual Orientation?

Individuals who believe that homosexual people choose their sexual orientation tend to think that homosexuals can and should change their sexual orientation. While about 2% of U.S. adults report that there has been a change in their sexual orientation (Mock and Eibach 2012), various forms of reparative therapy or **conversion therapy** are focused on changing homosexuals' sexual orientation. Some religious organizations sponsor "ex-gay ministries," which claim to "cure" homosexuals and transform them into heterosexuals by encouraging them to ask for "forgiveness for their sinful lifestyle" through prayer and other forms of "therapy." Serovich et al. (2008) reviewed 28 empirically based, peer-reviewed articles and found them methodologically problematic, which threatens the validity of interpreting available data on this topic.

**Conversion therapy** also called reparative therapy, focused on changing the sexual orientation of homosexuals.

Parelli (2007) attended private as well as group therapy sessions to change his sexual orientation from homosexuality to heterosexuality and noted that, for him, it did not work. He identified that reparative therapy focuses on outward behavioral change and does not acknowledge the inner yearnings:

> *By my mid-forties, I was experiencing a chronic need for appropriately affectionate male touch. It was so acute I could think of nothing else. Every cell of my body seemed relationally isolated and emotionally starved. Life was so completely and fatally ebbing out of my being that my internal life-saving system kicked in and put out a high-alert call for help. I desperately needed to be held by loving, human, male arms (p. 32).*

Other research brings into the question that reparative therapies "never work." Karten and Wade (2010) reported that the majority of 117 men who were dissatisfied with their sexual orientation and who sought sexual orientation change efforts (SOCE) were able to reduce their homosexual feelings and behaviors and increase their heterosexual feelings and behaviors. The primary motivations for their seeking change were religion ("homosexuality is wrong") and emotional dissatisfaction with the homosexual lifestyle. Being married, feeling disconnected from other men prior to seeking help, and feeling able to express nonsexual affection toward other men were the factors predictive of greatest change. Developing nonsexual relationships with same-sex peers was also identified as helpful in change. Jones and Yarhouse (2011) also provided data which suggest that religiously mediated sexual orientation change is possible. Better understanding the causes of homosexuality and one's homosexuality and one's emotional needs and issues as well as developing nonsexual relationships with same-sex peers were also identified as most helpful in change.

Critics of reparative therapy and ex-gay ministries emphasize that gay people are not the problem; social disapproval is the problem. The National Association for the Research and Therapy of Homosexuality (NARTH) has been influential in moving public opinion from "gays are sick" to "society is judgmental." The American Psychiatric Association, the American Psychological Association, the American Academy of Pediatrics, the American Counseling Association, the National Association of School Psychologists, the National Association of Social Workers, and the American Medical Association agree that homosexuality is not a mental disorder and needs no cure.

*Be who you are and say what you feel, because those who mind don't matter and those who matter don't mind.*

**DR. SEUSS**

# Heterosexism, Homonegativity, Homophobia, Biphobia, Binegativity, and Transphobia

The United States, along with many other countries throughout the world, is predominantly heterosexist. **Heterosexism** refers to "the institutional and societal reinforcement of heterosexuality as the privileged and powerful norm." Heterosexism is based on the belief that heterosexuality is superior to homosexuality. Of 4,549 undergraduates, 26% agreed (men more than women) with the statement "It is better to be heterosexual than homosexual" (Hall and Knox 2013). Heterosexism results in prejudice and discrimination against homosexual and bisexual people. The word *prejudice* refers to negative attitudes, whereas *discrimination* refers to behavior that denies equality of treatment for individuals or groups.

**Heterosexism** based on the belief that heterosexuality is superior to homosexuality.

Before reading further, you may wish to complete the Self-Assessment on page 188, which assesses the degree to which you are prejudiced toward gay men and lesbians.

*We struggled against apartheid because we were being blamed and made to suffer for something we could do nothing about. It is the same with homosexuality. The orientation is a given, not a matter of choice. It would be crazy for someone to choose to be gay, given the homophobia that is present.*

**DESMOND TUTU, SOUTH AFRICAN SOCIAL ACTIVIST**

## Attitudes toward Homosexuality, Homonegativity, and Homophobia

Adolfsen et al. (2010) noted that there are multiple dimensions of attitudes about homosexuality and identified five:

**1.** General attitude—Is homosexuality considered to be normal or abnormal? Do people think that homosexuals should be allowed to live their lives just as freely as heterosexuals?

**2.** Equal rights—Should homosexuals be granted the same rights as heterosexuals in regard to marriage and adoption?

**3.** Close quarters—Feelings in regard to having a gay neighbor or a lesbian colleague.

## SELF-ASSESSMENT | Sexual Prejudice Scale*

### Directions

The items below provide a way to assess one's level of prejudice toward gay men and lesbians. For each item identify a number from one to six which reflects your level of agreement and write the number in the space provided.

1 = strongly disagree
2 = disagree
3 = mildly disagree
4 = mildly agree
5 = agree
6 = strongly agree

### Gay Men Scale

_____ 1. You can tell a man is gay by the way he walks.
_____ 2. I think it's gross when I see two men who are clearly "together."
_____ 3. * Retirement benefits should include the partners of gay men.
_____ 4. Most gay men are flamboyant.
_____ 5. It's wrong for men to have sex with men.
_____ 6. *Family medical leave rules should include the domestic partners of gay men.
_____ 7. Most gay men are promiscuous.
_____ 8. Marriage between two men should be kept illegal.
_____ 9. *Health care benefits should include partners of gay male employees.
_____ 10. Most gay men have HIV/AIDS.
_____ 11. Gay men are immoral.
_____ 12. * Hospitals should allow gay men to be involved in their partners' medical care.
_____ 13. A sexual relationship between two men is unnatural.
_____ 14. Most gay men like to have anonymous sex with men in public places.
_____ 15. *There's nothing wrong with being a gay man.

### Scoring

Reverse score items 3, 6, 9,12, and 15. For example, if you selected a 6, replace the 6 with a 1. If you selected a 1, replace it with a 6, etc. Add each of the 15 items (from 1 to 6). The lowest possible score is 15, suggesting a very low level of prejudice against gay men; the highest possible score is 90, suggesting a very high level of prejudice against gay men. The midpoint between 15 and 90 is 52. Scores lower than this would reflect less prejudice against gay men; scores higher than this would reflect more prejudice against gay men.

### Participants

Both undergraduate and graduate students enrolled in social work courses comprised a convenience sample ($N = 851$). The sample was predominantly women (83.1%), white (65.9%), heterosexual (89.8%), single (81.3%), non-parenting (81.1%), 25 years of age or under (69.3%), and majoring in social work (80.8%).

### Results

The range of scores for the gay men scale was 15 to 84. The M = 31.53, SD=15.30. The sample had relatively low levels of prejudice against gay men.

### Lesbian Scale

_____ 1. Most lesbians don't wear makeup.
_____ 2. Lesbians are harming the traditional family.
_____ 3. *Lesbians should have the same civil rights as straight women.
_____ 4. Most lesbians prefer to dress like men.
_____ 5. *Being a lesbian is a normal expression of sexuality.
_____ 6. Lesbians want too many rights.
_____ 7. Most lesbians are more masculine than straight women.
_____ 8. It's morally wrong to be a lesbian.
_____ 9. *Employers should provide retirement benefits for lesbian partners.
_____ 10. Most lesbians look like men.
_____ 11. I disapprove of lesbians.
_____ 12. *Marriage between two women should be legal.
_____ 13. Lesbians are confused about their sexuality.
_____ 14. Most lesbians don't like men.
_____ 15. *Employers should provide health care benefits to the partners of their lesbian employees.

### Scoring

Reverse score items 3, 5, 9, 12 and 15. For example, if you selected a 6, replace the 6 with a 1. If you selected a 1, replace it with a 6, etc. Add each of the 15 items (from 1 to 6). The lowest possible score is 15, suggesting a very low level of prejudice against lesbians; the highest possible score is 90, suggesting a very high level of prejudice against lesbians. The midpoint between 15 and 90 is 52. Scores lower than this would reflect less prejudice against lesbians; scores higher than this would reflect more prejudice against lesbians.

### Participants

Both undergraduate and graduate students enrolled in social work courses comprised a convenience sample ($N = 851$). The sample was predominantly women (83.1%), white (65.9%), heterosexual (89.8%), single (81.3%), non-parenting (81.1%), 25 years of age or under (69.3%), and majoring in social work (80.8%).

### Results

The range of the scores for the lesbian scale was 15 to 86. M = 30.41, SD = 15.60. The sample had relatively low levels of prejudice against lesbians.

**Validity and Reliability**

Details are provided in the 2013 reference below.

*Scale is used with the permission of Jill Chonody, School of Psychology, Social Work and Public Policy. University of South Australia, Magill, South Australia jill.chonody@unisa.edu.au The scale is used with the permission of the *Journal of Homosexuality*.

**Reference**

Chonody, J. M. 2013. Measuring sexual prejudice against gay men and lesbian women: Development of the Sexual Prejudice Scale (SPS). *Journal of Homosexuality* 60: 895–926.

**4.** Public display—Reactions to a gay couple kissing in public.

**5.** Modern homonegativity—Feeling that homosexuality is accepted in society and that all kinds of special attention are unnecessary.

The term **homophobia** is commonly used to refer to negative attitudes and emotions toward homosexuality and those who engage in homosexual behavior. Homophobia is not necessarily a clinical phobia (that is, one involving a compelling desire to avoid the feared object despite recognizing that the fear is unreasonable). Other terms that refer to negative attitudes and emotions toward homosexuality include **homonegativity** (attaching negative connotations to being gay) and antigay bias.

*Do not follow where the path may lead. Go, instead, where there is no path and leave a trail.*

RALPH WALDO EMERSON, AMERICAN POET

**Homophobia** negative attitudes and emotions toward homosexuality and those who engage in homosexual behavior.

**Homonegativity** attaching negative connotations to being gay.

## International **Data**

Garrido et al. (2012) noted that Spain was one of the first countries to legalize same-sex marriage. Same-sex marriages are also recognized among citizens of Spain when one partner is from another country.

The Sex Information and Education Council of the United States (SIECUS) notes that "individuals have the right to accept, acknowledge, and live in accordance with their sexual orientation, be they bisexual, heterosexual, gay or lesbian. The legal system should guarantee the civil rights and protection of all people, regardless of sexual orientation" (SIECUS 2014). Nevertheless, negative attitudes toward homosexuality continue (Rutledge et al. 2012).

Characteristics associated with positive attitudes toward homosexuals and gay rights include younger age, advanced education, no religious affiliation, liberal political party affiliation, and personal contact with homosexual individuals (Lee & Hicks 2011). Women also tend to have more positive attitudes toward homosexuality (Wright and Bae 2013).

In one study, heterosexual women who kissed other women had more positive attitudes toward homosexuality (see Applying Social Research feature).

Negative social meanings associated with homosexuality can affect the self-concepts of LGBT individuals. **Internalized homophobia** is a sense of personal failure and self-hatred among lesbians and gay men resulting from social rejection and the stigmatization of being gay and has been linked to increased risk for depression, substance abuse and addiction, anxiety, and suicidal thoughts (Oswalt & Wyatt 2011; Ricks 2012; Rubinstein 2010). Being gay and growing up in a religious context that is antigay may be particularly difficult (Lalicha & McLaren 2010). Todd et al. (2013) conducted focus groups of LGBTQ individuals who confirmed that religion had a negative impact on them. One group member said, "Religion made me feel that something was wrong with me . . . it

**Internalized homophobia** a sense of personal failure and self-hatred among lesbians and gay men resulting from social rejection and the stigmatization of being gay.

## APPLYING SOCIAL RESEARCH | "I Kissed a Girl": Heterosexual Women who Report Same-Sex Kissing*

After a few drinks, it is not unusual that college-aged heterosexual females at a bar will kiss each other. Their motives are typically geared toward "entertainment for the guys," experimentation, or curiosity. The goal of this research was to identify correlations of such heterosexual female same-sex kissing on attitudes toward same-sex relationships. Specifically, are heterosexual females who kiss other females more likely to approve of same-sex marriage, approve of same-sex couples having their own children, and approve of same-sex couples adopting children?

### Methodology

Data for this study came from a sample of 436 undergraduate female student volunteers at two large southeastern universities who completed a 50-item questionnaire (approved by the Institutional Review Board) on "Attitudes toward Homosexuality of College Students." Respondents completed the questionnaire over the Internet.

The mean age of the respondents was 19.83. Over 40% of the respondents were freshmen (44.3%) with sophomores representing 33%, juniors, 13.2% and seniors, 9.5%. In comparing the responses of those heterosexual women who reported having kissed another woman (out of sexual experimentation/curiosity) with women who had not had this experience, cross-classification was conducted to determine any relationships by using Chi Square to assess statistical significance.

### Findings—Same-Sex Kissing

Almost half (47.9%) of the self-identified heterosexual women reported that they had kissed another woman out of sexual experimentation/curiosity. When the heterosexual women who reported same-sex kissing were compared with those who reported no such kissing, several statistically significant findings emerged in regard to pro gay attitudes.

1. *HSSK (heterosexual same-sex kissers) were less likely to think that homosexuality is immoral ($p < .000$).*
2. *HSSK were less likely to report getting sick if they watched two gay people being intimate ($p < .000$).*

made me feel worse about myself and become depressed." In contrast, McLaren et al. (2013) confirmed in a study of lesbian and gay men that involvement in the gay and lesbian community was related to fewer depressive symptoms.

Homophobia also results in gays staying in the closet (hiding their sexual orientation for fear of negative reactions/consequences). Such hiding may also have implications for gay interpersonal relationships. Easterling et al. (2012) found that gays were significantly more likely to keep a secret from their romantic partner than straights. The researchers hypothesized that gays learn early to keep secrets and that this skill slides over into their romantic relationships.

Gonzales (2014) noted that many LGBT people report "worse physical and mental health conditions than heterosexual and non-transgender populations, largely as a result of the stress caused by being a member of a stigmatized minority group or because of discrimination due to sexual orientation or gender nonconformity. Discriminatory environments and public policies stigmatize LGBT people and engender feelings of rejection, shame, and low self-esteem, which can negatively affect people's health-related behavior as well as their mental health. LGBT people living in states that ban same-sex marriage, for instance, are more likely than their counterparts in other states to report symptoms of depression, anxiety, and alcohol use disorder" (p. 1374).

## Biphobia/Binegativity/Transphobia

**Biphobia** a parallel set of negative attitudes toward bisexuality.

**Binegativity** same as biphobia.

Just as the term *homophobia* is used to refer to negative attitudes toward homosexuality, gay men, and lesbians, **biphobia** (also referred to as **binegativity**) refers to a parallel set of negative attitudes toward bisexuality/those identified as bisexual. Just as bisexuals are often rejected by heterosexuals (men are more rejecting than women) they are also rejected by many homosexual individuals (Yost & Thomas 2012). Thus, bisexuals experience "double discrimination."

3. *HSSK were less likely to report hating gay people ($p < .000$).*
4. *HSSK were more likely to be in favor of same-sex marriage ($p < .000$).*
5. *HSSK were more likely to be in favor of gay people having their own children ($p < .000$).*

### Discussion

Psychologist Gordon Allport (1954) asserted that contact between groups is necessary for the reduction of prejudice—an idea known as the **contact hypothesis**. In general, heterosexuals have more favorable attitudes toward gay men and lesbians if they have had prior contact with or know someone who is gay or lesbian (Mohipp & Morry 2004). Contact with openly gay individuals reduces negative stereotypes and ignorance and increases support for gay and lesbian equality (Wilcox and Wolpert 2000).

### References


Allport, G. W. 1954. *The nature of prejudice.* Cambridge, MA: Addison-Wesley.

Mohipp, C., and M. M. Morry. 2004. "Relationship of symbolic beliefs and prior contact to heterosexuals' attitudes toward gay men and lesbian women." *Canadian Journal of Behavioral Science* 36(1): 36–44.

Schiappa, E., P. B. Gregg, and D. E. Hewes. 2005. The parasocial contact hypothesis. *Communication Monographs* 72(1): 92–115.

Wilcox, C., and R. Wolpert. 2000. "Gay rights in the public sphere: Public opinion on gay and lesbian equality." In C. A. Rimmerman, Wald, K. D., & Wilcox, C. (eds.), *The politics of gay rights* (pp. 409–432). Chicago: University of Chicago Press.


*Abridged from a poster by Tiffany Beaver, David Knox, and Vaiva Kiskute, Southern Sociological Society Annual Meeting, Atlanta, April 2010. Complete article available in Knox, D, T. Beaver, and V. Kriskute. 2011. "I Kissed a Girl": Heterosexual women who report same-sex kissing (and more). *Journal of GLBT Family Studies* 7: 217–225.

Lesbians are a major source of negative views toward bisexuals. In a study of 346 self-identified lesbians, Rust (1993) found that lesbians view bisexuals as less committed to other women than are lesbians, perceive bisexuals as disloyal to lesbians, and resent bisexual women who have close relationships with the "enemy" of lesbians—men. Some negative attitudes toward bisexual individuals

**Contact hypothesis** contact between groups is necessary for the reduction of prejudice.

This bisexual couple feel acceptance of their diversity with each other.

Chelsea Curry

"are based on the belief that bisexual individuals are really lesbian or gay individuals who are in transition or in denial about their true sexual orientation" (Israel and Mohr 2004, p.121).

**Transphobia** negative attitudes toward transsexuality or those who self-identify as transsexual.

Transsexuals are targets of **transphobia**, a set of negative attitudes toward transsexuality or those who self-identify as transsexual. Singh et al. (2011) interviewed 21 transsexuals and found various strategies of coping with the discrimination/being resilient—a positive definition of self ("I am proud of who I am"), connecting with a supportive trans community, and cultivating hope for the future. In regard to the latter, one transsexual noted, "I have a right to be here and hope for a better future" (p. 24).

## Effects of Antigay/Trans Bias and Discrimination on Heterosexuals

The antigay and heterosexist social climate of our society is often viewed in terms of how it victimizes the gay population. However, heterosexuals are also victimized by heterosexism and antigay prejudice and discrimination. Some of these effects follow:

**Hate crimes** violence against an individual on the basis of the persons sexual orientation or gender identification.

1. ***Heterosexual victims of hate crimes.*** Extreme homophobia contributes to instances of violence against homosexuals—acts known as **hate crimes**. Such crimes include verbal harassment (the most frequent form of hate crime experienced by victims), vandalism, sexual assault and rape, physical assault, and murder. Hate crimes also target transsexuals (moreso than gay individuals).

Because hate crimes are crimes of perception, victims may not be homosexual; they may just be perceived as being homosexual. The National Coalition of Anti-Violence Programs (2012) reported that, in 2011, heterosexual individuals in the United States were victims of antigay hate crimes, representing 15% of all antigay hate crime victims.

2. ***Concern, fear, and grief over well-being of gay or lesbian family members and friends.*** Many heterosexual family members and friends of homosexual people experience concern, fear, and grief over the mistreatment of their gay or lesbian friends and/or family members. In 2011, there were 30 murders of LGBTQH (lesbian, gay, bisexual, transgender, queer, HIV-infected individuals) (National Coalition of Anti-Violence Programs 2012). Heterosexual parents who have a gay or lesbian teenager often worry about how the harassment, ridicule, rejection, and violence experienced at school might affect their gay or lesbian child. Will their child be traumatized/make bad grades or drop out of school to escape the harassment, violence, and alienation they endure there? Will the gay or lesbian child respond to the antigay victimization by turning to drugs or alcohol? Newcomb et al. (2014) compared national samples of gay and straight youth and found higher drug use among sexual minorities. Van Bergen et al. (2013) also found higher suicide rates. Higher rates of anxiety, depression, and panic attacks are associated with being gay (Oswalt and Wyatt 2011).

To heterosexuals who have lesbian and gay family members and friends, lack of family protections such as health insurance and rights of survivorship for same-sex couples can also be cause for concern. Finally, heterosexuals live with the painful awareness that their gay or lesbian family member or friend is a potential victim of antigay hate crime. Jamie Nabozny was a gay high school student in Ashland, Wisconsin, who was subjected to relentless harassment and abuse over a four-year period. He attempted suicide and dropped out of school. With his parents' help he sued the school, lost, and appealed in federal court. The school administration was held liable for his mistreatment, and the case was settled for close to a million dollars (see the documentary *Bullied*). Transsexuals also have no protection by the federal government in the workplace.

3. ***Restriction of intimacy and self-expression.*** Because of the antigay social climate, heterosexual individuals, especially males, are hindered in their own

National **Data**

Only twenty-nine states have laws banning discrimination based on sexual orientation (Human Rights Campaign 2014- see http://www.hrc.org/laws-and-legislation/federal-legislation/employment-non-discrimination-act).

self-expression and intimacy in same-sex relationships. Males must be careful about how they hug each other so as not to appear gay. Homophobic scripts also frighten youth who do not conform to gender role expectations, leading some youth to avoid activities—such as arts for boys, athletics for girls—and professions such as elementary education for males.

4. ***Dysfunctional sexual behavior.*** Some cases of rape and sexual assault are related to homophobia and compulsory heterosexuality. For example, college men who participate in gang rape, also known as "pulling a train," entice each other into the act "by implying that those who do not participate are unmanly or homosexual" (Sanday 1995, p. 399). Homonegativity also encourages early sexual activity among adolescent men. Adolescent male virgins are often teased by their male peers, who say things like "You mean you don't do it with girls yet? What are you, a fag or something?" Not wanting to be labeled and stigmatized as a "fag," some adolescent boys "prove" their heterosexuality by having sex with girls.

5. ***School shootings.*** Antigay harassment has also been a factor in many of the school shootings in recent years. For example, 15-year-old Charles Andrew Williams fired more than 30 rounds in a San Diego, California, suburban high school, killing two and injuring 13 others. A woman who knew Williams reported that the students had teased him and called him gay.

# GLBT and Mixed-Orientation Relationships

Research suggests that gay and lesbian couples tend to be more similar to than different from heterosexual couples (Kurdek 2005; 2004). However, Muraco et al. (2012) studied sexual orientation and romantic relationship quality in a sample of 15,534 adolescents and found that relationship quality was influenced by gender and age with older gay women reporting higher relationship quality. In the following section we look more closely at the similarities as well as the differences between heterosexual, gay male, and lesbian relationships in regard to relationship satisfaction, conflict and intimate partner violence, and monogamy and sexuality. We also look at relationship issues involving bisexual individuals and mixed-orientation couples.

Chelsea Curry

This lesbian couple is in a stable monogamous relationship.

## Relationship Satisfaction

In a review of literature on relationship satisfaction and sexual orientation, Kurdek (1994) concluded, "The most striking finding regarding the factors linked to relationship

## SELF-ASSESSMENT | Scale for Assessing Same-Gender Couples

### Directions

Couples often have good and not-so-good moments in their relationship. This measure was developed to get an objective view of your relationship. Thinking about your relationship with your partner, put a number between 0 and 6 in the black space before each item to indicate your level of agreement with the item.

| Strongly Disagree | Disagree | Somewhat Disagree | Neutral | Somewhat Agree | Agree | Strongly Agree |
|---|---|---|---|---|---|---|
| 0 | 1 | 2 | 3 | 4 | 5 | 6 |

_____ *1. There are some things about my partner that I do not like.
_____ *2. I wish my partner enjoyed more of the activities that I do.
_____ 3. My mate has the qualities I want in a partner.
_____ 4. My partner and I share the same values and goals in life.
_____ 5. My partner and I have an active social life.
_____ 6. My partner's sociability adds a positive aspect to our relationship.
_____ 7. If there is one thing that my partner and I are good at, it's talking about our feelings with each other.
_____ 8. Our differences of opinion lead to shouting matches.
_____ *9. I would lie to my partner if I thought it would "keep the peace."
_____ 10. During our arguments, I never put down my partner's point of view.
_____ 11. When there is a difference of opinion, we try to talk it out rather than fight.
_____ 12. We always do something to mark a special day in our relationship, like our anniversary.
_____ 13. I often tell my partner than I love him/her.
_____*14. Sometimes sex with my partner seems more like work than play to me.
_____ 15. I always seem to be in the mood for sex when my partner is.
_____*16. My partner sometimes turns away from my sexual advances.
_____*17. My family accepts my relationship with my partner.
_____ 18. My partner's family accepts our relationship.
_____ 19. My family would support our decision to adopt or have children.
_____ 20. My partner's family would support our decision to adopt or have children.
_____ 21. I feel as though my relationship is generally accepted by my friends.
_____ 22. I have a strong support system that accepts me as I am.
_____ 23. I have told my co-workers about my sexual orientation/attraction.
_____ 24. Most of my family members know about my sexual orientation/attraction.

### Scoring

Reverse score items **1, 2, 9, 14, 16, and 17**. For example, if you wrote down a 0 for item one, replace it with a 6. If you wrote a 1, replace it with a 5, etc. After reverse scoring the items, add them. Scores range from 0 to 144 with lower scores reflecting less contentment and more public disapproval and higher scores reflecting greater contentment and higher public approval.

### Norms

107 was the average score of individuals who took the scale.

### Source

Developed by Christopher K. Belous, Ph.D. and used with his permission. Dr. Belous is Assistant Professor and Clinic Director, Marriage and Family Therapy Program, Mercer Family Therapy Center, Mercer University-Atlanta, 1938 Peachtree Road, Suite 107, Atlanta, GA 30309. For additional use of this scale, contact Dr. Belous at Belous_ck@mercer.edu.

satisfaction is that they seem to be the same for lesbian couples, gay couples, and heterosexual couples" (p. 251). These factors include having equal power and control, being emotionally expressive, perceiving many attractions and few alternatives to the relationship, placing a high value on attachment, and sharing decision making. Kamen and colleagues (2011) also noted that commitment, trust, and support from one's partner were related to relationship satisfaction in same-sex relationships. In a comparison of relationship quality of cohabitants over a 10-year period involving both partners from 95 lesbian, 92 gay male, and 226 heterosexual couples living without children, and both partners from 312 heterosexual couples living with children, the researcher found that lesbian couples showed the highest levels of relationship quality averaged over all assessments (Kurdek 2008). This finding is not surprising. Females love deeply and are more focused on developing and maintaining good relationships. Issues unique to gay couples include if, when, and how to disclose their relationships to others and how to develop healthy intimate relationships in the absence of same-sex relationship models.

Individuals who identify as bisexual have the ability to form intimate relationships with both sexes. However, research has found that the majority of bisexual women and men tend toward primary relationships with the other sex (McLean 2004). Contrary to the common myth that bisexuals are, by definition, nonmonogamous, some bisexuals prefer monogamous relationships (especially in light of the widespread concern about HIV). Some gay and bisexual men have monogamous relationships in which both men have agreed that any sexual activity with casual partners must happen when both members of the couple are present and involved (e.g., three-ways or group sex) (Parsons et al. 2013).

## Conflict and Intimate Partner Violence

Gay and lesbian couples tend to disagree about the same issues that heterosexual couples argue about—finances, affection, sex, being overly critical, driving style, and household tasks (Kurdek 2004). However, same-sex couples and heterosexual couples tend to differ in how they resolve conflict. Gay and lesbian partners begin their discussions more positively and maintain a more positive tone throughout the discussion than do partners in heterosexual marriages (Gottman et al. 2003). One explanation is that same-sex couples value equality more and are more likely to have equal power and status in the relationship than are heterosexual couples. In addition, same-sex partners may have a higher motivation to reduce conflict because dissolution may involve losing a tight friendship network (Van Eeden-Moorefield et al. 2011).

Some same-sex relationships involve interpersonal violence. Finneran and Stephenson (2014) analyzed national Internet data from 1,575 men who had had sex with men. Nine percent reported that they had experienced some form of physical violence from a male partner in the previous 12 months. However, only 4.32% reported perpetrating physical violence. Sexual violence was less prevalent, with 4% of the men reporting such an experience in the past year (initiated by a partner). Less than 1% said that they had initiated sexual violence.

## Sexuality

### National **Data**

Oswalt and Wyatt (2013) reported data on 25,553 heterosexual, gay, lesbian, bisexual, and unsure college students throughout the United States and found that being uncertain about one's sexual orientation was significantly associated with having more sexual partners than gay, bisexual, and heterosexual men. Heterosexual men had significantly fewer partners than gay, bisexual, and unsure men with bisexual women reporting having had significantly more partners than females of other sexual orientations.

Like many heterosexual women, most gay women value stable, monogamous relationships that are emotionally as well as sexually satisfying. Gay and heterosexual women in U.S. society are taught that sexual expression should occur in the context of emotional or romantic involvement.

There are stereotypes and assumptions about what sexual behaviors various categories of lesbians engage in. Walker et al. (2012) studied a sample of 214 women who self-identified as lesbian to explore the relationship between lesbian labels (butch, soft butch, butch/femme, femme, and high femme) and attraction to sexual behavior (being on top, etc.). They found no relationship between the label and the sexual behavior. The researchers emphasized that sexual behaviors in the lesbian community are variable across labels.

A common stereotype of gay men is that they prefer casual sexual relationships with multiple partners versus monogamous, long-term relationships. However, although gay men report greater interest in casual sex than do heterosexual men (Gotta et al. 2011), most gay men prefer long-term relationships, and sex outside the primary relationship is usually infrequent and does not involve emotions (Green et al. 1996).

*The whole world goes on and on about love. Poets spend their lives writing about it. Everyone thinks it's the most wonderful thing. But, when you mention two guys in love, they forget all that and freak out.*

MARK A. ROEDER, *OUTFIELD MENACE*

The degree to which gay males engage in casual sexual relationships is better explained by the fact that they are male than by the fact that they are gay. In this regard, gay and straight men have a lot in common: They both tend to have fewer barriers to engaging in casual sex than do women (heterosexual or lesbian). However, Starks and Parsons (2014) found that males who reported secure adult attachment with their partners also reported higher levels of sexual communication with their partners, higher frequencies of sex with their partners, and fewer outside sexual partners.

One way that uninvolved gay men meet partners is through the Internet. Blackwell and Dziegielewski (2012) studied men who seek men for sex on the Internet and noted that these sites promote higher-risk sexual activities. "Party and play" (PNP) is one such activity and involves using crystal methamphetamine and having unprotected anal sex. More scholarly inquiry is needed on the extent of this phenomenon.

Such nonuse of condoms results in the high rate of human immunodeficiency virus (HIV) infection and acquired immunodeficiency syndrome (AIDS). Approximately 50,000 new cases of HIV are reported annually, and male-to-male sexual contact is the most common mode of transmission in the United States (29,700 infections annually) (Centers for Disease Control and Prevention 2012). Women who have sex exclusively with other women have a much lower rate of HIV infection than do men (both gay and straight) and women who have sex with men. Many gay men have lost a love partner to HIV infection or AIDS; some have experienced multiple losses.

*Homophobia is a social disease.*

BUMPER STICKER

## Love and Sex

Rosenberger et al. (2014) analyzed data from 24,787 gay and bisexual men who were members of online websites facilitating social or sexual interactions with other men. Over half of those who completed the questionnaire (61.4%) reported that they did not love their sexual partner during their most recent sexual encounter (28.3% reported being in love with their most recent sexual partner). Hence, most of these men did not require love as a context to having sex. However, they were mostly accurate in identifying their own feelings and those of their partners. Over 90% (91%) were matched by presence ("I love him/he loves me"), absence ("I don't love him/he doesn't love me"), or uncertainty ("I don't know if I do/I don't know if he does") of feelings of love with their most recent sexual partner.

## Division of Labor

Sutphina (2010) studied the division of labor patterns reported by 165 respondents in same-sex relationships. Partners with greater resources (e.g., income, education) performed fewer household tasks. Satisfaction with division of labor and sense of being appreciated for one's contributions to household tasks were positively correlated with overall relationship satisfaction. Esmaila (2010) noted that lesbians struggle to maintain a sense of fairness in regard to the division of labor. This includes invoking how heterosexual couples often slip into an unequal division of labor and how gay couples try to raise the bar on this issue.

## Mixed-Orientation Relationships

Mixed-orientation couples are those in which one partner is heterosexual and the other partner is lesbian, gay, or bisexual. Reinhardt (2011) studied a sample of bisexual women in heterosexual relationships and found that these women maintained a satisfactory relationship with the male partner. One half of these women were maintaining sexual relationships with other women while in primary heterosexual relationships. Their average sexual contact with another female was

1½ times per month. The majority of the women were satisfied with their sexual experiences with their male partner, and they were having sexual intercourse, on average, 3 times per week. Moss (2012) interviewed a sample of married bisexual women in a polyamorous relationship. These women recognized the non-normativeness of their relationships but were able to negotiate their needs with a supportive partner.

Gay and bisexual men are also in heterosexual relationships. Peter Marc Jacobson is gay and was married to Fran Drescher (actress in *The Nanny*) for 21 years. They report a wonderful relationship/friendship during and after the marriage. In a study of 20 gay or bisexual men who had disclosed their sexual orientation to their wives, most of the men did not intentionally mislead or deceive their future wives with regard to their sexuality. Rather, they did not fully grasp their feelings toward men, although they had a vague sense of their same-sex attraction prior to the marriage (Pearcey 2004). The majority of the men in this study (14 of 20) attempted to stay married after disclosure of their sexual orientation to their wives, and nearly half (9 of 20) stayed married for at least three years. The Straight Spouse Network (www.straightspouse.org) provides support to heterosexual spouses or partners, current or former, of GLBT mates.

Chelsea Curry

Sometimes a person of one sexual orientation becomes involved with a person of another sexual orientation. This woman is straight and the male is gay.

Some males are on the down low. Men on the **down low** (non-gay-identifying men who have sex with men and women) meet their partners out of town, not in predictable contexts—e.g., park—or on the Internet (Schrimshaw et al. 2013). There is considerable disruption in the marriage if their down low behavior is discovered by the straight spouse.

**Down low** non-gay-identifying men who have sex with men and women meet their partners out of town, not in predictable contexts, or on the Internet.

## Transgender Relationships

While a great deal is known about transgender individuals, little is known about transgender relationships. The exception is the research by Iantiffi and Bockting (2011) who reported on an Internet study of 1,229 transgender individuals over the age of 18 living in the United States. Fifty-seven percent were transwomen (MtF); 43%, transmen (FtM). The median age was 33 and most were white (80%). The average income was $32,000. Almost 70% (69.2) of the transwomen were living together; 57% of the transmen. While those who were living together were in a relationship, another 15% of the transwomen were in a couple relationship, but not living together; 29% of the transmen—hence around 85% of the transwomen and transmen were involved in a relationship. In regard to sexual fidelity 65% of the transwomen were monogamous (26% were not); 70% of the transmen (21% were not). Some of the partners of the trans individuals did not know their partner was trans—8% of transwomen and 2% of transmen. Transgender individuals live in fear they will not be accepted. Twenty-five percent of transwomen and 9% of transmen reported that they were afraid to tell their partner they were trans (which is required to a future marriage partner in Europe) (Sharpe 2012). Having sex in the dark/giving oral sex were strategies used to avoid discovery that one's genitals do not match their

*Let's get one thing straight, I'm not.*

BUMPER STICKER

presentation of self (if they had not had surgery). Indeed, the whole subject of the feminine/masculine parts of one's body is a subject to be avoided (56% of transwomen; 51% of transmen). Sexual orientation attractions were predictable. Transgendered individuals who were mostly lesbian, gay, or bisexual in sexual orientation were mostly attracted to women, men, and both sexes.

What is it like to be a spouse of a partner who is transitioning? Chase (2011) interviewed partners of transgendered individuals. One wife commented on what it was like to watch her male to female transgendered husband transition . . .

> *He's more girly than I am! Which [chuckles] is kind of frustrating at times. You know, he'll want to wear all these bracelets and you know jewelry and stuff. Oh, I just think sometimes he kind of goes overboard on it. And he'll always ask, I mean he asks me a lot my opinion on what he wears, you know, if I think this is going to look good or not and things like that. So, he's always asking about that, does this go with this and everything. [And] he's probably bought more makeup through the years than I have (p. 436).*

## Coming Out to Parents, Peers, at School, and at Work

> *My feelings for Ellen overrode all of my fear about being out as a lesbian. I had to be with her, and I just figured I'd deal with the other stuff later.*
>
> PORTIA DE ROSSI, PARTNER OF ELLEN DEGENERES

Coming out is a major decision with which GLBT individuals struggle. Svab and Kuhar (2014) studied the coming out process of 443 gay men and lesbians. Over three-fourths (77%) came out first to their friends in comparison to 7% who first came out to their mother, 5% to their brother/sister, and 3% to their father. Seventy-four percent of the surveyed gay men and lesbians reported experiencing a positive and supportive reaction to their first coming out, 18% reported neutral, and 4% negative reactions. However, some respondents interpreted silence and non-reaction as a form of a positive or at least a neutral reaction because they had expected a much worse response.

Parents of gay children often discover very positive outcomes from their child's "coming out." Gonzalez et al. (2013) interviewed 142 parents of LGVTQ children and identified five primary themes: personal growth (open

Chelsea Curry

Parents vary in their reaction to the knowledge that their son or daughter is gay. These parents embraced the openness of their son and made clear their love for him.

mindedness, new perspectives, awareness of discrimination, and compassion), positive emotions (pride and unconditional love), activism, social connection, and closer relationships (closer to child and family closeness).

## PERSONAL CHOICES

### Risks and Benefits of "Coming Out"

In a society where heterosexuality is expected and considered the norm, heterosexuals do not have to choose whether or not to tell others that they are heterosexual. However, decisions about "**coming out**," or being open and honest about one's sexual orientation and identity (particularly to one's parents), are agonizing for LGBT individuals. As noted above, friends, mothers, siblings, and fathers, in that order, are those to whom disclosure is first made (Svab and Kuhar 2014).

**Coming out** being open and honest about one's sexual orientation and identity.

### Risks of Coming Out

Whether GLBT individuals come out is influenced by the degree to which they are tired of hiding their sexual orientation, the degree to which they feel more "honest" about being open, their assessment of the risks of coming out, and their prediction of how others will respond. Some of the risks involved in coming out include disapproval and rejection by parents and other family members, harassment and discrimination at school, discrimination and harassment in the workplace, and hate crime victimization.

**1. *Parental and family members' reactions.*** Rothman et al. (2012) studied 177 LBG individuals who reported that two-thirds of the parents to whom they first came out responded with social and emotional support. Their research is in contrast to that of Mena and Vaccaro (2013), who interviewed 24 gay and lesbian youth about their coming out experience to their parents. All reported a less than 100% affirmative "we love you"/"being gay is irrelevant" reaction which resulted in varying degrees of sadness or depression (3 became suicidal).

Svab and Kuhar (2014) identified the concept of the "transparent closet" to describe a situation in which parents are informed about a child's homosexuality but do not talk about it . . . a form of rejection. The "family closet" refers to the wider kinship system having knowledge of a child's homosexuality but "keeping it quiet" (a form of rejection).

Padilla et al. (2010) found that parental reaction to a son or daughter coming out had a major effect on the development of their child. Acceptance had an enormous positive effect. When GLBT individuals in their study come out to their parents, parental reactions range from "I already knew you were gay and I'm glad that you feel ready to be open with me about it" to "get out of this house, you are no longer welcome here." We know of a father who responded to the disclosure of his son, "I'd rather have a dead son than a gay son." Parental rejection of GLBT individuals is related to suicide ideation and suicide attempts (Van Bergen et al. 2013).

Because black individuals are more likely than white individuals to view homosexual relations as "always wrong," African Americans who are gay or lesbian are more likely to face disapproval from their families than are white lesbians and gays (Glass 2014). The Resource Guide to Coming Out for African Americans (2011) is a useful model. Because most parents are heavily invested in their children, they find a way not to make an issue of their son or daughter being gay. "We just don't talk about it," said one parent.

Parents and other family members can learn more about homosexuality from the local chapter of Parents, Families, and Friends of Lesbians and Gays (PFLAG) and from books and online resources, such as those found at Human Rights Campaign's National Coming Out Project (http://www.hrc.org/). Mena and Vaccaro (2013) emphasized the importance of parents educating themselves about gay/lesbian issues and to know the importance of their loving and accepting their son or daughter at this most difficult time.

**2.** ***Harassment and discrimination at school.*** LGBT students are more vulnerable to being bullied, harassed, and discriminated against. The negative effects are predictable including "a wide range of health and mental health concerns, including sexual health risk, substance abuse, and suicide, compared with their heterosexual peers" (Russell et al. 2011). The U.S. Department of Health (2012) published new guidelines to be sensitive to the needs of and to protect GLBT youth. GLBT individuals are often targets of discrimination and bullying. Gilla et al. (2010) found evidence that gays are subjected to negative comments and name calling in contexts such as physical education classes or physical activities where they are perceived to be small or gay. Some communities offer charter schools which are "GLBT friendly" and which promote tolerance. Google "charter schools for GLBT youth" for examples of such schools offering a safe haven from the traditional bullying. About a third of the students at the Arts and College Preparatory Academy in Ohio identify with the GLBT community. Parental acceptance of a child who is LGBTQ is also important in buffering the negative effects on a child who is bullied at school due to their sexual orientation and difference (Russell 2013).

**3.** ***Discrimination and harassment at the workplace.*** The workplace continues to be a place where the 8 million LGBT individuals experience discrimination and harassment. Specifically, gay men are paid less than heterosexual men, LGBT individuals feel their potential for promotion is less than the heterosexual majority, and many remain closeted for fear of retribution. There is no federal law that explicitly prohibits sexual orientation and gender identity discrimination against LGBTs (Pizer et al. 2012). McIntyre et al. (2014) noted that being a racial minority and a sexual minority is associated with more stigma and psychological distress than just being a sexual minority.

**4.** ***Hate crime victimization.*** Another risk of coming out is being victimized by antigay hate crimes against individuals or their property that are based on bias against the victim because of his/her perceived sexual orientation. Such crimes include verbal threats and intimidation, vandalism, sexual assault and rape, physical assault, and murder. "Homosexuals are far more likely than any other minority group in the United States to be victimized by violent hate crime" (Potok 2010, p. 29).

## *Benefits of Coming Out*

Coming out to parents is associated with decided benefits. D'Amico and Julien (2012) compared 111 gay, lesbian, and bisexual youth who disclosed their sexual orientation to their parents with 53 who had not done so. Results showed that the former reported higher levels of acceptance from their parents, lower levels of alcohol and drug consumption, and fewer identity and adjustment problems. Similarly, Rothman et al. (2012) noted that for lesbian and bi females (not males), higher levels of illicit drug use, poorer self-reported health status, and being more depressed were associated with non-disclosure to parents.

One of the strongest contexts against coming out is that of professional football. Defensive lineman Michael Sam is an example of an openly gay male in this macho context. Since U.S culture is becoming more accepting, gays in all sports, including football, will eventually (and increasingly) come out.

The basic trajectory of reacting to a partner's homosexuality is not unlike learning of the death of one's beloved. Life as one knew it is altered. The stages of this transition are shock, disbelief, numbness, and mourning for the partner/life that was, readjustment, and moving on. The last two stages may involve staying in the relationship with the gay partner (the choice of about 15% of straight spouses) or divorcing/ending the relationship (the choice of 85%).

# Same-Sex Marriage

By a 5-4 decision, the Supreme Court declared DOMA (The Defense of Marriage Act) unconstitutional on equal protection grounds, thus giving same-sex married couples federal recognition and benefits (Wolf and Heath 2013). Over half (51%) of U.S. adults favor same-sex marriage. Almost three-quarters (72%) of 1,504 adults in a Pew Research Center poll say that legal same-sex marriage is "inevitable." This percent includes 85% of gay marriage supporters, as well as 59% of its opponents (Pew Research Center 2013). Hence, even those against gay marriage feel that it will be a legal option for gay couples. With the Supreme Court ruling in 2013, 800,000 legally married same-sex couples were no longer denied federal access to marriage benefits.

*I've just concluded that for me personally it is important for me to go ahead and affirm that I think same-sex couples should be able to get married.*

BARACK OBAMA

As of April 2014, 16 states (Connecticut, California, Delaware, Iowa, Maine, Maryland, Massachusetts, Minnesota, New Hampshire, New York, Rhode Island, Vermont, Washington D.C., New Jersey, Hawaii, New Mexico) have legalized same-sex marriage. Thirty-three states have expressly prohibited same-sex marriage by amending their constitution: "Marriage between one man and one woman is the only domestic legal union that shall be valid or recognized in this state." Hence, a same sex-couple married in one of the 16 states is no longer married when they move to one of the states which has not legalized same-sex marriage.

With the Supreme Court ruling in 2013 (U.S. versus Windsor), the federal government recognizes same-sex marriages (e.g., the survivor of a legal same-sex marriage can collect Social Security benefits as is true of a spouse in a traditional heterosexual marriage). No longer in effect is the **Defense of Marriage Act** (DOMA) (passed in 1996), which states that marriage is a "legal union between one man and one woman" and denies federal recognition of same-sex marriage. In 2014 the Justice Department affirmed that it will extend equal benefits to spouses in same-sex marriages. For example, spouses are not required to testify against each other in court, and spouses now have visitation rights for a spouse in prison, etc. (Doaring 2014).

**Defense of Marriage Act** states that marriage is a "legal union between one man and one woman" and denies federal recognition of same-sex marriage.

However, same-sex marriage remains a hotly contested issue in many states and political candidates are careful about their position in regard to same-sex relationships for fear of losing votes. After Obama announced his support for gay marriage, a quarter (25%) of Americans reported that they felt less favorable toward him because of his approval; 19% percent felt more favorable toward him (Pew Research Center 2012a).

**DIVERSITY IN OTHER COUNTRIES**

Same-sex marriage has been legal in Canada since 2005. Humble (2013) interviewed 28 individuals (including lesbians, gay men, bisexuals) aged 26 to 72 about their wedding experience. Overwhelmingly, the respondents reported a positive response from family, friends, and the community. The one caveat is that they felt that gay weddings were absent in wedding magazines.

## Same-Sex Marriage Legislation

In 2013, 53% of U.S. adults supported gay marriage. Where disapproval exists, the primary reason is morality. Gay marriage is viewed as "immoral, a sin, against the Bible." However, support for gay marriage varies by age; about half of young adults (18 to 29) support gay marriage. Moskowitz et al. (2010) also found that men are more supportive of lesbian marriage than gay male marriage.

## Arguments in Favor of Same-Sex Marriage

A major argument for same-sex marriage is that it would promote relationship stability among gay and lesbian couples. "To the extent that marriage provides status, institutional support, and legitimacy, gay and lesbian couples, if allowed to marry, would likely experience greater relationship stability" (Amato 2004, p. 963).

*I did want to get married while I was still in office. I think it's important that my colleagues interact with a married gay man.*

BARNEY FRANK, RETIRING DEMOCRATIC CONGRESSMAN, ON MARRYING HIS LONGTIME PARTNER JIM READY

Indeed, same-sex relationships, like cohabitation relationships, end at a higher rate than marriage relationships (Wagner 2006). This higher rate is attributed to the lack of institutional support. Supreme Court recognition of same-sex marriage will likely result in an increase of more stable unions.

Advocates of same-sex marriage argue that banning or refusing to recognize same-sex marriages granted in some states is a violation of civil rights that denies same-sex couples the many legal and financial benefits that are granted to heterosexual married couples. Rights and benefits accorded to married people include the following:

- The right to inherit from a spouse who dies without a will;
- The benefit of not paying inheritance taxes upon the death of a spouse;
- The right to make crucial medical decisions for a spouse and to take care of a seriously ill spouse or a parent of a spouse under current provisions in the federal Family and Medical Leave Act;
- The right to collect Social Security survivor benefits; and
- The right to receive health insurance coverage under a spouse's insurance plan.

Other rights bestowed on married (or once-married) partners include assumption of a spouse's pension, bereavement leave, burial determination, domestic violence protection, reduced-rate memberships, divorce protections (such as equitable division of assets and visitation of partner's children), automatic housing lease transfer, and immunity from testifying against a spouse. All of these advantages are now available to same-sex married couples because of the Supreme Court decision.

Positive outcomes for being married as a gay couple have been documented. Ducharme and Kollar (2012) evaluated a sample of 225 lesbian married couples in Massachusetts who reported physical, psychological, and financial well-being in their relationships. The researchers noted that these data support the finding in the heterosexual marriage literature that healthy marriage is associated with distinct well-being benefits for lesbian couples. Wright et al. (2013) found that same-sex married lesbian, gay, and bisexual persons were significantly less distressed than lesbian, gay, and bisexual persons not in a legally recognized relationship. However, Ocobock (2013) studied 32 gay men who were married to same-sex partners and found that while the legitimacy of marriage often led to positive family outcomes, it also commonly had negative consequences, including new and renewed experiences of family rejection. Hence, negative attitudes toward gay marriage may continue even when the marriage is legal.

Children of same-sex parents also benefit from legal recognition of same sex-marriage. Children living in gay- and lesbian-headed households are denied a range of securities that protect children of heterosexual married couples. These include the right to get health insurance coverage and Social Security survivor benefits from a nonbiological parent (Chonody et al. 2012).

Finally, there are religion-based arguments in support of same-sex marriage. Although many religious leaders teach that homosexuality is sinful and prohibited by God, some religious groups, such as the Quakers and the United Church of Christ (UCC), accept homosexuality, and other groups have made reforms toward increased acceptance of lesbians and gays.

## Arguments Against Same-Sex Marriage

Whereas advocates of same-sex marriage argue that they will not be regarded as legitimate families by the larger society so long as same-sex couples cannot be legally married, opponents do not want to legitimize same-sex couples and families. Opponents of same-sex marriage who view homosexuality as unnatural, sick, and/or immoral do not want their children to view homosexuality as socially acceptable.

Opponents of gay marriage also suggest that gay marriage leads to declining marriage rates, increased divorce rates, and increased nonmarital births. However, data in Scandinavia reflect that these trends were occurring 10 years before Scandinavia adopted registered same-sex partnership laws, liberalized alternatives to marriage (such as cohabitation), and expanded exit options (such as no-fault divorce; Pinello 2008).

Opponents of same-sex marriage commonly argue that such marriages would subvert the stability and integrity of the heterosexual family. However, Sullivan (1997) suggests that homosexuals are already part of heterosexual families:

> *[Homosexuals] are sons and daughters, brothers and sisters, even mothers and fathers, of heterosexuals. The distinction between "families" and "homosexuals" is, to begin with, empirically false; and the stability of existing families is closely linked to how homosexuals are treated within them (p. 147).*

Many opponents of same-sex marriage base their opposition on their religious views (viewing homosexuality as immoral). However, churches have the right to deny marriage for gay people in their congregations. Legal marriage is a contract between the spouses and the state; marriage is a civil option that does not require religious sanctioning.

In previous years, opponents of gay marriage have pointed to public opinion polls that suggested that the majority of Americans are against same-sex marriage. However, public opposition to same-sex marriage is decreasing. A 2012 Pew Research Center national poll found that 43% of U.S. adults oppose legalizing gay marriage, down from 51% in 2008 (Pew Research Center 2012). We also noted that support for gay marriage is higher among young adults.

### State Legal Recognition of Same-Sex Couples

A number of states allow same-sex couples legal status that entitles them to many of the same rights and responsibilities as married couples. There has been a relentless battle in regard to which entity has the power—the legislature, the courts, or the people. For example, Proposition 8 (defining marriage as a man/woman union), which by popular vote was overturned in 2012, paved the way for the Supreme Court decision which legalized same-sex marriage.

#### International **Data**

In 2001, the Netherlands became the first country in the world to offer full legal marriage to same-sex couples. Same-sex and other-sex married couples in the Netherlands are treated identically, with two exceptions. Unlike other-sex marriages, same-sex couples married in the Netherlands are unlikely to have their marriages recognized as fully legal abroad (while the United States recognizes same-sex marriages from other countries, not all countries do). Regarding children, parental rights will not automatically be granted to the nonbiological spouse in gay couples. To become a fully legal parent, the spouse of the biological parent must adopt the child. Fifteen countries have legalized same sex marriage—Argentina, Belgium, Brazil, Canada, Denmark, France, Iceland, The Netherlands, New Zealand, Norway, Portugal, South Africa, Spain, Sweden, and Uruguay.

*We're trained to see only male or female and to plot people into those categories when they actually don't fit neatly at all. But if we pause, watch and listen closely we'll see the multiplicity of ways in which people are sexed and gendered. There exists a range of personal identifications around woman, man, in-between—we don't even have names or pronouns that reflect that in- between place but people certainly live in it.*

MINNIE BRUCE PRATT, AMERICAN EDUCATOR

## GLBT Parenting Issues

About 90,000 same-sex households (15% of the 800,000 same-sex households) have children. This is a low estimate of children who have gay or lesbian parents, as it does not count children in same-sex households who did not identify their relationship in the census, those headed by gay or lesbian single parents, or those whose gay parent does not have physical custody but is still actively involved in the child's life. Regardless of the number, Becker and Todd (2013) noted that the children of gay and lesbian couples tend to face more challenges than children from other types of family arrangements.

### Gay Families—Lesbian Mothers and Gay Fathers

Many gay and lesbian individuals and couples have children from prior heterosexual relationships or marriages. Some of these individuals married as a "cover" for their homosexuality; others discovered their interest in same-sex relationships after they married. Children with mixed-orientation parents may be raised by a gay or lesbian parent, a gay or lesbian stepparent, a heterosexual parent, and a heterosexual stepparent. A gay or lesbian individual or couple may have children through the use of assisted reproductive technology, including in vitro fertilization, surrogate mothers, and donor insemination. Other gay couples adopt or become foster parents. Less commonly, some gay fathers are part of an emergent family form known as the hetero-gay family. In a hetero-gay family, a heterosexual mother and a gay father conceive and raise a child together but reside separately. While the struggle for acceptance of gay parenting has been difficult, data on lesbian mothers has revealed that they tend to have high levels of shared decision making, parenting, and family work, all reflecting an egalitarian ideology. When compared to heterosexual mothers, gay mothers tended to spend more time with their children and expressed more warmth and affection (Biblarz and Stacy 2010). Data on gay fathers has also revealed that they are more likely to co-parent equally and compatibly than fathers in heterosexual relationships. Gay fathers are also less likely to spank their children as a method of discipline (Biblarz and Savci 2010). Bergman et al. (2010) also found that gay men who became parents had heightened self-esteem as a result of becoming parents and raising children. Hence, the act of becoming a father has a very positive outcome on gay men's sense of self-worth. Part of this effect may be due to the fact that some gay men think that gay fatherhood is an unattainable role. While both gay females and gay males report increases in individual happiness during the first year of having a baby/adopting a child, relationship happiness decreases (Goldberg et al. 2010). This drop in relationship satisfaction after a child arrives in the gay relationship is the same as what happens in heterosexual relationships.

Antigay views concerning gay parenting include the belief that homosexual individuals are unfit to be parents and that children of lesbians and gays will not develop normally and/or that they will become homosexual. The Beliefs about Children's Adjustment on Same-Sex Families Scale is designed to measure the beliefs individuals have about the effects of the child-raising and education practices of same-sex parents on their children's adjustment.

### Bisexual Parents

Power et al. (2013) surveyed 48 bisexuals who were parenting inside a variety of family structures (heterosexual relationships, same-sex relationships, coparenting with ex-partners or nonpartners, sole parenting) and revealed issues relevant to all parents—discipline, combining work/parenting, etc. The dimension of bisexuality rarely surfaced. When it did, it was in the form of being closeted to help prevent their child being subjected to prejudice and dealing with prejudiced ex partners, in-laws, and grandparents.

### Development and Well-Being of Children with Gay or Lesbian Parents

A growing body of research on gay and lesbian parenting supports the conclusion that children of gay and lesbian parents are just as likely to flourish as are children of heterosexual parents.

## SELF-ASSESSMENT | Beliefs about Children's Adjustment on Same-Sex Families Scale (SBCASSF)*

Instruction: Please indicate your level of agreement with the following statements by circling your answer using the scale from 1 = *completely disagree* to 5 = *completely agree.*

| Item | 1 | 2 | 3 | 4 | 5 |
|---|---|---|---|---|---|
| 1. In general, the social development of a child is better when he/she is raised by a heterosexual father and a heterosexual mother, and not by a gay/lesbian couple. | 1 | 2 | 3 | 4 | 5 |
| 2. In general, children raised by gay/lesbian parents will have more problems than those who are raised by a heterosexual father and a heterosexual mother. | 1 | 2 | 3 | 4 | 5 |
| 3. It is more likely that the child will experience social isolation if his/her friends know that his/her parents are gay/lesbian person. | 1 | 2 | 3 | 4 | 5 |
| 4. If children are raised by a gay or lesbian couple, they will have more problems with their own sexual identity than when they are raised by a heterosexual father and a heterosexual mother. | 1 | 2 | 3 | 4 | 5 |
| 5. If we want to defend the interests of the child, only heterosexual couples should be able to adopt. | 1 | 2 | 3 | 4 | 5 |
| 6. A child adopted by a gay or lesbian couple will be the butt of jokes and rejection by his/her classmates. | 1 | 2 | 3 | 4 | 5 |
| 7. If a child is adopted by a gay or lesbian couple, s/he will surely have psychological problems in the future. | 1 | 2 | 3 | 4 | 5 |
| 8. Surely, the classmates will reject a child whose father or mother is gay/lesbian person. | 1 | 2 | 3 | 4 | 5 |
| 9. A child who is raised by a gay or lesbian couple will be teased by his/her classmates. | 1 | 2 | 3 | 4 | 5 |
| 10. The child raised by gay or lesbian parents will probably not be chosen as a leader by his/her class. | 1 | 2 | 3 | 4 | 5 |
| 11. When a child manifests same-sex sexual orientation behaviors, it would be wise to take him/her to a psychologist. | 1 | 2 | 3 | 4 | 5 |
| 12. If the parents are gay/lesbian people, it will be difficult for the child to be invited to friends' parties. | 1 | 2 | 3 | 4 | 5 |
| 13. A boy raised by lesbian mothers will be an effeminate child. | 1 | 2 | 3 | 4 | 5 |
| 14. The child usually hides the sexual orientation of his parents from his friends out of fear of being rejected | 1 | 2 | 3 | 4 | 5 |

### Scoring

Add the numbers you circled. The lowest possible score is 14 and the highest possible score is 70. The lower the score, the more positive the view of children being reared by same-sex parents; the higher the score the more negative the view of children being reared by same-sex parents. The midpoint is 28 so scores below that tend to reflect negative attitudes; scores above that tend to reflect positive attitudes.

### Source

Frias-Navarro, D. 2009. *Scale on Beliefs about Children's Adjustment in Same-Sex Families.* University of Valencia, Spain. http://www.uv.es/friasnav/Scale_Items14.pdf (http://www.uv.es/friasnav/Scale_Items14.pdf) Dr. Frias-Navarro is at the University of Valencia, Spain.

Frias-Navarro, D. and H. Monterde-i-Bort. 2012. A scale on beliefs about children's adjustment in same-sex families: Reliability and validity. *Journal of Homosexuality* 59: 1273–1288.

There is the belief that for children to develop totally, they need both a mother and father together. Biblarz and Stacey (2010) compared children from two-parent families with same or different sex co-parents and single-mother with single-father families and found that the strengths typically associated with married mother-father families appeared to the same extent in families with two mothers and potentially in those with two fathers. Hence, children seem to benefit when there are two parents in the household (rather than a single mom or dad) but the gender of the parents is irrelevant—the structure can be a woman and a man, two women, or two men. Crowl et al. (2008) also reviewed 19 studies on the developmental outcomes and quality of parent-child relationships among children raised by gay and lesbian parents. Results confirmed previous studies that children raised by same-sex parents fare equally well to children raised by heterosexual parents. Regardless of

Children who grow up with loving/nurturing parents flourish. The degree to which the child is loved and cared for, not the sexual orientation of the parents, is the important variable in the child's development.

Chelsea Curry

family type, adolescents were more likely to show positive adjustment when they perceived more caring from adults and when parents described having close relationships with them. Thus, the qualities of adolescent-parent relationships rather than the sexual orientation of the parents were significantly associated with adolescent adjustment.

What do children who are reared in a home with lesbian parents say? Gartrell et al. (2012) provided data from 78 adolescents who were being reared in lesbian families. Results revealed that these 17-year-old adolescents were academically successful in supportive school environments, had active social networks, and close family bonds. Almost all considered their mothers good role models and reported their overall well-being an average of 8.1 on a 10-point-scale.

In a series of interviews, daughters and sons (32) of lesbian parents reported that, in general, the reactions from their peers was largely positive. Most reported experiencing stigma at some point in their lives due to their mothers' lesbianism. A range of coping mechanisms were employed, including confrontation, secrecy, and seeking outside support (counseling) (Leddy et al. 2012). Gay parents may be particularly sensitive to potential stigmatization and seek gay- friendly neighborhoods to rear their children (Goldberg et al. 2012).

Some children of lesbian parents report appreciation for being connected to a unique group: "I've always been grateful for the queer community I was raised in, which provided me with lots of adults to be close to and lots of different models for how to be a person" (Leddy et al. 2012, p. 247). Bos and Sandfort (2010) studied children in lesbian and heterosexual families and found that children in lesbian families felt less parental pressure to conform to gender stereotypes, were less likely to experience their own gender as superior, and were more likely to be uncertain about future heterosexual romantic involvement. The last finding does not suggest that the children were more likely to feel that they were gay, but that they were living in a social context which allowed for more than the heterosexual adult model. One of our former female students who was reared by two gay moms (in a same-sex relationship with each other) said, "My experience was variable. It was a challenge since there is a taboo against having gay parents. But other times family life felt normal and I certainly feel that I was taken care of and loved just as much as if I had had heterosexual parents."

### Development and Well-Being of Children in Transgender Families

Research on children in transgender families is virtually nonexistent (Biblarz and Savci 2010). What is known focuses on the stress transgender children experience as they try to "fit in" to please parents and society at the expense of personal depression and loss of well-being. Meanwhile, children of transgender parents struggle with new definitions of who their parents are and how this affects them. Male children of male-to-female transsexuals may have a particularly difficult time adapting.

### Heterosexual Parental Concern for African American Gay Male

Heterosexual parents also have concerns/fears for their gay children, and, being African American and gay adds another layer of concern. LaSala and Frierson (2012) interviewed African Americans and their parents who talked about the issue of racism and homophobia. A theme of the parents was that black males are expected to live up to the macho stereotype of being with a woman and procreating. Having sex with other men is counter to the stereotype and makes the gay male more vulnerable to prejudice/discrimination.

### Discrimination in Child Custody, Visitation, Adoption, and Foster Care

A student in one of our classes reported that, after she divorced her husband, she became involved in a lesbian relationship. She explained that she would like to be open about her relationship to her family and friends, but she was afraid that if her ex-husband found out that she was in a lesbian relationship, he might take her to court and try to get custody of their children. Although several respected national organizations including the American Academy of Pediatrics, the Child Welfare League of America, the American Bar Association, the American Medical Association, the American Psychological Association, the American Psychiatric Association, and the National Association of Social Workers have gone on record in support of treating gays and lesbians without prejudice in parenting and adoption decisions, lesbian and gay parents are often discriminated against in child custody, visitation, adoption, and foster care.

Lehman (2010) identified the three issues the courts have used to deny homosexuals custody of their children—per se, presumption, and nexus. Per se takes the position that if the parent is homosexual, he or she is unfit to parent. The presumption position is similar to the per se position except the parent has the right to prove otherwise. The nexus approach focuses on the connection between being homosexual resulting in negative outcomes for the child. These outcomes include that the child will be ridiculed, that the parent will turn the child gay, and that homosexuality is immoral. Each of these positions is rarely taken seriously by the courts.

This chapter's Family Policy section asks whether gay and lesbian individuals and couples should be prohibited from adopting.

Most adoptions by gay people are second-parent adoptions. A **second-parent adoption** (also called co-parent adoption) is a legal procedure

**Second-parent adoption** legal procedure that allows individuals to adopt their partner's biological or adoptive child without terminating the first parent's legal status as parent.

## FAMILY POLICY | Should GLBT Individuals/Couples Be Prohibited from Adopting Children

Thousands of children in the U.S. child welfare system are waiting to be adopted. The American Psychological Association (2012) reviewed the research and noted that lesbian and gay parents are as likely as heterosexual parents to provide supportive and healthy environments for their children. In addition, the adjustment, development, and psychological well-being of children are unrelated to parental sexual orientation.

Despite the research confirming positive outcomes for children raised by gay or lesbian parents, and despite the support for gay adoption by child advocacy organizations, placing children for adoption with gay or lesbian parents remains controversial. Of 4,539 undergraduates, 14% agreed (men more than women) with the statement "Children reared by same-sex parents are more likely to be homosexual as adults than children reared by two parents of different sexes" (Hall and Knox 2013).

Social policies that prohibit LGBT individuals and couples from adopting children result in fewer

that allows individuals to adopt their partner's biological or adoptive child without terminating the first parent's legal status as parent. Second-parent adoption gives children in same-sex families the security of having two legal parents. Second-parent adoption potentially benefits a child by:

- Placing legal responsibility on the parent to support the child;
- Allowing the child to live with the legal parent in the event that the biological (or original adoptive) parent dies or becomes incapacitated;
- Enabling the child to inherit and receive Social Security benefits from the legal parent;
- Enabling the child to receive health insurance benefits from the parent's employer; and
- Giving the legal parent standing to petition for custody or visitation in the event that the parents break up (Clunis and Green 2003)

Second-parent adoption is not possible when a parent in a same-sex relationship has a child from a previous heterosexual marriage or relationship, unless the former spouse or partner is willing to give up parental rights.

### When LGB Relationships End: Reaction of the Children

How do children react to the ending of the relationship of their LGB parents? Goldberg and Allen (2013) interviewed 20 young adults who experienced their LGB parents' relationship dissolution and/or the formation of a new LGB stepfamily. Almost all families negotiated relational transitions informally and without legal intervention; the relationship with one's biological mother was the strongest tie from break-up to repartnering and stepfamily formation; and geographic distance from their nonbiological parents created hardships in interpersonal closeness. Overall, "young people perceived their families as strong and competent in handling familial transitions" (p. 529).

# The Future of GLBT Relationships

Moral acceptance and social tolerance/acceptance of gays, lesbians, bisexuals, and transsexuals as individuals, couples, and parents will come slowly. Heterosexism, homonegativity, biphobia, and transphobia are entrenched in American society. However, as more states recognize same-sex marriage, more children being adopted. What happens to children who are not adopted? The thousands of children who age out of the foster care system annually experience high rates of homelessness, incarceration, early pregnancy, failure to graduate from high school, unemployment, and poverty (Howard 2006). Essentially, family policies that prohibit gay adoption are policies that deny thousands of children the opportunity to have a nurturing family.

**Your Opinion?**

1. Do you believe children who grow up with same-sex parents are disadvantaged?
2. Do you believe that children without homes should be prohibited by law from being adopted by lesbian or gay individuals or couples?
3. If you were a six-year-old child with no family, would you rather remain in an institutional setting or be adopted by a gay or lesbian couple?

GLBT individuals will come out, their presence will become more evident, and tolerant, acceptance, and support will increase...slowly.

The future will also include GLBT suicide prevention efforts such as identifying at-risk individuals early and intervening through national and state programs (Haas et al. 2011). Such programs will include "educating community gatekeepers including teachers and staff in youth programs, senior centers, aging services agencies and others who come in contact with at-risk individuals."

# Summary

***What are problems with identifying and classifying sexual orientation?***

To avoid the stigma, prejudice and discrimination associated with homosexuality, some individuals hide their sexual orientation. Others do not self-identify as gay/bisexual even though they are attracted to those of the same sex or have engaged in same-sex behavior. Still others may change their attitudes, attractions, behaviors, and identity across time.

***What are the origins of sexual-orientation diversity?***

Although a number of factors have been correlated with sexual orientation, including genetics, gender role behavior in childhood, and fraternal birth order, no single theory can explain diversity in sexual orientation. Most gay people believe that homosexuality is an inherited, inborn trait. In a national study of homosexual men, 90% reported that they believed that they were born with their homosexual orientation; only 4% believed that environmental factors were the sole cause.

Individuals who believe that homosexual people choose their sexual orientation tend to think that homosexuals can and should change this orientation. Various forms of reparative therapy are focused on changing homosexuals' sexual orientation, but most professional associations agree that homosexuality is not a mental disorder and needs no cure.

***What are heterosexism, homonegativity, homophobia, biphobia, and transphobia?***

Heterosexism refers to "the institutional and societal reinforcement of heterosexuality as the privileged and powerful norm." Heterosexism is based on the belief that heterosexuality is superior to homosexuality. The term homophobia, also known as homonegativity, refers to negative attitudes and emotions toward homosexuality and those who engage in it. Biphobia and transphobia are negative attitudes and emotions toward bisexual and trans individuals.

***What are the relationships of GLBT individuals like?***

While GLBT relationships tend to be more similar to than different from those of heterosexual couples, GLBT relationships tend to involve partners of equal power and control who are emotionally expressive and share decision making. Lesbian couples tend to have the highest levels of relationship quality. A major difference from heterosexual relationships is to decide when and how to disclose their relationships to others.

While bisexuals have the ability to form intimate relationships with both sexes, they tend to have primary relationships with the other sex. About half of bisexuals are in open relationships.

Resolution of conflicts by GLBT couples tend to begin their discussions more positively and maintain a more positive tone throughout the discussion than do partners in heterosexual marriages (this is due to their equal power and status). Same-sex partners may have a higher motivation to reduce conflict because dissolution may involve losing a tight friendship network.

In a study of transgender relationships, almost 70% (69.2) of the transwomen were living together; 57% of the transmen. In regard to sexual fidelity, 65% of the transwomen were monogamous (26% were not), as were 70% of the transmen (21% were not). Some of the partners of the trans individuals did not know their partner was trans—8% of transwomen and 2% of transmen. Transgender individuals live in fear they will not be accepted.

***What is coming out to parents, peers, colleagues, etc., like?***

Almost three-fourths (72%) of GLBT individuals report that they "sometimes or often avoided disclosing their GLBT status due to fear of negative consequences, harassment, or discrimination." Parental reaction to a son or daughter coming out has a major effect on the development of their child. Most white parents are accepting; black parents have more difficulty due to the impact of their traditional religous beliefs.

Rejection by peers at school can invite harassment and discrimination, bullying, etc. Workplace involves risk to promotion and advancement. Being the target of a hate crime permeates the decision to come out.

The outcome of coming out is more often positive than negative. Individuals feel relieved at no longer having to hide their orientation and are surprised by the unaccepted support. Partners who have the most difficult reaction are spouses. The basic trajectory of reacting to a partner's homosexuality is not unlike learning of the death of one's beloved. Life as one knew it is altered. The stages of this transition are shock, disbelief, numbness, and mourning for the partner/life that was, followed by readjustment and moving on. The last two stages may involve staying in the relationship with the gay partner (the choice of about 15% of straight spouses) or divorcing/ending the relationship (the choice of 85%).

***What are the pros and cons of same-sex marriage?***

The argument for same-sex marriage is that banning such unions is a violation of civil rights that denies same-sex couples the many legal and financial benefits that are granted to heterosexual married couples. Children of gay/lesbian individuals would also benefit by being brought up in a more stable, socially approved relationship. Arguments against same-sex marriage focus on the "pathology" of homosexuality, the "immorality," and the subversion of traditional marriage.

***What does research on gay and lesbian parenting conclude?***

A growing body of credible, scientific research on gay and lesbian parenting concludes that children raised by gay and lesbian parents adjust positively and their families function well. Lesbian and gay parents are as likely as heterosexual parents to provide supportive and healthy environments for their children, and the children of lesbian and gay parents are as likely as those of heterosexual parents to flourish.

***What is the future of GLBT relationships?***

Moral acceptance and social tolerance/acceptance of GLBT individuals, couples, and parents will come slowly. Heterosexism, homonegativity, biphobia, and transphobia are entrenched in American society. However, as more states recognize same-sex marriage, more GLBT individuals will come out, their presence will become more evident, and tolerant, acceptance and support will increase . . . slowly.

## Key Terms

Binegativity
Biphobia
Bisexuality
Coming out
Contact hypothesis
Conversion therapy
Cross-dresser
Defense of Marriage Act
Down low
Gay
Genderqueer
GLBT
Hate crime
Heterosexism
Heterosexuality
Homonegativity
Homophobia
Homosexuality
Internalized homophobia
Lesbian
Queer
Second-parent adoption
Sexual identity
Sexual orientation
Transphobia
Transgender
Transsexual

# Web Links

Advocate (Online Newspaper for LGBT News)
http://www.advocate.com/

Bisexual Resource Center
http://www.biresource.org

People with a Lesbian, Gay, Transgender or Queer Parent (COLAGE)
http://www.colage.org

Compatible Partners
http://www.compatiblepartners.net/

Gay and Lesbian Support Groups for Parents
http://www.gayparentmag.com/

Human Rights Campaign
http://www.hrc.org

Mixed sexual orientation marriages
http://www.straightspouseconnection.com/
http://www.straightspouse.org/

National Gay and Lesbian Task Force
http://www.thetaskforce.org/

Other Sheep (Christian organization for empowering sexual minorities)
http://www.othersheep.org/

PFLAG (Parents, Families, and Friends of Lesbians and Gays)
http://www.pflag.org

PlanetOut (Online News Source)
http://www.planetout.com

# Chapter 8 Communication and Technology in Relationships

## Learning Objectives

- Explain the importance of congruence in regard to verbal and nonverbal behavior.
- Identify the communication choices individuals make to enhance their relationships.
- Understand how self-disclosure, secrecy, and lying/cheating impact relationships.
- Describe how technology impacts romantic relationships.
- Review two theoretical frameworks which explain the communication process.
- Discuss the sources of conflict and how conflict is resolved.
- Reveal the future for communication in relationships.

## TRUE OR FALSE?

1. Spouses with poor communication are more likely to divorce.
2. Spouses mirror each other in communication—when one spouse swears, the other swears back.
3. A similar percentage of individuals report receiving "bad news" ("I'm breaking up with you") as well as "good news" ("I love you") in a text message.
4. Couples who report good communication report enhanced physical intimacy.
5. Males, whites, and heterosexuals keep more secrets from their romantic partners than females, blacks, and homosexuals.

*Answers:* 1. T 2. T 3. T 4. T 5. F

You have begun the most important chapter in this text. Nothing is more important to the quality and stability of your relationship than the choices you make about how to communicate with your partner. For example, arguments are loud while discussions are soft. You can make a deliberate choice to have a discussion versus an argument by altering the volume of your voice. Two researchers followed 136 newlyweds over four years and revealed that those who ended up getting divorced did not make communication choices to enhance their relationship—they expressed anger, they were critical, and they chose not to make positive remarks to and about each other (Lavner and Bradbury 2012). Those who stayed together chose to speak respectfully to each other and made positive, affirming statements to each other. The communication pattern you choose to have with your partner will influence whether you live in misery or enjoy life with your partner. Communication is one of the primary elements researchers use to identify marital quality (Allendorf and Ghimire 2013).

Individuals know that how they communicate confirms the quality of their relationship ("We can talk all night about anything and everything") or condemns their relationship ("We have nothing to say to each other"). Braun et al. (2010) noted the positive effect of positive marital communication on mental health. In addition, good communication is associated with enhanced physical intimacy (Yoo et al. 2011).

Communication between romantic partners is influenced by the technological explosion of cell phones, texting, Facebook, sexting, Skype, etc. Lovers now "stay connected all day." This chapter acknowledges the effect of technology on romantic relationships today. We begin by looking at the nature of interpersonal communication.

*Ultimately the bond of all companionship, whether in marriage or in friendship, is conversation.*

OSCAR WILDE, ENGLISH PLAYWRIGHT

*"Hanging out" = an agreement to meet and spend time together.*

*"Talking" = we are discovering each other and deciding if we want to continue a relationship*

*"Dating = we are "boyfriend" and "girlfriend" and are monogamous—but the future is uncertain*

*"Engaged" = we are committed to marry*

DEFINITIONS FROM A COLLEGE STUDENT

# Interpersonal Communication

**Communication** can be defined as the process of exchanging information and feelings between two or more people. Wickrama et al. (2010) studied 540 couples and found that spouses mirror each other in terms of how they relate to each other. When one spouse insults, swears, or shouts, the other is likely to engage in the same behavior.

**Communication** the process of exchanging information and feelings between two or more people.

Communication is both verbal and nonverbal. Although most communication is focused on verbal content, most (estimated to be as high as 80%) interpersonal communication is nonverbal. **Nonverbal communication** is the "message about the message," the gestures, eye contact, body posture, tone, and rapidity of

**Nonverbal communication** the gestures, eye contact, body posture, tone, and rapidity of speech.

"I am upset with you" and "I don't care what you say" are the nonverbal messages these two are communicating to each other.

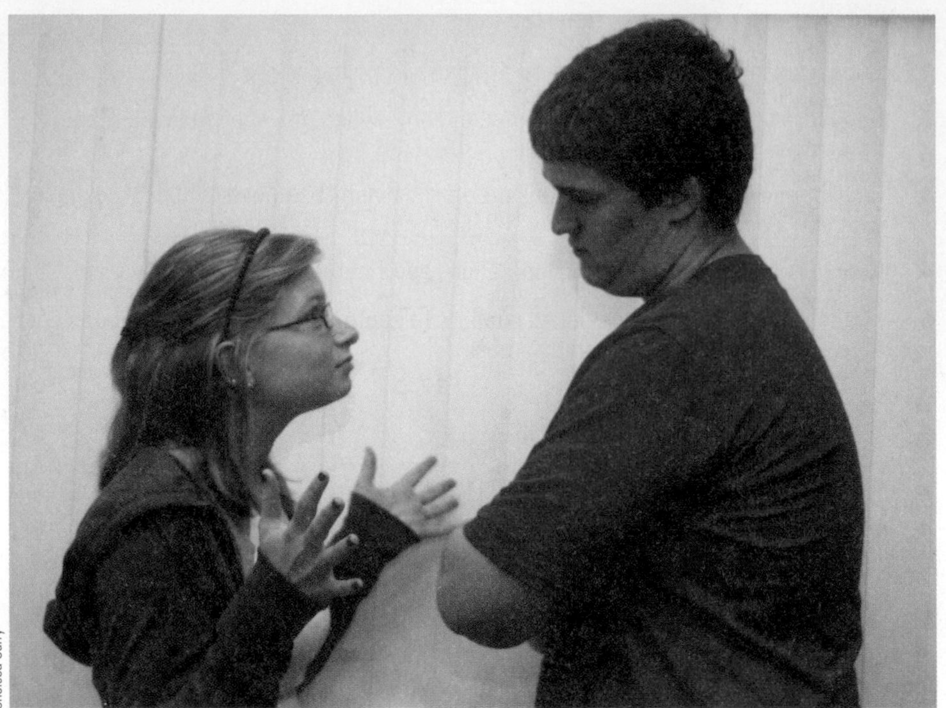

Chelsea Curry

speech. Even though a person says, "I love you and am faithful to you," crossed arms and lack of eye contact will convey a different meaning from the same words accompanied by a tender embrace and sustained eye-to-eye contact. The greater the congruence between verbal and nonverbal communication, the better. One aspect of nonverbal behavior is the volume of speech. The volume one uses when interacting with a partner has relevance for the relationship. When one uses a high frequency (yells) there are physiological (e.g., high blood pressure, higher heart rate, higher cortisol levels) and therapeutic (e.g., more limited gains in counseling—higher divorce rate) outcomes.

Flirting is a good example of both nonverbal and verbal behavior. One researcher defined **flirting** as showing another person romantic interest without serious intent (Moore 2010). Examples of how interest is shown include preening, such as stroking one's hair or adjusting one's clothing, and positional cues, such as leaning toward or away from the target person. Individuals go through a series of steps on their way to sexual intercourse, and the female is responsible for signaling the male that she is interested so that he moves the interaction forward. At each stage, she must signal readiness to move to the next stage. For example, if the male holds the hand of the female, she must squeeze his hand before he can entwine their fingers. Females generally serve as sexual gatekeepers and control the speed of the interaction toward sex.

**Flirting** showing another person romantic interest without serious intent.

Having a supportive communicative partner, one who listens and is engaged, is a plus for any relationship. The Supportive Communication Scale allows you to assess the degree to which your relationship is characterized by supportive communication.

*For most women, the language of conversation is primarily a language of rapport: a way of establishing connections and negotiating relationships.*

DEBORAH TANNEN, RESEARCHER

## Words versus Action

A great deal of social discourse depends on saying things that sound good but that have no meaning in terms of behavioral impact. "Let's get together" or "let's hang out" sounds good since it implies an interest in spending time with each other. But the phrase has no specific plan so that the intent is likely never to happen—just the opposite. So let's hang out really means "we won't be spending any time together." Similarly, come over anytime means "never come."

*Two monologues do not make a dialogue.*

JEFF DALY, CLINICAL PSYCHOLOGIST

## SELF-ASSESSMENT | Supportive Communication Scale

This scale is designed to assess the degree to which partners experience supportive communication in their relationships. After reading each item, circle the number that best approximates your answer.

0 = strongly disagree (SD)
1 = disagree (D)
2 = undecided (UN)
3 = agree (A)
4 = strongly agree (SA)

| | SD | D | UN | A | SA |
|---|---|---|---|---|---|
| 1. My partner listens to me when I need someone to talk to. | 0 | 1 | 2 | 3 | 4 |
| 2. My partner helps me clarify my thoughts. | 0 | 1 | 2 | 3 | 4 |
| 3. I can state my feelings without my partner getting defensive. | 0 | 1 | 2 | 3 | 4 |
| 4. When it comes to having a serious discussion, it seems we have little in common (reverse scored). | 0 | 1 | 2 | 3 | 4 |
| 5. I feel put down in a serious conversation with my partner (reverse scored). | 0 | 1 | 2 | 3 | 4 |
| 6. I feel discussing some things with my partner is useless (reverse scored). | 0 | 1 | 2 | 3 | 4 |
| 7. My partner and I understand each other completely. | 0 | 1 | 2 | 3 | 4 |
| 8. We have an endless number of things to talk about. | 0 | 1 | 2 | 3 | 4 |

### Scoring

Look at the numbers you circled. Reverse-score the numbers for questions 4, 5, and 6. For example, if you circled a 0, give yourself a 4; if you circled a 3, give yourself a 1, and so on. Add the numbers and divide by 8, the total number of items. The lowest possible score would be 0, reflecting the complete absence of supportive communication; the highest score would be 4, reflecting complete supportive communication. One-hundred-and-eighty-eight individuals completed the scale. Thirty-nine percent of the respondents were married, 38% were single, and 23% were living together. The average age was just over 24. The average score of 94 male partners who took the scale was 3.01; the average score of 94 female partners was 3.07.

### Source

Sprecher, S., S. Metts, B. Burelson, E. Hatfield, and A. Thompson. 1995. Domains of expressive interaction in intimate relationships: Associations with satisfaction and commitment. *Family Relations* 44: 203–10. Copyright © 1995 by the National Council on Family Relations.

Where there is behavioral intent, a phrase with meaning is "Let's meet Thursday night at seven for dinner" (rather than "let's hang out") and "Can you come over Sunday afternoon at four to play video games?" (rather than "come over any time"). With one's partner, "let's go camping sometime" means we won't ever go camping. "Let's go camping at the river this coming weekend" means you are serious about camping. A major theme of all communication is how emotionally close the partners want to be with each other. This issue is addressed in the Personal Choices section.

*Half the world is composed of people who have something to say and can't, and the other half who have nothing to say and keep on saying it.*

ROBERT FROST, AMERICAN POET

## PERSONAL CHOICES

### *How Close Do You Want to Be?*

Individuals differ in their capacity for, and interest in, an emotionally close and disclosing relationship. See Figure 8.1, which represents the various levels of closeness. These preferences may vary over time; the partners may want closeness at some times and distance at other times. Individuals frequently choose partners according to an "emotional fit"—agreement about the amount of closeness they desire in their relationship.

In addition to emotional closeness, some partners prefer a pattern of physical presence and complete togetherness (the current buzzword is *codependency*), in which they spend all of their leisure and discretionary time together. Others enjoy

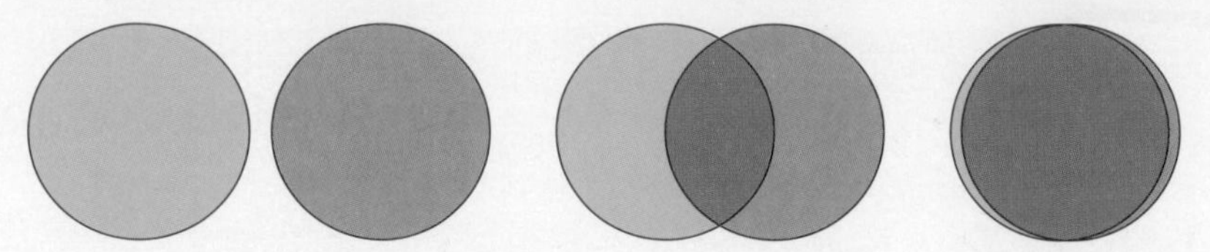

**Figure 8.1** The two circles on the left reflect a distant emotional relationship. The half and half overlapping circles in the middle reflect a moderate level of emotional closeness. The two mostly overlapping circles on the right reflect a very close emotional relationship. Any pattern is acceptable as long as both partners agree.

time alone and time with other friends and do not want to feel burdened by the demands of a partner with high companionship needs. Partners might consider their own choices and those of their partners in regard to emotional and spatial closeness.

## Making Choices for Effective Communication

*You can make more friends in two months by becoming really interested in other people than you can in two years by trying to get other people interested in you.*

DALE CARNEGIE, PUBLIC SPEAKER

*The single biggest problem with communication is the illusion that it has taken place.*

GEORGE BERNARD SHAW, PLAYWRIGHT

We continually make choices in how we communicate in our relationships. The following section identifies the various choices we can make to ensure that communication in our relationships has a positive outcome.

1. ***Make the choice that communication is a priority.*** Communicating effectively implies making communication an important priority in a couple's relationship. When communication is a priority, partners make time for it to occur in a setting without interruptions: they are alone; they are not texting or surfing the Internet; they do not answer the phone; and they turn the television off. Making communication a priority results in the exchange of more information between partners, which increases the knowledge each partner has about the other.

2. ***Avoid negative and hurtful statements to your partner.*** Because intimate partners are capable of hurting each other so intensely, it is important to avoid brutal statements to the partner. Such negativity is associated with vulnerability to divorce (Woszidlo and Segrin 2013). Indeed, be very careful how you give negative feedback or communicate disapproval to your partner. Markman et al. (2010)

Chelsea Curry

This couple is demonstrating their value for the relationship by sitting down together, looking into each other's eyes, and touching each other.

noted that couples in marriage counseling often will report "it was a bad week" based on one negative comment made by the partner.

**3.** ***Say positive things about your partner.*** Markman et al. (2010) also emphasized the need for partners to make positive comments to each other and that doing so was associated with more stable relationships. People like to hear others say positive things about them. These positive statements may be in the form of compliments ("You look terrific!") or appreciation ("Thanks for putting gas in the car"). Maatta and Uusiautti (2013) emphasized that compliments to the partner were associated with being involved in a romantic love relationship.

**4.** ***Establish and maintain eye contact.*** Shakespeare noted that a person's eyes are the "mirrors to the soul." Partners who look at each other when they are talking not only communicate an interest in each other but also are able to gain information about the partner's feelings and responses to what is being said. Not looking at your partner may be interpreted as lack of interest and prevents you from observing nonverbal cues.

**5.** ***Ask open-ended questions.*** When your goal is to find out your partner's thoughts and feelings about an issue, using **open-ended questions** is best. Such questions (e.g., "How do you feel about me?") encourage your partner to give an answer that contains a lot of information. **Closed-ended questions** (e.g., "Do you love me?"), which elicit a one-word answer such as *yes* or *no,* do not provide the opportunity for the partner to express a range of thoughts and feelings.

**Open-ended question** question which elicits a lot of information.

**Closed-ended question** question which elicits a one-word answer such as yes or no.

**6.** ***Use reflective listening.*** Effective communication requires being a good listener. One of the skills of a good listener is the ability to use the technique of **reflective listening**, which involves paraphrasing or restating what the person has said to you while being sensitive to what the partner is feeling. For example, suppose you ask your partner, "How was your day?" and your partner responds, "I felt exploited today at work because I went in early and stayed late and a memo from my new boss said that future bonuses would be eliminated because of a company takeover." Listening to what your partner is both saying and feeling, you might respond, "You feel frustrated because you really worked hard and felt unappreciated."

**Reflective listening** paraphrasing or restating what the person has said to you while being sensitive to what the partner is feeling.

*Women like silent men. They think they're listening.*

MARCEL ARCHARD, FRENCH PLAYWRIGHT

Reflective listening serves the following functions: (1) it creates the feeling for speakers that they are being listened to and are being understood; and (2) it increases the accuracy of the listener's understanding of what the speaker is saying. If a reflective statement does not accurately reflect what a speaker has just said, the speaker can correct the inaccuracy by saying it again.

An important quality of reflective statements is that they are nonjudgmental. For example, suppose two lovers are arguing about spending time with their respective friends and one says, "I'd like to spend one night each week with my friends and not feel guilty about it." The partner may respond by making a statement that is judgmental (critical or evaluative), such as those exemplified in Table 8.1. Judgmental responses serve to punish or criticize people for what they think, feel, or want and often result in an argument.

Table 8.1 also provides several examples of nonjudgmental reflective statements.

**7.** ***Use "I" statements.*** **"I" statements** focus on the feelings and thoughts of the communicator *without* making a judgment on others. Because "I" statements are a clear and nonthreatening way of expressing what you want and how you feel, they are likely to result in a positive change in the listener's behavior. Making "I" statements reflect being authentic. Impett et al. (2010) emphasized the need to be authentic when communicating. Being **authentic** means speaking and acting in a manner according to what one feels. Being authentic in a relationship means being open with the partner about one's preferences and feelings about the partner's behavior. Being authentic has positive consequences for the relationship in that one's thoughts and feelings are out in the open (in contrast to being withdrawn and resentful).

**"I" statements** words that reflect the thoughts and feelings of the communicator without being judgmental.

**Authentic** speaking and acting in a manner according to what one feels.

**TABLE 8.1 Judgmental and Nonjudgmental Responses to A Partner's Saying, "I'd like to go out with my friends one night a week."**

| Nonjudgmental, Reflective Statements | Judgmental Statements |
|---|---|
| You value your friends and want to maintain good relationships with them. | You only think about what you want. |
| You think it is healthy for us to be with our friends some of the time. | Your friends are more important to you than I am. |
| You really enjoy your friends and want to spend some time with them. | You just want a night out so that you can meet someone new. |
| You think it is important that we not abandon our friends just because we are involved. | You just want to get away so you can drink. |
| You think that our being apart one night each week will be good for our relationship. | You are selfish. |



**You statement** statement that blames or criticizes the listener and often results in increasing negative feelings and behavior in the relationship.

In contrast, **"you" statements** blame or criticize the listener and often result in increasing negative feelings and behavior in the relationship. For example, suppose you are angry with your partner for being late. Rather than say, "You are always late and irresponsible" (a "you" statement), you might respond with, "I get upset when you are late and ask that you call me when you will be delayed." The latter focuses on your feelings and a desirable future behavior rather than blaming the partner for being late.

**8.** ***Touch.*** Hertenstein et al. (2007) identified the various meanings of touch such as conveying emotion, attachment, bonding, compliance, power, and intimacy. The researchers also emphasized the importance of using touch as a mechanism of nonverbal communication to emphasize one's point or meaning.

**9.** ***Use soft emotions.*** Sanford and Grace (2011) identified "hard" emotions (e.g., angry or outraged) or "soft" emotions (sad or hurt) displayed during conflict. Examples of flat emotions are being bored or indifferent. The use of hard emotions resulted in an escalation of negative communication, whereas the display of soft emotions resulted in more benign communication and an increased feeling regarding the importance of resolving interpersonal conflict.

*Don't make decisions when you're angry and don't make promises when you're happy.*

BEN AFFLECK, ACTOR

**10.** ***Identify specific new behavior you want.*** Focus on what you want your partner to do rather than on what you don't want. Rather than say, "You spend too much time with your Xbox playing Halo, Call of Duty, and Gears of War," an alternative might be "When I come home, please help me with dinner, ask me about my day and turn on your Xbox after ten p.m.—after you have turned me on." Rather than say, "You never call me when you are going to be late," say, "Please call me when you are going to be late." Notice that you are asking your partner for what you want, not demanding. Demanding is associated with a demand/withdraw pattern in which one partner demands and the other withdraws. Couples with this pattern are more likely to be dissatisfied with their relationship and it is more likely for one of the partners to be having an affair (Balderrama-Durbin et al. 2012).

*Consider how hard it is to change yourself and you'll understand what little chance you have in trying to change others.*

JACOB M. BRAUDE, AUTHOR

**11.** ***Stay focused on the issue.*** Wheeler et al. (2010) studied conflict resolution strategies and found that couples who have a solution-oriented style in contrast to a confrontational (attack) or nonconfrontational (ignore) style report higher levels of marital satisfaction. Hence, couples who focus on the issue to get it resolved handle conflict with minimal marital fallout.

**Branching** going out on a different limb of an issue rather than staying focused on the issue.

**Branching** refers to going out on different limbs of an issue rather than staying focused on the issue. If you are discussing the overdrawn checkbook, stay focused on the checkbook. To remind your partner that he or she is equally irresponsible when it comes to getting things repaired or doing housework is to go off target. Stay focused.

**12.** ***Make specific resolutions to disagreements.*** To prevent the same issues or problems from recurring, agreeing on what each partner will do in similar circumstances in the future is important. For example, if going to a party together results in one partner's drinking too much and drifting off with someone else, what needs to be done in the future to ensure an enjoyable evening together? In this example, a specific resolution would be to decide how many drinks the partner will have within a given time period and make an agreement to stay together and dance only with each other.

**13.** ***Give congruent messages.*** **Congruent messages** are those in which the verbal and nonverbal behaviors match. A person who says, "Okay, you're right" and smiles while embracing the partner is communicating a congruent message. In contrast, the same words accompanied by leaving the room and slamming the door communicate a very different message.

**Congruent messages** those in which the verbal and nonverbal behaviors match.

*Never do something permanently stupid just because you are temporarily upset.*

ANONYMOUS

**14.** ***Share power.*** **Power** is the ability to impose one's will on the partner and to avoid being influenced by the partner. One of the greatest sources of dissatisfaction in a relationship is a power imbalance and conflict over power (Kurdek 1994). Over half (62%) of 4,419 undergraduates from two universities reported that they had the same amount of power in the relationship as their partner. Sixteen percent felt that they had less power; 22%, more power (men reported that they had more power than women in 25% of cases, versus 21%) (Hall and Knox 2013).

**Power** the ability to impose one's will on the partner and to avoid being influenced by the partner.

One way to assess power is to identify who has the least interest in the relationship. Waller and Hill (1951) observed that the person who has the least interest in continuing the relationship is in control of the relationship. This **principle of least interest** is illustrated by the woman who said, "He wants us to stay together more than I do so I am in control and when we disagree about something he gives in to me." Expressions of power in a relationship are numerous and include the following:

**Principle of least interest** the person who has the least interest in the relationship controls the relationship.

Withdrawal (not speaking to the partner)

Guilt induction ("How could you ask me to do this?")

Being pleasant ("Kiss me and help me move the sofa.")

Negotiation ("We can go to the movie if we study for a couple of hours before we go.")

Deception (running up credit card debts of which the partner is unaware)

Blackmail ("I'll find someone else if you won't have sex with me.")

Physical abuse or verbal threats ("You will be sorry if you try to leave me.")

Criticism ("You are stupid and fat.")

Dominance ("I make more money than you so I will decide where we go.")

Power may also take the form of love and sex. The person in the relationship who loves less and who needs sex less has enormous power over the partner who is very much in love and who is dependent on the partner for sex. This pattern reflects the principle of least interest we discussed earlier in the text.

**15.** ***Keep the process of communication going.*** Communication includes both content (verbal and nonverbal information) and process (interaction). It is important not to allow difficult content to shut down the communication process. To ensure that the process continues, the partners should focus on the fact that talking is important and reinforce each other for keeping the communication process alive. For example, if your partner tells you something that you do that bothers him or her, it is important to thank your partner for telling you rather than becoming defensive. In this way, your partner's feelings about you stay out in the open rather than being hidden behind a wall of resentment. If you punish such disclosure because you don't like the content, disclosure will stop. For example, a wife told her husband that she felt his lunches with a woman at work were becoming too frequent and wondered if it were a good idea. Rather than the husband becoming defensive and saying he could have lunch with whomever he wanted he might say, "I appreciate your telling me how you feel about this . . . you're right . . . maybe it would be best to cut back."

# Gender, Culture, and Communication

How individuals communicate with each depends on which gender is talking/listening and the society/culture in which they were socialized and live.

## Gender Differences in Communication

Numerous jokes address the differences between how women and men communicate. One anonymous quote on the Internet follows:

> *When a woman says, "Sure . . . go ahead," what she means is "I don't want you to." When a woman says, "I'm sorry," what she means is "You'll be sorry." When a woman says, "I'll be ready in a minute," what she means is "Kick off your shoes and start watching a football game on TV."*

Women and men differ in their approach to and patterns of communication. Women are more communicative about relationship issues, view a situation emotionally, and initiate discussions about relationship problems. Deborah Tannen (1990; 2006) is a specialist in communication. She observed that, to women, conversations are negotiations for closeness in which they try "to seek and give confirmations and support, and to reach consensus" (1990, p. 25). To men, conversations are about winning and achieving the upper hand.

The genders differ in regard to emotionality. Garfield (2010) reviewed men's difficulty with emotional intimacy. He noted that their emotional detachment stems from the provider role which requires them to stay in control. Being emotional is seen as weakness. Men's groups where men learn to access and express their feelings have been helpful in increasing men's emotionality.

In contrast, women tend to approach situations emotionally. For example, if a child is seriously ill, wives will want their husbands to be emotional, to cry, to show that they really care that their child is sick. But a husband might react to a seriously ill child by putting pressure on the wife to be "mature" about the situation and by encouraging stoicism.

Women disclose more in their relationships than men do (Gallmeier et al. 1997). In a study of 360 undergraduates, women were more likely to disclose information about previous love relationships, previous sexual relationships, their love feelings for the partner, and what they wanted for the future of the relationship. They also wanted their partners to reciprocate this level of disclosure, but such disclosure from their partners was, generally, not forthcoming.

## Cultural Differences in Communication

The meaning of a particular word varies by the country in which it is used. Xu (2013) emphasized the importance of being aware of cultural differences in communication. An American woman was dating a man from Iceland. When she asked him, "Would you like to go out to dinner?" he responded, "Yes, maybe." She felt confused by this response and was uncertain whether he wanted to eat out. It was not until she visited his home in Iceland and asked his mother, "Would you like me to set the table?"—to which his mother replied, "Yes, maybe"—that she discovered "Yes, maybe" means "Yes, definitely."

Individuals reared in France, Germany, Italy, or Greece regard arguing as a sign of closeness—to be blunt and argumentative is to keep the interaction alive and dynamic; to have a tone of agreement is boring. Asian cultures (e.g., Japanese, Chinese, Thai, and Pilipino) place a high value on avoiding open expression of disagreement and emphasizing harmony. Deborah Tannen (1998) observed the different perceptions of a Japanese woman married to a Frenchman:

> *He frequently started arguments with her, which she found so upsetting that she did her best to agree and be conciliatory. This only led him to seek another point on which to argue. Finally she lost her self-control and began to yell back. Rather than being angered, he was overjoyed. Provoking arguments was his way of showing interest in her, letting her know how much he respected her intelligence. To him, being able to engage in spirited disagreement was a sign of a good relationship (p. 211).*

# Self-Disclosure and Secrets

Shakespeare noted in *Macbeth* that "the false face must hide what the false heart doth know," suggesting that withholding information and being dishonest may affect the way one feels about oneself and relationships with others. All of us make choices, consciously or unconsciously, about the degree to which we disclose, are honest, and/or keep secrets.

## Self-Disclosure in Intimate Relationships

One aspect of intimacy in relationships is self-disclosure, which involves revealing personal information and feelings about oneself to another person.

Relationships become more stable when individuals disclose themselves to their partners (Tan et al. 2012). Areas of disclosure include one's formative

*A wonderful fact to reflect upon, that every human creature is constituted to be that profound secret and mystery to every other.*

**CHARLES DICKENS**

### PERSONAL CHOICES

### *How Much Do I Tell My Partner about My Past?*

Because of the fear of HIV infection and other sexually transmitted infections (STIs), some partners want to know the details of each other's previous sex life, including how many partners they have had sex with and in what contexts (e.g., hookups or stable relationships). Those who are asked will need to decide whether to disclose the requested information, which may include one's sexual orientation, present or past sexually transmitted infections, and any sexual preferences the partner might find bizarre (e.g., sadomachoism). Ample evidence suggests that individuals are sometimes dishonest with regard to the sexual information about their past they provide to their partners. The "number of previous sexual partners" is the most frequent lie undergraduates report telling each other. One female undergraduate who has had more partners than she wants to reveal said that when she is asked about her number, she smiles and says, "A lady never kisses and tells."

In deciding whether or not to talk honestly about your past to your partner, you may want to consider the following questions: How important is it to your partner to know about your past? Do you want your partner to tell you (honestly) about her or his past? What impact on your relationship will open disclosure have? What impact will withholding such information have on the level of intimacy you have with your partner?

years, previous relationships (positive and negative), experiences of elation and sadness, and goals (achieved and thwarted). We noted in the discussion of love in Chapter 2 that self-disclosure is a psychological condition necessary for the development of love. To the degree that you disclose yourself to another, you invest yourself in and feel closer to that person. People who disclose nothing are investing nothing and remain aloof. One way to encourage disclosure in one's partner is to make disclosures about one's own life and then ask about the partner's life.

## Secrets in Romantic Relationships

Most lovers keep a secret or two from their partners. Oprah Winfrey's biographer revealed a secret that Oprah kept from her long-term boyfriend Stedman Graham. When the couple was vacationing at a resort and Stedman left for a round of golf, Oprah promptly called room service and ordered two whole pecan pies. She called room service back a short time later to come and remove the empty tin plates (Kelley 2010).

College students also keep secrets from their partners. In a study of 431 undergraduates, Easterling et al. (2012) found the following:

1. ***Most kept secrets.*** Over 60% of the respondents reported having kept a secret from a romantic partner, and over one-quarter of respondents reported currently doing so.
2. ***Females kept more secrets.*** Sensitivity to the partner's reaction, desire to avoid hurting the partner, and desire to avoid damaging the relationship were the primary reasons why females were more likely than males to keep a secret from a romantic partner.
3. ***Spouses kept more secrets.*** Spouses have a great deal to lose if there is an indiscretion or if one partner does something the other will disapprove of (e.g., hook up). Partners who are dating or "seeing each other" have less to lose and are less likely to keep secrets.
4. ***Blacks kept more secrets.*** Blacks are a minority who live in a racist society and are still victimized by the white majority. One way to avoid such victimization is to keep one's thoughts to oneself—to keep a secret. This skill of deception may generalize to one's romantic relationships.
5. ***Homosexuals kept more secrets.*** Indeed, the phrase "in the closet" means "keeping a secret." Transgendered individuals in Europe are required to reveal their secret before marriage (Sharpe 2012).

Respondents were asked why they kept a personal secret from a romantic partner. "To avoid hurting the partner" was the top reason reported by 39% of the respondents. "It would alter our relationship" and "I feel so ashamed for what I did" were reported by 18% and 11% of the respondents, respectively.

Some secrets are technological—text messages, e-mails, and cell phone calls. Being deceptive with one's e-mails and cell phone is disapproved of by both women and men. In a study of 5,500 never-married individuals, 76% of the women and 53% of the men reported that being secretive with e-mails was a behavior they would not tolerate. Similarly, 69% of the women (47% of the men) said they would not put up with a partner who answered cell phone calls discretely (Walsh 2013).

## Family Secrets

Just as romantic partners have secrets, so do families. The family secret that takes the cake is the one Bernie Madoff kept from his wife and two adult sons—the information that he had defrauded almost 5,000 clients of over $50 billion. Madoff's wife and sons were adamant that their husband and father had acted alone and without their knowledge. Mark, the elder son, committed suicide because his father's behavior/secret had become known and lethal. Oprah

**TABLE 8.2 Family Secrets***

The following secrets were anonymously identified by the students in our marriage and family classes, who were asked to "identify a family secret (not necessarily your own)."

**(n = 70)**

**Abuse in the Family**

- Cousin molested by paternal uncle
- Sexual assault by parent
- Abusive grandfather
- Chopped off brother's finger on purpose but pretended it was an accident
- Brother threw a rock at me and knocked out a tooth

**Abuse from a partner in a relationship**

- Hidden abuse in marriage
- Sister's boyfriend beat her
- Dad's parents were abusive to each other in front of him

**Substance abuse**

- Dad smokes weed
- Aunt has marijuana plant
- Drug/alcohol abuse
- Alcoholism
- Cousin is a drug addict and a thief
- Best friend died of drug overdose and family claims it was a car accident
- Family member does heavy drugs

**Cheating/Adultery**

- infidelity/cheating
- Dad may have siblings in France from an affair when grandfather was in the war
- Friend's uncle had an affair, divorced her aunt after 20 years of marriage, and is dating the other lady
- Dad has three kids outside of marriage
- Husband is unfaithful
- Husband is cheating on wife
- Boyfriend's grandma left her fiancé for his grandfather
- Dad cheats on Mom frequently
- My dad cheated on my mom
- Aunt is pregnant; the baby's father is not her husband

**Homosexuality**

- Sister is a lesbian
- Gay cousin currently not out to family
- Neighbors are lesbians
- My cousin is a lesbian
- My uncle is gay

**Adoption**

- Child unaware that he is adopted
- Uncle gave child up for adoption because he wasn't married

**Heritage**

- Some of my relatives were Nazis
- I'm Polish and Cherokee
- Grandmother changed the spelling of our last name

**Illness/death**

- Grandmother has a serious mental illness
- Mother attempted suicide and is in therapy
- Best friend has an eating disorder
- Great grandfather committed suicide
- Grandpa committed suicide in 1995

**Other**

- Grandfather ran moonshine
- Brother's engaged
- A recipe I won't share
- I was in a wreck
- Family was placed under the Witness Protection Agency

** Unpublished data collected for this text. Courtship and Marriage/Marriage and Family classes, Department of Sociology, East Carolina University.*

Winfrey also had a family secret. At age 15 she gave birth to a son in her seventh month of pregnancy. The baby died a month later. When Oprah ran for Miss Black Tennessee she completed an application on which she stated that she had never had a child (Kelley 2010).

We asked our students to identify (anonymously) family secrets they were aware of. Table 8.2 reflects the family secrets revealed by 70 students.

**FAMILY POLICY**

## Should One Partner Disclose Human Immunodeficiency Virus (HIV)/Sexually Transmitted Infection (STI) Status to Another?

An estimated 25% of undergraduates report that they have or have had an STI. Individuals often struggle over whether or how to tell a partner if they have an STI, including HIV infection. If a person in a committed relationship acquires an STI, then that individual, or the partner, may have been unfaithful and have had sex with someone outside the relationship. Thus, disclosure about an STI may also mean confessing one's own infidelity or confronting the partner about his/her possible infidelity. (The infection may, however, have occurred prior to the current relationship but gone undetected.) Individuals who have an STI and who are beginning a new relationship face a different set of concerns. Will their new partner view them negatively? Will they want to continue the relationship? One Internet ad began, "I have herpes—Now that that is out of the way. . . ."

Although telling a partner about having an STI may be difficult and embarrassing, avoiding disclosure or lying about having an STI represents a serious ethical violation. The responsibility to inform a partner that one has an STI—before having sex with that partner—is a moral one. But there are also legal reasons for disclosing one's sexual health condition to a partner. If you have an STI and you do not tell your partner, you may be liable for damages if you transmit it to your partner.

# Dishonesty, Lying, and Cheating

Relationships are compromised by dishonesty, lying, and cheating.

## Dishonesty

*There are no degrees of honesty.*

AMISH PROVERB

Dishonesty and deception take various forms. One is a direct lie—saying something that is not true (e.g., telling your partner that you have had six previous sexual partners when, in fact, you have had 13). Not correcting an assumption is another form of dishonesty (e.g., your partner assumes you are heterosexual but you are bisexual).

## Lying in American Society

*When telling lies becomes not only easy but a part of you, what chance does love have?*

LARRY O'CONNER, *THE BELT*

Lying, a deliberate attempt to mislead, is pervasive. Lance Armstrong was stripped of his seven Tour de France titles for lying about doping. Journalist Mike Daisey admitted to fabricating stories about oppressed workers at an Apple contractor's factory in China. He said of his deception, "It's not journalism. It's theater." Sixty-two percent of 125 Harvard students admitted to cheating on either tests or papers (Webley 2012). Politicians routinely lie to citizens ("Lobbyists can't buy my vote"), and citizens lie to the government (via cheating on taxes). Teachers lie to students ("The test will be easy"), and students lie to teachers ("I studied all night"). Parents lie to their children ("It won't hurt"), and children lie to their parents ("I was at my friend's house"). Dating partners lie to each other ("I've had a couple of previous sex partners"), women lie to men ("I had an orgasm"), and men lie to women ("I'll call"). The price of lying is high—distrust and alienation. A student in class wrote:

> *At this moment in my life I do not have any love relationship. I find hanging out with guys to be very hard. They lie to you about anything and you wouldn't know the truth. I find that college dating is mostly about sex and having a good time before you really have to get serious. That is fine, but that is just not what I am all about.*

**Catfishing** a person makes up an online identity and an entire social façade to trick another person into becoming involved in an emotional relationship.

**Catfishing** refers to a process whereby a person makes up an online identity and an entire social façade to trick a person into becoming involved in an emotional relationship.

Reporting HIV infection and acquired immunodeficiency syndrome (AIDS) is mandatory in most states, although partner notification laws vary from state to state. New York has a strong partner notification law that requires health care providers to either notify any partners the infected person names or to forward the information about partners to the Department of Health, where public health officers notify the partners that they have been exposed to an STI and to schedule an appointment for STI testing. The privacy of the infected individual is protected by not revealing names to the partner being notified of potential infection. In cases where the infected person refuses to identify partners, standard partner notification laws require doctors to undertake notification without cooperation if they know of the sexual partner or spouse.

**Your Opinion?**

1. What percentage of undergraduates (who knowingly have an STI) would have sex with another undergraduate and not tell the partner?
2. What do you think the penalty should be for deliberately exposing a person to an STI?
3. What partner notification law do you recommend?

The catfish is the lonely person on the Internet who is susceptible to being seduced into this fake relationship. University of Notre Dame football player Manti Te'o reported that he was a victim of an online hoax, fooled into a relationship by someone pretending to be a woman named Lennay Kekua. The creator of the pretend Lennay Kekua then conspired with others to lead Te'o to believe that Kekua had died of leukemia.

## Lying and Cheating in Romantic Relationships

Lying is epidemic in college student romantic relationships. In response to the statement, "I have lied to a person I was involved with," 69% of 4,431 undergraduates (women more than men) reported "yes." Sixteen percent reported having lied to a partner about their previous number of sexual partners (Hall and Knox 2013).

*When she assured him that she had not kissed the other man, he knew she was lying—yet he was glad that she had taken the trouble to lie to him.*

**F. SCOTT FITZGERALD, *WINTER DREAMS***

Cheating may be defined as having sex with someone else while involved in a relationship with a romantic partner. When 4,558 undergraduates were asked if they had cheated on a partner they were involved with, almost a quarter (24%) reported that they had done so (Hall and Knox 2013). Even in monogamous relationships, there is considerable cheating. Vail-Smith et al. (2010) found that of 1,341 undergraduates, 27.2% of the males and 19.8% of the females reported having oral, vaginal, or anal sex outside of a relationship that their partner considered monogamous.

People most likely to cheat in these monogamous relationships were men over the age of 20, those who were binge drinkers, members of a fraternity, male NCAA athletes, and those who reported that they were nonreligious. White et al. (2010) also studied 217 couples where both partners reported on their own risk behaviors and their perceptions of their partner's behavior; 3% of women and 14% of men were unaware that their partner had recently had a concurrent partner. Eleven and 12%, respectively, were unaware that their partner had ever injected drugs; 10 and 12% were unaware that their partner had recently received an STI diagnosis; and 2 and 4% were unaware that their partner was HIV-positive. These data suggest a need for people in committed relationships to reconsider their risk of sexually transmitted infections and to protect themselves via condom usage. In addition, one of the ways in which college students deceive their partners is by failing to disclose that they have an STI. Approximately 25% of college students will contract an STI while they are in college (Purkett 2013). Because the potential to harm an unsuspecting partner is considerable, should we have a national social policy regarding such disclosure (see Family Policy feature)?

Strickler and Hans (2010) conceptualized infidelity (cheating) as both sexual and nonsexual. Sexual cheating was intercourse, oral sex, and kissing. Nonsexual cheating could be interpersonal (secret time together, flirting), electronic (text messaging, e-mailing), or solitary (sexual fantasies, pornography, masturbation). Of 400 undergraduates, 74% of the males and 67% of the females in a committed relationship reported that they had cheated according to their own criteria. Hence, in the survey, they identified a specific behavior as cheating and later reported whether they had engaged in that behavior.

WHAT IF?

### What If You Discover Your Partner Is Cheating on You?

Because cheating does occur rather frequently, to deny that this will ever happen in one's own relationship may be unrealistic. Reactions will vary from immediate termination of the relationship, to taking a break from the relationship, to revenge by cheating also. One scenario is to discuss the dishonesty with the partner to discover any relationship deficits that may be corrected. Another is to discuss the acceptability of the behavior in terms of frequency. Does everyone make a mistake sometime and this is to be overlooked, or is the dishonesty of a chronic variety that will continue? Most individuals are devastated to discover a betrayal and struggle to move beyond it. Continuing the relationship is functional only if the dishonesty is not chronic. Chronic unfaithfulness takes advantage of the forgiveness of the partner and permanently infuses the relationship with distrust and deceit—a recipe for disaster. Hannon et al. (2010) emphasized that getting beyond an act of betrayal requires the cooperation of the perpetrator who expresses sorrow, apologizes, and makes amends in exchange for the partner who forgives. Both work together to get beyond the impasse.

# Technology Mediated Communication in Romantic Relationships

The following scenario from one of our students reflects that technology has an enormous effect on romantic relationships.

*"I saw his photo on Match.com."*

*"I typed his name into Facebook and friend requested him."*

*"He accepted, messaged me and we began to text each other."*

*"We were long distance so we began to have long talks on Skype."*

*"To keep his interest I would sex text him ('What would you like me to do to you?')"*

*"After we moved in together, I snooped, checked his cell phone—discovered other women."*

*"I sent him a text message that it was over, and moved out."*

Taylor et al. (2013) surveyed 1,003 emerging adults (18-25) about the use of technology in romantic relationships. Both men and women reported that texting is appropriate with a potential romantic partner; however, neither agree with announcing a relationship on Facebook before having the "relationship talk" and becoming a committed couple.

Rappleyea et al. (2014) analyzed data on 1,003 young adults in regard to technology and relationship formation and found that the respondents believe that "talking," "hanging out," and "sharing intimate details" are more important when compared with using communication technologies to establish a relationship (p. 269).

Text messages have become a primary means for flirting (73% of 18–25 year olds; Gibbs 2012) and the initiation, escalation, and maintenance of romantic relationships (Bergdall et al. 2012). Some (to the disapproval of most) end a relationship with a text message (Gershon 2010). Text messages and communication technology are also used by divorced parents in the co-parenting of their children (Ganong et al. 2011).

Chelsea Curry

Some choose to end a relationship with a text message.

Chelsea Curry

Getting a break-up text message is one of the least welcome text messages.

## APPLYING SOCIAL RESEARCH | The Good, the Bad, and Technology Mediated Communication in Romantic Relationships*

Three-hundred-and-fifty-four undergraduates at a large southeastern university completed a 42-item Internet survey designed to reveal the degree to which electronic delivery (e.g., texting/e-mail versus face-to-face communication) was used to deliver good (e.g., "I love you," "Let's get married," "Are you down for sex tonight?") and bad (e.g., "I think we should break up," "I cheated," "I have an STI," "I got into graduate school and will be moving") news to a romantic partner.

**Findings**

1. Frequency – Nearly half of the respondents had experience receiving bad news via an electronic method—46% had been broken up with via text, e-mail, or another electronic method and 44% had been told that their partner had been unfaithful. Respondents not only received but also gave bad news—39% percent reported that they had ended a relationship and 25% had communicated an infidelity to a partner via electronic means.

   The respondents also revealed that technology-mediated communication (TMC) was used to convey good news to romantic partners. Over half (51%) reported that their partner communicated "I love you" for the first time via electronic methods such as a text or an e-mail. In addition, approximately 50% of the respondents noted that their partners had used text or e-mail to broach the subject of readiness to have sex. Thirty-five percent reported informing partners of good news about a job promotion and 6% reported receiving a marriage proposal by text or e-mail.
2. Gender – In regard to disclosing that one had been unfaithful, 15% more males than females used TMC. However, a similar percent of males and females responded that they had ended a relationship with a romantic partner electronically (35% and 39% respectively).
3. Media Ideologies and Emotional Response – While technology was used to convey both negative and positive content, it was not identified as preferable

### National and International **Data**

Based on a survey of 4,700 respondents in the United States and seven other countries (China, U.K., India, South Korea, South Africa, Indonesia, and Brazil), 9 in 10 carry a cell phone (one in four check it every 30 minutes; one in five check it every 10 minutes) (Gibbs 2012).

**Nomophobia** individual is dependent on virtual environments to the point of having a social phobia.

*I should just change my voicemail greeting to:*

*Please hang up and text me, thanks.*

UNIVERSITY STUDENT

## Texting and Interpersonal Communication

Personal, portable, wirelessly networked technologies in the form of iPhones, Droids, iPads, etc., have become commonplace in the lives of individuals (Gibbs 2012; Looi et al. 2010) as a way of staying connected with offline friends and partners (Reich et al. 2012). Bauerlein (2010) noted that today's youth are being socialized in a hyper-digital age where traditional modes of communication will be replaced by gadgets and texting will become the primary mode of communication. This shift to greater use of technology affects relationships in both positive and negative ways. On the positive side, it allows for instant and unabated connection—individuals can text each other throughout the day so that they are "in effect, together all the time." One of our students noted that on a regular day, she and her boyfriend will exchange 50 to 60 text messages. On the negative side is **nomophobia**, where the individual is dependent on virtual environments to the point of having a social phobia (King et al. 2013). He or she finds personal interaction difficult.

Coyne et al. (2011) examined the use of technology by 1,039 individuals in sending messages to their romantic partner. The respondents were more likely to use their cell phones to send text messages than any other technology. The most common reasons were to express affection (75%), discuss serious issues to face-to-face delivery. Almost 90 (88%) percent of the respondents listed face to face as their preferred method of receiving bad news and 71% listed face to face as their preferred method of delivering bad news. Similarly, face to face was the preferred method for delivering and receiving good news with both receiving a preference over 90%.

### Discussion and Theoretical Framework

Analysis of the data revealed that around half of the respondents reported using TMC in the delivery of good and bad news to romantic partners. While previous research emphasized that an advantage of TMC is that it helps keep romantic partners connected, individuals might be cautious in relying too heavily on electronic communication to communicate sensitive issues to partners. While the majority of respondents indicated that receiving good news via text or e-mail would not offend them, 76% would feel upset or have felt hurt by receiving bad news via such methods. Given that there can often be negative implications, partners should consider discussing media ideologies concerning appropriate use of technologies so as to avoid hurting each other or damaging their relationship.

Symbolic interactionism provided the theoretical framework for viewing the findings of this study. Symbolic interactionism is a micro-level theory that focuses on the meanings individuals attribute to phenomena. Symbolic interactionists focus on the importance of symbols, subjective versus objective reality, and the definition of social situations. The participants in this study revealed that they viewed a text message or e-mail as an undesirable delivery method in receiving bad news from a romantic partner when compared to receiving such news face to face.

### Source

Abridged and adapted from Faircloth, M., D. Knox, and J. Brinkley. 2012. The Good, the Bad and Technology Mediated Communication in Romantic Relationships. Paper, Southern Sociological Society, New Orleans, March 21–24.

(25%), and apologize (12%). There were no significant differences in use by gender, ethnicity, or religion. Pettigrew (2009) emphasized that **texting** or text messaging (short typewritten messages—maximum of 160 characters sent via cell phone) is used to "commence, advance, maintain" interpersonal relationships and is viewed as more constant and private than talking on a cell phone. Women text more than men. In a Nielsen State of the Media Survey conducted in the third quarter of 2010, it was reported that women spent 818 minutes texting 640 messages; men, 716 minutes texting 555 messages (Carey and Salazar 2011). On the negative side, these devices encourage the continued interruption of face-to-face communication between individuals and encourage the intrusion of one's work/job into the emotional intimacy of a couple and the family.

**Texting** short typewritten messages sent via a cell phone to commence, advance, and maintain interpersonal relationships.

Huang and Leung (2010) studied instant messaging and identified four characteristics of "addiction" in teenagers: preoccupation with IM, loss of relationships due to overuse, loss of control, and escape. Results also showed that shyness and alienation from family, peers, and school were significantly and positively associated with levels of IM addiction. As expected, both the level of IM use and level of IM addiction were significantly linked to teenagers' poorer academic performance.

Technology is also used to convey both good and bad news to a romantic partner. See the Applying Social Research that begins on page 228.

## When Texting/Facebook Become a Relationship Problem

Schade et al. (2013) studied the effects of technology on romantic relationships. They analyzed data from 276 adults ages 18–25 in committed relationships and found that male texting frequency was negatively associated with relationship satisfaction and stability scores for both partners while female texting frequency was positively associated with their own relationship stability scores. Hence, females thrived on texting, which had a positive relationship effect. Males tolerated it, which had the opposite/negative effect.

Cell phones/text messaging may be a source of conflict (e.g., partner text messages while the lover is talking to him or her). Over 60 percent of Chinese respondents and a quarter of U.S. respondents noted that a mobile device had come between them and their spouse (Gibbs 2012). But these devices may also help to reduce conflict. Perry and Werner-Wilson (2011) assessed the use of technology in conflict resolution in romantic relationships and found that text messaging allowed for de-escalation of conflict and provided time to construct ideas—to think about what they were going to say.

Norton and Baptist (2012) identified how social networking sites (e.g., Facebook, with over a billion users and 140 billion friendship connections) are problematic for couples—the sites are intrusive (e.g., partner surfs while lover is talking), and they encourage compulsive use (e.g., partner is always sending/receiving messages) and infidelity (e.g., flirting/cheating online). They studied how 205 married individuals mitigated the impact of technology on their relationship. Three strategies included openness (e.g., each spouse knew the passwords and online friends and had access to each other's online social networking accounts, e-mail, etc.), fidelity (e.g., flirting and online relationships were off limits), and appropriate people (e.g., knowing the friends of the partner and no former partners allowed).

## Sexting

**Sexting** sending erotic text and photo images via a cell phone.

Another way in which technology affects communication, particularly in romantic relationships, is **sexting** (sending erotic text and photo images via a cell phone). Sexting begins in high school. Strassberg et al. (2013) surveyed 606 high school students and found that almost 20% reported having *sent* a sexually explicit image of themselves via cell phone while almost twice as many reported that they had *received* a sexually explicit picture via cell phone (of these, over 25% indicated that they had *forwarded* this picture to others). Of those who sent a sexually explicit cell phone picture, over a third did so despite believing that there could be serious legal consequences attached to the behavior.

Burke-Winkelman et al. (2014) reported that 65% of 1,652 undergraduates reported sending sexually suggestive texts or photos to a current or potential partner (69% reported receiving). Almost a third (31%) reported sending the text messages to a third party. In regard to how they felt about sending nude photos, less than half were positive, and females were more likely to feel pressure to send nude photos.

Parker et al. (2011) analyzed data from 483 undergraduates at a large southeastern university who completed a 25-item Internet questionnaire. They found that about two-thirds (64%) reported sending a sex text message; 43%, a sex photo. There were no gender differences, but, as also occurred in the Burke-Winkelman research, African Americans reported higher frequencies of sending sex content to a romantic partner.

Dir et al. (2013) surveyed 278 undergraduates in regard to their receiving and sending sex text messages and sex photos. Gender and relationship status were significant predictors of specific sexting behaviors. Males reported sending more sex photos. In addition, those involved in a relationship (dating, serious relationship, cohabiting) sent more sexts than uninvolved singles. Males reported more positive outcomes of receiving sexts (e.g., sexual excitement); women were more likely to feel uncomfortable (e.g., embarrassed). Weisskirch and Delevi (2011) examined the role of attachment anxiety in sexting and found attachment anxiety predicted positive attitudes toward sexting such as accepting it as normal, believing that it would enhance the relationship, and stating that partners expected it. Temple et al. (2012) studied sexting in over

900 public high school students and found that doing so was associated with the likelihood of having sex, having multiple sex partners, and using alcohol before sex.

While undergraduates are not at risk as long as the parties are age 18 or older, sending erotic photos of individuals younger than 18 can be problematic. Sexting is considered by many countries as child pornography and laws related to child pornography have been applied in cases of sexting. Six high school students in Greensburg, Pennsylvania, were charged with child pornography after three teenage girls allegedly took nude or semi-nude photos of themselves and shared them with male classmates via their cell phones.

Some undergraduate females are under age 18. Having or sending nude images of underage individuals is a felony which can result in fines, imprisonment, and a record. We noted above that Strassberg et al. (2013) reported that a third of their high school respondents reported that they had forwarded a sexually explicit photo that could get them in serious legal trouble.

### Video-Mediated Communication (VMC)

Communication via computer between separated lovers, spouses, and family members is becoming more common. Furukawa and Driessnack (2013) assessed the use of VMC (**video-mediated communication**) in a sample of 341 online participants (ages 18 to 70-plus). Ninety-six percent reported that VMC was the most common method they used to communicate with their family and 60% reported doing so at least once a week. VMC allows the person to see and hear what is going on; for example, while the grandparents can't be present Christmas morning they can see the excitement of their grandchildren opening their gifts.

**Video-mediated communication** individuals are able to see and hear others they are separated from to simulate their presence and enjoy "being with" their beloved.

## Theories Applied to Relationship Communication

Symbolic interactionism and social exchange are theories that help to explain the communication process.

### Symbolic Interactionism

Interactionists examine the process of communication between two actors in terms of the meanings each attaches to the actions of the other. Definition of the situation, the looking-glass self, and taking the role of the other (discussed in Chapter 1) are all relevant to understanding how partners communicate. With regard to resolving a conflict over how to spend the semester break (e.g., vacation alone or go to see parents), the respective partners must negotiate their definitions of the situation (is it about their time together as a couple or their loyalty to their parents?). The looking-glass self involves looking at each other and seeing the reflected image of someone who is loved and cared for and someone with whom a productive resolution is sought. Taking the role of the other involves each partner's understanding the other's logic and feelings about how to spend the break.

### Social Exchange

Exchange theorists suggest that the partners' communication can be described as a ratio of rewards to costs. Rewards are positive exchanges, such as compliments, compromises, and agreements. Costs refer to negative exchanges, such as critical remarks, complaints, and attacks. When the rewards are high and the costs are low, the outcome is likely to be positive for both partners (profit).

When the costs are high and the rewards low, neither may be satisfied with the outcome (loss).

When discussing how to spend the semester break, the partners are continually in the process of exchange—not only in the words they use but also in the way they use them. If the communication is to continue, both partners need to feel acknowledged for their points of view and to feel a sense of legitimacy and respect. Communication in abusive relationships is characterized by the parties criticizing and denigrating each other, which usually results in a shutdown of the communication process and a drift toward ending the relationship.

## Conflict in Relationships

**Conflict** context in which the perceptions or behavior of one person is in contrast to or interferes with those of the other.

**Conflict** is the context in which the perceptions or behavior of one person is in contrast to or interferes with those of the other. A truism of relationships is, "If you haven't had a conflict with your partner, you haven't been going together long enough." This section explores the inevitability, benefits/drawbacks, and sources of conflict in relationships.

### Inevitability of Conflict

If you are alone this Saturday evening from six o'clock until midnight, you are assured of six conflict-free hours. But if you plan to be with your partner, roommate, or spouse during that time, the potential for conflict exists. Whether you eat out, where you eat, where you go after dinner, and how long you stay must be negotiated. Although it may be relatively easy for you and your companion to agree on one evening's agenda, marriage involves the meshing of desires on an array of issues for potentially 60 years or more. Indeed, conflict is inevitable in any intimate relationship.

### Benefits and Drawbacks of Conflict

*I'm not telling you it's going to be easy.*

*I'm telling you it is going to be worth it.*

ART WILLIAMS, BILLIONAIRE EXECUTIVE

Conflict can be healthy and productive for a couple's relationship. Ignoring an issue may result in the partners becoming increasingly resentful and dissatisfied with the relationship. Indeed, not talking about a concern may do more damage to a relationship (since resentment may build) than bringing up the issue and discussing it. Couples in trouble are not those who disagree but those who never discuss their disagreements. However, sustained conflict over chronic problems contributes to poor mental and physical health of the partners in the relationship (Wickrama et al. 2010).

### Sources of Conflict

Conflict has numerous sources, some of which are easily recognized, whereas others are hidden inside the web of relationship interaction.

**1. *Behavior.*** Money is a frequent issue over which couples conflict. The behavioral expression of a money issue might include how the partner spends money (excessively), the lack of communication about spending (e.g., does not consult the partner), and the target (e.g., items considered unnecessary by the partner).

*There is nothing either good or bad but thinking makes it so.*

SHAKESPEARE, *Hamlet*, II, II, 253

**2. *Cognitions, perceptions and attributions.*** How you perceive your partner's behavior and the motivations you attribute to your partner's behavior influence your behavior and your partner's behavior (Durtschi et al. 2011; Hall and Adam 2011). If you view your partner's being late as unavoidable (e.g., caught in traffic), you will behave kindly when your partner eventually shows up (as will your partner). But if you attribute your partner's being late as an attempt to make you angry, you will be angry, and the interaction will deteriorate. Claffey and Mickelson (2010) noted that it is not the actual division of labor but how the partners perceive it which has an impact on their lives and relationship.

Beliefs about marriage also influence conflicts in marriage. Spouses who believe that marriage is permanent and that sharing decisions is best report low conflict in marriage and higher marital happiness (Kamp Dush and Taylor 2012). Having an egalitarian relationship is also a context for continued positive communication (Jonathan and Knudson-Martin 2012).

**3. *Value differences.*** Because you and your partner have had different socialization experiences, you may also have different values—about religion (one feels religion is a central part of life; the other does not), money (one feels uncomfortable being in debt; the other has the buy-now-pay-later philosophy), in-laws (one feels responsible for parents when they are old; the other does not), and children (number, timing, discipline). The effect of value differences depends less on the degree of the difference than on the degree of rigidity with which each partner holds his or her values. Dogmatic and rigid thinkers, feeling threatened by value disagreement, may try to eliminate alternative views and thus produce more conflict. Partners who recognize the inevitability of difference may consider the positives of an alternative view and move toward acceptance. When both partners do this, the relationship takes priority and the value differences suddenly become less important.

**4. *Inconsistent rules.*** Partners in all relationships develop a set of rules to help them function smoothly. These unwritten but mutually understood rules include what time you are supposed to be home after work, whether you should call if you are going to be late, and how often you can see friends (same and other sex) alone. Conflict results when the partners disagree on the rules or when inconsistent rules develop in the relationship. For example, one wife expected her husband to take a second job so they could afford a new car, but she also expected him to spend more time at home with the family.

**5. *Leadership ambiguity.*** Unless a couple has an understanding about which partner will make decisions in which area (e.g., the wife may make decisions about money management; the husband will make decisions about disciplining the children), conflict may result. Couples may benefit from a clear understanding of which partner will function primarily in which role.

### Styles of Conflict Resolution

Conflict can be resolved by one or both partners changing positions to be more in line with the other partner, compromising, agreeing to disagree (no resolution), or forcing one's position. All are likely to be used over the course of a relationship with modifying one's position and compromising having the most positive outcome.

*In the end I can't explain why I didn't divorce any more that I can explain why I married.*

KATHLEEN AGUERO, WRITER

## Fighting Fair: Seven Steps in Conflict Resolution

Before reading about how to resolve conflict, you may wish to take the Communication Danger Signs Scale to assess the degree to which you may have a problem in this area.

*Beware of entrance to a quarrel.*

SHAKESPEARE, *HAMLET*, I, III, 65

When a disagreement begins, it is important to establish rules for fighting that will leave the partners and their relationship undamaged. Indeed, Lavner and Bradbury (2010) studied 464 newlyweds over a four-year period, noticed the precariousness of relationships (even those reporting considerable satisfaction divorced), and recommended that couples "impose and regularly maintain ground rules for safe and nonthreatening communication." Such guidelines for

## SELF-ASSESSMENT | Communication Danger Signs Scale

This scale is designed to assess the degree to which there is communication trouble in your relationship.

### Directions

After reading each sentence carefully, circle the number that best represents how often this happens in your relationship.

| 1 | 2 | 3 |
|---|---|---|
| Never/almost never | Occasionally | Frequently |

| | N | O | F |
|---|---|---|---|
| 1. Little arguments escalate into ugly fights with accusations, criticisms, name calling, or bringing up past hurts. | 1 | 2 | 3 |
| 2. My partner criticizes or belittles my opinions, feelings, or desires. | 1 | 2 | 3 |
| 3. My partner seems to view my words or actions more negatively than I mean them to be. | 1 | 2 | 3 |
| 4. When we have a problem to solve, it is like we are on opposite teams. | 1 | 2 | 3 |
| 5. I hold back from telling my partner what I really think and feel. | 1 | 2 | 3 |
| 6. I feel lonely in this relationship. | 1 | 2 | 3 |
| 7. When we argue, one of us withdraws, that is, doesn't want to talk about it anymore; or leaves the scene. | 1 | 2 | 3 |

### Scoring

Add the numbers you circled. 1 (never) is the response reflecting the ultimate safe context in which you communicate with your partner and a 3 (frequently) is the most toxic of communication contexts. The lower your total score (7 is the lowest possible score), the greater your communication context comfort. The higher your total score (21 is the highest possible score), the more you are likely to feel anxious when around or communicating with your partner. A score of 14 places you at the midpoint between being extremely comfortable communicating with your partner (a score of 7) and being extremely uncomfortable (a score of 21).

### Source

Dr. Howard Markman, Director, Center for Marital and Family Studies at the University of Denver. Used by permission. Contact Dr. Markman at http://loveyourrelationship.com/ for information about Communication/Relationship Retreats. Also see Markman, H. J., S. M. Stanley, and S. L. Blumberg. 2010. *Fighting for your marriage* (3rd ed.) San Francisco, CA: Jossey-Bass.

fair fighting/effective communication include not calling each other names, not bringing up past misdeeds, and not attacking each other.

Gottman (1994) identified destructive communication patterns to avoid which he labeled as "the four horsemen of the apocalypse"—criticism, defensiveness, contempt (the most damaging), and stonewalling. He also noted that being positive about the partner is essential—partners who said positive things to each other at a ratio of 5:1 (positives to negatives) were more likely to stay together. We have noted that "avoiding giving your partner a zinger" is also essential to maintaining a good relationship.

Fighting fairly also involves keeping the interaction focused and respectful, and moving toward a win-win outcome. If recurring issues are not discussed and resolved, conflict may create tension and distance in the relationship, with the result that the partners stop talking, stop spending time together, and stop being intimate. Developing and using skills for fair fighting and conflict resolution are critical for the maintenance of a good relationship.

Howard Markman is head of the Center for Marital and Family Studies at the University of Denver. He and his colleagues have been studying 150 couples at yearly intervals (beginning before marriage) to determine those factors most responsible for marital success. They have found that a set of communication skills that reflect the ability to handle conflict, which they call "constructive arguing," is the single biggest predictor of marital success over time (Marano 1992). According to Markman, "Many people believe that the causes of marital problems are the differences between people and problem areas such as money, sex, and children. However, our findings indicate it is not the differences that are important, but how these differences and problems are handled, particularly early in marriage" (Marano 1992, p. 53). Markman et al. (2010) provide details for constructive communication in their book *Fighting for Your Marriage*. The following sections identify standard steps for resolving interpersonal conflict.

### Address Recurring, Disturbing Issues

Addressing issues in a relationship is important. But whether partners do so is related to their level of commitment to the relationship. Partners who are committed to each other and to their relationship invest more time and energy to resolving problems. Those who feel stuck in relationships with barriers to getting out (e.g., children, economic dependence) avoid problem resolution (Frye 2011). The committed are intent on removing relationship problems.

### Identify New Desired Behaviors

Dealing with conflict is more likely to result in resolution if the partners focus on what they *want* rather than what they *don't want*. Tell your partner specifically what you want him or her to do. For example, if your partner routinely drives the car but never puts gas in it, rather than say, "Stop driving the gas out of the car," you might ask him or her to "always keep at least a fourth tank of gas in the car."

### Identify Perceptions to Change

Rather than change behavior, changing one's perception of a behavior may be easier and quicker. Rather than expect one's partner to always be on time, it may be easier to drop this expectation and to stop being mad about something that doesn't matter. South et al. (2010) emphasized the importance of perception of behavior in regard to marital satisfaction.

### Summarize Your Partner's Perspective

We often assume that we know what our partner thinks and why he or she does things. Sometimes we are wrong. Rather than assume how our partner thinks and feels about a particular issue, we might ask open-ended questions in an effort to learn our partner's thoughts and feelings about a particular situation. The answer to "How do you feel about me taking an internship abroad next semester?' will give you valuable information.

### Generate Alternative Win-Win Solutions

Looking for win-win solutions to conflicts is imperative. Solutions in which one person wins means that the other person is not getting needs met. As a result, the person who loses may develop feelings of resentment, anger, hurt, and hostility toward the winner and may even look for ways to get even. In this way, the winner is also a loser. In intimate relationships, one winner really means two losers.

Generating win-win solutions to interpersonal conflict often requires **brainstorming**. The technique of brainstorming involves suggesting as many alternatives as possible without evaluating them. Brainstorming is crucial to conflict resolution because it shifts the partners' focus from criticizing each other's perspective to working together to develop alternative solutions.

**Brainstorming** suggesting as many alternatives as possible without evaluating them.

With our colleagues (Knox et al. 1995), we studied the degree to which 200 college students in ongoing relationships were involved in win-win, win-lose, and lose-lose relationships. Descriptions of the various relationships follow:

**Win-win relationships** are those in which conflict is resolved so that each partner derives benefits from the resolution. For example, suppose a couple has a limited amount of money and disagrees on whether to spend it on eating out or staying at home. One possible win-win solution might be for the couple to eat out this time but eat in the next time. More than three-quarters (77.1%) of the students reported being involved in a win-win relationship, with men and women reporting similar percentages.

**Win-win relationship** relationship in which conflict is resolved so that each partner derives benefits from the resolution.

**Win-lose solution** outcome of a conflict in which one partner wins and the other loses.

**Lose-lose situation** a solution to a conflict in which neither partner benefits.

An example of a **win-lose solution** would be for one of the partners to get what he or she wanted (eat out or in with no promise of future exchange), with the other partner getting nothing of what he or she wanted. Of the respondents, 20% were involved in win-lose relationships.

A **lose-lose solution** is one in which both partners get nothing that they want—in the scenario presented, the partners would neither go out to eat nor stay in but spend time with their own friends . . . both would be mad at the other. Only 2% reported that they were involved in lose-lose relationships.

Of the students in win-win relationships, 85% reported that they expected to continue their relationship, in contrast to only 15% of students in win-lose relationships. No student in a lose-lose relationship expected the relationship to last.

After a number of solutions are generated, each solution should be evaluated and the best one selected. In evaluating solutions to conflicts, it may be helpful to ask the following questions:

1. Does the solution satisfy both individuals? (Is it a win-win solution?)
2. Is the solution specific? Does it specify exactly who is to do what, how, and when?
3. Is the solution realistic? Can both parties realistically follow through with what they have agreed to do?
4. Does the solution prevent the problem from recurring?
5. Does the solution specify what is to happen if the problem recurs?

*An honest disagreement is better than a dishonest agreement.*

VITHALDAS H. PATEL, AUTHOR

Kurdek (1995) emphasized that conflict-resolution styles that stress agreement, compromise, and humor are associated with marital satisfaction, whereas conflict engagement, withdrawal, and defensiveness styles are associated with lower marital satisfaction. In his own study of 155 married couples, the style in which the wife engaged the husband in conflict and the husband withdrew was particularly associated with low marital satisfaction for both spouses.

Communicating effectively and creating a context of win-win in one's relationship contributes to a high-quality marital relationship, which is good for one's health.

## Forgive

Too little emphasis is placed on forgiveness as an emotional behavior that can move a couple from a deadlock to resolution. Merolla and Zhang (2011) noted that offender remorse positively predicted forgiveness and that such forgiveness was associated with helping to resolve the damage. Hill (2010) studied forgiveness and emphasized that it is less helpful to try to "will" oneself to forgive the transgressions of another than to engage a process of self-reflection—that one has also made mistakes, hurt others, and is guilty—and to empathize with the fact that we are all fallible and need forgiveness. In addition, forgiveness ultimately means letting go of one's anger, resentment, and hurt and its power comes from offering forgiveness as an expression of love to the person who has betrayed him or her. Forgiveness also has a personal benefit—it reduces hypertension and feelings of stress. To forgive is to restore the relationship—to pump life back into it. Of course, forgiveness given too quickly may be foolish. A person who has deliberately hurt their partner without remorse may not deserve forgiveness.

*When you forgive, you in no way change the past—but you sure do change the future.*

BERNARD MELTZER, RADIO PERSONALITY

*Revenge and vengeance do more to the container in which they are held than to the person to whom they are intended.*

ANONYMOUS

It takes more energy to hold on to resentment than to move beyond it. One reason some people do not forgive a partner for a transgression is that one can use the fault to control the relationship. "I wasn't going to let him forget," said one woman of her husband's infidelity.

**Amae** expecting a close other's indulgence when one behaves inappropriately.

A related concept to forgiveness is **amae**. Marshall et al. (2011) studied the concept of amae in Japanese romantic relationships. The term means expecting a close other's indulgence when one behaves inappropriately. Thirty Japanese undergraduate romantic couples kept a diary for two weeks that assessed their amae behavior (requesting, receiving, providing amae). Results revealed that amae behavior was associated with greater relationship quality and less conflict. "Cutting one some slack" may be another way of expressing amae.

### Be Alert to Defense Mechanisms

Effective conflict resolution is sometimes blocked by **defense mechanisms**—techniques that function below the level of awareness to protect individuals from anxiety and to minimize emotional hurt. The following paragraphs discuss some common defense mechanisms.

**Defense mechanisms** techniques that function below the level of awareness to protect individuals from anxiety and to minimize emotional hurt.

**Escapism** is the simultaneous denial of and avoidance of dealing with a problem. The usual form of escape is avoidance. The spouse becomes "busy" and "doesn't have time" to think about or deal with the problem, or the partner may escape into recreation, sleep, alcohol, marijuana, or work. Denying and withdrawing from problems in relationships offer no possibility for confronting and resolving the problems.

**Escapism** simultaneous denial of and avoidance of dealing with a problem.

**Rationalization** is the cognitive justification for one's own behavior that unconsciously conceals one's true motives. For example, one wife complained that her husband spent too much time at the health club in the evenings. The underlying reason for the husband's going to the health club was to escape an unsatisfying home life. However, the idea that he was in a dead marriage was too painful and difficult for the husband to face, so he rationalized to himself and his wife that he spent so much time at the health club because he made a lot of important business contacts there. Thus, the husband avoided confronting his own miserable marital context.

**Rationalization** cognitive justification for one's own behavior that unconsciously conceals one's true motives.

**Projection** is attributing one's own thoughts, feelings, and desires to someone else while avoiding recognition that these are one's own thoughts, feelings, and desires. For example, the wife who desires to have an affair may accuse her husband of being unfaithful to her. Projection may be seen in such statements as "You spend too much money" (projection for "I spend too much money"), "You want to break up" (projection for "I want to break up"), and "I can't trust you" (projection for "You can't trust me"). Projection interferes with conflict resolution by creating a mood of hostility and defensiveness in both partners. The issues to be resolved in the relationship remain unchanged and become more difficult to discuss.

**Projection** attributing one's own thoughts, feelings, and desires to someone else while avoiding recognition that these are one's own thoughts, feelings, and desires.

**Displacement** involves shifting your feelings, thoughts, or behaviors from the person who evokes them onto someone else. The wife who is turned down for a promotion and the husband who is driven to exhaustion by his boss may direct their hostilities (displace them) onto each other rather than toward their respective employers. Similarly, spouses who are angry at each other may displace this anger onto someone else, such as the children.

**Displacement** shifting your feelings, thoughts, or behaviors from the person who evokes them onto someone else.

By knowing about defense mechanisms and their negative impact on resolving conflict, you can be alert to them in your own relationships. When a conflict continues without resolution, one or more defense mechanisms may be operating.

## The Future of Communication and Technology

The future of communication will increasingly involve technology in the form of texting, smart phones, Facebook, etc. Such technology will be used to initiate, enhance, and maintain relationships. Indeed, intimates today may text each other 60 times a day. Over 2,000 messages a month are not unusual. Parental communication with children will also be altered. Aponte and Pessagno (2010) noted that technology may have positive and negative effects on the family. A positive effect is that parents will be able to use technology to monitor content as

*Effective communication is 20% what you know and 80% how you feel about what you know.*

JIM ROHN, MOTIVATIONAL SPEAKER

their children surf the Internet, send text messages, and send/receive photos on their cell phone. Parents may also use technology to know where their children are by global tracking systems embedded in their cell phones. The downside is that children can use this same technology to establish relationships external to the family which may be nefarious (e.g., child predators).

# Summary

***What is communication, what are some communication choices we can make to enhance our relationships, and what are some gender and cultural differences?***

Communication is the exchange of information and feelings by two individuals. It involves both verbal and nonverbal messages. The nonverbal part of a message often carries more weight than the verbal part. "Good communication" is regarded as the primary factor responsible for a good relationship. Individuals report that communication confirms the quality of their relationship or condemns their relationship. Words have no meaning unless they are backed up by actions. Being told that one is loved is meaningless unless the person saying the words engages in behavior which shows love—being available for time together, being faithful, supporting the partner's interests, etc. In a study of newlyweds, those who ended up getting divorced had communication patterns where they were angry with each other, criticized each other, and were not supportive of each other. Those who stayed together spoke kindly to each other and were supportive of each other.

It is important that we make deliberate choices about how we communicate with our partner. Choosing to prioritize time to communicate, making positive statements to one's partner, and avoiding criticizing one's partner are essential to relationship enhancement. Other choices include maintaining eye contact, asking open-ended questions, using reflective listening, using "I" statements, and sharing power. Partners must also be alert to keeping the dialogue (process) going even when they don't like what is being said (content).

Men and women tend to focus on different content in their conversations. Men tend to focus on activities, information, logic, and negotiation and "to achieve and maintain the upper hand." To women, communication focuses on emotion, relationships, interaction, and maintaining closeness. A woman's goal is to preserve intimacy and avoid isolation. Women are also more likely than men to initiate discussion of relationship problems, and women disclose more than men.

***How do self-disclosure and secrets impact relationships?***

High levels of self-disclosure are associated with increased intimacy. Most individuals keep secrets to avoid hurting the partner and damaging the relationship.

***What is the extent of lying and cheating in relationships?***

Almost 60% of almost 300 undergraduates reported that they had lied to their partner. One in four reported that they had cheated on a partner. Even those in "monogamous" relationships had cheated—27% of males and 20% of females reported having had sex (oral, vaginal, or anal) with someone other than their partner.

***How is technology used in romantic relationships?***

Text messages are the primary way many individuals initiate, escalate, and maintain romantic relationships. Text messaging allows individuals to stay connected all day. Most text messages are related to expressing affection, discussing serious issues, and apologizing. Some individuals send a text message to break up with a partner.

Sexting is the sending of erotic text and photo images via a cell phone. Two thirds of 483 respondents reported sending a sexual text message, over 40% (42.9%) a sexual photo, and over 10% (12.7%) a sex video to a romantic partner. Men were the first to initiate sending sexual content and to perceive that sexting would have a positive effect on the couple's relationship (no such relationship was found).

***How are interactionist and exchange theories applied to relationship communication?***

Symbolic interactionists examine the process of communication between two actors in terms of the meanings each attaches to the actions of the other. Definition of the situation, the looking- glass self, and taking the role of the other are all relevant to understanding how partners communicate.

Exchange theorists suggest that the partners' communication can be described as a ratio of rewards to costs. Rewards are positive exchanges, such as compliments, compromises, and agreements. Costs refer to negative exchanges, such as critical remarks, complaints, and attacks. When the rewards are high and the costs are low, the outcome is likely to be positive for both partners (profit). When the costs are high and the rewards low, neither may be satisfied with the outcome (loss).

***What are the steps involved in conflict resolution?***

The sequence of resolving conflict includes deciding to address recurring issues rather than suppressing them, asking the partner for help in resolving issues, finding out the partner's point of view, summarizing in a nonjudgmental way the partner's perspective, and finding alternative win-win solutions. Defense mechanisms that interfere with conflict resolution include escapism, rationalization, projection, and displacement.

***What is the future of communication?***

The future of communication will increasingly involve technology to initiate, enhance, and maintain relationships. Parental communication with children will also be altered. Technology will have both positive and negative effects on communication in relationships.

## Key Terms

Amae
Authentic
Brainstorming
Branching
Catfishing
Closed-ended question
Communication
Conflict
Congruent messages
Defense mechanism
Displacement
Escapism
Flirting
"I" statements
Lose-lose solution
Nomophobia
Nonverbal communication
Open-ended question
Power
Principle of least interest
Projection
Rationalization
Reflective listening
Sexting
Texting
Win-lose solution
Win-win relationships
"You" statement
Video mediated communication

## Web Links

Association for Couples in Marriage Enrichment
http://www.bettermarriages.org/

Worldwide Marriage Encounter
http://www.episcopalme.com/

Love Your Relationship
http://loveyourrelationship.com/

# CHAPTER 9 Sexuality in Relationships

*Sex isn't good unless it means something. It doesn't necessarily need to mean "love" and it doesn't necessarily need to happen in a relationship, but it does need to mean intimacy and connection. . . . There exists a very fine line between being sexually liberated and being sexually used.*

LAURA SESSIONS STEPP, *UNHOOKED*

Chelsea Curry

## Learning Objectives

- Explain the various sexual values, sources, and choices.
- Identify the various sexual behaviors individuals engage in.
- Describe the various relationship contexts in which sexual behavior occurs.
- Review the importance of condom assertiveness.
- Discuss the various prerequisites to sexual fulfillment.
- Know the future of sexuality in relationships.

## TRUE OR FALSE?

1. Traditional gender roles tend to inhibit and interfere with a couple's sexual fulfillment.
2. Waiting to initiate sexual intimacy in a new unmarried relationship is associated with higher relationship satisfaction.
3. Researchers have found that having casual sex is negatively associated with psychological well-being and positively associated with psychological distress.
4. Of 5,500 never-married individuals, half of the women and 44% of the men said that "bad sex would be a deal-breaker."
5. Women who have a positive feeling/view of their body report higher sexual arousal, initiation of sex, and sexual satisfaction than women who view their bodies negatively.

*Answers:* 1. T 2. T 3. T 4. T 5. T

---

*Fifty Shades of Grey* (James 2011), a sexually explicit novel, captured the interest of readers (particularly women) in the United States and abroad. The fact that it was on the best-seller list for weeks reflects the cultural acceptance of female erotica into mainstream America. Just as the life of the central character, Anastasia Steele, was changed by her sexual experiences with billionaire Christian Grey, so sex in a new relationship changes the individuals and their relationship.

Gupta and Cacchioni (2013) emphasized that contemporary sex manuals demand that individuals have an active, fulfilling sex life. The rationale is that not only their individual and relationship happiness depend on it but that their physical health requires it. "Reflecting the healthicization of sex," there is increased pressure to "master, improve, and work on sex." The existence of this chapter in this text confirms the cultural value (for good or ill) for a good sexual relationship.

The choices we make in sexuality represent a major relationship game changer and are the subject of this chapter. We begin by discussing the sexual values lovers bring to the sexual encounter.

# Sexual Values and Sexual Choices

Think about the following situations:

*Two people are at a party, drinking and flirting. Although they met only two hours ago, they feel a strong attraction to each other. Each is wondering what level of sexual behavior will occur when they go back to one of their rooms later that evening. How much sexual involvement is appropriate in a first-time encounter?*

***

*Two students have decided to live together, but they know their respective parents would disapprove. If they tell their parents, the parents may cut off their money and both students will be forced to drop out of school. Should they tell?*

***

*While Maria was away for a weekend visiting her grandmother, her live-in partner hooked up with his ex-girlfriend. He regretted his behavior, asked for forgiveness, and promised never to be unfaithful again. Should Maria take him back?*

***

*A woman is married to a man whose career requires that he be away from home for extended periods. Although she loves her husband, she is lonely, bored, and sexually*

## FAMILY POLICY | Sex Education in the Public Schools

Sexuality education was introduced in the American public school system in the late nineteenth century with the goal of combating STIs (sexually transmitted infections) and instilling sexual morality (typically understood as abstinence until marriage). Over time, the abstinence agenda became more evident. Beginning in 1982 sex education programs that emphasized or promoted abstinence were eligible to qualify for federal funding. Programs that also discussed contraception and other means of pregnancy protection, referred to as **comprehensive sex education programs**, were not eligible. The Obama administration represented a change in favor of the latter. While these philosophical differences have been pervasive, a trend has emerged whereby schools and communities provide both abstinence education and contraception information (more comprehensive sex education). Karen (2013) noted that the Broward school district in Florida plans to require comprehensive sex education in every grade beginning in 2014 (parents could still opt their children out of such exposure).

**Comprehensive sex education programs** programs that discuss contraception and other means of pregnancy protection.

Chen et al. (2011) evaluated the cost effectiveness of such health education intervention programs which improve preadolescents' attitudes toward abstinence and pregnancy avoidance through contraceptive use. For each $1,000 spent in these programs, 13.67 unintended pregnancies among preadolescents were avoided. Other researchers have provided evidence of the positive outcomes of sex education. Erkut et al. (2013) examined whether a nine-lesson sex education intervention, "Get Real: Comprehensive Sex Education That Works," implemented in the sixth grade, could reduce the number of adolescents who might otherwise become "early starters" of sexual activity (defined as heterosexual intercourse) by the seventh grade. Participants were 548 boys and 675 girls who completed surveys in both sixth grade (baseline) and seventh grade (follow-up). Students randomly assigned to the control condition were 30% more likely to initiate sex at

*frustrated in his absence. She has been asked out by a colleague at work whose wife also travels. He, too, is in love with his wife but is lonely for emotional and sexual companionship. They are ambivalent about whether to hook up occasionally while their spouses are away. What would you tell them?*

**Sexual values** moral guidelines for making sexual choices in nonmarital, heterosexual, and homosexual relationships.

**Sexual values** are moral guidelines for making sexual choices in nonmarital, marital, heterosexual, and homosexual relationships. Attitudes and values sometimes predict sexual behavior. One's sexual values may be identical to one's sexual choices. For example, a person who values abstinence until marriage may choose to remain a virgin until marriage. One's behavior does not always correspond with one's values. Some who express a value of waiting until marriage have intercourse before marriage. One explanation for the discrepancy between values and behavior is that a person may engage in a sexual behavior, then decide the behavior was wrong, and adopt a sexual value against it.

At least three sexual values guide choices in sexual behavior: absolutism, relativism, and hedonism. See Table 9.1 for the respective sexual values of over 4,500 undergraduates. Individuals may have different sexual values at different stages of the family life cycle. For example, elderly individuals are more likely to be absolutist, whereas those in the middle years are more likely to be relativistic. Young unmarried adults are more likely than the elderly to be hedonistic.

*Lord give me abstinence... but not yet.*

ST. AUGUSTINE

**Absolutism** a sexual value system which is based on unconditional allegiance to tradition or religion.

### Absolutism

**Absolutism** is a sexual value system which is based on unconditional allegiance to tradition or religion (i.e., waiting until marriage to have sexual intercourse).

follow-up. Hence, exposure to the sex education program was associated with a delay in early sexual debut. In another study Vivancos et al. (2013) surveyed British university students about their sex education in school when they were 14 and subsequent sexual behavior. The researchers found that school-based sex education was effective at reducing the risk of unprotected intercourse and STIs in early adulthood.

Legislation was introduced in Congress in late 2013 (H.R. 725, the *Real Education for Healthy Youth Action*) which would provide comprehensive sexuality education to adolescents and young adults in public schools, communities, and institutions of higher education. The bill would also prevent federal funds from being spent on ineffective, medically inaccurate sex education programs. An April 2014 update on the bill is that it has a zero chance of being enacted into law.

**Your Opinion?**

1. To what degree do you support abstinence education in public schools?
2. Should condoms be made available for students already having sex?
3. Should parents control the content of sex education in public schools?

## Sources


Chen, C., T. Yamada, and E. M. Walker. 2011. Estimating the cost-effectiveness of a classroom-based abstinence and pregnancy avoidance program targeting preadolescent sexual risk behaviors. *Journal of Children & Poverty* 17: 87–109.

Erkut, S., J. M Grossman, A. A. Frye, I. Ceder, L. Charmaraman, and A. J. Tracy. 2013. Can sex education delay early sexual debut? *Journal of Early Adolescence* 33: 482–497.

Vivancos, R., I. Abubakar, P. Phillips-Howard, and P. R. Hunter. 2013. School-based sex education is associated with reduced risky sexual behavior and sexually transmitted infections in young adults. *Public Health* 127: 53–57.

Yi, K. "Broward school district plans to update sex ed," *Sun Sentinel*, November 16, 2013, accessed November 25, 2013, http://www.sun-sentinel.com/news/education/fl-school-sex-ed-policy-20131112,0,5848891.story


**TABLE 9.1 Sexual Values of 4,567 Undergraduates**

| Respondents | Absolutism | Relativism | Hedonism |
|---|---|---|---|
| Male students (N = 1,103) | 12% | 59% | 29% |
| Female students (N = 3,464) | 15% | 68% | 17% |

Source: Hall, S. and D. Knox. 2013. Relationship and sexual behaviors of a sample of 4,567 university students. Unpublished data collected for this text. Department of Family and Consumer Sciences, Ball State University and Department of Sociology, East Carolina University.

People who are guided by absolutism in their sexual choices have a clear notion of what is right and wrong.

The official creeds of fundamentalist Christian and Islamic religions encourage absolutist sexual values. Intercourse is solely for procreation, and any sexual acts that do not lead to procreation (masturbation, oral sex, homosexuality) are immoral and regarded as sins against God, Allah, self, and community. Waiting until marriage to have intercourse is also an absolutist sexual value. This value is often promoted in the public schools (see Family Policy feature).

Virginity loss for heterosexuals typically refers to vaginal sex (though some would say they are no longer virgins if there has been oral or anal sex). Lesbian and gay males typically refer to virginity loss if there has been oral or anal sex.

Individuals conceptualize their virginity in one of three ways—as a process, a gift, or a stigma. The process view regards first intercourse as a mechanism of learning about one's self and one's partner and sexuality. The gift view regards being a virgin as a valuable positive status wherein it is important to find the right person since sharing the gift is special. The stigma view considers virginity as something to be ashamed of, to hide, and to rid oneself of. When 215 undergraduates were asked their view, 54% classified themselves as process oriented, 38% as gift oriented, and 8.4% as stigma oriented at the time of first coitus (Humphreys 2013).

Sprecher (2014) examined data from 5,769 respondents over a 23-year period in regard to gender differences in pleasure, anxiety, and guilt in response to first intercourse. Men reported more pleasure and anxiety than women, and women reported more guilt than men. Anxiety decreased over the three decades for men; pleasure increased and guilt decreased for women. The result is that "although gender differences in emotional reactions to first intercourse have decreased over time, the first intercourse experience continues to be a more positive experience for men than for women." True Love Waits is an international campaign designed to challenge teenagers and college students to remain sexually abstinent until marriage. Under this program, created and sponsored by the Baptist Sunday School Board, young people are asked to agree to the absolutist position and sign a commitment to the following: "Believing that true love waits, I make a commitment to God, myself, my family, my friends, my future mate, and my future children to sexual purity including abstinence from this day until the day I enter a biblical marriage relationship" (True Love Waits 2014). How effective are the True Love Waits and virginity pledge programs in delaying sexual behavior until marriage? Data from the National Longitudinal Study of Adolescent Health revealed that, although youth who took the pledge were more likely than other youth to experience a later "sexual debut," have fewer partners, and marry earlier, most eventually engaged in premarital sex, were less likely to use a condom when they first had intercourse, and were more likely to substitute oral and/or anal sex in the place of vaginal sex. There was no significant difference in the occurrence of STIs between "pledgers" and "nonpledgers" (Brucker and Bearman 2005). The researchers speculated that the emphasis on virginity may have encouraged the pledgers to engage in noncoital (nonintercourse) sexual activities (e.g., oral sex), which still exposed them to STIs and to be less likely to seek testing and treatment for STIs. Similarly, Hollander (2006) collected national data on two waves of adolescents. Half of those who had taken the virginity pledge reported no such commitment a year later. Males and black individuals were particularly likely to retract their pledge. Landor and Simons (2010) studied 1,215 undergraduates and also found that male pledgers (in contrast to female pledgers) were less likely to remain virgins.

Most narratives about virginity refer to a heterosexual young women and an older experienced male. Caron and Hinman (2012) examined the experiences of undergraduate men who lost their virginity in an article, "I took his V-card." The researchers identified 237 individuals (women and men) who provided 195 stories about the male losing his virginity. In most cases (58%) the woman knew ahead of time that the partner was a virgin with a similar percentage (59%) of those reporting that it was a good first experience; 12% said it was "really bad." Gender roles were operative with 71% reporting pressure to conform to traditional roles (but 49% said they enjoyed the role reversal). A study of first intercourse experience of 475 young Canadian adults revealed orgasm for 6% of females and 62% of males with alcohol/drugs associated with fewer positive experiences and higher sexual regret (Reissing et al. 2012).

Boislard and Poulin (2011) were able to predict early virginity loss. In a sample of 402 youth, those engaging in more antisocial behaviors, having a lot of other sex friends, and being from a non-intact family predicted first intercourse at age 13 or less when compared to those who reported few antisocial behaviors, having few other sex friends, and being from an intact family.

Hans and Kimberly (2011) emphasized that traditional meanings of abstinence, sex, and virginity no longer hold. Some individuals still define themselves as virgins even though they have engaged in oral sex. Of 1,088 undergraduate males, 75% (70% of 3,427 undergraduate females) agreed with the statement "If you have oral sex, you are still a virgin." Hence, according to these undergraduates, having oral sex with someone is not really having sex (Hall and Knox 2013). Persons most likely to agree that "oral sex is not sex" are freshmen/sophomores and those self-identifying as religious (Dotson-Blake et al. 2012). Individuals may engage in oral sex rather than sexual intercourse to avoid getting pregnant, to avoid getting an STI, to keep their partner interested, to avoid a bad reputation, and to avoid feeling guilty over having sexual intercourse (Vazonyi and Jenkins 2010).

Rather than a dichotomous "one is or is not a virgin" concept that gets muddled by one's view of oral sex being "sex," a three-part view of virginity might be adopted—oral sex, vaginal sex, and anal sex. No longer might the term "virgin" be used to reveal sexual behaviors in these three areas. Rather, whether one has engaged in each of the three behaviors must be identified. Hence, an individual would not say "I am a virgin" but "I am an oral virgin" (or intercourse virgin or anal virgin as the case may be).

Carpenter (2010) discussed the concept of **secondary virginity**—a sexually experienced person's deliberate decision to refrain from intimate encounters for a set period of time and to refer to that decision as a kind of virginity (rather than "mere" abstinence). Secondary virginity may be a result of physically painful, emotionally distressing, or romantically disappointing sexual encounters. Of 61 young adults interviewed (most of whom were white, conservative, religious women), more than half believed that a person could, under some circumstances, be a virgin more than once. Fifteen people contended that people could resume their virginity in an emotional, psychological, or spiritual sense. Terence Duluca, a 27-year-old heterosexual, white Roman Catholic, explained:

**Secondary virginity** a sexually experienced person's deliberate decision to refrain from intimate encounters for a set period of time and to refer to that decision as a kind of virginity.

> *There is a different feeling when you love somebody and when you just care about somebody. So I would have to say if you feel that way, then I guess you could be a virgin again. Christians get born all the time again, so. . . . When there's true love involved, yes, I believe that.*

A subcategory of absolutism is **asceticism**. Ascetics believe that giving in to carnal lust is unnecessary and attempt to rise above the pursuit of sensual pleasure into a life of self-discipline and self-denial. Accordingly, spiritual life is viewed as the highest good, and self-denial helps one to achieve it. Catholic priests, monks, nuns, and some other celibate people have adopted the sexual value of asceticism.

**Asceticism** a belief that giving in to carnal lust is unnecessary and an attempt to rise above the pursuit of sensual pleasure into a life of self-discipline and self-denial.

## Relativism

Fifty-nine percent of 1,103 undergraduate males and 68% of 3,464 undergraduate females identified relativism as their sexual value (Hall and Knox 2013). **Relativism** is a value system emphasizing that sexual decisions should be made in reference to the emotional, security, and commitment aspects of the relationship. Whereas absolutists might feel that having intercourse is wrong for unmarried people, relativists might feel that the moral correctness of sex outside marriage depends on the particular situation. For example, a relativist might say that marital sex between two spouses who are emotionally and physically abusive to each other is not to

**Relativism** a value system emphasizing that sexual decisions should be made in reference to the emotional, security, and commitment aspects of the relationship.

be preferred over intercourse between two unmarried individuals who love each other, are kind to each other, and are committed to the well-being of each other.

*I had begun to think my ripening body would wither untasted on the vine.*

JACQUELINE CAREY, *KUSHIEL'S DART*

Relativists apparently do value having a good sex partner. About half of never-married individuals want to stay in a relationship only if the sex is good. In a study of 5,500 never marrieds, half of the women and 44% of the men said that "bad sex would be a deal-breaker" (Walsh 2013).

A disadvantage of relativism as a sexual value is the difficulty of making sexual decisions on a relativistic case-by-case basis. The statement "I don't know what's right anymore" reflects the uncertainty of a relativistic view. Once a person decides that mutual love is the context justifying intercourse, how often and how soon is it appropriate for the person to fall in love? Can love develop after some alcohol and two hours of conversation? How does one know that love feelings are genuine? The freedom that relativism brings to sexual decision making requires responsibility, maturity, and judgment. In some cases, individuals may convince themselves that they are in love so that they will not feel guilty about having intercourse.

## Script for Delaying Intercourse in a Relationship

The section below suggests a script for a female who wants to pace her partner to ensure that a relationship develops before sexual involvement.

Slowing the Sexual Speed of Our Relationship Down

*I need to talk about sex in our relationship. I'm not comfortable talking about this and don't know exactly what I want to say so I have written it down to try and help me get the words right. I like you, enjoy the time we spend together, and want us to continue seeing each other. The sex is something I need to feel good about to make it good for you, for me and for us.*

*I need for us to slow down sexually. I need to feel an emotional connection that goes both ways—we both have very strong emotional feelings for each other. I also need to feel secure that we are going somewhere—that we have a future and that we are committed to each other. We aren't there yet so I need to wait till we get there to be sexual (have intercourse) with you.*

*How long will this take? I don't know—the general answer is "longer than now." This may not be what you have in mind and you may be ready for us to increase the sex now. I'm glad that you want us to be sexual and I want this too but I need to feel right about it.*

*So, for now, let me drive the bus sexually . . . when I feel the emotional connection and that we are going somewhere, I'll be the best sex partner you ever had. But let me move us forward.*

*If this is too slow for you or not what you have in mind, maybe I'm not the girl for you. It is certainly OK for you to tell me you need more and move on. Otherwise, we can still continue to see each other and see where the relationship goes.*

*So . . . tell me how you feel and what you want . . . don't feel like you need to tell me you love me and want us to get married . . . ha! I'm not asking that . . . I'm just asking for time. . . .*

See the self-assessment to identify the degree to which you are conservative or liberal in your attitudes toward sexuality.

*The decision that saves you the most heartache is the first "NO." If you can make that one and not go along hoping things will get better you will save yourself a lot of trouble. When in doubt, DON'T.*

JOHNNY MERCER, LYRICIST

Absolutists and relativists have different views on whether or not two unmarried people should have intercourse. Whereas an absolutist would say that having intercourse is wrong for unmarried people and right for married people, a relativist would say, "It depends on the situation." Suppose, for example, that a married couple do not love each other and intercourse is an abusive, exploitative act. Suppose also that an unmarried couple love each other and their intercourse experience is an expression of mutual affection and respect. A relativist might conclude that, in this particular situation, having intercourse is "more right" for the unmarried couple than the married couple.

## SELF-ASSESSMENT | The Conservative – Liberal Sexuality Scale

This scale is designed to assess the degree to which you are conservative or liberal in your attitudes toward sex. There are no right or wrong answers.

**Directions**

After reading each sentence carefully, circle the number that best represents the degree to which you agree or disagree with the sentence.

| 1 | 2 | 3 | 4 | 5 |
|---|---|---|---|---|
| Strongly agree | Mildly agree | Undecided | Mildly disagree | Strongly disagree |

_____ 1. Abortion is wrong.
_____ 2. Homosexuality is immoral.
_____ 3. Couples should wait to have sexual intercourse until after they are married.
_____ 4. Couples who are virgins at marriage have more successful marriages.
_____ 5. Watching pornography is harmful.
_____ 6. Kinky sex is something to be avoided.
_____ 7. Having an extramarital affair is never justified.
_____ 8. Masturbation is something an individual should try to avoid doing.
_____ 9. One should always be in love when having sex with a person.
_____ 10. Transgender people are screwed up and "not right."
_____ 11. Sex is for youth, not for the elderly.
_____ 12. There is entirely too much sex on TV today.
_____ 13. The best use of sex is for procreation.
_____ 14. Any form of sex that is not sexual intercourse is wrong.
_____ 15. Our society is entirely too liberal when it comes to sex.
_____ 16. Sex education gives youth ideas about sex they shouldn't have.
_____ 17. Promiscuity is the cause of the downfall of an individual.
_____ 18. Too much sexual freedom is promoted in our country today.
_____ 19. The handicapped probably should not try to get involved in sex.
_____ 20. The movies in America are too sexually explicit.

**Scoring**

Add the numbers you circled. 1 (strongly agree) is the most conservative response and 5 (strongly disagree) is the most liberal response. The lower your total score (20 is the lowest possible score), the more sexually conservative your attitudes toward sex. The higher your total score (100 is the highest possible score), the more liberal your attitudes toward sex. A score of 60 places you at the midpoint between being the ultimate conservative and the ultimate liberal about sex. Of 191 undergraduate females, the average score was 69.85. Of 39 undergraduate males the average score was 72.89. Hence, both women and men tended to be more sexually liberal than conservative with men more sexually liberal than women.

**Source**

Knox, D. 2014. "The Conservative-Liberal Sexuality Scale" was developed for this text. The scale is intended to be thought provoking and fun. It is not intended to be used as a clinical or diagnostic instrument.

## Hedonism

**Hedonism** is the belief that the ultimate value and motivation for human actions lie in the pursuit of pleasure and the avoidance of pain. Twenty-nine percent of 1,103 undergraduate males and 17% of 3,464 undergraduate females identified hedonism as their primary sexual value (Hall and Knox 2013). Bersamin and colleagues (2014) analyzed data on single, heterosexual college students ($N = 3{,}907$) ages 18 to 25 from 30 institutions across the United States.

*Give a man free hands and you'll know where to find them.*

MAE WEST, ACTRESS

**Hedonism** belief that the ultimate value and motivation for human actions lie in the pursuit of pleasure and the avoidance of pain.

A greater proportion of men (18.6%) compared to women (7.4%) reported having had casual sex in the month prior to the study. The researchers also found that casual sex was negatively associated with psychological well-being (defined in reference to self-esteem, life satisfaction, and eudamonic well-being—having found oneself). Casual sex was also positively associated with psychological distress (e.g., anxiety, depression). There were no gender differences. Sandberg-Thoma and Kamp Dush (2014) also found suicide ideation and depressive symptoms associated with casual sexual relationships (sample of 12,401 adolescents). Fielder et al. (2014) studied hookups in first-year college women and found an association with experiencing depression, sexual victimization, and STIs.

Hedonism is sometimes viewed as sexual addiction. Levine (2010) noted that the term sexual addition has no professional agreement and is sometimes applied to those who watch pornography, have commercial sex, and engage in cybersex. More accurately, sex addition applies to those who have lost control over their sexual behavior which is often accompanied by spiraling psychological deterioration—i.e., depression.

## Sexual Double Standard

**Sexual double standard** view that encourages and accepts sexual expression of men more than women.

The **sexual double standard** is the view that encourages and accepts sexual expression of men more than women. Table 9.1 revealed that men are almost two times more hedonistic than women (Hall and Knox 2013). In another study of 1,004 undergraduates, 59% of the males and 24% of the females reported that they would "engage in casual sex" (Toews & Yazedjian 2011). Acceptance of the double standard is evident in the words used to describe hedonism—hedonistic men are thought of as "studs" but hedonistic women as "sluts." Indeed, Porter (2014) emphasized the double standard in her presentation on "slut-shaming," which she defined as "the act of making one feel guilty or inferior for engaging in certain sexual behaviors that violate traditional dichotomous gender roles." She pointed out that Charlie Sheen was a national celebrity for his flagrant debauchery but Kristen Stewart was shamed for her infidelity. Porter surveyed 240 undergraduates and found that 81% of the females reported having been slut-shamed (7.3% of the males).

*I consider sex a misdemeanor, the more I miss, de meaner I get.*

MAE WEST, ACTRESS

Some hedonists enjoy pornography and this becomes an issue in relationships (see What If?).

## Sources of Sexual Values

The sources of one's sexual values are numerous and include one's school, religion, and family, as well as technology, television, social movements, and the Internet. Previously we noted that public schools in the United States promote absolutist sexual values through abstinence education and that the effectiveness of these programs has been questioned.

WHAT IF?

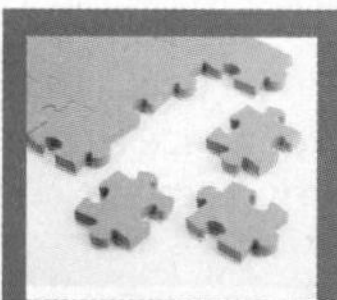

### What If My Partner Wants Me to Watch Pornography with Him?

Men watch pornography more frequently than women. Their motivations include emotional avoidance, excitement seeking, sexual pleasure, and sexual curiosity. The Pornography Consumption Inventory has been used to identify these uses (Reid et al. 2011). Maddox et al. (2011) studied the associations between viewing sexually explicit material (SEM) and relationship functioning in a random sample of 1,291 unmarried individuals in romantic relationships. As expected, more men (76.8%) than women (31.6%) reported having viewed SEM on their own—but almost half (45%) reported sometimes viewing SEM with their partner (44.8%). The researchers looked at communication, relationship adjustment, commitment, sexual satisfaction, and infidelity and found that those who never viewed SEM reported higher relationship quality on all indices than those who viewed SEM alone. In addition, those who viewed SEM only with their partners reported more dedication and higher sexual satisfaction than those who viewed SEM alone. Hence, while viewing pornography is a behavior individuals may feel very differently about, should they feel comfortable viewing SEM with their partner, there appear to be benefits.

## PERSONAL CHOICES

### *Deciding to Have Intercourse with a New Partner*

The following are issues to consider in making the decision to have sexual intercourse with a new partner:

**1.** ***Personal consequences.*** How do you predict you will feel about yourself after you have had intercourse with a new partner? An increasing percentage of college students are relativists and feel that the outcome will be positive if they are in love. The following quote is from a student in our classes:

> *I believe intercourse before marriage is OK under certain circumstances. I believe that when a person falls in love with another and the relationship is stable, it is then appropriate. This should be thought about very carefully for a long time, so as not to regret engaging in intercourse.*

Those who are not in love and have sex in a casual context often report sexual regret about their decision.

> *When I have the chance to experience sex with someone new, my first thought is to jump right in, ask questions later. However, my very second thought is, "If we do this, will he think I'm slutty?" followed by "Is it too early to be doing this? What if I regret it?" Looking back at my dating experiences over the past several years, it seems that earlier on, I let a guy talk me into doing something sexual, even though those questions were running circles in my head. However, in the most recent instances, I remained firm that sex wasn't going to happen so soon into dating someone, and I do not regret my decision. In the future, I don't plan on being intimate with anyone until we have gone from "going out on dates" to boyfriend/girlfriend status (Merrill and Knox 2010, p. 3).*

The effect intercourse will have on you will be influenced by your personal values, your religious values, and your emotional involvement with your partner. Some people prefer to wait until they are married to have intercourse and feel that this is the best course for future marital stability and happiness.

Strong personal and religious values against nonmarital intercourse may result in guilt and regret following an intercourse experience. In a sample of 270 unmarried undergraduates who had had intercourse, 71.9% regretted their decision to do so at least once (for example, they may have had intercourse more than once or with multiple partners and reported regret at least once). Higgins et al. (2010) analyzed data on first intercourse behavior from a cross-sectional survey of 1,986 non-Hispanic, white, and black 18- to 25-year-old respondents from four university campuses. Respondents were asked to rate the degree to which their first vaginal intercourse was physiologically and psychologically satisfying. Women were less likely than men to report a positive first vaginal intercourse experience, particularly physiological satisfaction. Being in a committed relationship was the condition associated with the most positive experience. Galperin et al. (2013) confirmed higher sexual regret among women—they lamented that they had not been more selective. In contrast, men regretted that they failed to act on some sexual opportunities.

**2.** ***Timing.*** In a sample of over 5,000 unmarried singles, about a third (28%) reported that they had sex by the third date; almost half (46%) had done so by the sixth date (Walsh 2013). Delaying intercourse with a new partner is important for achieving a positive outcome and avoiding regret. The table below reveals that a third of 429 undergraduates regretted having sexual intercourse "too soon" in a relationship (Merrill and Knox 2010, p. 2).

**Regret for Engaging in Behavior "Too Soon" or "Too Late"**
(N = 429)

| Behaviors | "Too Soon" | "Too Late" | "Perfect Timing" | "Did not do" |
|---|---|---|---|---|
| Sexual Intercourse | 33.3% | 3.3% | 48.4% | 15.0% |
| Spent the night | 26.6% | 3.7% | 58.8% | 11.4% |
| Saying "I love you" | 26.1% | 3.9% | 50.8% | 19.2% |
| Kissing | 11.1% | 3.1% | 82.8% | 3.1% |

Willoughby et al. (2014) confirmed the positive outcomes of delaying sex in a romantic relationship. The researchers analyzed data on a sample of 10,932 individuals in unmarried, romantic relationships and compared relationship outcomes from when sex was initiated—having sex prior to dating, having sex on the first date or shortly after, having sex after a few weeks of dating, and sexual abstinence. Results revealed that waiting to initiate sexual intimacy in unmarried relationships was associated with higher relationship satisfaction, more leisure time together, and greater affect (emotional feelings). This effect was strongly moderated by relationship length, with individuals who reported early sexual initiation reporting increasingly lower outcomes in relationships of longer than two years. Hence, the positive effects of delaying sex in a relationship continued the benefits over time.

3. ***Partner consequences.*** Because a basic moral principle is to do no harm to others, it is important to consider the effect of intercourse on your partner. Whereas intercourse may be a pleasurable experience with positive consequences for you, your partner may react differently. What is your partner's religious background, and what are your partner's sexual values? A highly religious person with absolutist sexual values will typically have a very different reaction (e.g., guilt/regret) to sexual intercourse than a person with low religiosity and relativistic or hedonistic sexual values.

4. ***Relationship consequences.*** What is the effect of intercourse on a couple's relationship? One's personal reaction to having intercourse may spill over into the relationship. Individuals might predict how they feel having intercourse will affect their relationship before including it in their relationship. One study of 209 undergraduates revealed that "feeling closer" was the most common outcome for having sexual intercourse (Vasilenko et al. 2012).

5. ***Contraception.*** Another potential consequence of intercourse is pregnancy. Once a couple decide to have intercourse, a separate decision must be made as to whether intercourse should result in pregnancy. Most sexually active undergraduates do not want children. People who want to avoid pregnancy must choose and plan to use a contraceptive method.

6. ***HIV and other sexually transmissible infections.*** Engaging in casual sex has potentially fatal consequences. Avoiding HIV infection and other STIs is an important consideration in deciding whether to have intercourse in a new relationship. The increase in the number of people having more partners results in the rapid spread of the bacteria and viruses responsible for numerous varieties of STIs. However, in a sample of 4,461 undergraduates, only 28% reported consistent condom use (Hall and Knox 2013). In one study, women with low GPAs who were binge drinkers reported the lowest condom use (Walsh et al. 2013). For those who believe they may have been exposed to the infection, quick and painless home tests such as OraQuick are available.

7. ***Influence of alcohol and other drugs.*** A final consideration with regard to the decision to have intercourse in a new relationship is to be aware of the influence of alcohol and other drugs on such a decision. Bersamin et al. (2014) confirmed that amount of alcohol consumed by undergraduates was associated with having sex with a stranger. LaBrie et al. (2013) analyzed data on 828 undergraduates and found that, among participants who consumed alcohol prior to their last

hookup, 30.7% of females and 27.9% of males indicated that they would likely not have hooked up with their partners had alcohol not been involved. Furthermore, females who were drinking before they hooked up were more likely to feel discontent with their hookup decision.

**8. *OK to change decision about including intercourse in a relationship.*** Although most couples who include intercourse in their relationship continue the pattern, some decide to omit it from their sexual agenda. One female student said, "Since I did not want to go on the pill and would be frantic if I got pregnant, we decided the stress was not worth having intercourse. While we do have oral sex, we don't even think about having intercourse any more."

### Sources


Bersamin, M. et al. 2014. Risky business: Is there an association between casual sex and mental health among emerging adults? *The Journal of Sex Research* 51: 43–51.

Galperin, A., M. G. Haselton, D.A. Frederick, J. Poore, W. Von Hippel, D. M. Buss, and G. C. Gonzaga. 2013. Sexual regret: Evidence for evolved sex differences. *Archives of Sexual Behavior* 42: 1145–1161.

Hall, S. and D. Knox. 2013. Relationship and sexual behaviors of a sample of 4,567 university students. Unpublished data collected for this text. Department of Family and Consumer Sciences, Ball State University and Department of Sociology, East Carolina University.

LaBrie, J. W. J. F. Hummer, T. M. Ghaidarov, A. Lac, and S. R. Kenney. 2014. Hooking up in the college context: Event-level effects of alcohol use and partner familiarity on hookup behaviors and contentment. *The Journal of Sex Research* 51: 62–73.

Merrill, J. and D. Knox. 2010. *When I fall in love again: A new study on finding and keeping the love of your life.* Santa Barbra, California: Praeger.

Vasilenko, S. A., E. S. Lefkowitz, and J. L. Maggs. 2012. Short-term positive and negative consequences of sex based on daily reports among college students. *Journal of Sex Research* 49: 558–569.

Willoughby, B. J., J. S. Carroll, and D. M. Busby. 2014. Differing relationship outcomes when sex happens before, on, or after first dates. *The Journal of Sex Research* 51: 52–61.


Religion is also an important influence. Thirty-eight percent of 1,107 undergraduate males (34% of 3,474 undergraduate females) self-identified as being "very" or "moderately" religious (Hall and Knox 2013). In regard to sexual behavior, researchers have found that religiously active young adults are more likely to agree that sexting and sexual intercourse are inappropriate activities to engage in before being in a committed dating relationship (Miller et al. 2011).

Parents are also influential. Wetherill et al. (2010) found that for individuals with parents who were aware of what their children were doing and cared about them, their behavior reported engaging in less frequent sexual behaviors. Purity Balls, which are events where fathers and daughters alike pledge purity—the fathers pledge that they will be faithful to their wives and the daughters pledge (both sign "pledge cards") that they will wait until marriage to have sexual intercourse—are held in 48 states. Purity Balls have their basis in evangelical conservative religious families. The documentary *Virgin Tales* focuses on one family in Colorado and their involvement in Purity Balls. As noted earlier, most pledgers end up having sexual intercourse before marriage and are more likely than nonpledgers to engage in oral sex. Pledgers are also less likely to use a condom than nonpledgers.

Reproductive technologies such as birth control pills, the morning-after pill, and condoms influence sexual values by affecting the consequences of behavior. Being able to reduce the risk of pregnancy and HIV infection with the pill and condoms allow one to consider a different value system than if these methods of protection did not exist.

Television also influences sexual values. A television advertisement shows an affectionate couple with minimal clothes on in a context where sex could

**TABLE 9.2 Sexual Values and Sexual Choices**

| Dilemma | Sexual Value | Sexual Choice |
|---|---|---|
| Have casual sex? | Absolutism | No |
| | Relativism | Depends on situation |
| | Hedonism | Yes |
| Cheat on partner? | Absolutism | No |
| | Relativism | Depends on situation |
| | Hedonism | Yes |
| Disclose STI infection? | Absolutism | Yes |
| | Relativism | Yes |
| | Hedonism | No |



occur. "Be ready for the moment" is the phrase of the announcer, and Levitra, the new quick-start Viagra, is the product for sale. The advertiser used sex to get the attention of the viewer and punch in the product to elicit buying behavior. Ballard et al. (2011) studied the sexual socialization of young adults. Some of the respondents noted the mixed messages in their socialization, e.g., the discrepancy between media messages and church messages regarding sexuality:

> *They [church members] say abstain, abstain, abstain, but it's so hard when sex is being thrown at you constantly from, you know, watching TV.*

In regard to magazines, exposure outcome is more positive. Walsh and Ward (2010) assessed sexual health behaviors and magazine reading among 579 undergraduate students. They found that more frequent reading of mainstream magazines was associated with greater sexual health knowledge, safe-sex self-efficacy, and consistency of using contraception (although results varied across sex and magazine genre).

The women's movement affects sexual values by empowering women with an egalitarian view of their sexuality. This view translates into encouraging women to be more assertive about their own sexual needs and giving them the option to experience sex in a variety of contexts (e.g., without love or commitment) without self-deprecation. The net effect is a potential increase in the frequency of recreational, hedonistic sex. The gay liberation movement has also been influential in encouraging values that are accepting of sexual diversity.

Finally, the Internet has an influence on sexual values. The Internet features erotic photos, videos, and "live" sex acts/stripping by webcam sex artists. Individuals can exchange nude photos, have explicit sex dialogue, arrange to have phone sex or meet in person, or find a prostitute. Indeed, the adult section of Craigslist was shut down because it featured blatant prostitution.

### Sexual Values and Sexual Choices

Table 9.2 shows how sexual values translate into sexual choices.

## Sexual Behaviors

We have been discussing the various sources of sexual values. We now focus on what people report that they do sexually.

> *My love life is terrible. The last time I was inside a woman was when I visited the Statue of Liberty.*
>
> WOODY ALLEN

### What Is Sex?

Horowitz and Spicer (2013) asked 124 emerging adults (40 male heterosexuals, 42 female heterosexuals, and 42 lesbians) to identify various sexual behaviors on a six-point scale (from "definitely" to "definitely not") as "having sex."

There was agreement that vaginal and anal sex were "definitely" sex while kissing was "definitely not" sex. Ratings of heterosexual males and females did not differ significantly but gays were more likely than heterosexuals to rate various forms of genital stimulation as "having sex."

Some individuals are **asexual**, which means there is an absence of sexual behavior with a partner and oneself (masturbation). About 4% of females and 11% of males reported being asexual in the last twelve months (DeLamater and Hasday 2007). In contrast, most individuals report engaging in various sexual behaviors. Penhollow et al. (2010) found that, for both male and female students, participation in recreational sexual behaviors (with or without a partner) enhanced their overall sexual satisfaction. The following discussion includes kissing, masturbation, oral sex, vaginal intercourse, and anal sex.

Chelsea Curry

Kissing is an outward physical behavior that often reflects intimacy between the couple. This couple is in love.

## Kissing

Kissing has been the subject of literature and science. The meanings of a kiss are variable—love, approval, hello, goodbye, or as a remedy for a child's hurt knee. There is also a kiss for luck, a stolen kiss, and a kiss to seal one's marriage vows. Kisses have been used to denote hierarchy. In the Middle Ages, only peers kissed on the lips; a person of lower status kissed someone of higher status on the hand, and a person of lower status showed great differential of status by kissing on the foot. Kissing also has had a negative connotation as in the "kiss of death," which reflects the kiss Judas gave Jesus as he was about to betray him. Kissing may be considered an aggressive act as some do not want to be kissed.

**Asexual** absence of sexual behavior with a partner and oneself (masturbation).

Although the origin of kissing is unknown, one theory posits that kissing is associated with parents putting food into their offspring's mouth . . . the bird pushes food down the throat of a chick in the nest. Some adult birds also exchange food by mouth during courtship.

The way a person kisses reflects the person's country, culture, and society. The French kiss each other once on each cheek or three times in the same region. Greeks tend to kiss on the mouth, regardless of the sex of the person. Anthropologists note that some cultures (e.g., Eskimos, Polynesians) upon meeting someone rub noses.

*Remember, we're madly in love, so it's all right to kiss me anytime you feel like it.*

SUZANNE COLLINS, *THE HUNGER GAMES*

## Masturbation

**Masturbation** involves stimulating one's own body with the goal of experiencing pleasurable sexual sensations. Ninety-seven percent of 1,092 undergraduate males and 65% of 3,421 undergraduate females reported having masturbated at some point (Hall and Knox 2013).

**Masturbation** stimulating one's own body with the goal of experiencing pleasurable sexual sensations.

Alternative terms for masturbation include *autoeroticism, self-pleasuring, solo sex,* and *sex without a partner.* An appreciation of the benefits of masturbation has now replaced various myths about it (e.g., it causes blindness). Most health care providers and therapists today regard masturbation as a normal and healthy sexual behavior. Masturbation is safe sex in that it involves no risk of transmitting diseases (such as HIV) or producing unintended pregnancy.

Masturbation is also associated with orgasm during intercourse. In a study by Thomsen and Chang (2000), 292 university undergraduates reported whether they had ever masturbated and whether they had an orgasm during their first intercourse experience. The researchers found that the strongest single predictor of orgasm and emotional satisfaction with first intercourse was previous masturbation.

## Oral Sex

**National Data**

Of 5,865 men and women ages 18 to 49, in the past year, more than half of women and men reported engaging in oral sex (Herbenick et al. 2010).

In a sample of 1,091 undergraduate males, 70% reported that they had given oral sex; 77% of 3,438 undergraduate women had done so (Hall and Knox 2013). Fellatio is oral stimulation of the man's genitals by his partner. In many states, legal statutes regard fellatio as a "crime against nature." "Nature" in this case refers to reproduction, and the "crime" is sex that does not produce babies. Nevertheless, most men have experienced fellatio. Cunnilingus is oral stimulation of the woman's genitals by her partner.

We noted earlier that, increasingly, some youth who have oral sex regard themselves as virgins, believing that only sexual intercourse constitutes "having sex." As noted previously, of 4,515 university students, 72.5% agreed that, "If you have oral sex, you are still a virgin." This perspective allows the person to avoid the social stigma of having had sexual intercourse, which puts them in a more positive light for subsequent partners ("I'm a virgin."). Dotson-Blake et al. (2012) developed a profile of those most likely to believe that "oral sex is sex": they tended to be first-year students, white, nonreligious, and hedonistic in sexual values with experience in hooking up, oral sex, and cohabitation.

**DIVERSITY IN OTHER COUNTRIES**

Malacad and Hess (2010) observed among Canadian adolescent females that oral sex has become an increasingly common and casual activity. In a study of 181 women aged 18–25 years, approximately three-quarters reported having engaged in oral sex, a prevalence rate almost identical to that for vaginal intercourse. The respondents were an average age of 17 (for both oral sex and intercourse) at their first experience. Most reported that their most recent oral sex experience was positive and in a committed relationship. Adolescent females who reported oral sex with a partner they were not in love with reported the most negative experience.

There is the mistaken belief that only intercourse carries the risk of contracting an STI. However, STIs as well as HIV can be contracted orally. Use of a condom or dental dam, a flat latex device that is held over the vaginal area, is recommended.

*There's nothing better than good sex. But bad sex? A peanut butter and jelly sandwich is better than bad sex.*

BILLY JOEL, SINGER

## Vaginal Intercourse

Vaginal intercourse, or **coitus**, refers to the sexual union of a man and woman by insertion of the penis into the vagina. In a study of 4,547 undergraduates, 72% reported that they had had sexual intercourse (men 67.6; women 73.7%) (Hall and Knox 2013). In regard to university students, each academic year, there are fewer virgins.

**Coitus** the sexual union of a man and woman by insertion of the penis into the vagina.

Vasilenko et al. (2012) reported on the consequences of vaginal sex for 209 first-year undergraduates at a large Northeastern university. Students kept a diary for 28 days on their sexual behavior and the personal and interpersonal consequences. The most commonly reported (81%) positive consequence of having vaginal sex was feeling physically satisfied; the most commonly reported negative consequence was worry about pregnancy (17%). In regard to interpersonal consequences, the most common consequence was feeling closer to the partner (89%); the most common negative consequence was worrying if the partner wanted more commitment. More positive outcomes resulted from dating rather than casual partners.

*Women complain more about sex than men. Their gripes fall into two major categories:*

*1) Not enough*

*2) Too much.*

ANN LANDERS, ADVICE COLUMNIST

## First Intercourse

Cavazos-Rehg et al. (2010) found that having first sexual intercourse at or before age 16 was associated with having an alcoholic parent, being black, and being born to a teenage mother. Intimacy was an important motivation for women to engage in first intercourse (Patrick and Lee 2010).

Hawes et al. (2010) studied first intercourse experiences in the United Kingdom and emphasized that **sexual readiness**, not age, is the more meaningful criteria. Such readiness can be determined in reference to contraception, autonomy of decision (not influenced by alcohol or peer pressure), consensuality (both partners equally willing), and the absence of regret (the right time for me). Using these criteria, the negative consequences of first intercourse are minimized.

**Sexual readiness** factors such as autonomy of decision (not influenced by alcohol or peers), consensuality (both partners equally willing), and absence of regret (the right time for me) are more important than age as meaningful criteria for determining when a person is ready for first intercourse.

## Anal Sex

### National **Data**

Of 5,865 men and women, more than 20% of men ages 25–49 and women ages 20–39 reported having had anal sex in the past year (Herbenick et al. 2010). Lifetime experience of heterosexuals with anal sex ranges from 6% to 40% with about 10% reporting having engaged in anal sex within the last year (McBride and Fortenberry 2010).

In a study of 4,533 undergraduates (primarily heterosexuals), 22% reported that they had had anal sex (men 24.3%; women 22%) (Hall and Knox 2013). Younger individuals, those with higher numbers of partners, and those with STIs are more likely to report having participated in anal sex. Motivations associated with anal sex include intimacy "anal sex is more intimate than regular sex," enjoyment in variety, domination by male, breaking taboos, and pain-pleasure enjoyment (McBride and Fortenberry 2010).

The greatest danger of anal sex is that the rectum might tear, in which case blood contact can occur; STIs (including HIV infection) may then be transmitted. Partners who use a condom during anal intercourse reduce their risk of not only HIV infection but also other STIs. Pain (physical as well as psychological) may occur for anyone of any sexual identity involved in receiving anal sex. Such pain is called **anodyspareunia**—frequent, severe pain during receptive anal sex. Of 505 women who experienced anal intercourse, 9 percent reported severe pain during every penetration (Stulhofer and Ajdukovic 2011).

**Anodyspareunia** frequent, severe pain during receptive anal sex.

## Cybersex

**Cybersex** is any consensual sexual experience mediated by a computer that involves at least two people. In this context, sexual experience includes sending text messages or photographic images that are sexual. Individuals typically send sex text and photos with the goal of arousal or looking at each other naked or masturbating when viewing each other on a web cam.

**Cybersex** any consensual sexual experience mediated by a computer.

# Sexuality in Relationships

Sexuality occurs in a social context that influences its frequency and perceived quality.

*We waste time looking for the perfect lover, instead of creating the perfect love.*

TOM ROBBINS

## Sexual Relationships among Never-Married Individuals

Never-married individuals and those not living together report more sexual partners than those who are married or living together. In one study, 9% of never-married individuals and those not living together reported having had five or more sexual partners in the previous 12 months; 1% of married people and 5% of cohabitants reported the same.

Unmarried individuals, when compared with married individuals and cohabitants, reported the lowest level of sexual satisfaction. One-third of a national sample of people who were not married and not living with anyone reported that they were emotionally satisfied with their sexual relationships. In contrast, 85% of the married and pair-bonded individuals reported emotional satisfaction in their sexual relationships. Hence, although never-married individuals have more sexual partners, they are less emotionally satisfied (Michael et al. 1994).

## Sexual Relationships among Married Individuals

*Personally I know nothing about sex because I have always been married.*

ZSA ZSA GABOR, ACTRESS

Marital sex is distinctive for its social legitimacy, declining frequency, and satisfaction (both physical and emotional).

1. ***Social legitimacy.*** In our society, marital intercourse is the most legitimate form of sexual behavior. Those who engage in homosexual, premarital, and extramarital intercourse do not experience as high a level of social approval as do those who engage in marital sex. It is not only okay to have intercourse when married, it is expected. People assume that married couples make love and that something is wrong if they do not.

*Money, it turned out, was exactly like sex, you thought of nothing else if you didn't have it and thought of other things if you did.*

JAMES ARTHUR BALDWIN, NOVELIST

**Satiation** repeated exposure to a stimulus results in the loss of its ability to reinforce.

2. ***Declining frequency.*** Sexual intercourse between spouses occurs about six times a month, which declines in frequency as spouses age. Pregnancy also decreases the frequency of sexual intercourse (Lee et al. 2010). In addition to biological changes due to aging and pregnancy, satiation also contributes to the declining frequency of intercourse between spouses and partners in long-term relationships. Psychologists use the term **satiation** to mean that repeated exposure to a stimulus results in the loss of its ability to reinforce. For example, the first time you listen to a new CD, you derive considerable enjoyment and satisfaction from it. You may play it over and over during the first few days. After a week or so, listening to the same music is no longer new and does not give you

David Knox

While the frequency of marital sex declines, it does not stop. This couple is in their eighties and report an active sex life.

the same level of enjoyment that it first did. So it is with intercourse between spouses or long-term partners. The thousandth time that a person has sex with the same partner is not as new and exciting as the first few times.

Polyamorists use the term **new relationship energy** (NRE) to refer to the euphoria of a new emotional/sexual relationship which dissipates over time. Polyamorists often talk with a long-term partner about the NRE they are feeling and both watch its eventual decline. Hence, polyamorists don't get upset when they see their partner experiencing NRE with a new partner since they view it as having a cycle that will not last forever (Wilson and Rodrigous 2010).

*For women the best aphrodisiacs are words. The G-spot is in the ears. He who looks for it below there is wasting his time.*

ISABEL ALLENDE, CHILEAN WRITER

**3.** ***Satisfaction (emotional and physical).*** Despite declining frequency and less satisfaction over time, marital sex remains a richly satisfying experience. Contrary to the popular belief that unattached singles have the best sex, married and pair-bonded adults enjoy the most satisfying sexual relationships. In the national sample referred to earlier, 88% of married people said they received great physical pleasure from their sexual lives, and almost 85% said they received great emotional satisfaction (Michael et al. 1994). Individuals least likely to report being physically and emotionally pleased in their sexual relationships are those who are not married, not living with anyone, or not in a stable relationship with one person.

**New relationship energy (NRE)** the euphoria of a new emotional/sexual relationship which dissipates over time.

## Sexual Relationships among Divorced Individuals

Of the almost 2 million people getting divorced, most will have intercourse within one year of being separated from their spouses. The meanings of intercourse for separated or divorced individuals vary. For many, intercourse is a way to reestablish—indeed, repair—their crippled self-esteem. Questions such as, "What did I do wrong?" "Am I a failure?" and "Is there anybody out there who will love me again?" loom in the minds of divorced people. One way to feel loved, at least temporarily, is through sex. Being held by another and being told that it feels good give people some evidence that they are desirable. Because divorced people may be particularly vulnerable, they may reach for sexual encounters as if for a lifeboat. "I felt that, as long as someone was having sex with me, I wasn't dead and I did matter," said one recently divorced person.

Because divorced individuals are usually in their mid-30s or older, they may not be as sensitized to the danger of contracting HIV as are people in their 20s. Divorced individuals should always use a condom to lessen the risk of an STI, including HIV infection and AIDS.

## Friends with Benefits

Friends with benefits is often part of the relational sexual landscape. Mongeau et al. (2013) defined **friends with benefits** as platonic friends (i.e., those not involved in a romantic relationship) who engage in some degree of sexual intimacy on multiple occasions. This sexual activity could range from kissing to sexual intercourse and is a repeated part of a friendship, not just a one-night stand. Forty-three percent of 1,099 undergraduate males reported that they had been in a FWB relationship (42% of 3,466 undergraduate females). These undergraduates were primarily first- and second-year students (Hall and Knox 2013). Mongeau et al. (2013) identified seven types of friends with benefits relationships (FWBR):

*Companionship without commitment suited him just fine.*

SAID OF EINSTEIN IN *EINSTEIN: HIS LIFE AND UNIVERSE*

**Friends with benefits** platonic friends who engage in some degree of sexual intimacy on multiple occasions.

**1.** True friends—close friends who have sex on multiple occasions (similar to but not labeled as romantic partner). The largest percent (26%) of the 258 respondents reported this type of FWBR.

**2.** Just sex—focus is sex, serial hookup with same person. Don't care about person other than as sexual partner. Twelve percent reported this type of FWBR.

**3.** Network opportunism—part of same social networks who just hang out and end up going home to have sex together since there is no better option; a sexual fail safe. Fifteen percent reported this type of FWBR.

**4.** Successful transition—intentional use of friends with benefits relationship to transition into a romantic relationship. Eight percent reported this type of FWBR.

**5.** Unintentional transition—sexual relationship morphs into romantic relationship without intent. Relationship results from regular sex, hanging out together, etc. Eight percent reported this type of FWBR.

**6.** Failed transition—one partner becomes involved while the other does not. The relationship stalls. The lowest percentage (7%) of the 258 respondents reported this type of FWBR.

**7.** Transition out—the couple were romantic, the relationship ended, but the sexual relationship continued. Eleven percent reported this type of FWBR. Advantages to having sex with one's former partner include having a "safe" sexual partner, having a predictably "good" sexual partner who knows likes/dislikes and does not increase the number of lifetime sexual partners, and "fanning sexual flames might facilitate rekindling partners' emotional connections" (Mongeau et al. 2013). Karlsen and Traeen (2013) asserted that the majority of partners in a living together relationship do not explicitly define the relationship or have explicit rules to regulate it.

Advantages of being in a FWBs relationship involve having a regular sexual partner, not feeling vulnerable to having to hook up with a stranger just to have sex, and feeling free to have other relationships (including sexual). Disadvantages include developing a bad reputation as someone who does not really care about emotional involvement, coping with discrepancy of becoming more or less involved than the partner, and losing the capacity to give oneself emotionally.

Chelsea Curry

This male is involved in three "friends with benefits" relationships. He says with a smile that these relationships are a man's dream—sex with no commitment. He also notes that some women hate him as a player.

Lehmiller et al. (2014) analyzed data on 376 individuals in a relationship and compared those who were in a friends with benefits relationship (50.5%) with those who were romantically involved but not in a FWB relationship (49.5%). Differences included that FWB partners were less likely to be sexually exclusive, were less sexually satisfied, more likely to practice safe sex, and generally communicated less about sex than romantic partners who were not in a FWB relationship.

## Concurrent Sexual Partnerships

**Concurrent sexual partnerships** are those in which the partners have sex relationships with several individuals with whom they have a relationship. In a study of 783 adults ages 18 to 59, 10% reported that both they and their partners had had other sexual partners during their relationship (Paik 2010). Men were more likely than women to have been nonmonogamous (17% versus 5%). A serious relationship was the context in which going outside the relationship for sex was least likely. However, if the partners had a casual or friends-with-benefits relationship, the chance of nonmonogamy rose 30% and 44%, respectively. Wester and Phoenix (2013) studied relationships at the "talking" phase and noted that men were much more likely to approve of all forms of sexual behavior in multiple relationships than women.

**Concurrent sexual partnerships** the partners have sex relationships with several individuals at the same time.

## Swinging

**Swinging relationships** (also called **open relationships**) consist of focused recreational sex. These relationships involve married/pair bonded individuals agreeing that they may have sexual encounters with others.

**Swinging relationships** individuals in a committed relationship agree to have recreational sex with other individuals, independently or as a group.

Kimberly and Hans (2012) reviewed the literature on swinging couples. There are about 3 million swinging couples in the United States with most reporting positive emotional and sexual relationships. Swinging for them is a mutually agreeable recreational behavior that has positive benefits for the relationship.

Almost a quarter (24%) of 960 undergraduate males reported that they were comfortable with their partner being emotionally and sexually involved with someone else; 13% of 2,931 undergraduate females agreed (Hall and Knox 2013). Couples who have open relationships often develop certain rules. The following are rules of a couple in an open relationship (author's files):

1. ***Honesty***—we tell each other everything we do with someone outside the relationship. If we flirt, we even tell that. Openness about our feelings is a must—if we get uncomfortable or jealous, we must talk about it.
2. ***Recreational sex***—sex with the other person will be purely recreational—it is not love and the relationship with the other person is going nowhere. The people we select to have sex with must know that we have a loving committed relationship with someone else.
3. ***Condom***—a requirement every time.
4. ***Approval***—every person we have sex with must be approved by the partner in advance. Each partner has the right to veto a selection. The person in question must not be into partner snatching, looking for romance, or jealous. Persons off the list are co-workers, family (he can't have sex with her sister or she, with his brother), old lovers, and old friends.
5. ***Online hunting***—prohibited. Each agrees not to go looking on the Internet for sex partners.

## Sexual Problems

Sexual relationships are not without sexual problems. Hendrickx and Enzlin (2014) analyzed data on an Internet survey of 35,132 heterosexual Flemish men and women (aged 16 to 74 years). In men, the most common sexual difficulties (impairment in sexual function regardless of level of distress) were hyperactive

## APPLYING SOCIAL RESEARCH | Sexual Correlates of Relationship Happiness*

"Sex is the icing on the relationship cake" is an accepted truth. Few believe that individuals can "screw themselves into a good relationship." Rather, it is the relationship which is more likely to have a positive influence on the couple's sex life. Nevertheless, there may be certain sexual factors associated with relationship satisfaction/ happiness—the focus of this study.

### Sample and Methods

The data for analysis were taken from a nonrandom sample of 1,319 undergraduate volunteers at a large southeastern university who responded to an online 100-item questionnaire. Since there were no identifying codes associated with the submitted questionnaires, the identity of the respondents was anonymous.

Almost three-fourths (73.1%) of the respondents were female; over a fourth (26.7%) were male. The average age of the respondents was 20.1 with a median age of 19 years. Regarding race, 74% self-identified as white; 17.1% self-identified as black (African-American, African black, and Caribbean black were combined); 3% Asian; 2.1% Hispanic; 1.7% biracial; and 1.7% other. Regarding relationship status of the respondents, almost half (49%) were emotionally involved with one person, 29.3% were not dating, and 21.7% were casually dating.

The dependent variable, relationship happiness, was measured by asking respondents to respond to the following statement: "In regard to my current relationship, on a scale of 1 to 10, with 1 being extremely unhappy and 10 being extremely happy, I rate my current level of happiness as:" by circling the number that described their level of relationship happiness. Various independent variables which were found to be statistically significant are included in the results section. Ordinal Least Squares (OLS) regression analysis was used to determine sexual predictors of relationship happiness among the respondents.

sexual desire (frequent sexual urges or activity) (27.7%), premature ejaculation (12.2%), and erectile difficulty (8.3%). For women, the most common sexual difficulties were absent or delayed orgasm (20.1%), hypoactive sexual desire (19.3%), and lack of responsive desire (13.7%).

# Condom Assertiveness

**Condom assertiveness** the unambiguous messaging that sex without a condom is unacceptable.

Wright et al. (2012) reviewed the concept of **condom assertiveness**—the unambiguous messaging that sex without a condom is unacceptable—and identified the characteristics of undergraduate women who are more likely to insist on condom use. More condom-assertive females have more faith in the effectiveness of condoms, believe more in their own condom communication skills, perceive that they are more susceptible to STIs, believe there are more relational benefits to being condom assertive, believe their peers are more condom assertive, and intend to be more condom assertive. Peasant et al. (2014) identified the various negotiation strategies for condom use.

Condom assertiveness is important not only for sexual and anal intercourse but also for oral sex. Not to use a condom or dental dam is to increase the risk of contracting a sexually transmitted infection (STI). Indeed, individuals think "I am on the pill and won't get pregnant" or "No way I am getting pregnant by having oral sex" only to discover HPV or another STI in their mouth or throat.

Also known as sexually transmitted disease, or *STD*, STI refers to the general category of sexually transmitted infections such as chlamydia, genital herpes, gonorrhea, and syphilis. The most lethal of all STIs is that caused by the human immunodeficiency virus (HIV), which attacks the immune system and can lead to autoimmune deficiency syndrome (AIDS).

Here we note the importance of lowering one's risk by assessing one's risk (see Self-Assessment feature).

**Results**

The results are summarized into those variables associated with relationship happiness (see the table below).

**Sexual Correlates of Relationship Happiness**

Emotional involvement—being involved with one person, being engaged, being married
Absolutist sexual value—waiting until marriage to have sexual intercourse
Sex with love—history of having had sex in the context of a love relationship
Open sexual communication—expressing preferences to one's sexual partner
Use of birth control—not allowing sexual intercourse to occur without it
No forced sex—absence of having been forced to have sex by a partner
No regrets—about timing of first sexual experience



Some unique findings of this study included that relationship happiness was unrelated to one's sexual orientation, having had sexual intercourse, and having engaged in various sexual behaviors—masturbation, oral sex, and anal sex.

*Abridged and adapted from an original paper, 2011, by Dantzler, T., D. Knox, V. Kriskute, and K. Vail-Smith, East Carolina University.

## The Illusion of Safety in a "Monogamous" Relationship

Most individuals in a serious "monogamous" relationship assume that their partner is faithful and that they are at zero risk for contracting an STI. In a study of 1,341 undergraduates at a large southeastern university, almost 30%

*Oh, what a tangled web we weave . . . when first we practice to deceive.*

WALTER SCOTT, *MARMION*

This female is "condom assertive." Indeed, most females today require a condom before they have intercourse.

## SELF-ASSESSMENT | Student Sexual Risks Scale

Safer sex means sexual activity that reduces the risk of transmitting STIs, including the AIDS virus. Using condoms is an example of safer sex. Unsafe, risky, or unprotected sex refers to sex without a condom, or to other sexual activity that might increase the risk of transmitting the AIDS virus. For each of the following items, check the response that best characterizes your option.

A = Agree
U = Undecided
D = Disagree

_____ 1. If my partner wanted me to have unprotected sex, I would probably give in.
_____ 2. The proper use of a condom could enhance sexual pleasure.
_____ 3. I may have had sex with someone who was at risk for HIV/AIDS.
_____ 4. If I were going to have sex, I would take precautions to reduce my risk for HIV/AIDS.
_____ 5. Condoms ruin the natural sex act.
_____ 6. When I think that one of my friends might have sex on a date, I remind my friend to take a condom.
_____ 7. I am at risk for HIV/AIDS.
_____ 8. I would try to use a condom when I have sex.
_____ 9. Condoms interfere with romance.
_____ 10. My friends talk a lot about safer sex.
_____ 11. If my partner wanted me to participate in risky sex and I said that we needed to be safer, we would still probably end up having unsafe sex.
_____ 12. Generally, I am in favor of using condoms.
_____ 13. I would avoid using condoms if at all possible.
_____ 14. If a friend knew that I might have sex on a date, the friend would ask me whether I was carrying a condom.
_____ 15. There is a possibility that I have HIV/AIDS.
_____ 16. If I had a date, I would probably not drink alcohol or use drugs.
_____ 17. Safer sex reduces the mental pleasure of sex.
_____ 18. If I thought that one of my friends had sex on a date, I would ask the friend if he or she used a condom.
_____ 19. The idea of using a condom doesn't appeal to me.
_____ 20. Safer sex is a habit for me.
_____ 21. If a friend knew that I had sex on a date, the friend wouldn't care whether I had used a condom or not.
_____ 22. If my partner wanted me to participate in risky sex and I suggested a lower-risk alternative, we would have the safer sex instead.
_____ 23. The sensory aspects of condoms (smell, touch, and so on) make them unpleasant.
_____ 24. I intend to follow safer sex guidelines within the next year.
_____ 25. With condoms, you can't really give yourself over to your partner.
_____ 26. I am determined to practice safer sex.
_____ 27. If my partner wanted me to have unprotected sex and I made some excuse to use a condom, we would still end up having unprotected sex.
_____ 28. If I had sex and I told my friends that I did not use condoms, they would be angry or disappointed.
_____ 29. I think safer sex would get boring fast.
_____ 30. My sexual experiences do not put me at risk for HIV/AIDS.
_____ 31. Condoms are irritating.
_____ 32. My friends and I encourage each other before dates to practice safer sex.
_____ 33. When I socialize, I usually drink alcohol or use drugs.
_____ 34. If I were going to have sex in the next year, I would use condoms.
_____ 35. If a sexual partner didn't want to use condoms, we would have sex without using condoms.
_____ 36. People can get the same pleasure from safer sex as from unprotected sex.
_____ 37. Using condoms interrupts sex play.
_____ 38. Using condoms is a hassle.

### Scoring

(to be read after completing the previous scale)
Begin by giving yourself 80 points. Subtract one point for every undecided response. Subtract two points every time that you disagreed with odd-numbered items or with item number 38. Subtract two points every time you agreed with even-numbered items 2 through 36.

### Interpreting Your Score

Research shows that students who score higher on the SSRS are more likely to engage in risky sexual activities, such as having multiple sex partners and failing to consistently use condoms during sex. In contrast, students who practice safer sex tend to endorse more positive attitudes toward safer sex, and tend to have peer networks that encourage safer sexual practices. These students usually plan on making sexual activity safer, and they feel confident in their ability to negotiate safer sex, even when a dating partner may press for riskier sex. Students who practice safer sex often refrain from using alcohol or drugs, which may impede negotiation of safer sex, and often report having engaged in lower-risk activities in the past. How do you measure up?

### (Below 15) Lower Risk

Congratulations! Your score on the SSRS indicates that, relative to other students, your thoughts and behaviors are more supportive of safer sex. Is there any room for improvement in your score? If so, you may want to examine items for which you lost points and try to build safer sexual strengths in those areas. You can help protect others from HIV by educating your peers about making sexual activity safer. (Of 200 students surveyed by DeHart and Berkimer, 16% were in this category.)

**(15 to 37) Average Risk**

Your score on the SSRS is about average in comparison with those of other college students. Though you don't fall into the higher-risk category, be aware that "average" people can get HIV, too. In fact, a recent study indicated that the rate of HIV among college students is 10 times that of the general heterosexual population. Thus, you may want to enhance your sexual safety by figuring out where you lost points and work toward safer sexual strengths in those areas (Of 200 students surveyed by DeHart and Berkimer, 68% were in this category.)

**(38 and Above) Higher Risk**

Relative to other students, your score on the SSRS indicates that your thoughts and behaviors are less supportive of safer sex. Such high scores tend to be associated with greater HIV-risk behavior. Rather than simply giving in to riskier attitudes and behaviors, you may want to empower yourself and reduce your risk by critically examining areas for improvement. On which items did you lose points? Think about how you can strengthen your sexual safety in these areas. Reading more about safer sex can help, and sometimes colleges and health clinics offer courses or workshops. You can get more information about resources in your area by contacting the Center on Disease Control's HIV/AIDS Information Line at 1-800-342-2437. (Of 200 students surveyed by DeHart and Birkimer, 16% were in this category.)

**Source**

DeHart, D. D. and J. C. Birkimer. 1997. The Student Sexual Risks Scale (modification of SRS for popular use; facilitates student self-administration, scoring, and normative interpretation). Developed for this text by Dana D. DeHart, College of Social Work at the University of South Carolina; John C. Birkimer, University of Louisville. Used by permission of Dana DeHart.

(27%) of the males and 20% of the females reported having oral, vaginal, or anal sex outside of a relationship that their partner considered monogamous. People most likely to cheat were men over the age of 20, those who were binge drinkers, members of a fraternity, male NCAA athletes, or nonreligious people (Vail-Smith et al. 2010). In another study of 227 couples who were involved in a monogamous relationship and made a specific agreement not to have sex with others, just under 30% reported cheating (Warren et al. 2012). These data suggest that individuals, even in committed monogamous relationships, cannot count on their partner being faithful…and that they themselves are not immune from infidelity. Hence, individuals might be aware that they are always at risk for contracting a STI.

*The worst part of writing fiction is the fear of wasting your life behind a keyboard. The idea that, dying, you'll realize you only lived on paper. Your only adventures were make-believe, and while the world fought and kissed, you sat in some dark room masturbating and making money.*

CHUCK PALAHNIUK, *STRANGER THAN FICTION*

# Sexual Fulfillment: Some Prerequisites

There are several prerequisites for having a good sexual relationship.

## Self-Knowledge, Body Image and Health

Sexual fulfillment involves knowledge about yourself and your body. Such information not only makes it easier for you to experience pleasure but also allows you to give accurate information to a partner about pleasing you. It is not possible to teach a partner what you don't know about yourself.

Sexual fulfillment also implies having a positive body image. To the degree that you have positive feelings about your body, you will regard yourself as a person someone else would enjoy touching, being close to, and making love with. If you do not like yourself or your body, you might wonder why anyone else would. Van den Brink et al. (2013) analyzed body image data on 319 Dutch undergraduates. Most reported neutral or mildly positive body evaluations, and in 30% of the sample these evaluations were clearly positive. Comparisons between

*Leave something to be desired, in order not through glut to become unhappy.*

BALTASAR GRACIAN, SPANISH WRITER IN 1600S

women who reported positive versus neutral body evaluations showed that the body-satisfied women had lower body mass indexes (BMIs) and reported less body image investment, less overweight preoccupation, and less body self-consciousness during sexual activity. These women also reported higher sexual self-esteem and better sexual functioning. Woertman and Van den Brink (2012) found sexual arousal, initiating sex, sexual satisfaction, and orgasm related to a positive body image in women.

*Commandment #1: Believe in yourself.*

*Commandment #2: Get over yourself.*

KRISTAN HIGGINS, AUTHOR

Effective sexual functioning also requires good physical and mental health. This means regular exercise, good nutrition, lack of disease, and lack of fatigue. Performance in all areas of life does not have to diminish with age—particularly if people take care of themselves physically.

Good health also implies being aware that some drugs may interfere with sexual performance. Alcohol is the drug most frequently used by American adults (including college students). Although a moderate amount of alcohol can help a person become aroused through a lowering of inhibitions, too much alcohol can slow the physiological processes and deaden the senses. Shakespeare may have said it best: "It [alcohol] provokes the desire, but it takes away the performance" (*Macbeth,* act 2, scene 3). The result of an excessive intake of alcohol for women is a reduced chance of orgasm; for men, overindulgence results in a reduced chance of attaining an erection.

*It's no good pretending that any relationship has a future if your music collections disagree violently or if your favorite films wouldn't even speak to each other if they met at a party.*

NICK HORNBY, ENGLISH NOVELIST

The reactions to marijuana are less predictable than the reactions to alcohol. Though some individuals report a short-term enhancement effect, others say that marijuana just makes them sleepy. In men, chronic use may decrease sex drive because marijuana may lower testosterone levels.

## A Committed, Loving Relationship

A guideline among therapists who work with couples who have sexual problems is to treat the relationship before focusing on the sexual issue. The sexual relationship is part of the larger relationship between the partners, and what happens outside the bedroom in day-to-day interaction has a tremendous influence on what happens inside the bedroom. Indeed, relationship satisfaction is associated with sexual satisfaction (Stephenson et al. 2013). The statement, "I can't fight with you all day and want to have sex with you at night" illustrates the social context of the sexual experience. Partners in committed relationships reported the highest sexual satisfaction (Galinsky and Sonenstein 2013).

*Well, I'm here to tell you that expecting that kind of love—that perfection—from a man is unrealistic. That's right, I said it—it's not gonna happen, no way, no how. Because a man's love isn't like a woman's love.*

STEVE HARVEY, COMEDIAN

In the chapter on love, we reviewed the concept of alexithymia or the inability to experience and express emotion. Scimeca et al. (2013) studied a sample of 300 university students who revealed that higher alexithymia scores were associated with lower levels of sexual satisfaction and higher levels of sexual detachment for females, and with sexual shyness and sexual nervousness for both genders. Conversely, being able to experience and express emotion has positive outcomes for one's sexual relationship.

## An Equal Relationship

*What has been eroticized by male dominant systems of all kinds is dominance and passivity. We need to eroticize equality. I always say to audiences of men: "Cooperation beats submission. Trust me."*

GLORIA STEINEM, FEMINIST

Sanchez et al. (2012) emphasized that traditional gender roles interfere with and inhibit a sexually fulfilling relationship. These roles dictate that the woman not initiate sex, be submissive, disregard her own pleasure, and not give accurate feedback to the male. By disavowing these roles, adopting an egalitarian perspective, and engaging in new behavior (initiating sex, taking the dominant role, insisting on her own sexual pleasure, and informing her partner about what she needs), the couple are on the path to a new and more fulfilling sexual relationship. Of course, such a change in the woman requires that the male give up that he must always initiate sex and belief in the double standard (e.g., the belief that women who love sex are sluts), and delight in not having to drive the sexual bus all the time.

## Open Sexual Communication (Sexual Self Disclosure) and Feedback

Sexually fulfilled partners are comfortable expressing what they enjoy and do not enjoy in the sexual experience. Unless both partners communicate their needs, preferences, and expectations to each other, neither is ever sure what the other wants. In essence, the Golden Rule ("Do unto others as you would have them do unto you") is *not* helpful, because what you like may not be the same as what your partner wants.

Sexually fulfilled partners take the guesswork out of their relationship by communicating preferences and giving feedback. Ali (2011) compared Australian and Malaysian couples and found that the former were sexually self-disclosing with their partners (telling the partner what he or she wanted sexually) while the latter were not. She noted that the norms of Malaysian parental socialization do not allow for talking about sex since doing so is thought to destroy childhood innocence. Even adult married couples do not discuss sex as it is not regarded as proper.

**DIVERSITY IN OTHER COUNTRIES**

Nagao et al. (2014) surveyed 5,665 Japanese women who were involved with a male partner and were interested in sexual activity. In general, the women reported a desire for longer foreplay and afterplay as well as more influence over when sex occurred and what sexual positions were experienced. Their partner's unilateral action during sexual activity negatively affected the quality of their sexual life with their partner. These findings emphasize the need for more sexual communication between the partners.

## Frequent Initiation of Sexual Behavior

Simms and Byers (2013) assessed the sexual initiation behaviors of 151 individuals (33% men and 66% women) who were 18–25 years of age, had been in an

WHAT IF?

### What If My Partner Does Not Like Sex or Is Hypersexual?

Sex is an important part of a couple's relationship, particularly in new relationships. However, individuals vary in their interest in, capacity for, and preference for different sexual behaviors. Although some need sex daily, are orgasmic, and enjoy a range of sexual behaviors, others never think of sex, have never had an orgasm, and want to get any sexual behavior over with as soon as possible. When two people of widely divergent views on sex end up in the same relationship, clear communication and choices are necessary. The person who has no interest must decide if developing an interest is a goal and be open to learning about masturbation and sexual fantasy. Where becoming interested in sex is not a goal, the partner must decide the degree to which this is an issue. Some will be pleased that the partner has no interest because this means no sexual demands, whereas others will bolt. Openness about one's sexual feelings and hard choices will help resolve the dilemma.

While hypersexuality (also known as hyperphilia, hypersexual disorder, and compulsive sexual disorder) has no agreed-upon definition in terms of how much sex is excessive, it may become a problem for the partner through its expression of being driven to and feeling a loss of control over spending hours viewing pornography over the Internet, having cybersex, visiting strip clubs, having affairs, etc. If there is a feeling of not having control over one's sexual appetite, behavioral sex therapy may be helpful (Kaplan and Krueger 2010). Alternatively, sex addiction is treated in the same way as any other addiction (Niven 2010).

exclusive, heterosexual, non-cohabitating dating relationship between 3 and 18 months, had seen their dating partners at least 3–4 days per week over the previous month, and had engaged in genitally focused sexual activities. The researchers found that both men and women who reported initiating sex more frequently and who perceived their partner as initiating more frequently reported greater sexual satisfaction.

## Having Realistic Expectations

To achieve sexual fulfillment, expectations must be realistic. A couple's sexual needs, preferences, and expectations may not coincide. It is unrealistic to assume that your partner will want to have sex with the same frequency and in the same way that you do on all occasions. It may also be unrealistic to expect the level of sexual interest and frequency of sexual interaction to remain consistently high in long-term relationships.

Sexual fulfillment means not asking things of the sexual relationship that it cannot deliver. Failure to develop realistic expectations will result in frustration and resentment. One's age, health (both mental and physical), sexual dysfunctions of self and partner, and previous sexual experiences will have an effect on one's sexuality and one's sexual relationship and sexual fulfillment (McCabe & Goldhammer 2012).

Each partner brings to a sexual encounter, sometimes unconsciously, a motive (pleasure, reconciliation, procreation, duty), a psychological state (love, hostility, boredom, excitement), and a physical state (tense, exhausted, relaxed, turned on). The combination of these factors will change from one encounter to another. Tonight one partner may feel aroused and loving and seek pleasure, but the other partner may feel exhausted and hostile and have sex only out of a sense of duty. Tomorrow night, both partners may feel relaxed and have sex as a means of expressing their love for each other.

## Sexual Compliance

**Sexual compliance** an individual willingly agrees to participate in sexual behavior without having the desire to do so.

Given that partners may differ in sexual interest and desire, Vannier and O'Sullivan (2010) identified the concept of **sexual compliance** whereby an individual willingly agrees to participate in sexual behavior without having the desire to do so. The researchers studied 164 heterosexual young (18–24) adult couples in committed relationships to assess the level of sexual compliance. Almost half (46%) of the respondents reported at least one occasion of sexual compliance with sexual compliance comprising 17% of all sexual activity recorded over a three-week period. Indeed, sexual compliance was a mechanism these individuals used in their committed relationships to resolve the issue of different levels of sexual desire that is likely to happen over time in a stable couple's relationship. Others felt guilty they did not desire sex and still others did it because their partner provided sex when the partner was not in the mood.

There were no gender differences in differences of sexual desire and no gender difference in providing sexual compliant behavior. The majority of participants reported enjoying the sexual activity despite not wanting to engage in it at first.

## Job Satisfaction

Stulhofer et al. (2013) analyzed data on job satisfaction and sexual health among a sample of over 2,000 males and found that negative mood resulting from job stress/unhappiness at work was associated with sexual difficulties. Having low job stress, high income, emotional intimacy with one's partner, and having children were associated with sexual health.

## Avoiding Spectatoring

One of the obstacles to sexual functioning is **spectatoring**, which involves mentally observing your sexual performance and that of your partner. When the researchers in one extensive study observed how individuals actually behave during sexual intercourse, they reported a tendency for sexually dysfunctional partners to act as spectators by mentally observing their own and their partners' sexual performance. For example, the man would focus on whether he was having an erection, how complete it was, and whether it would last. He might also watch to see whether his partner was having an orgasm (Masters and Johnson 1970). Just focusing on one's own body can have an effect. Van den Brink et al. (2013) confirmed that body image self-consciousness was negatively associated with sexual functioning and frequency of sexual activity with a partner. Spectatoring, as Masters and Johnson conceived it, interferes with each partner's sexual enjoyment because it creates anxiety about performance, and anxiety blocks performance. A man who worries about getting an erection reduces his chance of doing so. A woman who is anxious about achieving an orgasm probably will not. Montesi et al. (2013) confirmed the negative effect of anxiety on a couple's sexual relationship. The desirable alternative to spectatoring is to relax, focus on and enjoy your own pleasure, and permit yourself to be sexually responsive.

**Spectatoring** mentally observing your sexual performance and that of your partner.

**DIVERSITY IN OTHER COUNTRIES**

How does ethnicity influence sexual desire? Woo et al. (2012) compared Chinese women with European and Canadian women and found that the former reported higher sex guilt and lower sexual desire than the latter. The researchers identified the Confucian philosophy which viewed sex as positive as long as it was in reference to producing children to carry on the family line. Sex outside of marriage and sex for pleasure were discouraged.

## Female Vibrator Use, Orgasm, and Partner Comfort

It is commonly known that vibrators (also known as sex toys and novelties) are beneficial for increasing the probability of orgasmic behavior in women. During intercourse women typically report experiencing a climax 30% of the time; vibrator use increases orgasmic reports to over 90%. Herbenick et al (2010) studied women's use of vibrators within sexual partnerships. They analyzed data from 2,056 women aged 18–60 years in the United States. Partnered vibrator use was common among heterosexual-, lesbian-, and bisexual-identified women. Most vibrator users indicated comfort using them with a partner and related using them to positive sexual function. In addition, partner knowledge and perceived liking of vibrator use was a significant predictor of sexual satisfaction for heterosexual women.

That men are accepting of using a vibrator with their female partners was confirmed by Reece et al. (2010), who surveyed a nationally representative sample of heterosexual men in the U.S. Forty-three percent reported having used a vibrator. Of those who had done so, most vibrator use had occurred within the context of sexual interaction with a female partner. Indeed, 94% of male vibrator users reported that they had used a vibrator during sexual play with a partner, and 82% reported that they had used a vibrator during sexual intercourse. In another study which used a national sample, women and men held positive attitudes about female vibrator

Chelsea Curry

Vibrators are available in erotic adult shops or online.

use (Herbenick et al. 2011).These data support recommendations of therapists and educators who often suggest the incorporation of vibrators into partnered relationships.

## The Future of Sexual Relationships

The future of sexual relationships will involve continued individualism as the driving force in sexual relationships. Numerous casual partners (hooking up) with predictable negative outcomes—higher frequencies of STIs, unexpected pregnancies, sexual regret, and relationships going nowhere—will continue to characterize individuals in late adolescence and early twenties. As these persons reach their late twenties, the goal of sexuality begins to transition to seeking a partner not just to hook up and have fun with but to settle down with. This new goal is accompanied by new sexual behaviors such as delayed first intercourse in the new relationship, exclusivity, and movement toward marriage. The monogamous move toward the marriage context creates a transitioning of sexual values from hedonism to relativism to absolutism, where strict morality rules become operative in the relationship (expected fidelity).

## Summary

***How is sex defined and what are three main sexual values?***

Sex is defined variably but includes the elements of anatomy/physiology, behaviors, relationship context, and sexual orientation. The definition also includes sexual values which are moral guidelines for making sexual choices in nonmarital, marital, heterosexual, and homosexual relationships.

Three sexual values are absolutism (rightness is defined by official code of morality), relativism (rightness depends on the situation—who does what, with whom, in what context), and hedonism ("if it feels good, do it"). Relativism is the sexual value most college students hold, with women being more relativistic than men and men being more hedonistic than women.

Individuals most likely to have sexual intercourse early (age 13 or earlier) report engaging in more antisocial behaviors, having a lot of other sex friends, and being from a non-intact family. About 60% of college students believe that if they have oral sex, they are still virgins.

***What are various sexual behaviors?***

Some individuals are asexual—having an absence of sexual behavior with a partner and one's self (masturbation). About 4% of females and 11% of males report being asexual in the last twelve months. Kissing involves various meanings including love, approval, hello, goodbye; it is also used as an ointment such as when a parent kisses the hurt knee of a child, and so on. French kissing may be dangerous by increasing risk of a bacterial infection that may lead to meningitis.

Masturbation involves stimulating one's own body with the goal of experiencing pleasurable sexual sensations. Fellatio is oral stimulation of the man's genitals by his partner. In many states, legal statutes regard fellatio as a "crime against nature," in that the sex does not produce babies. Cunnilingus is oral stimulation of the woman's genitals by her partner. Anal (not vaginal) intercourse is the sexual behavior associated with the highest risk of HIV infection. The potential for the rectum to tear and blood contact to occur presents the greatest danger. AIDS is lethal. Partners who use a condom during anal intercourse reduce their risk of not only HIV infection but also other STIs.

Sexual behavior with a partner is associated with feelings of well-being. Indeed, researchers have found that experiencing affection and sex with a partner on one day predicted lower negative mood, higher positive mood, and lower stress the following day.

Cybersex refers to a computer-mediated sexual experience which may include sending sexual text or photos.

***What are the various relationship contexts of sexual behavior?***

Never-married and noncohabiting individuals report more sexual partners than those who are married or living with a partner. Marital sex is distinctive for its social legitimacy, declining frequency, and satisfaction (both physical and emotional). Marital sex is also the most emotionally and physically rewarding. Divorced individuals have a lot of sexual partners but are the least sexually fulfilled.

***What is condom assertiveness and which women are more likely to practice it?***

Condom assertiveness is the unambiguous message from one sexual partner to another that sex without a condom is unacceptable. The characteristics of undergraduate women who are most likely to insist on condom use are those who believe that condoms are effective in preventing STIs, believe in their own condom communication skills, perceive that they are susceptible to STIs, believe that there are relational benefits to being condom assertive, believe their peers are condom assertive, and have the intention to be more condom assertive.

The person most likely to get an STI has sexual relations with a number of partners or with a partner who has a variety of partners. Even if you are in a mutually monogamous relationship, you may be at risk for acquiring an STI, as 30% of male undergraduate students and 20% of female undergraduate students in monogamous relationships reported having oral, vaginal, or anal sex with another partner outside of the relationship.

***What are the prerequisites of sexual fulfillment?***

Fulfilling sexual relationships involve self-knowledge, positive body image, self-esteem, health, a good nonsexual relationship, open sexual communication, sexual self disclosure, safer sex practices, and making love with, not to, one's partner. Other variables include realistic expectations ("my partner will not always want what I want"), and avoiding spectatoring (not being self-conscious and observing one's "performance").

***What is the future of sexual relationships?***

Individualism will continue to be the driving force in sexual relationships. Numerous casual partners (hooking up) with predictable negative outcomes—higher frequencies of STIs, unexpected pregnancies, sexual regret, and relationships going nowhere—will continue to characterize individuals in late adolescence and early twenties. As these persons reach their late twenties, the goal of sexuality begins to transition to seeking a partner not just to hook up and have fun with but to provide an emotional connection. This new goal is accompanied by new sexual behaviors such as delayed first intercourse in the new relationship, exclusivity, and movement toward marriage.

## Key Terms

Absolutism
Anodyspareunia
Asceticism
Asexual
Coitus
Comprehensive sex education program
Concurrent sexual partnership
Condom assertiveness
Cybersex
Friends with benefits
Hedonism
Masturbation
New relationship energy (NRE)
Relativism
Satiation
Secondary virginity
Swinging relationships
Sexual compliance
Sexual double standard
Sexual readiness
STI (sexually transmitted infection)
Sexual values
Spectatoring

## Web Links

Body Health: A Multimedia AIDS and HIV Information Resource http://www.thebody.com

Centers for Disease Control and Prevention (CDC) http://www.cdc.gov

Go Ask Alice: Sexuality http://www.goaskalice.columbia.edu/

National Center for Health Statistics http://www.cdc.gov/nchs/

Kinsey Confidential http://kinseyconfidential.org/

Sex Education Library http://www.sexedlibrary.org/

Kinsey Institute http://www.kinseyinstitute.org/

Sexual Health Network http://www.sexualhealth.com/

Sexuality Information and Education Council of the United States (SIECUS) http://www.siecus.org

# CHAPTER 10 Violence and Abuse in Relationships

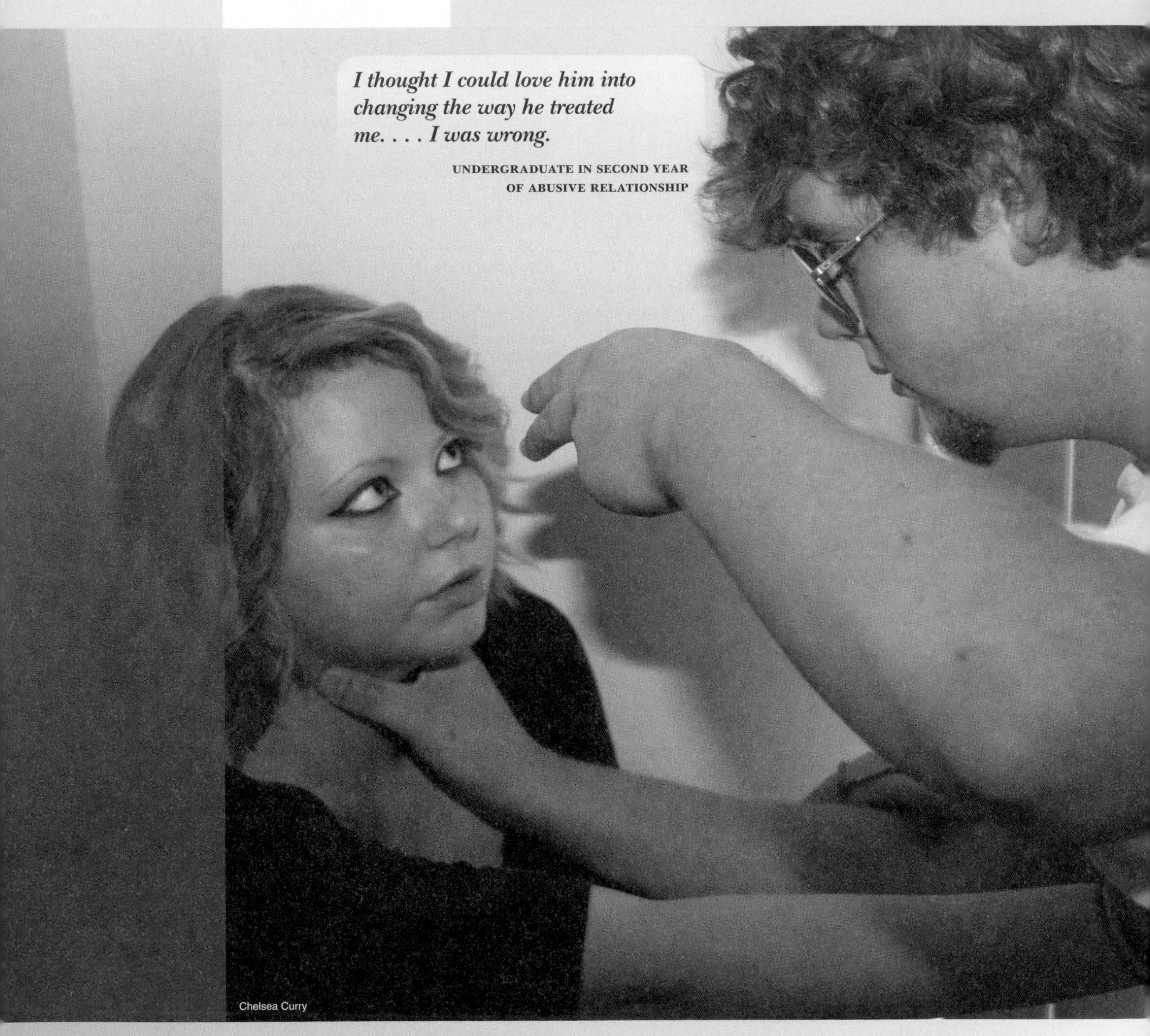

*I thought I could love him into changing the way he treated me. . . . I was wrong.*

UNDERGRADUATE IN SECOND YEAR OF ABUSIVE RELATIONSHIP

Chelsea Curry

## Learning Objectives

- Explain the various types of abuse.
- Identify the explanations for abusive behavior in intimate relationships.
- Describe abuse in undergraduate relationships and in marriage.
- Review the effects and the cycle of abuse.
- Discuss parent, sibling, and elder abuse.
- Know the future of abuse in intimate relationships.

## TRUE OR FALSE?

1. Violence in relationships is more likely to be mutual than unilaterial.
2. Females and males report similar motivations for being violent toward their partner.
3. There is more intimate partner violence in committed romantic relationships than in casual dating relationships.
4. Child development researchers have found that spanking a child results in the child's higher self-esteem, higher quality marriages, and higher life satisfaction.
5. Taking out a protective order against an abusive partner actually increases the number of assaultive acts.

*Answers:* 1. F 2. F 3. T 4. F 5. F

Lady Gaga, Christina Aguilera, Rihanna, and Halle Berry are celebrities who have made public that they were abused by their intimate partners. Halle Berry, for example, noted that her partner beat her to the point that she lost 80% of her hearing in her right ear. Television media (e.g., (*Dateline*, 20/20)) regularly feature horror stories of women who are beaten up or killed by their partners. What began as an intimate love relationship ends in the abuse of one's former partner or spouse.

*I have had abusive relationships. . . . I was never brave enough to be the person who I am today.*

LADY GAGA

# Nature of Relationship Abuse

In this chapter, we examine the other side of intimacy in one's relationships. The various terms and types of abuse are discussed in this section.

## Violence and Homicide

Also referred to as physical abuse, **violence** is defined as physical aggression with the purpose of controlling, intimidating, and subjugating another human being (Jacobsen and Gottman 2007). Examples of physical violence include pushing, throwing something at the partner, slapping, hitting, and forcing the partner to have sex.

**Violence** physical aggression with the purpose of controlling, intimidating, and subjugating another human being.

Eleven percent of 3,471 undergraduate females and 3% of 1,102 undergraduate males reported that they had been in a physically abusive relationship (Hall and Knox 2013). Women are more likely to suffer health consequences from being in a physically abusive relationship than men (Sillito 2012). **Intimate partner violence** (IPV) is an all-inclusive term that refers to crimes committed against current or former spouses, boyfriends, or girlfriends.

**Intimate partner violence** all-inclusive term that refers to crimes committed against current or former spouses, boyfriends, or girlfriends.

There are two types of violence (Brownridge 2010). One type is **situational couple violence** (SCV) where conflict escalates over an issue (e.g., money, sex) and one or both partners lose control. The person feels threatened and seeks to defend himself or herself. Control is lost and the partner strikes out. Both partners may lose control at the same time so it is symmetrical. A second type of violence, referred to as **intimate terrorism** (IT), is designed to control the partner. Hertzog et al. (2013) noted the similarity of interpersonal violence and bullying. Both involve forms of aggression with the former being specific to a two-person romantic relationship.

**Situational couple violence** conflict escalates over an issue and one or both partners lose control.

**Intimate Terrorism (IT)** behavior designed to control the partner.

Witte and Kendra (2010) found that individuals who had been victims of interpersonal violence were less likely to discern when a social encounter was becoming dangerous and moving toward violence. The authors emphasized that learning to recognize cues (e.g., the partner is overly aggressive or encourages alcohol consumption) is essential to avoiding such experiences.

**Battered woman syndrome** legal term used in court stating the person accused of murder was suffering from justifying his or her behavior.

**Battered woman syndrome** (also referred to as **battered woman defense**) is a legal term used in court that the person accused of murder was suffering from abuse so as to justify his or her behavior ("I shot him because he raped me"). While there is no medical or psychological term, the "syndrome" refers to frequent, severe maltreatment which often requires medical treatment. Therapists define battering as physical aggression that results in injury and is accompanied by fear and terror (Jacobsen and Gottman 2007).

Klipfel et al. (2014) assessed the occurrence of emotional, physical, and sexual interpersonal aggression reported by 161 individuals within various levels of relationships. The relationship and percent of reported intimate partner violence in the various relationships follow: committed romantic relationships (69%), casual dating relationships (33%), friends-with-benefits relationships (31%), booty calls (36%), and one-night stands (35%). The take home message is that the greater the commitment in the relationship, the greater the reported IPV (intimate partner violence).

**Uxoricide** the murder of a woman by a romantic partner.

Battering may lead to murder. **Uxoricide** is the murder of a woman by a romantic partner. The brutal first-degree murder of student-athlete Yeardley Love at Virginia Tech University by men's lacrosse player George Huguely was uxoricide. Thirty percent of homicides of women 18–24 are by an intimate partner (Teten et al. 2009). Most perpetrators are a male partner. Blacks and Hispanic women have higher rates of being killed by their intimate partners than non-Hispanic women (Azziz-Baumgartner et al. 2011). **Intimate partner homicide** is the murder of a spouse.

**Intimate partner homicide** murder of a spouse.

**Filicide** murder of an offspring by a parent.

**Parricide** murder of a parent by an offspring.

**Siblicide** murder of a sibling.

Other forms of murder in the family are **filicide** (murder of an offspring by a parent), **parricide** (murder of a parent by an offspring), and **siblicide** (murder of a sibling). Family homicide is rare—less than one in 100,000 (Diem and Pizarro 2010). Regarding filicide, parental mental illness, domestic violence, and alcohol/substance abuse are common contexts (Sidebotham 2013).

*Every woman who thinks she is the only victim of violence has to know that there are many more.*

SALMA HAYEK, CELEBRITY

## National **Data**

One in four women in the United States experiences interpersonal partner violence in her lifetime (Liu et al. 2013).

A great deal more abuse occurs than is reported. As few as 5% of rapes are reported to the authorities (Heath et al. 2013). The primary reasons rape victims do not report their experience to the police are not wanting others to know, nonacknowledgment of rape (e.g., "It wasn't really rape . . . I just gave in"), and criminal justice concerns ("It won't do any good" "I will be blamed") (Cohn et al. 2013). Yet the psychological damage which results from rape is enormous. Johnson et al. (2008) emphasized that IPV is associated with post-traumatic stress disorder (PTSD) and results in social maladjustment and personal or social resource loss.

Regardless of the type, violence is not unique to heterosexual couples (Porter & Williams 2011). In a study of violence in a sample of 284 gay and bisexual men, the researchers found that almost all reported psychological abuse, more than a third reported physical abuse, and 10% reported being forced to have sex. Abuse in gay relationships was less likely to be reported to the police because some gays did not want to be "outed" (Bartholomew et al. 2008). Differences between violence in other-sex couples versus same-sex couples include that the latter is more mild, the threat of outing is present, and the

This boyfriend disapproves of the way his girlfriend is dressed and tells her she looks like a slut and to change her clothes. His controlling behavior is an example of emotional abuse.

Chelsea Curry

violence is more isolated because it often occurs in a context of being in the closet. Finally, victims of same-sex relationship abuse lack legal protections and services (most battered women's shelters are not available to lesbians).

The Self-Assessment Abusive Behavior Inventory allows you to assess the degree of abuse in your current or most recent relationship. There are several types of abuse in relationships.

## SELF-ASSESSMENT | Abusive Behavior Inventory

Write the number that best represents your closest estimate of how often each of the behaviors have happened in the relationship with your current or former partner during the previous six months.

| 1 | 2 | 3 | 4 | 5 |
|---|---|---|---|---|
| Never | Rarely | Occasionally | Frequently | Strongly |

_____ 1. Called you a name and/or criticized you
_____ 2. Tried to keep you from doing something you wanted to do (for example, going out with friends or going to meetings)
_____ 3. Gave you angry stares or looks
_____ 4. Prevented you from having money for your own use
_____ 5. Ended a discussion and made a decision without you
_____ 6. Threatened to hit or throw something at you
_____ 7. Pushed, grabbed, or shoved you
_____ 8. Put down your family and friends
_____ 9. Accused you of paying too much attention to someone or something else
_____ 10. Put you on an allowance
_____ 11. Used your children to threaten you (for example, told you that you would lose custody or threatened to leave town with the children)
_____ 12. Became very upset with you because dinner, housework, or laundry was not done when or how it was wanted
_____ 13. Said things to scare you (for example, told you something "bad" would happen or threatened to commit suicide)
_____ 14. Slapped, hit, or punched you
_____ 15. Made you do something humiliating or degrading (for example, begging for forgiveness or having to ask permission to use the car or do something)
_____ 16. Checked up on you (for example, listened to your phone calls, checked the mileage on your car, or called you repeatedly at work)
_____ 17. Drove recklessly when you were in the car
_____ 18. Pressured you to have sex in a way you didn't like or want
_____ 19. Refused to do housework or child care
_____ 20. Threatened you with a knife, gun, or other weapon
_____ 21. Spanked you
_____ 22. Told you that you were a bad parent
_____ 23. Stopped you or tried to stop you from going to work or school

(*continued*)

## SELF-ASSESSMENT (continued)

_____ 24. Threw, hit, kicked, or smashed something
_____ 25. Kicked you
_____ 26. Physically forced you to have sex
_____ 27. Threw you around
_____ 28. Physically attacked the sexual parts of your body
_____ 29. Choked or strangled you
_____ 30. Used a knife, gun, or other weapon against you

### Scoring

Add the numbers you wrote down and divide the total by 30 to determine your score. The higher your score (five is the highest score), the more abusive your relationship.

The inventory was given to 100 men and 78 women equally divided into groups of abusers or abused and nonabusers or nonabused. The men were members of a chemical dependency treatment program in a veterans' hospital and the women were partners of these men. Abusing or abused men earned an average score of 1.8; abusing or abused women earned an average score of 2.3. Nonabusing abused men and women earned scores of 1.3 and 1.6, respectively.

### Source

M. F. Shepard and J. A. Campbell. The Abusive Behavior Inventory: A measure of psychological and physical abuse. *Journal of Interpersonal Violence* 7(3): 291–305. 

## Emotional Abuse

**Emotional abuse** nonphysical behavior designed to denigrate and control the partner.

**Emotional abuse** (also known as **psychological abuse, verbal abuse,** or **symbolic aggression**) is nonphysical behavior designed to denigrate and control the partner. Follingstad and Edmundson (2010) identified the top emotionally abusive behaviors of one's partner in a national sample of adults in the United States: refusal to talk to the partner as a way of punishing the partner, making personal decisions for the partner (e.g., what to wear, what to eat, whether to smoke), throwing a temper tantrum and breaking things to frighten the partner, criticizing/belittling the partner to make him or her feel bad, and acting jealous when the partner was observed talking or texting a potential romantic partner. Other emotionally abusive behaviors include:

Yelling and screaming as a way of intimidating
Staying angry/pouting until the partner gives in
Requiring an accounting of the partner's time
Treating the partner with contempt
Making the partner feel stupid
Withholding emotional and physical contact
Isolation—prohibiting the partner from spending time with friends, siblings, and parents
Threats—threatening the partner with abandonment or threats of harm to oneself, family, or partner's pets
Public demeaning behavior—insulting the partner in front of others
Restricting behavior—restricting the partner's mobility (e.g., being told not to spend time with friends, siblings, parents) or telling the partner what to do
Demanding behavior—requiring the partner to do as the abuser wishes (e.g., have sex)

Thirty-seven percent of 3,469 undergraduate females (and 21% of 1,101 undergraduate males) reported that they "had been involved in an *emotionally* abusive relationship" (Hall and Knox 2013). According to Follingstad and Edmundson (2010), while both partners in a relationship may report that they engage in emotionally abusive behavior, each partner tends to report that he or she engages in less frequent and less severe emotionally abusive behavior. In effect, the respondent creates "a picture of their own use of psychological abuse as limited in scope and not harmful in nature" (p. 506).

Jacobsen and Gottman (2007) revealed that batterers tend to fall into one of two categories, which they call "Pit Bulls" and "Cobras." Pit Bulls are merciless in their abuse, have little guilt, and are dangerous when the partner ends the relationship. Cobras are cool while inflicting pain, may experience guilt, and are less dangerous at the end of the relationship. There is nothing the battered can do to stop the battering but leave.

Chelsea Curry

While the stereotype is that only males physically and emotionally abuse their partners, women are also abusive.

## Mutual or Unilateral?

Marcus (2012) assessed the frequency of unilateral or mutual partner violence in a sample of 1,294 young adults. A quarter of the couples reported a mutually violent pattern compared to three-quarters who evidenced a unilateral pattern. Mathes (2013) noted that among those couples where violence is a part of their relationship, individuals select each other to play their game—to be violent. In all cases, the more violence, the lower the relationship quality.

*The one thing that is unforgivable is deliberate cruelty.*

**BLANCHE, *A STREETCAR NAMED DESIRE***

## Female Abuse of Partner

Swan et al. (2008) revealed that females who abuse their partners are more likely to be motivated by self-defense and fear, whereas male abusers are more likely to be driven by the desire to control. Hence, women who are violent toward their partners tend to strike back rather than throw initial blows. But abusive females may have more variable motives. Whitaker (2014) noted that women were more likely to express violence for reasons of losing their temper, to try to get their partner to listen, to make their partner do as they wanted, or to punish the partner. The motives may be mixed. One of our female students wrote the following:

> *When people think of physical abuse in relationships, they get the picture of a man hitting or pushing the woman. Very few people, including myself, think about the "poor man" who is abused both physically and emotionally by his partner—but this is what I do to my boyfriend. We have been together for over a year now and things have gone from really good to terrible.*
>
> *The problem is that he lets me get away with taking out my anger on him. No matter what happens during the day, it's almost always his fault. It started out as kind of a joke. He would laugh and say "oh, how did I know that this was going to somehow be my fault?" But then it turns into me screaming at him, and not letting him go out and be with his friends as his "punishment."*
>
> *Other times, I'll be talking to him or trying to understand his feelings about something that we're arguing about and he won't talk. He just shuts off and refuses to say anything but "alright." That makes me furious, so I start verbally and physically attacking him just to get a response.*
>
> *Or, an argument can start from just the smallest thing, like what we're going to watch on TV, and before you know it, I'm pushing him off the bed and stepping on his stomach as hard as I can and throwing the remote into the toilet. That really gets to him.*

## What If I'm Not Sure I Am Being Abused?

It is not always easy to know what constitutes being abused. Is a partner's moodiness, irritability, and criticism considered abuse? Is calling a partner stupid considered abuse? Is shoving a partner an abusive act? In general, any verbal or physical behavior that is designed to denigrate or control a partner is abuse.

How much abuse should a person tolerate? The answer is "very little," because the partner will be reinforced for being abusive and will continue/escalate the abuse. In only a short time, demeaning and pushing or shoving a partner can become routine acts considered normative in a couple's relationship. If you don't like the way your partner is treating you, it is important to register your disapproval quickly so that being abused stops. In Chapter 1, we discussed, "Not to decide is to decide." If you don't make a decision to stop abusive behavior, you have made a decision to stay in an abusive relationship.

When women are abusive, they are rarely arrested by the police. Men do not want others to know that they "aren't man enough" to stand up to an abusive woman.

## Stalking

**Stalking** unwanted following or harassment that induces fear in the victim.

Abuse may take the form of stalking. **Stalking** is defined as unwanted following or harassment that induces fear in the victim. This pathological state is characterized by obsessional thinking (Duntley and Buss (2012).

Five types of stalkers have been identified by Mullen et al. (1999):

**1.** ***Rejected stalkers***—ex-partners of a terminated romantic relationship seeking revenge. They often want the partner back. These individuals may be dangerous/kill the partner.

**2.** ***Resentful stalkers***—individuals who have a vendetta against the victim (e.g., jealous of the person for stealing a boyfriend). Their goal is to frighten or cause the person distress.

**3.** ***Intimacy stalker***—seeks to have an intimate relationship with the victim. Celebrities David Letterman and Jodi Foster have been stalked by persons who wanted a relationship with them.

**4.** ***Incompetent suitor***—stalker feels bested by a competitor but still seeks the attention of the victim. The stalker may feel entitled to be loved by the victim.

**5.** ***Predatory stalker***—sexual predator who is planning to attack the victim.

Twenty-eight percent of 3,460 undergraduate females and 18% of 1,102 undergraduate males reported that they had been stalked (followed and harassed) (Hall and Knox 2013). Most (77%) stalkers are male (Lyndon et al. 2012).

**Obsessive relational intrusion (ORI)** relentless pursuit of intimacy with someone who does not want it.

Less threatening than stalking is **obsessive relational intrusion** (ORI), the relentless pursuit of intimacy with someone who does not want it. The person becomes a nuisance but does not have the goal of harm as does the stalker. People who cross the line in terms of pursuing an ORI relationship or responding to being rejected (stalking) engage in a continuum of behavior. Spitzberg and Cupach (2007) have identified ORI behaviors:

1. *Hyperintimacy*—telling a person that they are beautiful or desirable to the point of making them uncomfortable.

2. *Relentless electronic contacts*—flooding the person with text or e-mail messages and phone calls. These contacts may reach the extent of **cybervictimization** which includes being sent threatening text/e-mail, unsolicited obscene e-mail, computer viruses, or junk mail (spamming). Short of cybervictimization is **cybercontol**, whereby individuals use communication technology, such as cell phones, e-mail, and social networking sites, to monitor or control romantic partners. In one study of 745 undergraduates, close to 25% of female college students monitored their partner's behavior by checking e-mails, even password-protected accounts, versus only 6% of males. In addition, more females than males thought it was appropriate to check e-mail and cell phone call histories of their partners. A sociobiological explanation can be used to explain that more females are involved in monitoring their partner's behavior than vice versa—males are more likely to cheat and women are protecting their relationship (Burke et al. 2011).

**Cybervictimization** being sent threatening text/e-mail, unsolicited obscene e-mail, computer viruses, or junk mail.

**Cybercontrol** using communication technology, such as cell phones, e-mail, and social networking sites, to monitor or control romantic partners.

3. *Interactional contacts*—showing up at the person's workplace or gym. The intrusion may also include joining the same volunteer groups as the pursued.

4. *Surveillance*—monitoring the movements of the pursued (e.g., by following the person or driving past the person's house).

5. *Invasion*—breaking into the person's house and stealing objects that belong to the person; stealing the person's identity; or putting Trojan horses (viruses) on the person's computer. One woman downloaded child pornography on her boyfriend's computer and called the authorities to arrest him. After a jury trial he was sent to prison.

6. *Harassment or intimidation*—leaving unwanted notes on the person's desk or a dead animal on the doorstep.

7. *Threat or coercion*—threatening physical violence or harm to the person's family or friends.

8. *Aggression or violence*—carrying out a threat by becoming violent (e.g., kidnapping or rape).

The following example of being stalked is from one of our students. It illustrates stalking in person and cyberstalking.

***My Fiancé Turned into a Stalker***

*I ended the engagement to my boyfriend/fiancé of three years and he did not take the news so well. He demanded that he wanted back "everything he ever bought me." So he drove to my apartment at 2 A.M. to collect what he thought was rightfully his. He left the place a wreck.*

*I also began receiving multiple text messages, multiple phone calls, and lengthy voice mails. He began posting personal information about our relationship on Facebook. And, he began contacting my friends via Facebook trying to keep tabs on what I was doing, who I was seeing, and where I was going.*

*He also randomly showed up at my apartment unannounced with roses and a card trying to get me back. And, he showed up unannounced at my parents' house trying to get them involved in getting me back. He was grasping at straws trying to appear that he had changed so I would l take him back.*

*To this day I am still afraid that he might show up somewhere I am or continue with the many text messages and phone calls. He never physically threatens me but he did threaten to kill himself if I ever broke up with him—I was terrified that he would actually go through with it. He had been emotionally, mentally, and sometimes physically abusive during our relationship but he had never physically threatened me.*

## Reacting to the Stalker

Being very controlling in an existing relationship is predictive that the partner will become a stalker when the relationship ends. Ending a relationship with a potential stalker in a way that does not spark stalking is important. Some guidelines include:

**1.** Make a direct statement to the person ("I am not interested in a relationship with you, my feelings about you will not change, and I know that you will respect my decision and direct your attention elsewhere").

**2.** Seek protection through formal channels (police or court restraining order).

**3.** Avoid the perpetrator (ignore text messages/e-mail, don't walk with or talk to, hang up if the person calls).

**4.** Use informal coping methods (block text messages/phone calls).

**5.** Stay away. Do not offer to be friends since the person may misinterpret this as a romantic overture.

*If he mistreated and abused his last girlfriend, why would you want to be his new girlfriend?*

KAREN E. QUINONES MILLER, AUTHOR

# Explanations for Violence and Abuse in Relationships

Research suggests that numerous factors contribute to violence and abuse in intimate relationships. These factors operate at the cultural, community, individual, and family levels.

## Cultural Factors

In many ways, American culture tolerates and even promotes violence (Walby 2013). Violence in the family stems from the acceptance of violence in our society as a legitimate means of enforcing compliance and solving conflicts at interpersonal, familial, national, and international levels. Violence and abuse in the family may be linked to such cultural factors as violence in the media, acceptance of corporal punishment, gender inequality, and the view of women and children as property.

**DIVERSITY IN OTHER COUNTRIES**

In 1979, Sweden passed a law that effectively abolished corporal punishment as a legitimate child-rearing practice. Fifteen other countries, including Italy, Germany, and Ukraine, have banned all corporal punishment in all settings, including the home.

**Violence in the Media** One need only watch boxing matches, football, or the evening news to see the violence in war, school shootings, and domestic murders. New films and TV movies and TV programs (e.g., "Dateline") regularly feature themes of violence.

**Corporal Punishment of Children** The use of physical force with the intention of causing a child pain, but not injury, for the purposes of correcting or controlling the child's behavior and/or making the child obedient is **corporal punishment**. In the United States, corporal punishment in the form of spanking, hitting, whipping, and paddling or otherwise inflicting pain on the child is legal in all states (as long as the corporal punishment does not meet the individual state's definition of child abuse). Seventy percent of U.S. respondents agreed that spanking is necessary to discipline children (Ma et al. 2012). Unlike Sweden, Italy, Germany, and 12 other countries which have banned corporal punishment in the home, violence against children has become a part of America's cultural heritage. Ma et al. (2012) found that the use of corporal punishment on adolescents by mothers and fathers was associated with a greater proportion of youth externalizing behaviors (e.g., breaking rules at home/school, hanging around kids who get into trouble, being mean to others, and threatening to hurt people).

**Corporal punishment** use of physical force with the intention of causing a child pain, but not injury, for the purposes of correcting or controlling the child's behavior and/or making the child obedient.

Other researchers have found that children who are victims of corporal punishment display more violence, have an increased incidence of depression as adults, report lower relationship quality as adult college students (Larsen et al. 2011), and are more likely to experience abuse in adult relationships (particularly women) (Maneta et al. 2012). Male exposure to harsh physical punishment (when combined with exposure to sexually explicit pornography) is associated with sexual aggression against women (Simons et al. 2012). Child development specialists recommend an end to corporal punishment to reduce the risk of physical abuse, harm to other children, and to break the cycle of abuse.

**Gender Inequality** Domestic violence and abuse may also stem from traditional gender roles. Traditionally, men have also been taught that they are superior to women and that they may use their aggression toward women, believing that women need to be "put in their place." The greater the inequality and the more the woman is dependent on the man, the more likely the abuse. Conversely, women with a higher income and education than their partner report more frequent abuse since these achievements may be a threat to the male's masculinity (Anderson 2010). Adams and Williams (2014) emphasized that violence among Mexican-American adolescents is sometimes normative in the context of jealousy. Hence the partner is seen as property and when who has access to this property is violated, violence may be legitimized.

Some occupations, such as police officers and military personnel, lend themselves to contexts of gender inequality. In military contexts, men notoriously devalue, denigrate, and sexually harass women. In spite of the rhetoric about gender equality in the military, women in the Army, Navy, and Air Force academies continue to be sexually harassed. In 2012, according to the Department of Defense, there were 26,000 sexual assaults in the military, yet only about 14% filed a complaint. Fear of reporting sexual harassment is a primary reason for not reporting such sexual harassment (Lawrence and Penaloza 2013). Male perpetrators may not separate their work roles from their domestic roles. One student in our classes noted that she was the ex-wife of a Navy Seal and that "he knew how to torment someone, and I was his victim."

## Community Factors

Community factors that contribute to violence and abuse in the family include social isolation, poverty, and inaccessible or unaffordable health-care, day-care, elder-care, and respite-care services and facilities.

**DIVERSITY IN OTHER COUNTRIES**

In cultures where a man's honor is threatened if his wife is unfaithful, the husband's violence toward her is tolerated by the legal system. Unmarried women in Jordan who have intercourse (even if through rape) are viewed as bringing shame on their parents and siblings and may be killed; this action is referred to as an **honor crime** or **honor killing**. The legal consequence is minimal to nonexistent. For example, a brother may kill his sister if she has intercourse with a man she is not married to and spend no more than a month or two in jail as the penalty.

**Social Isolation** Living in social isolation from extended family and community members increases the risk of being abused. Spouses whose parents live nearby are least vulnerable.

**Poverty** Abuse in adult relationships occurs among all socioeconomic groups. However, poverty and low socioeconomic status are a context of high stress which lends itself to expression of this stress by violence and abuse in interpersonal relationships.

**Honor crime/honor killing** unmarried woman who have intercourse (even if through rape) bring shame on their parents and siblings and may be killed.

**Inaccessible or Unaffordable Community Services** Failure to provide medical care to children and elderly family members sometimes results from the lack of accessible or affordable health-care services in the community. Without elder-care and respite-care facilities, families living in social isolation may not have any help with the stresses of caring for elderly family members and children.

## Individual Factors

Individual factors associated with domestic violence and abuse include psychopathology, personality characteristics, and alcohol or substance abuse. A number of personality characteristics have also been associated with people who are abusive in their intimate relationships. Some of these characteristics follow:

**1.** ***Dependency.*** Therapists who work with batterers have observed that they are overly dependent on their partners. Because the thought of being left by their partners induces panic and abandonment anxiety, batterers use physical aggression and threats of suicide to keep their partners with them.

**2.** ***Jealousy.*** Along with dependence, batterers exhibit jealousy, possessiveness, and suspicion. An abusive husband may express his possessiveness by isolating his wife from others; he may insist she stay at home, not work, and not socialize with others. His extreme, irrational jealousy may lead him to accuse his wife of infidelity and to beat her for her presumed affair.

**3.** ***Need to control.*** Abusive partners have an excessive need to exercise power over their partners and to control them. The abusers do not let their partners make independent decisions (including what to wear), and they want to know where their partners are, whom they are with, and what they are doing. Abusers like to be in charge of all aspects of family life, including finances and recreation.

**4.** ***Unhappiness and dissatisfaction.*** Abusive partners often report being unhappy and dissatisfied with their lives, both at home and at work. Many abusers have low self-esteem and high levels of anxiety, depression, and hostility. They may take out their frustration with life on their partner.

**5.** ***History of aggressiveness.*** Abusers often have a history of interpersonal aggressive behavior. They have poor impulse control and can become instantly enraged and lash out at the partner. Battered women report that episodes of violence are often triggered by minor events, such as a late meal or a shirt that has not been ironed.

**6.** ***Quick involvement.*** Because of feelings of insecurity, the potential batterer will rush his partner quickly into a committed relationship. If the woman tries to break off the relationship, the man will often try to make her feel guilty for not giving him and the relationship a chance.

**7.** ***Blaming others for problems.*** Abusers take little responsibility for their problems and blame everyone else. For example, when they make mistakes, they will blame their partner for upsetting them and keeping them from concentrating on their work. A man may become upset because of what his partner said, hit her because she smirked at him, and kick her in the stomach because she poured him too much (or not enough) alcohol.

**8.** ***Jekyll-and-Hyde personality.*** Abusers have sudden mood changes so that a partner is continually confused. One minute an abuser is nice, and the next minute angry and accusatory. Explosiveness and moodiness are the norm.

**9.** ***Isolation.*** An abusive person will try to cut off a partner from all family, friends, and activities. Ties with anyone are prohibited. Isolation may reach the point at which an abuser tries to stop the victim from going to school, church, or work.

**10.** ***Alcohol and other drug use.*** Whether alcohol reduces one's inhibitions to display violence, allows one to avoid responsibility for being violent, or increases one's aggression, it is associated with violence and abuse (even if the partner is pregnant; Eaton et al. 2012). Younger individuals with more severe drug addictions (e.g., a history of overdose) are more likely to be violent (Fernandez-Montalvo et al. 2012).

**11.** ***Criminal/Psychiatric background.*** Eke et al. (2011) examined the characteristics of 146 men who murdered or attempted to murder their intimate partner. Of these, 42% had prior criminal charges, 15% had a psychiatric history, and 18% had both. Shorey et al. (2012) identified the mental health problems

Chelsea Curry

Alcohol is a common precursor to abuse in a relationship.

in men arrested for domestic violence and found high rates of PTSD, depression, generalized anxiety disorder (GAD), panic disorder, and social phobia.

**12. *Impulsive.*** Miller et al. (2012) identified one of the most prominent personality characteristics associated with aggression/abuse. "Impulsive behavior in the context of negative affect" was consistently related to aggression across multiple indices.

## Relationship Factors

Halpern-Meekin et al. (2013) studied a sample of 792 relationships and found that "relationship churning" was associated with both physical and emotional violence. The researchers compared relationships which had ended, those which were still together, and those which had an on-and-off pattern (churners). Individuals in the latter type relationship were twice as likely to report physical violence and half again as likely to report emotional abuse compared to the other two patterns.

## Family Factors

Family factors associated with domestic violence and abuse include being abused as a child, having parents who abused each other, and not having a father in the home.

**Child Abuse in Family of Origin** Individuals who were abused as children are more likely to be abusive toward their intimate partners as adults.

**Family Conflict** Children learn abuse from their family context. Children whose fathers were not affectionate were more likely to be abusive to their own children.

**Parents Who Abuse Each Other** Parents who are aggressive toward each other create a norm of aggression in the family and are more likely to be aggressive toward their children (Graham et al. 2012). While most parents are not aggressive toward each other in front of their children (Pendry et al. 2011), Dominguez et al. (2013) reconfirmed the link between witnessing parental abuse (either father or mother abusive toward the other) and being in an abusive relationship as an adult. However, a majority of children who witness abuse do not continue the pattern—a family history of violence is only one factor out of many associated with a greater probability of adult violence.

## Sexual Abuse in Undergraduate Relationships

*I just want to sleep. A coma would be nice. Or amnesia. Anything, just to get rid of this, these thoughts, whispers in my mind. Did he rape my head, too?*

LAURIE HALSE ANDERSON, *SPEAK*

Katz and Myhr (2008) noted that some women experience sexual abuse in addition to a larger pattern of physical abuse and that the combination is associated with less life satisfaction in general, less sexual satisfaction, more conflict in one's relationship, and more psychological abuse from the partner. In effect, women in these co-victimization relationships are miserable.

In a sample of 3,468 undergraduate females at a large southeastern university, 33% reported being pressured by a partner they were dating to have sex (Hall and Knox 2013). Men are also pressured to have sex (19% of 1,101 undergraduate males) (Hall and Knox 2013). In other research, in a sample of 1,400 men ages 18–24, 6% reported that they were coerced to have vaginal sex with a female; 1% by a male to have oral or anal sex (Smith et al. 2010).

**DIVERSITY IN OTHER COUNTRIES**

In 2012, a 23-year-old college student was gang raped on a bus in New Delhi, India. She and her fiancé were beaten and thrown off the bus and left for dead. "A woman [here] is a possession, like a piece of land" said Amon Deol, general secretary of a Punjab women's rights group (Mahr 2013, p. 12).

### Acquaintance and Date Rape

The word *rape* often evokes images of a stranger jumping out of the bushes to attack an unsuspecting victim. However, most rapes are perpetrated not by strangers but by people who have a relationship with the victim. About 85% of rapes are perpetrated by someone the woman knows (12% of victims have been raped by both an acquaintance and a stranger—double victims) (Hall & Knox 2012). The type of rape by someone the victim knows is referred to as **acquaintance rape**, which is defined as nonconsensual sex between adults (of the same or other sex) who know each other. The behaviors of sexual coercion occur on a continuum from verbal pressure and threats to use of physical force to obtain sexual acts, such as oral sex, sexual intercourse, and anal sex.

**Acquaintance rape** nonconsensual sex between adults who know each other.

The perpetrator of a rape is likely to believe in various rape myths. **Female rape myths** are beliefs about the female that deny that she was raped or cast blame on the woman for her own rape. These beliefs are false, widely held, and justify male aggression. Examples include: "women deserve to be raped" (particularly when they drink too much and are provocatively dressed); "women fantasize about and secretly want to be raped" and "women who really don't want to be raped resist more—they could stop a guy if they really wanted to." McMahon (2010) found that females rape myths were more likely to be accepted by males, those pledging a fraternity/sorority, athletes, those without previous rape education, and those who did not know someone who had been sexually assaulted. Mouilso and Calhoun (2013) noted that being mentally disturbed is associated with belief in these myths.

**Female rape myths** beliefs about the female that deny that she was raped or cast blame on the woman for her own rape.

There are also **male rape myths** such as "men can't be raped" and "men who are raped are gay." Not only is there a double standard of perceptions that only women can be raped but a double standard is operative in the stereotypical perception of the gender of the person engaging in sexual coercion (e.g., men who rape are aggressive; women who rape are promiscuous) (Oswald and Russell 2006).

**Male rape myths** "men can't be raped" and "men who are raped are gay".

One type of acquaintance rape is **date rape**, which refers to nonconsensual sex between people who are dating or on a date. The woman (while both women and men may be raped on a date, we will refer to the female since male-female rape is more prevalent) has no idea that rape is on the agenda. She may have

**Date rape** nonconsensual sex between people who are dating or on a date.

gone out to dinner, had a pleasant evening, and gone back to her apartment with her date only to be raped.

Women are also vulnerable to repeated sexual force. Daigle et al. (2008) noted that 14% to 25% of college women experience repeat sexual victimization during the same academic year—they are victims of rape or other unwanted sexual force more than once, often in the same month. The primary reason is that they do not change their context; they may stay in the relationship with the same person and continue to use alcohol or drugs.

Rape also occurs in same-sex relationships.

## Sexual Abuse in Same-Sex Relationships

Gay, lesbian, bisexual individuals are not immune to experiencing sexual abuse in their relationships. Rothman et al. (2011) reviewed studies which involved a total sample of 139,635 gay, lesbian, bisexual women and men and found that the highest estimates reported for lifetime sexual assault were for lesbian and bisexual women (85%). In regard to childhood sexual assault of lesbian and bisexual women the percentage was 76%. For childhood sexual assault of gay and bisexual men, 59%.

## Alcohol and Rape

Alcohol is the most common rape drug. A person under the influence cannot give consent. Hence, a person who has sex with someone who is impaired is engaging in rape. In a study of 340 college rape victims, 41% reported being impaired and 21% reported being incapacitated—hence over 60% were in an altered state (Littleton et al. 2009). About two-thirds of all three groups—impaired, incapacitated, and not impaired—reported moderate physical force by the rapist. The nonimpaired group reported the highest percent of verbal threats (27%) and severe physical force (9%). The greatest resistance (56% nonverbal; 53% verbal; 45% physical) came from the nonimpaired. Farris et al. (2010) also found that alcohol dose was related to men interpreting a woman's friendliness as sexual interest.

## Rophypnol—The Date Rape Drug

Of those occasions in which a woman is incapacitated due to alcohol or drugs, 85% are voluntary (the woman is willingly drinking or doing drugs). In 15% of the cases, however, drugs are used against her will (Lawyer et al. 2010).

**Rophypnol**—also known as the date rape drug, rope, roofies, Mexican Valium, or the "forget (me) pill"—causes profound, prolonged sedation and short-term memory loss. Similar to Valium but ten times as strong, Rophypnol is a prescription drug used as a potent sedative in Europe. It is sold in the United States for about $5, is dropped in a drink (where it is tasteless and odorless), and causes victims to lose their memory for 8 to 10 hours. During this time, victims may be raped yet be unaware until they notice signs of it (e.g., blood in panties) the next morning.

**Rophypnol** date rape drug that causes profound, prolonged sedation and short-term memory loss.

The Drug-Induced Rape Prevention and Punishment Act of 1996 makes it a crime to give a controlled substance to anyone without his or her knowledge and with the intent of committing a violent crime (such as rape). Violation of this law is punishable by up to 20 years in prison and a fine of $250,000.

The effects of rape include loss of self-esteem, loss of trust, and the inability to be sexual. Zinzow et al. (2010) studied a national sample of women and found that forcible rape was more likely to be associated with PTSD and with major depression than was drug- or alcohol-facilitated or incapacitated rape.

Acknowledging that one is a rape victim (rather than keeping it a secret) is associated with fewer negative psychological symptoms (e.g., depression) and increased coping (e.g., moving beyond the event to a new relationship)

*Rape is a culturally fostered means of suppressing women. Legally we say we deplore it, but mythically we romanticize and perpetuate it, and privately we excuse and overlook it.*

VICTORIA BILLINGS, AUTHOR

## APPLYING SOCIAL RESEARCH | Double Victims: Sexual Coercion by a Dating Partner and a Stranger

Whether one is a victim of rape or sexual coercion by an acquaintance or a stranger, unwanted sexual advances are traumatizing. This study attempted to identify the characteristics of victims of sexual coercion where the perpetrator was a dating partner and a stranger, with emphasis on victims of both types of perpetrators. Who are these victims? What is a profile of the respective groups and to what degree are the profiles alike or different? These profiles may help to identify individuals who may be vulnerable to sexual coercion or rape by either a dating partner or by a stranger.

**Sample and Methods**

The sample consisted of 2,747 undergraduates from two relatively large universities. Age of respondents ranged from 18 to 30. Females made up 72.4% of the combined sample; males, 27.6%. In regard to racial identity, 85% were white, 5.3% black, and 4.5% Hispanic. The participants completed a 100-item online survey, "Sexual Attitudes and Behaviors of College Students Questionnaire," which included questions on gender, age, race, religiosity, cohabitation, substance use, child abuse, emotional abuse, and physical abuse.

The dependent variable was created based on the responses to whether the respondent had been pressured to have sex by a stranger and whether the respondent had been pressured to have sex by a dating partner. Four levels of the *pressure/coercion* variable were created to represent 1) those who answered "no" to both questions (n = 1,516; 58.5%) referred to as the "no sexual pressure group," 2) those who answered "yes" to the *stranger* item and "no" to the *dating* item (*n* = 166; 6.4%) referred to as "yes-stranger/no dating pressure," 3) those who answered "no" to the *stranger* item and "yes" to the *dating* item (*n* = 620; 23.9%) referred to as "no-stranger/yes dating pressure," and 4) those who answered "yes" to both items (n = 291; 11.2%) referred to as "double victims."

(Clements and Ogle 2009). Denial, not discussing it with others, and avoiding the reality that one has been raped prolongs one's recovery.

# Abuse in Marriage Relationships

The chance of abuse in a relationship increases with marriage. Indeed, the longer individuals know each other and the more intimate their relationship, the greater the abuse. Below is an example of an abusive marriage experienced by the mother of one of our students.

*My 13 Years in an Abusive Marriage**

*I was verbally, physically, and sexually abused for 13 years in my marriage from ages 22-34. I never had any hint during the two years we were dating that the man I loved (he was my soul mate) would end up making my life a living hell. He was a little jealous but I thought this was cute and that he cared about me. In fact he made me feel special . . . he made me feel LOVED . . . We could talk for hours and hours.*

> *I would do anything for him. . . . I worshipped the ground he walked on . . . the sun rose and set on him. . . . we had the same interests . . . same hobbies. . . . . same music. . . . we were both clingy. . . . where you saw one. . . . you saw the other..... when we went somewhere we stayed together . . . the entire time . . . we actually lived together for about 6 months before we were married.......life was perfect!*

*But the very next day after we were married, when I came home from work he had put all my dresses and skirts on the bed with the hem taken out of them. . . . I wondered what in the world I had gotten myself into.*

#### Findings

The percent of undergraduates experiencing sexual coercion in the respective categories: 23.9% date, 6.4% stranger, 11.2% both, and 58.5% neither. A multinomial logistic regression was used to predict membership in the various groups of the dependent variable. Analysis revealed some distinct differences between each of the victim groups and the non-victim group. Those who reported pressure to have sex from both a stranger and a date—**double victims**—were more likely to be a white female who had been abused as a child, been emotionally abused by a partner, cohabited, used alcohol/drugs, used the Internet to find a partner, dated interracially, hooked up, been in a friends with benefits relationship, and lied to a partner. What emerged may be an early victimized female (child abuse, possible prior emotional partner abuse) who lowers inhibitions via alcohol/drug use and engages in sex in non-committed relationships (e.g., hooking up, friends with benefits). Lying to one's partner suggests a less than ideal relationship.

#### Implication

Universities that sponsor rape prevention programs for undergraduates might consider a unit which focuses on these double victims as a unique group. Such an emphasis may increase awareness on the part of those who may be particularly vulnerable to sexual coercion, which may lead to choices by the individual to avoid such exposure/victimization.

#### Source

Adapted and abridged from Hall, S. and D. Knox. 2011. Double Victims: Sexual Coercion by a Dating Partner and a Stranger. Poster, National Council on Family Relations Annual Meeting, Orlando, November.

**Double victims** those who reported pressure to have sex from both a stranger and a date.

*The abuse occurred mostly when he had been drinking*

- *Accusations . . . accusing me of sleeping with anyone breathing*
- *Fear . . . he jumped at me . . . got into my face. . . . he threatened me with bodily harm. . . . anything to intimidate me*
- *Shame . . . . he belittled me, called me names, cussed at me*
- *Control . . . he got mad at me for visiting my family*
- *Physical abuse . . . he pushed, pinched, squeezed, spit and pulled a gun on me*
- *Rape . . . he forced me to have sex with him (when he was drunk he would last forever*

  - *He would stay out all night; come home and wake me up. . . . and accuse me of going out and being with someone else. . . . (looked for stamps on my hand from a club).*
  - *I can't tell you how many black eyes he gave me . . . I kept sunglasses on for about 5-6 years. . . . I will never forget one night when he was on top of me. . . . he had me pinned down just hitting me in the face with his fist just as hard as he could.*
  - *He once backhanded me and knocked four of my front teeth loose (they later died and had to be pulled). I now have a plate in my mouth.*
  - *There was another time when he picked me up and slung me across the room with both of my legs hitting the coffee table and seriously bruising the front shins of my legs. Today, I still have problems . . . broken veins and poor circulation in my lower legs.*
  - *I have left him so many times. . . . I would not even dare take a guess at how many time we separated . . . sometimes he stayed gone for 3-4 months. . . . he would always talk me into coming back. . . . buying me back. . . . he would start back to church with me and we would go to church for months . . . then he would start back drinking again.*

- *I once called the police and he was arrested. . . . but I dropped the charges.*
- *I can't remember how many cars he totaled . . . After one of his accidents he was out of work for over six years. I went to see a lawyer about leaving him and was told I would have to pay him alimony.*
    - *The reason I stayed is because I loved him and when he wasn't drinking he was perfect and treated me right. But his abuse was changing my personality and one time I lost it and pulled a gun on him. . . . so I knew it was time to go . . . No man is worth going to jail for . . . even though I cried for months every day and still loved him. . . . . I knew that I had enough. I have since forgiven him and moved on. My message to my daughter and other young people is to make a deliberate choice to get out immediately after the abuse starts.*

*Used with the permission of Teresa Carol Wimberly

## General Abuse in Marriage

Abuse in marriage may differ from unmarried abuse in that the husband may feel ownership of the wife and feel the need to control her. But the behaviors of abuse are the same—belittling the spouse, controlling the spouse, physically hurting the spouse, etc. Deciding to end a marriage in which there is abuse is considered in the Personal Choices section that follows.

## PERSONAL CHOICES

### *Would You End an Abusive Marital Relationship?*

Students in our classes were asked whether they would end a marriage if the spouse hit or kicked them. Many felt that marriage was too strong a commitment to end if the abuse could be stopped. Some of their comments follow.

> *I would not divorce my spouse if she hit or kicked me. I'm sure that there's always room for improvement in my behavior, although I don't think it's necessary to assault me. I recognize that under certain circumstances, it's the quickest way to draw my attention to the problems at hand. I would try to work through our difficulties with my spouse.*
>
> ***
>
> *The physical contact would lead to a separation. During that time, I would expect him to feel sorry for what he had done and to seek counseling. My anger would be so great, it's quite hard to know exactly what I would do.*
>
> ***
>
> *I wouldn't leave him right off. I would try to get him to a therapist. If we could not work through the problem, I would leave him. If there was no way we could live together, I guess divorce would be the answer.*
>
> ***
>
> *I would not divorce my husband, because I don't believe in breaking the sacred vows of marriage. But I would separate from him and let him suffer!*
>
> ***
>
> *I would tell him I was leaving but that he could keep me if he would agree for us to see a counselor to ensure that the abuse never happened again.*

Some said, "Seek a divorce." Those opting for divorce (a minority) felt they couldn't live with someone who had abused or might abuse them again.

*I abhor violence of any kind, and since a marriage should be based on love, kicking is certainly unacceptable. I would lose all respect for my husband and I could never trust him again. It would be over.*

Marriage therapists emphasize that a pattern of abuse develops and continues if such behavior is not addressed and stopped. The first time abuse occurs, the couple should seek therapy. The second time it occurs, they should separate. The third time, they might consider a divorce. Later in the chapter we discuss disengaging from an abusive relationship.

### Men Who Abuse

Henning and Connor-Smith (2011) studied a large sample of men who were recently convicted of violence toward a female intimate partner (N = 1,130). More than half of the men (59%) reported that they were continuing or planning to continue the relationship with the partner they had abused. Reasons included being older (e.g., too late to start over), being married to the victim, having children together, attributing less blame to the victim for the recent offense, and having a childhood history of family violence.

### Rape in Marriage

**Marital rape**, now recognized in all states as a crime, is forcible rape (includes vaginal, oral, anal) by one's spouse. Some states (Washington) recognize a marital defense exception for third-degree rape in which force is not used even though there is no consent. Over 30 countries (e.g., China, Afghanistan, Pakistan) have no laws against marital rape.

**Marital rape** forcible rape (includes vaginal, oral, anal) by one's spouse.

## Effects of Abuse

Abuse has devastating effects on the physical and psychological well-being of victims. Abuse between parents also affects the children.

### Effects of Partner Abuse on Victims

Effects of abuse are in reference to perception. Rhatigan and Nathanson (2010) found in a study of 293 college women that women took into account their own behavior in evaluating their boyfriends' abusive behavior. If they felt they had set him up by their own behavior, they were more understanding of his aggressiveness. Additionally, those with low self-esteem were more likely to take some responsibility for their boyfriend's abusive behavior. While some partners will rationalize or overlook their partner's abusive behavior, others are hurt by it and the effect is negative.

In general, abuse has devastating consequences. Becker et al. (2010) confirmed that being a victim of intimate partner violence is associated with symptoms of PTSD—loss of interest in activities/life in general, feeling detached from others, inability to sleep, irritability, etc. Sarkar (2008) confirmed that IPV (intimate partner violence) affected the woman's physical (increased the risk for unintended pregnancy/multiple abortions) and mental health (higher levels of anxiety and drug abuse).

### Effects of Partner Abuse on Children

The most dramatic effects of abuse occur on pregnant women, which include increased risk of miscarriage, birth defects, low birth weight, preterm delivery,

WHAT IF?

## What If My Close Friend Is Being Abused and Keeps Going Back?

An abused wife will leave and return to her husband an average of seven times before she finally makes the break (most end up leaving eventually). Hence, an abused individual must reach a point where staying is more aversive than leaving (or leaving more attractive than staying). Even after staying away, the abused may still return to the abuser. Rihanna, the singer who was severely beaten by her partner recording artist/actor Chris Brown, renewed the relationship with him.

Until your friend makes a decision to leave, you can do little but make it clear that you are available and have a place for your friend to stay when she is ready to make the break. Without a place to go, some will remain in an abusive relationship until they are, literally, beaten to death. Your promise of a safe place to stay is a stimulus to action . . . you just won't know when your friend will take you up on your offer. In addition, you might consider suggesting reasons to leave "to save your children" and to empower them with your confidence that they are "strong enough to leave . . . to actually go through with it." Such a strong motivation (children's safety) and a confident peer have been associated with leaving an abusive relationship (Baly 2010).

*Even if it's a mistake [getting back with Chris Brown who beat her three years ago] it's my mistake.*

RIHANNA

and neonatal death. Negative effects may also accrue to children who witness domestic abuse. Russella et al. (2010) found that children who observed parental domestic violence were more likely to be depressed as adults. Hence, a child need not be a direct target to abuse, but merely a witness to incur negative effects. However, Kulkarni et al. (2011) compared the effects of witnessing domestic violence and actually experiencing it on PTSD in later life and found that witnessing alone was not associated with PTSD.

# The Cycle of Abuse

The following reflects the cycle of abuse.

*I Got Flowers Today*

*I got flowers today. It wasn't my birthday or any other special day. We had our first argument last night, and he said a lot of cruel things that really hurt me. I know he is sorry and didn't mean the things he said, because he sent me flowers today.*

*I got flowers today. It wasn't our anniversary or any other special day. Last night, he threw me into a wall and started to choke me. It seemed like a nightmare. I couldn't believe it was real. I woke up this morning sore and bruised all over. I know he must be sorry, because he sent me flowers today.*

*Last night, he beat me up again. And it was much worse than all the other times. If I leave him, what will I do? How will I take care of my kids? What about money? I'm afraid of him and scared to leave. But I know he must be sorry, because he sent me flowers today.*

*I got flowers today. Today was a very special day. It was the day of my funeral. Last night, he finally killed me. He beat me to death.*

*If only I had gathered enough courage and strength to leave him, I would not have gotten flowers today.*

Paulette Kelly, Allen Dowell

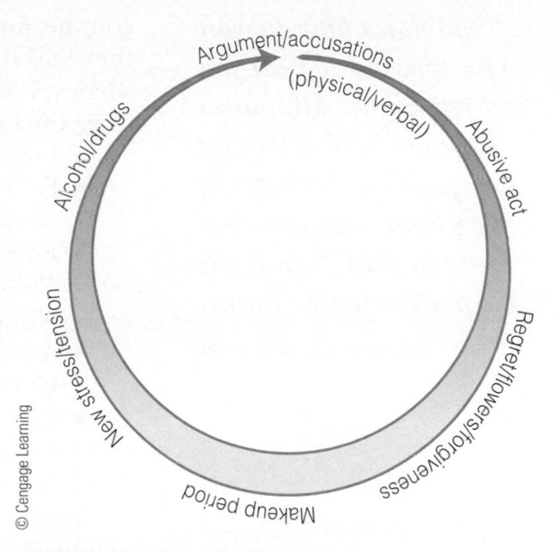


**Figure 10.1** The Cycle of Abuse.

The cycle of abuse begins when a person is abused and the perpetrator feels regret, asks for forgiveness, and engages in positive behavior (gives flowers). The victim, who perceives few options and feels anxious terminating the relationship with the abusive partner, feels hope for the relationship at the contriteness of the abuser and does not call the police or file charges.

After the forgiveness, couples usually experience a period of making up or honeymooning, during which the victim feels good again about the partner and is hopeful for a nonabusive future. However, stress, anxiety, and tension mount again in the relationship, which the abuser relieves by violence toward the victim. Such violence is followed by the familiar sense of regret and pleadings for forgiveness, accompanied by a new round of positive behavior (flowers and candy).

As the cycle of abuse reveals, some victims do not prosecute their partners who abuse them. In response to this problem, Los Angeles has adopted a zero tolerance policy toward domestic violence. Under the law, an arrested person is required to stand trial and his victim required to testify against the perpetrator. The sentence in Los Angeles County for partner abuse is up to six months in jail and a fine of $1,000.

*Tho prun'st a rotten tree.*

**SHAKESPEARE, *AS YOU LIKE IT***

Figure 10.1 illustrates this cycle, which occurs in clockwise fashion. In the rest of this section, we discuss reasons why people stay in an abusive relationships and how to get out of them.

*There is an ache in my heart for the imagined beauty of a life I haven't had, from which I had been locked out, and it never goes away.*

**ROBERT GOOLRICK, *THE END OF THE WORLD AS WE KNOW IT: SCENES FROM A LIFE***

## Why Victims Stay in Abusive Relationships

One of the most frequently asked questions to people who remain in abusive relationships is, "Why do you stay?" Alexander et al. (2009) noted that the primary reason a woman returns again and again to an abusive relationship is the emotional attachment to her partner—she is in love. While someone who criticizes ("you're ugly/stupid/pitiful"), is dishonest (sexually unfaithful), or physically harms a partner will create a context of interpersonal misery, the victim might still love that person.

Another explanation for why some people remain with abusive partners is that the abuse is only one part of the relationship. When such partners are not being abusive, they are kind, caring, and loving. It is these positive behaviors which keep the victim hooked The psychological term is a **periodic reinforcement** which means that every now and then the abusive person "floods" the partner with strong love/positives which keep the partner in the relationship. In effect, the person who is abused stays in the relationship since it includes positives such as flowers, gifts, or declarations of love to entice a partner to stay in the relationship.

**A periodic reinforcement** every now and then the abusive person floods the partner with strong love/positives which keep the partner in the relationship.

Whether the abused has children may have a positive or negative effect in terms of staying in an abusive relationship (Rhodes et al. 2010). Some mothers will leave their abusive partners so as not to subject their children to the abuse. But others will stay in an abusive marriage since the mothers want to keep the family together (and keep a father close for their children).

In earlier research, Few and Rosen (2005) interviewed 25 women who had been involved in abusive dating relationships from three months to nine years (average = 2.4 years) to find out why they stayed. The researchers conceptualized the women as **entrapped**—stuck in an abusive relationship and unable to extricate oneself from the abusive partner. Indeed, these women escalated their

**Entrapped** stuck in an abusive relationship and unable to extricate oneself from the abusive partner.

*. . . you don't have to wait for someone to treat you bad repeatedly. All it takes is once, and if they get away with it that once, if they know they can treat you like that, then it sets the pattern for the future.*

JANE GREEN, *BOOKENDS*

commitment to stay in hopes that doing so would eventually pay off. In effect, they had invested time with a partner they were in love with and wanted to do whatever was necessary to maintain the relationship. Some of the factors which drive the entrapment of these women include:

- Love ("I love him.")
- Fear of loneliness ("I'd rather be with someone who abuses me than alone.")
- Emotional dependency ("I need him.")
- Commitment to the relationship ("I took a vow 'for better or for worse.'")
- Hope ("He will stop being abusive—he's just not himself lately.")
- A view of violence as legitimate ("All relationships include some abuse.")
- Guilt ("I can't leave a sick man.")
- Fear for one's life ("He'll kill me if I leave him.")
- Economic dependence ("I have no money and no place to go.")
- Isolation ("I don't know anyone who can help me.")

Halligan et al. (2013) noted the impact of technology on the maintenance of abusive relationships. She and her colleagues found that undergraduates in abusive relationships could not stop themselves from checking their text messages—even if the message was from an abusive partner. Some of the messages from the abusive partners included an apology, a promise to never be abusive again, and a request to resume the relationship.

Battered women also stay in abusive relationships because they rarely have escape routes related to educational or employment opportunities, they do not want to disrupt the lives of their children, and they may be so emotionally devastated by the abuse (anxious, depressed, or suffering from low self-esteem) that they feel incapable of planning and executing their departure.

Disengaging from an abusive relationship is a process that takes time. One should not be discouraged when they see a friend return to an abusive context but recognize that progress in disengagement is occurring (she did it once) and that positive movement is predictive of eventually getting out.

### Fighting Back: What Is the Best Strategy?

How might a partner respond when being physically abused? First, the goal is to avoid serious physical injury. Such injury is increased if the partner is drunk, on drugs, or has a weapon, so forceful physical resistance (pushing, striking, struggling) to the mind-altered partner should be avoided. Verbal resistance (pleading, crying, or trying to assuage the offender) or nonforceful physical resistance (fleeing or hiding) might be more helpful in avoiding injury. Fortunately, the frequency of severe injury is low (7%) with most (69%) abusive incidents resulting in no injury (Powers and Simpson 2012). Of course, any abuse in an intimate relationship may be too much and should be avoided or the relationship ended.

*I've heard that people stay in bad situations because a relationship like that gets turned up by degrees. It is said that a frog will jump out of a pot of boiling water. Place him in a pot and turn it up a little at a time, and he will stay until he is boiled to death. Us frogs understand this.*

DEB CALETTI, *STAY*

### How to Leave an Abusive Relationship

Couple therapy to reduce abuse in an abusive relationship is hampered by the fact of previous abuse. Being a victim of previous psychological aggression (female or male) is related to fear of speaking in front of the partner and fear of being in therapy with the partner (O'Leary et al. 2013). Hence, gains may be minimal.

Leaving an abusive partner begins with the person making a plan and acting on the plan—packing clothes/belongings, moving in with a sister, mother, or friend, or going to a homeless shelter. If the new context is better than being in the abusive context, the person will stay away. Otherwise, the person may go back and start the cycle all over. As noted previously, this leaving and returning typically happens seven times.

Sometimes the woman does not leave while the abuser is at work but calls the police and has the man arrested for violence and abuse. While the abuser is in jail, she may move out and leave town. In either case, disengagement from the abusive relationship takes a great deal of courage. Calling the National Domestic Violence Hotline (800-799-7233 [SAFE]), available 24 hours, is a point of beginning.

Kress et al. (2008) noted that involvement with an intimate partner who is violent may be life threatening. Particularly if the individual decides to leave the violent partner, the abuser may react with more violence and murder the person who has left. Indeed, a third of murders that occur in domestic violence cases occur shortly after a breakup. Specific actions that could be precursors to murder of an intimate partner are stalking, strangulation, forced sex, physical abuse, gun ownership, and drug or alcohol use on the part of the abuser. Moving quickly to a safe context (e.g., parents) is important.

Taking out a protective order whereby the accused is prohibited from being within close proximity of the victim partner is one option some abused partners take. While Kothari et al. (2012) confirmed that taking out such an order against a partner who has engaged in intimate partner violence was associated with a reduced frequency of repeated violence against the victim, other research is less clear. Ward et al. (2014) studied the effect of involving the police when interpersonal abuse occurs and found a different answer. When there was a history of abuse, there was an increased likelihood of future violent offending subsequent to police contact. When there was no history of abuse, there seemed to be no change, no increase in violence. So what is a person in an abusive relationship to do? The best answer seems to be to take the position that the abusive behavior will continue and to removing oneself from the abusive context and stay in a safe context.

## Healing after Leaving an Abusive Relationship

Healing from being in an abusive relationship takes time. Allen and Wozniak (2011) discouraged repetitive disclosure of one's abuse and history. Instead they focused on "holistic, integrative, and alternative healing approaches such as prayer, meditation, yoga, creative visualization, and art therapy." They provided positive quantitative and qualitative results for this approach. Healing is also facilitated by involvement in a new, positive, nurturing relationship where the person is loved, respected, and cherished. Such a context creates a new set of relationship experiences that removes one's anxiety, depression, and self-loathing.

## Treatment of Abusers

LeCouteur and Oxlad (2011) interviewed men who revealed why they abused their partners and found that they constructed a justification that their partner had "breached the normative moral order." Successful therapy for abusers requires that they acknowledge that their violent/abusive behavior was wrong, to feel remorse, and to commit to finding other ways to cope with their frustration. Shamai and Buchbinder (2010) surveyed men who had participated in a treatment program for partner-violent men. Most experienced therapy as positive and meaningful and underwent personal changes, especially the acquisition of self-control. Taking a time out, counting to ten, and reassessing the situation before reacting were particularly helpful techniques. While gains were made, the men still tended to create dominant relationships with women, which left open the potential for them to be abusive again.

The involvement of the abused partner in the rehabilitation of the abuser may also be important. Antunes-Alves and De Stefano (2014) noted that joint couple therapy is replacing group treatments as a primary strategy to reduce/eliminate interpersonal violence. Such therapy must address the equality of the female in the relationship. Some men stop abusing their partners only when their partners no longer put up with their abuse.

This concludes our discussion of abuse in adult relationships. In the following pages, we discuss other forms of abuse.

# Child, Parent, Sibling, and Elder Abuse

In addition to intimate partner abuse, child, parent, sibling, and elder abuse occurs.

## Child Abuse

Child abuse may take many forms, including physical, neglect, sexual, and murder. The percentages of various types of child abuse in substantiated victim cases are illustrated in Figure 10.2. Notice that neglect is the largest category of abuse.

**Child abuse** any behavior or lack of behavior by parents or caregivers that results in deliberate harm to a child's physical or psychological well-being.

**Child abuse** is any behavior or lack of behavior by parents or caregivers that results in deliberate harm to a child's physical or psychological well-being. Child abuse includes physical abuse, such as beating and burning; psychological abuse, such as insulting or demeaning; and neglect, such as not providing a child needed medical care. Abusive acts may be acts of commission where the child is deliberately harmed by the parents or acts of omission where the parents fail to look out for the child (e.g., medical treatment for a severe injury).

Various forms of child psychological maltreatment and neglect were identified by Nash et al. (2012) and include lack of emotional involvement by parents (e.g., the child does not feel loved), terrorizing/spurning (e.g., the child is frightened or made to cry for no reason), isolating (e.g., confining the child to a closet or other small place), abandonment (e.g., throwing child out of house after an argument), absence of monitoring (e.g., parents don't require child to attend school), and medical neglect (e.g., parents do not feed the child/provide hygiene, or take child to doctor).

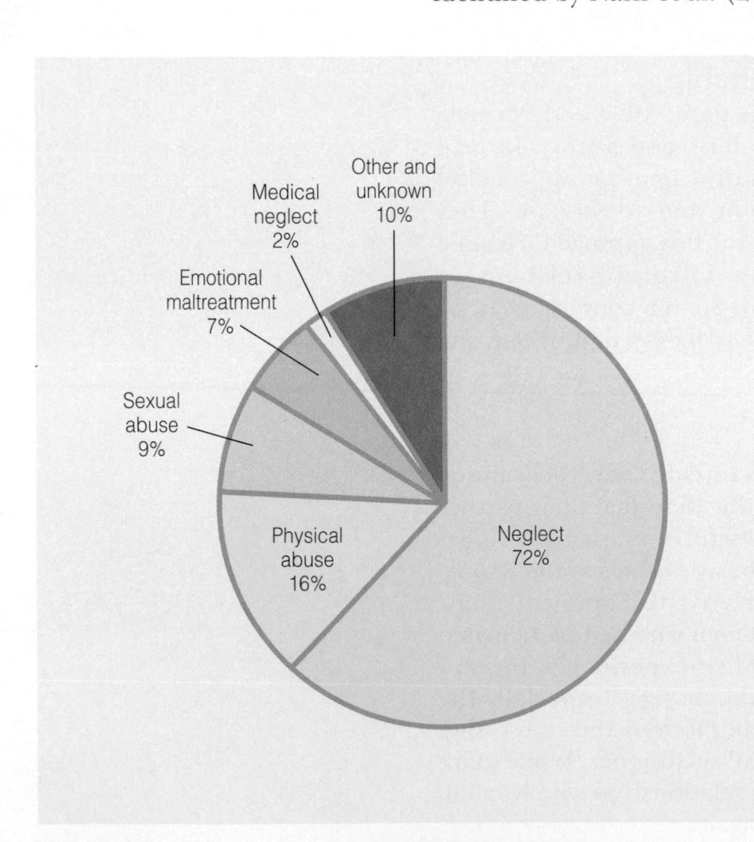


**Figure 10.2** Child Abuse and Neglect Cases: 2009.
Source: *Statistical Abstract of the United States*, 2012–2013, 131th ed. Washington, DC. U.S. Bureau of the Census, Table 342.

Factors contributing to child abuse:

1. ***Parental psychopathology.*** Symptoms of parental psychopathology that may predispose a parent to abuse or neglect children include low frustration tolerance, inappropriate expression of anger, and alcohol or substance abuse.

2. ***Unrealistic expectations.*** Abusive parents often have unrealistic expectations of their child's behavior. For example, a parent might view the crying of a baby as a deliberate attempt on the part of the child to irritate the parent so the abusive parent strikes back.

3. ***History of abuse.*** Individuals who were abused as children are more likely to report abusing their own children (Kerley et al. 2010). The researchers observed that the absence of positive parenting carries over into the next generation. Although parents who were physically or verbally abused or neglected as children are

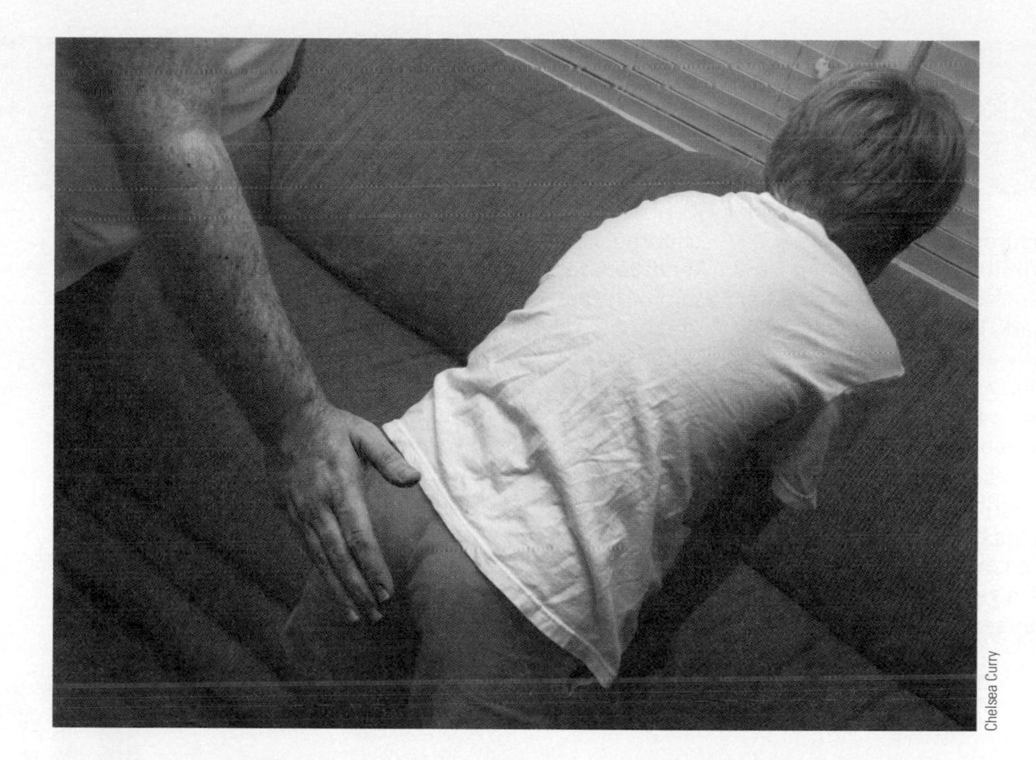

Chelsea Curry

Physical abuse of a child is sometimes excused by parents as spanking and discipline.

somewhat more likely to repeat that behavior than parents who were not abused, the majority of parents who were abused do *not* abuse their own children. Indeed, many parents who were abused as children are dedicated to ensuring nonviolent parenting of their own children precisely because of their own experience of abuse.

**4.** ***Displacement of aggression.*** One cartoon shows several panels consisting of a boss yelling at his employee, the employee yelling at his wife, the wife yelling at their child, and the child kicking the dog, who chases the cat up a tree. Indeed, frustration may spill over from the adults to the children, the latter being less able to defend themselves.

**5.** ***Social isolation.*** The saying "it takes a village to raise a child" is relevant to child abuse. Unlike inhabitants of most societies of the world, many Americans rear their children in closed and isolated nuclear units. In extended kinship societies, other relatives are always present to help with the task of child rearing. Isolation means parents have no relief from the parenting role as well as no supervision by others who might interrupt observed child abuse.

**6.** ***Disability of a child.*** Jaudesa and Mackey-Bilaverb (2008) found that children who were developmentally delayed were twice as likely to be victims of abuse and neglect.

**7.** ***Young/poor/no partner.*** Camperio Ciani and Fontanesi (2012) studied 110 mothers who killed 123 of their babies (**neonaticide**) within the first day of life. All fit the same profile—young, poor, and no partner.

**Neonaticide** mothers who kill babies within the first day of life.

**8.** ***Other factors.*** In addition to the factors just mentioned, the following factors are also associated with child abuse and neglect: premarital/unplanned pregnancy, mother-infant attachment is lacking, spousal abuse, and adopted/foster children.

The effects of child abuse are devastating. There are more than 1,500 deaths resulting from child abuse in the United States annually (Paluscia et al. 2010), and there are also profound effects for abused children who are not killed. These effects include (Reyome, 2010):

**1.** Few close social relationships and an inability to love or trust.

**2.** Lower quality of subsequent relationships (Larsen et al. 2011).

## FAMILY POLICY | Megan's Law and Beyond

In 1994, Jesse Timmendequas lured seven-year-old Megan Kanka into his Hamilton Township house in New Jersey to see a puppy. He then raped and strangled her and left her body in a nearby park. Timmendequas had two prior convictions for sexually assaulting girls. Megan's mother, Maureen Kanka, argued that she would have kept her daughter away from her neighbor if she had known about his past sex offenses. She campaigned for a law, now known as **Megan's Law**, requiring that communities be notified of a neighbor's previous sex convictions. Forty-five other states have enacted laws similar to New Jersey's Megan's Law.

The 1994 Jacob Wetterling Act requires states to register individuals convicted of sex crimes against children. The law requires that convicted sexual offenders register with local police in the communities in which they live. It also requires the police to go out and notify residents and certain institutions (such as schools) that a previously convicted sex offender has moved into the area. This provision of the law has been challenged on the belief that individuals should not be punished forever for past deeds. Critics of the law argue that convicted child molesters who have been in prison have paid for their crime. To stigmatize them in

**Megan's Law** requires that communities be notified of a neighbor's previous sex convictions.

*Children are often the silent victims of drug abuse.*

RICK LARSEN, CONGRESSMAN

**3.** Communication problems and learning disabilities, particularly in children who have been neglected. These children lag in language development (Sylvestrea and Mérettec 2010).

**4.** Aggression, low self-esteem, depression, and low academic achievement.

**5.** Increased risk of alcohol or substance abuse and suicidal tendencies as adults.

**6.** Adolescent delinquency (true for all racial groups) (McNeil et al. 2012).

Child sexual abuse is a specific type of child abuse. Perpetrators may be both in (**intrafamilial**) and outside (**extrafamilial**) the family. A neighbor is an example of an extrafamilial perpetrator for whom Megan's Law was passed (see the Family Policy feature). McElvaney et al. (2014) interviewed 22 young people who experienced child sexual abuse and revealed ambivalence about disclosing. They feared not being believed, being asked questions about their well-being, felt ashamed of what happened, and blamed themselves for the abuse and for not telling.

**Intrafamilial** inside the family

**Extrfamilial** external to the family

## Parent Abuse

Some people assume that, because parents are typically physically and socially more powerful than their children, they are immune from being abused by their children. However, parents are often targets of their children's anger, hostility, and abuse. Routt and Anderson (2011) emphasized the prevalence of adolescent to parent violence and provided details from interviews with these youth and their parents. Teenage and even younger children physically and verbally lashed out at their parents. Children have been known to push parents down stairs, set the house on fire while their parents were in it, and use weapons such as guns or knives to inflict serious injuries or even to kill a parent.

## Sibling Abuse

Observe a family with two or more children and you will likely observe some amount of sibling abuse. Even in "well-adjusted" families, some degree of fighting and name calling among children is expected. Most incidents of sibling violence communities as sex offenders may further alienate them from mainstream society and increase their vulnerability for repeat offenses.

In many states, Megan's Law is not operative because it is on appeal. Parents ask, "Would you want a convicted sex offender, even one who has completed his prison sentence, living next door to your eight-year-old daughter?" However, the reality is that little notification is afforded parents in most states. Rather, the issue is tied up in court and will likely remain so until the Supreme Court decides it. A group of concerned parents (Parents for Megan's Law) are trying to implement the law nationwide. See the Web Links section at the end of this chapter for the group's website address.

**Your Opinion?**

1. To what degree do you believe the Supreme Court should uphold Megan's Law?
2. To what degree do you believe that convicted child molesters who have served their prison sentence should be free to live wherever they like without neighbors being aware of their past?
3. Independent of Megan's Law, how can parents protect their children from sex abuse?

consist of slaps, pushes, kicks, bites, and punches. What passes for normal, acceptable, or typical behavior between siblings would often be regarded as violent and abusive behavior outside the family. Inside the family, sibling rivalry is the label used for playful abusive acts . . . but when these acts become secret or one sibling is always the aggressor or there is a considerable age difference, the label of sibling abuse applies.

Sibling abuse may include sexual exploitation, whereby an older brother will coerce younger female siblings into nudity or sex. Though some sex between siblings is consensual, often it is not. "He did me *and* all my sisters," reported one woman. Another woman reported that, as a child and young adolescent, she performed oral sex on her brother three times a week for years because he told her that ingesting a man's semen was the only way she would be able to have babies as an adult. Morrill (2014) noted that sibling sexual abuse may be particularly disruptive to one's individual development in that it can last for years.

In addition to physical and sexual sibling abuse, there is **sibling relationship aggression**. This is behavior of one sibling toward another sibling that is intended to induce social harm or psychic pain in the sibling. Examples include social alienation or exclusion (e.g., not asking the sibling to go to a movie when a group is going), telling secrets or spreading rumors (e.g., revealing a sibling's sexual or drug past), and withholding support or acceptance (e.g., not acknowledging a sibling's achievements in school or sports). Tucker et al. (2013) provided data on 8th- and 12th-grade sibling aggression and found greater reactive than proactive aggression to their closest-aged sibling.

**Sibling relationship aggression** behavior of one sibling toward another sibling that is intended to induce social harm or psychic pain in the sibling.

## Elder Abuse

As increasing numbers of the elderly end up in the care of their children, abuse, though infrequent, may occur.

Elder abuse occurs across various ethnic and socioeconomic groups. DeLiema et al. (2012) assessed the prevalence of abuse among the elderly (age 66 and older) in a sample of low-income immigrant Latinos: 40% reported having experienced some form of abuse or neglect within the previous year. Nearly

25% reported psychological abuse, 10.7% physical assault, 9% sexual abuse, and 16.7% financial exploitation; 11.7% were neglected by their caregivers. These percentages are higher in all mistreatment domains than findings from previous research, suggesting that low-income Latino immigrants are highly vulnerable to elder mistreatment.

Various examples of elder abuse include:

1. ***Neglect.*** Also known as passive abuse, examples include failing to buy or give the elderly needed medicine, failing to take them for necessary medical care, or failing to provide adequate food, clean clothes, and a clean bed. Neglect is the most frequent type of domestic elder abuse.
2. ***Physical abuse.*** This abuse includes inflicting injury or physical pain or sexual assault.
3. ***Psychological abuse.*** Examples of psychological abuse include verbal abuse, deprivation of mental health services, harassment, and deception.
4. ***Social abuse.*** Unreasonable confinement and isolation, lack of supervision, and abandonment are examples of social abuse.

**Granny dumping** refers to taking one's invalid grandmother to the emergency room of a hospital and leaving her there with no identification.

Another type of elder abuse is **granny dumping**. Adult children or grandchildren who feel burdened with the care of their elderly parent or grandparent leave an elder at the emergency entrance of a hospital with no identification. If the hospital cannot identify responsible relatives, it is required by state law to take care of the abandoned elder or transfer the person to a nursing home facility, which is paid for by state funds. Relatives of the dumped elder, hiding from financial responsibility, never visit or see the elder again.

Adult children who are most likely to dump or abuse their parents tend to be under a great deal of stress and to use alcohol or other drugs. In some cases, parent abusers are getting back at their parents for mistreating them as children. In other cases, the children are frustrated with the burden of having to care for their elderly parents.

The ultimate form of elder abuse is murder. Karch and Nunn (2011) provided data on 68 homicides of dependent elderly by a caregiver and found that they were mostly (97%) non-Hispanic or white (88%) women (63%) killed in their own homes (92.6%) with a firearm (35.3%) or by intentional neglect (25.0%) by a husband (30.9%) or a son (22.1%). Nearly half of the victims were aged 80 years and older (48.5%). "Many homicide by caregiver incidents are precipitated by physical illness of the victim or caregiver, opportunity for perpetrator financial gain, mental illness of the caregiver, substance use by the caregiver, or an impending crisis in the life of the caregiver not related to illness."

# The Future of Violence and Abuse in Relationships

Violence and abuse will continue to occur behind closed doors, in private contexts where social constraint is minimal. Reducing such violence and abuse will depend on prevention strategies focused at three levels: the general population, specific groups thought to be at high risk for abuse, and families who have already experienced abuse. Public education and media campaigns aimed at the general population will continue to convey the criminal nature of domestic assault, suggest ways the abuser might learn to prevent or escape from abuse, and identify where abuse victims and perpetrators can get help. MacMillan et al. (2013) emphasized that children need to be socialized early that violence, even that between their parents, is not acceptable.

Preventing or reducing family violence through education necessarily involves altering aspects of American culture that contribute to such violence.

For example, violence in the media must be curbed (not easy, with nightly news clips of bombing assaults in other countries, gun shootings, violent films, etc.). Chavis et al. (2013) documented that parental exposure to a brief multimedia program designed to suggest alternatives to physical punishment resulted in parents making more frequent choices not to discipline their 6-to-24-month-old infants via spanking. Bennett et al. (2014) emphasized the need for bystander intervention programs which sensitized individuals to be aware of abuse and to intervene. Katz and Moore (2013) provided some data on the efficacy of such intervention programs. Finally, traditional gender roles and views of women and children as property must be replaced with egalitarian gender roles and respect for women and children. Children should also be socialized to value their bodies and react as necessary to protect them. *Inside Out: Your body is amazing inside and out and belongs only to you* (Podgurski 2013) is a resource to help children accomplish this goal.

Another important cultural change is to reduce violence-provoking stress by reducing poverty and unemployment and by providing adequate housing, nutrition, medical care, and educational opportunities for everyone. Integrating families into networks of community and kin would also enhance family well-being and provide support for families under stress.

# Summary

***What is the nature of relationship abuse?***

Violence or physical abuse is defined as physical aggression with the purpose of controlling, intimidating, and subjugating another human being. Emotional abuse (also known as psychological abuse, verbal abuse, or symbolic aggression) is designed to denigrate the partner, thereby giving the abuser more control.

***What are explanations for violence in relationships?***

Cultural explanations for violence include violence in the media, corporal punishment in childhood, gender inequality, and stress. Community explanations involve social isolation of individuals and spouses from extended family, poverty, inaccessible community services, and lack of violence prevention programs. Individual factors include psychopathology of the person (antisocial), personality (dependency or jealousy), and alcohol abuse. Family factors include child abuse by one's parents, observing parents who abuse each other, and not having a father in the home.

***How does sexual abuse in undergraduate relationships manifest itself?***

About a third of undergraduate females (and 20% of undergraduate males) report having been pressured to have sex. Acquaintance rape is the most frequent context with the perpetrator believing in male or female sex myths.

***Is there abuse in marriage?***

The longer individuals know each other and the more intimate the relationship, the greater the abuse. All states have laws prohibiting marital rape (though there are exceptions to the law in Washington state). Over 30 countries provide no protection for wives against rape by their husbands.

***What are the effects of abuse?***

The effects of IPV include physical harm, mental harm (depression, anxiety, low self-esteem, lost of trust in others, sexual dysfunctions), unintended pregnancy, and multiple abortions. High levels of anxiety and depression often lead to alcohol and drug abuse. Violence on pregnant women significantly increased the risk for infants of low birth weight, preterm delivery, and neonatal death.

***What is the cycle of abuse and why do people stay in an abusive relationship?***

The cycle of abuse begins when a person is abused and the perpetrator feels regret, asks for forgiveness, and starts acting nice (for example, gives flowers). The victim, who perceives few options and feels guilty terminating the relationship with the partner who asks for forgiveness, feels hope for the relationship at the contriteness of the abuser and does not call the police or file charges. The couple usually experiences a period of making up or honeymooning, during which the victim feels good again about the abusing partner. However, tensions mount again and are released in the form of violence. Such violence is followed by the familiar sense of regret and pleadings for forgiveness, accompanied by being nice (a new bouquet of flowers, and so on).

The reasons people stay in abusive relationships include love, emotional dependency, commitment to the relationship, hope, view of violence as legitimate, guilt, fear, economic dependency, and isolation. The catalyst for breaking free combines the sustained aversiveness of staying, the perception that they and their children will be harmed by doing so, and the awareness of an alternative path or of help in seeking one.

***What is the nature of child, parent, sibling, and elder abuse?***

Child abuse includes physical abuse, such as beating and burning; verbal abuse, such as insulting or demeaning the children; and neglect, such as failing to provide adequate food, hygiene, medical care, or adult supervision for children. Children can also experience emotional neglect by their parents and sexual abuse from both within (intrafamilial) and outside (extrafamilial) the family.

Parent abuse is the deliberate harm (physical or verbal) of parents by their children. Sibling abuse is a frequent form of abuse that occurs with limited detection. The difference between sibling rivalry and sibling abuse is that the latter is secret and there is usually one perpetrator.

Elder abuse is another form of abuse in relationships. Granny dumping occurs by children or grandchildren who feel burdened with the care of their elderly parents or grandparents and leave them at the emergency entrance of a hospital. If the relatives of the elderly patient cannot be identified, the hospital will put the patient in a nursing home at state expense.

# Key Terms

A periodic reinforcement
Acquaintance rape
Battered-woman syndrome
Child abuse
Corporal punishment
Cybercontrol
Cybervictimization
Date rape
Double victims
Emotional abuse
Entrapped
Extrafamilial
Female rape myths
Filicide
Granny dumping
Honor crime/Honor killing
Intimate partner homicide
Intimate partner violence (IPV)
Intimate terrorism (IT)
Intrafamilial
Male rape myths
Marital rape
Megan's Law
Neonaticide
Obsessive relational intrusion (ORI)
Parricide
Rape myths
Rophypnol
Siblicide
Sibling relationship aggression
Situational couple violence (SCV)
Stalking
Uxoricide
Violence

# Web Links

Childabuse.com
http://www.childabuse.com/

Male Survivor
http://www.malesurvivor.org/

Minnesota Center against Violence and Abuse
http://www.mincava.umn.edu

National Sex Offender Data Base
http://www.nationalalertregistry.com

Parents for Megan's Law and the Crime Victims Center
http://www.parentsformeganslaw.com

Rape, Abuse & Incest National Network (RAINN)
http://www.rainn.org/

Stop and Recover from Abuse
www.chooserespect.org
www.safeplace.org

Stop It Now! The Campaign to Prevent Child Sexual Abuse
http://www.stopitnow.com/

V-Day (movement to stop violence against women and girls)
http://www.vday.org/

# CHAPTER 11 Planning for Children

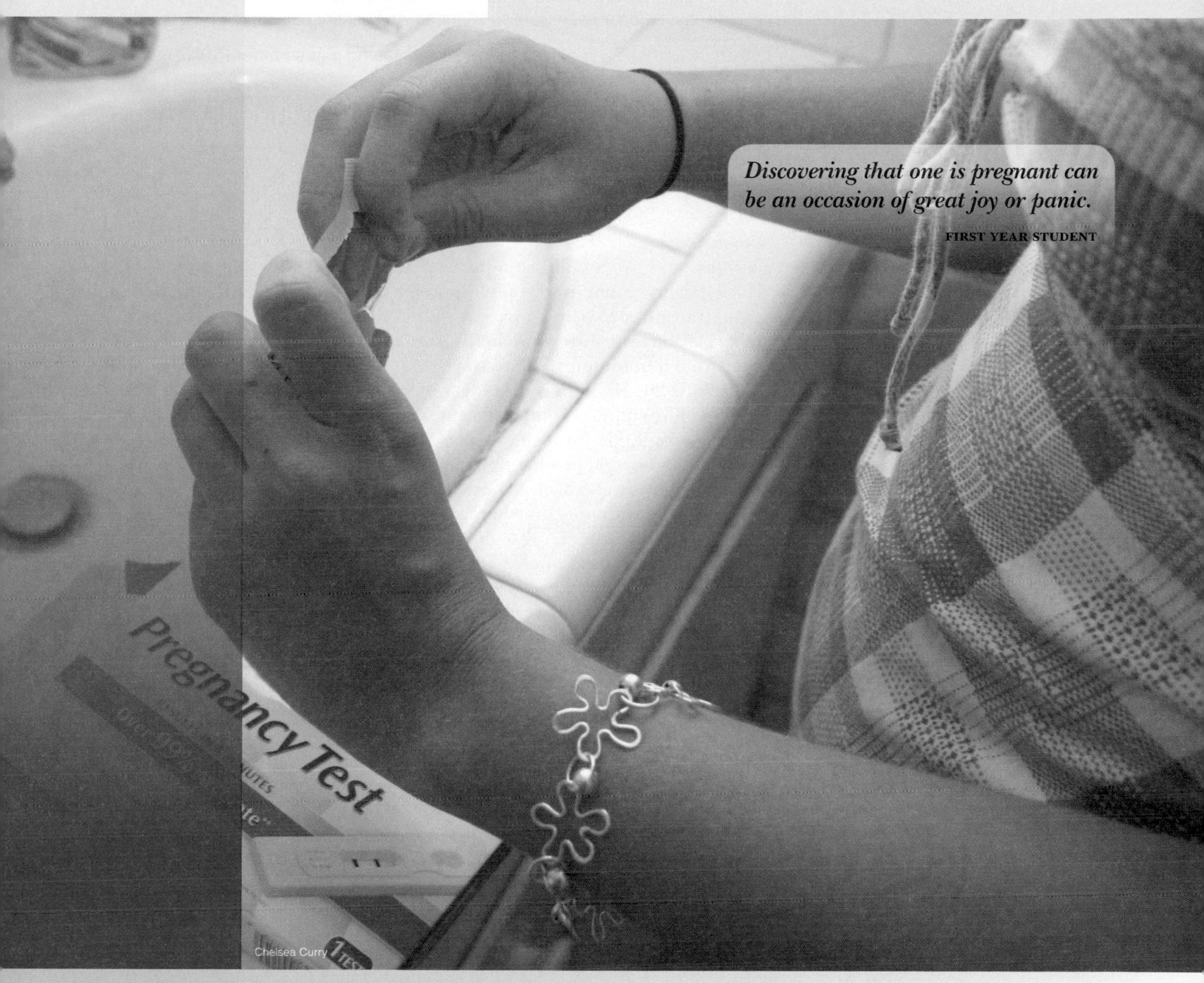


*Discovering that one is pregnant can be an occasion of great joy or panic.*

FIRST YEAR STUDENT

## Learning Objectives

- Discuss the social influences motivating an individual to have a child.
- Review the individual motivations for having a child.
- Identify the various categories/motivations of individuals who decide to remain childfree.
- Know the causes of infertility and technology available to help induce a pregnancy.
- Be aware of the motives for adoption and the demographics of those who adopt.
- Discuss the types of abortion and the outcome of having an abortion.
- Predict the future of planning for children.

## TRUE OR FALSE?

1. Over half of all pregnancies in the United States are unintended.
2. Most women obtaining abortions are able to rely on their male partners for social support.
3. Women who have had a tubal ligation are less likely to experience high levels of sexual satisfaction, relationship satisfaction, and sexual pleasure.
4. In a study on 18–29 year olds, most reported "being a good parent" was more important than "having a good marriage."
5. Obesity is related to problems of infertility—difficulty getting pregnant and having the number of children the woman wants.

*Answers:* 1. T 2. T 3. F 4. T 5. T

Deciding to have a child is important not only for the child but also is a life-changing event for the adult (and for the couple) making the decision. The transition to being a parent changes the life of the person and that person's relationships forever. Ninety-percent (an equal percent for women and men) of 4,536 undergraduates reported that they wanted to have a child (Hall and Knox 2013). Among 18–29 year olds, most reported that "being a good parent" was more important than "having a successful marriage" (52% to 30%) (Wang and Taylor 2011). Women who plan their pregnancies are more likely to be educated, employed, and in a happy relationship than those who have unplanned pregnancies. They are also likely to be psychologically healthy (e.g., not depressed) compared to those with unplanned pregnancies (Yanikerem et al. 2013).

Planning children, or failing to do so, is a major societal issue. Planning when to become pregnant has benefits for both the mother and the child. Having several children at short intervals increases the chances of premature birth, infectious disease, and death of the mother or the baby. Would-be parents can minimize such risks by planning fewer children with longer intervals in between.

Women who plan their pregnancies can also modify their behaviors and seek preconception care from a health care practitioner to maximize their chances of having healthy pregnancies and babies. For example, women planning pregnancies can make sure they eat properly and stop drinking alcohol and smoking cigarettes that could harm the developing fetus. Partners who plan their children also benefit financially by pacing the economic needs of their offspring. Having children four years apart helps to avoid having more than one child in college at the same time. Conscientious family planning will also help to reduce the number of unwanted pregnancies.

### National **Data**

Just over a third (34.4%) of adolescent and young women reported using a condom, 29.1% used hormonal contraception or an intrauterine device alone, and 15.8% used another method or no method during their last sexual intercourse (Tyler et al. 2014).

**Pregnancy coercion** coercion by a male partner for the woman to become pregnant.

**Birth control sabotage** partner interference with contraception.

Fifty-one percent of all pregnancies in the United States are not intended (Finer and Zolna 2014). In addition to "forgetting" to use contraception or assuming one will not become pregnant with just "one" exposure, Miller et al. (2010) noted physical or sexual partner force/violence including **pregnancy coercion** (coercion by a male partner for the woman to become pregnant) and **birth control sabotage** (partner interference with contraception—e.g., the woman stops taking the pill without her partner's knowledge) as reasons for unintended pregnancies. The researchers studied 1,278 females ages 16–29 years and found that 19% reported pregnancy coercion, and 15% reported birth control sabotage.

Your choices in regard to whether you want to have children and the use of contraception have important effects on your happiness, lifestyle, and resources. These choices, in large part, are influenced by social and cultural factors that may operate without your awareness.

*Mother Nature, in her infinite wisdom, has instilled within each of us a powerful biological instinct to reproduce; this is her way of assuring that the human race, come what may, will never have any disposable income.*

DAVE BARRY, COMEDIAN

# Do You Want to Have Children?

Beyond a biological drive to reproduce (which not all adults experience), societies socialize their members to have children. This section examines the social influences that motivate individuals to have children, the lifestyle changes that result from such a choice, and the costs of rearing children.

## Social Influences Motivating Individuals to Have Children

Our society tends to encourage childbearing, an attitude known as **pronatalism**. Our family, friends, religion, and government encourage positive attitudes toward parenthood. Cultural observances also function to reinforce these attitudes.

**Pronatalism** attitude that encourages childbearing.

**Family** Our experience of being reared in families encourages us to have families of our own. Our parents are our models. They married; we marry. They had children; we have children. We also expect to have a "happy family."

**DIVERSITY IN OTHER COUNTRIES**

*In a cross-national comparison of industrialized nations, the United States ranked virtually at the top in the percentage of those disagreeing with this statement: "The main purpose of marriage is having children." Nearly 70 percent of Americans believe the main purpose of marriage is something else, compared, for example, to 51 percent of Norwegians and 45 percent of Italians.*

*State of Our Unions, 2012*

**Friends** Our friends who have children influence us to do likewise. After sharing an enjoyable weekend with friends who had a little girl, one husband wrote to the host and hostess, "Lucy and I are always affected by Lisa—she is such a good child to have around. We haven't made up our minds yet, but our desire to have a child of our own always increases after we leave your home." This couple became parents 16 months later.

**Religion** Religion is a strong influence on an individual's decision to have children. Catholics are taught that having children is the basic purpose (procreation) of marriage and gives meaning to the union. Mormons and ultra-Orthodox Jews also have a strong interest in having and rearing children.

**Race** Hispanics have the highest fertility rate of any racial/ethnic category.

**Government** The taxes (or lack of them) that our federal and state governments impose support parenthood. Although individuals have children for more emotional than financial reasons, married couples with children pay lower taxes than married couples without children. Assume there are two couples (one married couple with two children and one childfree married couple), each making $100,000. The couple with children would pay $1,950 less in federal tax than the childfree couple.

**Cultural Observances** Our society reaffirms its value for having children by identifying special days for Mom and Dad. Each year on Mother's Day and Father's Day (and now Grandparents' Day), parenthood is celebrated across the nation with cards, gifts, and embraces. People choosing not to have children have no cultural counterpart (e.g., Childfree Day). In addition to influencing individuals

## FAMILY POLICY | How Old Is Too Old to Have a Child?

Rajo Devi was age 70 when she gave birth to a baby girl in Calcutta. Dougall et al. (2013) noted that women are not aware of the steep rate of fertility decline with age. Popular media encourages young women to feel that they have forever to get pregnant. Yet questions are now being asked about the appropriateness of elderly individuals becoming parents. Should social policies on this issue be developed?

There are advantages and disadvantages of having a child as an elderly parent. The primary developmental advantage for the child of retirement-aged parents is the attention the parents can devote to their offspring. Not distracted by their careers, these parents have more time and interest to nurture, play with, and teach their children. Although they may have less energy, their experience and knowledge are doubtless better. However, elderly fathers are more likely to die sooner, leaving their offspring without a father. Abundant data reveal the value of a father for the emotional and physical development of a child (Dearden et al. 2013).

The primary disadvantage of having a child in the later years is that the parents are likely to die before, or early in, the child's adult life. Steve Martin was 67 when his baby was born in 2012. He will need to live until his mid-eighties to experience the high school graduation and even later to attend the marriage of his child.

There are also medical concerns for both the mother and the baby during pregnancy in later life. They include an increased risk of morbidity (chronic illness and disease) and mortality (death) for the mother. These risks are typically a function of chronic disorders that go along with aging, such as diabetes, hypertension, and cardiac disease. Stillbirths, miscarriages, ectopic pregnancies, multiple births, and congenital malformations are also more frequent for women with advancing age. However, prenatal testing can identify some potential problems such as the risk of Down syndrome and any chromosome abnormality—negative neonatal outcomes are not inevitable. Because an older woman can usually have a healthy baby, government regulations on the age at which a woman can become pregnant are not likely. Aasheim et al. (2012) studied age of mother and outcomes and found that women age 32 and older compared to women age 25–31 reported more

to have children, society and culture also influence feelings about the age parents should be when they have children. Recently, couples have been having children at later ages. Is this a good idea? The Family Policy feature discusses this issue.

## Individual Motivations for Having Children

Individual motivations, as well as social influences, play an important role in the decision to have children. Some of these inducements are obvious as in the desire to love and to be loved by one's own child, companionship, and the desire to be personally fulfilled as an adult by having a child. Some people also want to recapture their own childhood and youth by having a child. Motives that are less obvious may also be operative—wanting a child to avoid career tracking (e.g., "I'd rather have a baby than tenure") and to gain the acceptance and approval of one's parents and peers. Teenagers sometimes want to have a child to have someone to love them (discussed later in the chapter). Goldberg et al. (2012) interviewed 35 pre-adoptive gay male couples (70 men) and examined their reasons for pursuing parenthood. While most reasons were similar to heterosexual couples, a reason unique to their minority status was "desire to teach a child tolerance."

*IF I HAD MY LIFE TO LIVE OVER*

*I would have talked less and listened more…I would have sat on the lawn with my children and not worried about grass stains.*

—ERMA BOMBECK

## Lifestyle Changes and Economic Costs of Parenthood

**Lifestyle Changes** Becoming a parent often involves changes in lifestyle. Daily living routines become focused around the needs of the children. Living arrangements change to provide space for another person in the household. Some parents change their work schedule to allow them to be home more. Food shopping and menus change to accommodate the appetites of children. A major lifestyle change is the loss of freedom of activity and flexibility in one's personal schedule. Lifestyle changes are particularly dramatic for women. The time and psychological distress from early pregnancy to 18 months postpartum. This outcome was not catastrophic.

Age of the father may also be an issue in older parenting. Puleo et al. (2012) found that children born to older fathers had an increased risk of autism. Similarly, Krishnaswamy et al. (2011) found that Malaysian children born to fathers aged 50 or above had an increased risk of having CMD (common mental disorders) compared to children who were fathered by young men. Indeed, males in their 20s might consider having some of their sperm frozen since they may wish to have a child in later life (after 40).

### Your Opinion?

1. How old do you think is "too old" to have a child?
2. Do you think the government should attempt to restrict people from having a biological child after the age of 60?
3. Who do you feel benefits most and least from having a child in later life?

### Sources


Aasheim, V., U. Waldenstrom, A. Hjelmstedt, S. Rasmussen, H. Pettersson, and E. Schytt. 2012. Associations between advanced maternal age and psychological distress in primiparous women, from early pregnancy to 18 months postpartum. *BJOG: An International Journal of Obstetrics & Gyneaecology* 119: 1108–1116.

Dearden, K., B. Crookston, H. Madanat, J. West, M. Penny, and Santiago Cueto. 2013. What difference can fathers make? Early parental absence compromises Peruvian children's growth. *Maternal & Child Nutrition* 9: 143–154.

Dougall, K. M., Y. Beyene, and R. D. Nachtigall. 2013. Age shock: Misperceptions of the impact of age on fertility before and after IVF in women who conceived after age 40. *Human Reproduction* 28: 350–356.

Krishnaswamy, S., K. Subramaniam, P. Ramachandran, T. Indran, and J. Abdul Aziz. 2011. Delayed fathering and risk of mental disorders in adult offspring. *Early Human Development* 87: 171–175.

Puleo, C. M. et al. 2012. Advancing paternal age and simplex autism. *Autism: The International Journal of Research & Practice* 16: 367–380.

David Knox

One motivation for having children is the joy of teaching them new things.

## APPLYING SOCIAL RESEARCH | Gender Differences in Attitudes toward Children*

Although undergraduate life includes parties, alcohol, and sex, the larger personal agenda of pair-bonding, procreation, and socialization of one's offspring remain major life goals. This study examined how undergraduates feel about having children.

### Sample

Data for the study involved a sample of 293 undergraduate student volunteers at a large southeastern university who completed a 50-item questionnaire. Of the respondents, 73.4% were female and 26.6% were male. The median age of the respondents was 20 with a range of 17 to 46. Of the respondents, 40.8% were freshmen, 20.5% sophomores, 17.5% juniors, and 21.2% seniors. Racial background of the respondents was 82% whites and 17.9% blacks (respondent self-identified as African American black, African black, or Caribbean black). In comparing the responses of women and men, cross-classification was conducted to determine any relationships with chi-square utilized to assess statistical significance.

### Findings

Analysis of the data revealed the following 10 statistically significant gender differences in regard to attitudes toward children:

1. ***Females were more likely to feel having children is important*** (94.3% versus 90.5%) ($p < .007$).
2. ***Females were more likely to view children as providing a reason to live*** (40.1% versus 23.2%) ($p < .008$).
3. ***Females were more likely to enjoy being around toddlers or young children*** (93.2% versus 74.1%) ($p < .001$).
4. ***Females were more likely to have taken care of an infant.*** The pro-child value females have is related to the fact that they have almost three times more experience of taking care of an infant than a male. Almost 60% of the female respondents, compared to less than a fourth of male respondents (59.7% versus 23.1%) ($p < .000$) reported that they had taken care of an infant.
5. ***Females were less likely to be annoyed by crying babies.*** In response to the statement, "crying babies drive me crazy," 32.4% of the female respondents compared to 56.2% ($p < .004$) of the male respondents agreed.
6. ***Females were more likely to view "wanting children" as an important criteria for a mate*** (91.7% versus 70.5%) ($p < .001$).

effort required to be pregnant and rear children often compete with the time and energy needed to finish one's education or build one's career. Parents learn quickly that being both an involved parent and climbing the career ladder are difficult. The careers of women may suffer most.

To give our students a simulated exposure to the effect of a baby on their lifestyle we asked them to take care of a "fake baby" for a week. Baby Think It Over (BTIO) is a life-sized computerized infant simulation doll that has been used in pregnancy prevention programs with adolescents. The following is part of the write-up of one of our students:

> *The whole idea of the electronic baby was to see if I was ready to be a mother. I am sad to say that I failed the test. I am not ready to be a mother. This whole experience was extremely difficult for me because I am a full-time student and I have a job. It was really hard to get the things that I needed to get done. Suddenly, I couldn't just think about myself but I also had a little one to think about.*
>
> *The baby seemed to cry a lot, even if I had just changed her, she still cried. The experience really hit me when I had to wake up four and five times in the night to feed and to change the baby. I learned that when the baby sleeps, I need to sleep as well. I also learned that if I had to take care of the baby by myself, I just couldn't do it. What is sad is that single moms do it every day. If I had a supportive boyfriend or husband to help, the whole idea of having a baby wouldn't be so bad. But my boyfriend told me to call him when the project was over and I had given the baby back.*

Other students had a very positive experience with the baby and did not want to part with it at the end of the week. Indeed, in one case the "fake baby" became

7. ***Females were less likely to marry a man who did not want children.*** Consistent with the previous finding, females were more likely to eliminate from consideration marrying a person who could not or would not have children (44.3% versus 21.5%) ($p < .007$).
8. ***Females were more likely to divorce a spouse if the spouse turned against children.*** Were a marriage to occur with the woman assuming that her husband wanted children and she were to find out that he had changed his mind, she would divorce him. Almost three times as many female as male (29.5% versus 10.8%) respondents agreed that they would divorce their spouse if "I was married and my spouse turned against having children."
9. ***Females were more likely to consider adoption if the spouses were sterile*** (89.3% versus 76.2%) ($p < .001$).
10. ***Females were more likely to be open to having a child of either sex.*** In our sample, 46.2% of the female respondents compared to 33.8% of the male respondents reported that they "couldn't care less" whether they had a male or a female child.

### Theoretical Explanation for the Findings

Sociobiology is helpful in explaining the findings of this study. The data emphasized that women evidenced significantly more interest in having children, in selecting a mate who wanted children, and in divorcing a husband who turned against children. Sociobiology (social behavior is influenced by biology) emphasizes the fact that women carrying their babies to term and providing milk for their survival are a reflection of a biological genetic wiring that predisposes women to greater interest in having a baby and in bonding with them. Indeed the species demands that at least one parent take responsibility for ensuring the survival of the species. That men do not have such a strong biological link but more often derive their reward from social approval for economic productivity in the workplace may help to explain the discrepancy in female and male attitudes and behaviors.

*Abridged from B. Bragg, D. Knox, and M. Zusman. 2008. The little ones: Gender differences in attitudes toward children among university students. Poster, Southern Sociological Society, April, Richmond, VA.

a part of the family. One student said that she awakened early one morning to find her father rocking the baby.

**Financial Costs** Meeting the financial obligations of parenthood is difficult for many parents. The costs begin with prenatal care and continue at childbirth. For an uncomplicated vaginal delivery, with a two-day hospital stay, the cost may total $10,000, whereas a Cesarean section birth may cost $14,000. The annual cost of a child less than two years old for middle-income parents ($57,600 to $99,730)—which includes housing ($3,870), food ($1,350), transportation ($1,540), clothing ($740), health care ($820), child care ($2,740), and miscellaneous ($890)—is $11,950. For a 15- to 17-year-old, the cost is $13,830 (*Statistical Abstract of the United States*, 2012–2013, Table 689). These costs do not include the wages lost when a parent drops out of the workforce to provide child care. Indeed, a woman with a child (particularly a heterosexual woman) is less likely to be hired (Baumle 2013).

Most parents value their children attending college. See the website at the end of this chapter in regard to the College Cost Calculator. The price varies depending on whether a child attends a public or a private college. The annual cost for a son or daughter attending a four-year public college in the state of residence in 2014 was $21,447 (including tuition,

Chelsea Curry

Being responsible for an electronic baby that cries and needs "feeding" and "diapering" can give the childfree a glimpse of the responsibility to come.

## National **Data**

The estimated cost to parents for rearing a child born in 2012 to age 18 will be $241,080 (United States Department of Agriculture 2013).

dorm, meals); for a private college the cost was $42,224 annually. Collegeboard .com will identify the cost of a specific college.

# How Many Children Do You Want?

*A two year old is kind of like having a blender, but you don't have a top for it.*

JERRY SEINFELD, COMEDIAN

**Procreative liberty** the freedom to decide whether or not to have children.

Individuals no longer feel required to have children to avoid being stigmatized or to experience personal fulfillment. Couples can now choose not to have children or to have them in or out of marriage. Most decide to have children, and most inside marriage.

**Procreative liberty** is the freedom to decide whether or not to have children. More women are deciding not to have children or to have fewer children (Department of Commerce et al. 2011).

## Childfree Marriage?

*I hate kids.*

BILL MAHER, COMEDIAN

## National **Data**

About 20% of American women over the age of 40 do not have children. The percent is expected to increase (Allen and Wiles 2013). Of those who do not have children, 44% are voluntarily childfree, 40% are involuntarily childfree, and 16% plan a child in the future (Smock and Greenland 2010).

*The reproductive imperative of marriage is over.*

PAMELA HAAG, MARRIAGE CONFIDENTIAL

Just as cultural forces help to ensure that a person will marry (90%) and select someone of the same race (85%), there are similar influences to have children. The intentionally childfree may be viewed with suspicion ("they are selfish"), avoidance ("since they don't have children they won't like us or support our family values"), discomfort ("what would I have in common with these people?"), rejection ("they are wrong not to want children/I don't want to spend time with these people"), and pity ("they don't know what they are missing") (Scott 2009). Indeed "the mere existence of a growing childless by choice population is a challenge to people who believe procreation is instinctive, intrinsic, biological, or obligatory" (p. 191).

Stereotypes about couples who deliberately elect not to have children include that they don't like kids, are immature, and are not fulfilled because they don't have a child to make their lives "complete." The reality is that such individuals may enjoy children and some deliberately choose careers to work with them (e.g., elementary school teacher). But they don't want the full-time emotional and economic responsibility of having their own children (see insert on Our Decision to be Childfree). Allen and Wiles (2013) interviewed 38 older (ages 63–93) childfree individuals and found an array of meanings (from grief to relief, from making a deliberate choice to do what felt "natural"—not have a child—to breaking a family of violence cycle). Pelton and Hertlein (2011) emphasized that voluntary childfree couples do not progress through the traditional family life cycle. Indeed they pass through a different set of stages specific to these couples. And they do so without lamenting the fact that they do not have children.

John Martin

Vivian and Jeff (see their "Our Decision to be Childfree" below)

**Childlessness concerns** is a concept which refers to the idea that holidays and family gatherings may be difficult because of not having children or feeling left out or sad that others have children. McQuillan et al. (2012) studied a representative sample of 1,180 women without children from the National Survey of Fertility Barriers and found that the degree to which women report these concerns is related to their reason for being childless; women who were voluntarily childless reported few concerns. Those most reactive were the childless due to biomedical causes or having delayed parenthood until they could no longer get pregnant. The feature on "Our Decision to be Childfree" reflects a happy childfree couple who chose not to have children.

**Childlessness concerns** the idea that holidays and family gatherings may be difficult because of not having children or feeling left out or sad that others have children.

**Our Decision to be Childfree***

*I am 51 years old and my husband, Jeff, is 56 years old. We both came from stable families and have been married over 25 years. We have no children by choice. I have never had what most of my female friends describe as a desire to have children. I never thought about having children growing up, nor did I think of names for my future children as young girls often do. I do not recall ever having much of a "maternal instinct." I babysat one time, and I called my mother to come over because I was very bored with the children and didn't really know what to do with them, nor did I want to entertain them any longer! My husband and I discussed not having children, and we had a premarital agreement, though not a written one, not to have children. We used birth control strictly to ensure that I would not get pregnant. We did not discuss having children a lot after marriage, though we did not make any decision to permanently prevent it until after I turned 40 and we had been married for about 15 years. I never had a fear of having children; I just had no desire to have them. Neither of our families gave us any pressure to have children, though we did hear the occasional, "You two would make great parents." And, the oft said, "You'll feel differently when they are your own." Having worked in education for 25 years, I have seen first-hand that for many people, it wasn't different even with their own as many parents of the children I taught didn't seem to give them the care they needed. I trusted my instincts and my feelings and I have no regrets. It was helpful for both of our families that older siblings did have children, thus there were grandchildren for our parents. I was raised Catholic and my husband was raised Baptist, though neither of us practice these religions. I had no issue reconciling my decision with any religious beliefs or tenets.*

*When we are asked if we have children we both respond, "No, we do not and by choice." I learned this after saying no numerous times and seeing people awkwardly change the subject as they always seem to wonder if perhaps we couldn't have children. I do not feel that we have an "empty marriage" as is often said about marriages without children. My husband and I have many friends with whom we socialize. We have worked hard on these relationships over the years. As our friends had children, their lives changed drastically, and we had to respect that. We went to a lot of birthday parties for their small children, bought a lot of 16th birthday presents, and gave monetary gifts for their high school and college graduations. However, now that their children are grown, we are free to travel and socialize as much with these individuals as we did before they had children. This was important to us and we worked hard to stay connected. If we wanted our friends to respect our choice to not have children, we knew we had to respect their choice to have them as well.*

*There is one typical societal comment that I have grown to dislike. It is the well-meaning, "Who will take care of you when you get older?" Having to care for my mother for two weeks after complications from hip surgery, I couldn't help but wonder what she would do if she had only my older brother to reply upon? Even though he is her child, he would not care for her. There are no guarantees that if you have children, they will care for you. I would take that even further, and say they shouldn't have to take care of you. Not all people are comfortable being caregivers, especially to loved ones. And most people have their own families for whom to care, and they have to work to make ends meet. My husband and I have done as much preplanning as we can for our future and our future healthcare needs.*

*Both my husband and I feel that we have given back to society in many ways, and that we have helped other children in our lives. We give to scholarships to help young people get ahead, we donate to organizations that assist young people, and I have worked in education for 25 years, with high school and college age students. We live a full life, with friends and family and no regrets.*

*Dr. Vivian Martin Covington, College of Education, East Carolina University. Written for this text and used by permission.

**Antinatalism** a perspective against children.

Some people simply do not like children. Aspects of our society reflect **antinatalism** (a perspective against children). Indeed, there is a continuous fight for corporations to implement or enforce any family policies (from family leaves to flex time to on-site day care). Profit and money—not children—are priorities. In addition, although people are generally tolerant of their own children, they often exhibit antinatalistic behavior in reference to the children of others. Notice the unwillingness of some individuals to sit next to a child on an airplane. AsiaAir has responded to this need by offering the first 12 rows in coach only for individuals ages 12 and above.

Laura Scott (2009), a childfree wife, set up the Childless by Choice Project and surveyed 171 childfree adults (ages 22 to 66, 71% female, 29% male) to identify their motivations for not having children. The categories of her respondents and the percentage of each follow:

Early articulators (66%)—these adults knew early that they did not want children

Postponers (22%)—adults who kept delaying when they would have children and remained childless

Acquiescers (8%)—those who made the decision to remain childless because their partner did not want children

Undecided (4%)—those who are childless but still in the decision-making process

## SELF-ASSESSMENT | Childfree Lifestyle Scale

The purpose of this scale is to assess your attitudes toward having a childfree lifestyle. After reading each statement, select the number that best reflects your answer, using the following scale:

| 1 | 2 | 3 | 4 | 5 | 6 | 7 |
|---|---|---|---|---|---|---|
| Strongly Disagree | | | | | | Strongly Agree |

_____ 1. I do not like children.
_____ 2. I would resent having to spend all my money on kids.
_____ 3. I would rather enjoy my personal freedom than have it taken away by having children.
_____ 4. I would rather focus on my career than have children.
_____ 5. Children are a burden.
_____ 6. I have no desire to be a parent.
_____ 7. I am too "into me" to become a parent.
_____ 8. I lack the nurturing skills to be a parent.
_____ 9. I have no patience for children.
_____ 10. Raising a child is too much work.
_____ 11. A marriage without children is empty.
_____ 12. Children are vital to a good marriage.
_____ 13. You can't really be fulfilled as a couple unless you have children.
_____ 14. Having children gives meaning to a couple's marriage.
_____ 15. The happiest couples that I know have children.
_____ 16. The biggest mistake couples make is deciding not to have children.
_____ 17. Child-free couples are sad couples.
_____ 18. Becoming a parent enhances the intimacy between spouses.
_____ 19. A house without the "pitter patter" of little feet is not a home.
_____ 20. Having a child means your marriage is successful.

### Scoring

Reverse score items 11 through 20. For example if you wrote a 1 for item 20, change this to a 7. If you wrote a 2, change it to a 6, etc. Add the numbers. The higher the score (140 is the highest possible score), the greater the value for a childfree lifestyle. The lower the score (20 is the lowest possible score), the less the desire to have a childfree lifestyle. The midpoint between the extremes is 80. Scores below 80 suggest a preference for having children; scores above 80 suggest a desire for a childfree lifestyle. The average score of 52 male and 138 female undergraduates at Valdosta State University was below the midpoint (M = 68.78, SD = 17.06), suggesting a desire for a lifestyle with children. A significant difference was found between males, who scored 72.94 (*SD* = 16.82), and females, who scored 67.21 (*SD* = 16.95), suggesting that males are more approving of a childfree lifestyle. There were no significant differences between whites and blacks or between students in different ranks (freshmen, sophomore, junior, or senior).

### Source

"The Childfree Lifestyle Scale" 2010 by Mark A. Whatley, Ph.D., Department of Psychology, Valdosta State University, Valdosta, GA 31698-0100. Used by permission. Other uses of this scale only by written permission of Dr. Whatley (mwhatley@valdosta.edu). Information on the reliability and validity of this scale is available from Dr. Whatley.

The top five reasons her respondents gave for wanting to remain childfree were (Scott 2009):

**1.** Current life/relationship satisfaction was great and they feared that parenthood would only detract from it.

**2.** Freedom and independence were strong values that they feared would be affected by children.

**3.** Avoidance of the responsibility for rearing a child.

**4.** No maternal/paternal instinct.

**5.** Accomplishing career and travel goals would be difficult as a parent.

Some of Scott's interviewees also had an aversion to children, having had a bad childhood or concerns about childbirth. But other individuals love children, enjoy them, and benefit from having them. Hoffnung and Williams (2013) analyzed data on 200 women and found that being a mother was associated with higher life satisfaction than being childfree.

*Having children is like living in a frat house—nobody sleeps, everything's broken, and there's a lot of throwing up.*

RAY ROMANO, COMEDIAN

## One Child?

Only 3% of adults view one child as the ideal family size (Sandler 2010). Those who have only one child may do so because they want the experience of parenthood without children markedly interfering with their lifestyle and careers. Still others have an only child because of the difficulty in pregnancy or

> **DIVERSITY IN OTHER COUNTRIES**
>
> China has a One Child Policy, which has led to forced abortions, sterilizations, and economic penalties for having more than one child. National visibility for this policy came in 2012 when Feng Jianmei could not pay the $6,300 fine for a second child and was forced to have an abortion to terminate her seven-month pregnancy. She was literally tied up, forced into a vehicle, and dragged to the hospital where she was given an injection to end the pregnancy (MacLeod 2012). Feng et al. (2013) reviewed China's one-child policy and considered it an example of "extreme state intervention in reproduction and a serious and harmful policy mistake."

birthing the child. One mother said, "I threw up every day for nine months including on the delivery table. Once is enough for me." There are also those who have only one child because they can't get pregnant a second time.

## Two Children?

The most preferred family size in the United States (for non-Hispanic white women) is the two-child family (1.9 to be exact!). Reasons for this preference include feeling that a family is "not complete" without two children, having a companion for the first child, having a child of each sex, and repeating the positive experience of parenthood enjoyed with their first child. Some couples may not want to "put all their eggs in one basket." They may fear that, if they have only one child and that child dies, they will not have another opportunity to enjoy parenting (unless they have a second child).

*It is not altogether remarkable that parents may have one child, if only in error or because of confused expectations of bliss. What is truly remarkable is that most parents have more than one child.*

LIONEL TIGER, ANTHROPOLOGIST

## Three Children?

Couples are more likely to have a third child, and to do so quickly, if they already have two girls rather than two boys. They are least likely to bear a third child if they already have a boy and a girl. One male said, "I have 12 older sisters . . . my parents kept having children till they had me."

Having a third child creates a middle child. This child is sometimes neglected because parents of three children may focus more on the baby and the firstborn than on the child in between. However, an advantage to being a middle child is the chance to experience both a younger and an older sibling. Each additional child also has a negative effect on the existing children by reducing the amount of parental time available to the other children. The economic resources available for each child are also affected by each subsequent child.

Most couples have two children and stop. Sometimes a third child is a surprise.

David Knox

## Four Children

More than three children are often born to parents who are immersed in a religion which encourages procreation. Catholics, Mormons, and ultra Orthodox Jews typically have more children than Protestants. When religion is not a factor, **competitive birthing** may be operative whereby individuals have the same number (or more) of children as their peers.

The addition of each subsequent child dramatically increases the possible relationships in the family. For example, in a one-child family, four interpersonal relationships are possible: mother-father, mother-child, father-child, and father-mother-child. In a family of four, eleven relationships are possible.

*I wanted to have a child at every window, waving to me when I came home. I got my wish.*

CHICO MARX, FAMOUS MARX BROTHERS

**Competitive birthing** individuals have the same number (or more) of children as their peers.

## Contraception

To achieve the desired family size and to prevent unwanted pregnancies, a good beginning is to be sober, plan whether or not sexual intercourse will be part of your relationship, and discuss contraception. Having a conversation about birth control is a good way to begin sharing responsibility for it; one can learn of the partner's interest in participating in the choice and use of a contraceptive method (see Table 11.1). Men can also share responsibility by purchasing and using condoms, paying for medical visits and the pharmacy bill, reminding their partner to use the method, assisting with insertion of barrier methods, checking contraceptive supplies, and having a vasectomy if that is what the couple decide on. The partners should also remember that contraception provides no protection against STIs and to use condoms/dental dams.

*Happiness is having a large, loving, caring, close-knit family in another city.*

GEORGE BURNS, COMEDIAN

**TABLE 11.1 Methods of Contraception and Sexually Transmitted Infections Protection**

| Method | Typical Use[1] Effectiveness Rates | STI Protection | Benefits | Disadvantages | Cost[2] |
|---|---|---|---|---|---|
| Oral contraceptive (The Pill) | 92% | No | High effectiveness rate, 24-hour protection, and menstrual regulation | Daily administration, side effects possible, medication interactions | $10–42 per month |
| Nexplanon/Implanon NXT® or Implanon® (3-year implant) | 99.95% | No | High effectiveness rate, long-term protection | Side effects possible, menstrual changes | $400–600 insertion |
| Depo-Provera® (3-month injection) or Depo-subQ Provera 104® | 97% | No | High effectiveness rate, long-term protection | Decreases body calcium, not recommended for use longer than 2 years for most users, side effects likely | $45–75 per injection |
| Ortho Evra® (transdermal patch) | 92% | No | Same as oral contraceptives except use is weekly, not daily | Patch changed weekly, side effects possible | $15–32 per month |
| NuvaRing® (vaginal ring) | 92% | No | Same as oral contraceptives except use is monthly, not daily | Must be comfortable with body for insertion | $15–48 per month |
| Male condom | 85% | Yes | Few or no side effects, easy to purchase and use | Can interrupt spontaneity | $2–10 a box |
| Female condom | 79% | Yes | Few or no side effects, easy to purchase | Decreased sensation and insertion takes practice | $4–10 a box |

*Continued*

**TABLE 11.1 (Continued)**

| Method | Typical Use[1] Effectiveness Rates | STI Protection | Benefits | Disadvantages | Cost[2] |
|---|---|---|---|---|---|
| Spermicide | 71% | No | Many forms to choose, easy to purchase and use | Can cause irritation, can be messy | $8–18 per box/tube/can |
| Today® Sponge[3] | 68–84% | No | Few side effects, effective for 24 hours after insertion | Spermicide irritation possible | $3–5 per sponge |
| Diaphragm & Cervical cap[3] | 68–84% | No | Few side effects, can be inserted within 2 hours before intercourse | Can be messy, increased risk of vaginal/UTI infections | $50–200 plus spermicide |
| Intrauterine device (IUD): *Paraguard or Mirena* | 98.2%–99% | No | Little maintenance, longer term protection | Risk of PID increased, chance of expulsion | $150–300 |
| Withdrawal | 73% | No | Requires little planning, always available | Pre-ejaculatory fluid can contain sperm | |
| Periodic abstinence | 75% | No | No side effects, accepted in all religions/cultures | Requires a lot of planning, need ability to interpret fertility signs | $0 |
| Emergency contraception | 75% | No | Provides an option after intercourse has occurred | Must be taken within 72 hours, side effects likely | $10–32 |
| Abstinence | 100% | Yes | No risk of pregnancy or STIs | Partners both have to agree to abstain | |

[1]Effectiveness rates are listed as percentages of women not experiencing an unintended pregnancy during the first year of typical use. Typical use refers to use under real-life conditions. Perfect use effectiveness rates are higher.

[2]Costs may vary. The Affordable Care Act, health care legislation passed by Congress and signed into law by President Obama on March 23, 2010, requires health insurance plans to cover preventive services and eliminate cost sharing for some services, including "All Food and Drug Administration approved contraceptive methods, sterilization procedures, and patient education and counseling for all women with reproductive capacity." (U.S. Department of Health and Human Services Health Resources and Services Administration, Women's Preventive Services Guidelines, http://www.hrsa.gov/womensguidelines/ January 9, 2014.)

[3]Lower percentages apply to women who have already given birth. Higher rates apply to nulliparous women (women who have never given birth).

Source: Developed and used by permission of Beth Credle Burt, MAEd, CHES, a health education specialist. Education Services Project Manager, Siemens Healthcare.

## Emergency Contraception

**Emergency contraception** various types of combined estrogen-progesterone morning-after pills or post-coital IUD insertion.

**Emergency contraception** (also called **postcoital contraception**) refers to various types of combined estrogen-progesterone morning-after pills or post-coital IUD insertion used primarily in three circumstances: when a woman has unprotected intercourse, when a contraceptive method fails (such as condom breakage, which occurs 7% of the time, or slippage), and when a woman is raped. "Better safe than sorry" requires immediate action because the sooner the EC pills are taken, the lower the risk of pregnancy—twelve hours is best, and 72 hours at the latest (Tal et al., 2014).

While emergency contraception medication is available over the counter—no prescription is necessary (and no pregnancy test is required) for women age 17 and above—some parents feel their parental rights are being undermined. For females under age 17, a prescription is required. Although side effects (nausea, vomiting, and so on) may occur, they are typically over in a couple of days and the risk of being pregnant is minimal.

## PERSONAL CHOICES

### *Is Genetic Testing for You?*

Because each of us may have flawed genes that carry increased risk for diseases such as cancer and Alzheimer's, the question of whether to have a genetic test before becoming pregnant becomes relevant. The test involves having a blood sample drawn. About 900 genetic tests are available. The advantage is the knowledge of what defective genes you may have and what diseases you may pass to your children. McKee and Blow (2010) emphasized that genetic testing is a collective decision since it impacts one's partner and parents and others might be consulted.

The disadvantage is stress or anxiety (for example, what do you do with the information?) as well as discrimination from certain health insurance companies (who may deny coverage). The validity of the test is also problematic as some tests may provide inaccurate information. And, because no treatment may be available if the test results are positive, the knowledge that they might pass diseases to their children can be unsettling to a couple. However, Aspinwall et al. (2013) studied individuals who had a family history of melanoma and pancreatic cancer and who completed genetic testing. The researchers noted "benefits to both carriers and noncarriers without inducing distress in general or worry about melanoma or pancreatic cancer." Genetic testing is expensive, ranging from $200 to $2,400. The National Society of Genetic Counselors (http://www.nsgc.org) offers information about this testing.

### Sex Selection

Some couples use sex selection technologies to help ensure a boy or a girl. MicroSort is the new preconception sperm-sorting technology, which allows parents to increase the chance of having a girl or a boy baby. The procedure is also called "family balancing" since couples that already have several children of one sex often use it. The eggs of a woman are fertilized and the sex of the embryos three to eight days old is identified. Only embryos of the desired sex are then implanted in the woman's uterus.

Puri and Nachtigall (2010) examined ethical considerations in regard to sex selection. One perspective held by sex-selection technology providers argues that sex selection is an expression of reproductive rights and a sign of female empowerment to prevent unwanted pregnancies and abortions. A contrasting view is that sex selection contributes to gender stereotypes that could result in neglect of children of the lesser-desired sex.

# Teenage Motherhood

### National **Data**

Forty-one percent of all births are to unmarried women, most of whom are teenagers. Seventy-two percent of these births are to black women; 28% to white women (Pew Research Center 2010b).

Reasons for teenagers having a child include not using contraception, having limited parental supervision, and perceiving few alternatives to parenthood. Indeed, motherhood may be one of the only remaining meaningful roles available to them. Some teenagers feel lonely and unloved and have a baby to create

**DIVERSITY IN OTHER COUNTRIES**

Teens in countries such as France and Germany are as sexually active as U.S. teenagers, but the teenage birth rate in these countries is very low. Teens in France and Germany grow up in a society that promotes responsible contraceptive use. In the Netherlands, for example, individuals are taught to use both the pill and the condom (an approach called **double Dutch**) for prevention of pregnancy and sexually transmitted infections (STIs).

**Double dutch** using both the pill and the condom for prevention of pregnancy and sexually transmitted infections (STIs).

a sense of being needed and wanted. Teenage moms may also have had a teenager for a mother, thus making having a child as a teenager normative.

### Problems Associated with Teenage Motherhood

Teenage parenthood is associated with various negative consequences, including the following:

1. ***Stigmatization and marginalization.*** Because teen mothers resist the typical life trajectory of their middle-class peers, they are often stigmatized and marginalized. Not only are they more likely to drop out of school, they are also more likely to be economic burdens to the larger society in terms of welfare payments.

2. ***Poor health habits.*** Teenage unmarried mothers are less likely to seek prenatal care and are more likely than older women and married women to smoke, drink alcohol, and take other drugs. These factors have an adverse effect on the health of the baby. Indeed, babies born to unmarried teenage mothers are more likely to have low birth weights (less than five pounds, five ounces) and to be born prematurely. Children of teenage unmarried mothers are also more likely to be developmentally delayed. These outcomes are largely a result of the association between teenage unmarried childbearing and persistent poverty.

3. ***Lower academic achievement/income/satisfaction for child.*** Children born to teen mothers are more likely to have lower educational achievement, income, and life satisfaction (Lipman et al. 2011).

4. ***Compromised personal health and psychosocial adjustment.*** Teen mothers, when compared to older mothers, are more likely to have poorer general health, be more depressed, and have lower self-esteem. Patel and Sen (2012) also found that teen mothers had poorer physical health outcomes than sexually active teenagers who did not get pregnant. However, teen mothers are less likely to drink alcohol than their same age counterparts (Amato and Kane 2011). The latter emphasizes a positive outcome of teenage motherhood often ignored.

5. ***Lower sense of adequacy in role of mother.*** Uzun et al. (2013) compared 100 adolescent mothers (average age 17.8) with 100 older mothers (average age 26) and found the former more likely to feel inadequate in the role of mother.

6. ***Child absent from father.*** In only 20% of the cases does the biological father live with the teenage mother and their child. When this occurs, the child seems to benefit. Martin et al. (2013) studied 509 children born to fathers who lived with the teenage mother and found that they (the children) were more likely to be securely attached to their mother and had fewer externalizing problems than other children at age three.

**Infertility** the inability to achieve a pregnancy after at least one year of regular sexual relations without birth control, or the inability to carry a pregnancy to a live birth.

*He'd sensed the strength she'd called on to haul her sexuality out from under the weight of infertility. In his experience childlessness in women either warped into a dedication to self-hating sexual expertise or formed a subsonic noise of sadness and loss.*

GLEN DUNCAN, *A DAY AND A NIGHT AND A DAY: A NOVEL*

## Infertility

**Infertility** is the inability to achieve a pregnancy after at least one year of regular sexual relations without birth control, or the inability to carry a pregnancy to a live birth. Different types of infertility include the following:

1. ***Primary infertility.*** The woman has never conceived even though she wants to and has had regular sexual relations for the past twelve months.

2. ***Secondary infertility.*** The woman has previously conceived but is currently unable to do so even though she wants to and has had regular sexual relations for the past twelve months.

3. ***Pregnancy wastage.*** The woman has been able to conceive but has been unable to produce a live birth.

Shapiro (2012) emphasized the importance of conceiving in one's twenties rather than delaying pregnancy until ones thirties or forties. The chance of conceiving per month in one's thirties is 20%; in one's forties, it is 10%. Dougall et al. (2013) interviewed women who delayed getting pregnant until age 40 or later—44% reported that they were "shocked" to learn that the difficulty in getting pregnant declines so steeply as a woman moves into her mid to late thirties. Schmidt et al. (2012) also noted that delaying pregnancy until 35 and beyond is associated with higher risk of preterm births and stillbirths.

## Causes of Infertility

Although popular usage does not differentiate between the terms *fertilization* and the *beginning of pregnancy*, fertilization or **conception** refers to the fusion of the egg and sperm, whereas **pregnancy** is not considered to begin until five to seven days later, when the fertilized egg is implanted (typically in the uterine wall). Hence, not all fertilizations result in a pregnancy. An estimated 30% to 40% of conceptions are lost prior to or during implantation. Forty percent of infertility problems are attributed to the woman, 40% to the man, and 20% to both of them. Some of the more common causes of infertility in men include low sperm production, poor semen motility, effects of STIs (such as chlamydia, gonorrhea, and syphilis), and interference with passage of sperm through the genital ducts due to an enlarged prostate. Additionally, there is some association between high body mass index in men and sperm that is problematic in impregnating a female (Sandlow 2013).

Infertility in the woman is related to her age, not having been pregnant before, blocked fallopian tubes, endocrine imbalance that prevents ovulation, dysfunctional ovaries, chemically hostile cervical mucus that may kill sperm, and effects of STIs (Van Geloven et al. 2013). Obesity in the woman is also related to her infertility. Frisco and Weden (2013) analyzed national data on 1,658 females at two different time periods and found that young women who were obese at baseline had higher odds of remaining childless and increased odds of underachieving fertility intentions than young women who were normal weight at baseline. The researchers concluded that obesity has long-term ramifications for women's childbearing experiences with respect to whether and how many children women have in general relative to the number of children they want. Brandes et al. (2011) noted that unexplained infertility is one of the most common diagnoses in fertility care and is associated with a high probability of achieving a pregnancy, most spontaneously.

Difficulty conceiving and carrying a fetus to term is strenuous for both individuals and couples. Teskereci and Oncel (2013) reported data on 200 couples undergoing infertility treatment and found that the quality of life of the women in the study was lower than that of the men. Advanced age, low education level, unemployment status, lower income, and long duration of infertility were also associated with lower quality of life. Baldur-Felskov et al. (2013) studied a large sample of Danish women who were infertile and found a higher incidence of hospitalizations due to psychiatric disorders. Pritchard and Kort-Butler (2012) found involuntarily childless women less happy than biological or adoptive mothers.

An at-home fertility kit, Fertell, allows women to measure the level of their follicle-stimulating hormone on the third day of their menstrual cycles. An abnormally high level means that egg quality is low. The test takes 30 minutes and involves a urine stick. The same kit allows men to measure the concentration of motile sperm. Men provide a sample of sperm (e.g., masturbation) that swim through a solution similar to cervical mucus. This procedure takes about 80 minutes. Fertell has been approved by the Food and Drug Administration (FDA;, no prescription is necessary and the cost for the kit is about $100).

*If you want to know what's in motherhood for you, as a woman, then—in truth—it's nothing you couldn't get from, say, reading the 100 greatest books in human history; learning a foreign language well enough to argue in it; climbing hills; loving recklessly; sitting quietly, alone, in the dawn; drinking whisky with revolutionaries; learning to do close-hand magic; swimming in a river in winter; growing foxgloves, peas and roses; calling your mum; singing while you walk; being polite; and always, always helping strangers. No one has ever claimed for a moment that childless men have missed out on a vital aspect of their existence, and were the poorer, and crippled by it.*

CAITLIN MORAN, *HOW TO BE A WOMAN*

**Conception** the fusion of the egg and sperm.

**Pregnancy** five to seven days after conception, when the fertilized egg is implanted, typically in the uterine wall.

WHAT IF?

## What If Your Partner Is Infertile—Would You Marry the Partner?

Being infertile (for the woman) may have a negative lifetime effect. Wirtberg et al. (2007) interviewed 14 Swedish women 20 years after their infertility treatment and found that childlessness had had a major impact on all the women's lives and remained a major life theme. The effects were both personal (sad) and interpersonal (half were separated and all reported negative effects on their sex lives). The effects of childlessness were especially increased at the time the study was conducted, as the women's peer group was entering the grandparent phase. The researchers noted that infertility has lifetime consequences for the individual woman and her relationships. One can feel the emotional pain of Celine Dion, who at age 41 revealed her fourth failed IVF attempt to get pregnant—"I'm going to try till it works." It worked on the sixth try–with twins.

## Assisted Reproductive Technology (ART)

### International **Data**

Worldwide about five million babies are born annually in developed countries with the assistance of medical interventions (ART). Between 1% and 4% of all births in developed countries are conceived by in vitro fertilization (Kondapalli and Perales-Puchalt 2013).

A number of technological innovations are available to assist women and couples in becoming pregnant. These include hormonal therapy, artificial insemination, ovum transfer, in vitro fertilization, gamete intrafallopian transfer, and zygote intrafallopian transfer.

**Hormone Therapy** Drug therapies are often used to treat hormonal imbalances, induce ovulation, and correct problems in the luteal phase of the menstrual cycle. Frequently used drugs include Clomid, Pergonal, and human chorionic gonadotropin (HCG), a hormone extracted from human placenta. These drugs stimulate the ovary to ripen and release an egg. Although they are fairly effective in stimulating ovulation, hyperstimulation can occur, which may result in permanent damage to the ovaries.

Hormone therapy also increases the likelihood that multiple eggs will be released, resulting in multiple births. The increase in triplets and higher order multiple births over the past decade in the United States is largely attributed to the increased use of ovulation-inducing drugs for treating infertility. Infants of higher order multiple births are at greater risk of having low birth weight and their mortality rates are higher. Mortality rates have improved for these babies, but these low birth weight survivors may need extensive neonatal medical and social services.

**Artificial Insemination** When the sperm of the male partner are low in count or motility, sperm from several ejaculations may be pooled and placed directly into the cervix. This procedure is known as *artificial insemination by husband* (AIH). When sperm from someone other than the woman's partner are used to fertilize a woman, the technique is referred to as *artificial insemination by donor* (AID).

Women who want to become pregnant obtain sperm from a friend or from a sperm bank. Regardless of the source of the sperm, it should be screened for genetic abnormalities and STIs, quarantined for 180 days, and retested for human immunodeficiency virus (HIV); also, the donor should be younger than 50 to diminish hazards related to aging. These precautions are not routinely taken—let the buyer beware.

**Artificial Insemination of a Surrogate Mother** In some instances, artificial insemination does not help a woman get pregnant. (Her fallopian tubes may be blocked, or her cervical mucus may be hostile to sperm.) The couple that still wants a child and has decided against adoption may consider parenthood through a surrogate mother. There are two types of surrogate mothers. One is the contracted surrogate mother who supplies the egg, is impregnated with the male partner's sperm, carries the child to term, and gives the baby to the man and his partner. Alternatively, donor sperm may also be used to fertilize the surrogate mother. A second type is the surrogate mother who carries to term a baby to whom she is not genetically related. Sarah Jessica Parker and Matthew Broderick contracted with a surrogate mother to carry their embryo (they provided both sperm and egg) to term; they now have twin daughters. Or a donor egg can be fertilized by donor sperm and the embryo can be implanted in the surrogate mother's uterus.

Although some American women are willing to "rent their wombs," women in India also provide this service. For $5,000, an Indian wife who already has a child will carry a baby to term for an infertile couple (for a fraction of the cost of an American surrogate). *Google Baby* is the name of a documentary showing how infertile couples with a credit card can submit an order for a baby—the firm will find donor egg and donor sperm, fertilize the egg, and implant it in a surrogate mother in India. The couple need only fly to India, pick up their baby, and return to the States.

When a donor sperm is used, the question of the importance of a father in the life of a child is brought into question. The provision of sperm from an anonymous donor allows the father to walk away with no financial obligations to the child. Countries such as Britain, Sweden, Norway, and Switzerland have banned anonymity in sperm donation, which sends the message that fathers are important.

Some data are available on social fathers of children who were conceived via donor sperm to their mothers. Casey et al. (2013) examined the psychological well-being of fathers and father–child relationships in families with a seven-year-old child conceived by donor insemination. These relationships were compared with 25 egg donations and 32 unassisted-conception families. Findings revealed the fathers whose child was conceived with donor sperm reported psychological well-being, parenting, and absence of conflict equal to the control groups. The researchers concluded that commitment to parenthood may be more important than genetic relatedness for positive father-child relationships.

California is one of 12 states in which entering into an arrangement with a surrogate mother is legal. The fee to the surrogate mother is around $30,000. Other fees (travel, hospital, lawyers, and so on) can add another $70,000. Surrogate mothers typically have their own children which may make giving up a child that they carry easier.

**In Vitro Fertilization** About 2 million couples cannot have a baby because the woman's fallopian tubes are blocked or damaged, preventing the passage of eggs to the uterus. In some cases, blocked tubes can be opened via laser surgery or by inflating a tiny balloon within the clogged passage. When these procedures are not successful (or when the woman decides to avoid invasive tests and exploratory surgery), *in vitro* (meaning "in glass") *fertilization* (IVF), also known as test-tube fertilization, is an alternative.

Using a laparoscope (a narrow, telescope-like instrument inserted through an incision just below the woman's naval to view tubes and ovaries), the physician is able to see a mature egg as it is released from the woman's ovary. The time of release can be predicted accurately within two hours. When the egg emerges, the physician uses an aspirator to remove the egg, placing it in a small tube containing stabilizing fluid. The egg is taken to the laboratory, put in a culture petri dish, kept at a certain temperature and acidity level, and surrounded by sperm from the woman's partner (or donor). After one of these sperm fertilizes the egg, the egg divides and is implanted by the physician in the wall of the woman's uterus. Usually, several eggs are implanted in the hope one will survive. Sometimes, several eggs survive as in the case of Nadya Suleman, who ended up giving birth to 8 babies.

**Cryopreservation** fertilized eggs are frozen and implanted at a later time, if necessary.

Occasionally, some fertilized eggs are frozen and implanted at a later time, if necessary. This procedure is known as **cryopreservation**. Separated or divorced couples may disagree over who owns the frozen embryos, and the legal system is still wrestling with the fate of their unused embryos, sperm, or ova after a divorce or death.

**Ovum transfer** an egg is donated, fertilized in vitro with the husband's sperm, and then transferred to his wife.

**Ovum Transfer** In conjunction with in vitro fertilization is **ovum transfer**, also referred to as embryo transfer. In this procedure, an egg is donated, fertilized in vitro with the husband's sperm, and then transferred to his wife. Alternatively, a physician places the sperm of the male partner in a surrogate woman. After about five days, her uterus is flushed out (endometrial lavage), and the contents are analyzed under a microscope to identify the presence of a fertilized ovum.

The fertilized ovum is then inserted into the uterus of the otherwise infertile partner. Although the embryo can also be frozen and implanted at another time, fresh embryos are more likely to result in successful implantation. Infertile couples that opt for ovum transfer do so because the baby will be biologically related to at least one of them (the father) and the partner will have the experience of pregnancy and childbirth.

**Success Using Reproductive Technologies** The cost of treating infertility is enormous. Katz et al. (2011) examined the costs for 398 women in eight infertility practices over an 18-month period. For the half who pursued IVF the median per-person medication costs ranged from $1,182 for medications only to $24,373 and $38,015 for IVF and IVF–donor egg groups, respectively. In regard to the costs of successful outcomes (delivery or ongoing pregnancy by 18 months) for IVF—$61,377. Within the time frame of the study, costs were not significantly different for women whose outcomes were successful and women whose outcomes were not. Only 28% of couples who invest in a fertility clinic end up with a live birth (Lee 2006). The sooner an infertility problem is suspected the more successful the intervention.

Researchers continue to investigate the reason for low birth weight in babies which result from ART. It is not known if such is due to ART procedures or to the fact of infertility itself (Kondapalli and Perales-Puchalt 2013).

# Adoption

**National Data**

Adoption by U.S. parents is rare. Just over 1% of 18- to 44-year-old women reported having adopted a child (Smock and Greenland 2010).

Angelina Jolie and Brad Pitt are celebrities who have given national visibility to adopting children. They are not alone in their desire to adopt children.

The various routes to adoption are public (children from the child welfare system), private agency (children placed with nonrelatives through agencies), independent adoption (children placed directly by birth parents or through an intermediary such as a physician or attorney), kinship (children placed in a family member's home), and stepparent (children adopted by a spouse). Motives for adopting a child include an inability to have a biological child (infertility), a desire to give an otherwise unwanted child a permanent loving home, or a desire to avoid contributing to overpopulation by having more biological children. Some couples may seek adoption for all of these motives. Fifteen percent of adoptions will be children from other countries.

### Characteristics of Children Available for Adoption

Adoptees in the highest demand are healthy, white infants. Those who are older than three, of a racial or ethnic group different from that of the adoptive parents, of a sibling group, or with physical or developmental disabilities, have been difficult to place. Sankar (2012) noted that only 1% and 2% of 200 individuals reported a preference for a child ages 6 to 10 and ages 11 to 14, respectively. She emphasized the importance of providing opportunities for potential parents to learn from adults who had been adopted as an older child and to provide competency parental training for parenting older children.

### Children Who Are Adopted

Children who are adopted have an enormous advantage over those who are not adopted. Juffer et al. (2011) examined 270 research articles including more than 230,000 children to compare the physical growth, attachment, cognitive development, school achievement, self-esteem, and behavioral problems of adopted and non-adopted children. Results revealed that adopted children out-performed their non-adopted peers who remained in institutions and they showed a dramatic recovery in practically all areas of development.

"Who are your real parents?", "Why did your mother give you up?", and "Are those your real parents?" are questions children who are adopted must sometimes cope with. W.I.S.E. UP is a tool provided to adopted children to help them cope with these intrusive, sometimes uncomfortable questions (Singer 2010). W.I.S.E. is an acronym for: **W**alk away, **I**gnore or change the subject, **S**hare what you are comfortable sharing, and **E**ducate about adoption in general. The tool emphasizes that adopted children are wiser about adoption than their peers and can educate them or remove themselves from the situation.

### Costs of Adoption

Adopting from the U.S. foster care system is generally the least expensive type of adoption, usually involving little or no cost, and states often provide subsidies to adoptive parents. However, before the adoption process begins, a couple who are foster care parents to a child and who become emotionally bonded with the child, risk that the birth parents might reappear and request their child back.

**National Data**

There are almost half a million children in foster care (Stinhart et al. 2012).

Stepparent and kinship adoptions are also less costly and have less risk of the child being withdrawn. Agency and private adoptions can range from $5,000 to $40,000 or more, depending on travel expenses, birth mother expenses, and requirements in the state. International adoptions can range from $7,000 to $30,000 (see http://costs.adoption.com/).

## Transracial Adoption

**Transracial adoption** adopting a child of a race different from that of the parents.

**Transracial adoption** is defined as the practice of adopting children of a race different from that of the parents—for example, a white couple adopting a Korean or an African American child. The actress Sandra Bullock adopted a child of another race (African American).

**National Data**

Annually, 50,000 children are adopted transracially (Malott et al. 2012).

Transracial adoptions are controversial. The motivations of persons wanting to adopt cross racially is sometimes questioned...are they trying to make a political statement? Another controversy is whether it is beneficial for children to be adopted by parents of the same racial background. In regard to the adoption of African-American children by same-race parents, the National Association of Black Social Workers (NABSW) passed a resolution against transracial adoptions, citing that such adoptions prevented black children from developing a positive sense of themselves "that would be necessary to cope with racism and prejudice that would eventually occur" (Hollingsworth 1997, p. 44).

The counterargument is that healthy self-concepts, an appreciation for one's racial heritage, and coping with racism or prejudice can be learned in a variety of contexts. Legal restrictions on transracial adoptions have disappeared, and social approval for transracial adoptions is increasing.

## Open versus Closed Adoptions

Another controversy is whether adopted children should be allowed to obtain information about their biological parents. In general, there are considerable benefits for having an open adoption, especially the opportunity for the biological parent to stay involved in the child's life. Adoptees learn early that they are adopted and who their biological parents are. Birth parents are more likely to avoid regret and to be able to stay in contact with their child. Adoptive parents have information about the genetic background of their adopted child. Goldberg et al. (2011) studied lesbian, gay, and heterosexual couples who were involved in an open adoption. While there were some tensions with the birth parents over time, most of the 45 adoptive couples reported satisfying relationships.

## Internet Adoption

Some couples use the Internet to adopt a baby. The Donaldson Adoption Institute surveyed 2,000 adoptees, adoptive parents, birth parents and adoption professionals. The survey revealed the Internet's value of being able to make connections and provide information for parents seeking to adopt (Healy, 2013). But Roby and White (2010) noted the Internet's use to adopt a baby can pose serious problems of potential fraud, exploitation, and, most important, lack of professional consideration of the child's best interest. Couples should proceed with great caution. Policy makers should also be aware of the practice of "re-homing" where parents who have adopted a child use the Internet to place unwanted adopted children in new families. There is no monitoring or regulation of this practice (Healy 2013).

## Foster Parenting

**Foster parent** neither a biological nor an adoptive parent but a person who takes care of and fosters a child taken into custody.

Some individuals seek the role of parent via foster parenting. A **foster parent**, also known as a family caregiver, is neither a biological nor an adoptive parent but is one who takes care of and fosters a child taken into custody. Foster care may be temporary or long term. A foster parent has made a contract with the state

David Knox

This couple are foster parents who decided to adopt their foster child and make him a permanent family member.

for the service, has judicial status, and is reimbursed by the state. Foster parents are screened for previous arrest records and child abuse and neglect. They are licensed by the state; some states require a "foster parent orientation" program.

Children placed in foster care have typically been removed from parents who are abusive, who are substance abusers, and/or who are mentally incompetent. Parents who are incarcerated or have mental health problems are more likely to have their parental rights terminated (Meyer et al. 2010). Although foster parents are paid for taking care of children in their home, they may also be motivated by their love of children. The goal of placing children in foster care is to improve their living conditions and then either return them to their family of origin or find a more permanent adoptive or foster home for them. Some couples become foster parents in the hope of being able to adopt a child who is placed in their custody.

Due to longer delays for foreign adoptions (e.g., it typically takes three years to complete a foreign adoption) and less availability of American infants, increasingly, couples are adopting a foster child. Waterman et al. (2011) studied behavior problems and parenting stress for children adopted from foster care homes and found that parents reported high adoption satisfaction despite ongoing behavioral problems with a third of the children in the sample.

# Sterilization

## National **Data**

Worldwide over 220 million couples have used sterilization as their method of birth control—nearly 43 million men and 180 million women. The lifetime chance of a pregnancy after sterilization is one in 200 (Beerthuizen 2010).

**Sterilization** is a permanent surgical procedure that prevents reproduction. Sterilization may be the contraceptive method of choice when the woman should not have more children for health reasons or when individuals are certain about their desire to have no more children or to remain childfree. Most couples complete their intended childbearing in their late twenties or early thirties, leaving more than fifteen years of continued risk of unwanted pregnancy.

**Sterilization** permanent surgical procedure that prevents reproduction.

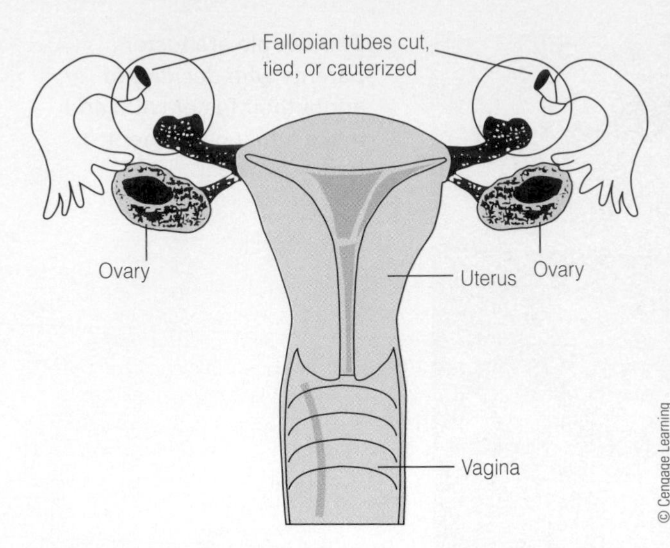


**Figure 11.1** Female Sterilization: Tubal Sterilization.

Because of the risk of pill use at older ages and the lower reliability of alternative birth control methods, sterilization has become the most popular method of contraception among married women who have completed their families.

## Female Sterilization

Although a woman may be sterilized by removal of her ovaries (**oophorectomy**) or uterus (**hysterectomy**), these operations are not normally undertaken for the sole purpose of sterilization because the ovaries produce important hormones (as well as eggs) and because both procedures carry the risks of major surgery. Sometimes, however, another medical problem requires hysterectomy.

The usual procedures of female sterilization are the salpingectomy and a variant of it, the laparoscopy. **Salpingectomy**, also known as tubal ligation or tying the tubes (see Figure 11.1), is performed under a general anesthetic, often while the woman is in the hospital just after she has delivered a baby. An incision is made in the lower abdomen, just above the pubic line, and the fallopian tubes are brought into view one at a time. A part of each tube is cut out, and the ends are tied, clamped, or cauterized (burned). The operation takes about 30 minutes. Smith et al. (2010) studied 3,448 women (ages 16 to 64) to assess if having a tubal ligation was related to the frequency of sexual problems and ratings of sexual satisfaction, relationship satisfaction, and sexual pleasure. Results revealed that having had a tubal ligation was not associated with any specific sexual problem, such as physical pain during sex or an inability to reach orgasm. In fact, sterilized women were more likely to experience extremely high levels of sexual satisfaction, relationship satisfaction, and sexual pleasure.

**Oophorectomy** removal of a woman's ovaries.

**Hysterectomy** removal of a woman's uterus.

**Salpingectomy** tubal ligation or tying the tubes.

A less expensive and quicker (about 15 minutes) form of salpingectomy, which is performed on an outpatient basis, is **laparoscopy**. Often using local anesthesia, the surgeon inserts a small, lighted viewing instrument (laparoscope) through the woman's abdominal wall just below the navel, through which the uterus and the fallopian tubes can be seen. The surgeon then makes another small incision in the lower abdomen and inserts a special pair of forceps that carry electricity to cauterize the tubes. The laparoscope and the forceps are then withdrawn, the small wounds are closed with a single stitch, and small bandages are placed over the closed incisions. (Laparoscopy is also known as "the Band-Aid operation.") As an alternative to reaching the fallopian tubes through an opening below the navel, the surgeon may make a small incision in the back of the vaginal barrel (vaginal tubal ligation).

**Laparoscopy** a form of tubal ligation that involves a small incision through the woman's abdominal wall just below the navel.

Essure is a permanent sterilization procedure that requires no cutting and only a local anesthetic in a half-hour procedure that blocks the fallopian tubes. Women typically may return home within forty-five minutes (and to work the next day). These procedures for female sterilization are greater than 95% effective. Although some female sterilizations may be reversed, a woman should become sterilized only if she does not want to have a biological child.

## Male Sterilization

Vasectomies are the most frequent form of male sterilization. They are usually performed in the physician's office under a local anesthetic. Michielsen and Beerthuizen (2010) reviewed the data and noted that it convincingly demonstrates that **vasectomy** is a safe and cost-effective intervention for permanent male contraception. They recommend the no-scalpel vasectomy under local anesthesia. Sperm are still produced in the testicles, but because there is no tube to the penis, they remain in the epididymis and eventually dissolve.

**Vasectomy** most frequent form of male sterilization.

The procedure takes about 15 minutes. The patient can leave the physician's office within a short time. Because sperm do not disappear from the ejaculate immediately after a vasectomy (some remain in the vas deferens above the severed portion), a couple should use another method of contraception until the man has had about 20 ejaculations. In about 1% of the cases, the vas deferens grows back and the man becomes fertile again. A vasectomy does not affect the man's desire for sex, ability to have an erection or an orgasm, amount of ejaculate (sperm account for only a minute portion of the seminal fluid), health, or chance of prostate cancer. Although a vasectomy may be reversed (with a 30 to 60% success rate), a man should get a vasectomy only if he does not want to have a biological child.

**DIVERSITY IN OTHER COUNTRIES**

There are wide variations in the range of cultural responses to the abortion issue. On one end of the continuum is the Kafir tribe in Central Asia, where an abortion is strictly the choice of the woman. In this preliterate society, there is no taboo or restriction with regard to abortion, and the woman is free to exercise her decision to terminate her pregnancy. One reason for the Kafirs' approval of abortion is that childbirth in the tribe is associated with high rates of maternal mortality. Because birthing children may threaten the life of significant numbers of adult women in the community, women may be encouraged to abort. Such encouragement is particularly strong in the case of women who are viewed as too young, too old, too sick, or too small to bear children.

A tribe or society may also encourage abortion for a number of other reasons, including practicality, economics, lineage, and honor. Abortion is practical for women in migratory societies. Such women must control their pregnancies, because they are limited in the number of children they can nurse and transport. Economic motivations become apparent when resources are scarce—the number of children born to a group must be controlled. Abortion for reasons of lineage or honor involves encouragement of an abortion in those cases in which a woman becomes impregnated in an adulterous relationship. To protect the lineage and honor of her family, the woman may have an abortion.

# Abortion

Abortion remains a controversial issue in America. Among American women, half will have an unintended pregnancy and 30% will have an abortion. About 60% of women having an abortion are in their 20s are unmarried (Guttmacher Institute 2012). An abortion may be either an **induced abortion**, which is the deliberate termination of a pregnancy through chemical or surgical means, or a **spontaneous abortion (miscarriage)**, which is the unintended termination of a pregnancy. Miscarriages often represent a significant loss which is associated with depression/anxiety (Geller et al. 2010) and marital unhappiness (Sugiura-Ogasawara et al. 2013).

In this text we will use the term *abortion* to refer to induced abortion. In general, abortion is legal in the United States but it was challenged under the Bush administration. Specifically, federal funding was withheld if an aid group offered abortion or abortion advice. However, the Obama administration restored governmental approval for abortion. Obama said that denying such aid undermined "safe and effective voluntary family planning in developing countries."

*No woman can call herself free until she can choose consciously whether she will or will not be a mother.*

MARGARET SANGER, EARLY CONTRACEPTION ADVOCATE

**Induced abortion** the deliberate termination of a pregnancy through chemical or surgical means.

**Spontaneous abortion (miscarriage)** the unintended termination of a pregnancy.

## Incidence of Abortion

When this country was founded abortion was legal and accepted by the public up until the time of "quickening"—the moment when a woman can feel the fetus inside her. By the time of the Civil War, one in five pregnancies was terminated by an abortion. Opposition to abortion grew in the 1870s, led by the American Medical Association launching a fierce campaign to make abortion illegal unless authorized and performed by a licensed physician. Abortion was made illegal in 1880, which did not stop the practice; a million abortions were performed illegally by the 1950s. In 1973, the Supreme Court upheld in Roe v. Wade the right of a woman to have a legal abortion. About 1.2 million abortions are performed annually in the United States. Although the number of abortions has been increasing among the poor (lower access to health care and health education), there has been a decrease among higher income women (due to increased acceptability of having a child without a partner and increased use of contraception). Ninety percent of abortions occur within the first three months of pregnancy (Guttmacher Institute 2012).

*I feel about Photoshop the way some people feel about abortion. It is appalling and a tragic reflection on the moral decay of our society . . . unless I need it, in which case, everybody be cool.*

TINA FEY, COMEDIAN

**Abortion rate** number of abortions per 1,000 women ages 15 to 44.

**Abortion ratio** the number of abortions per 1,000 live births.

**Parental consent** a woman needs permission from a parent to get an abortion if she is under a certain age, usually 18.

**Parental notification** a woman has to tell a parent she is getting an abortion if she is under a certain age, usually 18, but she doesn't need parental permission.

The **abortion rate** (the number of abortions per 1,000 women ages 15 to 44) increased 1% between 2005 and 2008, from 19.4 to 19.6 abortions; the total number of abortion providers was virtually unchanged (Jones and Kooistra 2011). About 40% of abortions are repeat abortions (Ames and Norman 2012).

The **abortion ratio** refers to the number of abortions per 1,000 live births. Abortion is affected by the need for parental consent and parental notification. **Parental consent** means that a woman needs permission from a parent to get an abortion if she is under a certain age, usually 18. **Parental notification** means that a woman has to tell a parent she is getting an abortion if she is under a certain age, usually 18, but she doesn't need parental permission. Laws vary by state. Call the National Abortion Federation Hotline at 1-800-772-9100 to find out the laws in your state.

The Self-Assessment provides a way for you to assess your abortion views.

## Reasons for an Abortion

In a survey of 1,209 women who reported having had an abortion, the most frequently cited reasons were that having a child would interfere with a woman's education, work, or ability to care for dependents (74%); that she could not afford a baby now (73%); and that she did not want to be a single mother or

## SELF-ASSESSMENT | Abortion Attitude Scale

This is not a test. There are no wrong or right answers to any of the statements, so answer as honestly as you can. The statements ask your feelings about legal abortion (the voluntary removal of a human fetus from the mother during the first three months of pregnancy by a qualified medical person). Tell how you feel about each statement by giving only one response. Use the following scale for your answers:

| Strongly Agree | Slightly Agree | Disagree | Slightly Disagree | Strongly Disagree |
|---|---|---|---|---|
| 5 | 4 | 3 | 2 | 1 |

_____ 1. The Supreme Court should strike down legal abortions in the United States.

_____ 2. Abortion is a good way of solving an unwanted pregnancy.

_____ 3. A mother should feel obligated to bear a child she has conceived.

_____ 4. Abortion is wrong no matter what the circumstances are.

_____ 5. A fetus is not a person until it can live outside its mother's body.

_____ 6. The decision to have an abortion should be the pregnant mother's.

_____ 7. Every conceived child has the right to be born.

_____ 8. A pregnant female not wanting to have a child should be encouraged to have an abortion.

_____ 9. Abortion should be considered killing a person.

_____ 10. People should not look down on those who choose to have abortions.

_____ 11. Abortion should be an available alternative for unmarried pregnant teenagers.

_____ 12. People should not have the power over the life or death of a fetus.

_____ 13. Unwanted children should not be brought into the world.

_____ 14. A fetus should be considered a person at the moment of conception.

### Scoring and Interpretation

As its name indicates, this scale was developed to measure attitudes toward abortion. Sloan (1983) developed the scale for use with high school and college students. To compute your score, first reverse the point scale for items 1, 3, 4, 7, 9, 12, and 14. For example, if you selected a 5 for item one, this becomes a 1; if you selected a 2, this becomes a 4, etc. After reversing the scores on the seven items specified, add the numbers you circled for all the items. Sloan provided the following categories for interpreting the results:

| | |
|---|---|
| 70–56 | Strong pro-abortion |
| 54–44 | Moderate pro-abortion |
| 43–27 | Unsure |
| 26–16 | Moderate pro-life |
| 15–0 | Strong pro-life |

### Reliability and Validity

The Abortion Attitude Scale was administered to high school and college students, Right to Life group members, and abortion service personnel. Sloan (1983) found that the mean score for Right to Life members was 16.2; the mean score for abortion service personnel was 55.6; and student' scores fell between these values.

### Source

"Abortion Attitude Scale" by L. A Sloan. *Journal of Health Education* Vol. 14, No. 3, May/June 1983. *The Journal of Health Education* is a publication of the American Allegiance for Health, Physical Education, Recreation and Dance, 1900 Association Drive, Reston, VA 20191. Reprinted by permission.

was having relationship problems (48%). Nearly four in ten women said they had completed their childbearing, and almost one-third of the women were not ready to have a child (Finer 2005). Falcon et al. (2010) confirmed that the use of drugs was related to unintended pregnancy and the request for an abortion. Clearly, some women get pregnant when high on alcohol or other substances, regret the pregnancy, and want to reverse it.

Abortions performed to protect the life or health of the woman are called **therapeutic abortions**. However, there is disagreement over this definition. Garrett et al. (2001) noted, "Some physicians argue that an abortion is therapeutic if it prevents or alleviates a serious physical or mental illness, or even if it alleviates temporary emotional upsets. In short, the health of the pregnant woman is given such a broad definition that a very large number of abortions can be classified as therapeutic" (p. 218).

**Therapeutic abortions** abortions performed to protect the life or health of the woman.

Some women with multifetal pregnancies (a common outcome of the use of fertility drugs) may have a procedure called *transabdominal first-trimester selective termination*. In this procedure, the lives of some fetuses are terminated to increase the chance of survival for the others or to minimize the health risks associated with multifetal pregnancy for the woman. For example, a woman carrying five fetuses may elect to abort three of them to minimize the health risks of the other two.

## Pro-Life Abortion Position

A dichotomy of attitudes toward abortion is reflected in two opposing groups of abortion activists. Individuals and groups who oppose abortion are commonly referred to as "pro-life" or "antiabortion."

*Abortion is murder.*

JACK WRIGHT JR., ATTORNEY

Of 4,540 undergraduates at two large universities, 24% reported that abortion was not acceptable under certain conditions (Hall and Knox 2013). Pro-life groups favor abortion policies or a complete ban on abortion. They essentially believe the following:

1. The unborn fetus has a right to live and that right should be protected.
2. Abortion is a violent and immoral solution to unintended pregnancy.
3. The life of an unborn fetus is sacred and should be protected, even at the cost of individual difficulties for the pregnant woman.

Foster et al. (2013) studied the effect of pro-life protesters outside of abortion clinics on those individuals who came to the clinic to get an abortion. The researchers interviewed 956 women, 16% of whom said that they were very upset by the protesters. However, exposure to the protesters was not associated with differences in emotions one week after the abortion.

## Pro-Choice Abortion Position

In the sample of 4,540 undergraduates referred to above, 63% reported that "abortion is acceptable under certain conditions" (Hall and Knox 2013). Pro-choice advocates support the legal availability of abortion for all women. They essentially believe the following:

1. Freedom of choice is a central value—the woman has a right to determine what happens to her own body.
2. Those who must personally bear the burden of their moral choices ought to have the right to make these choices.
3. Procreation choices must be free of governmental control.

*Unless pro-lifers are committed to supporting and raising all the children they don't want aborted, they are being egregiously morally dishonest.*

MARY-HOWELL MARTENS, PENN YAN, NY

Although many self-proclaimed feminists and women's organizations, such as the National Organization for Women (NOW), have been active in promoting abortion rights, not all feminists are pro-choice.

## Confidence in Making an Abortion Decision

To what degree do women considering abortion feel confident of their decision? Foster et al. (2012) studied 5,109 women who sought 5,387 abortions at one U.S.

clinic and found that 87% of the women reported having high confidence in their decision before receiving counseling. Variables associated with uncertainty included being younger than 20, being black, not having a high school diploma, having a history of depression, having a fetus with an anomaly, having general difficulty making decisions, having spiritual concerns, believing that abortion is murder and fearing not being forgiven by God.

## Physical Effects of Abortion

Part of the debate over abortion is related to its presumed effects. In regard to the physical effects, legal abortions, performed under safe medical conditions, are safer than continuing the pregnancy. The earlier in the pregnancy the abortion is performed, the safer it is. Vacuum aspiration, a frequently used method in early pregnancy, does not increase the risks to future childbearing. However, late-term abortions do increase the risks of subsequent miscarriages, premature deliveries, and babies of low birth weight.

Weitz et al. (2013) compared the outcome of 11,487 early aspiration abortions depending on who performed the abortion—nurse, certified nurse midwife, physician's assistant, and physician—and found a complication rate of 1.8% with insignificant variation. These data support the adoption of policies to allow these providers to perform early aspirations to expand access to abortion care.

## Psychological Effects of Abortion

Of equal concern are the psychological effects of abortion. The American Psychological Association reviewed all outcome studies on the mental health effects of abortion and concluded, "Based on our comprehensive review and evaluation of the empirical literature published in peer-reviewed journals since 1989, this Task Force on Mental Health and Abortion concludes that the most methodologically sound research indicates that among women who have a single, legal, first-trimester abortion of an unplanned pregnancy for nontherapeutic reasons, the relative risks of mental health problems are no greater than the risks among women who deliver an unplanned pregnancy" (Major et al. 2008, p. 71).

In contrast, Fergusson et al. (2013) surveyed eight publications looking at five outcome domains—anxiety, depression, alcohol misuse, illicit drug use/misuse, and suicidal behavior. Results revealed that abortion was associated with small to moderate increases in risks of anxiety, alcohol misuse, illicit drug use/misuse, and suicidal behavior. The researchers concluded that there is no available evidence to suggest that abortion has therapeutic effects in reducing the mental health risks of unwanted or unintended pregnancy.

However, in rebuttal, Steinberg (2013) disagreed with Fergusson's conclusions "because there are many poorly conducted studies in the literature and they cannot be combined with well-conducted studies" and noted that "there are some high-quality studies that provide evidence that abortion relative to birth of an unwanted pregnancy is not associated with higher rates of subsequent mental health problems, suggesting that abortion is not a cause of mental health problems."

Independent of the Fergusson and Steinberg standoff, Canario et al. (2013) analyzed data on 50 women (and 15 partners) one and six months after abortion and found a decrease in emotional disorder for all etiologies of abortion and an increase in perceived quality of a couple's relationship in therapeutic abortion over time. The researchers concluded that the psychological adjustment of an individual after abortion seems to be influenced by factors such as the couple's relationship.

## Knowledge and Support of Partners of Women Who Have Abortion

Jones et al. (2011) examined data from 9,493 women who had obtained an abortion to find out the degree to which their male partners knew of the abortion and their feelings about it. The overwhelming majority of women reported that the men by whom they got pregnant knew about the abortion, and most perceived these men to be supportive. Cohabiting men were particularly supportive. The researchers concluded that that most women obtaining abortions are able to rely on male partners for social support.

## PERSONAL CHOICES

### Should You Have an Abortion?

The decision to have an abortion continues to be a complex one. Women who are faced with the issue may benefit by considering the following guidelines:

1. ***Consider all the alternatives available to you, realizing that no alternative will have only positive consequences and no negative consequences.*** As you consider each alternative, think about both the short-term and the long-term outcomes of each course of action, what you want, and what you can live with.
2. ***Obtain information about each alternative course of action.*** Inform yourself about the medical, financial, and legal aspects of abortion, childbearing, parenting, and placing the baby up for adoption.
3. ***Talk with trusted family members, friends, or unbiased counselors.*** Consider talking with the man who participated in the pregnancy. If possible, also talk with women who have had abortions as well as with women who have kept and reared a baby or placed a baby for adoption. If you feel that someone is pressuring you in your decision making, look for help elsewhere.
4. ***Consider your own personal and moral commitments in life.*** Understand your own feelings, values, and religious beliefs concerning the fetus and weigh those against the circumstances surrounding your pregnancy.

# The Future of Planning for Children

While having children will continue to be the option selected by most couples, fewer will select this option and being childfree will lose some of its stigma. Indeed, the pregnancy rate for women ages 15–29 has dropped steadily since 1990 (Curtin et al. 2013). Individualism and economics are the primary factors responsible for reducing the obsession to have children. To quote Laura Scott, "having children will change from an assumption to a decision." Once the personal, social, and economic consequences of having children come under close scrutiny, the automatic response to have children will be tempered.

# Summary

***What are the social influences and individual motivations for having children?***

Having children continues to be a major goal of most individuals (women more than men). Social influences to have a child include family, friends, religion, government, favorable economic conditions, and cultural observances. The reasons people give for having children include love and companionship with one's own offspring, the desire to be personally fulfilled as an adult by having a child, and the desire to recapture one's youth. Having a child (particularly for women) reduces one's educational and career advancement. The cost for housing, food, transportation, clothing, health care, and child care for a child up to age 18 (college not included) is over $240,000.

***How many children do individuals want?***

About 20% of women aged 40 to 44 do not have children. Whether these women will remain childfree or eventually have children is unknown. Categories of those who do not have children include early articulators who knew early that they did not want children, postponers who kept delaying when they would have children and remained childfree, acquiescers who made the decision to remain childless because their partner did not want children, and undecided, who are childfree but still in the decision-making process.

The top five reasons individuals give for wanting to remain childfree are a high level of current life satisfaction, being free/independent, avoiding the responsibility of rearing a child, the absence of maternal/paternal instinct, and a desire to accomplish/experience things in life which would be difficult as a parent.

About 3% of individuals have one child. The most preferred family size is two children—a boy and a girl. To get this sex balance, some couples use sex selection. Contrasting views include that sex selection is an expression of reproductive rights and a sign of female empowerment to prevent unwanted pregnancies and abortions. A contrasting view, more likely to be held by primary care physicians, is that sex selection contributes to gender stereotypes that could result in neglect of children of the lesser-desired sex.

***What are the causes of teenage motherhood?***

Forty-one percent of all births are to unmarried women, most of whom are teenagers. Reasons for teenagers having a child include not being socialized as to the importance of contraception, having limited parental supervision, and perceiving few alternatives to parenthood. Indeed, motherhood may be one of the only remaining meaningful roles available to them. In addition, some teenagers feel lonely and unloved and have a baby to create a sense of being needed and wanted. In contrast, in Sweden, eligibility requirements for welfare payments make it almost necessary to complete an education and get a job before becoming a parent.

***What are the causes of infertility and the technology available to help?***

Infertility is defined as the inability to achieve a pregnancy after at least one year of regular sexual relations without birth control, or the inability to carry a pregnancy to a live birth. Forty percent of infertility problems are attributed to the woman, 40% to the man, and 20% to both of them. The causes of infertility in women include blocked fallopian tubes, endocrine imbalance that prevents ovulation, dysfunctional ovaries, chemically hostile cervical mucus that may kill sperm, and effects of STIs.

A number of technological innovations are available to assist women and couples in becoming pregnant. These include hormonal therapy, artificial insemination, ovum transfer and in vitro fertilization. Being infertile (for the woman) may have a negative lifetime effect, both personal and interpersonal (half the women in one study were separated or reported a negative effect on their sex lives).

***What are the motives for adoption?***

Just over 1% of 18- to 44-year-old women reported having adopted a child. Motives for adoption include a couple's inability to have a biological child (infertility), their desire to give an otherwise unwanted child a permanent loving home, or their desire to avoid contributing to overpopulation by having more biological children.

***How prevalent is sterilization?***

Sterilization is a permanent surgical procedure that prevents reproduction. Worldwide over 220 million couples have used sterilization as their method of birth control—nearly 43 million men and 180 million women. The lifetime chance of a pregnancy after sterilization is one in 200.

***What are the motives and outcomes for an abortion?***

The most frequently cited reasons for induced abortion were that having a child would interfere with a woman's education, work, or ability to care for dependents (74%); that she could not afford a baby now (73%); and that she did not want to be a single mother or was having relationship problems (48%). In regard to the psychological effects of abortion, the American Psychological Association reviewed the literature and concluded that "among women who have a single, legal, first-trimester abortion of an unplanned pregnancy for nontherapeutic reasons, the relative risks of mental health problems are no greater than the risks among women who deliver an unplanned pregnancy."

## Key Terms

Abortion rate
Abortion ratio
Antinatalism
Birth control sabotage
Childlessness concerns
Competitive birthing
Conception
Cryopreservation
Emergency contraception
Foster parent
Double dutch
Hysterectomy
Induced abortion
Infertility
Laparoscopy

Oophorectomy
Ovum transfer
Parental consent
Parental notification
Pregnancy
Pregnancy coercion
Procreative liberty
Pronatalism
Salpingectomy
Spontaneous abortion (miscarriage)
Sterilization
Therapeutic abortion
Transracial adoption
Vasectomy

## Web Links

Alan Guttmacher Institute
http://www.guttmacher.org/

Childfree by Choice
http://www.childfreebychoice.com/
http://www.childlessbychoiceproject.com/

College Cost Calculator
http://apps.collegeboard.com/fincalc/college_cost.jsp

Contraception
http://bedsider.org/

EngenderHealth
http://www.engenderhealth.org/

The Evan B. Donaldson Adoption Institute
http://www.adoptioninstitute.org/

Fetal Fotos (bonding with your fetus)
http://www.fetalfotosusa.com/

National Right to Life
http://www.nrlc.org/

No Kidding!
http://www.nokidding.net/

MicroSort
http://www.microsort.net/

Planned Parenthood Federation of America, Inc.
http://www.plannedparenthood.org

NARAL Pro-Choice America (reproductive freedom and choice)
http://www.naral.org/

Selecting a Method of Contraception
http://www.mayoclinic.com/health/birth-control/BI99999

CHAPTER 12

# Parenting

*To be in your children's memories tomorrow, You have to be in their lives today.*

BARBARA JOHNSON, LITERARY CRITIC

David Knox

**Learning Objectives**

- Identify the various roles involved in parenting children.
- Review the nature of parenting choices.
- Identify how women, men, and couples transition to parenthood.
- Summarize the facts of parenthood.
- Review the various principles of effective parenting.
- Know the unique challenges of single parents.
- Review the future of parenting.

## TRUE OR FALSE?

1. Adult children whose parents pay their bills, party more (drink/binge drinking) than adult children with jobs who take care of themselves.
2. New mothers who feel that they must be perfect with their newborn are more likely to report postpartum depression.
3. New research confirms that co-sleeping with one's infant has positive outcomes for the infants learning to self-soothe and self-regulate.
4. New mothers who are cohabiting with the father of their baby report less relationship satisfaction and a poorer transition to motherhood than mothers who are married to the father.
5. The more parents monitor/supervise their children's behavior the higher the rate of delinquent behavior in their adolescent.

*Answers:* 1. T 2. T 3. F 4. T 5. F

Michele Obama emphasizes that motherhood is her priority. She noted in an interview, "Like any mother, I am just hoping that I don't mess them up," which for her "means getting them out of the White House and paying attention to their lives given the fact that they are in the public spotlight." Mrs. Obama notes that their father's job is the last topic on the menu. "It's sitting down at the dinner table and having Barack's day be the last thing anyone really cares about." Parents everywhere have the same values, goals, and fears of Mrs. Obama. In this chapter we review the role of parenting with the goal as facilitating happy, economically independent, socially contributing adult offspring. We begin by looking at the various roles of parenting in a photo essay.

*Regardless of your relationship with your parents, you'll miss them when they're gone from your life.*

MAYA ANGELOU, POET

*You give your children enough money to do something but not enough to do nothing.*

GEORGE CLOONEY AS MATT KING, *THE DESCENDANTS*

*There is an expiration date on blaming your parents for steering you in the wrong direction; the moment you are old enough to take the wheel, responsibility lies with you.*

J.K. ROWLING, NOVELIST

*The moment a child is born, the mother is also born. She never existed before. The woman existed, but the mother, never. A mother is something absolutely new.*

BHAGWAN SHREE RAJNEESH

# The Choices Perspective of Parenting

Although both genetic and environmental factors are at work, the choices parents make have a profound impact on their children. In this section, we review the nature of parental choices and some of the basic choices parents make.

## Nature of Parenting Choices

Parents might keep the following points in mind when they make choices about how to rear their children:

**1. *Not to make a parental decision is to make a decision.*** Parents are constantly making choices even when they think they are not doing so. When a child is impolite and the parent does not provide feedback and encourage polite behavior, the parent has chosen to teach the child that being impolite is acceptable. When a child makes a promise ("I'll text you when I get to my friend's house") and does not do as promised, unless the parent addresses the issue the parent has chosen to allow the child to not take commitments seriously. Hence, parents cannot choose not to make choices in their parenting, because their inactivity is a choice that has as much impact as a deliberate decision to require politeness and responsibility.

**2. *All parental choices involve trade-offs.*** Parents are also continually making trade-offs in the parenting choices they make. The decision to take on a second job or to work overtime to afford the larger house will come at the price of having less time to spend with one's children and being more exhausted when such time is available. The choice to enroll a younger child in the highest-quality day care will mean less money for an older child's karate lessons. Parents should

**PHOTO ESSAY** | Parenting Roles

Although finding one definition of **parenting** is difficult, there is general agreement about the various roles parents play in the lives of their children. New parents assume at least seven roles:

**Parenting** defined in terms of roles including caregiver, emotional resource, teacher, and economic resource.

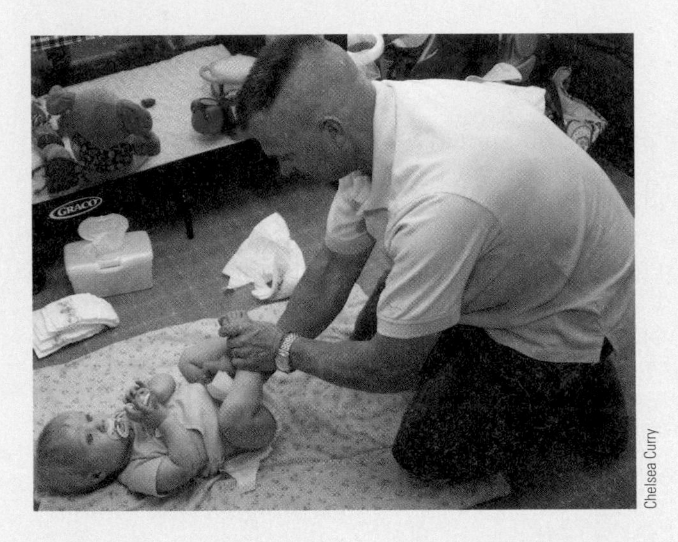

Chelsea Curry

Chelsea Curry

## Caregiver

A major role of parents is the physical care of their children. From the moment of birth, when infants draw their first breath, parents stand ready to provide nourishment (milk), cleanliness (diapers), and temperature control (warm blanket). The need for such sustained care in terms of a place to live/eat continues into adulthood as one-fourth of 18 to 34 year olds have moved back in with their parents after living on their own (Parker 2012). These **boomerang generation** children are back primarily for economic reasons.

**Boomerang generation** adult children who return to live with their parents.

## Emotional Resource

Beyond providing physical care, parents are sensitive to the emotional needs of children in terms of their need to belong, to be loved, and to develop positive self-concepts. In hugging, holding, and kissing an infant, parents not only express their love for the infant but also reflect awareness that such displays of emotion are good for the child's sense of self-worth.

Parents also provide "emotion work" for children—listening to their issues, helping them figure out various relationships they are struggling with, etc. Minnottea et al. (2010) studied the emotion work of parents with their children in a sample of 96 couples and found that women did more of it. Indeed, the greater the number of labor hours by men, the fewer the number of emotion work hours for their children and the higher the number of hours for women.

## Teacher

All parents think they have a philosophy of life or set of principles their children will benefit from. Parents soon discover that their children may not be interested in their religion or philosophy and, indeed, may rebel against it. This possibility does not deter parents from their role as teacher.

An array of self-help parenting books provide parents with ideas about the essentials children need that parents must teach. Galinsky (2010) identified

Chelsea Curry

seven skills all parents should be responsible for teaching their children: focus and self-control, seeing some else's point of view, communicating, making connections, critical thinking, taking on challenges, and self-directed engaged learning.

## Economic Resource

New parents are also acutely aware of the costs for medical care, food, and clothes for infants, and seek ways to ensure that such resources are available to their children. Working longer hours, taking second jobs, and cutting back on leisure expenditures are attempts to ensure that money is available to meet the needs of the children.

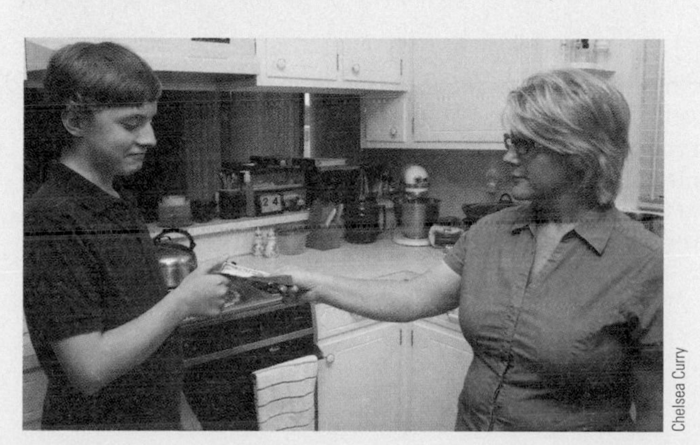
Chelsea Curry

Sometimes the pursuit of money for the family has a negative consequence for children. Rapoport and Le Bourdais (2008) investigated the effects of parents' working schedules on the time they devoted to their children and confirmed that the more parents worked, the less time they spent with their children. In view of extensive work schedules, parents are under pressure to spend quality time with their children, and it is implied that putting children in day care robs children of this time. However, Booth et al. (2002) compared children in day care with those in home care in terms of time the mother and child spent together per week. Although the mothers of children in day care spent less time with their children than the mothers who cared for their children at home, the researchers concluded that the "groups did not differ in the quality of mother-infant interaction" and that the difference in the "quality of the mother-infant interaction may be smaller than anticipated" (p. 16).

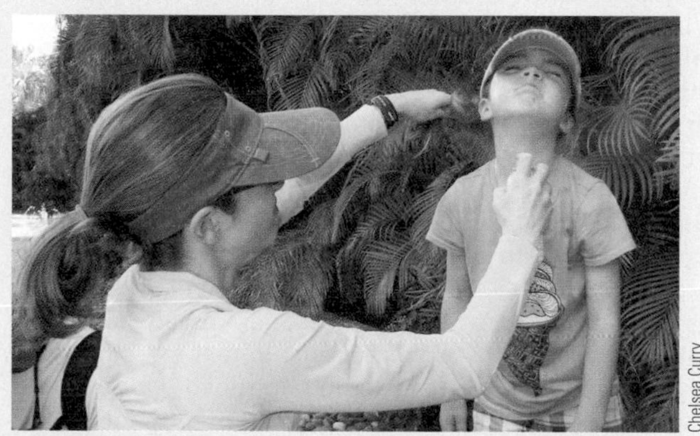
Chelsea Curry

## Protector

Parents also feel the need to protect their children from harm. This role may begin in pregnancy. Castrucci et al. (2006) interviewed 1,451 women about their smoking behavior during their pregnancy—of those who smoked, 89% reduced their smoking during pregnancy, while 24.9% stopped smoking completely during pregnancy.

Other expressions of the protective role include insisting that children wear seat belts in cars and jackets in cold weather, protecting them from violence or nudity in the media, and protecting them from strangers.

Taubman-Ben-Ari and Noy (2011) found that while parents of young children may initially become safer drivers, over time most revert back to their driving behaviors prior to becoming a parent.

Diamond et al. (2006) studied 40 middle-class mothers of young children and identified 15 strategies they used to protect their children. Their three principal strategies were to educate, control, and remove risk. The strategy used depended on the age and temperament of the child. For example, some parents felt protecting their children from certain television content was important. TV monitoring ranges from families who do not allow a television in their home to monitoring everything their children watch on television.

Some parents feel that protecting their children from harm implies appropriate discipline for inappropriate behavior. Galambos et al. (2003) noted that "parents' firm behavioral control seemed to halt the upward trajectory in externalizing problems among adolescents with deviant peers." In other words, parents who intervened when they saw a negative context developing were able to help their children avoid negative peer influences. Kolko et al. (2008) compared clinically referred boys and girls (ages 6 to 11) diagnosed with **oppositional defiant disorder** (children do not comply with requests of authority figures) to a matched sample of healthy control children and found that the former had greater exposure to delinquent peers. Hence, parents who monitor their children's peer relationships minimize their children's exposure to delinquent models, making it a wise time investment.

Increasingly, parents are joining the technological age and learning how to text message. In their role as protector, this ability allows parents to text message their children to tell them to come home, to phone home, or to work out a logistical problem—"meet me at the food court in the mall." Children can also use text messaging to let parents know that they arrived safely at a destination, when they need to be picked up, or when they will be home.

Protection also becomes relevant to teens who may be asked to sign a parent-teen driving contract which specifies details about use of the family car—no texting, drinking, or drugs while driving; keep a quarter of a tank of gas in the car; meet curfew; etc. Such contracts may also include the child paying for car insurance with penalties for late payments (Copeland 2010).

**Oppositional defiant Disorder**
children do not comply with requests of authority figures.

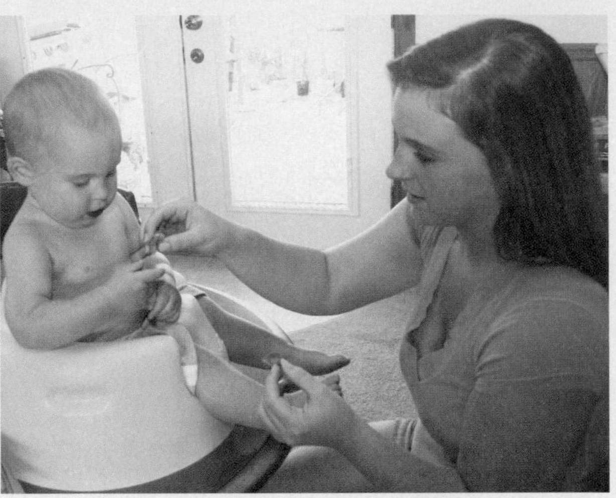
David Knox

## Health Promoter

The family is a major agent for health promotion—not only in promoting healthy food choices, responsible use of alcohol, nonuse of drugs, and safe driving skills, but also in ending smoking behavior. Knog et al. (2012) observed that parents who were most successful in getting their adolescents to stop smoking were positive models (they did not smoke themselves) and disapproved of their adolescent smoking. Baltazar et al. (2011) found that having a close relationship with 7th–9th grade children was associated with less smoking, drinking, and using inhalants by those children.

## Ritual Bearer

To build a sense of family cohesiveness, parents often foster rituals to bind members together in emotion and in memory. Prayer at meals and before bedtime, birthday and holiday celebrations, and vacations at the same place (beach, mountains, and so on) provide predictable times of togetherness and sharing.

increase their awareness that no choice is without a trade-off and should evaluate the costs and benefits of making such decisions.

**3. *Reframe "regretful" parental decisions.*** All parents regret a previous parental decision (e.g., they should have held their child back a year in school or not done so; they should have intervened in a bad peer relationship; they should have handled their child's drug use differently). Whatever the issue, parents chide themselves for their mistakes. Rather than berate themselves as parents, they might emphasize the positive outcome of their choices: not holding the child back made the child the first to experience some things among his or her peers; they made the best decision they could at the time; and so on.

**4. *Parental choices are influenced by society and culture.*** Parents are continually assaulted by commercial interests to get them to buy products for their children. Corporations regularly market to young parents to get them to buy the latest learning aid for their child, which promises a genius by age five.

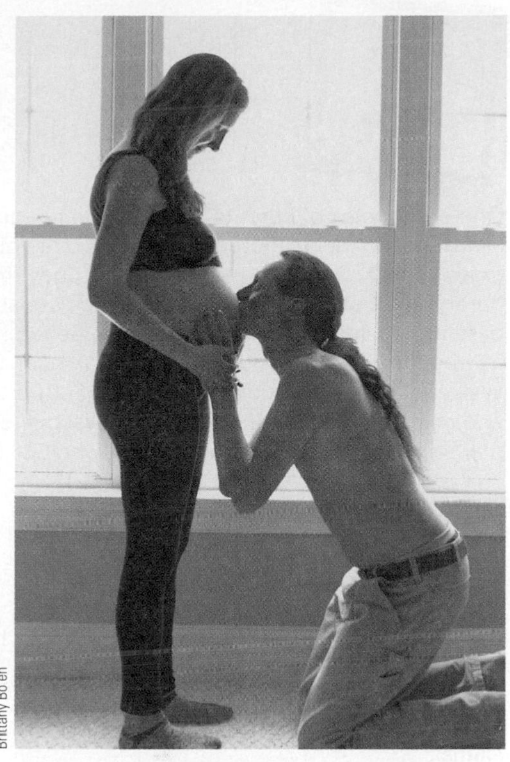
Brittany Bo en

The transition to parenthood begins with managing the pregnancy as a couple.

### Six Basic Parenting Choices

The six basic choices parents make include deciding (1) whether or not to have a child, (2) the number of children to have, (3) the interval between children, (4) the method of discipline, (5) the degree to which they will invest time with their children, and (6) whether or not to co-parent (the parents cooperate in the development of the lives of their children).

Though all of the above decisions are important, the relative importance one places on parenting as opposed to one's career will have implications for the parents, their children, and their children's children.

**Transition to parenthood** that period from the beginning of pregnancy through the first few months after the birth of a baby.

## Transition to Parenthood

The **transition to parenthood** refers to that period from the beginning of pregnancy through the first few months after the birth of a baby. The mother, father, and two of them as a couple undergo changes and adaptations during this period.

*The birth of my daughter changed everything forever. The moment I held her sweet self in my arms, I became a different person.*

SUSAN CHEEVER, AUTHOR

### Transition to Motherhood

Being a mother, particularly a biological or adoptive mother (in contrast to a stepmother), is a major social event for a woman (Pritchard and Kort-Butler 2012). Indeed, a baby changes a woman's life forever. Her initial reaction is influenced by whether the baby was intended. Su (2012) studied 825 women and found that unintended births were associated with decreased happiness among mothers. Nevertheless, whatever her previous focus, her life will now include her baby.

One woman expressed: "Our focus shifts from our husbands to our children, who then take up all our energy. We become angry and lonely and burdened . . . and that's if we LIKE it." Of course, the new role of parent can become exhausting. Kotila et al. (2013) asked 182 dual-earner parents to keep diaries on their time involvement with their children. While both parents were highly involved with their infants, mothers were more involved than fathers in routine child care. The Self-Assessment feature on page 338 examines one's view of traditional motherhood.

*My children have always existed at the deepest center of me, right there in the heart/hearth, but I struggled with the powerful demands of motherhood, chafing sometimes at the way they pulled me away from my separate life, not knowing how to balance them with my unwieldy need for solitude and creative expression.*

SUE MONK KIDD, *TRAVELING WITH POMEGRANATES: A MOTHER-DAUGHTER STORY*

Sociobiologists suggest that the attachment between a mother and her offspring has a biological basis (one of survival). The mother alone carries the

## APPLYING SOCIAL RESEARCH | Providing Financial Support for Emerging Adult Offspring

One of the roles of parents is to provide financial assistance to their children. Indeed, 62% of young adults (ages 19–22) report that they receive money from their parents—mostly to pay bills. How much assistance are parents providing? A great deal, is the general answer—an average of $12,185 annually (Wightman et al. 2012). Padilla-Walker et al. (2012) studied a sample of 401 undergraduates and at least one of their parents and discovered that over half reported paying between $5,000 and $30,000 (30% paid $30,000 or more and 20% paid less than $5,000) yearly. Most felt it was their obligation as parents to help their children financially, particularly while their children were "emerging adults" (between 18–late 20s).

When the parents were asked how long they should provide financial support, about half (49%) said until their offspring gets a job, about a quarter (23%) said when they graduate from college, and 6% said never. It seems that that "lower levels of financial support may indeed

**Oxytocin** a hormone from the pituitary gland that has been associate with the onset of maternal behavior in lower animals.

**Baby blues** transitory symptoms of depression 24 to 48 hours after the baby is born.

**Postpartum depression** a more severe reaction following the birth of a baby which occurs in reference to a complicated delivery as well as numerous physiological and psychological changes occurring during pregnancy, labor, and delivery; usually in the first month after birth but can be experienced after a couple of years have passed.

*I feel like a defective model, like I came off the assembly line flat-out fucked and my parents should have taken me back for repairs before the warranty ran out.*

ELIZABETH WURTZEL, *PROZAC NATION*

**Postpartum psychosis** a reaction in which a woman wants to harm her baby.

fetus in her body for nine months, lactates to provide milk, and, during the expulsive stage of labor, produces **oxytocin**, a hormone from the pituitary gland that has been associated with the onset of maternal behavior in lower animals. Most new mothers become emotionally bonded with their babies and resist separation.

Not all mothers feel joyous after childbirth. Naomi Wolf uses the term "the conspiracy of silence" to note motherhood is "a job that sucks 80 percent of the time" (quoted in Haag 2011, p. 83). Some mothers don't bond immediately and feel overworked, exhausted, mildly depressed, and irritable; they cry, suffer loss of appetite, and have difficulty in sleeping. Many new mothers experience **baby blues**—transitory symptoms of depression 24 to 48 hours after the baby is born.

About 10% to 15% experience **postpartum depression**, a more severe reaction than baby blues (usually occurs within a month of the baby's birth). Postpartum depression is more likely in the context of a low quality relationship when the partner is not supportive (Wickel 2012). A complicated delivery is also associated with postpartum depression. Gelabert et al. (2012) looked at various personality traits that were associated with postpartum depression. They found that women who were perfectionistic (characterized by concern over mistakes, personal standards, parental expectations, parental criticism, doubt about actions and organization) were more likely to report postpartum depression. To minimize baby blues and postpartum depression, antidepressants such as Zoloft and Prozac are used. Most women recover within a short time. With prolonged negative feelings, a clinical psychologist should be consulted.

**Postpartum psychosis**, a reaction in which a woman wants to harm her baby, is even more rare than postpartum depression. While having misgivings about a new infant is normal, the parent who wants to harm the infant should make these feelings known to the partner, a close friend, and a professional.

## Transition to Fatherhood

Just as some mothers experience postpartum depression, fathers, too, may become depressed following the birth of a baby. Quing et al. (2011) examined mothers' and fathers' postnatal (378 pairs) reactions and found that some parents of both genders reported postnatal depression (15% for mothers and 13% for fathers). The personal reaction of the male to the role of becoming a new father is related to whether the birth of the baby was intended. Su (2012) facilitate a greater perception of oneself as an adult and promote more adult-like behaviors including fewer risky behaviors (i.e., drinking/binge drinking), greater identity development (at least in the domain of occupational identity), and higher numbers of work hours per week" (Padilla-Walker et al. 2012, p. 56). Hence, underwriting children completely seemed to interfere with the transition of those children to more responsible adult behavior.

**Sources**

Padilla-Walker, L. M., L. J. Nelson, and J. S. Carroll. 2012. Affording emerging adulthood: Parental financial assistance of their college-aged children. *Journal of Adult Development* 12: 50–58.

Wightman, P, R. Schoeni, and K. Robinson. 2012. Familial financial assistance to young adults. Paper, Annual Meeting of the Population Association of America. May 3.

studied 889 men and found that unintended births were associated with depressive symptoms among fathers, particularly where financial strain was involved.

Regardless of whether the father gets depressed or the birth was intended, there is an economic benefit to the family for the husband becoming a father. Killewald (2013) noted that fathers who live with their wives and biological children become more focused and committed to work and their wage gains increase 4%. Regardless, mothers are typically disappointed in the amount of time the father helps with the new baby (Biehle and Michelson 2012). Nomaguchi et al. (2012) confirmed that fathers' spending time with children, engagement, and cooperative coparenting were related to less parenting stress for mothers who were married to, cohabiting with, or dating the father.

Ian and Dakota

This father enjoys time with his son.

## SELF-ASSESSMENT | The Traditional Motherhood Scale

The purpose of this survey is to assess the degree to which you possess a traditional view of motherhood. Read each item carefully and consider what you believe. There are no right or wrong answers, so please give your honest reaction and opinion. After reading each statement, select the number that best reflects your level of agreement, using the following scale:

| 1 | 2 | 3 | 4 | 5 | 6 | 7 |
|---|---|---|---|---|---|---|
| Strongly Disagree | | | | | | Strongly Agree |

_____ 1. A mother has a better relationship with her children than a father does.
_____ 2. A mother knows more about her child than a father, thereby being the better parent.
_____ 3. Motherhood is what brings women to their fullest potential.
_____ 4. A good mother should stay at home with her children for the first year.
_____ 5. Mothers should stay at home with the children.
_____ 6. Motherhood brings much joy and contentment to a woman.
_____ 7. A mother is needed in a child's life for nurturance and growth.
_____ 8. Motherhood is an essential part of a female's life.
_____ 9. I feel that all women should experience motherhood in some way.
_____ 10. Mothers are more nurturing than fathers.
_____ 11. Mothers have a stronger emotional bond with their children than do fathers.
_____ 12. Mothers are more sympathetic to children who have hurt themselves than are fathers.
_____ 13. Mothers spend more time with their children than do fathers.
_____ 14. Mothers are more lenient toward their children than are fathers.
_____ 15. Mothers are more affectionate toward their children than are fathers.
_____ 16. The presence of the mother is vital to the child during the formative years.
_____ 17. Mothers play a larger role than fathers in raising children.
_____ 18. Women instinctively know what a baby needs.

### Scoring

After assigning a number from 1 (strongly disagree) to 7 (strongly agree), add the numbers and divide by 18. The higher your score (7 is the highest possible score), the stronger the traditional view of motherhood. The lower your score (1 is the lowest possible score), the less traditional the view of motherhood.

### Norms

The norming sample of this self-assessment was based upon 20 male and 86 female students attending Valdosta State University. The average age of participants completing the scale was 21.72 years (SD 2.98), and ages ranged from 18 to 34. The ethnic composition of the sample was 80.2% white, 15.1% black, 1.9% Asian, 0.9% American Indian, and 1.9% other. The classification of the sample was 16.0% freshmen, 15.1% sophomores, 27.4% juniors, 39.6% seniors, and 1.9% graduate students.

The most traditional score was 6.33; the score reflecting the least support for traditional motherhood was 1.78. The midpoint (average score) between the top and bottom score was 4.28; thus, people scoring above this number tended to have a more traditional view of motherhood and people scoring below this number have a less traditional view of motherhood.

There was a significant difference ($p$ .05) between female participants' scores (mean = 4.19) and male participants' scores (mean = 4.68), suggesting that males had more traditional views of motherhood than females.

### Source

"Attitudes Toward Motherhood Scale," 2004 by Mark Whatley, Ph.D., Department of Psychology, Valdosta State University, Valdosta, GA 31698-0100. Used by permission. Other uses of this scale only by written permission of Dr. Whatley (mwhatley@valdosta.edu). Information on the reliability and validity of this scale is available from Dr. Whatley.

Schoppe-Sullivan et al. (2008) emphasized that mothers are the gatekeepers of the father's involvement with his children. A father may or may not be involved with his children to the degree that a mother encourages or discourages his involvement. The **gatekeeper role** is particularly pronounced after a divorce in which the mother receives custody of the children (the role of the father may be severely limited).

**Gatekeeper role** term used to refer to the influence of the mother on the father's involvement with his children.

The Self-Assessment on traditional fatherhood examines one's view of traditional fatherhood.

## SELF-ASSESSMENT | The Traditional Fatherhood Scale

The purpose of this survey is to assess the degree to which you have a traditional view of fatherhood. Read each item carefully and consider what you believe. There are no right or wrong answers. After reading each statement, select the number that best reflects your level of agreement, using the following scale:

| 1 | 2 | 3 | 4 | 5 | 6 | 7 |
|---|---|---|---|---|---|---|
| Strongly Disagree | | | | | | Strongly Agree |

_____ 1. Fathers do not spend much time with their children.
_____ 2. Fathers should be the disciplinarians in the family.
_____ 3. Fathers should never stay at home with the children while the mother works.
_____ 4. The father's main contribution to his family is giving financially.
_____ 5. Fathers are less nurturing than mothers.
_____ 6. Fathers expect more from children than their mothers do.
_____ 7. Most men make horrible fathers.
_____ 8. Fathers punish children more than mothers do.
_____ 9. Fathers do not take a highly active role in their children's lives.
_____ 10. Fathers are very controlling.

### Scoring

After assigning a number from 1 (strongly disagree) to 7 (strongly agree), add the numbers and divide by 10. The higher your score (7 is the highest possible score), the stronger the traditional view of fatherhood. The lower your score (1 is the lowest possible score), the less traditional the view of fatherhood.

### Norms

The norming sample was based upon 24 male and 69 female students attending Valdosta State University. The average age of participants completing the Traditional Fatherhood Scale was 22.15 years, and ages ranged from 18 to 47. The ethnic composition of the sample was 77.4% white, 19.4% black, 1.1% Hispanic, and 2.2% other. The classification of the sample was 16.1% freshmen, 11.8% sophomores, 23.7% juniors, 46.2% seniors, and 2.2% graduate students.

The most traditional score was 5.50; the score representing the least support for traditional fatherhood was 1.00. The average score of the respondents 3.33, suggesting a less-than-traditional view.

There was a significant difference ($p$ .05) between female participants' attitudes (average 3.20) and male participants' attitudes toward fatherhood (average 3.69), suggesting that males had more traditional views of fatherhood than females. There were no significant differences between ethnicities.

### Source

"Traditional Fatherhood Scale," 2004 by Mark Whatley, Ph.D., Department of Psychology, Valdosta State University, Valdosta, GA 31698-0100. Used by permission. Other uses of this scale only by written permission of Dr. Whatley (mwhatley@valdosta.edu). Information on the reliability and validity of this scale is available from Dr. Whatley.

The importance of the father in the lives of his children is enormous and goes beyond his economic contribution (Dearden et al. 2013; McClain 2011). While the role of father is not clearly defined and positive models are lacking (Ready et al. 2011), children who have a father who maintains active involvement in their lives tend to:

| | |
|---|---|
| Make good grades | Have higher incomes as adults |
| Be less involved in crime | Have higher education levels |
| Have good health/self-concept | Have higher cognitive functioning |
| Have a strong work ethic | Have stable jobs |
| Have durable marriages | Have fewer premarital births |
| Have a strong moral conscience | Have lower incidences of child sex abuse |
| Have higher life satisfaction | Exhibit healthier/on time physical development |

Daughters may have an extra benefit of a close relationship with fathers. Byrd-Craven et al. (2012) noted that such a relationship was associated with the daughters' having lower stress levels which assisted them in coping with problems and in managing interpersonal relationships. To foster the relationship between

*You signed no contract to become a parent, but the responsibilities were written in invisible ink. There was a point when you had to support your child, even if no one else would. It was your job to rebuild the bridge, even if your child was the one who burned it in the first place.*

JODI PICOULT, *THE TENTH CIRCLE*

fathers and daughters, the Indian Princess program has emerged. Google Indian Princess Program for an organization in your community.

### Transition from a Couple to a Family

*There's no such thing as fun for the whole family.*

**JERRY SEINFELD, COMEDIAN**

The birth of a baby is usually a stressor event for each parent and for their relationship regardless of whether the child is biological or adopted and regardless of whether the parents are heterosexual or homosexual (Goldberg et al. 2010). However, not all couples report problems. Holmes et al. (2013) studied 125 couples as they transitioned to parenthood and found that although many parents reported declines in love and increases in conflict, 23% of mothers and 37% of fathers reported equal or increased love; 20% of mothers and 28% of fathers reported equal or lower conflict. Durtschi and Soloski (2012) found that coparenting was associated with less parental stress and greater relationship quality. Similarly, salivary cortisol (stress hormone) levels of parents with babies was lower in those parents who worked together at night to get the children into bed (McDaniel et al. 2012). Hence, coparenting was related to lower stress levels.

Do women who are cohabiting experience the same negative change in relationship satisfaction as do married women? Yes. Mortensen et al. (2012) examined the data on 71,504 women who transitioned into motherhood over a two-year period and observed a similar negative change in relationship satisfaction during the transition to parenthood. However, cohabiting women started off and stayed less satisfied throughout the transition period.

*We may not be able to prepare the future for our children, but we can at least prepare our children for the future.*

**FRANKLIN D. ROOSEVELT, 32ND PRESIDENT**

Regardless of how children affect relationship satisfaction, spouses report more commitment to their relationship once they have children (Stanley and Markman 1992). Figure 12.1 illustrates that the more children a couple has, the more likely the couple will stay married. A primary reason for this increased commitment is the desire on the part of both parents to provide a stable family context for their children. In addition, parents of dependent children may keep their marriage together to maintain continued access to and a higher standard of living for their children. Finally, people (especially mothers) with small children feel more pressure to stay married (if the partner provides sufficient economic resources) regardless of how unhappy they may be. Hence, though children may decrease happiness, they increase stability, because pressure exists to stay together (the larger the family is, the more difficult the economic post-divorce survival is).

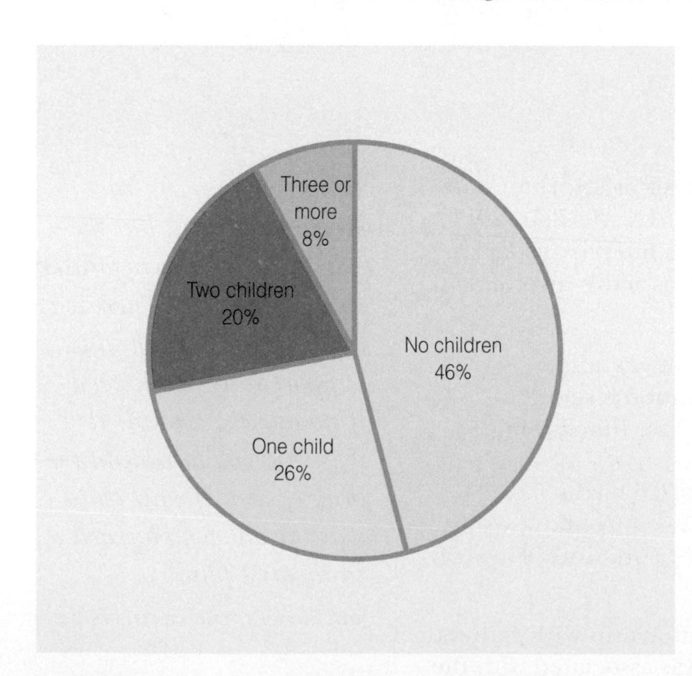


**Figure 12.1** Percentage of Couples Getting Divorces by Number of Children. The greater the number of children, the higher the chance of divorce.

# Parenthood: Some Facts

Parenting is only one stage in an individual's or couple's life (children typically live with an individual 30% of that person's life and with a couple 40% of their marriage). Parenting involves responding to the varying needs of children as they grow up, and parents require help from family and friends in rearing their children.

Some additional facts of parenthood follow.

### Views of Children Differ Historically

Whereas children of today are thought of as dependent, playful, and adventurous, they were historically viewed quite differently (Beekman 1977). Indeed, the concept of childhood, like gender, is a social construct rather than a fixed life stage. From the 13th through the 16th centuries, children were viewed as innocent, sweet, and a source of amusement for adults. From the 16th through the 18th centuries, they were viewed as being in need of discipline and moral training. In the 19th century, whippings were routine as a means of breaking a child's spirit and bringing the child into submission. Although remnants of both the innocent and moralistic views of children exist today, the lives of children are greatly improved. Child labor laws in the United States protect children from early forced labor; education laws ensure a basic education; and modern medicine has been able to increase the life span of children.

**DIVERSITY IN THE UNITED STATES**

Parents are diverse. Amish parents rear their children in homes without electricity (no television, cell phones, or video games). Charismatics and evangelicals (members of the conservative religious denominations) frequently send their children to Christian schools which emphasize a Biblical and spiritual view of life. More secular parents tend to bring up their children with individualistic liberal views.

Children of today are the focus of parental attention. In some families, everything children do is "fantastic" and deserves a gold star. The result is a generation of young adults who feel that they are special and should be catered to; they are entitled. Nelson (2010) noted that some parents have become "helicopter parents" (also referred to as hovercrafts and PFH—Parents From Hell) in that they are constantly hovering overprotectively at school and elsewhere with excessive interest to ensure their child's success.

*Parents are not interested in justice, they're interested in peace and quiet.*

BILL COSBY, COMEDIAN

## Parents Are Only One Influence in a Child's Development

Although parents often take the credit and the blame for the way their children turn out, they are only one among many influences on child development. Although parents are the first significant influence, peer influence becomes increasingly important during adolescence. For example, Ali and Dwyer (2010) studied a nationally representative sample of adolescents and found that having peers who drank alcohol was related to the adolescent drinking alcohol.

Siblings also have an important and sometimes lasting effect on each other's development. Siblings are social mirrors and models (depending on the age) for each other. They may also be sources of competition and can be jealous of each other.

Teachers are also significant influences in the development of a child's values. As noted above, some parents send their children to religious schools to ensure that they will have teachers with conservative religious values. Selecting this structure for a child's education may continue into college.

Media in the form of television, replete with MTV and "parental discretion advised" movies are a major source of language, values, and lifestyles for children that may be different from those of the parents. Parents are also concerned about the sexuality and violence that their children are exposed to through television.

Another influence of concern to parents is the Internet. Though parents may encourage their children to conduct research and write term papers using the Internet, they may fear their children are also looking at pornography. Parental supervision of teenagers and their privacy rights on the Internet remain potential sources of conflict (see the Family Policy feature). Parents are also concerned about **sextortion** (online sexual extortion). Sextortion often takes the form of teenage girls with a webcam at a party who will visit an Internet chat room and flash their breasts. A week later they will get an e-mail from a stranger who informs them that he has captured their video from the Internet and will post their photos on Facebook if they do not send him explicit photos (Wilson et al. 2010).

**Sextortion** online sexual extortion.

## Each Child Is Unique

Children differ in their genetic makeup, physiological wiring, intelligence, tolerance for stress, capacity to learn, comfort in social situations, and interests. Parents soon

## FAMILY POLICY | Government or Parental Control of Internet Content for Children?

Governmental censoring of content on the Internet is the focus of an ongoing public debate. At issue is what level of sexual content children should be exposed to and should parents or the government make the decision? This issue has already been decided in China, where the government ensures that all major search engines filter what is made available to Internet users. In the United States, Congress passed the Communications Decency Act in 1996, which prohibited sending "indecent" messages over the Internet to people under age 18. However, the Supreme Court struck down the law in 1997, holding that it was too broadly worded and violated free speech rights by restricting too much material that adults might want to access. Other laws to restrict content access on the Internet have been passed but struck down on the basis that they violate First Amendment rights. However, The Protection of Children from Sexual Predators Act of 1998 required Internet Service Providers (ISPs) to report any knowledge of child victimization or child pornography to law enforcement.

Many object to government control of sexual content on the Internet on the grounds of First Amendment rights. Many others believe government restrictions are necessary to protect children from inappropriate sexual content.

An overriding question is, "Should parents or the government be in control of what children are exposed to?" In the United States, parents, not the government, prefer to be responsible for regulating their children's use of the Internet. Beyond exposure to Internet pornography, children sometimes post information about themselves on the Internet that encourages pedophiles to e-mail them. Some of these profiles may be more revealing than intended. Software products, such as Net Nanny, Surfwatch, CYBERsitter, CyberPatrol, and Time's Up are being marketed to help parents control what their children view on the Internet. These

become aware of the uniqueness of each child—of her or his difference from every other child they know. Parents of two or more children are often amazed at how children who have the same parents can be so different. Children also differ in their mental and physical health. Solomon (2012) interviewed 300 families in which the children were very different from their parents—autistic, gay, deaf, schizophrenic, etc. He entitled his book *Far from the Tree.*

*An entire generation of parents has spent years panicking about the effects of hookup culture on girls—making it all too easy to ignore the emotional lives of boys. But it's boys who often lack the skills to adapt. And it's boys who are falling behind.*

ROSALIND WISEMAN, CHILD DEVELOPMENT RESEARCHER

### Each Gender Is Unique

Wiseman (2013) reminds parents that the genders are also different and encourages them not to assume that boys are always emotionally resilient. Just because girls are socialized to be more open about their feelings (e.g., sad and depressed), boys may have these same feelings without a way to cope with them. She recommends that parents stay close to the emotional worlds of boys too—their falling in love, having their hearts broken, being bullied.

### Parenting Styles Differ

**Responsiveness** the extent to which parents respond to and meet the emotional needs of their children.

**Demandingness** the manner in which parents place demands on children in regard to expectations and discipline.

Diana Baumrind (1966) developed a typology of parenting styles that has become classic in the study of parenting. She noted that parenting behavior has two dimensions: responsiveness and demandingness. **Responsiveness** refers to the extent to which parents respond to and meet the emotional needs of their children. In other words, how supportive are the parents? Warmth, reciprocity, person-centered communication, and attachment are all aspects of responsiveness. **Demandingness**, on the other hand, is the manner in which parents place demands on children in regard to expectations and discipline. How much control do they exert over their children? Monitoring and confrontation are also aspects of demandingness. Categorizing parents in terms of their responsiveness and their demandingness creates four categories of parenting styles: permissive (also known as indulgent), authoritarian, authoritative, and uninvolved:

software programs allow parents to block unapproved websites and categories (such as pornography), block transmission of personal data (such as address and telephone numbers), scan pages for sexual material before they are viewed, and track Internet usage.

Another alternative is for parents to use the Internet with their children both to monitor what their children are viewing and to teach their children values about what they believe is right and wrong. Some parents believe that children must learn how to safely surf the Internet. Several Internet sites have been established in regard to this issue, including BlogSafety.com, a forum for parents to discuss blogging and other aspects of social networking; NetSmartz.org service, which teaches kids 5 to 17 how to be safe on the Internet; and WiredSafety .org, which posts information on Internet safety.

Monitoring Internet usage by children has another function. In a Dutch study of 1,796 adolescents, 17% had real life contacts with someone they had met on the Internet; their parents were aware of this contact only 30% of the time (Van den Heuvel et al. 2012).

**Your Opinion?**

1. To what degree do you believe the government should control Internet content?
2. To what degree do you believe that parents should monitor what their children are exposed to on the Internet?
3. What can parents do to teach their children responsible use of the Internet?

### Source

Van den Heuvel, A. R., J. J. M. Van den Eijnden, A. Van Rooij, and D. Van de Mheen. 2012. Meeting online contacts in real life among adolescents: The predictive role of psychosocial well-being and internet-specific parenting. *Computers in Human Behavior* 28: 465–472.

1. ***Permissive parents are high on responsiveness and low on demandingness.*** These parents are very lenient and allow their children to largely regulate their own behavior. Walcheski and Bredhoft (2010) found that the permissive parenting style is associated with overindulgence. For example, these parents state punishments but do not follow through; they give in to their children. These parents act out of fear that disciplining the child will cause the child to dislike his or her parents. The parents are controlled by the potential disapproval of the child.

2. ***Authoritarian parents are high on demandingness and low in responsiveness.*** They feel that children should obey their parents no matter what. Surjadi et al. (2013) confirmed that harsh, inconsistent discipline from parents was related to poor relationship quality with romantic partners during cohabitation or the early years of marriage.

3. ***Authoritative parents are both demanding and responsive.*** They impose appropriate limits on their children's behavior but emphasize reasoning and communication. Authoritative parenting offers a balance of warmth and control and is associated with the most positive outcome for children. Examples of this style include parents telling the child their expectations of the child's behavior before the child engages in the activity, giving the child reasons why rules should be obeyed, talking with the child when he or she misbehaves, and explaining consequences (Walcheski and Bredehoft 2010).

*That's the nature of being a parent, Sabine has discovered. You'll love your children far more than you ever loved your parents, and—in the recognition that your own children cannot fathom the depth of your love—you come to understand the tragic, unrequited love of your own parents.*

**URSULA HEGI, WRITER**

Panetta et al. (2014) studied outcomes of various parenting styles of mothers and fathers on 195 7th to 11th grade adolescents. Temperament of adolescent (e.g., mood, flexibility, rigidity) explained half of the variance in adolescent outcomes. Parenting styles contributed a smaller but significant role. When both parents were authoritative, it was associated with more optimal outcomes in adolescents' personal adjustment than any other parenting style combination. In addition, when both parents were permissive and neglectful, these parenting styles were associated with poorer adolescent outcomes. Similar positive outcomes for authoritative parenting were observed in Japanese children by Uji et al. (2014).

WHAT IF?

## What If You and Your Partner Disagree over Parenting Styles?

Because spouses grow up in families where different parenting philosophies are operative, it is not unusual for them to disagree over how to rear or discipline their own children. Some parents divorce because they argue constantly over how to rear their children. One alternative is for parents to agree that the most involved parent's child-rearing style will prevail. Such an agreement is obviously a concession on the part of one parent.

Another alternative is to use the parenting philosophy that research suggests has the best outcome. As previously noted, authoritative parenting, which is a blend of compassion and discipline, seems to have the best outcome in terms of fewer problems with children and their development of higher levels of social competence.

Laukkanen et al. (2014) characterized parenting styles of affection, behavioral and psychological. They noted that the child's temperament impacted the parenting style used by mothers. In a Finnish study of first-grade children, the more active and the less positive a child, the lower the mother's well-being and, consequently, the more psychological control (e.g., use of guilt and withdrawal of love) she applied. In effect, the mother felt inadequate as a parent since the child offered few emotional rewards and the mother was exhausted trying to manage the child's high activity level.

**4.** ***Uninvolved parents are low in responsiveness and demandingness.*** These parents are not invested in their children's lives. Panetta et al. (2014) confirmed that permissive and neglectful parenting has negative outcomes for adolescents.

McKinney and Renk (2008) identified the differences between maternal and paternal parenting styles, with mothers tending to be authoritative and fathers tending to be authoritarian. Mothers and fathers also use different parenting styles for their sons and daughters, with fathers being more permissive with their sons than with their daughters. Overall, this study emphasized the importance of examining the different parenting styles on adolescent outcome and suggested that having at least one authoritative parent may be a protective factor for older adolescents. In summary, the authoritative parenting style is the combination of warmth, guidelines, and discipline—"I love you but you need to be in by midnight or you'll lose the privileges of having a cell phone and a car."

*Children have never been very good at listening to their elders, but they have never failed to imitate them.*

JAMES BALDWIN, NOVELIST

# Principles of Effective Parenting

**Parenting self-efficiency** feeling competent as a parent.

Dumka et al. (2010) noted that **parenting self-efficiency** (feeling competent as a parent) was associated with parental control. In effect, parents who viewed themselves as good teachers felt confident in their role which, in turn, translated into their ability to effectively manage the behavior of their adolescents. Numerous principles are involved in being effective parents (Keim & Jacobson 2011).

We begin with the most important of these principles, which involves giving time/love to your children as well as praising and encouraging them.

## Give Time, Love, Praise, Encouragement, and Acceptance

Children need to feel that they are worth spending time with and that they are loved. Because children depend first on their parents for the development of their sense of emotional security, it is critical that parents provide a warm emotional context in which the children develop. Feeling loved as an infant also affects one's capacity to become involved in adult love relationships. Abundant evidence from children reared in institutions where nurses attended only to their physical needs testify to the negative consequences of early emotional neglect. **Reactive attachment disorder** is common among children who were taught as infants that no one cared about them. Such children have no capacity to bond emotionally with others since they have no learning history of the experience and do not trust adults/caretakers/parents. A five to one ratio of positive to negative statements to the child is also associated with positive outcomes for the child (Armstrong and Clinton 2012).

*Parents are like God because you wanna know they're out there, and you want them to think well of you, but you really only call when you need something.*

CHUCK PALAHNIUK, *INVISIBLE MONSTERS*

**Reactive attachment disorder** common among children who were taught as infants that no one cared about them; these children have no capacity to bond emotionally with others since they have no learning history of the experience and do not trust adults, caretakers, or parents.

## Avoid Overindulgence

**Overindulgence** may be defined as more than just giving children too much; it includes overnurturing and not providing enough structure. Using this definition, a study of 466 participants revealed that those who are overindulged tend to hold materialist values for success, are not able to delay gratification, and are less grateful for things and to others. Another result of overindulgence is that children grow up without consequences so they avoid real jobs where employers expect them to show up at 8:00 A.M. Indeed, not being overindulged promotes the ability to delay gratification, be grateful, and experience a view (nonmaterialism) that is associated with happiness (Slinger and Bredehoft 2010).

Parents typically overindulge their children because they feel guilty for not spending time with them or because they (the parents) did not have certain material goods in their own youth. In a study designed to identify who overindulged, Clarke (2004) found that mothers were four times more likely to overindulge than fathers. Mothers spend more hours in child care so they are more vulnerable to being worn down—they overindulge their children as a coping mechanism for stress (give them expensive toys, etc. to entertain them).

*Educate your children to self-control, to the habit of holding passion and prejudice and evil tendencies subject to an upright and reasoning will, and you have done much to abolish misery from their future and crimes from society.*

BENJAMIN FRANKLIN, FOUNDING FATHER OF UNITED STATES

**Overindulgence** overnurturing and not providing enough structure.

## Monitor Child's Activities/Drug Use

Abundant research suggests that parents who monitor their children and teens and know where their children are, who they are with, and what they are doing are less likely to report that their adolescents receive low grades, or are engaged in early sexual activity, delinquent behavior, and alcohol or drug use (Crosnoe and Cavanagh 2010). Crutzen et al. (2012) assessed the effects of parental approval of children's drinking alcohol at home on their subsequent drinking behavior. Of 1,500 primary school children, those who were allowed to drink alcohol around their parents were less likely to consume alcohol when they were away from their parents.

Parents who drank alcohol under age and who used marijuana or other drugs wonder how to go about encouraging their own children to be responsible alcohol users and drug free. Drugfree.org has some recommendations for parents, including being honest with their children about previous alcohol and drug use, making clear that they do not want their children to use alcohol or drugs, and explaining that although not all alcohol or drug use leads to negative consequences, staying clear of alcohol/drug use is the best course of action.

*We must view young people not as empty bottles to be filled, but as candles to be lit.*

ROBERT SHAFFER, COLLEGE STUDENT ADVISOR

## Monitor Television and Pornography Exposure

Some parents monitor the amount and content of media exposure their children experience. A Kaiser Foundation (2010) study of media use of 8 to 18 year olds found that they were watching television an average of

4.5 hours per day. Nonschool computer use was an hour and a half a day—much of which was unsupervised (two-thirds of the students said there were no rules about Internet use).

Children and youth watching pornography is not unusual. Flood (2009) noted the negative effects of pornography exposure among children including their developing sexist and unhealthy notions of sex and relationships. In addition, among boys and young men, frequent consumption of pornography is associated with supportive attitudes of sexual coercion and increases their likelihood of perpetrating assault. Pornography teaches unrealistic expectations (e.g., females are expected to have bodies like porn stars, to enjoy "facials," and anal sex) and is devoid of integrating intimacy and sexual expression (Knox et al. 2004).

## Use Technology to Monitor Cell Phone/Text Messaging Use

**National Data**

Thirteen is the average age at which parents feel that it is appropriate for their child to own a cell phone (Gibbs 2012).

"Technological advances" was identified by 2,000 parents as the top issue that makes parenting children today tougher than in previous years. Sixty-one percent identified this issue; 57% noted increased violence (Payne and Trapp 2013). While media technology can be used to enhance the bond between parents and children (Coyne 2011), some parents are concerned about their child's use of a cell phone and texting. Not only may predators contact children/teenagers without the parents' awareness, but also 39% of teens report having sent sexually suggestive text messages and 20% have sent nude or semi-nude photos or videos of themselves. New technology called SMS Tracker for Android permits parents to effectively take over their child's mobile phone. A parent can see all incoming and outgoing calls, text messages, and photos. Another protective device on the market is MyMobileWatchdog (known as MMWD), which monitors a child's cell phone use and instantly alerts the parents online if their son or daughter receives unapproved e-mail, text messages, or phone calls. SecuraFone can reveal how fast the car in which the cell phone of the user is moving and alert parents. If a teen is speeding, the parents will know. One version shuts the texting capability down if the phone is going faster than five miles an hour. Mobiflock for Android allows parents to lock their child's phone for a predetermined time (e.g., 7 to 10 P.M. when studying is scheduled). In effect, these smart phones have web filters (which can block inappropriate websites), app filters (which make sure apps are kid-friendly), and contact filters (which can prevent harassing calls or texts by blocking certain numbers), which allow parents to monitor their child's location, texts, calls, browsing histories, app downloads, and photos they send and receive.

## Set Limits and Discipline Children for Inappropriate Behavior

The goal of setting limits and disciplining children is their self-control. Parents want their children to be able to control their own behavior and to make good decisions on their own.

Parental guidance involves reinforcing desired behavior and punishing undesirable behavior. Unless parents levy negative consequences for lying, stealing, and hitting, children can grow up to be dishonest, to steal, and to be inappropriately aggressive. **Time-out** (a noncorporal form of punishment that involves removing the child from a context of reinforcement to a place of isolation for one minute for each year of the child's age) has been shown to be an effective consequence for inappropriate behavior (Morawska and Sanders

**Time-out** a noncorporal form of punishment that involves removing the child from a context of reinforcement to a place of isolation for one minute for each year of the child's age.

2011). Withdrawal of privileges (use of cell phone, watching television, being with friends), pointing out the logical consequences of the misbehavior ("you were late; we won't go"), and positive language ("I know you meant well but . . .") are also effective methods of guiding children's behavior.

Physical punishment is less effective in reducing negative behavior (see the Personal Choices section on spanking); it teaches the child to be aggressive and encourages negative emotional feelings toward the parents. When using time-out or the withdrawal of privileges, parents should make clear to the child that they disapprove of the child's behavior, not the child. It is also important to reinforce the child for engaging in positive behavior by such statements as "I appreciate your being honest with us," "Thank you for texting me when you got to the party," and "The kindness you showed your sister is wonderful."

*It wasn't uncommon for the Hemingway children to be razor-strapped and then commanded to get down on their knees and ask God for forgiveness.*

**PAUL HENDRICKSON, *HEMINGWAY'S BOAT***

## PERSONAL CHOICES

### *Should Parents Use Corporal Punishment?*

Parents differ in the type of punishment they feel is appropriate for children. Some parents use corporal punishment as a means of disciplining their children. In a study on 1,580 mothers of infants at 14, 24, and 36 months, 18% reported consistent physical discipline (spanking) (Mayer and Blome 2013). Parents who spanked tended to have younger rather than older children and to be black rather than white or Latino (Grogan-Kaylor and Otis 2007; Jordan and Curtner-Smith 2011).

The decision to choose a corporal or noncorporal method of punishment should be based on the consequences of use. In general, the use of time-out and withholding of privileges is more effective than corporal punishment in stopping undesirable behavior (Paintal 2007). Though beatings and whippings will temporarily decrease negative verbal and nonverbal behaviors, they may have major side effects. First, punishing children by inflicting violence teaches them that it is acceptable to physically hurt someone you love. Hence, parents may be inadvertently teaching their children to use violence in the family. Children who are controlled by corporal punishment are more likely to grow up to be violent toward their own children, spouses, and friends (Paintal 2007).

Second, parents who beat their children should be aware that they are teaching their children to fear and to avoid them. The result is that children who grow up in homes where corporal punishment is used have more distant relationships with their parents (Paintal 2007). Jordan and Curtner-Smith (2011) studied a sample of 374 undergraduates and found that mothers (the primary disciplinarians) who relied on corporal punishment as a method of discipline incurred more negative feelings from their children, which affected the children's attachment to them. African-American children were at greater risk for corporal punishment and negative attachment.

Third, children who grow up in homes in which corporal punishment was used are more likely to feel helpless, to have low self-esteem, and to withdraw (Paintal 2007). While there is disagreement (Ferguson 2013), some children who are spanked evidence more externalization of behavior in terms of being aggressive, fighting, and bullying (Gromoske and Maguire-Jack 2012). In recognition of the negative consequences of corporal punishment, the law in Sweden forbids parents to spank their children.

A review of some of the alternatives to corporal punishment includes the following (Crisp and Knox 2009):

**1. *Be a positive role model.*** Children learn behaviors by observing their parents' actions, so parents must model the ways in which they want their children to behave. If a parent yells or hits, the child is likely to do the same.

**2. *Set rules and consequences.*** Make rules that are fair, realistic, and appropriate to a child's level of development. Explain the rules and the consequences of not following them. If children are old enough, they can be included in establishing the rules and consequences for breaking them.

3. ***Encourage and reward good behavior.*** When children are behaving appropriately, give them verbal praise and reward them with tangible objects (occasionally), privileges, or increased responsibility.

4. ***Use charts.*** Charts to monitor and reward behavior can help children learn appropriate behavior. Charts should be simple and focus on one behavior at a time, for a certain length of time.

5. ***Use time-out.*** Time-out involves removing children from a situation following a negative behavior. This can help children calm down, end an inappropriate behavior, and reenter the situation in a positive way. Explain what the inappropriate behavior is, why the time-out is needed, when it will begin, and how long it will last. Set an appropriate length of time for the time-out based on age and level of development, usually one minute for each year of the child's age (see the following Self-Assessment on Spanking versus Time-Out). Awareness of these alternatives is associated with a reduction in the use of spanking. Mothers exposed to child development information report a 30% reduction in the use of spanking (Mayer and Blome 2013).

### Sources


Crisp, B., and D. Knox. 2009. *Behavioral family therapy: An evidence based approach.* Durham, NC: Carolina Academic Press.

Ferguson, C. J. 2013. Spanking, corporal punishment and negative long-term outcomes: A meta-analytic review of longitudinal studies. *Clinical Psychology Review* 33: 196–208.

Grogan-Kaylor, A. and M. D. Otis. 2007. The predictors of parental use of corporal punishment *Family Relations* 56: 80–91.

Jordan, E. F. and M. E. Curtner-Smith. 2011. Multiple indicators of corporal punishment and current attachment to mother. Paper, 73rd Annual Meeting of National Council on Family Relations, Orlando, November.

Mayer, L. and W. W. Blome. 2013. The importance of early, targeted intervention: The effect of family, maternal, and child characteristics on the use of physical discipline. *Journal of Human Behavior in the Social Environment.* 23: 144–158.

Paintal, S. 2007. Banning corporal punishment of children. *Childhood Education* 83: 410–21.


## Have Family Meals

Parents who stay emotionally connected with their children build strong relationships with them and report fewer problems. Bisakha (2010) found that families who have regular meals with their adolescents report fewer behavioral problems, such as less substance use and running away from home for females. For males, there was also less running away as well as less drinking, less physical violence, less property-destruction, and less stealing. The researcher recommends that family meals be made a regular family ritual.

## Encourage Responsibility

Giving children increased responsibility encourages the autonomy and independence they need to be assertive and self-governing. Giving children more responsibility as they grow older can take the form of encouraging them to choose healthy snacks and letting them decide what to wear and when to return from playing with a friend (of course, the parents should praise appropriate choices).

Children who are not given any control over, and responsibility for, their own lives remain dependent on others. A dependent child is a vulnerable child. Successful parents can be defined in terms of their ability to rear children who can function as independent adults.

Parents also recognize that there is a balance they must strike between helping their children and impeding their own growth and development. One

## SELF-ASSESSMENT | Spanking versus Time-Out Scale

Parents discipline their children to help them develop self-control and correct misbehavior. Some parents spank their children; others use time-out. Spanking is a disciplinary technique whereby a mild slap (that is, a "spank") is applied to the buttocks of a disobedient child. Time-out is a disciplinary technique whereby, when a child misbehaves, the child is removed from the situation. The purpose of this survey is to assess the degree to which you prefer spanking versus time-out as a method of discipline. Please read each item carefully and select a number from 1 to 7, which represents your belief. There are no right or wrong answers.

| 1 | 2 | 3 | 4 | 5 | 6 | 7 |
|---|---|---|---|---|---|---|
| Strongly Disagree | | | | | | Strongly Agree |

_____ 1. Spanking is a better form of discipline than time-out.
_____ 2. Time-out does not have any effect on children.
_____ 3. When I have children, I will more likely spank them than use a time-out.
_____ 4. A threat of a time-out does not stop a child from misbehaving.
_____ 5. Lessons are learned better with spanking.
_____ 6. Time-out does not give a child an understanding of what the child has done wrong.
_____ 7. Spanking teaches a child to respect authority.
_____ 8. Giving children time-outs is a waste of time.
_____ 9. Spanking has more of an impact on changing the behavior of children than time-out.
_____ 10. I do not believe "time-out" is a form of punishment.
_____ 11. Getting spanked as a child helps you become a responsible citizen.
_____ 12. Time-out is used only because parents are afraid to spank their kids.
_____ 13. Spanking can be an effective tool in disciplining a child.
_____ 14. Time-out is watered-down discipline.

### Scoring

If you want to know the degree to which you approve of spanking, reverse the number you selected for all odd-numbered items (1, 3, 5, 7, 9, 11, and 13). For example, if you selected a 1 for item 1, change this number to a 7 (1 = 7; 2 = 6; 3 = 5; 4 = 4; 5 = 3; 6 = 2; 7 = 1). Now add these seven numbers. The lower your score (7 is the lowest possible score), the lower your approval of spanking; the higher your score (49 is the highest possible score), the greater your approval of spanking. A score of 21 places you at the midpoint between being very disapproving of or very accepting of spanking as a discipline strategy.

If you want to know the degree to which you approve of using time-out as a method of discipline, reverse the number you selected for all even-numbered items (2, 4, 6, 8, 10, 12, and 14). For example, if you selected a 1 for item 2, change this number to a seven (that is, 1 = 7; 2 = 6; 3 = 5; 4 = 4; 5 = 3; 6 = 2; 7 = 1). Now add these seven numbers. The lower your score (7 is the lowest possible score), the lower your approval of time-out; the higher your score (49 is the highest possible score), the greater your approval of time-out. A score of 21 places you at the midpoint between being very disapproving of or very accepting of time-out as a discipline strategy.

### Scores of Other Students Who Completed the Scale

The scale was completed by 48 male and 168 female student volunteers at East Carolina University. Their ages ranged from 18 to 34, with an average of 19.65. The ethnic background of the sample included 73.1% white, 17.1% African American, 2.8% Hispanic, 0.9% Asian, 3.7% from other ethnic backgrounds; 2.3% did not indicate ethnicity. The college classification level of the sample included 52.8% freshman, 24.5% sophomore, 13.9% junior, and 8.8% senior. The average score on the spanking dimension was 29.73, and the time-out dimension was 22.93, suggesting greater acceptance of spanking than time-out.

### Source

"The Spanking vs. Time-Out Scale," 2004 by Mark Whatley, Ph.D., Department of Psychology, Valdosta State University, Valdosta, GA 31698-0100. Used by permission. Other uses of this scale only by written permission of Dr. Whatley (mwhatley@valdosta.edu). Information on the reliability and validity of this scale is available from Dr. Whatley.

example is making decisions on how long to provide free room and board for an adult child. See the following Personal Choices section, which details this issue.

## Establish Norm of Forgiveness

Carr and Wang (2012) emphasized the importance of forgiveness in family relationships and the fact that forgiveness is a complex time-involved process rather than a one-time cognitive event (e.g., "I forgive you"). Respondents revealed in interviews with the researchers that forgiveness involved head over heart (finding ways to explain the transgression), time (sometimes months and years), and distance (giving the relationship a rest and coming back with renewed understanding).

Increasingly, adult children are staying at home . . . and some parents wonder when they are going to leave and go out on their own.

Chelsea Curry

## PERSONAL CHOICES

### *Adult children living with parents.*

Forty-two percent of adult children ages 20–24 live with their parents (El Nasser 2012). The recession (and resulting unemployment), college debt, and divorce all point to the primary reason—it's cheaper at mom's house. And, based on an Americatrade Survey of 1,000 Gen Z members, adult children's level of embarrassment at living with their parents varies based on their own age—49% at age 25, 88% at age 30, and 91% at age 35 report hanging their head at living at mom's (Payne and Trap 2013). Parents vary widely in how they view and adapt to adult children living with them. Some parents prefer that their children live with them, enjoy their company, and hope the arrangement continues indefinitely (this type of arrangement is often the norm in Italy). They have no rules about their children living with them and expect nothing from them. The adult children can come and go as they please, pay for nothing, and have no responsibilities or chores.

Other parents develop what is essentially a rental agreement, whereby their children are expected to pay rent, cook and clean the dishes, mow the lawn, and service the cars. No overnight guests are allowed, and a time limit is specified as to when the adult child is expected to move out.

A central issue from the point of view of young adults who live with their parents is whether their parents perceive them as adults or children. This perception has implications for whether they are free to come and go as well as to make their own decisions (Sassler et al. 2008). Males are generally left alone whereas females are often under more scrutiny.

Adult children also vary in terms of how they view the arrangement. Some enjoy living with their parents, volunteer to pay rent (though most do not), take care of their own laundry, and participate in cooking and housekeeping. Others are depressed that they are economically forced to live with their parents, embarrassed that they do so, pay nothing, and do nothing to contribute to the maintenance of the household. In a study of 30 young adults who had returned to the parental home, two-thirds paid nothing to live there and wanted to keep it that way. Those who did pay contributed considerably less than what rent would cost on the open market. Most paid for their cell phones, long-distance charges, and personal effects like clothing and toiletries (Sassler et al. 2008).

Whether parents and adult children discuss the issues of their living together will depend on the respective parents and adult children. Although there is no best way, clarifying expectations will help prevent misunderstandings. For example, what is the norm about bringing new pets into the home? Take, for example, a divorced son who moved back in with his parents along with his six-year-old son *and* a dog. His parents enjoyed being with them but were annoyed that the dog chewed on the furniture. Parental feelings eventually erupted that dismayed their adult child. He moved out, and the relationship with his parents became very strained. However, about three-fourths of the respondents in the Sassler et al. (2008) study reported generally satisfactory feelings about returning to their parental home. Most looked forward to moving out again but, in the short run, were content to stay to save money. In the sample of 1,094 undergraduate males, 14% (and 21% of 3,447 females) said that they would likely live with their parents after college (Hall and Knox 2013).

### Sources

El Nasser, H. 2012. No place like home for adult kids. *USA Today.* August 1. 1A.

Hall, S. and D. Knox. 2013. Relationship and sexual behaviors of a sample of 4,541 university students. Unpublished data collected for this text. Department of Family and Consumer Sciences, Ball State University and Department of Sociology, East Carolina University.

Payne, C. and P. Trap 2013. When is it embarrassing to live with parents? *USA Today.* Aug 28, 1B.

Sassler, S., D. Ciambrone, and G. Benway. 2008. Are they really mamma's boys/daddy's girls? The negotiation of adulthood upon returning home to the parental home. *Sociological Forum* 23: 670–698.

## Teach Emotional Competence

Wilson et al. (2012) emphasized the importance of teaching children **emotional competence**—experiencing, expressing, and regulating emotion. Being able to label when one is happy or sad (experiencing emotion), expressing emotion ("I love you"), and regulating emotion (e.g., reducing anger) assists children in getting in touch with their feelings and being empathetic with others. Wilson et al. (2012) reported on "Tuning in the Kids," a training program for parents to learn how to teach their children to be emotionally competent. Follow-up data on parents who took the six-session, two-hour-a-week program revealed positive outcomes/changes.

**Emotional competence** experiencing, expressing, and regulating emotion.

## Provide Sex Education

Hyde et al. (2013) interviewed 32 mothers and 11 fathers about the sexuality education of their children. Most gave little to no explicit information about safe sex—they assumed that the school had done so and that their son or daughter was not in a romantic relationship. Some felt that talking to their children about sex might be viewed as encouraging sexual behavior.

## Express Confidence

"One of the greatest mistakes a parent can make," confided one mother, "is to be anxious all the time about your child, because the child interprets this as your lack of confidence in his or her ability to function independently." Rather, this mother noted that it is best to convey to one's child that you know that he or she will be fine because you have confidence in him/her. "The effect on the child," said this mother, "is a heightened sense of self-confidence." Another way to conceptualize this parental principle is to think of the self-fulfilling prophecy as a mechanism that facilitates self-confidence. If parents show their child that they

*When your children are teenagers, it's important to have a dog so that someone in the house is happy to see you.*

—NORA EPHRON, *I FEEL BAD ABOUT MY NECK: AND OTHER THOUGHTS ON BEING A WOMAN*

have confidence in him or her, the child begins to accept these social definitions as real and becomes more self-confident.

## Respond to the Teen Years Creatively

Parenting teenage children presents challenges that differ from those in parenting infants and young children. Teenagers literally have altered brains that have lower amounts of dopamine, which may disrupt their reward function and make them less responsive to social stimuli (Forbes et al. 2012). Teenagers are more likely to engage in a high rate of risky behavior—smoking, alcohol consumption, hazardous driving, drug use, delinquency, dares, sporting risks, rebellious behavior, and sexual intercourse (Becker 2010). Seeking novelty, peer influences, genetic factors, and brain function are among the elements accounting for the vulnerability of adolescents.

### National **Data**

Sometimes strain between a teen and parent(s) ends in the teen running away. The National Runaway Switchboard receives more than 100,000 calls annually, 72% from females. The most frequent reason (29%) given for running away is "family dynamics" (National Runaway Switchboard 2010).

Conflicts between parents and teenagers often revolve around money and independence. The desired cell phone, designer clothes, and car can outstrip the budgets of many parents. Teens also want increasingly more freedom. However, neither of these issues needs result in conflict. When they do, the effect on the parent-child relationship may be inconsequential. One parent tells his children, "I'm just being the parent, and you're just being who you are; it is okay for us to disagree but you can't go." The following suggestions can help to keep conflicts with teenagers at a lower level:

1. ***Catch them doing what you like rather than criticizing them for what you don't like.*** Adolescents are like everyone else—they don't like to be criticized, but they do like to be noticed for what they do that is examplary.

2. ***Be direct when necessary.*** Though parents may want to ignore some behaviors of their children, addressing certain issues directly is necessary. Regarding the avoidance of STIs or HIV infection and pregnancy, parents should inform their teenagers of the importance of using a condom or dental dam before sex.

3. ***Provide information rather than answers.*** When teens are confronted with a problem, try to avoid making a decision for them. Rather, provide information on which they may base a decision. What courses to take in high school and what college to apply to are decisions that might be made primarily by the adolescent. The role of the parent might best be to listen.

*The first half of our lives is ruined by our parents, the second half by our children.*

CLARENCE DARROW, AMERICAN LAWYER

4. ***Be tolerant of high activity levels.*** Some teenagers are constantly listening to loud music, going to each other's homes, and talking on cell phones for long periods of time. Parents often want to sit in their easy chairs in peace. Recognizing that it is not realistic to always expect teenagers to be quiet and sedentary may be helpful in tolerating their disruptions.

5. ***Engage in leisure activity with your teenagers.*** Whether renting a DVD, eating a pizza, or taking a camping trip, structuring some activities with your teenagers is important. Such activities permit a context in which to communicate with them and their emotional well-being. Offer (2013) found that having family meals was associated with higher emotional well-being of adolescents.

Sometimes teenagers present challenges with which their parents feel unable to cope. Aside from monitoring their behavior closely, family therapy may be helpful. Two major goals of such therapy are to increase the emotional bond between the parents and the teenagers and to encourage positive consequences for desirable behavior (e.g., concert tickets for good grades) and negative consequences for undesirable behavior (e.g., loss of cell phone/car privileges).

**6. *Use technology to encourage safer driving.*** GPS devices are now available which can tell a parent where their teenager is, record any sudden stops, and record speed. Some teenagers scream foul and accuse their parents of not trusting them. One answer is that of a father who required his daughter to pay for her car insurance but told her she could get a major discount if the GPS device was installed. She then thought the GPS was a good idea (Copeland 2012).

## Ten Steps to Talking With Teens about Sex

Karen Rayne, Ph.D., is a specialist in talking with teens about sex. The mother of two teenagers, she recommends the following:

**1. *Know yourself.*** What are your expectations, your hopes, and your fears about your teenager's sexual and romantic development? You will have far more control over yourself and your interactions if you have a full understanding of these things.

**2. *Remember that it's not about you.*** Your teenager is, in fact, discovering sex for the first time. They want to talk about their current exciting, overwhelming path. So let them! That's how you'll find out what you can do to help your teenager walk this path—and remember, that's what matters most.

**3. *Stop talking!*** As the parent of a teenager, you are in the business of getting to know your teenager, not to give information. If you're talking, you can't hear anything your teenager is trying to tell you.

**4. *Start listening!*** Stop talking. Start listening. Remember what's most important about your role as a parent? And that can't happen if you don't really, really listen.

**5. *You get only one question.*** Since there's only one, you had better make it a good one that can't be answered with a yes or a no. Spend some time mulling over it. You can ask it when you're sure it's a good one.

**6. *Do something else.*** Anything else. Many teenagers, especially boys, will have an easier time talking about sexuality and romance if you're doing something "side by side" like driving, walking, or playing a game rather than sitting and looking at each other.

**7. *About pleasure and pain.*** You have to talk about both. If you don't acknowledge the physical and emotional pleasure associated with sexuality, your teenager will think you're completely out of touch, and so you will be completely out of touch.

**8. *Be cool like a cucumber.*** It is only when you manage to have a calm, loving demeanor that your teenager will feel comfortable talking with you. And remember—you're in the business of getting to know your teenager—and the only way to do that is if your teenager keeps talking.

**9. *Bring it on!*** Your teenagers have tough questions, some of them quite specific and technical. If you're able to answer these questions with honesty, humor, and without judgment, your teenager will feel much more at home coming to you with the increasingly difficult emotional questions that touch their heart.

**10. *Never surrender.*** There may be times you feel like quitting. Like the millionth time when you've tried to have an actual conversation with your teenager—about anything, much less sex!—and your teenager has once again completely avoided eye contact and has not even acknowledged your existence. But you can't. You're still doing some good, so keep going. Trust me. ©Karen Rayne 2013.

*Before I got married I had six theories about bringing up children; now I have six children and no theories.*

JOHN WILMONT, EARL OF ROCHESTER
1647–1680

## Child Rearing Theories

Parenting professionals have offered parents conflicting advice across time. For example, in 1914, parents who wanted to know what to do about their child's thumb sucking were told to try to control such a bad impulse by pinning the sleeves of the child to the bed if necessary. Today, parents are told that thumb sucking meets an important psychological need for security and they should not try to prevent it. If a child's teeth become crooked as a result, an orthodontist should be consulted.

Advice may also be profit driven. Much of what passes for "research-based advice" to provide parents with information about specific toys or learning programs turns out to be child development research funded by corporations who are intent on selling parents merchandise for their children. In her book *Buy Buy Baby: How Consumer Culture Manipulates Parents and Harms Young Minds*, Thomas (2007) identified the dangerous economic and cultural norm that encourages parents to buy things for their toddlers and infants that may actually harm them.

There are several theories about rearing children. These are reviewed in the following pages and summarized in Table 12.1. In examining these approaches, it is important to keep in mind that no single approach is superior to another. What works for one child may not work for another. Any given approach may not even work with the same child at two different times.

It is important for new parents to use a cafeteria approach when examining parenting advice from health care providers, family members, friends, parenting educators, and other well-meaning individuals. Take what makes sense, works, and feels right as a parent and leave the rest behind. Parents know their own child better than anyone else and should be encouraged to combine different approaches to find what works best for them and their child.

### Developmental-Maturational Approach

For the past 60 years, Arnold Gesell and his colleagues at the Yale Clinic of Child Development have been known for their ages-and-stages approach to childrearing. The developmental-maturational approach has been widely used in the United States. The basic perspective, considerations for childrearing, and criticisms of the approach follow.

**Basic Perspective** Gesell views what children do, think, and feel as being influenced by their genetic inheritance. Although genes dictate the gradual unfolding of a unique person, every individual passes through the same basic pattern of growth. This pattern includes four aspects of development: motor behavior (sitting, crawling, walking), adaptive behavior (picking up objects and walking around objects), language behavior (words and gestures), and personal-social behavior (cooperativeness and helpfulness). Through the observation of hundreds of normal infants and children, Gesell and his coworkers identified norms of development. Although there may be large variations, these norms suggest the ages at which an average child displays various behaviors. For example, on the average, children begin to walk alone (although awkwardly) at age 13 months and use simple sentences between the ages of two and three.

**Considerations for Childrearing** Gesell suggests that if parents are aware of their children's developmental clock, they will avoid unreasonable expectations. For example, a child cannot walk or talk until the neurological structures necessary for those behaviors have matured. Also, the hunger of a 4-week-old must be immediately appeased by food, but at 16 to 28 weeks, the child has some capacity to wait because the hunger pains are less intense. In view of this and other developmental patterns, Gesell suggested that infants need to be cared for on a demand schedule; instead of having to submit to a schedule imposed by parents, infants are fed, changed, put to bed, and allowed to play when they want. Children are likely to be resistant to a hard-and-fast schedule because they may be developmentally unable to cope with it.

**Criticisms of the Developmental-Maturational Approach** Gesell's work has been criticized because of (1) its overemphasis on the idea of a biological clock, (2) the deficiencies of the sample he used to develop maturational norms, (3) his insistence on the merits of a demand schedule, and (4) the idea that environmental influences are weak.

Most of the children who were studied to establish the developmental norms were from the upper middle class. Children in other social classes are exposed to different environments, which influence their development. Norms established on upper-middle-class children may not adequately reflect the norms of children from other social classes.

Gesell's suggestion that parents do everything for the infant when the infant desires has also been criticized. Rearing an infant on the demand schedule can drastically interfere with the parents' personal and marital interests. In the United States, with its emphasis on individualism, many parents feed their infants on a demand schedule but put them to bed to accommodate the parents' schedule.

## Behavioral Approach

The behavioral approach to childrearing, also known as the social learning approach, is based on the work of B. F. Skinner and the theme of *Behavioral Family Therapy* (Crisp and Knox 2009). Behavioral approaches to child behavior have received the most empirical study. Public health officials and policymakers are encouraged to promote programs with empirically supported treatments. The basic perspective, considerations, and criticisms of a behavioral approach to child rearing follow:

**Basic Perspective** Behavior is learned through classical and operant conditioning.

**Classical conditioning** involves presenting a stimulus with a reward. For example, infants learn to associate the faces of their parents with food, warmth, and comfort. Although initially only the food and feeling warm will satisfy the infant, later just the approach of the parent will soothe the baby. This may be observed when a father hands his infant to a stranger. The infant may cry because the stranger is not associated with pleasant events. But when the stranger hands the infant back to the parent, the crying may subside because the parent represents positive events and the stimulus of the parent's face is associated with pleasurable feelings and emotional safety.

**Classical conditioning** presenting a stimulus with a reward.

**Operant conditioning** focuses on the consequences of behavior.

Other behaviors are learned through **operant conditioning**, which focuses on the consequences of behavior. Two principles of learning are basic to the operant explanation of behavior—reward and punishment. According to the reward principle, behaviors that are followed by a positive consequence will increase. If the goal is to teach the child to say "please," doing something the child likes after he or she says "please" will increase the use of "please" by the child. Rewards may be in the form of attention, praise, desired activities, or privileges. Whatever consequence increases the frequency of an occurrence is, by definition, a reward. If a particular consequence does not change the behavior in the desired way, a different consequence needs to be tried.

*Find out what a child will work for and will work to avoid. Systematically manipulate the consequences and you can change behavior.*

JACK TURNER, CLINICAL PSYCHOLOGIST

The punishment principle is the opposite of the reward principle. A negative consequence following a behavior will decrease the frequency of that behavior; for example, the child could be isolated for 5 or 10 minutes following an undesirable behavior. The most effective way to change behavior is to use the reward and punishment principles together to influence a specific behavior. Praise children for what you want them to do and provide negative consequences for what they do that you do not like.

**Considerations for Childrearing** Parents often ask, "Why does my child act this way, and what can I do to change it?" The behavioral approach to childrearing suggests the answer to both questions. The child's behavior has been learned through being rewarded for the behavior, and it can be changed by eliminating the reward for or punishing the undesirable behavior and rewarding the desirable behavior.

The child who cries when the parents are about to leave home to see a movie is often reinforced for crying by the parents' staying home longer. To teach the child not to cry when they leave, the parents should reward the child for not crying when they are gone for progressively longer periods of time. For example, they might initially tell the child they are going outside to walk around the house and they will give the child a treat when they get back if he or she plays until they return. The parents might then walk around the house and reward the child for not crying on their return. If the child cries, they should be out of sight for only a few seconds and gradually increase the amount of time they are away. The essential point is that children learn to cry or not to cry depending on the consequences of crying. Because children learn what they are taught, parents might systematically structure learning experiences to achieve specific behavioral goals.

**Criticisms of the Behavioral Approach** Professionals and parents have attacked the behavioral approach to childrearing on the basis that it is deceptively simple and does not take cognitive issues into account. Although the behavioral approach is often presented as an easy-to-use set of procedures for child management, many parents do not have the background or skill to implement the procedures effectively. What constitutes an effective reward or punishment, how it should be presented, in what situation and with what child, to influence what behavior are all decisions that need to be made before attempting to increase or decrease the frequency of a behavior. Parents often do not know the questions to ask or lack the training to make appropriate decisions in the use of behavioral procedures. One parent locked her son in the closet for an hour to punish him for lying to her a week earlier—a gross misuse of learning principles.

Behavioral childrearing has also been said to be manipulative and controlling, thereby devaluing human dignity and individuality. Some professionals feel that humans should not be manipulated to behave in certain ways through the use of rewards and punishments. Behaviorists counter that rewards and punishments are already involved in any behavior the child engages in and parents are simply encouraged to arrange them in a way that benefits the child in terms of learning appropriate behavior.

Finally, the behavioral approach has been criticized because it de-emphasizes the influence of thought processes on behavior. Too much attention, say the critics, has been given to rewarding and punishing behavior and not enough attention has been given to how the child perceives a situation. For example, parents might think they are rewarding a child by giving her or him a bicycle for good behavior. But the child may prefer to upset the parents by rejecting the bicycle and may be more rewarded by their anger than by the gift. Behaviorists counter by emphasizing they are very much focused on cognitions and always include the perception of the child in developing a treatment program (Crisp and Knox 2009).

## Parent Effectiveness Training Approach

Thomas Gordon (2000) developed a model of childrearing based on parent effectiveness training (PET).

**Basic Perspective** Parent effectiveness training focuses on what children feel and experience in the here and now—how they see the world. The method of trying to understand what the child is experiencing is active listening, in which the parent reflects the child's feelings. For example, the parent who is told by the child, "I want to quit taking piano lessons because I don't like to practice" would reflect, "You're really bored with practicing the piano and would rather have fun doing something else." PET also focuses on the development of the child's positive self-concept.

To foster a positive self-concept in their child, parents should reflect positive images to the child—letting the child know he or she is loved, admired, and approved of.

**Considerations for Childrearing** To assist in the development of a child's positive self-concept and in the self-actualization of both children and parents, Gordon (2000) recommended managing the environment rather than the child, engaging in active listening, using "I" messages, and resolving conflicts through mutual negotiation.

An example of environmental management is putting breakables out of reach of young children. Rather than worry about how to teach children not to touch breakable knickknacks, it may be easier to simply move the items out of their reach.

The use of active listening becomes increasingly important as the child gets older. When Joanna is upset with her teacher, it is better for the parent to reflect the child's thoughts than to take sides with the child. Saying "You're angry that Mrs. Jones made the whole class miss play period because Becky was chewing gum" rather than saying "Mrs. Jones was unfair and should not have made the whole class miss play period" shows empathy with the child without blaming the teacher.

Gordon also suggested using "I" rather than "you" messages. Parents are encouraged to say "I get upset when you are late and don't text me" rather than "You're an insensitive, irresponsible brat for not texting me when you said you would." The former avoids damaging the child's self-concept but still expresses the parent's feelings and encourages the desired behavior.

Gordon's fourth suggestion for parenting is the no-lose method of resolving conflicts. Gordon rejects the use of power by parent or child. In the authoritarian home, the parent dictates what the child is to do and the child is expected to obey. In such a system, the parent wins and the child loses. At the other extreme is the permissive home, in which the child wins and the parent loses. The alternative, recommended by Gordon, is for the parent and the child to seek a solution that is acceptable to both and to keep trying until they find one. In this way, neither parent nor child loses and both win.

**Criticisms of the Parent Effectiveness Training Approach** Although much is commendable about PET, parents may have problems with two of Gordon's suggestions.

First, he recommends that because older children have a right to their own values, parents should not interfere with their dress, career plans, and sexual behavior. Some parents may feel they do have a right (and an obligation) to interfere. Second, the no-lose method of resolving conflict is sometimes unrealistic.

Suppose a 16 year old wants to spend the weekend at the beach with her boyfriend and her parents do not want her to. Gordon advises negotiating until a decision is reached that is acceptable to both. But what if neither the daughter nor the parents can suggest a compromise or shift their position? The specifics of how to resolve a particular situation are not always clear.

## Socioteleological Approach

Alfred Adler, a physician and former student of Sigmund Freud, saw a parallel between psychological and physiological development. When a person loses her or his sight, the other senses (hearing, touch, taste) become more sensitive—they compensate for the loss. According to Adler, the same phenomenon occurs in the psychological realm. When individuals feel inferior in one area, they will strive to compensate and become superior in another. Rudolph Dreikurs, a student of Adler, developed an approach to childrearing that alerts parents as to how their children might be trying to compensate for feelings of inferiority (Soltz and Dreikurs 1991). Dreikurs's socioteleological approach is based on Adler's theory.

**Basic Perspective** According to Adler, it is understandable that most children feel they are inferior and weak. From the child's point of view, the world is filled with strong giants who tower above him or her. Because children feel powerless in the face of adult superiority, they try to compensate by gaining attention (making noise, becoming disruptive), exerting power (becoming aggressive, hostile), seeking revenge (becoming violent, hurting others), and acting inadequate (giving up, not trying). Adler suggested that such misbehavior is evidence that the child is discouraged or feels insecure about her or his place in the family. The term *socioteleological* refers to social striving or seeking a social goal. In the child's case, the goal is to find a secure place within the family—the first "society" the child experiences.

**Considerations for Childrearing** When parents observe misbehavior in their children, they should recognize it as an attempt to find security. According to Dreikurs, parents should not fall into playing the child's game by responding to a child's disruptiveness with anger but should encourage the child, hold regular family meetings, and let natural consequences occur. To encourage the child, the parents should be willing to let the him or her make mistakes. If John wants to help Dad carry logs to the fireplace, rather than say, "You're too small to carry the logs," Dad should allow John to try to carry whatever he can pick up and learn for himself.

As well as being constantly encouraged, the child should be included in a weekly family meeting. During this meeting, such family issues as bedtimes, the appropriateness of between-meal snacks, assignment of chores, and family fun are discussed. The meeting is democratic; each family member has a vote. Participation in family decision making is designed to enhance the self-concept of each child. By allowing each child to vote on family decisions, parents respect the child as a person as well as the child's needs and feelings.

Resolutions to conflicts with the child might also be framed in terms of choices the child can make. "You can go outside and play only in the backyard, or you can play in the house" gives the child a choice. If the child strays from the backyard, he or she can be brought in and told, "You can go out again later." Such a framework teaches responsibility for and consequences of one's choices.

Finally, Driekurs suggested that the parents let natural consequences occur for their child's behavior. If a daughter misses the school bus, she walks or is charged taxi fare out of her allowance. If she won't wear a coat and boots in bad weather, she gets cold and wet. Of course, parents are to arrange logical consequences when natural consequences will not occur or would be dangerous if they did. For example, if a child leaves the television on overnight, access might be taken away for the next day or so.

**Criticisms of the Socioteleological Approach** The socioteleological approach is sometimes regarded as impractical, since it teaches the importance of letting children take the natural consequences for their actions. Such a principle may be interpreted to let the child develop a sore throat if he or she wishes to go out in the rain without a raincoat. In reality, advocates of the method would not let the child make a dangerous decision. Rather, they would give the child a choice with a logical consequence such as "You can go outside wearing a raincoat, or you can stay inside—it is your choice."

## Attachment Parenting

Dr. William Sears and his wife, Martha Sears, developed an approach to parenting called attachment parenting (Sears and Sears 1993). This common sense parenting approach focuses on parents' emotional connecting with their baby.

**Basic Perspective** The emotional attachment process between mother and child is thought to begin prior to birth and continues to be established during the next three years. Sears identified three parenting goals: to know your child, to help your child feel right, and to enjoy parenting. He also suggested five concepts or tools (identified in the following section) that comprise attachment parenting that will help parents to achieve these goals. Overall, the ultimate goal is for parents to get connected with their baby. Once parents are connected, it is easy for parents to figure out what works for them and to develop a parenting style that fits them and their baby. Meeting a child's needs early in life will help him or her form a secure attachment with parents. This secure attachment will help the child to gain confidence and independence as he or she grows up. Attachment Parenting International is a nonprofit organization committed to educating society and parents about the critical emotional and psychological needs of infants and children.

**Considerations for Childrearing** The first attachment tool is for parents to connect with their baby early. The initial months of parenthood are a sensitive time for bonding with the baby and starting the process of attachment.

The second tool is to read and respond to the baby's cues. Parents should spend time getting to know their baby and learn to recognize his or her unique cues. Once a parent gets in tune with the baby's cues, it is easy to respond to the child's needs. Sears encourages parents to be open and responsive and to pick their baby up if he or she cries. Responding to a baby's cries helps the baby to develop trust and encourages good communication between child and parent. Eventually, babies who are responded to will internalize their security and will not be as demanding.

The third attachment tool is for mothers to breast feed their babies and to do this on demand rather than trying to follow a schedule. He emphasizes the important role that fathers also play in successful breast feeding by helping to create a supportive environment.

The fourth concept of attachment parenting is for parents to carry their baby in a sling or carrier. The closeness is good for the baby and it makes life easier for the parent. Wearing your baby in a sling or carrier allows parents to engage in regular day-to-day activities and also makes it easier to leave the house.

Finally, Sears advocates that parents let the child sleep in their bed with them since it allows parents to stay connected with their child throughout the

**TABLE 12.1 Theories of Child Rearing**

| Theory | Major Contributor | Basic Perspective | Focal Concerns | Criticisms |
|---|---|---|---|---|
| Developmental-Maturational | Arnold Gesell | Genetic basis for child passing through predictable stages | Motor behavior<br>Adaptive behavior<br>Language behavior<br>Social behavior | Overemphasis on biological clock<br>Inadequate sample to develop norms<br>Demanding schedule questionable<br>Upper-middle-class bias |
| Behavioral | B. F. Skinner<br>Charles Madsen<br>Bryan Crisp | Behavior is learned through operant and classical conditioning | Positive reinforcement<br>Negative reinforcement<br>Punishment<br>Extinction<br>Stimulus response | De-emphasis on cognitions of child<br>Theory too complex for parent to accurately or appropriately apply<br>Too manipulative or controlling<br>Difficult to know reinforcers and punishers in advance |
| Parent Effectiveness Training | Thomas Gordon | The child's world view is the key to understanding the child | Change the environment before attempting to change the child's behavior<br>Avoid hurting the child's self-esteem<br>Avoid win-lose solutions | Parents must sometimes impose their will on the child's<br>How to achieve win-win solutions is not specified |
| Socioteleological | Alfred Adler | Behavior is seen as attempt of child to secure a place in the family | Insecurity<br>Compensation<br>Power<br>Revenge<br>Social striving<br>Natural consequences | Limited empirical support<br>Child may be harmed taking "natural consequences" |
| Attachment | William Sears | Goal is to establish a firm emotional attachment with child | Connecting with baby<br>Responding to cues<br>Breast-feeding<br>Wearing the baby<br>Sharing sleep | May result in spoiled, overly dependent child<br>Exhausting for parents |

night. However, some parents and babies often sleep better if the baby is in a separate crib, and Sears recognizes that wherever parents and their baby sleep best is the best policy. Wherever you choose to have your baby sleep, Sears is clear on one thing—it is never acceptable to let your baby cry when he or she is going to sleep. Parents need to parent their child to sleep rather than leaving him or her to cry.

**Criticisms of Attachment Parenting** Some parents feel that responding to their baby's cries, carrying or wearing their baby, and sharing sleep with their baby will lead to a spoiled baby who is overly dependent. Some parents may feel more tied down using this parenting approach and may find it difficult to get their child on a schedule. Many women return to work after the baby is born and find some of the concepts difficult to follow. Some women choose not to breast feed their children for a variety of reasons. Finally, Treat et al. (2013) collected data on mothers who co-slept with their infants and found that doing so did *not* help the infants self-soothe or self-regulate. Co-sleeping

Kelly and Minnie

This single mother is also a medical student and knows the meaning of exhaustion and stress.

was also associated with elevated SIDS risk scores, greater infant emotionality, and sleeping problems.

## Single-Parenting Issues

Forty percent of births in the United States are to unmarried mothers. Distinguishing between a single-parent "family" and a single-parent "household" is important. A **single-parent family** is one in which there is only one parent—the other parent is completely out of the child's life through death or complete abandonment or as a result of sperm donation, and no contact is ever made. In contrast, a **single-parent household** is one in which one parent typically has primary custody of the child or children, but the parent living out of the house is still a part of the child's life. This arrangement is also referred to as a binuclear family. In most divorce cases in which the mother has primary physical custody of the child, the child lives in a single-parent household because the child is still connected to the father, who remains part of the child's life. In cases in which one parent has died, the child, or children live with the surviving parent in a single-parent family because there is only one parent.

**Single-parent family** one in which there is only one parent.

**Single-parent household** one parent has primary custody of the child/children with the other parent living outside of the house but still being a part of the child's family; also called binuclear family.

### Single Mothers by Choice

Single parents enter their role though divorce or separation, widowhood, adoption, or deliberate choice to rear a child or children alone. Actress Diane Keaton, never married, has two adopted children. An organization for women who want children and who may or may not marry is Single Mothers by Choice.

## Challenges Faced by Single Parents

The single-parent lifestyle involves numerous challenges. See singleparent.lifetips.com for some interesting advice. Challenges associated with being a single parent include the following.

**1.** ***Responding to the demands of parenting with limited help.*** Perhaps the greatest challenge for single parents is taking care of the physical, emotional, and disciplinary needs of their children alone. Depression and stress are common among single parents (Hong and Welch 2013). One single mother said, "I'm working two jobs, taking care of my kids, and trying to go to school. Plus my mother has cancer. Who am I going to talk with to help me get through all of this?" Many single women solve the dilemma with a network of friends.

**2.** ***Resolving the issue of adult sexual needs.*** Some single parents regard their parental role as interfering with their sexual relationships. They may be concerned that their children will find out if they have a sexual encounter at home or be frustrated if they have to go away from home to enjoy a sexual relationship. Some choices with which they are confronted include, "Do I wait until my children are asleep and then ask my lover to leave before morning?" or "Do I openly acknowledge my lover's presence in my life to my children and ask them not to tell anybody?" and "Suppose my kids get attached to my lover, who may not be a permanent part of our lives?"

**3.** ***Coping with lack of money.*** Single-parent families, particularly those headed by women, report that money is always lacking.

### National **Data**

The median income of a single-woman householder is $25,269, much lower than that of a single-man householder ($36,611) or a married couple ($71, 830) (*Statistical Abstract of the United States 2012-2013,* Table 692).

**4.** ***Ensuring guardianship.*** If the other parent is completely out of the child's life, the single parent needs to appoint a guardian to take care of the child in the event of the parent's death or disability.

**5.** ***Obtaining prenatal care.*** Single women who decide to have a child have poorer pregnancy outcomes than married women. Their children are likely to be born prematurely and to have low birth weight (Mashoa et al. 2010). The reason for such an association may be the lack of economic funds (no male partner with economic resources available) as well as the lack of social support for the pregnancy or the working conditions of the mothers.

**6.** ***Coping with the absence of a father.*** Another consequence for children of single-parent mothers is that they often do not have the opportunity to develop an emotionally supportive relationship with their father. Barack Obama noted, "I know what it is like to grow up without a father." The late comedian Rodney Dangerfield said that in his entire life he spent an average of two hours a year with his father. We earlier detailed the value of an involved father for a child's life.

**7.** ***Avoiding negative life outcomes for the child in a single-parent family.*** Researcher Sara McLanahan, herself a single mother, set out to prove that children reared by single parents were just as well off as those reared by two parents. McLanahan's data on 35,000 children of single parents led her to a different conclusion: children of only one parent were twice as likely as those reared by two married parents to drop out of high school, get pregnant before marriage, have drinking problems, and experience a host of other difficulties, including getting divorced themselves (McLanahan and Booth 1989; McLanahan 1991). In addition,

Freeman and Temple (2010) found that adolescents from single-parent homes were more likely to be raped than those from two-parent homes. Lack of supervision, fewer economic resources, and less extended family support were among the culprits.

**8.** ***Perpetuating a single family structure.*** Growing up in a single-family home increases the likelihood that the adult child will have a first child while unmarried and in a cohabitation relationship, thus perpetuating the single-family structure.

Though the risk of negative outcomes is higher for children in single-parent homes, most are happy and well-adjusted. Maier and McGeorge (2013) emphasized that single parents (both mothers and fathers) are victims of negative stereotyping. The reality is that there are numerous positives associated with being a single parent. These include having a stronger bonding experience with one's children since they "are" the family, a sense of pride and self-esteem for being independent, and being a strong role model for offspring who observe their parent being able to "wear many hats."

## The Future of Parenting

The future of parenting will involve new contexts for children and new behaviors that children learn and parents tolerate. While parents will continue to be the primary context in which their children are reared, because the financial need for both parents to earn an income will increase, children will, increasingly, end up in day care, afterschool programs, and day camps during the summer. There will also be an increasing number of children reared in single-family contexts. These changed contexts will result in new parental norms where a wider range of behaviors on the part of their children will be accepted. Hence, since parents will be increasingly preoccupied with their job/careers, the norms their children are learning in day care and other contexts will be more readily accepted since parents will have less time and energy trying to reverse them.

Hence, children may be less polite, less obedient, and less compliant to authority resulting more often in parental acceptance than addressing or correcting each behavior. Teachers in the public school system often comment that "lack of parental involvement" is showing. We are not suggesting that children will become wild hellions, only that the behaviors they learn in contexts other than the home will be less restrictive.

## Summary

***What are the basic roles of parents?***

Parenting includes providing physical care for children, loving them, being an economic resource, providing guidance as a teacher or model, protecting them from harm, promoting their health, and providing meaningful family rituals. Seven skills all children must have (that parents are responsible for teaching) are: focus and self-control, seeing some else's point of view, communicating, making connections, critical thinking, taking on challenges, and self-directed engaged learning. One of the biggest problems confronting parents today is the societal influence on their children. These include drugs and alcohol; peer pressure; TV, Internet, and movies; and crime or gangs.

***What is a choices perspective of parenting?***

Although both genetic and environmental factors are at work, the choices parents make have a dramatic impact on their children. Parents who don't make a choice about parenting have already made one. The six basic choices parents make include deciding (1) whether to have a child, (2) the number of children, (3) the interval between children, (4) one's method of discipline and guidance, (5) the degree to which one will be invested in the role of parent, and (6) the degree to which the parents will co-parent. The choices perspective also emphasizes that not to make a choice is to make a choice, all decisions are trade-offs, reframing regretful decisions is important, and one's society/culture has an impact on choices.

***What is the transition to parenthood like for women, men, and couples?***

Transition to parenthood refers to that period of time from the beginning of pregnancy through the first few months after the birth of a baby. The mother, father, and couple all undergo changes and adaptations during this period. Most mothers relish their new role; some may experience the transitory feelings of baby blues; a few report postpartum depression.

A summary of almost 150 studies involving almost 50,000 respondents on the question of how children affect marital satisfaction revealed that parents (both women and men) reported lower marital satisfaction than nonparents. In addition, the higher the number of children, the lower the marital satisfaction; the factors that depressed marital satisfaction were conflict and loss of freedom. The new couple also gradually withdraws from friends and spends increasing amounts of time at home.

***What are several facts about parenthood?***

The concept of childhood, like gender, has been socially constructed rather than being a fixed life stage. Parenthood will involve about 40% of the time a couple live together, parents are only one influence on their children, each child is unique, and parenting styles differ. Research suggests that an authoritative parenting style, characterized as both demanding and warm, is associated with positive outcomes. In addition, being emotionally connected to a child, respecting the child's individuality, and monitoring the child's behavior to encourage positive contexts have positive outcomes.

***What are some of the principles of effective parenting?***

Giving time, love, praise, and encouragement; monitoring the activities of one's child; setting limits; encouraging responsibility; and providing sexuality education are aspects of effective parenting.

***What are the issues of single parenting?***

About 40% of all children will spend one-fourth of their lives in a female-headed household. The challenges of single parenthood for the parent include taking care of the emotional and physical needs of a child alone, meeting one's own adult emotional and sexual needs, money, and rearing a child without a father (the influence of whom can be positive and beneficial).

***What is the future of parenting?***

Focused on income getting, parents will depend more on secondary/nonfamily resources to take care of and rear their children. The result will be a wider set of norms that children learn and behaviors they engage in with less parental correction. Parents will remain in control of their children but less so than in the past.

## Key Terms

Baby blues
Boomerang generation
Demandingness
Emotional competence
Classical conditioning
Gatekeeper role
Operant conditioning
Oppositional defiant disorder
Overindulgence
Oxytocin
Parenting
Parenting self-efficiency
Postpartum depression
Postpartum psychosis
Reactive attachment disorder
Responsiveness
Sextortion
Single-parent family
Single-parent household
Time-out
Transition to parenthood

# Web Links

Attachment Parenting International
http://www.attachmentparenting.org/

Family Wellness Workshops
http://www.familywellness.com/

The Partnership for a Drug-Free America
http://www.drugfree.org/

The Children's Partnership Online
http://www.childrenspartnership.org

National Fatherhood Initiative
http://www.fatherhood.org/

Single Parenting
http://singleparent.lifetips.com/

Single Parenting—Grants to go back to school
http://www.schoolgrantsblog.com/grants-for-single-mothers/

Un/Hushed—Talking with Adolescents about Sex
http://www.unhushed.net/

# CHAPTER 13 Money, Work, and Relationships

*I slept and dreamed that life was beauty;*
*I woke and found that life was duty.*

ELLEN STURGIS HOOPER, AMERICAN POET

Chelsea Curry

## Learning Objectives

- Discuss how money, and the lack of it—poverty—affects relationships.
- Review money as power, office romance, and working wives.
- Identify the effect of a wife's employment on her children.
- Review the various strategies of balancing work and family life.
- Predict the future of the effect of work on family life.

## TRUE OR FALSE?

1. Working mothers rate themselves higher on their job as parent than non-working mothers.
2. The more hours a couple work, the greater the emotional distance they feel from each other.
3. Due to a new corporate climate, women who drop out of the work force to have their babies often find that the corporation will not only hold their jobs but also promote them more quickly on their return.
4. The more traditional the marriage, the higher the sex frequency.
5. A woman who is happy in her marriage but who takes a lucrative job becomes vulnerable to divorcing her husband.

*Answers:* 1. T 2. T 3. F 4. T 5. F

Marissa Mayher was the 37-year-old CEO of Yahoo when she announced her pregnancy. The news generated a mixed reaction from "she can't have it all" to "good for her for showing that women can balance career and children." There is certainly a precedent for women continuing to be career involved once children arrive. Model Heidi Klum gave birth in October and was back on the runway in November. When fashion designer Victoria Beckham had her fourth child, she expressed no intention of taking maternity leave. These women have all of the important factors for combining work and family—excellent child care, a flexible work schedule, and a supportive partner who has a flexible work schedule. What is interesting in the whole debate of career women and children is that career CEO men who have children never experience the same scrutiny (Petrecca 2012, p. 2). We never see a news article on a male "CEO Manager Having First Child." In this chapter, we discuss work and family and its impact on individuals, relationships, and children. We begin with a discussion of how money and poverty affect relationships.

# Money and Relationships

Money is a concern for most individuals. When 487 men and 513 women were asked what area of their life they would most want to change (study conducted by Kelton Research), finances was at the top of the list (selected by 75% of men and 81% of women) over health, appearance, and self-esteem (Healy and Trap 2012). High inflation creates economic uncertainty, which reduces the marriage rate (Schellekens and Gliksberg 2013). A recession, which is typically associated with job layoffs and housing foreclosures, often means more negative interaction between the partners (Masarik et al. 2012). Intimate partner homicides (murder of a spouse) also increase (Diem and Pizarro 2010). Arguing about money is a gateway to more problems in the relationship, such as spending less time together and having disagreements about sex (Wheeler and Kerpelman 2013).

*It isn't so much that hard times are coming; the change observed is mostly soft times going.*

GROCHO MARX, COMEDIAN

Difficult economic times are also associated with positive consequences such as causing individuals to become less consumer oriented, more engaged in their relationships, and more involved in transcendental activities (religious, contemplative) (Etzioni 2011). Indeed, the entrenched value of **consumerism**—to buy everything and to have everything now—has come under fire. The real stress that money inflicts on relationships is the result of internalizing the

*He who has no money is poor, he who has nothing but money is even poorer.*

AMISH PROVERB

*We all think we're going to get out of debt.*

LOUIE ANDERSON, COMEDIAN

societal expectations of who one is, or should be, in regard to the pursuit of money. A wife whose husband had just bought a second McDonald's franchise said, "It's not fun anymore. We have money but I never see him. It isn't worth it." (The husband subsequently sold both stores, and while the couple had less money, their marriage recovered.)

*Get as much pleasure from your saving as from your spending.*

SUZE ORMON, INVESTMENT GURU

WHAT IF?

## What If You and Your Partner Feel Differently about Being in Debt?

Being in debt is a dating liability. In a study of 5,500 never-married individuals, almost two-thirds (65%) reported that they would not date someone with a credit card debt over $5,000 (Walsh 2013). Some individuals have always been in debt, regard it as a given in life, and have no anxiety about it. Others have never been in debt, view it as something to avoid, and become depressed when thinking about their debt. These different philosophies may create conflict. Resolution of this dilemma usually comes through deferring to the preferences of the partner with the most anxiety. For example, the partner who can't sleep if in debt will be in greater distress and will create more stress in the relationship than the partner who can tolerate a great deal of debt. Hence, the least negative consequences for the couple are in favor of having minimal debt.

## Money as Power in a Couple's Relationship

*Use it up, wear it out, make it do or do without.*

AMISH PROVERB

Money is a central issue in relationships because of its association with power, control, and dominance. Generally, the more money a partner makes, the more power that person has in the relationship. Males in general make considerably more money than females and generally have more power in relationships.

### National **Data**

The average annual income of a male with some college education who is working full time is $52,580 compared with $36,553 for a female with the same education, also working full time (*Statistical Abstract of the United States, 2012–2013,* Table 703).

*It is pretty hard to tell what does bring happiness; poverty and wealth have both failed.*

KIN HUBBARD, AMERICAN HUMORIST

When a wife earns an income, her power in the relationship increases. We (the authors of your text) know of a married couple in which the wife recently began to earn an income. Before doing so, her husband's fishing boat was stored

### National **Data**

A record 40% of all households with children under the age of 18 include mothers who are either the sole or primary source of income for the family. These "breadwinner moms" are made up of two very different groups: 5.1 million (37%) are married mothers who have a higher income than their husbands, and 8.6 million (63%) are single mothers (Wang et al. 2013).

in the couple's protected carport. Her new job/income resulted in her having more power in the relationship, as reflected in her parking her car in the carport and her husband putting his fishing boat underneath the pine trees to the side of the house.

*I'd like to live as a poor man with lots of money.*

PABLO PICASSO

To some individuals, money means love. While admiring the engagement ring of her friend, a woman said, "What a big diamond! He must really love you." The cultural assumption is that a big diamond equals expense and a lot of sacrifice and love. Similar assumptions are often made when gifts are given or received. People tend to spend more money on presents for the people they love, believing that the value of the gift symbolizes the depth of their love. People receiving gifts may make the same assumption. "She must love me more than I thought," mused one man. "I gave her a Blu-Ray movie for Christmas, but she gave me a Blu-Ray player. I felt embarrassed."

## Effects of Poverty on Marriages and Families

In 2014, a two-person household with an income below $15,370 was defined as living in poverty (Poverty line 2014). Such individuals have poorer physical and mental health, report lower personal and relationship satisfaction, and die at younger ages. The anxiety over lack of money may result in relationship conflict. Hardie and Lucas (2010) analyzed data from over 4,000 respondents in both married and cohabitation contexts and found that economic hardship was associated with more conflict in both sets of relationships. Predictably, money (the lack of it, disagreement over how it is spent) is a frequent problem reported by couples who become involved in marriage counseling.

*We pretend to work because they pretend to pay us.*

AUTHOR UNKNOWN

Parenting is also negatively affected by economic pressure (e.g., can't cover expenses, financial cutbacks, material needs). Parents experiencing economic

Chelsea Curry

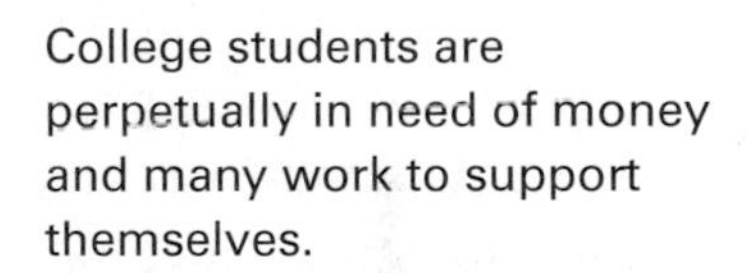

College students are perpetually in need of money and many work to support themselves.

*Closet space is like money. You use up as much as you have.*

RAY MICHEL, FOOTBALL PLAYER

*If you would be wealthy, think of saving as well as getting.*

BEN FRANKLIN

pressure report more emotional distress and harsher parenting (Neppl 2012). Mesarik et al. (2012) emphasized that conflict resolution skills may help to reduce the negative effects of economic pressures.

Kahneman and Deaton (2010) defined emotional well-being as the quality of an individual's everyday experience—the frequency and intensity of experiences of joy, stress, sadness, anger, and affection that make one's life pleasant or unpleasant. They found that reporting emotional well-being rises with income and tops out at $75,000 a year (adjusting for inflation through 2015, the figure is around $80,000). Increases in income above this figure are not associated with increases in satisfaction.

# Work and Marriage: Effects of Work on Spouses

A couple's marriage is organized around the work of each spouse. Where the couple live is determined by where the spouses can get jobs. Jobs influence what time spouses eat, which family members eat with whom, when they go to bed, and when, where, and for how long they vacation. In this section, we examine some of the various influences of work on a couple's relationship. We begin by looking at the skills identified by dual earner spouses to manage their work/job/career so as to provide income for the family but minimal expense to their relationship.

*It doesn't interest me what you do for a living. I want to know what you ache for, and if you dare to dream of meeting your heart's desire.*

ORIAH MOUNTAIN DREAMER, *THE INVITATION*

## Basic Rules for Managing One's Work Life to Have a Successful Marriage

Ma a tta and Uusiautti (2012) analyzed data from 342 married couples who explained their secrets for maintaining a successful relationship in the face of the demands of work. These secrets included turning a negative into a positive, being creative, tolerating dissimilarity (accepting the partner), and being committed to the relationship. The latter focused on accepting as a premise that the partners will work through whatever difficulties they encounter.

A wife who earns an income has more power in the relationship than a wife who does not earn an income.

Chelsea Curry

## PERSONAL CHOICES

### *Work or Relationships?*

People who consistently choose their work over their relationships either have partners who have also made such choices or partners who are disenchanted. Traditionally, men have chosen their work over their relationships; women have chosen their relationships over work. One exception is Katy Perry, singer, who noted in her documentary the toll her exhaustive concert schedule/tour had on her marriage. Most people try to balance their work and relationship so as to experience a life they enjoy. Professions or careers that inherently provide the opportunity for balance include elementary school teaching, where one is home relatively early in the day with a couple of months off in the summer. The role of college or university teacher is even better, with greater flexibility during the day, week, and year. Most university faculty do not teach classes in the summer. While they may conduct research or prepare and publish articles in the summer, they are not burdened by punching a nine-to-five clock. Some individuals may work from home full time. A **momprenuer** is a mother who has a successful at-home business.

Some individuals require very little income and have no interest in material wealth. We, the authors, have a friend who has little to no regard for material wealth. He has been homeless and now runs a homeless shelter where he and his staff feed two meals a day to over 100 individuals, paid for through private funding and donations. At night, he goes back to his $200 a month loft where he reads and paints. We asked him how his lack of concern for money affects the interest women have in establishing a long-term relationship with him. "It kills it," he noted. "They simply have no interest in being pair-bonded to someone living this vagabond existence."

Work may also have a positive effect on an individual and his or her relationships. Danner-Vlaardingerbroek et al. (2013) noted that while work can have a negative effect on parent-child interaction (mood, rumination about work, exhaustion) it can also provide a source of positive mood since individuals can enjoy their work and be proud of what they do.

*I wanted the gold, and I got it—*
*Came out with a fortune last fall,—*
*Yet somehow life's not what I thought it,*
*And somehow the gold isn't all.*

ROBERT W. SERVICE, SPELL OF THE YUKON

**Mompreneur** a mother who has a successful at-home business.

## Employed Wives

Driven primarily by the need to provide income for the family, 77% of all U.S. wives with children are in the labor force. The time wives are most likely to be in the labor force is when their children are teenagers (between the ages of 14 and 17), the time when food and clothing expenses are the highest (*Statistical Abstract of the United States, 2012–2013,* Tables 599 and 600).

*He who marries for love without money has good nights and sorry days.*

ANONYMOUS

Some women prefer to be employed part time rather than full time (Parker and Wang 2013). One option is the teaching profession, which allows employees to work about ten months a year and to have two months free in the summer. Although many low-wage earners need two incomes to afford basic housing and a minimal standard of living, others have two incomes to afford expensive homes, cars, vacations, and educational opportunities for their children. Treas et al. (2011) found that homemakers (in 28 countries) are happier than full-time working wives. However, to be clear, many women enjoy their jobs and careers, and staying at home with children/being a full-time homemaker is not a role in which these women are interested.

Hoffnung and Williams (2013) assessed 200 female college seniors, most of whom wanted both a career and children, and followed up on them 16 years later. Though they could be divided into three role-status outcome groups—Have It All (mothers, employed full time), Traditional (mothers, employed part time or not at all), and Employed Only (childfree, employed full time)—most of

Some women prefer to be at home with their babies than to be in the work force.

Chelsea Curry

the women still wanted to "have it all." Many traditional women looked forward to returning to employment, and many of the employed only women wanted to have children. Some parents wonder if the money a wife earns by working outside the home is worth the sacrifices to earn it. Not only is the mother away from her children, but she must pay for others to care for the children. Sefton (1998) calculated that the value of a stay-at-home mother is $36,000 per year in terms of what a dual-income family spends to pay for all the services that she provides (domestic cleaning, laundry, meal planning and preparation, shopping, providing transportation to activities, taking the children to the doctor, and running errands). Adjusting for changes in the consumer price index, this figure was over $62,000 in 2015. The value of a househusband would be the same. However, because males typically earn higher incomes than females, the loss of income would be greater than for a female.

Of interest is that working mothers give themselves slightly higher ratings than non-working mothers for the job they are doing as parents. Among mothers with children under age 18 who work full or part time, 78% say they are doing an excellent or very good job as parents. Among mothers who are not employed, 66% say the same (Parker and Wang 2013).

**Mommy track** stopping paid employment to spend time with young children.

The **mommy track** (stopping paid employment to spend time with young children) is another potential cost to those women who want to build a career. Taking time out to rear children in their formative years can derail a career. Unless a young mother has a supportive partner with a flexible schedule, family who live close and who can take care of her children, or money to pay for child care, she will discover that corporations need the work done and don't care about the kids. Kahn et al. (2014) reconfirmed that motherhood is costly to women's careers, but mostly to women who have three children. Those who have one or two children experience the greatest expense when the woman is young but this disadvantage is eliminated by the 40s and 50s.

## Office Romance

The office or workplace is where people earn money to pay bills and pay down their debt. It is also a place where they meet and establish relationships,

Chelsea Curry

The office is a prime context for a love relationship to develop.

including love and sexual relationships. In a survey of over a thousand employees in the workplace, almost 60% (56%) said that they had been involved in an office romance (Vault Office Romance Study 2014). In CareerBuilder's annual romance survey (2014), 38% reported having dated someone who worked in the same company. Jane Merrill interviewed 70 adults who had been involved in an office romance. Of these interviews she reported that "...the office is still a scene of seduction and amorality. The hallmark of office romances is secrecy—either hiding a steamy short or long love relationship between two single people or where one or both is married or in a serious relationship" (Merrill and Knox 2010). Undergraduates have also experienced romance on the job (see Applying Social Research feature).

## Types of Dual-Career Marriages

A **dual-career marriage** is defined as one in which both spouses pursue careers and may or may not include dependents. A career is different from a job in that the former usually involves advanced education or training, full-time commitment, long hours/night/weekend work "off the clock," and a willingness to relocate. Dual-career couples typically operate without a person (e.g., a "wife") who stays home to manage the home and children. However, some couples hire a nanny.

Types of dual-career marriages are those in which the husband's career takes precedence (**HIS/her career marriage**), the wife's career takes precedence (**HER/his career marriage**), both careers are equally important (**HIS/HER career marriage**), or both spouses share a career or work together (**THEIR career marriage**).

When couples hold traditional gender role attitudes, the husband's career is likely to take precedence (HIS/her career). This situation translates into the wife being willing to relocate and to disrupt her career for the advancement of her husband's career. Davis et al. (2012) examined a national sample of married couples during the "great recession" and found that women were more willing to move in support of their husbands' careers than vice versa.

**Dual-career marriage** one in which both spouses pursue careers and may or may not include dependents.

**HIS/her career marriage** dual-career marriage in which the husband's career takes precedence.

**HER/his career marriage** dual-career marriage in which the wife's career takes precedence.

**HIS/HER career marriage** dual-career marriage in which both careers are equally important.

**THEIR career marriage** dual-career marriage in which both spouses share a career or work together.

## APPLYING SOCIAL RESEARCH | Undergraduate Love and Sex on the Job

Seven-hundred seventy-four undergraduates in North Carolina, Florida, and California completed an Internet survey on office romance. The purpose was to identify the percent of the sample who had fantasized about love and sex on the job and the percent who actually experienced their fantasies. Other objectives included the percent reporting having kissed a coworker, whether the involvement was with someone of equal or higher status, having disclosed a relationship to others, having been sexually harassed, etc.

Over three-fourths (78%) of the respondents were female; 22% were male. Forty-seven percent were employed in a service job such as sales, fast food, or retail; 9% worked in an academic context, 12% in an office, and 4% in a medical context. Most (73%) regarded where they worked as a job, not a career, and 13% saw it as a place to meet a future spouse.

Over 40% of the student employees reported fantasies of both love and sex with a coworker in contrast to a quarter who *actually experienced* love and sex with a coworker (see Table 13.1).

**TABLE 13.1 Fantasies and Realities on the Job**
(N = 774)

| | Fantasized About | Actually Experienced |
|---|---|---|
| Love Relationship at the Office | 43% | 25% |
| Sexual Relationship at the Office | 41% | 24% |

Other findings included:

### Kissing

Almost 30% (28.6%) reported that they had kissed a fellow employee at work.

### Rank of Person Undergraduate Had Sex With

Almost 80% (79%) reported having had sex with a peer/co-worker; 11% of those surveyed had sex with a boss, supervisor, or someone above them.

Wives who move in reference to their husbands' careers or who have children in the middle of their careers often find their climb to the top is slow. Couples often drift into traditional roles because the husband's income is

Chelsea Curry

Some couples agree that the husband will be Mr. Mom.

### Telling Others about the Office Romance

About 30% (28.1%) reported that they told someone else about their office romance.

### Sexual Harassment

About 20% (19%) reported that they had been physically touched at work in a way that made them uncomfortable. A higher percentage (30%) said that a person at work said something sexual to them that made them uncomfortable (only 3% filed a formal complaint of sexual harassment).

### Duration of Office Romance

Most (18%) of the office romances lasted less than six months; 5% lasted six months to a year; 5% lasted between one and two years; and 2% lasted five years and continued. The remaining respondents said they had not been involved in an office romance.

### After the Office Romance Ends

Of the workplace romances that had ended, almost three-quarters (73%) ended positively with over half (54%) remaining friends. Thirteen percent still see each other, 5% are married to each other, and 1% live together.

### Losing One's Job

Less than 2% (1.7%) ended up losing their job at the place where they had the office romance.

The Society for Human Resources Management conducted a survey in which they interviewed 380 human resource professionals about office romance policies. Over half (54%) had no policy; 36% had a written policy; and 6% had a verbal policy (Carey and Trapp 2013).

### Sources


Carey, A. R. and P. Trapp. 2013. Companies workplace policies. *USA Today*, Oct. 20, A1.

Merrill, J. and D. Knox. 2010. *Finding Love from 9 to 5: Trade Secrets of an Office Love.* Santa Barbara, CA: Praeger.


higher. Some women avoid getting into a context where they cannot pursue their careers. In a study of educated Papua New Guinean women, Spark (2011) found them reluctant to marry Papua New Guinean men. Indeed, these women avoided marriage because they feared it would destroy their career prospects.

### National **Data**

About six in ten wives work today, nearly double the percentage in 1960. More than 62% of survey respondents endorse the modern marriage (HIS/HER career marriage) in which the husband and wife both work and both take care of the household and children (Parker and Wang 2013).

For couples who do not have traditional gender role attitudes, the wife's career may take precedence (HER/his career). In such marriages, the husband is willing to relocate and to subordinate his career for his wife's. Such a pattern is also likely to occur when a wife earns considerably more money than her husband. In some cases, the husband who is downsized by his employer or who prefers the role of full-time parent becomes "Mr. Mom." Not only may the wife benefit in terms of her career advancement, the relationship of the father with his children will be closer than that of fathers in traditional marriages who spend less time with their children.

Some dual-career marriages are those in which the careers of both the wife and husband are given equal status in the relationship (HIS/HER career). Deutsch et al. (2007) surveyed 236 undergraduate senior women to assess their views on husbands, managing children, and work. Most envisioned two egalitarian scenarios in which both spouses would cut back on their careers and/or both would arrange their schedules to devote time to child care. Still other couples would hire domestic child care help so that neither spouse functions in the role of traditional parent. In reality, equal status to HIS and HER careers is not a dominant pattern (Stone 2007).

Finally, some couples have the same career and may work together (THEIR career). Some news organizations hire both spouses to travel abroad to cover the same story. These careers are rare.

*Money often costs too much.*

RALPH WALDO EMERSON, AMERICAN ESSAYIST

In the following sections, we look at the effects on women, men, their marriage, and their children when a wife is employed outside the home. The Self-Assessment on dual-employed couples permits a way to examine ways of coping with this choice.

## SELF-ASSESSMENT | Dual Employed Coping Scales (DECS)

### Purpose

The DECS is designed to record what spouses find helpful to them in managing family and work roles when both spouses are employed outside the home. Coping is defined as personal or collective (with other individuals or programs) efforts to manage the demands associated with the dual-employed family.

### Directions

Read each of the following "Coping behaviors" and write the number which describes your coping as a spouse in a dual employed family:

1 = Strongly disagree
2 = Moderately disagree
3 = Neither agree nor disagree
4 = Moderately agree
5 = Strongly agree
NC = No child

I "cope" with the demands of our dual employed family by:

_____ 1. Becoming more efficient, making better use of my time "at home"
_____ 2. Using modern equipment (e.g., microwave oven, etc.) to help out at home
_____ 3. Believing that we have much to gain financially by our both working
_____ 4. Working out a "fair" schedule of household tasks for all family members
_____ 5. Getting by on less sleep than I'd ideally like to have
_____ 6. Ignoring comments of how we "should" behave as men and women (e.g., women shouldn't work; men shouldn't clean house)
_____ 7. Deciding I will do certain housekeeping tasks at a regular time each week
_____ 8. Buying convenience foods which are easy to prepare at home
_____ 9. Believing that my working has made me a better parent than I otherwise would be
_____ 10. Leaving some things undone around the house (even though I would like to have them done)
_____ 11. Getting our children to help out with household tasks
_____ 12. Ignoring criticisms of others about parents who both work outside the home
_____ 13. Making friends with other couples who are both employed outside the home
_____ 14. Specifically planning "family time together" into our schedule; planning family activities for all of us to do together
_____ 15. Hiring outside help to assist with our housekeeping and home maintenance
_____ 16. Overlooking the difficulties and focusing on the good things about our lifestyle
_____ 17. Planning for various family relations to occur at a certain regular time each day or week (e.g., from the time we get home until their bedtime is the "children's time")
_____ 18. Eating out frequently
_____ 19. Believing that my working has made me a better spouse
_____ 20. Hiring help to care for the children
_____ 21. Relying on extended family members for encouragement
_____ 22. Covering household family responsibilities for each other when one spouse has extra work
_____ 23. Leaving work and work-related problems at work when I leave at the end of the day
_____ 24. Having friends at work whom I can talk to about how I feel
_____ 25. Planning for time alone with my spouse
_____ 26. Modifying my work schedule (e.g., reducing amount of time at work or working different hours)
_____ 27. Relying on extended family members for financial help when needed
_____ 28. Negotiating who stays home with an ill child on a case-by-case basis
_____ 29. Planning work changes (e.g., transfer, promotion, shift change) around family needs
_____ 30. Relying on extended family members for childcare help

____ 31. Identifying one partner as primarily responsible for household tasks
____ 32. Believing that we are good "role models" for our children by both working
____ 33. Identifying one partner as primarily responsible for household tasks
____ 34. Planning time for myself to relieve tensions (e.g., jogging, exercising, mediating, etc.)
____ 35. Buying more goods and services (as opposed to "do-it-yourself" projects)
____ 36. Encouraging our children to help each other out when possible (e.g., homework, rides to activities, etc.)
____ 37. Trying to be flexible enough to fit in special needs and events (e.g., child's concert at school, etc.)
____ 38. Planning ahead so that major changes at home (e.g., having a baby) will not disturb our work requirements
____ 39. Making better use of time at work
____ 40. Having good friends whom I talk to about how I feel
____ 41. Limiting our home entertaining to only our close friends
____ 42. Believing that, with time, our lifestyle will be easier
____ 43. Planning schedules out ahead of time (e.g., who takes kid(s) to the doctor; who works late)
____ 44. Sticking to an established schedule of work and family-related activities
____ 45. Believing that I must excel at both my work and my family roles
____ 46. Cutting down on the number of outside activities in which I can be involved
____ 47. Establishing whose role responsibility it is to stay home when child(ren) are ill
____ 48. Identifying one partner as primarily responsible for bread winning
____ 49. Believing that working is good for my personal growth
____ 50. Believing that, overall, there are more advantages than disadvantages to our lifestyle
____ 51. Limiting job involvement in order to have time for my family
____ 52. Lowering my standards for "how well" household tasks must be done
____ 53. Encouraging our child(ren) to be more self-sufficient, where appropriate
____ 54. Eliminating certain activities (home entertaining, volunteer work, etc.)
____ 55. Frequent communication among all family members about individual schedules, needs, and responsibilities
____ 56. Maintaining health (eating right, exercising, etc.)
____ 57. Believing that I need a lot of stimulation and activity to keep from getting bored
____ 58. Limiting my involvement on the job—saying "no" to some of the things I could be doing

### Scoring

Reverse score item 45 so that so that 1 = 5, 2 = 4, 3 = 3, 4 = 2, and 5 = 1. Then add each of the 58 items for a total score. The range of possible scores is from 58 to 290 with 174 the midpoint. The higher one's total score, the more one is able to cope with the demands of being a spouse in a dual income family. The scale was completed by several families including those with parents of children with serious illness. Overall means for wives were 173.9 and for husbands, 162.3.Wives had significantly higher scores on the use of coping behaviors than husbands.

### Source

McCubbin, H. I. and A. I. Thompson (eds). 1991. Family assessment inventories for research and practice. Madison, WI: University of Wisconsin. Call the University of Wisconsin, Family Stress Coping, Coping and Health Project at 608-262-5070 for information.

## Effects of the Wife's Employment on the Wife

While some new mothers enjoy their work role and return to work soon after their children are born, others anguish over leaving their baby to return to the work force. One new mother who went back to work after the birth of her daughter said

> *You go through periods of guilt for leaving her, sadness, missing her, worrying to death and even a slight bit of anger at your spouse. "Hey, if you made more money, we could afford for me to stay home."*

The new mother discovers that there are now two spheres to manage—work and family—which may result in **role overload**—not having the time or energy to meet the demands of their responsibilities in the roles of wife, parent, and worker. Because women have traditionally been responsible for most of the housework and child care, employed women come home from work to what Hochschild (1989) calls the second shift, housework and child care that have to be done after work. According to Hochschild, the **second shift** has the following result:

**Role overload** not having the time or energy to meet the demands of responsibilities in the roles of wife, parent, and worker.

**Second shift** housework and child care that employed women do when they return home from their jobs.

> *. . . women tend to talk more intently about being overtired, sick, and "emotionally drained." Many women could not tear away from the topic of sleep. They talked about how much they could "get by on" . . . six and a half, seven, seven and a half, less, more. . . . Some apologized for how much sleep they needed. . . . They talked about how to avoid fully waking up when a child called them at night, and how to get back to sleep. These women talked about sleep the way a hungry person talks about food (p. 9).*

Part of this role overload may be that women have more favorable attitudes toward housework and childcare than men. Two researchers assessed the attitudes of 732 spouses and found that women had more favorable attitudes toward cleaning, cooking, and child care than did men—women enjoyed these activities more, set higher standards for them, and felt more responsible for them. The researchers also noted that these favorable and men's unfavorable attitudes were associated with women's greater contribution to household labor (Poortman and Van der Lippe 2009).

**Role conflict** being confronted with incompatible role obligations.

Another stressful aspect of employment for employed mothers in dual-earner marriages is **role conflict**, being confronted with incompatible role obligations. For example, the role of a career woman is to stay late and prepare a report for the following day. However, the role of a mother is to pick up her child from day care at 5 P.M. When these roles collide, there is role conflict. Although most women resolve their role conflicts by giving preference to the mother role, some give priority to the career role and feel guilty about it.

**Role strain** anxiety that results from being able to fulfill only a limited number of role obligations.

**Role strain**, the anxiety that results from being able to fulfill only a limited number of role obligations, occurs for both women and men in dual-earner marriages. No one is at home to take care of housework and children while they are working, and they feel strained at not being able to do everything.

## Effects of the Wife's Employment on Her Husband

Husbands also report benefits from their wives' employment. These include being relieved of the sole responsibility for the financial support of the family and having more freedom to quit their job, change jobs, or go to school. Since men traditionally had no options but to work full time, they now benefit by having a spouse with whom to share the daily rewards and stresses of employment.

## Effects of the Wife's Employment on the Marriage

*The difference between a job and a career is the difference between forty and sixty hours a week.*

ROBERT FROST

Helms et al. (2010) analyzed the relationship between employment patterns and marital satisfaction of 272 dual-earner couples and found that co-provider couples (in contrast to those where there were distinct primary and secondary providers) reported the highest marital satisfaction. In addition, these couples reported the most equitable division of housework. Lam et. al. (2012) confirmed that employment of the wife was associated with greater household participation on the part of the husband. However, Minnotte et al. (2013) found reduced marital satisfaction for more egalitarian husbands and wives, suggesting that husbands may have resented the intrusion of family life into their work roles. More traditional relationships were also associated with higher sex frequency (Kornrich et al. 2013).

Are marriages in which the wife has her own income more vulnerable to divorce? Not if the wife is happy. But if she is unhappy, her income will provide her a way to take care of herself when she leaves. Schoen et al. (2002) wrote, "Our results provide clear evidence that, at the individual level, women's employment does not destabilize happy marriages but increases the risk of disruption in unhappy marriages" (p. 643). Hence, employment won't affect a happy marriage but it can effect an unhappy one. Further research by Schoen et al. (2006) revealed that full-time employment of the wife is actually associated with marital stability.

In one-fourth of marriages, wives earn higher incomes than their husbands. In a survey of 4,606 adult workers, 59% of the men agreed that "I am comfortable if my spouse earns more than I do" (Yang and Ward 2010). Meisenbach (2010) interviewed 15 U.S. female breadwinners (FBWs) and identified their issues, concerns, and worries. They enjoyed having control in their relationships but were ambivalent about whether the control was good for the relationship.

What is the effect of control over one's time devoted to work? Olsen and Dahl (2010) confirmed that having control over the hours one works is related to positive family functioning. Sandberg et al. (2013) analyzed data on 281 couples and found that negative couple interaction was associated with less work satisfaction and elevated depression scores; hence, a marriage-to-work spillover can be costly for families, organizations, and governments.

# Work and Family: Effects on Children

Independent of the effect on the wife, husband, and marriage, what is the effect on the children of the wife earning an income outside of the home? Individuals disagree on the effects of maternal employment on children. The Self-Assessment provides a way to assess your beliefs in this regard.

## Effects of the Wife's Employment on the Children

Mothers with young children are the least likely to be in the labor force. Mothers most likely to be in the work force are those with children between the ages of 14 and 17—the teen years. Teenagers are no longer dependent on the physical care of their mother and they create more expenses, often requiring a second income.

Dual-earner parents want to know how children are affected by maternal employment. Johnson et al. (2013) found that the number of hours worked by the mother was unrelated to the behavior of middle school children. However, the externalizing behavior of boys was related to fathers working more than 55 hours a week. Nevertheless, there is an effect of maternal employment on eating healthy food. Bauer et al. (2012) confirmed that the more hours a mother worked outside the home the less often the family shared family meals and the less often adolescents were encouraged to eat fruits/vegetables. And, there is a positive effect of wife's employment on the relationship of the father with his children. Meteyer and Perry-Jenkins (2012) confirmed that the more hours the wife worked, the greater the father was involved with his children. However, a disadvantage for children of two-earner parents is that they receive less supervision. Letting children come home to an empty house is particularly problematic. The issue of self-care children is discussed in the next section.

**Self-Care/Latchkey Children** Long work hours, not being able to control their work schedules, and leaving their children unsupervised result in considerable concern for parents about the well-being of their children (Barnett et al. 2010). Although these self-care, or latchkey, children often fend for themselves very well, some are at risk. Over 4 million children are alone at home for an average of 6.5 hours a week—they are vulnerable to a lack of care in case of an accident or emergency. Children who must spend time alone at home should know the following:

1. How to reach their parents at work (the phone number, extension number, and name of the person to talk to if the parent is not there)
2. Their home address and phone number in case information must be given to the fire department or an ambulance service
3. How to call emergency services, such as the police and fire departments

## SELF–ASSESSMENT | Maternal Employment Scale

### Directions

Using the following scale, mark a number on the blank next to each statement to indicate how strongly you agree or disagree.

| 1 | 2 | 3 | 4 | 5 | 6 |
|---|---|---|---|---|---|
| Disagree Very Strongly | Disagree Strongly | Disagree Slightly | Agree Slighty | Agree Strongly | Agree Very Strongly |

____ 1. Children are less likely to form a warm and secure relationship with a mother who works full time outside the home.
____ 2. Children whose mothers work are more independent and able to do things for themselves.
____ 3. Working mothers are more likely to have children with psychological problems than mothers who do not work outside the home.
____ 4. Teenagers get into less trouble with the law if their mothers do not work full time outside the home.
____ 5. For young children, working mothers are good role models for leading busy and productive lives.
____ 6. Boys whose mothers work are more likely to develop respect for women.
____ 7. Young children learn more if their mothers stay at home with them.
____ 8. Children whose mothers work learn valuable lessons about other people they can rely on.
____ 9. Girls whose mothers work full time outside the home develop stronger motivation to do well in school.
____ 10. Daughters of working mothers are better prepared to combine work and motherhood if they choose to do both.
____ 11. Children whose mothers work are more likely to be left alone and exposed to dangerous situations.
____ 12. Children whose mothers work are more likely to pitch in and do tasks around the house.
____ 13. Children do better in school if their mothers are not working full time outside the home.
____ 14. Children whose mothers work full time outside the home develop more regard for women's intelligence and competence.
____ 15. Children of working mothers are less well nourished and don't eat the way they should.
____ 16. Children whose mothers work are more likely to understand and appreciate the value of a dollar.
____ 17. Children whose mothers work suffer because their mothers are not there when they need them.
____ 18. Children of working mothers grow up to be less competent parents than other children because they have not had adequate parental role models.
____ 19. Sons of working mothers are better prepared to cooperate with a wife who wants both to work and to have children.
____ 20. Children of mothers who work develop lower self-esteem because they think they are not worth devoting attention to.
____ 21. Children whose mothers work are more likely to learn the importance of teamwork and cooperation among family members.
____ 22. Children of working mothers are more likely than other children to experiment with alcohol, other drugs, and sex at an early age.
____ 23. Children whose mothers work develop less stereotyped views about men's and women's roles.
____ 24. Children whose mothers work full time outside the home are more adaptable; they cope better with the unexpected and with changes in plans.

### Scoring

Items 1, 3, 4, 7, 11, 13, 15, 17, 18, 20, and 22 refer to "costs" of maternal employment for children and yield a Costs Subscale score. High scores on the Costs Subscale reflect strong beliefs that maternal employment is costly to children. Items 2, 5, 6, 8, 9, 10, 12, 14, 16, 19, 21, 23, and 24 refer to "benefits" of maternal employment for children and yield a Benefits Subscale score. To obtain a total score, reverse the score of all items in the Benefits Subscale so that 1 = 6, 2 = 5, 3 = 4, 4 = 3, 5 = 2, and 6 = 1. The higher one's total score, the more one believes that maternal employment has negative consequences for children.

### Source

E. Greenberger, W. A. Goldberg, T. J. Crawford, and J. Granger. 1988. Beliefs about the consequences of maternal employment for children. *Psychology of Women Quarterly,* Maternal Employment Scale 12:35–59. Used by permission of Blackwell Publishing.

**4.** The name and number of a relative or neighbor to call if the parent is unavailable

**5.** To keep the door locked and not let anyone in

**6.** How to avoid telling callers their parents are not at home; instead, they should tell the caller that their parents are busy or can't come to the phone

**7.** How to avoid playing with appliances, matches, or the fireplace

Parents should also consider the relationship of the children they leave alone. If the older one terrorizes the younger one, the children should not be left alone. Also, if the younger one is out of control, it is inadvisable to put the older one in the role of being responsible for the child. If something goes wrong (such as a serious accident), the older child may be unnecessarily burdened with guilt.

## Quality Time

The term *quality time* has become synonymous with good parenting. Dual-income parents struggle with not having enough quality time with their children. Snyder (2007) studied 220 parents from 110 dual-parent families and found that quality time was defined in different ways. Some parents (structured-planning parents) saw quality time as planning and executing family activities. Mormons typically set aside "Monday home evenings" as a time to bond, pray, and sing together. Other parents (child-centered parents) noted that quality time occurred when they were having heart-to-heart talks with their children. Still other parents believed that all the time they were with their children was quality time. Whether they were having dinner together or riding to the post office, quality time was occurring if they were together. As might be expected, mothers assumed greater responsibility for quality time.

*Well, it's 1 A.M. Better go home and spend some quality time with the kids.*

**HOMER SIMPSON, *THE SIMPSONS***

## Day Care Considerations

Parents going into or returning to the workforce are intent on finding high-quality day care. Rose and Elicker (2008) surveyed the various characteristics of day care that were important to 355 employed mothers of children under six years of age and found that warmth of caregivers, a play-based curriculum, and educational level of caregivers emerge as the first-, second-, and third-most important factors in selecting a day-care center.

**Quality of Day Care** While most mothers prefer relatives (spouse or partner or another relative) for the day-care arrangement for their children, more than half of U.S. children are in center-based child-care programs. Forty-three percent of 3 year olds and 69% of 5 year old children are in center-based day-care programs (*Statistical Abstract of the United States 2012–2013*, Table 578).

Employed parents are concerned that their children get good-quality care. Their concern is warranted. Cortisol is a steroid hormone that plays an important role in adaptation to stress. Higher levels of cortisol reflect higher levels of stress experienced by the individual. Gunnar et al. (2010) assessed the cortisol stress levels of 151 children in full-time, home-based day-care centers and compared these levels to those of children the same age who were cared for in their own home. Increases were noted in the majority of children (63%) at day care, with 40% classified as a stress response. These increased cortisol levels began in the morning and continued throughout the afternoon.

Vandell et al. (2010) examined the effects of early child care 15 years later and found that higher quality care predicted higher cognitive–academic achievement at age 15, with escalating positive effects at higher levels of quality. These findings do not mean that day care is bad for children, but suggest the need for caution in selecting a day-care center.

Wagner et al. 2013 confirmed the exhaustion/stress experienced by full-time day-care workers. However, De Schipper et al. (2008) noted that day-care workers who engage in high-frequency positive behavior engender secure attachments with the children they work with. Hence, children of such workers don't feel they are on an assembly line but bond with their caretakers. Parents concerned about the quality of day care their children receive might inquire about the availability of webcams. Some day-care centers offer full-time webcam access so that parents or grandparents can log onto their computers and see the interaction of the day-care worker with their children.

Ahnert and Lamb (2003) emphasized that attentive, sensitive, loving parents can mitigate any potential negative outcomes in day care. "Home remains the center of children's lives even when children spend considerable amounts of time in child care. . . . [A]lthough it might be desirable to limit the amount of time children spend in child care, it is much more important for children to spend as much time as possible with supportive parents" (pp. 1047–1048).

**Cost of Day Care** Day-care costs are a factor in whether a low-income mother seeks employment, because the cost can absorb her paycheck. Even for dual-earner families, cost is a factor in choosing a day-care center. Day-care costs vary widely from nothing, where friends trade off taking care of the children, to very expensive institutionalized day care in large cities.

The cost of high quality infant care can be as high as $2,000 per month. Day care for children ages two and over ranged from $300 to $1,600 a month in 2014. These costs are for one child; most spouses have two children. Do the math.

# Balancing Work and Family Life

Work is definitely stressful on individuals, spouses, and relationships. Not only is working longer hours (50 or more hours a week) associated with increased alcohol use for both the male and female worker, spouses report feeling greater emotional distance in their relationship (Lavee and Ben-Ari 2007). Failure to harness one's work stress may have consequences for one's sex life. Stulhofer et al. (2013) analyzed data on 2,112 men and found that the chance of experiencing one or more sexual health difficulties in the past 12 months were about 1.8 times higher among men who reported the highest levels of workplace difficulties than among men who experienced no such difficulties.

**Spillover thesis** one's work role impacts one's family life in terms of working overtime, being "on call" on weekends, etc.

Hoser (2012) examined work and family life and found that the **spillover thesis** worked in only one direction—work spread into family life in the form of doing overtime, taking work home, attending seminars organized by the company, and being "on call" on the weekend/during vacation. Rarely did one's family life dictate the work role.

## National **Data**

Based on a survey of 2,500 adults age 18 and older conducted by Ricoh Americas and Harris International, 64% of the respondents reported a preference for working "virtually" compared to 34% who preferred to work in an office (Yang and Gonzalezz 2013).

One of the major concerns of employed parents and spouses is how to juggle the demands of work and family. Many family-friendly policies are not family friendly (Schlehofer 2012). When there is conflict between work and family, various strategies are employed to cope with the stress of role overload and role conflict, including (1) the superperson strategy, (2) cognitive restructuring, (3) delegation of responsibility, (4) planning and time management, and (5) role compartmentalization (Stanfield 1998).

## Superperson Strategy

The superperson strategy involves working as hard and as efficiently as possible to meet the demands of work and family. A superperson often skips lunch and cuts back on sleep and leisure to have more time available for work and family. Women are particularly vulnerable because they feel that if they give too much attention to child-care concerns, they will be sidelined into lower-paying jobs with no opportunities.

Hochschild (1989) noted that the terms **superwoman** or **supermom** are cultural labels that allow a woman to regard herself as very efficient, bright, and confident. However, Hochschild noted that this is a "cultural cover-up" for an overworked and/or frustrated woman. As noted earlier, not only does the woman have a job in the workplace (first shift), she comes home to another set of work demands in the form of house care and child care (second shift). Finally, she has a third shift (Hochschild 1997).

**Superwoman/supermom** a cultural label that allows a woman to regard herself as very efficient, bright, and confident; usually a cultural cover-up for an overworked and frustrated woman.

The **third shift** is the expense of emotional energy by a spouse or parent in dealing with various issues in family living. Although young children need time and attention, responding to conflicts and problems with teenagers also involves a great deal of emotional energy. Minnottea et al. (2010) studied 96 couples and found that women perform more "emotion work." Opree and Kalmijn (2012) noted that employed women who also take care of an adult child or aging parents are more likely to report negative changes in their own mental health.

**Third shift** the expense of emotional energy by a spouse or parent in dealing with various issues in family living.

Mothers who try to escape the supermom trap through eliciting the help of their husbands may experience a downside. Sasaki et al (2010) studied 78 dual career couples with an 8-month-old infant and found that a greater husband's contribution to care giving was associated with the wife's lower self-competence. The authors concluded that "despite increasingly egalitarian sex roles, employed mothers seem to be trapped between their desire for help with childrearing and the threat to their personal competence posed by failure to meet socially constructed ideals of motherhood."

## Cognitive Restructuring

Another strategy used by some women and men experiencing role overload and role conflict is cognitive restructuring, which involves viewing a situation in positive terms. Exhausted dual-career earners often justify their time away from their children by focusing on the benefits of their labor: their children live in a nice house in a safe neighborhood and attend the best schools. Whether these outcomes offset the lack of quality time may be irrelevant; the beliefs serve simply to justify the two-earner lifestyle.

## Delegation of Responsibility and Limiting Commitments

A third way couples manage the demands of work and family is to delegate responsibility to others for performing certain tasks. Because women tend to bear most of the responsibility for child care and housework, they may choose to ask their partner to contribute more or to take responsibility for these tasks.

Another form of delegating responsibility involves the decision to reduce one's current responsibilities and not take on additional ones. For example, women and men may give up or limit commiting to volunteer responsibilities. One woman noted that her life was being consumed by the responsibilities of her church; she had to change churches because the demands were relentless. In the realm of paid work, women and men can choose not to become involved in professional activities beyond those that are required.

## Time Management

While two-thirds of women prefer to work part time as opposed to full time, Vanderkam (2010) argues that women who work part time end up spending just

## FAMILY POLICY | Government and Corporate Work-Family Policies and Programs

Family satisfaction is impacted by the work context of individuals. Leach and Butterworth (2012) analyzed data on a large representative community sample of midlife (aged 40–44 at baseline) employed persons in marriage-like relationships (n = 2,054) to investigate the relationship between psychosocial job characteristics and relationship quality. The researchers found that high job demands, low job control, and job insecurity were independently associated with lower levels of positive support from partners for both men and women. Desrochers et al. (2012) also confirmed that a family-unfriendly work culture was negatively associated with family satisfaction.

These results identify the broader social costs of adverse psychosocial characteristics at work and implore government agencies and corporations to facilitate family-friendly policies. One law on the books is the family leave option. Under the Family and Medical Leave Act, all companies with 50 or more employees are required to provide each worker with up to 12 weeks of unpaid leave for reasons of family illness, birth, or adoption of a child. In a subsequent amendment, the Family Leave Act permits states to provide unemployment pay to workers who take unpaid time off to care for a newborn child or a sick relative.

Currently, the United States is the only industrialized country that does not provide paid child leave. Over 160 countries provide paid leave for mothers to birth their babies; 45 countries provide the same benefits for fathers. Australia guarantees a year of leave to all new mothers. Germany provides 14 weeks off with 100% salary. Aside from government-mandated work-family policies, corporations and employers have begun to initiate policies and programs that address the family concerns of their employees. Fifteen percent of large corporations in the United States provide employer-sponsored child care benefits (4 percent of small businesses) (Totenhagen et al. 2012).

Employer-provided assistance with elderly parent care, options in work schedules, and job relocation assistance are becoming more common. Widener (2008) noted that some U.S. companies have advanced strategies to balance work and life and gender equity, and that these family-friendly policies "can garner a competitive edge, attracting and retaining young, early career professionals." Galea et al. (2014) noted that the more family responsibilities workers have, "the more they tend to perceive flexible working hours as a necessity rather than an extra benefit." Such flexibility "creates a situation which is advantageous for both employer and employee."

While the various family friendly options are commendable, Stone (2007) noted that one of the reasons women opt out of their career paths is the inflexibility of corporations. Her interviewees noted that, when a woman drops out to have or to take care of her children, her clients are assigned to someone else and, when she returns, she never seems able to get back on the same standing as those who did not drop out. She suggested family-friendly policies are simply window dressings and that, in fact, corporations view families as interfering with production. Schlehofer (2012) agreed and said that women who drop out to have their babies often find that they have been marginalized on their return.

41 more minutes daily on child care and 10 minutes more per day playing with the child than if they worked full time. By working full-time, she says the woman affords high-quality day care and can focus on the child when she is not at work. The full-time worker is not exhausted from being with the children all day but may be more "emotionally ready" to spending dinner, bath, and reading time with their children at night.

Other women use time management by prioritizing and making lists of what needs to be done each day. This method involves trying to anticipate stressful periods, planning ahead for them, and dividing responsibilities with the spouse. Such division of labor allows each spouse to focus on an activity that needs to be done (grocery shopping, picking up children at day care) and results in a smoothly functioning unit.

Having flexible jobs and/or careers is particularly beneficial for two-earner couples. Being self-employed, telecommuting, or working in academia permits flexibility of schedule so that individuals can cooperate on what needs

The Obama administration has emphasized informing businesses about the benefits of flexible work schedules, helping businesses create flexible work opportunities, and increasing federal incentives for telecommuting. Moen (2011) pointed out that corporations need to "break open the time clocks around paid work—the tacit, taken-for-granted beliefs, rules, and regulations about the time and timing of work days, work weeks, work years, and work lives."

Similarly, Soo Jung and Zippay (2011) emphasized the tensions associated with the demands of employment and home life which may have negative effects on mental and physical health and called for employer-based policies to facilitate work-life balance and well-being. Over half (52%) of 2,035 employed adults in a Harris survey noted that their employer offered flextime (37% telecommuting/working from home) (Carey and Trapp 2013). Not all corporations are jumping on board—Yahoo CEO Marissa Mayer sent out an e-mail in 2013 banning telecommuting, stating that "We need to be one Yahoo and that starts with physically being together."

**Your Opinion?**

1. Argue for and against the fact that businesses benefit from having family-friendly policies.
2. Argue for and against the theory that childfree workers should work later and on holidays so that parents can be with their children.
3. Why do you think that the United States lags behind other industrialized nations in terms of paid leave for parents?

## Sources


Carey, A. and P. Trapp. 2013. Fewer hours at the office. *USA Today.* October 17. A1.

Desrochers, S., L. D. Sargent, and A. J. Hostetler. 2012 Boundary-spanning demands, personal mastery, and family satisfaction: Individual and crossover effects among dual-earner parents. *Marriage and Family Review* 48: 443–464.

Galea, C., I. Houkes, and A. DeRijk. 2014. An insider's point of view: How a system of flexible working hours helps employees to strike a proper balance between work and personal life. *International Journal of Human Resource Management* 25: 1011–1111.

Leach, L. and P. Butterworth. 2012. Psychosocial adversities at work are associated with poorer quality marriage-like relationships. *Journal of Population Research* 29: 351–372.

Moen, P. 2011. From "work-family" to the "gendered life course" and "fit": five challenges to the field. *Community, Work & Family* 14: 81–96.

Schlehofer, M. 2012. Practicing what we teach? An autobiographical reflection on navigating academia as a single mother. *Journal of Community Psychology* 40: 112–128.

Soo Jung, J. and A. Zippay. 2011. Juggling act: Managing work-life conflict and work-life balance. *Families in Society* 92: 84–90.

Stone, P. 2007. *Opting out?* Berkley: University of California Press.

Totenhagen, C. J., S. A. Hawkins, W. Arias Jr., and L. Borden. 2012. Employer-sponsored child care benefits and employeed outcomes. Paper, National Council on Family Relations, Phoenix, November.

Widener, A. J. 2008. Family-friendly policy: Lessons from Europe-Part II Public Manager. Winter 2007/2008. 36: 44–50.


to be done. Alternatively, some dual-earner couples attempt to solve the problem of child care by having one parent work during the day and the other parent work at night so that one of them can always be with the children. Shift workers often experience sleep deprivation and fatigue, which may make fulfilling domestic roles as a parent or spouse difficult for them. Similarly, shift work may have a negative effect on a couple's relationship because of their limited time together.

Presser (2000) studied the work schedules of 3,476 married couples and found that recent husbands (married fewer than five years) who had children and who worked at night were six times more likely to divorce than husbands or parents who worked days.

**DIVERSITY IN OTHER COUNTRIES**

France, Denmark, and Germany have experimented with fewer required work hours per week. France tried the 35-hour workweek with eight weeks of vacation. The goal was to increase employment. However, over the ten years during which the 35-hour workweek was in place, the desired gains did not occur. In 2008, France moved to increase its workweek, with the result that the French now work an average of 41 hours a week; in America, the average is 41.7 hours (Keller 2008a). The Fair Labor Standards Act in 1938 established the 40-hour week and overtime payment laws. This law was a way to allow companies to get employees to work more rather than hire new employees (Farr 2012).

## Role Compartmentalization

**Role compartmentalization** separating the roles of work and home so that spouses do not think about or dwell on the problems of one when they are at the other.

Some spouses use **role compartmentalization** separating the roles of work and home so that they do not think about or dwell on the problems of one when they are at the other. Spouses unable to compartmentalize their work and home feel role strain, role conflict, and role overload, with the result that their efficiency drops in both spheres. Some families look to the government and their employers for help in balancing the demands of family and work.

# Balancing Work and Leisure Time

Finding a balance between work and leisure is challenging.

## Intrusive Technology

*Do not live at too great pace. To know how to spread things out is to know how to enjoy them.*

BALTASAR GRACIAN, SPANISH WRITER IN 1600S

*We give up leisure in order that we may have leisure, just as we go to war in order that we may have peace.*

ARISTOTLE

The workplace has become the home place. Because of the technology of the workplace, such as laptops, iPads, and iPhones/smart phones, some spouses work all the time, wherever they are. Seven in ten U.S. workers say that technology has resulted in work becoming a part of their personal time (Petrecca 2013). One dual-career couple noted that they are sending text messages or working on their iPads when they are at home so that the time they are not working and are communicating directly with each other is becoming less frequent.

Even on vacation, some spouses insist on Internet access. "Do you have wireless Internet?" is one of the first questions asked by individuals making a motel reservation. Seventy-seven percent of adults in one study noted a preference for Internet access during vacation. Specifically they wanted to "check personal e-mail," "find trip information," "online banking," "read the news," "social media profiles," and "work e-mail" (Carey and Trapp 2010).

## Definition and Importance of Leisure

**Leisure** the use of time to engage in freely chosen activities perceived as enjoyable and satisfying, including exercise.

**Leisure** refers to the use of time to engage in freely chosen activities perceived as enjoyable and satisfying, including exercise. Tucker et al. (2008) identified three conditions of leisure—quiet leisure at home (reading a book), active leisure (playing basketball), and more work. Parry and Light (2014) suggested that reading erotic literature, particularly for women, is becoming more normative. Ratings of rest, recuperation, and satisfaction were lowest in the additional work condition.

*The real dividing line between things we call work and the things we call leisure is that in leisure, however active we may be, we make our own choices and our own decisions. We feel for the time being that our life is our own.*

UNKNOWN

The value of family leisure was confirmed by West and Merriam (2009). They interviewed 306 families who camped at the St. Croix State Park, Pine County, Minnesota, and found that cohesiveness as was measured by the amount of intimate communication of troubles, secrets, and mood among family members was created and maintained by families who engaged in outdoor recreation. In addition, analysis of data on 898 families from throughout the United States revealed a positive relationship between all family leisure satisfaction variables and satisfaction with family life (Agate et al. 2009). The take-home message of these two studies is that parents might recognize the benefits of family leisure and make it happen as often as possible. Leisure also has benefits for the individual.

## View of Work/Leisure by Millennials

**Millennials** the 80 million workers born between 1980 and 1995.

Corporations are learning that individuals are no longer interested in being consumed by their work. This new work ethic is particularly operative among **millennials**, the 80 million workers born between 1980 and 1995. Having been told that they are special, that performance is not required for praise, and that fun, leisure, and lifestyle come before giving one's life to a job or career, these millennials are socializing their bosses to be more flexible or to find someone

Leisure provides excitement and a break from the routine of work. This retired husband is enjoying his time off the grid.

Chelsea Curry

else to work for them. Indeed, corporations today are hiring consultants to socialize management how to cope with a workforce that wants a job on their own terms. They are told that they must be not just a boss but also a life coach and a psychiatrist, and that they must motivate rather than demand.

## Functions of Leisure

Leisure fulfills important functions in our individual and interpersonal lives. Leisure activities may relieve work-related stress and pressure; facilitate social interaction and family togetherness; foster self-expression, personal growth, and skill development; and enhance overall social, physical, and emotional well-being.

*No one needs a vacation so much as the person who has just had one.*

ELBERT HUBBARD, AMERICAN PHILOSOPHER

Though leisure represents a means of family togetherness and enjoyment, it may also represent an area of stress and conflict. Some couples function best when they are busy with work and child care so that they have limited time to interact or for the relationship. Indeed, some prefer not to have a lot of time alone together.

## Individual and Relationship Problems Related to Leisure

Individual or relationship problems may occur in the context of leisure. Excessive drinking may occur in leisure contexts with friends. Grekin et al. (2007) analyzed the alcohol consumption of 3,720 undergraduates (mean age 17.96) who spent spring break with either friends or parents and found that students who vacationed with friends during spring break dramatically increased their alcohol use. In contrast, students who stayed home or vacationed with parents during spring break were at low risk for excessive alcohol use.

*Americans have an inability to relax into sheer pleasure. Ours is an entertainment seeking-nation, but not necessarily a pleasure-seeking one.... This is the cause of that great sad American stereotype—the overstressed executive who goes on vacation, but who cannot relax.*

ELIZABETH GILBERT, *EAT, PRAY, LOVE*

Relationship problems may also occur in reference to leisure. Video games are a major source of relationship stress (particularly among young couples) (see the "Applying Social Research" feature).

## Family Vacations and Vacation Stress

Farr (2012) reviewed the history and value of family vacations. Physicians in the late 1800s prescribed vacations for health benefits and taking a vacation was normative and socially approved. Around 1900, the norms changed in reference

## APPLYING SOCIAL RESEARCH | Anger over Romantic Partner's Video Game Playing*

Playing Call of Duty, Halo, and Gears of War on one's Xbox are major sources of enjoyment for college students, particularly males. But relationships with their partners may pay a price—the focus of this study.

### Sample and Methods

The data consisted of a nonrandom sample of 148 undergraduate volunteers at a large southeastern university who responded to a 45-item questionnaire on "Video Game Survey" (approved by the Institutional Review Board of the university). The objective of the study was to identify current uses and interpersonal consequences of playing video games.

### General Findings

Of the respondents, 97.6% of the men and 92.6% of the women reported having played a video game. Although previous researchers (Rau et al. 2006; Wan and Chiou 2006) reported the capacity of video games to become addictive, these respondents were *not* obsessed by or addicted to playing video games; they rarely skipped class or played relentlessly for days. Indeed, only two males and one female (less than 3% of the respondents) used the term *obsessive* to describe their video game playing. The average GPA of these students was 2.96, suggesting that their academic performance was standard for their year in school. However, while some researchers suggest that there is little negative impact of video game playing/intrusion into one's daily living (Billieux et al. 2013), other researchers suggest that both stress and aggression are associated with video game playing (Hansan et al. 2013).

### Significant Gender Differences

In addition to the previous general findings, three significant sex differences emerged.

1. ***Women viewed video games as more experimental than recreational.*** Women and men differed in why they played video games. Female respondents were significantly more likely than male respondents (33.7% versus 7.5%) to report that playing video games was experimental in the sense of curiosity—just seeing what it was like. In contrast, male respondents were significantly more likely to report that playing video games was recreational (65% versus 41.3%; $p < .000$).
2. ***Male partners of females were more likely to play video games.*** The data reflected that females were much more likely to have a partner who played video games than vice versa. Of the undergraduate females in this study,

## WHAT IF? What If Partner Sends/Receives Text Messages throughout Your Vacation?

Doing business via texting, smart phones, and e-mail while on vacation is not unusual. Where both partners do so or the other partner is indifferent, there is no problem. However, some partners are angered that their vacation partner never detaches from work and is always texting and always "on." Of course, a partner legitimizes texting, talking on the cell phone, and e-mailing as necessary "business" for income. One solution is a discussion before leaving town about the definition of the "vacation"—whether it is a working vacation, a pure vacation, or what? Although couples will vary in their requests, have the discussion and reach an agreement about how to spend vacation time. Otherwise, the vacation might become not only "all about business," but also about a great deal of arguing.

75.6% reported having been involved in a relationship with a partner who played video games (in contrast to 42.4% of male undergraduates involved with a female who played video games) ($p < .001$).

3. ***Females were more likely to be angry over a partner playing video games.*** Of the female respondents, 32.9% reported that they had become upset with their partners' relentless video game playing. Of the males, only 4.5% reported having told his partner that she spent too much time playing video games. The difference was statistically significant ($p < .005$). We asked an undergraduate female who reported that her partner's game playing was a problem in their relationship to explain the source of her frustration:

*Not only are video games expensive (not to speak of the equipment) but he hogs the TV for an entire evening and acts like I don't exist. Quality time—what is that?*

**Implications**

Three implications emerged from the data. First, in spite of the stereotype that video game players are geeks who are focused on machines and have no relationships, 74.5% of the sample reported that they were dating or involved with someone. Indeed, over half were emotionally involved, engaged, or married. Second, these respondents were neither addicted to nor obsessed with playing video games. Missing class and being ensconced in their room playing video games for days on end was virtually nonexistent. Third, video game playing was a problem in relationships for about a third of the female respondents. Feeling neglected, disrespected ("doesn't care what I want to watch on TV"), and wasting money were among the dissatisfactions.

**Sources**


Billieux, J., M. Van der Linden, S. Achab, Y. Khazaal, L. Paraskevopoulos, D. Zullino, and G. Thorens. 2013. Why do you play World of Warcraft? An in-depth exploration of self-reported motivations to play online and in-game behaviors in the virtual world of Azeroth. *Computers in Human Behavior* 29: 103–109.

Hansan, Y., L. Begue, and B. J. Bushman. 2013. Violent video games stress people out and make them more aggressive. *Aggressive Behavior* 39: 64–70.

Rau, P. L. P., S. Y. Peng, and C. C. Yang. 2006. Time distortion for expert and novice online game players. *CyberPsychology & Behavior* 9: 396–403.

Wan, C. S., and W. B. Chiou. 2006. Psychological motives and online games addiction: A test of flow theory and humanistic needs theory of Taiwanese adolescents. *CyberPsychology & Behavior* 9: 317–24.


*Updated and abridged from D. Knox, M. Zusman, A. White, and G. Haskins. 2009. "Coed anger over romantic partner's video game playing." *Psychology Journal* 6: 10–16.

to the immigrant work ethic (e.g, full- time work is good), religion ("Protestant ethic"), and the belief that idleness was bad. The 1950s and 1960s experienced a resurgence of interest in the family vacation with a "see America" campaign. However, "during the last 30 years, family vacation has become a distant memory" (Farr 2012). Due to increasing fuel costs, Stevens et al. (2014) confirmed that family visitation of national parks has continued to decline.

Vacations can be stressful. Not only must individuals plan and pack, put their work on hold and arrange for pet care, but they can also be exhausted with air travel. Delayed flights and packed planes do not contribute to enjoying one's vacation. Car travel has its own challenge with children who squabble in the backseat and ask every five minutes, "When will we be there?"

# The Future of Work and Family Life

Families will continue to be stressed by work. Employers will, increasingly, ask employees to work longer and do more without the commensurate increases in salary or benefits. Businesses are struggling to stay solvent and workers will take the brunt of the instability.

*I don't care too much for money, for money can't buy me love.*

THE BEATLES

The number of wives who work outside the home will increase—the economic needs of the family will demand that they do so. Husbands will adapt, most willingly, some reluctantly. Children will become aware that budgets are tight, tempers are strained, and leisure with the family in the summer may not be as expansive as previously. As children go to college they can benefit from exposure to financial management information—being careful about credit card debt and how they spend money (Fiona et al. 2012).

While the percent of wives in the work force may increase, the percent of mothers who do not work outside the home is increasing. The percent in 2012 was 29%, up from 23% in 1999 (Cohn et al. 2014). Difficulty finding employment and concerns about the employment effects on children are explanations.

# Summary

***What is the effect of money on relationships?***

Having economic security is a top concern of individuals and spouses, more important than health, appearance, and self-esteem. Poverty is devastating to couples and families. Those living in poverty suffer from poorer physical and mental health, report lower personal and marital satisfaction, and die sooner. The issue of money (the lack of it, disagreement over how it is spent) is a frequent problem that couples report in marriage counseling.

Emotional well-being rises with income and tops out at $80,000 a year. Increases in income above this figure are not associated with increases in satisfaction. The reason money is such a profound issue in marriage is its symbolic significance as well as the fact that individuals do something daily in reference to money—work, spend, save, or worry about money.

***How does work affect the marriage?***

Money is a central issue in relationships because of its association with power, control, and dominance. Generally, the more money a partner makes, the more power that person has in the relationship. Males make considerably more money than females and generally have more power in relationships. When a wife earns an income, her power increases in the relationship. Money also provides an employed woman the power to be independent and to leave an unhappy marriage.

Driven primarily by the need to provide income for the family, about 70% of all U.S. wives are in the labor force. The time wives are most likely to be in the labor force is when their children are teenagers (between the ages of 14 and 17), the time when the family's food and clothing expenses are the highest.

The office or workplace is often a place where people meet and establish relationships—including love and sexual connections. About 60% of workers report having been involved in an office romance. Whether a wife is satisfied with her job is related to the degree to which the job takes a toll on her family life. When the wife's employment interferes with her family life, she reports less job satisfaction. Husbands also report benefits from their wives' employment. These include being relieved of the sole responsibility for the financial support of the family and having more freedom to quit jobs, change jobs, or go to school.

In regard to marriage, a woman's high income at her job does not destabilize a happy marriage but increases the risk of disruption in an already unhappy marriage. Couples may be particularly vulnerable when a wife earns more money than her husband. Thirty percent of working wives earn more than their husbands.

***What is the effect of parents' work decisions on the children?***

Children do not appear to suffer cognitively or emotionally from their parents working as long as positive, consistent child-care alternatives are in place. However, having working parents (particularly working mothers) translates into fewer family meals and a less nutritious diet for offspring. In addition, less supervision of children by parents is an outcome of having dual-earner parents. Letting children come home to an empty house is particularly problematic. Parents view quality time as structured, planned activities, talking with their children, or just hanging out with them.

***What are the various strategies for balancing the demands of work and family?***

Strategies used for balancing the demands of work and family include the superperson strategy, cognitive restructuring, delegation of responsibility, planning and time management, and role compartmentalization. Government and corporations have begun to respond to the family concerns of employees by implementing work-family policies and programs. These policies are typically inadequate and cosmetic.

***What is the importance of leisure and what are its functions?***

Corporations are learning that the new millenials value leisure and will not forego it for their work life. Leisure helps to relieve stress, facilitate social interaction and family togetherness, and foster personal growth and skill development. However, leisure time may also create conflict over how to use it. Females sometimes become angry at their romantic partners for playing video games.

***What is the future of work and the family?***

Families will continue to be stressed by work. Employers will, increasingly, ask employees to work longer and do more. Businesses are struggling to stay solvent and the workers will take the brunt of the instability.

## Key Terms

Consumerism
Dual-career marriage
HER/his career marriage
HIS/her career marriage
HIS/HER career marriage
Leisure
Millenials
Mommy track
Momprenuer
Role compartmentalization
Role conflict
Role overload
Role strain
Second shift
Spillover thesis
Superwoman/supermom
THEIR career marriage
Third shift

## Web Links

At Home Dads
http://www.angelfire.com/zine2/athomedad/index.blog

Identity Theft: Prevention and Survival
http://www.identitytheft.org/

Ms. Money (budgeting documents)
http://www.msmoney.com

Network on the Family and the Economy
http://www.olin.wustl.edu/macarthur/

# CHAPTER 14 Stress and Crisis in Relationships

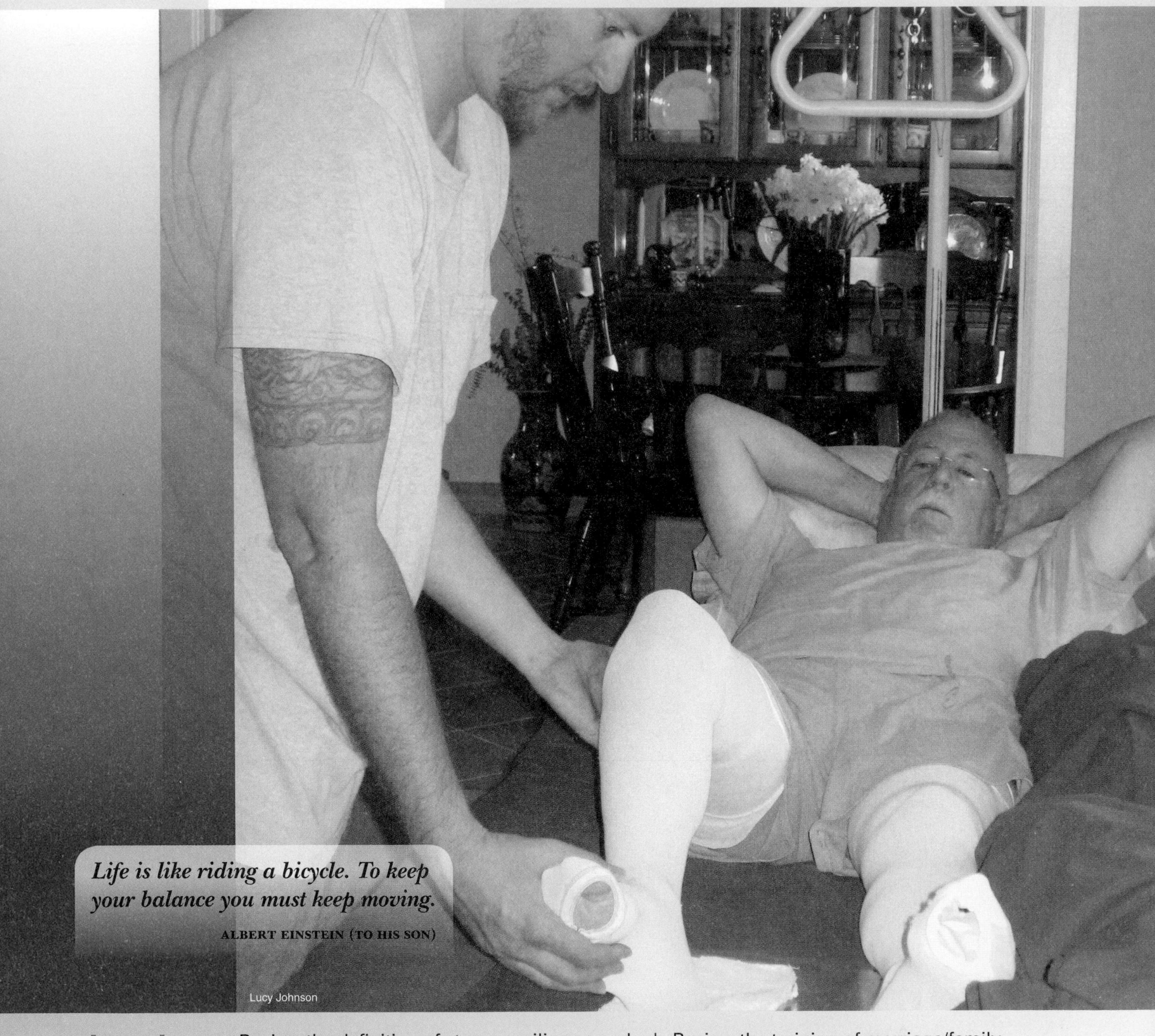

*Life is like riding a bicycle. To keep your balance you must keep moving.*

ALBERT EINSTEIN (TO HIS SON)

Lucy Johnson

## Learning Objectives

- Review the definition of stress, resilience, and the family stress model.
- Identify positive stress management strategies (and what doesn't work).
- Summarize five major crisis events.
- Review the training of marriage/family therapists.
- Know the caveats about becoming involved in marriage and family therapy.
- Predict the future of stress and crisis in the relationships.

## TRUE OR FALSE?

1. Data on telerelationship therapy reveal that only face-to-face (not Skype) therapy is effective in the resolution of individual, couple, and family problems.
2. The primary reason individuals confess an affair is their unresolved guilt.
3. High energy drinks are associated with divorce.
4. Exercise is the most important of all behaviors for good mental and physical health.
5. Men and women agree on the behaviors that constitute both offline and online cheating.

*Answers:* 1. F  2. F  3. T  4. T  5. F

Newlyweds invariably start out happy and hopeful. Researcher Gian Gonzaga followed 602 newlyweds in southern California and noted that stress was the test for whether they would endure. Almost 40% (39% to be exact) reported one area of high chronic stress the first year of marriage. In-laws, work, and money were the top stressors (Jayson 2012). Gonzaga also pointed out that stress can end a relationship as well as bring a couple closer together. This chapter is about choices individuals, spouses, and families make in coping with stress and crisis.

*My mind is troubled, like a fountain stirred;*
*And I myself see not the bottom of it.*

SHAKESPEARE, *TROILUS AND CRESSIDA*, III, III., 310

# Personal Stress and Crisis Events

In this section, we review the definitions of stress and crisis events, the characteristics of resilient families, and how individuals and couples can choose to view and respond to stress and crisis events.

## Definitions of Stress and Crisis Events

**Stress** is a reaction of the body to substantial or unusual demands (physical, environmental, or interpersonal). Stress is often accompanied by irritability, high blood pressure, and depression (Barton & Kirtley 2012). Stress is also associated with reducing relationship satisfaction (Bodenmann et al. 2010) and sexual desire (Stephenson and Meston 2012).

**Stress** reaction of the body to substantial or unusual demands (physical, environmental, or interpersonal).

### National **Data**

Cohen and Janicki-Deverts (2012) assessed psychological stress in three national surveys and found stress higher among women than men; persons who were younger, less educated, and had lower incomes also reported more stress.

*Sometimes you need a little crisis to get your adrenaline flowing and help you realize your potential."*

—JEANNETTE WALLS, THE GLASS CASTLE

Stress is a process rather than a state. For example, a person will experience different levels of stress throughout a divorce—acknowledging that one's marriage is over, telling the children, leaving the family residence, getting the final decree, and seeing one's ex with a new partner may all result in varying levels of stress.

A **crisis** is a situation that requires changes in normal patterns of response behavior. A family crisis is a situation that upsets the normal functioning of the family and requires a new set of responses to the stressor. Sources of stress and crises can be external, such as the hurricanes that annually hit our coasts

**Crisis** situation that requires changes in normal patterns of responsible behavior.

or devastating tornadoes in the spring (such as the "force 5" 16-minute tornado in Moore, Oklahoma, in 2013 that killed 158 people). Other examples of an external crisis are economic recession, downsizing, or military deployment. The source of stress and crisis events may also be internal (e.g., alcoholism, an extramarital affair, or Alzheimer's disease of one's spouse or parents). Another internal source of a crisis event is inheriting money from one's deceased parents. One spouse reported a dramatic change in the relationship when the partner suddenly "became rich" and did not "share the wealth."

Stressors or crises may also be categorized as expected or unexpected. Examples of expected family stressors include the need to care for aging parents and the death of one's parents. Unexpected stressors include contracting a human immunodeficiency virus (HIV), having a miscarriage, or experiencing the suicide of one's teenager.

Both stress and crisis events are normal aspects of family life and sometimes reflect a developmental sequence. Pregnancy, childbirth, job change or loss, children's leaving home, retirement, and widowhood are predictable for most couples and families. Crisis events may have a cumulative effect: the greater the number in rapid succession, the greater the stress.

*In this sad world, sadness comes to all . . . perfect relief is not possible, except with time . . . you are sure to be happy again. I have experienced enough to know what I say.*

ABRAHAM LINCOLN

## Resilient Families

Just as the types of stress and crisis events vary, individuals and families vary in their abilities to respond successfully to crisis events. **Resiliency** refers to a family's strengths and ability to respond to a crisis in a positive way. Lane et al. (2012) identified the various aspects of **family resilience** as beliefs (e.g., finding meaning in diversity, a positive outlook, spirituality), organizational patterns (e.g., flexibility, connectedness), and communication process (e.g., clarity, open emotional sharing). A family's ability to bounce back from a crisis (from negative health diagnosis, loss of one's job to the death of a family member) reflects its level of resiliency. Resiliency is related to the degree to which a family is hardy. The Self-Assessment section measures Family Hardiness.

**Resiliency** a family's strengths and ability to respond successfully to crisis events.

**Family resilience** beliefs, organizational patterns, and communication process.

## A Family Stress Model

Various theorists have explained how individuals and families experience and respond to stressors. The ABCX model of family stress was developed by Reuben Hill in the 1950s. The model can be explained as follows:

A = stressor event
B = family's management strategies, coping skills
C = family's perception, definition of the situation
X = family's adaptation to the event

A is the stressor event, which interacts with B, the family's coping ability or crisis-meeting resources. Both A and B interact with C, the family's appraisal or perception of the stressor event. X is the family's adaptation to the crisis. Thus, a family that experiences a major stressor (e.g., a spouse with a spinal cord injury) but has great coping skills (e.g., religion or spirituality, love, communication, and commitment) and perceives the event to be manageable will experience a moderate crisis. However, a family that experiences a less critical stressor event (e.g., their child makes Cs and Ds in school) but has minimal coping skills (e.g., everyone blames everyone else) and perceives the event to be catastrophic will experience an extreme crisis.

*Bob was there. He sat next to me in the doctor's office. He came with me when I shaved my head. He curled up with me and cried with me on the couch, and made love to me, bald and with a bandage on my breast.*

JEAN TROUNSTINE, AUTHOR

*Things without remedy should be without regard: what's done is done.*

*MACBETH*, III, II, 11

The importance of how one views his or her situation is crucial in determining how one responds to difficulty. Movie star Glenn Ford became despondent at the end of his life. He spoke into a tape recorder . . .

## SELF-ASSESSMENT | Family Hardiness Scale

This scale is designed to identify the degree to which your family has the characteristics of hardiness, which is defined as resistance to stress, having internal strength, and having a sense of control over life events and hardships.

### Directions

Read each statement and decide to what degree each describes your family. Choices include false, mostly false, mostly true, or totally true about your family. Write a 0 to 3 next to each statement.

0 = False
1 = Mostly false
2 = Mostly true
3 = Totally true
NA = Not applicable

In our family . . .

_____ 1. Trouble results from mistakes we make.
_____ 2. It is not wise to plan ahead and hope because things do not turn out anyway.
_____ 3. Our work and efforts are not appreciated no matter how hard we try and work.
_____ 4. In the long run, the bad things that happen to us are balanced by the good things that happen.
_____ 5. We have a sense of being strong even when we face big problems.
_____ 6. Many times I feel I can trust that even in difficult times things will work out.
_____ 7. While we don't always agree, we can count on each other to stand by us in times of need.
_____ 8. We do not feel we can survive if another problem hits us.
_____ 9. We believe that things will work out for the better if we work together as a family.
_____ 10. Life seems dull and meaningless.
_____ 11. We strive together and help each other no matter what.
_____ 12. When our family plans activities we try new and exciting things.
_____ 13. We listen to each other's problems, hurts and fears.
_____ 14. We tend to do the same things over and over . . . it's boring.
_____ 15. We seem to encourage each other to try new things and experiences.
_____ 16. It is better to stay at home than go out and do things with others.
_____ 17. Being active and learning new things are encouraged.
_____ 18. We work together to solve problems.
_____ 19. Most of the unhappy things that happen are due to bad luck.
_____ 20. We realize our lives are controlled by accidents and luck.

### Scoring

Reverse score items 1, 2, 3, 8, 10, 14, 16, 19, and 20. For example, if for number 1 you wrote down a 0, replace the 0 with a 3 and vice versa. Now add all the numbers from 1 to 20.

### Norms

A "low" score indicating very low hardiness is zero and a "high" score indicating very high hardiness is 60. The overall average of 304 families who took the scale scored 47.4 (SD = 6.7).

### Source

The authors of the scale are Marilyn A. McCubbin, Hamilton I. McCubbin, and Anne I. Thompson. See McCubbin, H. I. and A. I. Thompson (eds.) 1991. *Family assessment inventories for research and practice.* Madison, WI: University of Wisconsin. For permission call the University of Wisconsin, Family Stress Coping, Coping and Health Project at 608-262-5070.

*Is this the way my life is going to end up, in deep sadness? I don't know if I'm even capable of working. I'm very insecure. I think I've lost it. I don't know if I can handle memorizing lines. My mind is so shook up, I don't know. . . . I've thought about taking my own life. It's a drastic thing to do—a horrible thing to do. But I don't think I can go on living like I am now . . . I'm lost. I'm just completely washed up. I'm finished (Ford 2011, p. 296).*

# Positive Stress-Management Strategies

Researchers Burr and Klein (1994) administered an 80-item questionnaire to 78 adults to assess how families experiencing various stressors such as bankruptcy, infertility, a disabled child, and a troubled teen used various coping strategies

*If we are lucky, we spend our lives in a fool's paradise, never knowing how close we skirt the abyss.*

ROGER EBERT, FILM CRITIC

*I have learned to seek my happiness by limiting my desires, rather than in attempting to satisfy them.*

JOHN STUART MILL, BRITISH PHILOSOPHER

*I make the most of all that comes and the least of all that goes.*

SARA TEASDALE, POET

and how they felt these strategies worked. In the following sections, we detail some helpful stress-management strategies.

## Scaling Back and Restructuring Family Roles

Higgins et al. (2010) studied 1,404 men and 1,623 women in dual-earner families. These spouses evidenced two stress-reducing strategies: scaling back and restructuring family roles. Men were more likely to scale back than women. Scaling back means being selective in volunteering for various committees or activities so as not to overextend oneself. Restructuring family roles means abandoning rigid roles and recruiting children to help.

## Choosing a Positive Perspective

The strategy that most respondents reported as being helpful in coping with a crisis was choosing a positive view of the crisis situation. Survivors of hurricanes, tornados, and earthquakes routinely focus on the fact that they and their loved ones are alive rather than the loss of their home or material possessions. Buddhists have the saying, "Pain is inevitable; suffering is not." This perspective emphasizes that how one views a situation, not the situation itself, determines its impact on you. One Chicago woman said after a pipe burst and caused $30,000 worth of damage to her home: "If it's not about your health, it's irrelevant." A team of researchers (Baer et al. 2012) confirmed the importance of selecting positive cognitions for 87 adults who enrolled/participated in an 8-week program on mindfulness-based stress reduction (MBSR). Mindfulness can be thought of

Chelsea Curry

College students are sometimes overwhelmed with their coursework. Exercise helps to cut their stress level.

as choosing how you view a phenomenon. These adults were coping with chronic illness and chronic pain and reduced their stress by being mindful to frame their situation positively: "we are alive," "we are coping," and "we have each other."

Regardless of the crisis event, one can view the crisis positively. A betrayal can be seen as an opportunity for forgiveness, unemployment as a stage to spend time with one's family, and ill health as a chance to appreciate one's inner life. Positive cognitive functioning is associated with keeping the crisis in perspective and moving on (Phillips et al. 2012).

*Do you not see how necessary a world of pains and troubles is to school an intelligence and make it a soul?*

JOHN KEATS, POET

### Exercise

The Centers for Disease Control and Prevention (CDC) and the American College of Sports Medicine (ACSM) recommend that people ages six years and older engage regularly, preferably daily, in light to moderate physical activity for at least 30 minutes at a time. These recommendations are based on research that has shown the physical, emotional, and cognitive benefits of exercise. Researchers (Mata et al. 2013, Silveira et al. 2013, and Vina et al. 2012) emphasized the psychological benefits of exercise (including its role in reducing depression and viewing exercise as a proactive drug). Exercise is the most important of all behaviors for good physical and mental health.

*The only way you can hurt your body is not use it. Inactivity is the killer—it is never too late.*

JACK LALANNE, FITNESS GURU WHO DIED AT 96

### Family Cohesion, Friends, and Relatives

Mendenhall et al. (2013) found that individuals who are imbedded in a network of close family relationships tend to cope well with stress. A network of friendships and relatives is also associated with coping with various life transitions. Such relationships provide both structural and emotional support.

### Love

A love relationship also helps individuals cope with stress. Being emotionally involved with another and sharing the crisis experience with that person helps to insulate individuals from being devastated by a traumatic event. Love is also viewed as helping resolve relationship problems. Over 86% of 1,087 undergraduate males and 89% of 3,446 undergraduate females agreed with the statement, "A deep love can get a couple through any difficulty or difference" (Hall and Knox 2013).

*What is to give light must endure burning.*

VICTOR FRANKL, HOLOCAUST SURVIVOR

### Religion and Spirituality

A strong religious belief is associated with coping with stress (Mendenhall et al. 2013). Goodman et al. (2013) conducted interviews with spouses representing different religions throughout the United States and found that religion was functional in assisting the spouses cope with stress and crisis events. Ellison et al. (2011) previously examined the role of religion and marital satisfaction/coping with stress. They found that **sanctification** (viewing the marriage as having divine character or significance) was associated both with predicting positive marital quality and providing a buffer for financial and general stress on the marriage. Green and Elliott (2010) also noted that those who identify as religious report better health and marital satisfaction. Not only does religion provide a rationale for one's plight ("It is God's will"), but it also offers a mechanism to ask for help. Religion is a social institution which connects one to others who may offer both empathy and physical assistance.

**Sanctification** viewing the marriage as having divine character or significance.

*It is the set of the sails and not the gales that determines the path you go.*

AMISH PROVERB

### Laughter and Play

A sense of humor is related to lower anxiety (Grases et al. 2012). Forty-one undergraduates took part in a study in which they watched a 25-minute drama video and a 25-minute humor video. The latter was related to reducing their level of anxiety.

*A person without a sense of humor is like a wagon without springs—jolted by every pebble in the road.*

HENRY WARD BEECHER, AMERICAN CLERGYMAN

Having a pet is positively associated with one's physical and mental well being.

Chelsea Curry

### Sleep

Getting an adequate amount of sleep is associated with lower stress levels. Midday naps are also associated with positive functioning, particularly memory and cognitive function (Pietrzak et al. 2010).

### Pets

Hughes (2011) emphasized that having an animal as a pet is associated with reducing blood pressure and stress, preventing heart disease, and fighting depression. Veterinary practices are encouraged to increase the visibility of this connection so that more individuals might benefit.

### Harmful Stress-Management Strategies

*The fool doth think he is wise, but the wise man knows himself to be a fool.*

WILLIAM SHAKESPEARE, *AS YOU LIKE IT*

Some coping strategies not only are ineffective for resolving family problems but also add to the family's stress by making the problems worse. Respondents in the Burr and Klein (1994) research identified several strategies they regarded as harmful to overall family functioning. These included keeping feelings inside, taking out frustrations on or blaming others, and denying or avoiding the problem.

Burr and Klein's research also suggested that women and men differ in their perceptions of the usefulness of various coping strategies. Women were more likely than men to view as helpful such strategies as sharing concerns with relatives and friends, becoming more involved in religion, and expressing emotions. Men were more likely than women to use potentially harmful strategies such as using alcohol, keeping feelings inside, or keeping others from knowing how bad the situation was.

## Family Crisis Examples

Some of the more common crisis events that spouses and families face include physical illness, mental illness, an extramarital affair, unemployment, substance abuse, and death.

## Physical Illness and Disability

When one partner has a debilitating illness there are profound changes in the roles of the respective partners and their relationship. Mutch (2010) interviewed eight partners who took care of a spouse with multiple sclerosis after 20 years of marriage. The partners reported experiencing a range of feelings, including a sense of duty and a sense of loss, as they prioritized the health and needs of their spouse above their own. Partners reported losing their sense of identity as a spouse as their caretaking role became "the career." Some partners also felt out of control due to the unpredictable and progressive nature of MS and because it consumed their life 24 hours every day. Some felt guilt at not being satisfied with their life and wanting some freedom and independence.

This couple continues to cope with Huntington's disease.

While 24 million adults in the United States report that they have a disability, only 6.8 million of these use a visible assistive device. Hence, individuals may have a hidden disability such as chronic back pain, multiple sclerosis, rheumatoid arthritis, or chronic fatigue syndrome (Lipscomb 2009). These illnesses are particularly invasive in that conventional medicine has little to offer besides pain medication. For example, spouses with chronic fatigue syndrome may experience financial consequences ("I could no longer meet the demands of my job so I quit"), gender role loss ("I couldn't cook for my family" or "I was no longer a provider"), and changed perceptions by their children ("They have seen me sick for so long they no longer ask me to do anything").

Some couples cope with a debilitating illness such as Huntington's disease. Below is the experience of a couple who know the meaning of "In Sickness and in Health."

**In Sickness and in Health—The Effects of Huntington's Disease in a Marriage***

*My wife, Tina, and I met in New York City. We had both traveled to New York from Florida and North Carolina, respectively. We were both in school and met while waiting tables in an Upper East Side sports bar. We dated for approximately a year and then married a year later. Our marriage was an elopement. Our time in New York was idyllic and magical. The city offered nonstop energy and cultural opportunities, especially for an aspiring actor as myself.*

*Tina has a warm and inviting personality. When we met, she was the consummate social butterfly. She comes from a family of five children. The family has an affinity for playing silly, practical jokes, which she just loves. Tina has a huge heart and a strong faith. She is always thinking of other people.*

*Wanting to start a family, we decided to relocate closer to her family in Florida. We also reaffirmed our vows in Tina's faith, the Catholic Church. Once in Florida, we set out looking for employment. It was known at that time that Tina's mother had medical issues that had not been concretely diagnosed. After extensive testing, it was determined that her mother suffered from Huntington's Disease. Huntington's is an inherited disease of the central nervous system characterized by progressive dementia, unsteady balance, involuntary movements, and emotional outbursts. Typically, HD's symptoms will manifest themselves in adults in their mid 30s to mid 40s. There is no cure and the slide in one's life is slow and steadily downward. Because it is inherited, any offspring of the carrier has a 50% chance of carrying the gene. The disease went unchecked, because both grandparents died rather early in their 50s. A DNA test can determine if an individual carries the gene.*

*Tina and I already had our first daughter before we knew the medical ramifications of the HD. At first we were adamant about not having more children; however, after weighing the chances of improved genetic engineering, we had a second daughter. Tina was 32 when our second daughter was born. She had already completed training as an LPN and was working in the mental health field, for which she had an intense passion.*

*During our time in Florida, I became aware of mood swings in Tina's behavior, characterized by strong outbursts of emotion, sometimes at the simplest distractions or occurrences. They could sometimes be brutally offensive, especially toward me. Sometimes these occurrences would take place in front of the children. I found out early that the best solution was to disengage. I know all couples go through issues, but this was different. There was no rational way of dealing with Tina in those moments.*

*Tina and I relocated to my home state of North Carolina. We settled in Apex. Tina obtained an LPN position at the state mental health hospital, Dorothea Dix. I became employed with Wake County Human Services as a social worker.*

*We both were happy about our move to North Carolina. Tina's commitment to her field of work was impressive. She had a real passion for her patients. Some of the same issues were continuing to occur as far the mood swings. About five years after the move, other small signs such as facial tics and difficulty with balance appeared. Soon after, Tina tested positive for HD. After several falls at work, she went on disability with the hospital and Social Security. She exhibits classic symptoms of the HD today. She has an unbalanced gait, her thought processes are compromised to the point that she deliberately has to search for thoughts and words, and she has stopped driving.*

*Our oldest daughter is a freshman in college and more emotionally mature/understanding of her mother's disease. Our youngest who is 14 is more affected by the lack of emotional contact from her mother. As the father and husband, I have observed a devastating change in my wife. This woman, who was such a social whirlwind, is much less comfortable in those social settings now. I have tremendous empathy for her, knowing that she is slowly losing her mind and what it must feel like day in day out for her. Despite that, Tina has a strong faith and has found a real home with a small group in our church.*

*Tina feels remorse sometimes about her relationship with our youngest. She knows that if tempers flare, she has the capability of losing her temper and she does not want to do that. She fully understands the disease. She says, "I talk like a drunk and I walk like a drunk"! Tina spends her days reading on the back porch. Even though she has issues with her thoughts and verbalization, she still comprehends reading material.*

*We are coping with HD and Tina will get worse, not better. We signed up for "in sickness and in health" and are doing the best that we can. Support from friends and family helps immeasurably.*

** Jason Weeks, written for this text and used with permission.*

**Palliative care** health care for the individual who has a life-threatening illness (focusing on relief of pain and suffering) and support for the individual and his or her loved ones.

In those cases in which the illness is fatal, **palliative care** is helpful. This term describes the health care for the individual who has a life-threatening illness (focusing on relief of pain and suffering) and support for the individual and his or her loved ones. Such care may involve the person's physician or a palliative care specialist who works with the physician, nurse, social worker, and chaplain.

Pharmacists or rehabilitation specialists may also be involved. The goals of such care are to approach the end of life with planning (how long should life be sustained on machines?) and forethought to relieve pain (medication) and provide closure.

We have been discussing reacting to the crisis of a spouse with a disability. Smith and Grzywacz (2013) focused on the crisis of having a child with special health care needs and found negative physical and mental health effects on parents. Their study covered a 10-year period.

## Mental Illness

*Turn your wounds into wisdom.*

OPRAH WINFREY

Mental illness is defined as "alternations in thinking, mood or behavior that are associated with distress and impaired function" (Marshall et al. 2010). Eight percent of adults 18–29 have a serious mental illness. This percentage drops as individuals age to 1.4% of those over the age of 65 (Hudson 2012). Depression is common among college students. Field et al. (2012) found an unusually high percentage (52%) of depression among the 238 respondents who also reported anxiety, intrusive thoughts, and sleep disturbances.

The toll of mental illness on a relationship can be immense. A major initial attraction of partners to each other includes intellectual and emotional qualities. Butterworth and Rodgers (2008) surveyed 3,230 couples to assess the degree to which mental illness of a spouse or spouses affects divorce and found that couples in which either men or women reported mental health problems had higher rates of marital disruption than couples in which neither spouse experienced mental health problems. For couples in which both spouses reported mental health problems, rates of marital disruption reflected the additive combination of each spouse's separate risk. See The Black Hole for insight into depression.

**Black Hole—A Spouse Talks about Being Depressed***

*If you have experienced life in the Black Hole, then you don't need an explanation of it.*

*If you have never experienced life in the Black Hole, it is impossible for anyone to explain it to you, and even if someone could explain it to you, you still wouldn't understand it.*

*The Black Hole is, by definition, an irrational state.*

*The only thing you can comprehend about it is that you cannot comprehend it.*

*Offering ANY advice, judgmental comments, suggestions like—"If you would only . . .", "You've got to want to help yourself . . .", "I know you are depressed, BUT . . .", "You are not being rational . . ."—to someone who is in the Black Hole is not helpful.*

*On the contrary, it is very destructive. It may temporarily relieve YOUR frustration with the person, but all it does for them is to give them a serving of guilt to deal with as they wait for the Black Hole to pass.*

*If you really want to help somebody who is in the Black Hole, there IS one thing you can do—and that one thing is NOTHING.*

*After living in the Black Hole for a lifetime, a person has pretty much heard everything that you plan to tell them about it. Eventually, people who visit the Black Hole learn that it is a Monster, which comes without warning or invitation, it stays for a while, and it leaves when it is ready. The Monster doesn't time its visits to avoid holidays, vacations, or rainy days. It just barges in.*

*A person also knows what works for them while they are in the Black Hole. For some it may be exercise, or fishing, or sex (if they are still physically able), or music, or going to the beach, or talking about it, or employing logic to deal with it, or prayer, or reciting positive affirmations.*

*For others, like me, there are only two things that help—complete solitude and sleep.*

*Those two things do not provide a cure, but they do help you cope with the Monster until he leaves.*

*Nobody asks why you wear glasses. They just assume that you wear glasses because you need to, they don't ask questions about it, they don't offer suggestions and they don't try to fix it. And they don't assume that you wear glasses because of something they did. What a blessing it would be, if people treated those who struggle with clinical depression with the same respect.*

*Source: Former student of the authors. Name withheld by request.*

## Middle Age Crazy (Midlife Crisis)

*Failure is the condiment that gives success its flavor.*

TRUMAN CAPOTE, AUTHOR

The stereotypical explanation for a 45-year-old male who buys a convertible, has an affair, or marries a 20-year old is that he is "having a midlife crisis." The label conveys that such individuals feel old, think that life is passing them by, and seize one last great chance to do something they have always wanted to do. Indeed, one father (William Feather) noted, "Setting a good example for your children takes all the fun out of middle age."

However, a 10-year study of close to 8,000 U.S. adults ages 25 to 74 by the MacArthur Foundation Research Network on Successful Midlife Development revealed that, for most respondents, the middle years brought no crisis at all but a time of good health, productive activity, and community involvement. Less than a quarter (23%) reported a crisis in their lives. Those who did experience a crisis were going through a divorce. Two-thirds were accepting of getting older; one-third did feel some personal turmoil related to the fact that they were aging (Goode 1999).

Of those who initiated a divorce in midlife, 70% had no regrets and were confident that they did the right thing. This fact is the result of a study of 1,147 respondents ages 40 to 79 who experienced a divorce in their 40s, 50s, or 60s. Indeed, midlife divorcers' levels of happiness or contentment were similar to those of single individuals their own age and those who remarried (Enright 2004).

Some people embrace middle age. The Red Hat Society (http://www.redhatsociety.com/) is a group of women who have decided to "greet middle age with verve, humor, and élan. We believe silliness is the comedy relief of life [and] share a bond of affection, forged by common life experiences and a genuine enthusiasm for wherever life takes us next." The society traces its beginning to when Sue Ellen Cooper bought a bright red hat because of a poem Jenny Joseph wrote in 1961, titled the "Warning Poem." The poem reads:

*When I am an old woman I shall wear purple*
*With a red hat which doesn't go and doesn't suit me.*

Cooper then gave red hats to friends as they turned 50. The group then wore their red hats and purple dresses out to tea, and that's how it got started. Now there are over 1 million members worldwide.

In the rest of this chapter, we examine how spouses cope with the crisis events of an extramarital affair, unemployment, drug abuse, and death. Each of these events can be viewed either as devastating and the end of meaning in one's life or as an opportunity and challenge to rise above.

*After all the things you told me*

*And the promises you made*

*Why can't you behave?*

COLE PORTER, *WHY CAN'T YOU BEHAVE?*

## Extramarital Affair

Affairs are not unusual. Of spouses in the United States, about one-fourth of husbands and one-fifth of wives report ever having had intercourse with someone to whom they were not married (Russell et al. 2013).

## International **Data**

Fifteen percent of the husbands and 5% of the wives in China have engaged in extramarital sex (Zhang 2010).

**Types of Extramarital Affairs** The term **extramarital affair** refers to a spouse's sexual involvement with someone outside the marriage. Affairs are of different types, which may include the following:

**Extramarital affair** a spouse's sexual involvement with someone outside the marriage.

1. ***Brief encounter.*** A spouse meets and hooks up with a stranger. In this case, the spouse is usually out of town, and alcohol is often involved.
2. ***Paid sex.*** A spouse seeks sexual variety with a prostitute who will do whatever he wants (as happened in the case of former New York governor Eliot Spitzer). These encounters usually go undetected unless there is an STI, the person confesses, or the prostitute exposes the client.
3. ***Instrumental or utilitarian affair.*** This is sex in exchange for a job or promotion, to get back at a spouse, to evoke jealousy, or to transition out of a marriage.
4. ***Coping mechanism.*** Sex can be used to enhance one's self-concept or feeling of sexual inadequacy, compensate for failure in business, cope with the death of a family member, or test one's sexual orientation.
5. ***Paraphiliac affairs.*** In these encounters, the on-the-side sex partner acts out sexual fantasies or participates in sexual practices that the spouse considers bizarre or abnormal, such as sexual masochism, sexual sadism, or transvestite fetishism.
6. ***Office romance.*** Two individuals who work together may drift into an affair. David Petraeus (former CIA director) and John Edwards (former presidential candidate) became involved in affairs with women they met on the job.
7. ***Internet use.*** Internet usage now tops 1.6 billion people (Hertlein and Piercy 2012). Although, legally, an extramarital affair does not exist unless two people (one being married) have intercourse, Internet use can be disruptive to a marriage or a couple's relationship.

*After being married for over 37 years, I showed extremely poor judgment by engaging in an extramarital affair.*

**DAVID PETRAEUS, FORMER CIA DIRECTOR**

While men and women agree that actual kissing, touching breasts/genitals, and sexual intercourse constitute infidelity, they disagree about the degree to which online behaviors constitute cheating. Based on data collected by Hines (2012), men are less likely than women to define as cheating e-mailing a person online for relationship advice (27% versus 51%), having a friendly conversation with someone in a chat room called "Married and Lonely" (64% versus 84%), and creating a pet name for a person he/she met in an Internet chat room (40% versus 70%). Males also are less likely to view online pornography as cheating (27% versus 64%).

Computer friendships may move to feelings of intimacy, involve secrecy (one's partner does not know the level of involvement), include sexual tension (even though there is no overt sex), and take time, attention, energy, and affection away from one's partner. Cavaglion and Rashty (2010) noted the anguish embedded in 1,130 messages on self-help chat boards from female partners of males involved in cybersex relationships and pornographic websites. The females reported distress and feelings of ambivalent loss that had an individual, couple, and sexual relationship impact. Cramer et al. (2008) also noted that women become more upset when their man was *emotionally* unfaithful with another woman (although men become more upset when their partner was *sexually* unfaithful with another man). The Self-Assessment section allows you to measure your attitude toward infidelity.

Cybersex can also be problematic. Examples of cybersex include " . . . participating in sexual acts through the use of webcams, playing sexual computer

## SELF-ASSESSMENT | Attitudes toward Infidelity Scale

Infidelity can be defined as unfaithfulness in a committed monogamous relationship. Infidelity can affect anyone, regardless of race, color, or creed; it does not matter whether you are rich or attractive, where you live, or how old you are. The purpose of this survey is to gain a better understanding of what people think and feel about issues associated with infidelity. There are no right or wrong answers to any of these statements. Please read each statement carefully, and respond by using the following scale:

| 1 | 2 | 3 | 4 | 5 | 6 | 7 |
|---|---|---|---|---|---|---|
| Strongly Disagree | | | | | | Strongly Agree |

_____ 1. Being unfaithful never hurt anyone.
_____ 2. Infidelity in a marital relationship is grounds for divorce.
_____ 3. Infidelity is acceptable for retaliation of infidelity.
_____ 4. It is natural for people to be unfaithful.
_____ 5. Online/Internet behavior (for example, visiting sex chat rooms, porn sites) is an act of infidelity.
_____ 6. Infidelity is morally wrong in all circumstances, regardless of the situation.
_____ 7. Being unfaithful in a relationship is one of the most dishonorable things a person can do.
_____ 8. Infidelity is unacceptable under any circumstances if the couple is married.
_____ 9. I would not mind if my significant other had an affair as long as I did not know about it.
_____ 10. It would be acceptable for me to have an affair, but not my significant other.
_____ 11. I would have an affair if I knew my significant other would never find out.
_____ 12. If I knew my significant other was guilty of infidelity, I would confront him/her.

### Scoring

Selecting a 1 reflects the least acceptance of infidelity; selecting a 7 reflects the greatest acceptance of infidelity. Before adding the numbers you selected, reverse the scores for item numbers 2, 5, 6, 7, 8, and 12. For example, if you responded to item 2 with a "6," change this number to a "2"; if you responded with a "3," change this number to "5," and so on. After making these changes, add the numbers. The lower your total score (12 is the lowest possible), the less accepting you are of infidelity; the higher your total score (84 is the highest possible), the greater your acceptance of infidelity. A score of 48 places you at the midpoint between being very disapproving and very accepting of infidelity.

### Scores of Other Students Who Completed the Scale

The scale was completed by 150 male and 136 female student volunteers at Valdosta State University. The average score on the scale was 27.85. Their ages ranged from 18 to 49, with a mean age of 23.36. The ethnic backgrounds of the sample included 60.8% white, 28.3% African American, 2.4% Hispanic, 3.8% Asian, 0.3% American Indian, and 4.2% other. The college classification level of the sample included 11.5% freshmen, 18.2% sophomores, 20.6% juniors, 37.8% seniors, 7.7% graduate students, and 4.2% post baccalaureates. Male participants reported more positive attitudes toward infidelity (mean = 31.53) than did female participants (mean = 23.78; $p < .05$). White participants had more negative attitudes toward infidelity (mean = 25.36) than did non-white participants (mean = 31.71; $p < .05$). There were no significant differences in regard to college classification.

### Source

"Attitudes toward Infidelity Scale" 2006 by Mark Whatley, Ph.D., Department of Psychology, Valdosta State University, Valdosta, GA 31698-0100. Used by permission. Other uses of this scale only by written permission of Dr. Whatley (mwhatley@valdosta.edu). Information on the reliability and validity of this scale is available from Dr. Whatley.

games, and participating in sexual dialogue in chat rooms. These activities may or may not end in sexual climax" (Jones and Tuttle 2012, p. 275). Spouses who discover their partner spending increasing amounts of time engaging in cybersex feel angry, betrayed, and depressed.

**Extradyadic involvement** the sexual involvement of a pair-bonded invididual with someone other than the partner.

**Extradyadic involvement** or extrarelational involvement refers to the sexual involvement of a pair-bonded individual with someone other than the partner. Extradyadic involvements are not uncommon. Of 1,099 undergraduate males, 21% agreed with the statement, "I have cheated on a partner I was involved with" (25% of 3,459 undergraduate females). When the statement was "A partner I was involved with cheated on me," 39% of the males and 52% of the females agreed (Hall and Knox 2013).

Men are more upset if their wife has a heterosexual than a homosexual affair while women are equally upset if their spouse has a homosexual or a heterosexual

**TABLE 14.1 Till Death Do us Part?**

| Monogamy | Cheating | Swinging | Polyamory |
|---|---|---|---|
| Spouse is only sex partner | Husband and/ or wife cheats | Spouses agree multiple sex partners | Spouses agree multiple love and sex partners |

affair (Confer & Cloud 2011). Characteristics associated with spouses who are more likely to have extramarital sex include male gender, a strong interest in sex, permissive sexual values, low subjective satisfaction in the existing relationship, employment outside the home, low church attendance, greater sexual opportunities, higher social status (power and money), and alcohol abuse (Hall et al. 2008).

Traditional marriage scripts fidelity. Traditional wedding vows state, "Hold myself only unto you as long as we both should live." Table 14.1 identifies the alternatives for resolving the transition from multiple to one sexual partner till death.

Unless a couple have a polyamorous or swinging relationship, if they are to have sex outside their marriage, it is via cheating.

**DIVERSITY IN OTHER COUNTRIES**

Researcher Pam Druckerman (2007) wrote *Lust in Translation*, in which she reflects on how affairs are viewed throughout the world. First, terms for having an affair vary; for the Dutch, it is called, "pinching the cat in the dark"; in Taiwan, it is called "a man standing in two boats"; and in England, "playing off sides." Second, how an affair is regarded differs by culture. In America, the script for discovering a partner's affair involves confronting the partner and ending the marriage. In France, the script does not involve confronting the partner and does not assume that the affair means the end of the marriage; rather, "letting time pass to let a partner go through the experience without pressure or comment is the norm."

**Reasons for Extramarital Affairs** Reasons spouses give for becoming involved with someone other than their mate include (Jeanfreau et al. 2014; MaddoxShaw et al. 2013; Omarzu et al. 2012):

**1. *Variety, novelty, and excitement.*** Most spouses enter marriage having had numerous sexual partners. Extradyadic sexual involvement may be motivated by the desire for continued variety, novelty, and excitement. One of the characteristics of sex in long-term committed relationships is the tendency for it to become routine. Early in a relationship, the partners cannot seem to have sex often enough. However, with constant availability, partners may achieve a level of satiation, and the attractiveness and excitement of sex with the primary partner seem to wane.

The **Coolidge effect** is a term used to describe this waning of sexual excitement and the effect of novelty and variety on sexual arousal:

**Coolidge effect** waning of sexual excitement over time and the positive effect of novelty and variety on sexual arousal.

> *One day President and Mrs. Coolidge were visiting a government farm. Soon after their arrival, they were taken off on separate tours. When Mrs. Coolidge passed the chicken pens, she paused to ask the man in charge if the rooster copulated more than once each day. "Dozens of times," was the reply. "Please tell that to the President," Mrs. Coolidge requested. When the President passed the pens and was told about the rooster, he asked, "Same hen every time?" "Oh no, Mr. President, a different one each time." The President nodded slowly and then said, "Tell that to Mrs. Coolidge." (Bermant 1976, pp. 76–77)*

Whether or not individuals are biologically wired for monogamy continues to be debated. Monogamy among mammals is rare (from 3% to 10%), and monogamy tends to be the exception more often than the rule (Morell 1998). Pornography use, which involves viewing a variety of individuals in sexual contexts, is associated with extramarital sex (Wright 2013). Even if such biological wiring for plurality of partners does exist, it is equally debated whether such wiring justifies nonmonogamous behavior—that individuals are responsible for their decisions.

*My second husband was married and I was engaged to be married when we met at a Sunday lunch party in Rockland County. Our connection was so electrifying. . . . As we spoke, the other people at the party fell away.*

SUSAN CHEEVER, AUTHOR

**2. *Workplace friendships.*** Drifting from being friends to lovers is not uncommon in the workplace (Merrill and Knox 2010). We noted in the last chapter that about 60% of persons in the work force reported having become involved with someone at work. Coworkers share the same world eight to ten hours a day and, over a period of time, may develop good feelings for each other that eventually lead to a sexual relationship. Tabloid reports regularly reflect that romances develop between married actors making a movie together (e.g., Brad Pitt and Angelina Jolie met on a movie set).

**3. *Relationship dissatisfaction.*** It is commonly believed that people who have affairs are not happy in their marriage. Spouses who feel misunderstood, unloved, and ignored sometimes turn to another who offers understanding, love, and attention. An affair is a context where a person who feels unloved and neglected can feel loved and important.

**4. *Sexual dissatisfaction.*** Some spouses engage in extramarital sex because their partner is not interested in sex. Others may go outside the relationship because their partners will not engage in the sexual behaviors they want and enjoy. The unwillingness of the spouse to engage in oral sex, anal intercourse, or a variety of sexual positions sometimes results in the other spouse's looking elsewhere for a more cooperative and willing sexual partner.

**5. *Revenge.*** Some extramarital sexual involvements are acts of revenge against one's spouse for having an affair. When partners find out that their mate has had or is having an affair, they are often hurt and angry. One response to this hurt and anger is to have an affair to get even with the unfaithful partner.

**6. *Homosexual relationship.*** Some individuals marry as a front for their homosexuality. Cole Porter, known for such songs as "I've Got You Under My Skin," "Night and Day," and "Every Time We Say Goodbye," was a homosexual who feared no one would buy his music if his sexual orientation were known. He married Linda Lee Porter (alleged to be a lesbian), and their marriage lasted until Porter's death 30 years later.

Other gay individuals marry as a way of denying their homosexuality. These individuals are likely to feel unfulfilled in their marriage and may seek involvement in an extramarital homosexual relationship. Other individuals may marry and then discover later in life that they desire a homosexual relationship. Such individuals may feel that (1) they have been homosexual or bisexual all along, (2) their sexual orientation has changed from heterosexual to homosexual or bisexual, (3) they are unsure of their sexual orientation and want to explore a homosexual relationship, or (4) they are predominantly heterosexual but wish to experience a homosexual relationship for variety. The term **down low** refers to African-American married men who have sex with men and hide this behavior from their spouse.

**Down low** African-American men who have sex with men and hide this behavior from their spouse.

**7. *Aging.*** A frequent motive for intercourse outside marriage is the desire to return to the feeling of youth. Ageism, which is discrimination against the elderly, promotes the idea that being young is good and being old is bad. Sexual attractiveness is equated with youth, and having an affair may confirm to older partners that they are still sexually desirable. Also, people may try to recapture the love, excitement, adventure, and romance associated with youth by having an affair.

**8. *Absence from partner.*** One factor that may predispose a spouse to an affair is prolonged separation from the partner. Some wives whose husbands are away for military service report that the loneliness can become unbearable. Some husbands who are away say that remaining faithful is difficult. Partners in commuter relationships may also be vulnerable to extradyadic sexual relationships.

**Revealing One's Affair by Confession/Partner Snooping** Walters and Burger (2013) identified how individuals revealed their infidelity—in person (38%), over the phone (38%), by a third partner (12%), via e-mail (6%), and through text messaging (6%). A primary motivation for disclosure was respect either for the primary partner or for the history of the primary relationship. While some felt the need to confess because they were guilty or felt the need to be honest, others felt the relationship would benefit from openness/honesty.

In other cases, the cheating was discovered by snooping. **Snooping**, also known as covert intrusive behavior, is defined as investigating (without the partner's knowledge or permission) a romantic partner's private communication (such as text messages, e-mail, and cell phone use) motivated by concern that the partner may be hiding something. Derby et al. (2012) analyzed snooping behavior in 268 undergraduates and found that almost two-thirds (66%) reported that they had engaged in snooping behavior, most often when the partner was taking a shower. Primary motives were curiosity and suspicion that the partner was cheating. Being female, being jealous, and having cheated were associated with higher frequencies of snooping behavior.

**Snooping** investigating (without the partner's knowledge or permission) a romantic partner's private communication motivated by concern that the partner may be hiding something.

WHAT IF?

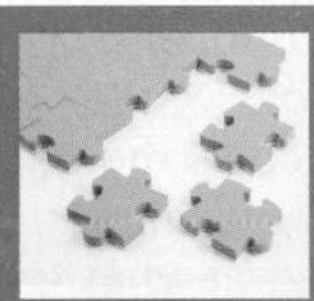

### What Is the Meaning of Cheating to You?

There are multiple meanings associated with cheating (Walters and Burger 2013). While some are ashamed, guilty, and driven to confess, others regard cheating as a confirmation that their relationship is deteriorating, confirmation that they are not cut out to be monogamous, and confirmation that they are transitioning out of the relationship.

**Other Effects of an Affair** Reactions to the knowledge that one's spouse has been unfaithful vary. Some relationships end. Negash et al. (2014) found that 36% of 539 young adult females reported emotional or sexual extradyadic involvement (EDI) in the last two months. Such behavior was particularly predictive of ending one's primary relationship if the partner thought the relationship was of high quality and felt particularly disillusioned and betrayed by the EDI.

Even if the relationship does not end, the awareness of unfaithfulness may be difficult. The following is an example of a wife's reaction to her husband's affairs:

*Those who are faithful know only the trivial side of love; it is the faithless who know of love's tragedies.*

OSCAR WILDE, *THE PICTURE OF DORIAN GRAY*

> *My husband began to have affairs within six months of our being married. Some of the feelings I experienced were disbelief, doubt, humiliation and outright heart-wrenching pain! When I confronted my husband he denied any such affair and said that I was suspicious, jealous and had no faith in him. In effect, I had the problem. He said that I should not listen to what others said because they did not want to see us happy but only wanted to cause trouble in our marriage. I was deeply in love with my husband and knew in my heart that he was guilty as sin; I lived in denial so I could continue our marriage.*
>
> *Of course, my husband continued to have affairs. Some of the effects on me included:*
>
> 1. *I lost the ability to trust my husband and, after my divorce, other men.*
> 2. *I developed a negative self-concept—the reason he was having affairs is that something was wrong with me.*
> 3. *He robbed me of the innocence and my "VIRGINITY"—clearly he did not value the opportunity to be the only man to have experienced intimacy with me.*
> 4. *I developed an intense hatred for my husband.*

## APPLYING SOCIAL RESEARCH | CHEATING: Reactions to Discovery of a Partner's Cheating

"How could you?" asked an angry wife who had discovered her husband in a secluded parked car with a woman from his office. The scene was on *Cheaters*, a television program which features spouses caught in the act of cheating on their mates. Cheating also occurs in undergraduate relationships. The purpose of this research was to identify how undergraduates react to the knowledge that one's romantic partner has cheated—had sexual intercourse with someone else.

### The Survey

A 47-item questionnaire was posted on the Internet and completed by 244 undergraduates (83% of the survey respondents were female, 69% were white, and 52% were in their first year). Over 60% were emotionally involved with one person.

### Results—Gender Differences in Reactions

Over half (51%) reported they had been cheated on by a partner with whom they were in a romantic relationship (17% of the sample reported that they had cheated on a romantic partner). Significantly more women than men (55% versus 31%) reported that a romantic partner had cheated on them.

In addition to a significant gender difference with men cheating more than women, there were significant gender differences in the reactions to a partner's cheating. Women were more likely to cry, put their partner under surveillance, confront their partner, and get tested for an STI.

### Results—Differences in "Unhealthy" and "Healthy" Reactions by Gender

A subset of 19 "Unhealthy" and 9 "Healthy" reactions to a partner's cheating were identified to ascertain if there were gender differences. For example, increasing one's alcohol consumption, becoming suicidal, and having an affair out of revenge were identified as "unhealthy" reactions while confronting the partner, forgiving the partner, and seeing a therapist were regarded as more "healthy" reactions. For each item, participants were assigned a score of 1 point if they agreed (somewhat agree, agree, strongly agree) with the reaction (e.g., "I drank more alcohol" or "I forgave my partner"). Scores on the unhealthy and healthy reactions were summed to create an overall count of demonstrated unhealthy and healthy reactions.

When unhealthy reactions were considered there were no significant gender reactions. Women and men were just as likely to have unhealthy reactions to a partner's infidelity. However, in regard to healthy reactions, females reported a significantly higher percentage of

*It took years for me to recover from this crisis. I feel that through faith and religion I have emerged "whole" again. Years after the divorce my husband made a point of apologizing and letting me know that there was nothing wrong with me, that he was just young and stupid and not ready to be serious and committed to the marriage.*

**Alienation of affection** lawsuits which give a spouse the right to sue a third party for taking the affections of a spouse away.

Seven states (Hawaii, Illinois, North Carolina, Mississippi, New Mexico, South Dakota, and Utah) recognize **alienation of affection** lawsuits which give a spouse the right to sue a third party for taking the affections of a spouse away. Alienation of affection claims evolved from common law, which considered women property of their husbands. The reasoning was if another man was accused of stealing his "property," a husband could sue him for damages. The law applies to both women and men so a woman who steals another woman's man can be sued for taking her property away. Such was the case of Cynthia Shackelford who sued Anne Lundquist in 2010 for "alienating" her husband from her and breaking up her 33-year marriage. A jury awarded Cynthia Shackelford $4 million in punitive damages and $5 million in compensatory damages. The decision has been appealed. In North Carolina, about 200 alienation of affection lawsuits are filed annually. The infraction must have occurred while the couple was still married (not during the separation period) and there is a three-year statute of limitations.

healthy reactions/behaviors than males; females averaged 3.71 healthy behaviors (SD = 1.46) compared to males, who averaged 2.44 healthy behaviors (SD = 1.58).

### Theoretical Framework and Discussion

Symbolic interaction theory and social exchange theory provide frameworks for understanding reactions to the knowledge that one's partner has cheated. Symbolic interaction posits that a couple's relationship is created and maintained on agreed-upon meanings of various behaviors. A major concept inherent in this theoretical framework is definition of the situation. In reference to cheating, partners in a relationship have definitions about the meaning of one partner having sexual intercourse with someone outside the dyad. Among the undergraduates in the current study, most of whom expected fidelity, the definition of the situation was a monogamous relationship with the expectation of fidelity. As such, a feeling of betrayal occurred when fidelity was breeched.

The social exchange framework views the interaction between partners in a romantic relationship in terms of profit and cost. Both partners enter the relationship with promised love and fidelity and expect the same in return. When one partner does not exchange fidelity for fidelity, there is a significant cost to the faithful partner for remaining in a relationship where the partner has been unfaithful. Indeed a common reaction among the sample of undergraduates in this study to the knowledge that their partner had cheated was to terminate the relationship; almost half (47%) ended the relationship with the partner who cheated.

### Implications of the Study

There are three implications of this study. One, cheating in romantic relationships among undergraduates is not uncommon. Over half (51%) reported having been cheated on. Two, the knowledge that one's partner has cheated is traumatic. Feeling betrayed (6.24 out of 7) was the most common reaction and was often accompanied by crying (5.81 out of 7), depression (4.65 out of 7), and increased drinking (2.92 out of 7). Three, the range of alternative reactions to the knowledge of infidelity was extensive, including many healthy alternatives including forgiveness, exercise, and seeing a therapist (with women more likely than men to select healthy alternatives).

### Source

Abridged and adapted from Barnes, H., D. Knox, and J. Brinkley. 2012. CHEATING: Gender differences in reactions to discovery of a partner's cheating. Paper, Southern Sociology annual meeting. New Orleans, March.

Beyond emotional and economic consequences for an affair, there are medical consequences. Fisher et al. (2012 and 2012A) found that involvement in an extramarital affair is associated with MACE (major adverse cardiovascular events, including a fatal heart attack) for both the person having the affair and for the person being betrayed.

**Successful Recovery from Infidelity** When an affair is discovered, a sense of betrayal pervades the nonoffending spouse or partner (Barnes et al. 2012). While an affair is a high frequency cause of a couple deciding to divorce, keeping the relationship together (with forgiveness and time) is the most frequent outcome. Abrahamson et al. (2012) interviewed seven individuals who had experienced an affair in their relationship and who were still together two years later. The factors involved in rebuilding their relationship included:

1. ***Motivation to stay together.*** Having been together several years, having children, having property jointly, not wanting to "fail," and fearing life alone were factors which motivated the partners to stay together. The basic feeling is that we have a lot to gain by working this out.

2. ***Taking joint responsibility.*** The betrayed partner found a way to acknowledge she or he had contributed to the affair so that there was joint responsibility for the affair.

*Life is short, Break the Rules.*

*Forgive quickly, Kiss SLOWLY.*

*Love truly. Laugh uncontrollably*

*And never regret ANYTHING*

*That makes you smile.*

MARK TWAIN

**3.** ***Forgiveness, counseling, and not referring to the event again.*** Forgiveness involved letting go of one's resentment, anger, and hurt; accepting that we all need forgiveness; and moving forward (Hill 2010). Bagarozzi (2008) noted that the offending spouse must take responsibility for the affair, agree not to repeat the behavior, and grant their partner the right to check up on them to regain trust. The personal choices section focuses on the decision to end a relationship rather than work through it when a partner has an affair.

**4.** ***Vicarious learning***—noting that others who ended a relationship over an affair were not necessarily happier/better off.

**5.** ***Feeling pride in coming through a difficult experience.*** One wife whose husband had had an affair for seven years (with her best friend) was told by their counselor, "You're going to need to resolve this in a way that it does not wreck your life whether or not you stay married . . . . one option is to resolve it AND stay married . . . . to keep your family together." The couple worked it out, improved their relationship, and are still together.

## PERSONAL CHOICES

### *Should You Seek a Divorce If Your Partner Has an Affair?*

About 20% of spouses face the decision of whether to stay with a mate who has had an extramarital affair, whether the affair was physical or online.

*Before you embark on a journey of revenge, dig two graves.*

CONFUCIUS

Regardless of whether the indiscretion is physical or emotional, one alternative for the partner is to end the relationship immediately on the premise that trust has been broken and can never be mended. People who take this position regard fidelity as a core element of the marriage that, if violated, necessitates a divorce. College students disapprove of an extramarital affair. In a sample of 4,539 undergraduates, 67% (higher percentage of men than women) said that they would "divorce a spouse who had an affair" (Hall and Knox 2013). Americans in general tend to be unforgiving about an affair. In a national survey, 64% would not forgive their spouse for having an extramarital affair and almost as high a percentage (62%) say they would leave their spouse and get a divorce if they found out their spouse was having an affair; 31% would not divorce (Jones 2008). For some, emotional betrayal is equal to sexual betrayal.

Other couples build into their relationship the fact that each will have external relationships. In Chapter 3, we discussed polyamory, in which partners are open and encouraging of multiple relationships at the same time. The term *infidelity* does not exist for polyamorous couples.

Even for traditional couples, infidelity need not be the end of a couple's marriage but the beginning of a new, enhanced, more understanding/communicative relationship. Healing takes time; the relationship may require a commitment on the part of the straying partner not to repeat the behavior, forgiveness by the partner (and not bringing it up again), and a new focus on keeping the relationship. In spite of the difficulty of adjusting to an affair, most spouses are reluctant to end a marriage. Not one of fifty successful couples said that they would automatically end their marriage over adultery (Wallerstein and Blakeslee 1995).

The spouse who chooses to have an affair is often judged as being unfaithful to the vows of the marriage, as being deceitful to the partner, and as inflicting enormous pain on the partner (and children). When an affair is defined in terms of giving emotional energy, time, and economic resources to something or someone outside the primary relationship, other types of "affairs" may be equally as devastating to a relationship. Spouses who choose to devote their lives to their careers, parents, friends, or recreational interests may deprive the partner of significant amounts of emotional energy, time, and money and create a context in which the partner may choose to become involved with a person who provides more attention and interest.

Another issue in deciding whether to take the spouse back following extramarital sex is the concern over HIV. One spouse noted that, though he was willing to forgive and try to forget his partner's indiscretion, he required that she be tested for HIV and that they use a condom for six months. She tested negative, but their use of a condom was a reminder, he said, that sex outside one's bonded relationship in today's world has a life-or-death meaning. Related to this issue is that wives are more likely to develop cervical cancer if their husbands have other sexual partners.

There is no single way to respond to a partner who has an extramarital relationship. Most partners are hurt and think of ways to work through the crisis. Some succeed and come to regard themselves as being in a great marriage (Tulane et al. 2011).

### Sources

Hall, S. and D. Knox. 2013. Relationship and sexual behaviors of a sample of 4,590 university students. Unpublished data collected for this text. Department of Family and Consumer Sciences, Ball State University and Department of Sociology, East Carolina University.

Jones, J. M. 2008. Most Americans not willing to forgive unfaithful spouse. *The Gallup Poll Briefing,* March. Washington.

Tulane, S., L. Skogrand and J. DeFrain. 2011. Couples in great marriages who considered divorcing. *Marriage and Family Review* 47: 289–310.

Wallerstein, J. S., and S. Blakeslee. 1995. *The good marriage*. Boston: Houghton Mifflin.

Positive outcomes of having experienced and worked through infidelity include a closer marital relationship, placing higher value on each other, and realizing the importance of good marital communication.

Spouses who remain faithful to their partners have made a conscious decision to do so. They avoid intimate conversations with members of the other sex and a context (e.g., alcohol/being alone in a hotel room) that is conducive to physical involvement. The best antidote to an affair is a strong emotional/sexual connection with one's spouse and avoiding contexts conductive to extradyadic involvement.

**Prevention of Infidelity** Allen et al. (2008) identified the premarital factors predictive of future infidelity. The primary factor for both partners was a negative pattern of interaction. Partners who ended up being unfaithful were in relationships where they did not connect emotionally, they argued, and they criticized each other. Hence, spouses least vulnerable are in loving, nurturing, communicative relationships where each affirms the other. Neuman (2008) also noted that avoiding friends who have affairs and establishing close relationships with married couples who value fidelity further insulate individuals from having an affair. Hertlein and Piercy (2012) studied the treatment of Internet affairs and emphasized the importance of the individual being true to various boundaries to protect the primary relationship.

Some couples feel that an open marriage is the answer to their being unfaithful. Rather than be dishonest, cheat, and see another partner in secret, the partners decide that they will remain committed to each other but have other sexual partners as well. Bergstrand and Williams (2000) reported online data from 1,092 swingers (85% married with a committed partner), mostly white, middle class, middle aged, divorced, and in the current marriage over 10 years. Most (62%) reported that swinging improved their relationship; only 6% reported that jealousy was a problem.

Some issues partners in open relationships might discuss include: What sexual behaviors with others are acceptable (e.g., OK to kiss)? Do both agree a condom will always be worn? Will the partners discuss their liaisons with each

## FAMILY POLICY | Alcohol Abuse on Campus

While some college students do not drink (Herman-Kinney and Kinney 2013), most do. And while most college students who drink do not have a problem with alcohol, some do. Twelve percent of 1,084 and 7% of 3,438 female undergraduates reported that "I have a problem with alcohol" (Hall and Knox 2013). Individuals with personalities characterized by urgency and sensation seeking also drink more, have more binges, and more alcohol problems (Shin et al. 2012). Alcohol and other drug use (Pedrelli et al. 2013) including high energy drinks (Snipes and Benotsch 2013) are associated as is the greater likelihood of cohabitation, early marriage, and subsequent divorce (Williams et al. 2012).

Campus policies throughout the United States include alcohol-free dorms; alcohol bans, enforcement, and sanctions; peer support; and alerting parents and providing education. Most colleges and universities do not ban alcohol or its possession on campus. Administrators fear that students will attend other colleges where they are allowed to drink. Alumni may want to drink at football games and view such university banning as intrusive. Some attorneys think colleges and universities can be held liable for not stopping dangerous drinking patterns, but others argue that college is a place for students to learn how to behave responsibly. Should sanctions be used (e.g., expelling a student or closing down a fraternity)? If policies are too restrictive, drinking may go underground, where detecting use may be more difficult.

other or keep them a secret? Can the same person be seen more than once? Is emotional attachment OK? Are mutual friends OK to have sex with? Who else will be told about the open relationship? (Zimmerman 2012).

Some individuals who have cheated and feel out of control seek help for their "addiction" by entering rehab (e.g., Tiger Woods). Whether their attending a sex rehabilitation clinic is a marital ploy to be forgiven, do public penance, and resume the marriage or a call for help is unknown. Levine (2010) studied a sample of 30 married men who had been discovered to have violated monogamy rules with their wife via pornography, cybersex, commercial sex involvement, paraphilic pursuits, or affairs and who went to a sex addiction clinic. According to Levine (2010) only 25% of the sample had issues (spiraling psychological deterioration) which could reasonably be described as having a sexual addiction, which is a primary criterion for being labeled as a sex addict.

## Unemployment

The unemployment rate is defined as the number of people actively looking for a job divided by the labor force. Changes in unemployment depend mostly on inflows made up of non-employed people starting to look for jobs, of employed people who lose their jobs and look for new ones, and of people who stop looking for employment. In early 2014, the U.S. unemployment rate was 6.6. African-Americans, Latinos, young workers and less-educated workers are the most vulnerable to being chronically unemployed, which will affect their economic, social, and personal health (Schmitt and Jones 2012).

Unemployment/economic stress is associated with a decline in relationship satisfaction (Williamson et al. 2013), child maltreatment (Euser et al. 2010), and divorce (Eliason 2012). The personal effects of unemployment are more severe for men than for women (Backhans and Hemmingsson 2012). Our society expects men to be the primary breadwinners in their families and equates masculine self-worth and identity with job and income.

**Your Opinion?**

1. To what degree do you believe drinking on your campus is an issue and policies should be developed to control it?
2. What policies might universities develop to reduce binge drinking?
3. Under what conditions should a student who abuses alcohol be expelled?

## Sources


Hall, S. and D. Knox. 2013. Relationship and sexual behaviors of a sample of 4,590 university students. Unpublished data collected for this text. Department of Family and Consumer Sciences, Ball State University and Department of Sociology, East Carolina University.

Herman-Kinney, N. J. and D. A. Kinney. 2013. Sober as deviant: The stigma of sobriety and how some college students "stay dry" on a "wet" campus. *Journal of Contemporary Ethnography* 42: 64–103.

Pedrelli, P., K. Bentley, M. Vitali, A. J. Clain, M. Nyer, M. Fava and A. H. Farabaugh. 2013. Compulsive use of alcohol among college students. *Psychiatry Research* 205: 95–102.

Shin, S. H., H. G. Hong, and S. M. Jeon. 2012. Personality and alcohol use: The role of impulsivity. *Addictive Behaviors* 37: 102–107.

Snipes, D. J. and E. G. Benotsch. 2013. High-risk cocktails and high-risk sex: Examining the relation between alcohol mixed with energy drink consumption, sexual behavior, and drug use in college students. *Addictive Behaviors* 38: 1418–1423.

Williams, L. R., L. Wray-Lake, E. Loken, and J. L. Maggs. 2012. The effects of adolescent heavy drinking on the timing and stability of cohabitation and marriage. *Families in Society* 93: 181–188.


When spouses or parents lose their jobs as a result of physical illness or disability, the family experiences a double blow—loss of income combined with higher medical bills. Unless an unemployed spouse is covered by the partner's medical insurance, unemployment can result in loss of health insurance for the family. Insurance for both health care and disability is very important to help protect a family from an economic disaster.

## Alcohol/Substance Abuse

A person has a problem with alcohol or a substance if it interferes with their health, job, or relationships. Spouses, parents, and children who abuse alcohol and/or drugs contribute to the stress and conflict experienced in their respective marriages and families. Although some individuals abuse drugs to escape from unhappy relationships or the stress of family problems, substance abuse inevitably adds to the individual's marital and family problems when it results in health and medical problems, legal problems, loss of employment, financial ruin, school failure, emotionally distant relationships (Cox et al. 2013; Lotspeich-Younkin and Bartle-Haring, 2012), and divorce. Table 14.2 reflects substance abuse at various age categories.

The Family Policy section deals with alcohol abuse on campus.

**TABLE 14.2 Current Drug Use by Type of Drug and Age Group**

| Type of Drug Used | Age 12 to 17 | Age 18 to 25 | Age 26 to 34 |
|---|---|---|---|
| Marijuana and hashish | 6.7% | 18.5% | 8.8% |
| Cocaine | .4% | 1.5% | 1.5% |
| Alcohol | 14.8% | 61.2% | no data |
| Cigarettes | 9.1% | 36.7% | no data |

Source: Adapted from *Statistical Abstract of the United States 2012–2013*, 131th ed. Washington, DC: U.S. Bureau of the Census, Table 207.

Chelsea Curry

Alcohol is the most frequently used drug in college.

## What If Your Partner Is an Alcoholic?

If your partner is an alcoholic, it is important to recognize that you can do nothing to stop your partner from drinking. Your partner must "hit bottom" or have an epiphany and make a personal decision to stop. In the meantime, you can make choices to join **Al-Anon** (a support group) to help you cope with your partner's alcoholism, regard the alcoholism as a sickness and decide to stay with your partner, or end the relationship. Most individuals stay with their partner because there is an emotional cost of leaving. However, some pay the price. Only you can decide.

**Al-Anon** organization that provides support for family members and friends of alcohol abusers.

## Death of Family Member

Even more devastating than drug abuse are family crises involving death—of one's child, parent, or loved one (we discuss the death of one's spouse in Chapter 16 on Relationships in the Later Years). The crisis is particularly acute when the death is a suicide.

**Death of One's Child** A parent's worst fear is the death of a child. Most people expect the death of their parents but not the death of their children. Amy Winehouse died at the age of 27 in 2011. Her distraught parents, Mitch and Janis, said that they were "left bereft" at her death. Grief feelings may be particularly acute on the anniversary of the death of an individual with whom one was particularly close.

Maple et al. (2013) interviewed 22 parents following the death of a young adult child and found that the parents needed to maintain a relationship with their child including public and private memorials to internal dialogues. Mothers and fathers sometimes respond to the death of their child in different ways. When they do, the respective partners may interpret these differences in negative ways, leading to relationship conflict and unhappiness. To deal with these differences, spouses need to be patient and practice tolerance in allowing each to grieve in her or his own way. Men typically become work focused and women become focused on the grieving sibling (Alam et al. 2012).

**Death of One's Parent** Terminally ill parents may be taken care of by their children. Such care over a period of years can be emotionally stressful, financially draining, and exhausting. Hence, by the time the parent dies, a crisis has already occurred.

Reactions to the death of a loved one (whether parent or partner) is not something one "gets over." Burke et al. (1999) noted that grief is not a one-time experience that people adjust to and move on. Rather, for some, there is **chronic sorrow**, where grief-related feelings occur periodically throughout the lives of those left behind. Grief feelings may be particularly acute on the anniversary of the death or when the bereaved individual thinks of what might have been had the person lived. Burke et al. (1999) noted that 97% of the individuals in one study who had experienced the death of a loved one 2 to 20 years earlier met the criteria for chronic sorrow.

**Chronic sorrow** grief-related feelings occur periodically throughout the lives of those left behind.

Boss (2013) also noted that closure may not be a realistic goal but to learn to cope with the death of a loved one by finding meaning and dropping the expectation that one should "get over it." Actor Liam Neeson noted that three years after the death of his wife, actress Natasha Richardson, he still experienced waves of grief. Particularly difficult are those cases of ambiguous loss where the definitions of death become muddled, as when a person disappears or has a disease such as Alzheimers.

**Suicide of Family Member** In 2013, 37-year-old country-western singer Mindy McCready shot and killed herself. She did so a month after her boyfriend had committed suicide. Previously she had written, "I call my life a beautiful mess." Suicide is a devastating crisis event for families. Annually there are 31,000 suicides (750,000 attempts), and each suicide immediately affects at least six other people in that person's life. These effects include depression (e.g., grief), physical disorders (e.g., shingles due to stress), and social stigma (e.g., the person is viewed as weak and the family as a failure for not having been able to help the suicide victim) (De Castro and Guterman 2008). Miers et al. (2012) interviewed six parental units (mother and dad) whose teenager had committed suicide. Critical needs of the couple included that they needed support from another suicide survivor, support when viewing the deceased teen, and support in remembering the teen. People who are between the ages of 15 and 19, homosexuals, males, those with a family history of suicide, mood disorders, or substance abuse and those with a past history of child abuse and parental sex abuse are more vulnerable to suicide than others (Melhem et al. 2007).

Suicide is viewed as a rational act in that the person feels that taking his or her life is the best option available at the time. Therapists view suicide as a "permanent solution to a temporary problem" and routinely call 911 to have people hospitalized or restrained who threaten suicide or who have been involved in an attempt. Suicide can sometimes be predicted (Stefansson et al. 2012).

Adjustment to the suicide of a family member takes time. Part of the recovery process is accepting that one cannot stop the suicide of another who is adamant about taking his or her own life and that one is not responsible for the suicide of another. Indeed, family members often harbor the belief that they could have done something to prevent the suicide.

*The undiscovered country, from whose bourn,No traveler returns.*

WILLIAM SHAKESPEARE, *HAMLET*, III, I, 79

# Marriage and Family Therapy

University students have a positive view of marriage therapy. In a study of 288 undergraduate and graduate students, 93% of the females and 82% of the males agreed that, "I would be willing to see a marriage counselor before I got a divorce" (Dotson-Blake et al. 2010).

Undergraduate couples might consider consulting a marriage therapist (who specializes in relationships . . . being married is irrelevant) about their relationship rather than remaining in an unhappy relationship. Signs to look for in your own relationship that suggest you might consider seeing a therapist include feeling distant and not wanting or being unable to communicate with your partner, avoiding each other, feeling depressed, drifting into a relationship with someone else, increased drinking, and privately contemplating separation or breaking up. If you are experiencing one or more of these symptoms, it may be wise to intervene early so as to stop the spiral toward an estranged relationship before there is no motivation to do so. A relationship is like a boat. A small unattended leak can become a major problem and sink the boat. Marriage therapy sometimes serves to reverse relationship issues early by helping the partners to sort out values, make decisions, and begin new behaviors so that they can start feeling better about each other.

*Love seems the swiftest, but it is the slowest of all growths. No man or woman really knows what perfect love is until they have been married a quarter of a century.*

**MARK TWAIN, AMERICAN AUTHOR AND HUMORIST**

## Availability of Marriage and Family Therapists

There are around 50,000 marriage and family therapists in the United States. Their professional roles include medical doctors (MDs), nurses, psychologists, social workers, professional counselors, and those who focus exclusively on marriage and family therapy. About 40% of "marriage counselors" are clinical members of the American Association for Marriage and Family Therapy (AAMFT). Currently there are 57 masters, 19 doctoral, and 16 postgraduate programs accredited by the AAMFT. Forty-eight states require a license to practice marriage and family therapy. Bordoloi et al. (2013) recommended renaming the organization American Association for Couples and Family Therapy to be more inclusive of same-sex relationships.

Therapists holding membership in AAMFT have had graduate training in marriage and family therapy, two years of post-graduate experience including 1,000 hours of direct client contact concurrently completed with 200 hours of supervision with an AAMFT Approved Supervisor (100 of these hours must be individual hours in supervision).

Whatever marriage therapy costs, it can be worth it in terms of improved relationships. If divorce can be averted, both spouses and children can avoid the trauma and thousands of dollars will be saved. Managed care has resulted in some private therapists making marriage therapy affordable. Effective marriage therapy usually involves seeing the spouses together (conjoint therapy). However, there are occasions where individual therapy is necessary and helpful. Issues of depression, alcoholism/substance abuse, chronic infidelity, and PTSD may best be dealt with in individual sessions.

## Variety of Approaches to Couple Therapy

Members of AAMFT use more than 20 different treatment approaches, most of which attempt to be evidence based (Lindblad-Goldberg and Northey 2013). Dattilio et al. (2014) emphasized that while evidence-based approaches to therapy continue to grow in the field of marital and family therapy, some therapists suggest a broader view of evaluating effective therapy. If achieving the behavioral goals identified by the client/s is the definition of evidence based, a broader view must be considered carefully.

If you and your partner are seeing a marriage therapist, the two of you should clearly identify your goal to the therapist: "Our goal is to feel better about each other, improve our communication with each other, and spend more time together doing things of mutual enjoyment" and ask the therapist if and how he or she can help you to achieve your goals. If the answer you receive is not satisfactory, find another therapist.

Most therapists (31%) report that they use either a behavioral or cognitive-behavioral approach. A behavioral approach (also referred to as **behavioral couple therapy**) means that the therapist focuses on behaviors the respective spouses want increased or decreased, initiated or terminated, and then negotiates behavioral exchanges between the partners.

**Behavioral couple therapy** the therapist focuses on behaviors the respective spouses want increased or decreased, initiated or terminated, and then negotiates behavioral exchanges between the partners.

Some therapists use behavior contracts which are agreements partners make of new behaviors to engage in between sessions. The following is an example and assumes that the partners argue frequently, never compliment each other, no longer touch each other, and do not spend time together. The contract calls for each partner to make no negative statements to the other, give two compliments per day to the other, hug or hold each other at least once a day, and allocate Saturday night to go out to dinner alone with each other. On the contract, under each day of the week, the partners check that they did what they agreed to; the contract is given to the therapist at the next appointment. Partners who change their behavior toward each other often discover that the partner changes also and there is a new basis for each to feel better about each other and their relationship.

*Behavior Contract for Partners**

Name of Partners ____________ Date:_Week of *June 8–14*

Behaviors each partner agrees to engage in and Days of Week

| | Mon. | Tues. | Wed. | Thurs. | Fri. | Sat. | Sun. |
|---|---|---|---|---|---|---|---|
| 1. *No negative statements to partner* | □ | □ | □ | □ | □ | □ | □ |
| 2. *Compliment partner twice each day* | □ | □ | □ | □ | □ | □ | □ |
| 3. *Hug or hold partner once a day* | □ | □ | □ | □ | □ | □ | □ |
| 4. *Out to dinner Saturday night* | □ | □ | □ | □ | □ | □ | □ |

*From B. Crisp and D. Knox. 2009. *Behavioral family therapy*. Durham, NC: Carolina Academic Press.

Sometimes clients do not like behavior contracts and say to the behavior therapist, "I want my partner to compliment me and hug me because my partner wants to, not because you wrote it down on this silly contract." The behavior therapist acknowledges the desire for the behavior to come from the heart of the partner and points out that the partner is making a choice to engage in new behavior to please the partner.

Cognitive-behavioral therapy, also referred to as **integrative behavioral couple therapy** (IBCT), emphasizes a focus on cognitions. South et al. (2010) found that spouses were most happy when they perceived the positive and negative behaviors occurred at the desirable frequency—high frequency of positives and low frequency of negatives.

**Integrative Behavioral Couple Therapy** (IBCT) therapy which focuses on the cognitions or assumptions of the spouses, which impact the way spouses feel and interpret each other's behavior.

Whether a couple in therapy remain together will depend on their motivation to do so, how long they have been in conflict, the severity of the problem, and whether one or both partners are involved in an extramarital affair. Two moderately motivated partners with numerous conflicts over several years are

This therapist is conducting a session with a couple over the Internet (referred to as telerelationship therapy or Skype therapy).

David Knox

less likely to work out their problems than a highly motivated couple with minor conflicts of short duration. Severe depression or alcoholism on the part of either spouse is a factor that will limit positive marital and family gains. In general, these issues must be resolved individually before the spouses can profit from marital therapy.

## Telerelationship (Skype) Therapy

**Telerelationship therapy** therapy sessions conducted online, often through Skype, where both therapist and couple can see and hear each other.

An alternative to face-to-face therapy is **telerelationship therapy**, which uses the Internet (Skype). Both therapist and couple log on to Skype, where each can see and hear the other while the session is conducted online. Terms related to telerelationship therapy are telepsychology, telepsychiatry, virtual therapy, and video interaction guidance (VIG) (Magaziner 2010; Doria et al. 2014). Telerelationship therapy allows couples to become involved in marriage/family therapy independent of where they live (e.g., isolated rural areas), the availability of transportation, and time (i.e., sessions can be scheduled outside the 9 to 5 block). While the efficacy of telerelationship therapy compared with face-to-face therapy continues to be researched, evidence regarding its value in individual (Nelson and Bui 2010) and family therapy (Doria et al. 2014) has been documented. Regarding the latter, Doria et al. (2014) provided content analysis of 15 therapeutic sessions (by three therapists in Europe) which improved family happiness, parental self-esteem and self-efficacy, and attitude–behavior change. These data emphasize that therapy can be effectively conducted over the Internet.

## Some Caveats about Marriage and Family Therapy

In spite of the potential benefits of marriage therapy, some valid reasons exist for not becoming involved in such therapy. Not all spouses who become involved in marriage therapy regard the experience positively. Some feel that their marriage is worse as a result. Reasons some spouses cite for negative outcomes include saying things a spouse can't forget, feeling hopeless at not being able to resolve a problem "even with a counselor," and feeling resentment over new demands a spouse makes in therapy.

Therapists may also give clients an unrealistic picture of loving, cooperative, and growing relationships in which partners always treat each other with respect and understanding, share intimacy, and help each other become whomever each wants to be. In creating this idealistic image of the perfect relationship, therapists may inadvertently encourage clients to focus on the shortcomings in their relationship and to expect more of their marriage than is realistic. Psychiatrist Robert Sammons (2014) calls this his first law of therapy: "That spouses always focus on what is missing rather than what they have . . . indeed the only thing that is important to couples in therapy is that which is missing. A new focus of what the person does that pleases them rather than what they are missing is needed."

Couples and families in therapy must also guard against assuming that therapy will be a quick and easy fix. Changing one's way of viewing a situation (cognitions) and behavior requires a deliberate, consistent, relentless commitment to ensure that these changes occur. Talking and hoping mean nothing and change nothing. Only by changing one's perceptions and behaviors does a relationship change.

## The Future of Stress and Crisis in Relationships

Stress and crisis will continue to be a part of relationships. No spouse, partner, marriage, family, or relationship is immune. A major source of stress will be economic—the difficulty in securing and maintaining employment and sufficient income to take care of the needs of the family.

Most relationship partners will also show resilience to rise above whatever crisis happens. The motivation to do so is strong, and having a partner to share one's difficulties reduces the sting. As noted, it is always one's perception of an event, not the event itself, which will determine the severity of a crisis and the capacity to cope with and overcome it.

## Summary

***What is stress and what is a crisis event?***

Stress is a reaction of the body to substantial or unusual demands (physical, environmental, or interpersonal). Stress is associated with irritability, high blood pressure, and depression. A crisis is a situation that requires changes in normal patterns of behavior. A family crisis is a situation that upsets the normal functioning of the family and requires a new set of responses to the stressor. Sources of stress and crises can be external (for example, hurricane, tornado, downsizing, military separation) or internal (for example, alcoholism, extramarital affair, Alzheimer's disease, inherited wealth).

Family resilience is when family members successfully cope under adversity, which enables them to flourish with warmth, support, and cohesion. Key factors include positive outlook, spirituality, flexibility, communication, financial management, family-shared recreation, routines or rituals, and support networks.

***What are positive stress management strategies?***

Changing one's view is the most helpful strategy in reacting to a crisis. Viewing ill health as a challenge, bankruptcy as an opportunity to spend time with one's family, and infidelity as an opportunity to improve communication are examples. Other positive coping strategies are exercise, adequate sleep, love, religion, friends or relatives, and humor. Some harmful strategies include keeping feelings inside, taking out frustrations on one's partner, and denying or avoiding the problem.

***What are several of the major family crisis events?***

Some of the more common crisis events that spouses and families face include physical illness, mental illness, an extramarital affair, unemployment, substance abuse, and the death of one's spouse or children. Surviving an affair involves forgiveness on the part of the offended spouse to grant a pardon to the offending spouse, to give up feeling angry, and to relinquish the right to retaliate against the offending spouse. In exchange, an offending spouse must take responsibility for the affair, agree not to repeat the behavior, and grant the partner the right to check up on the offending partner to regain trust. The occurrence of a midlife crisis is reported by less than a quarter of adults in the middle years. Those who did experience a crisis were going through a divorce.

***What help is available from marriage and family therapists?***

There are over 50,000 marriage and family therapists in the United States. About 40% are clinical members of the American Association for Marriage and Family Therapy (AAMFT). Whether a couple in therapy remain together will depend on their motivation to do so, how long they have been in conflict, and the severity of the problem. Two moderately motivated partners with numerous conflicts over several years are less likely to work out their problems than a highly motivated couple with minor conflicts of short duration. Couples tend to benefit from early intervention. About 70% of couples involved in integrative couple behavior therapy (ICBT) (which involves behavior change and cognitions) report continued significant gains two years after the end of treatment. Telerelationship therapy is an alternative to face-to-face therapy.

***What caveats should be kept in mind for becoming involved in marriage/family therapy?***

Not all spouses who become involved in marriage therapy regard the experience positively. Some feel that their marriage is worse as a result. Reasons some spouses cite for negative outcomes include saying things a spouse can't forget, feeling hopeless at not being able to resolve a problem "even with a counselor," and feeling resentment over new demands a spouse makes in therapy.

## Key Terms

Al-Anon
Alienation of affection
Behavioral couple therapy (BCT)
Chronic sorrow
Coolidge effect
Crisis
Down low
Extradyadic involvement
Extramarital affair
Family resilience
Integrative behavioral couple therapy (IBCT)
Palliative care
Resiliency
Sanctification
Snooping
Stress
Telerelationship therapy

## Web Links

American Association for Marriage and Family Therapy
http://www.aamft.org

Association for Applied and Therapeutic Humor
http://www.aath.org

Association for Couples in Marriage Enrichment
http://www.bettermarriages.org/

Couples in Trouble
http://www.couplesintrouble.com/

Mental Health
http://www.bringchange2mind.org/

Red Hat Society
http://www.redhatsociety.com/

CHAPTER 15

# Divorce and Remarriage

David Knox

*Divorce is a snakebite. One's next thought should not be, "Where am I going to find another snake?"*

— SUZANNE FINNAMORE, *SPLIT: A MEMOIR OF DIVORCE*

## Learning Objectives

- Explain the factors involved in predicting whether a couple will divorce.
- Identify macro and micro factors associated with divorce.
- Understand the consequences of divorce for spouses, parents, and children.
- Describe the conditions under which one can have a successful divorce.
- Review what is involved in preparing for a remarriage and the issues involved.
- Discuss the differences between first marriages and remarriages.
- Suggest what the future holds for divorce and remarriage.

## TRUE OR FALSE?

1. Deal breakers for continuing a relationship for women and men are the same.
2. The top reason 478 undergraduates gave for breaking up was betrayal by the partner.
3. There is no universal agreement on child custody by parents and half the agreements change over time.
4. Most undergraduates prefer that their partner break up with them in a text message than face to face.
5. Income drops for both women and men who divorce but women suffer more.

*Answers:* 1. F 2. F 3. T 4. F 5. T

Divorce is a choice. Indeed, spouses who end a marriage have made choices which have resulted in the continuation of a negative relationship and one or both have chosen to stop the pain. This chapter is about choices which lead to divorce as well as creating an awareness of how American society and culture create a context where divorce is an option. As we will note, individualistic societies such as the United States have much higher divorce rates than familistic societies such as India.

Undergraduates know about divorce since over a third have experienced the divorce of their parents. In a sample of 4,531 undergraduates, 35% reported that their parents were divorced (Hall and Knox 2013). In this chapter, we look at the social and individual causes of divorce, the consequences for the spouses and their children, and ways to make the ending of one's marriage as civil and forward-looking as possible. We begin by examining whether to end a relationship.

# Deciding Whether to End a Relationship/Marriage

*There's a word for couples who believe that the feelings they share now are the feelings they'll share forever: delusional. If you must enter a relationship that's bound to turn sour and is almost impossible to get out of, look into a time-share.*

BILL MAHR, COMEDIAN

Relationships follow a trajectory including a decision about whether to continue.

## Factors Predictive of Continuing or Ending a Relationship

Walsh (2013) reported on Match.com's survey of 5,500 singles who identified "deal breakers." Deal breakers for women were "dating someone who was secretive with their texts" (77%), dating someone who was lazy (72%), dating someone who was shorter than them (71%), or dating someone who was a virgin (51%). Deal breakers for men were dating someone who was disheveled/unclean (63%), who was lazy (60%), or who did not care about her career (46%).

Rhoades et al. (2010) identified four factors involved in whether a person continues or ends a relationship.

1. ***Dedication***—motivation to build/maintain a high-quality relationship, have a long-term relationship.

2. ***Perceived constraints***—factors operating to keep the couple together. For example: social pressure to stay together from parents/friends/church group, intangible investments (e.g., years together) lost if relationship ends, belief that one's quality of life would deteriorate should the relationship end, concern for the welfare of one's partner, fear that taking the steps (e.g., talking to partner) to end the relationship would be difficult, or fear of finding a suitable replacement if the relationship ends.

**3.** ***Material constraints***—observable changes which would occur if the relationship ends. For example: dealing with shared debt, apartment, furniture, pets, etc.

**4.** ***Feeling Trapped***—individuals rate the degree to which they feel trapped by the investment in the relationship. The more trapped, the less likely they are to stay in the relationship.

The researchers analyzed data on 1,184 unmarried adults in a romantic relationship and found that each of the four factors predicted whether the relationship would continue or end. Basically the partners choose to end a relationship when the rewards of leaving outweigh the costs of staying so that there is little profit for staying in the relationship.

## End an Unsatisfactory Relationship?

All relationships have difficulties, and all necessitate careful consideration of various issues before a decision is made to divorce. Before you make the decision to divorce, consider the following:

*I love the way you're breaking my heart. Although you're gonna ruin it, it's Heaven while you're doing it. So darling just keep playing your part, cause I love the way you're breaking my heart.*

PEGGY LEE, TORCH SINGER OF THE 50S

**1.** ***Consider improving the relationship rather than ending it.*** In some cases, people end relationships and later regret having done so. Setting unrealistically high standards may end a marriage prematurely. Particularly in our individualistic society with the fun/love/sex focus of relationships, anything that deviates is considered for dumping. Contacting a marriage therapist before an attorney may be an important choice. Of course, we do not recommend giving an abusive relationship more time, as abuse, once started, tends to increase in frequency and intensity.

The new buzzword for divorce is "conscious uncoupling," which was used by Gwyneth Paltrow and Chris Martin of their divorce in 2014 (the term was developed by psychotherapist Katherine Woodward Thomas). Of course, couples may also choose to continue their coupling (and improve their relationship).

**2.** ***Acknowledge and accept that terminating a relationship will be difficult and painful.*** Yazedjian and Toews (2010) studied college breakups and found that they were associated with depression and a drop in grades. Both partners are usually hurt, although the person with the least interest in maintaining the relationship will suffer less.

**3.** ***Select your medium of breaking up.*** While some break up face to face, others do so in a text message, in an e-mail, or on Facebook. People have different ideologies and values about the use of various technologies in breaking up (Gershon 2010). Most prefer that their partner break up with them face to face, not in a text message (Faircloth et al. 2012). Be sensitive to how your partner will view getting the bad news.

**4.** ***In talking with your partner, blame yourself for the breakup.*** One way to end a relationship is to blame yourself by giving a reason that is specific to you (e.g., "I need more freedom," "I want to go to graduate school," or "I'm not ready to settle down"). If you blame your partner, the relationship may continue because your partner may promise to change and you may feel obligated to give your partner a second chance.

**5.** ***Cut off the relationship completely.*** If you are the person ending the relationship, it will probably be easier for you to continue to see the other person without feeling too hurt. However, the other person will probably have a more difficult experience and will heal faster if you stay away completely. Alternatively, some people are skilled at ending love relationships and turning them into friendships. Though this is difficult and infrequent, it can be rewarding across time for the respective partners.

**6.** ***Learn from the terminated relationship.*** Included in the reasons a relationship may end are behaviors of being too controlling; being oversensitive, jealous, or too picky; cheating; fearing commitment; and being unable to compromise and negotiate conflict. Some of the benefits of terminating a relationship are recognizing one's own contribution to the breakup and working on any behaviors that might be a source of problems. Otherwise, one might repeat the process in the next relationship.

*Divorce is failure. But it is better to fail than to continue in an unhappy marriage.*

BETTE DAVIS, ACTRESS

**7. *Allow time to grieve over the end of the relationship.*** Ending a love relationship is painful. Allowing yourself time to experience such grief will help you heal for the next relationship. Recovering from a serious relationship can take 12 to18 months.

**8. *Clean your Facebook page.*** Angry spouses sometimes post nasty notes about their ex on their Facebook page, which can be viewed by the ex's lawyer. Also, if there are any incriminating photos of indiscreet encounters, drug use, wild parties or the like, these can be used in court and should be purged. Twitter and blog postings should also be scrutinized.

## Definition and Prevalence of Divorce

**Divorce** the legal ending of a valid marriage contract.

**Divorce** is the legal ending of a valid marriage contract. The lifetime probability of couples getting divorced today is between 40% and 50% (Cherlin 2010). But the individual's age at marriage, education, race, and religion influence probability of divorce for that person. For college graduates who wed in the 1980s at age 26 or older, 82% were still married 20 years later. For college grads who married when the partners were less than age 26, 65% were still married 20 years later. If the couple were high school graduates and married when the partners were below age 26, 49% were still married. Hence, education and age at marriage influence one's chance of divorce. In addition, race (whites are less likely to divorce), religion (less devout are more likely to divorce), and previous marriage (previously married are more likely to divorce) also influence one's chance of divorce (Parker-Pope 2010).

Divorce rates have been relatively stable in recent years. The principal factor for the lack of increase in the divorce rate is that people are delaying marriage so that they are older at the time of marriage. Indeed, the older a person at the time of marriage, the less likely the person is to divorce.

*When somebody leaves he's making a statement that goes far beyond the personal: It's like (a) I can't stand you anymore, and (b) there isn't any redeeming social value in these human connections. I want to disconnect so badly that I'm willing to sell all my human colleagues in this "connection" enterprise down the river! Or (c) if that isn't plain enough for you, I don't love you anymore. The Velcro of love on your shirt doesn't stick anymore!*

CAROLINE SEE, *BEULAH LAND*

**National Data**

About 900,000 divorces occur each year (*Statistical Abstract of the United States, 2012–2013,* Table 78).

## High- and Low-Risk Occupations for Divorce

Some occupations are more prone to divorce than others. The highest-risk occupations are those of being a dancer/choreographer (43%) or bartender (38%); among the lowest are clergy and optometrists at 5% and 4% respectively. Other examples include nurses (29%), sociologists (23%), psychologists (19%), authors/teachers (15%), and lawyers (12%) (Shawn and Aamodt 2009). Regardless of one's occupation, there are numerous reasons for divorce (Ducanto 2013). These can be conceptualized as both macro and micro factors which help to explain divorce in U.S. society.

# Macro Factors Contributing to Divorce

Sociologists emphasize that social context creates outcome. This concept is best illustrated by the fact that from 1639 to 1760, the Puritans in Massachusetts averaged only one divorce per year (Morgan 1944). The social context of that

Chelsea Curry

Being a bartender is an occupation with a high risk of divorce.

era involved strong pro-family values and strict divorce laws, with the result that divorce was almost nonexistent for over 100 years. In contrast, divorce occurs more frequently today as a result of various structural and cultural factors, also known as macro factors.

## Increased Economic Independence of Women

In the past, an unemployed wife was dependent on her husband for food and shelter. No matter how unhappy her marriage was, she stayed married because she was economically dependent on her husband. Her husband was her lifeline. But finding gainful employment outside the home made it possible for a wife to afford to leave her husband if she wanted to. Now that three-fourths of wives are employed, fewer women are economically trapped in unhappy marriages. A wife's employment does not increase the risk of divorce in a happy marriage. However, it does provide an avenue of escape for women in unhappy or abusive marriages (Kesselring and Bremmer 2006).

Employed wives are also more likely to require an egalitarian relationship; although some husbands prefer this role relationship, others are unsettled by it. Another effect of a wife's employment is that she may meet someone new in the workplace and become aware of an alternative to her current partner. Finally, unhappy husbands may be more likely to divorce if their wives are employed and able to be financially independent (requiring less alimony and child support).

## Changing Family Functions and Structure

Many of the protective, religious, educational, and recreational functions of the family have been largely taken over by outside agencies. Family members may now look to the police for protection, the church or synagogue for meaning, the school for education, and commercial recreational facilities for fun rather than to each other within the family for fulfilling these needs. The result is that, although meeting emotional needs remains an important and primary function of the family, fewer reasons exist to keep a family together.

In addition to the change in functions of the family brought on by the Industrial Revolution, family structure has changed from that of the larger extended family in a rural community to a smaller nuclear family in an urban community. In the former, individuals could turn to a lot of people in times of stress; in the latter, more stress necessarily falls on fewer shoulders. Also, with marriages more isolated and scattered, kin may not live close enough to express their disapproval for the breakup of a marriage. With fewer social consequences for divorce, Americans are more willing to escape unhappy unions.

## Liberal Divorce Laws/Social Acceptance

**No-fault divorce** neither party is identified as the guilty party or the cause of the divorce.

All states recognize some form of **no-fault divorce** in which neither party is identified as the guilty party or the cause of the divorce (e.g., committing adultery). In effect, divorce is granted after a period of separation (typically 12 months). Nevada requires the shortest waiting period of six weeks. Most other states require from 6 to 12 months. The goal of no-fault divorce is to make divorce less acrimonious. However, this objective has not been achieved as spouses who divorce may still fight over custody of the children, child support, spouse support, and division of property. Nevertheless, social acceptability as well as legal ease may affect the frequency of divorce. Frimmel et al. (2013) noted that in Austria the increase in divorce is greatly influenced by its acceptance.

**DIVERSITY IN OTHER COUNTRIES**

Young Maasai wives of Arusha, Tanzania (East Africa), do not have the option of divorce. Rather, they remain married because divorce would result in their return to their parents' home where they would bring shame on their parents. In effect, they have no other role than that of wife and mother—the role of divorcée is not an option.

## Prenuptial Agreements and the Internet

New York family law attorney Nancy Chemtob notes that those who have prenuptial agreements are more likely to divorce, since one can cash out without economic devastation. In addition, she suggested that the Internet contributes to divorce since a bored spouse can go online to various dating sites and see what alternatives are out there. Spinning up a new relationship online before dumping the spouse of many years is not uncommon. Brown and Lin (2012) documented that divorce is becoming more common (1 in 4) in adults age 50 and older.

## Fewer Moral and Religious Sanctions

While previously some churches denied membership to the divorced, today many priests and clergy recognize that divorce may be the best alternative in particular marital relationships. Churches increasingly embrace single and divorced or separated individuals, as evidenced by divorce adjustment groups.

## More Divorce Models

The prevalence of divorce today means that most individuals know someone who is divorced. The more divorced people a person knows, the more normal divorce will seem to that person. The less deviant the person perceives divorce to be, the greater the probability the person will divorce if that person's own marriage becomes strained. Divorce has become so common that numerous websites for the divorced are available.

## Mobility and Anonymity

When individuals are highly mobile, they have fewer roots in a community and greater anonymity. Spouses who move away from their respective families and friends often discover that they are surrounded by strangers who don't care if they stay married or not. Divorce thrives when pro-marriage social expectations are not operative. In addition, the factors of mobility and anonymity also result in the removal of a consistent support system to help spouses deal with the difficulties they may encounter in marriage.

**DIVERSITY IN OTHER COUNTRIES**

Compared with other countries, the United States has one of the highest divorce rates in the world. Only Belarus of Eastern Europe comes close to the United States in its crude divorce rate. The society in which a person lives may have more to do with whether a person gets a divorce than the personality of the person. Prior to 1997, there were no divorces in Ireland; Ireland is predominately Catholic and has historically been staunchly against divorce. While annulments are possible, only about 25 per year are approved.

### Social Class, Ethnicity, and Culture

Charles Murray (2012) argues that social class influences who stays married and points out that educated white Americans are the least likely to divorce. Indeed, the less educated with less income are more likely to divorce.

Asian Americans and Mexican Americans also have lower divorce rates than European Americans or African Americans because they consider the family unit to be of greater value (familism) than their individual interests (individualism). Unlike familistic values in Asian cultures, individualistic values in American culture emphasize the goal of personal happiness in marriage. When spouses stop having fun (when individualistic goals are no longer met), they sometimes feel no reason to stay married. Of 4,537 undergraduates, only 9% agreed that "I would not divorce my spouse for any reason" (Hall and Knox 2013).

## Micro Factors Contributing to Divorce

Macro factors are not sufficient to cause a divorce. One spouse must choose to divorce and initiate proceedings. Such a view is micro in that it focuses on individual decisions and interactions within a specific relationship. A partner whose behavior does not meet one's expectations (infidelity, abuse, value parents over spouse) spells marital dissatisfaction (Dixon et al. 2012). The section on Applying Social Research reveals the top reasons undergraduates gave for the end of their last romantic relationship.

### Growing Apart/Differences

The top reason for seeking divorce given by a sample of 886 divorcing individuals was "growing apart" (55%) (Hawkins et al. 2012). The individuals found that they no longer had anything in common.

### Falling Out of Love

Benjamin et al. (2010) noted that the absence of love in a relationship was associated with an increased chance of divorce. Indeed, almost half (44%) of 4,581 undergraduates reported that they would divorce a spouse they no longer loved (Hall and Knox 2013). No couple is immune to falling out of love and getting divorced. Lavner and Bradbury (2010) studied 464 newlyweds over a four-year period and found that, even in those cases of reported satisfaction across the four years, some couples abruptly divorced. Whether the divorce was triggered by a personal indiscretion (e.g., infidelity) or crisis (e.g., death of child), the point is that years of satisfaction do not make a couple immune to divorce.

*Love and hate are two horns on the same goat.*

CHARLOTTE PHELAN IN *THE HELP*

### Limited Time Together

Some spouses do not make time to be together. Time devoted to children and career interferes with couple time. Partners who spend little time together doing things they mutually enjoy often feel estranged from each other and have little motivation to stay together.

*It is not a lack of love, but a lack of friendship that makes unhappy marriages.*

FRIEDRICH NIETZSCHE, PHILOSOPHER

## APPLYING SOCIAL RESEARCH | Saying Goodbye in Romantic Relationships*

Most undergraduates have experienced the end of a romantic relationship. The purpose of this study was to investigate the strategies they used or were subjected to in regard to the end of their last romantic relationship and to identify the outcomes for the individuals and their relationship. Basically we wanted to know why relationships ended (e.g., infidelity, boredom, etc.), how (e.g., face to face or text message), and the post breakup relationship outcomes (e.g., enemies, friends, etc.).

### Background

Previous research has been conducted on relationship dissolution. Bullock et al. (2012) asked whether couples could remain friends after a breakup. They found that the more satisfied the individuals were during the dissolved romance, the more likely they were to remain friends when the relationship ended. Larson and Sweeten (2012) focused on substance abuse and found that it increased after a breakup. Park et al. (2012) emphasized that one's reaction to a breakup depended on how much one's self-worth was tied to being in a romantic relationship (the greater the connection, the more difficult the adjustment). Rhoades et al. (2011) studied how life satisfaction/mental health were impacted by a romantic breakup. They found that if the couple had discussed a future/were planning to marry, the breakup was more difficult. While these studies are insightful, none provide an overview trajectory from why and how a couple broke up and the consequences of the breakup.

### Sample and Methods

A convenience sample from a large southeastern university completed a voluntary, anonymous 25-item online survey on relationship breakups. The sample (n = 478) was predominately female (70%), white (71%), and heterosexual (80%). Descriptive statistics as well as bivariate correlations and logistic regression were utilized to address various research questions.

### Research Questions and Findings

Two major questions guided this research.

**Research Question 1:** What reasons did undergraduates give for ending their last romantic relationship? See Table 1 for the reasons given by 478 respondents. The top two reasons were the same as those identified by a sample of Norwegian individuals over the age of 50 (Traeen and Thuen 2013). There was no significant difference between sex of the respondent and the reason given for ending the relationship.

**Table 1**
**Main Reasons for Breaking Up**
**N = 478**

| | Male | Female | Total |
|---|---|---|---|
| Bored in the relationship/not happy | 32% | 22% | 23% |
| Betrayal of partner | 9% | 17% | 14% |
| Different interests | 11% | 11% | 11% |
| I met someone new | 12% | 10% | 10% |
| Different values | 13% | 7% | 8% |
| Moved away | 6% | 9% | 8% |

**Research Question 2:** Following the ending of the romantic relationship, what were the effects and feelings associated with the breakup (i.e., depression, guilt, relief, etc.)? And to what degree did these vary by sex of respondent? (See Table 2.)

"Feeling initially upset but recognizing the breakup was for the best" was the primary reaction to breaking up. "I was glad it was over" was the second most frequent response. Chi-square analysis showed that for females there was a significant difference between who initiated the breakup and the reactions to those breakups ($\chi^2 = 55.51$, df = 8, $p < .001$). For example, if the woman ended the relationship she was more likely to report "I'm glad it was over" and "I was initially upset but felt it was for the best." In contrast, males showed no significant difference in effect regardless if they were the initiators of the breakup or not.

### Theoretical Framework

Symbolic interactionism provided the theoretical framework for viewing the findings of this study. Symbolic interactionism is a micro-level theory that focuses on the meanings individuals attribute to phenomena. Symbolic interactionists focus on the importance of symbols, subjective versus objective reality, and the definition of social situations. The respondents identified "feeling bored/unhappy" and "betrayal of partner" as reasons to terminate a

**Table 2**
**Reactions to the Breakup**

| | I ended the relationship | | My partner ended the relationship | | It was mutual | | |
|---|---|---|---|---|---|---|---|
| | Male | Female | Male | Female | Male | Female | Total |
| I was glad it was over | 16 | 44 | 2 | 4 | 9 | 9 | 84 |
| I was initially upset but feel it was for the best | 27 | 78 | 11 | 43 | 16 | 48 | 223 |
| I was depressed | 5 | 11 | 3 | 30 | 5 | 7 | 61 |
| I saw a counselor to help with the breakup | 0 | 0 | 1 | 2 | 0 | 0 | 3 |
| I felt suicidal | 0 | 2 | 0 | 1 | 0 | 1 | 4 |
| Total | 48 | 135 | 17 | 80 | 30 | 65 | 375 |

relationship. The fact that Americans are socialized to think in individualist rather than Asian familistic terms emphasizes the cultural backdrop on which romantic breakup decisions are made. In addition, American youth are taught to get upset and end a romantic relationship in response to a partner's cheating. French lovers are less quick to end a relationship over an indiscretion or an affair.

Similarly, these undergraduates viewed the ending of their romantic relationships as an undesirable event which resulted in the culturally scripted response—being sad, being upset, being depressed. Older individuals might take another view—that the end of an unfulfilling romantic relationship is an opportunity to meet a new partner and create a more fulfilling relationship.

### Implications

There are three implications of the data. First, the ending of romantic relationships is filled with angst. Respondents spoke of feeling unhappy, betrayed, and replaced by another lover. They also revealed feelings of anger and jealousy.

Second, the aftermath of a romantic relationship may have negative consequences. Thirty percent of the females reported feeling depressed when their partner ended the relationship. Over 40% reported that they were initially upset.

Third, romantic breakups are not serious enough to induce thoughts of suicide or to seek counseling. Only 4 of the 478 respondents revealed that they felt suicidal; only three sought counseling. Indeed, most undergraduates were resilient and moved on. Almost half (46%) said that while they were initially upset, they believed the ending of the romantic relationship was for the best.

### Sources


Arkes, J. 2013. The temporal effects of parental divorce on youth substance use. *Substance Use & Misuse* 48: 290–297.

Bullock, M., J. Hackathorn, E. M. Clark, and B. A. Mattingly. 2012. Can we be (and stay) friends? Remaining friends after dissolution of a romantic relationship. *Journal of Social Psychology* 151: 662–666.

Larson, M. and G. Sweeten. 2012. Breaking up is hard to do: Romantic dissolution, offending and substance use during the transition to adulthood. *Criminology* 50: 605–636.

Park, L. E., D. T. Sanchez, and K. Brynildsen. 2012. Maladaptive responses to relationship dissolution. The role of relationship contingent self-worth. *Journal of Applied Social Psychology* 41: 1749–1773.

Rhoades, G. K., C. M. Kamp Dush, D. C. Atkins, S. M. Stanley, and H. J. Markman. 2011. Breaking up is hard to do: The impact of unmarried relationship dissolution on mental health and life satisfaction. *Journal of Family Psychology* 25: 366–374.

Traeen, B. and F. Thuen. 2013. Relationship problems and extradyadic romantic and sexual activity in a web-sample of Norwegian men and women. *Scandinavian Journal of Psychology* 54: 137–145.


*Updated and adapted from "Saying goodbye in romantic relationships: Strategies and outcomes" by Brackett, A., J. Fish, and D. Knox. 2013. Poster, Southeastern Council on Family Relations, Birmingham, AL, February 21–23.

## Low Frequency of Positive Behavior

People marry because they anticipate greater rewards from being married than from being single. During courtship, each partner engages in a high frequency of positive verbal (compliments) and nonverbal (eye contact, physical affection) behavior toward the other. The good feelings the partners experience as a result of these positive behaviors encourage them to marry to "lock in" these feelings across time. Mitchell (2010) interviewed 390 married couples and found that intimacy was associated with marital happiness. Just as love feelings are based on partners making the choice to engage in a high frequency of positive behavior toward each other, defeatist feelings are created when these positive behaviors stop and negative behaviors begin. Thoughts of divorce then begin (to escape the negative behavior).

## Having an Affair

In a survey of U.S individuals, half of those who reported having participated in extramarital sex also reported that they were either divorced or separated (Allen and Atkins 2012). In an Oprah.com survey of 6,069 adults, an affair was the top reason respondents said they would seek a divorce (Healy and Salazar 2010). Thirty-three percent said an affair would be the deal breaker. Other top responses were chronic fighting (28%), no longer being in love (24%), boredom (8%), and sexual incompatibility (4%).

*He announces that lately he keeps losing things. "Like your wife and child," I want to say, but don't. At forty, I've learned not to say everything clever, not to score every point.*

**SUZANNE FINNAMORE, *SPLIT: A MEMOIR OF DIVORCE***

## Poor Communication/Conflict Resolution Skills

The second most frequent reason for seeking divorce given by a sample of 886 divorcing individuals was "not able to talk together" (Hawkins et al. 2012). Not only do individuals distance themselves from each other by not talking, they further complicate their relationship since they have no way to reduce conflict.

Talking issues out is a major strength for spouses.

David Knox

Managing differences and conflict in a relationship helps to reduce the negative feelings that develop in a relationship. Some partners respond to conflict by withdrawing emotionally from their relationship; others respond by attacking, blaming, and failing to listen to their partner's point of view.

### Changing Values

Both spouses change throughout the marriage. "He's not the same person I married" is a frequent observation of people contemplating divorce. One minister married and decided seven years later that he did not like the confines of his religious or marital role. He left the ministry, earned a Ph.D., and began to drink and have affairs. His wife now found herself married to a clinical psychologist who spent his evenings at bars with other women. The couple divorced.

Because people change throughout their lives, the person selected at one point in life may not be the same partner one would select at another point. Margaret Mead, the famous anthropologist, noted that her first marriage was a student marriage; her second, a professional partnership; and her third, an intellectual marriage to her soul mate, with whom she had her only child. At each of several stages in her life, she experienced a different set of needs and selected a mate who fulfilled those needs.

*A good divorce is better than a bad marriage . . . but not much.*

ERIC MICHAEL MOBERG, AUTHOR

### Satiation

**Satiation**, also referred to as habituation, refers to the state in which a stimulus loses its value with repeated exposure. Spouses may tire of each other. Their stories are no longer new, their sex is repetitive, and their presence for each other is no longer exciting as it was at the beginning of the relationship. Some people who feel trapped by the boredom of constancy decide to divorce and seek what they believe to be more excitement by returning to singlehood and new partners. One man said, "I traded something good for something new." A developmental task of marriage is for couples to enjoy being together and not demand a constant state of excitement (which is not possible over a 50-year period). The late comedian George Carlin said, "If all of your needs are not being met, drop some of your needs." If spouses did not expect so much of marriage, maybe they would not be disappointed.

**Satiation** the state in which a stimulus loses its value with repeated exposure.

### Perception That One Would Be Happier If Divorced

Women file most divorce applications. Their doing so may be encouraged by their view that they will achieve greater power over their own life. They feel that by getting a divorce they will have their own money (in the form of child support and/or alimony) without having a man they don't want in the house. In addition, they will have greater control over their children, since women are more often awarded custody.

### Top 20 Factors Associated with Divorce

Researchers have identified the characteristics of those most likely to divorce (Park and Raymo 2013; Djamba et al. 2012; Amato 2010; Nunley and Alan Seals 2010; Chiu and Busby 2010). Some of the more significant associations include the following:

**1.** Less than two years of hanging out together (partners know little about each other)

**2.** Having little in common (similar interests serve as a bond between people)

**3.** Marrying at age 17 and younger (associated with low education and income and lack of maturity)

**4.** Being different in race, education, religion, social class, age, values, and libido (widens the gap between spouses)

**5.** Not being religiously devout (less bound by traditional values)

**6.** Having a cohabitation history with different partners (pattern of establishing and breaking relationships)

**7.** Having been previously married (less fearful of divorce)

**8.** Having no children or fewer children (less reason to stay married)

**9.** Having limited education (associated with lower income, more stress, less happiness)

**10.** Falling out of love (spouses have less reason to stay married)

**11.** Being unfaithful (broken trust, emotional reason to leave relationship)

**12.** Growing up with divorced parents (models for ending rather than repairing relationship; may have inherited traits such as alcoholism that are detrimental to staying married)

**13.** Having poor communication skills (issues go unresolved and accumulate)

**14.** Having mental problems (bipolar, depression, anxiety) or physical disability (chronic fatigue syndrome)

**15.** Having seriously ill child (impacts stress, finances, couple time)

**16.** Having premarital pregnancy or unwanted child (spouses may feel pressure to get married; stress of parenting unwanted child)

**17.** Emotional/physical abuse (relationship is aversive)

**18.** Lacking commitment (for nontraditional spouses, divorce is seen as an option if the marriage does not work out)

**19.** Unemployment (finances decrease, stress increases)

**20.** Alcoholism/substance abuse (partner no longer dependable)

The more of these factors that exist in a marriage, the more vulnerable a couple is to divorce. Regardless of the various factors associated with divorce, there is debate about the character of people who divorce. Are they selfish, amoral people who are incapable of making good on a commitment to each other and who wreck the lives of their children? Or are they individuals who care a great deal about relationships and won't settle for a bad marriage? Indeed, they may divorce precisely because they value marriage and want to rescue their children from being reared in an unhappy home.

*So if you are a reasonably well-educated person with a decent income, come from an intact family and are religious, and marry after age 25 without having a baby first, your chances of divorce are very low indeed.*

STATE OF OUR UNIONS, 2012

Caryl Rusbult's investment model of commitment may also be used to identify why a relationship ends. When the emotional satisfaction for being in the relationship, an attractive available alternative, and a willingness to give up one's investments (e.g., material goods, friendships, etc.) converge, the person is vulnerable to divorce (Finkenauer 2010). In effect spouses see little reason to stay. Hawkins et al. (2012) in their study of 886 divorcing spouses found that couples were particularly vulnerable to divorce if they were growing apart, had differences in tastes, and had money problems.

## Consequences of Divorce for Spouses/Parents

For both women and men, divorce is often an emotional and financial disaster. After death of a spouse, separation and divorce are among the most difficult of life's crisis events (notice that all three top life crisis events are in reference to the end of a love relationship).

*Revenge is like drinking poison and hoping that it will kill your enemy.*

— NELSON MANDELA, IMPRISONED FOR 27 YEARS.

### Recovering from Ending a Relationship

A sample of 410 undergraduates revealed how their relationships ended and they recovered (Knox et al. 2000):

**1.** ***Women were more likely to end the relationship.*** Women were significantly more likely than men to report that they initiated the breakup (50% versus 40%). One female student recalled, "I got tired of his lack of ambition—I just thought I could do better. He's a nice guy but living in a trailer is not my idea of a life."

**2.** ***Sex differences in relationship recovery.*** Though recovery was not traumatic for either men or women, men reported more difficulty than women did in adjusting to a breakup. When respondents were asked to rate their level of difficulty from "no problem" (0) to "complete devastation" (10), women scored 4.35 and men scored 4.96. In explaining why men might have more difficulty adjusting to terminated relationships, some of the female students said, "Men have such inflated egos, they can't believe that a woman would actually dump them." Others said, "Men are oblivious to what is happening in a relationship and may not have a clue that it is over. When it does end, they are in shock."

Siegler and Costa (2000) also noted that women fare better emotionally after separation or divorce than do men. They noted that women are more likely than men not only to have a stronger network of supportive relationships but also to profit from divorce by developing a new sense of self-esteem and confidence, because they are thrust into more independent roles. However, whether one is the "divorcee" or "divorcer" is relevant—the person who was terminated (divorcee) has a more difficult time adjusting to divorce than the divorcer who ended the relationship.

*And I felt like my heart had been so thoroughly and irreparably broken that there could be no real joy again. . . . Everyone wanted me to get help and rejoin life, pick up the pieces and move on, and I tried to, but I just had to lie in the mud with my arms wrapped around myself, eyes closed, grieving, until I didn't have to any more.*

ANNE LAMOTT

**3.** ***Recovering took time and a new partner.*** The passage of time and involvement with a new partner were identified as the most helpful factors in getting over a love relationship that ended. Though the difference was not statistically significant, men more than women reported a new partner was more helpful in relationship recovery (34% versus 29%). Similarly, women more than men reported that time was more helpful in relationship recovery (34% versus 29%). However, a study of Jewish women who were adjusting to divorce emphasized that a new romantic partner was helpful in their adjustment (Kulik and Heine-Cohen 2011).

**4.** ***Other findings.*** Other factors associated with recovery for women and men were "moving to a new location" (13% versus 10%) and recalling that "the previous partner lied to me" (7% versus 5%). Men were much more likely than women to use alcohol to help them get over a previous partner (9% versus 2%). Neither men nor women reported using therapy to help them get over a partner (1% versus 2%). These data suggest that breaking up was not terribly difficult for these undergraduates (but more difficult for men than women) and that both time and a new partner enabled their recovery.

*Divorce isn't such a tragedy. A tragedy's staying in an unhappy marriage, teaching your children the wrong things about love. Nobody ever died of divorce.*

JENNIFER WEINER, *FLY AWAY HOME*

Maintaining one's positive self-identity is associated with a positive recovery from a romantic relationship that has ended (Mason et al. 2012). Individuals who define themselves solely in reference to their ex-partner experience more difficulty in getting over and moving on (e.g., more likely to remain in love with the ex). These individuals may also use Facebook to find out current information about the ex-partner in regard to their involvement in a new relationship (Tong 2013).

Is breaking up harder on spouses or cohabitants? Tavares and Aassve (2013) compared those who were married with those who were living together to assess the level of psychological distress due to a breakup. When children are involved, the breakup is more difficult . . . but controlling for children, the distress is similar. Is a divorce always traumatic? Not if having a party is an indication. In Las Vegas, Mari-Rene Alue, married 13 years when her husband had an affair, burned her wedding dress and tied her wedding ring to a balloon to let it float away (Trejos 2013).

*Whether life finds us guilty or not guilty, we ourselves know we are not innocent.*

— SÁNDOR MÁRAI, *JUDIT . . . ÉS AZ UTÓHANG*

## Financial Consequences

Both women and men experience a drop in income following divorce, but women may suffer more (Warrener et al. 2013). Because men usually have greater financial resources, they may take all they can with them when they leave. The only money they may continue to give to an ex-wife is court-ordered child support or spousal support (alimony). Republican Senator John McCain pays his former wife $17,000 in alimony annually. While most states do not provide alimony, they provide for an equitable distribution of property.

*Divorce is an embarrassing public admission of defeat.*

TRACY LETTS, PULITZER PRIZE-WINNING PLAYWRIGHT AND ACTOR

Remarriage generally restores a woman's economic stability. When remarried mothers and fathers are compared there is a "matrilineal tilt" in terms of money transfers to children (Clark and Kenny 2010). In effect, while single divorced fathers give more money to their children than single divorced mothers, remarried mothers give more money to their children than remarried men. According to their data, 21% of divorced mothers gave financial transfers to their biological children over the past two years as compared to 16% of divorced fathers doing so.

Remarried mothers prefer to give their resources to their biological children with whom they are more likely to have maintained a relationship since the divorce. In contrast, remarried fathers are less likely to make money transfers since they have a less close relationship to their children since the divorce and they have a new wife monitoring their spending behavior. The new wife of a remarried father said to him when his biological son asked him for money, "We don't want to encourage his economic dependence . . . do we?" The son did not get the money.

**Postnuptial agreement** an agreement about how money is to be divided should a couple later divorce, which is made after the couple marry.

How money is divided at divorce depends on whether the couple had a prenuptial agreement or a **postnuptial agreement**. Appendix C at the end of this text provides an example of a prenuptial agreement for a couple getting remarried. Such agreements are most likely to be upheld if an attorney insists on four conditions—full disclosure of assets by both parties, independent representation by separate counsel, absence of coercion or duress, and terms that are fair and equitable.

## Effect of Divorce on Friendships

What is the effect of divorce on the friendships of the spouses? Based on a study of 58 divorced individuals and 123 couples, individual friendships were maintained and strengthened but friendships with both members of a couple were rarely maintained. Hence, divorce will cost at least one member of a couple with whom the divorcing individual is friends (Greif and Dea, 2012).

*You're best off if you can put your emotions off to the side, and really realize that, in the end, marriage is a business transaction.*

JEAN CHATZY, FINANCIAL ADVISOR

## Fathers' Separation from Children?

Mercadante et al. (2014) studied fathers going through divorce in Western Australia and found that they were "at an emotional disadvantage during separation, not only grieving the loss of their former marital relationship, but also their simultaneous loss of contact with their children, their fathering role, and their former family routine." While fathers in the United States are also disadvantaged, they have a choice of how involved they want to be with their children post divorce. Most judges today recognize the value of the father for children and provide joint or full custody if warranted. But fathers must be aggressive and get a legal contract which gives them equal joint and physical custody. Otherwise, they may be cut out of the lives of their children. Trusting the former spouse to be amicable about access to the children may be a mistake.

**Parental alienation syndrome** an alleged disturbance in which children are obsessively preoccupied with deprecation and/or criticism of a parent, denigration that is unjustified and/or exaggerated.

## Parental Alienation

More involved fathers reduces parental alienation syndrome. **Parental alienation syndrome** (PAS) is an alleged disturbance in which children are obsessively preoccupied with deprecation and/or criticism of a parent, denigration that is unjustified and/or exaggerated (Gardner 1998). Meier (2009) reviewed the history of parental alienation syndrome and parental alienation (PA) and found the former to be questioned by most researchers (e.g., PAS is not a medical psychosis with specific criteria). However, **parental alienation** can be defined as an alliance between a parent and a child that isolates the other parent (Godbout and Parent 2012).

**Parental alienation** an alliance between a parent and a child that isolates the other parent.

The following are examples of behaviors that either parent may engage in to alienate a child from the other parent (Godbout & Parent 2012; Schacht 2000; Teich 2007).

**1.** Minimizing the importance of contact and the relationship with the other parent, including moving far away with the child so as to make regular contact difficult.

**2.** Exhibiting excessively rigid boundaries; rudeness or refusal to speak to or inability to tolerate the presence of the other parent, even at events important to the child; refusal to allow the other parent near the home for visitation drop-off or pick-up.

**3.** Having no concern about missed visits with the other parent (e.g., taking the child out of town when the other parent is supposed to have legal access).

**4.** Showing no positive interest in the child's activities or experiences during visits with the other parent and withholding affection if the child expresses positive feelings about the absent parent.

**5.** Granting autonomy to the point of apparent indifference ("It's up to you if you want to see your dad. I don't care.").

**6.** Overtly expressing dislike of a visitation ("If you loved me you would not want to see your mom.").

**7.** Refusing to discuss anything about the other parent ("I don't want to hear about . . . .") or showing selective willingness to discuss only negative matters.

**8.** Using innuendo and accusations against the other parent, including statements that are false, and blaming the other parent for the divorce.

**9.** Portraying the child as an actual or potential victim of the other parent's behavior.

**10.** Demanding that the child keep secrets from the other parent.

**11.** Destroying gifts from or memorabilia of the other parent.

**12.** Promoting loyalty conflicts (e.g., offering an opportunity for a desired activity that conflicts with scheduled visitation).

The most telling sign that children have been alienated from a parent is the irrational behavior of the children, who for no properly explained reason say that they want nothing further to do with one of the parents. Indeed, such children have a lack of ambivalence toward the alienation, lack of guilt or remorse about the alienation, and always take the alienating parent's side in the conflict (Baker and Darnall 2007).

Children who are alienated from one parent are sometimes unable to see through the alienation process and regard their negative feelings as natural. Such children are similar to those who have been brainwashed by cult leaders to view outsiders negatively. Godbout and Parent (2012) interviewed children who had been alienated from a parent and observed "difficulties at school, internal and external behavior problems, and a search for identity after reaching adulthood." Toren et al. (2013) reported on a 16-session parallel group therapy program for 22 children with parental alienation and their parents. The children's level of anxiety and depression decreased significantly following the therapeutic intervention.

**DIVERSITY IN OTHER COUNTRIES**

Spouses from different countries who divorce may discover that their children are taken to another country that does not recognize domestic law in the United States. Over 1,600 children were taken out of the United States in 2009 in violation of a court order or over the objections of the other parent (Filisko 2010). The two children of Christopher Savoie of Nashville were taken by their mother to Japan. Savoie went to Japan to get his children but Japanese law supported the children staying with their mother. He has not seen his children since they were abducted by his ex-wife.

WHAT IF?

## What If Your Former Spouse Tries to Turn Your Children against You?

The best antidote to parental alienation syndrome is to spend time with your children so that they can discover for themselves who you are as a parent, how you feel about them, and how much you love and value them. Regardless of what your former spouse says to your children about you, the reality of how you treat them is what will determine how your children feel about you. However, should your spouse not allow you to see your children, it is imperative to hire a lawyer (and go to court if necessary) to ensure that you are awarded time with your children. You cannot have a relationship with your children if you don't spend time with them.

*Within the year, my mother did indeed tell us of their plans to live apart—maybe by then she was even using the word "divorce"—and my only question, through deep gasping sobs, was: "Well, why did you even give birth to me if you were going to do this?"*

NELL CASEY, WRITER

# Consequences of Divorce for Children

Over a million children annually experience the divorce of their parents. Like all life experiences, there are both negative and positive outcomes. Negative outcomes include lower psychological well-being, more likely to drop out of first year of college or make lower grades (Soria and Linder 2014), have earlier sexual behavior (Carr and Springer 2010), have higher rates of alcohol/marijuana use (Arkes 2013), exhibit depressive symptoms or antisocial behavior (Strochschein 2012), and have lower commitment in romantic relationships (Cui 2011).

But there are also positive aspects of divorce. Halligan et al. (2014) analyzed data from 336 undergraduates who were asked to identify positive outcomes they experienced from the divorce of their parents. Table 15.1 provides the percentages "agreeing" or "strongly agreeing" with various outcomes.

Table 15.1 reflects that children of divorce may choose to notice the positive aspects of divorce rather than buy into the cultural script that "divorce is terrible and my life will be ruined because of my parent's divorce." South (2013) also found that some children of divorced parents are very deliberate in trying not to repeat the mistakes of their parents. Some of their respondents reported that "they were hard on their romantic partners," meaning that were unwilling to let issues slide but addressed them quickly even though doing so might cause conflicts.

Although a major factor that determines the effect of divorce on children is the degree to which the divorcing parents are civil, legal and physical custody are important issues. The following section details how judges go about making this decision.

## Who Gets the Children?

### National **Data**

Twenty-three percent of children live with single mothers, 3% with single fathers (Carr and Springer 2010).

**TABLE 15.1 Positive Effects of Divorce: Percentage of Respondents Agreement**

| | % |
|---|---|
| Since my parents' divorce, I am more compassionate for people who are going through a difficult time. | 65.63% |
| I have greater tolerance for people with different viewpoints since my parents' divorce. | 63.16% |
| Since my parents' divorce, I have been exposed to different family values, tradition and life styles. | 60.01% |
| I have liked spending time alone with my mother since my parents' divorce. | 57.71% |
| My mother is happier since the divorce. | 57.20% |
| I rely less on my parents for making decisions since my parents' divorce. | 53.51% |
| I have liked spending time alone with my father since my parents' divorce. | 45.61% |
| My mother has made a greater effort to spend quality time with me since the divorce. | 45.61% |
| I can spend more time with the parent I prefer since my parents' divorce. | 45.37% |
| Since my parents' divorce, I have felt closer to my mother. | 44.98% |
| My relationship with my mother has improved since my parents' divorce. | 44.71% |
| My father is happier since the divorce. | 43.85% |
| My parents' divorce has made me closer to my friends. | 42.54% |
| Since my parents' divorce, I have greater appreciation for my siblings. | 40.78% |
| My father has made a greater effort to spend quality time with me since the divorce. | 38.60% |
| Since my parents' divorce, I have felt closer to my siblings. | 35.96% |
| After my biological parents' divorce, I noticed that I was exposed to less conflict between my parents on a daily basis. | 34.93% |
| My relationship with my father has improved since my parents' divorce. | 34.65% |
| I think my parents have a "good" divorce. | 34.35% |
| My parents are more civil to each other since the divorce. | 34.21% |
| My parents' divorce has resulted in a closer relationship with my grandparents. | 32.20% |
| My standard of living improved since my parents' divorce. | 30.70% |
| Since my parents' divorce, I have felt closer to my father. | 30.13% |

Judges assigned to hear child custody cases make a judicial determination regarding whether one or both parents will have decisional authority on major issues affecting the children (called **legal custody**), and the distribution of parenting time (called visitation or **physical custody**). Toward this end, judges in all states are guided by the statutory dictum called "best interests of the child." Some of the factors involved in deciding custody include (Lewis 2008):

**Legal custody** a judicial determination regarding whether one or both parents will have decisional authority on major issues affecting the children.

**Physical custody** the distribution of parenting time, also called "visitation".

**1.** The child's age, maturity, sex, and activities, including culture and religion—all relevant information about the child's life—keeping the focus of custody on the child's best interests.

**2.** The wishes expressed by the child, particularly the older child (judges will often interview children six years old or older in chambers).

**3.** Each parent's capacity to care for and provide for the emotional, intellectual, financial, religious, and other needs of the child (including the work schedules of the parents).

**4.** The parents' ability to agree, communicate, and cooperate in matters relating to the child.

**5.** The nature of the child's relationship with the parents, which includes the child's relationships with other significant people, such as members of the child's extended family.

**6.** The need to protect the child from physical and psychological harm caused by abuse or ill treatment.

**7.** Past and present parental attitudes and behavior (e.g., parenting skills and personalities).

**8.** The proposed plans for caring for the child (the judge will want to know how each parent proposes to raise the child, including proposed parenting times for the other parent).

The judge may assign a custody evaluator to interview the child/assess the child's situation in reference to the eight items and write a recommendation. These and other custody factors will become the basis for the judge's custody determination. In some cases joint custody may be awarded.

## Joint Custody?

**Family relations doctrine** belief that even nonbiological parents may be awarded custody or visitation rights if they have been economically and emotionally involved in the life of the child.

As men have become more involved in the role of father, the courts no longer assume that "parent" means "mother." Indeed, a new **family relations doctrine** has emerged which suggests that even nonbiological parents (such as stepparents) may be awarded custody or visitation rights if they have been economically and emotionally involved in the life of the child.

Hartenstein and Markham (2013) studied a sample of 30 parents who were going through divorce to identify the process of their decision about custody for their children. The most frequent decision (76%) was made by the parents without court involvement. There was no universal arrangement and about half of the arrangements changed over time.

New terminology is making its way into the lives of divorcing spouses and in the courts. The term *joint custody,* which implies ownership, is being replaced with *shared parenting,* which implies cooperation in taking care of children.

There are several advantages of joint custody or shared parenting. Ex-spouses may fight less if they have joint custody because there is no inequity in their involvement in their children's lives. Children will benefit from the resultant decrease in hostility between parents who have both "won" them. Unlike sole-parent custody, in which one parent wins (usually the mother) and the other parent loses, joint custody allows children to continue to benefit from the love and attention of both parents. Children in homes where joint custody has been awarded might also have greater financial resources available to them than children in sole-custody homes.

Joint physical custody may also be advantageous in that the stress of parenting does not fall on one parent but, rather, is shared (Markham and Coleman 2011). One mother who has a joint custody arrangement with her ex-husband said, "When my kids are with their dad, I get a break from the parenting role, and I have a chance to do things for myself." Another joint-parenting father said, "When you live with your kids every day, you can get very frustrated and are not always happy to be with them. But after you haven't seen them for three days, it feels good to see them again (more quality time)."

A disadvantage of joint custody is that it tends to put hostile ex-spouses in more frequent contact with each other, and the marital war continues. In a study (Markham and Coleman 2010) of 20 divorced women with shared physical custody, almost half (45%) were categorized as "continuously contentious," reflecting incessant anger. Factors fueling such anger/contentiousness were irresponsible fathers (e.g., they did not show up, drank too much, and/or did not provide structure or physical care, like bathing the children), lack of money such as child support, and trying to control the ex-spouse.

Depending on the level of hostility between the ex-partners, their motivations for seeking sole or joint custody, and their relationship with their children, any arrangement can have positive or negative consequences for the ex-spouses as well as for the children. In those cases in which the spouses exhibit minimal hostility toward each other, have strong emotional attachments to their children, and want to remain an active influence in their children's lives, joint custody may be the best of all possible choices. Bauserman (2012) reviewed the literature on the custody condition which results in the highest satisfaction for mothers and fathers and found that sole custody is associated with the highest satisfaction, followed by joint custody. The least satisfactory outcome is when there is a noncustodial parent who feels shut out of the child's life.

## Minimizing Negative Effects of Divorce on Children

Researchers have identified the following conditions under which a divorce has the fewest negative consequences for children:

**1.** ***Healthy parental psychological functioning.*** Children of divorced parents benefit to the degree that their parents remain psychologically fit and positive, and socialize their children to view the divorce as a "challenge to learn from." Parents who nurture self-pity, abuse alcohol or drugs, and socialize their children to view the divorce as a tragedy from which they will never recover create negative outcomes for their children. Some divorcing parents can benefit from therapy as a method for coping with their anger or depression and to ensure that they make choices in the best interests of their children (e.g., encouraging the child's relationship with the other parent). Some parents become involved in divorce education programs. Crawford (2012) observed a reduction in conflict between parents who completed a divorce education program.

**2.** ***A cooperative relationship between the parents.*** The most important variable in a child's positive adjustment to divorce is when the child's parents continue to maintain a cooperative relationship throughout the separation, divorce, and post-divorce period. It is not divorce, but the post-divorce relationship with the parents which has a negative effect on children. Bitter parental conflict places the children in the middle. One daughter of divorced parents said, "My father told me, 'If you love me, you would come visit me,' but my mom told me, 'If you love me, you won't visit him.'" Numerous states mandate parenting classes as part of the divorce process.

**3.** ***Parental attention to the children and allowing them to grieve.*** Children benefit when both the custodial and the noncustodial parent continue to spend time with

## PERSONAL CHOICES

### *Choosing to Maintain a Civil Relationship with Your Former Spouse*

Spouses who divorce must choose the kind of new relationship they will have with each other. This choice is crucial in that it affects not only their own but also their children's lives. What kind of relationship do former spouses have? Drapeau et al. (2009) interviewed children ages 8 to 11 and their parents at 1 and 2.5 years after the separation and identified various patterns.

The typical pattern was for a high degree of conflict between the parents that lessened progressively over time. About a quarter of parents displayed a low level of conflict that remained stable over time. Sometimes referred to as "sustained adjustment," these couples were committed to get along as well as they could, given that they were divorced. Another 25% of the former spouses were successful in achieving a level of cooperation after separation. This pattern was the most beneficial to children. A fourth pattern (a third of the divorces) reflected a sustained conflict that lasted for years after the divorce. This pattern was the most devastating for children since it continually exposed them to conflict between the parents despite their physical separation. We might call these four patterns typical, sustainers, cooperatives, and conflictors.

A final pattern (slow ragers) that characterized 10% of the sample was that the early low level of conflict that set the tone for the relationship after divorce tended to rise over time. A factor associated with low-conflict was high income—if there was plenty of money after the divorce, the ex-spouses tended to get along. In addition, couples who had a high degree of initial agreement tended to fare better.

Clearly, spouses make a choice in regard to how they relate to each other following their divorce. Only those choices to get along benefit their children.

---

them and to communicate to them that they love them and are interested in them. Parents also need to be aware that their children do not want the divorce and to allow them to grieve over the loss of their family as they knew it.

4. ***Encouragement to see noncustodial parent.*** Children benefit when custodial parents (usually mothers) encourage their children to maintain regular and stable visitation schedules with the noncustodial parent.

5. ***Attention from the noncustodial parent.*** Children benefit when they receive frequent and consistent attention from noncustodial parents, usually the fathers. Noncustodial parents who do not show up at regular intervals exacerbate their children's emotional insecurity by teaching them, once again, that parents cannot be depended on. Parents who show up consistently teach their children to feel loved and secure.

6. ***Assertion of parental authority.*** Children benefit when both parents continue to assert their parental authority and continue to support the discipline practices of each other.

7. ***Regular and consistent child support payments.*** Support payments (usually from the father to the mother) are associated with economic stability for the child.

8. ***Stability.*** Not moving and keeping children in the same school system is beneficial to children. Some parents—called **latchkey parents**—spend every other week with the children in the family home so the children do not have to alternate between residences (Luscombe 2011).

**Latchkey parents** they spend every other week with the children in the family home so the children do not have to alternate between residences.

Moving to a new location causes children to be cut off from their friends, neighbors, and teachers. It is important to keep their life as stable as possible during a divorce.

9. ***Children in a new marriage.*** Manning and Smock (2000) found that divorced noncustodial fathers who remarried and who had children in the new marriage were more likely to shift their emotional and economic resources to the new family unit than were fathers who did not have a new marriage and new biological children. Fathers might be alert to this potential and consider each child, regardless of when or with whom the child was born, as worthy of continued love, time, and support.

10. ***Age and reflection on the part of children of divorce.*** Sometimes children whose parents are divorced benefit from growing older and reflecting on their parents' divorce as adults rather than as children. Nielsen (2004) emphasized that daughters who feel distant from their fathers can benefit from examining the divorce from the viewpoint of the father (was he alienated by the mother?), the cultural bias against fathers (they are maligned as "deadbeat dads" who "abandon their families for a younger woman"), and the facts about divorced dads (they are more likely than mothers to be depressed and suicidal following divorce). In general, age and temperament impact the adjustment of a child to the divorce of his or her parents. Young children have fewer cultural negative labels than teenagers, and highly sensitive children react more negatively than those who seem unaffected by whatever.

# Prerequisites for Having a "Successful" Divorce

*No divorce is ever a friendly divorce.*

BETTE DAVIS, ACTRESS

Although the concept of a "successful" divorce has been questioned (Amato et al. 2011), some divorces may cause limited damage to the partners and their children. Indeed, most people are resilient and "are able to adapt constructively to their new life situation within two to three years following divorce, a minority being defeated by the marital breakup, and a substantial group of women being enhanced" (Hetherington 2003, p. 318). The following are some of the behaviors spouses can engage in to achieve a successful divorce:

1. ***Mediate rather than litigate the divorce.*** Divorce mediators encourage a civil, cooperative, compromising relationship while moving the couple toward an agreement on the division of property, custody, and child support. In contrast, attorneys make their money by encouraging hostility so that spouses will prolong the conflict, thus running up higher legal bills. In addition, the couple cannot divide money spent on divorce attorneys (average is $17,500 for *each* side so a litigated divorce cost will start at $35,000). Benton (2008) noted that the worst thing divorcing spouses can do is for each to hire the "meanest, nastiest, most expensive yard dog lawyer in town" because doing so will only result in a protracted expensive divorce in which neither spouse will win. Because the greatest damage to children from a divorce is caused by a continuing hostile and bitter relationship between their parents, some states require **divorce mediation** as a mechanism to encourage civility in working out differences and to clear the court calendar of protracted legal battles. Research confirms there are enormous benefits of mediation versus litigation (Amato 2010). The following Family Policy section focuses on divorce mediation.

**Divorce mediation** a mechanism to encourage civility in working out differences and to clear the court calendar of protracted legal battles.

2. ***Coparent with your ex-spouse.*** Setting aside negative feelings about your ex-spouse so as to cooperatively coparent not only facilitates parental adjustment but also takes children out of the line of fire. Such coparenting translates into being cooperative when one parent needs to change a child-care schedule, sitting together during a performance by the children, and showing appreciation for the other parent's skill in responding to a crisis with the children.

*She got the gold mine, I got the shaft.*

JERRY REED

3. ***Take some responsibility for the divorce.*** Because marriage is an interaction between spouses, one person is seldom totally to blame for a divorce. Rather, both spouses share reasons for the demise of the relationship. Take some responsibility for what went wrong. What did *you* do wrong that you could correct in a subsequent relationship?

*Divorce is a game played by lawyers.*

CARY GRANT, ACTOR

4. ***Create positive thoughts.*** Divorced people are susceptible to feeling as though they are failures—they see themselves as Divorced with a capital D, a situation sometimes referred to as "hardening of the categories" disease. Improving self-esteem is important for individuals going through divorce. They can do this by systematically thinking positive thoughts about themselves.

One technique (called the stop-think technique) for improving one's self-concept, is to write down 21 positive statements about yourself ("I am honest," "I have strong family values," "I am a good parent") and transfer these to three-by-five cards, each containing three statements. Take one of the cards with you each day and read the thoughts at three regularly spaced intervals (e.g.,7:00 A.M., 1:00 P.M., and 7:00 P.M.). This ensures that you are thinking positive thoughts about yourself throughout the day and are not allowing yourself to drift into a negative state (e.g., "I am a failure" or "no one wants to be with me").

*Divorce lawyers stoke anger and fear in their clients, knowing that as long as the conflicts remain unresolved the revenue stream will keep flowing.*

CRAIG FERGUSON, *AMERICAN ON PURPOSE*

5. ***Avoid alcohol and other drugs.*** The stress and despair that some people feel during and following the divorce process sometimes make them vulnerable to the use of alcohol or other drugs. These should be avoided because they produce an endless negative cycle. For example, stress is relieved by alcohol; alcohol produces a hangover and negative feelings; the negative feelings are relieved by more alcohol, producing more negative feelings, etc.

*Nothing erases unpleasant thoughts more effectively than concentration on pleasant ones.*

HANS SELYE, STRESS RESEARCHER

6. ***Engage in aerobic exercise.*** Exercise helps one to not only counteract stress but also avoid it. Jogging, swimming, riding an exercise bike, or engaging in other similar exercise for 30 minutes every day increases oxygen to the brain and helps facilitate clear thinking. In addition, aerobic exercise produces endorphins in the brain, which create a sense of euphoria (runner's high).

# FAMILY POLICY | Should Divorce Mediation Be Required before Litigation?

Divorce mediation is a process in which spouses who have decided to separate or divorce meet with a neutral third party (mediator) to negotiate four issues: (1) how they will parent their children, which is referred to as child custody and visitation; (2) how they are going to financially support their children, referred to as child support; (3) how they are going to divide their property, known as property settlement; and (4) how each one is going to meet their financial obligations, referred to as spousal support.

## Benefits of Mediation

There are enormous benefits from avoiding litigation and mediating one's divorce:

1. ***Better relationship***. Spouses who choose to mediate their divorce have a better chance for a more civil relationship because they cooperate in specifying the conditions of their separation or divorce. Mediation emphasizes negotiation and cooperation between the divorcing partners. Such cooperation is particularly important if the couple has children, in that it provides a positive basis for discussing issues in reference to the children and how they will be parented across time.
2. ***Economic benefits***. Mediation is less expensive than litigation. The combined cost (total cost to both spouses) of hiring attorneys and going to court over issues of child custody and division of property is around $35,000. A mediated divorce typically costs less than $5,000.
3. ***Less time-consuming process***. Whereas a litigated divorce can take two to three years, a mediated divorce takes two to three months; for highly motivated individuals, a "mediated settlement conference" ". . . can take place in one session from 8:00 A.M. until both parties are satisfied with the terms" (Amato).
4. ***Avoidance of public exposure***. Some spouses do not want to discuss their private lives and finances in open court. Mediation occurs in a private and confidential setting.
5. ***Greater overall satisfaction***. Mediation results in an agreement developed by the spouses, not one imposed by a judge or the court system. A comparison of couples who chose mediation with couples who chose litigation found that those who mediated their own settlement were more satisfied with the conditions of their agreement. In addition, children of mediated divorces were exposed to less marital conflict, which may facilitate their long-term adjustment to divorce.

## Basic Mediation Guidelines

Divorce mediators conduct mediation sessions with certain principles in mind:

1. ***Children***. What is best for a couple's children should be the major concern of the parents because they know their children far better than a judge or a mediator. Children of divorced parents adjust best under three conditions: (1) that both parents have regular and frequent access to the children; (2) that the children see the parents relating in a polite and positive way; and (3) that each parent talks positively about the other parent and neither parent talks negatively about the other to the children. Sometimes children are included in the mediation.

   They may be interviewed without the parents present to provide information to the mediator about their perceptions and preferences. Such involvement of the children has superior outcomes for both the parents and the children (McIntosh et al. 2008).
2. ***Fairness***. It is important that the agreement between the soon-to-be ex-spouses be fair, with neither party being exploited or punished. It is fair for both parents to contribute financially to the children and to have regular access to their children.
3. ***Open disclosure***. The spouses will be asked to disclose all facts, records, and documents to ensure an informed and fair agreement regarding property, assets, and debts.
4. ***Other professionals***. During mediation, spouses may be asked to consult an accountant regarding tax laws. In addition, each spouse is encouraged to consult an attorney throughout the mediation and to have the attorney review the written agreements that result from the mediation. However, during the mediation sessions, all forms of legal action by the spouses against each other should be stopped.

Another term for involving a range of professionals in a divorce is **collaborative practice**, a process that brings a team of professionals (lawyer, psychologist, mediator, social worker, financial counselor) together to help a couple separate and divorce in a humane and cost-effective way.

5. ***Confidentiality.*** The mediator will not divulge anything spouses say during the mediation sessions without their permission. The spouses are asked to sign a document stating that, should they not complete mediation, they agree not to empower any attorney to subpoena the mediator or any records resulting from the mediation for use in any legal action. Such an agreement is necessary for spouses to feel free to talk about all aspects of their relationship without fear of legal action against them for such disclosures.

Divorce mediation is not for every couple. It does not work where there is a history of spouse abuse, where the parties do not disclose their financial information, where one party is controlled by someone else (e.g., a parent of one of the divorcing spouses), or where there is the desire for revenge. Mediation should be differentiated from **negotiation** (where spouses discuss and resolve the issues themselves), **arbitration** (where a third party, an arbitrator, listens to both spouses and makes a decision about custody, division of property, and so on), and **litigation** (where a judge hears arguments from lawyers representing the respective spouses and decides issues of custody, child support, division of property, and spousal support).

The following chart identifies a continuum of consequences from negotiation to litigation.

### Your Opinion?

1. To what degree do you believe the government should be involved in mandating divorce mediation?
2. How can divorce mediation go wrong? Why should a couple not want to mediate their divorce?
3. What are the advantages for children when parents mediate their divorce?

### Source

Appreciation is expressed to Mike Haswell for contributing to this section. See www.haswellmeditation.com. Costs of litigation versus mediation have been updated for 2015.

**Collaborative practice** process that brings a team of professionals (lawyer, psychologist, mediator, social worker, financial counselor) together to help a couple separate and divorce in a humane and cost-effective way.

**Negotiation** spouses discuss and resolve the issues themselves.

**Arbitration** a third party, an arbitrator, listens to both spouses and makes a decision about custody, division of property, and so on.

**Litigation** a judge hears arguments from lawyers representing the respective spouses and decides issues of custody, child support, division of property, and spousal support.

| Negotiation | Mediation | Arbitration | Litigation |
|---|---|---|---|
| Cooperative | | | Competitive |
| Low Cost | | | High Cost |
| Private | | | Public |
| Protects Relationships | | | Damages Relationships |
| Focus on the Future | | | Focus on the Past |
| Parties in Control | | | Parties Lose Control |

Clyndell Godfrey

This remarried couple are a stepfamily; the child is from the bride's first marriage.

**7. *Continue interpersonal connections.*** Adjustment to divorce is facilitated by continuing relationships with friends and family. These individuals provide emotional support and help buffer the feeling of isolation and aloneness. First Wives World (www.firstwivesworld.com) is a new interactive website that provides an Internet social network for women transitioning through divorce.

**8. *Let go of the anger for your ex-partner.*** Former spouses who stay negatively attached to an ex by harboring resentment and trying to get back at the ex prolong their adjustment to divorce. The old adage that "you can't get ahead by getting even" is relevant to divorce adjustment.

**9. *Allow time to heal.*** Because self-esteem usually drops after divorce, a person is often vulnerable to making commitments before working through feelings about the divorce. Most individuals need between 12 and 18 months to adjust to the end of a marriage. Although being available to others may help to repair one's self-esteem, getting remarried during this time should be considered cautiously. Two years between marriages is recommended.

# Remarriage

Divorced spouses are not sour on marriage. Although they may want to escape from the current spouse, they are open to having a new spouse. In the past, two-thirds of divorced females and three-fourths of divorced males remarried (Sweeney 2010), with the new unions occurring between two and four years of the previous divorce (Brown and Porter 2013). When comparing divorced individuals who remarried and divorced individuals who have not remarried, the remarried individuals reported greater personal and relationship happiness.

*Some people think that it's holding on that makes one strong; sometimes it's letting go.*

UNKNOWN

The majority of divorced people remarry for many of the same reasons as those for a first marriage—love, companionship, emotional security, and a regular sex partner. Other reasons are unique to remarriage and include financial security (particularly for a wife with children), help in rearing one's children, the desire to provide a "social" father or mother for one's children, escape from the stigma associated with the label "divorced person," and legal threats regarding the custody of one's children. With regard to the latter, the courts view a parent seeking custody of a child more favorably if the parent is married. The religiously devout remarry, in part because religion is pro-family and they seek the context which reflects those values (Brown and Porter 2013).

*Remarriage is the triumph of hope over experience.*

SAMUEL JOHNSON, ENGLISH WRITER

**National Data**

Forty percent of weddings in the U.S. involve a spouse who has been married before (Maggie Scarf, The Remarriage Blueprint).

*But I'm throwing myself back in because I like being married. I don't want to end this whole fabulous journey alone. I want someone by my side who I love and who loves me.*

NEIL DIAMOND, AGE 71, OF HIS THIRD MARRIAGE (TO A WOMAN 29 YEARS YOUNGER)

## Preparation for Remarriage

Since remarried spouses show a steeper decline in marital satisfaction than first married spouses (Reck and Higginbotham 2012), preparation for remarriage is particularly important. As most remarriages involve children, taking a stepfamily education program may be particularly valuable. Higginbotham and Skogrand (2010) studied 356 adults who attended a 12-hour stepfamily relationship education course. The participants were either remarried, cohabiting, or seriously dating someone who had children from a previous relationship. Results revealed that regardless of their race or marital status, the individuals benefited from the stepfamily relationship education—commitment to the relationship, agreement on finances, relationships with ex-partners, and parenting all improved over time.

Children also benefit when parents participate in stepfamily education programs. Higginbotham et al. (2010) interviewed 40 parents and 20 facilitators who were part of a stepfamily education program. They found that children were perceived to benefit by their parents' increased empathy, engagement in family time, and enhanced relationship skills.

Self-help books on remarriage and stepfamily issues are lacking. One researcher (Shafer 2010) noted that such books tend to lack empirical research and are unable to provide practical solutions for blended families and parents. *The Remarriage Blueprint* by Maggie Scarf (2013) is a welcomed researched based guide for those the second time around.

Trust is a major issue for people getting remarried. Brimhall et al. (2008) interviewed 16 remarried individuals and found that most reported their first marriage ended over trust issues—the partner betrayed them by having an affair or by hiding or spending money without their knowledge. Each was intent on ensuring a foundation of trust in the new marriage.

*I am happy by myself. I am not going to marry again.*

DEBBIE REYNOLDS, AGE 81

## Issues of Remarriage for the Divorced

Several issues challenge people who remarry (Ganong & Coleman 1994; Goetting 1982; Kim 2011; Martin-Uzzi and Duval-Tsioles 2013; Scarf 2013).

*Divorce used to be considered a sign of failure; today it is often deemed the first step toward true happiness.*

HELEN FISHER, ANTHROPOLOGIST

**Boundary Maintenance** Ghosts of the first marriage, in terms of the ex-spouse, must be dealt with. A parent must decide how to relate to an ex-spouse to maintain a good parenting relationship for the biological children while an emotional distance to prevent problems from developing with the new partner.

Some spouses continue to be emotionally attached to and have difficulty breaking away from an ex-spouse. These former spouses have what Masheter (1999) terms a **negative commitment** whereby such individuals "have decided to remain [emotionally] in this relationship and to invest considerable amounts of time, money, and effort in it . . . [T]hese individuals do not take responsibility

*I think I like living alone. I'd be happy to have some fella who came over, just as long as he was out of there by 9 P.M.*

BARBARA EHRENREICH, WHEN ASKED IF SHE WAS INTERESTED IN FINDING ANOTHER PARTNER

**Negative commitment** such individuals have decided to remain [emotionally] in this relationship and to invest considerable amounts of time, money, and effort in it.

WHAT IF?

### What If Your Ex-Spouse or Ex-Partner Wants to Get Back Together with You?

The dilemma of an ex-spouse or ex-partner wanting to get back together is not unusual and has been the subject of classic novels. Rhett Butler said of getting back with Scarlett O'Hara in *Gone with the Wind*, "I'd rather remember it as it was at its best than mend it and look at the broken pieces as long as I live." His decision is worthy of duplicating. Once an intense love relationship has been broken by divorce, mending it and returning it to a durable, happy relationship is not likely. The reason is that the factors that ended the relationship earlier may resurface if the partners get back together. "If it didn't work the first time, it is not likely that it will work the second time" is the take home message.

for their own feelings and actions, and often remain 'stuck,' unable to move forward in their lives" (p. 297).

In some cases an uncooperative ex-spouse can be a source of bonding for the remarried spouses—but they may also be frustrated in dealing with the ex (Martin-Uzzi and Duval-Tsioles 2013). For example, the newly remarried couple may have planned an adult couple vacation and the ex might call the night before saying she or he can't take the children for their regular weekend.

**Emotional Remarriage** Remarriage involves beginning to trust and love another person in a new relationship. Such feelings may come slowly as a result of negative experiences in a previous marriage.

**Psychic Remarriage** Divorced individuals considering remarriage may find it difficult to give up the freedom and autonomy of being single and to develop a mental set conducive to pairing. This transition may be particularly difficult for people who sought a divorce as a means to personal growth and autonomy. These individuals may fear that getting remarried will put unwanted constraints on them.

**Community Remarriage** This aspect involves a change in focus from single friends to a new mate and other couples with whom the new pair will interact. The bonds of friendship established during the divorce period may be particularly valuable because they have given support at a time of personal crisis. Care should be taken not to drop these friendships.

**Parental Remarriage** Because most remarriages involve children, people must work out the nuances of living with someone else's children. Mothers are usually awarded primary physical custody, and this circumstance translates into a new stepfather adjusting to the mother's children and vice versa. The late film critic Roger Ebert married a woman with two children and said of his new stepchildren.

> *...I took joy in the role and loved the children. They represented children I believed I might never have. I never saw them as competing with me for their mother's attention, but as sharing their family with me Ebert (2011), (p. 568).*

**Economic and Legal Remarriage** A second marriage may begin with economic responsibilities to a first marriage. Alimony and child support often threaten the harmony and sometimes even the economic survival of second marriages. Although the income of a new wife is not used legally to decide the amount her new husband is required to pay in child support for his children of a former marriage, his ex-wife may petition the court for more child support. The ex-wife may do so, however, on the premise that his living expenses are reduced with a new wife and that, therefore, he should be able to afford to pay more child support. Although an ex-wife is not likely to win, she can force the new wife to go to court and to disclose her income (all with considerable investment of time and legal fees for a newly remarried couple).

Economic issues in a remarriage may become evident in another way. A remarried woman who receives inadequate child support from an ex-spouse and needs money for her child's braces, for instance, might wrestle with how much money to ask her new husband for.

There may also be a need for a marriage contract to be drawn up before the wedding. Suppose a wife moves into the home of her new husband. If he has a will stating that his house goes to his children from a former marriage at his death and no marriage contract that either gives his wife the house or allows her to stay in the house rent free until her death, his children can legally throw her out of the house. The same is true for their beach house which he brought into the marriage. If his will leaves the beach house to his children, his wife may have no place to live.

## Remarriage for Widowed Individuals

Only 10% of remarriages involve widows or widowers. Nevertheless, remarriage for widowed individuals is usually very different from remarriage for those who are divorced. Unlike divorced individuals, widowed individuals are usually much older and their children are grown.

Brimhall and Engblom-Deglmann (2011) interviewed 24 remarried individuals about the death of a previous spouse, either theirs or their partner's, and how this previous marriage was affecting the new marriage. Participants were interviewed individually and as a couple. Several themes emerged including putting the past spouse on a pedestal, comparing the current and past spouses, insecurity of the current spouse, curiosity about the past spouse and relationship, the new partner's response to this curiosity, and its impact on the current relationship. The best new relationship outcomes seemed to happen when the spouse of a deceased partner talked openly about the past relationship and reassured the current partner of his or her love for the new partner and the new relationship.

A widow or widower may marry someone of similar age or someone who is considerably older or younger. Marriages in which one spouse is considerably older than the other are referred to as May-December marriages (discussed in Chapter 7, Marriage Relationships). Here we will discuss only **December marriages**, in which both spouses are elderly.

**December marriages** marriages in which spouses are elderly.

A study of 24 elderly couples found that the primary motivation for remarriage was the need to escape loneliness or the need for companionship (Vinick 1978). Men reported a greater need to remarry than did the women.

Most of the spouses (75%) met through a mutual friend or relative and married less than a year after their partner's death (63%). Increasingly, elderly individuals are meeting online. Some sites cater to older individuals seeking partners, including www.seniorfriendfinder.com and www.ourtime.com.

The children of the couples in Vinick's study had mixed reactions to their parent's remarriage. Most of the children were happy that their parent was happy and felt relieved that someone would now meet the companionship needs of their elderly parent on a more regular basis. However, some children disapproved of the marriage out of concern for their inheritance. "If that woman marries Dad," said a woman with two children, "she'll get everything when he dies. I love him and hope he lives forever, but when he's gone, I want the house I grew up in." Though children may be less than approving of the remarriage of their widowed parent, adult friends of the couple, including the kin of the deceased spouses, are usually very approving (Ganong and Coleman 1999).

## Stages of Involvement with a New Partner

After a legal separation or divorce (or being widowed), a parent who becomes involved in a new relationship passes through various transitions. These include deciding on the level of commitment to the new partner, introducing one's children to the new partner, and allowing time for the children to adjust to the new partner. Progressing through the stages may not be easy.

## Stability of Remarriages

In a comparison of first and remarried individuals (Whitton 2013), the latter had more positive attitudes toward divorce and a weaker commitment to marriage. Indeed, when marital quality was low in both first and second marriages, the latter were much more likely to feel divorce was an option. This finding was true independent of whether or not the remarried adults brought children into the new marriage. Sweeney (2010) confirmed that remarriages are more likely than first marriages to end in divorce in the early years of remarriage. Reck and Higginbotham (2012) found that remarried men and women reported significant difficulties within the first three years of remarriage. For example, men reported their role as husband, father, and stepfather was particularly challenging due to the difficulties in gaining new familial expectations and parenting their own biological and new stepchildren (e.g., disciplining, establishing trust). Women reported similar difficulties in being a wife, mother, and stepmother; however, women generally reported higher levels of difficulty in these roles. Mirecki et al. (2013) confirmed lower marital satisfaction of spouses in second marriages. Since 65 percent of second marriages include the presence of stepchildren, integrating the various individuals into a functioning family is challenging. Higher education on the part of the spouses seemed to help (higher education is associated with higher income). That second marriages, in general, are more susceptible to divorce than first marriages may also be related to the fact that divorced individuals are less fearful of divorce (e.g., they know they can survive divorce) than individuals who have never divorced.

*Love and magic have a great deal in common. They enrich the soul, delight the heart. And they both take practice.*

NORA ROBERTS, AMERICAN AUTHOR

Though remarried people are more vulnerable to divorce in the early years of their subsequent marriage, they are less likely to divorce after 15 years of staying in the second marriage than those in first marriages (Clarke and Wilson 1994). Hence, these spouses are likely to remain married because they want to, not because they fear divorce.

*Divorce is a journey that the children involved in do not ask to take. They are forced along for a ride where the results are dictated by the road their parents decide to travel.*

DIANE GREENE, GOOGLE BOARD OF DIRECTORS

## Stepfamilies

### National **Data**

More than half of all children will spend some time in a family arrangement other than the traditional family (Crosnoe and Cavanagh 2010). At any given time 5.3 million children under age 18 are living with a biological parent and a married or cohabiting stepparent (Sweeney 2010).

Stepfamilies, also known as blended, binuclear, remarried, or reconstituted families, represent the fastest-growing type of family in the United States. A **blended family** is one in which spouses in a new marriage relationship blend their respective children from at least one other spouse from a previous marriage. The term **binuclear** refers to a family that spans two households; when a married couple with children divorce, their family unit typically spreads into two households. There is a movement away from the use of the term *blended* because stepfamilies really do not blend. The term **stepfamily** (sometimes referred to as step relationships) is the term currently in vogue. This section examines how stepfamilies differ from nuclear families; how they are experienced from the viewpoints of women, men, and children; and the developmental tasks that must be accomplished to make a successful stepfamily.

**Blended family** one in which spouses in a new marriage relationship blend their respective children from at last one other spouse from a previous marriage.

**Binuclear family** family that lives in two households as when parents live in separate households following a divorce.

**Stepfamily** family in which spouses in a new marriage bring children from previous relationships into the new home.

Although there are various types of stepfamilies, the most common is a family in which the partners bring children from previous relationships into the new home, where they may also have a child of their own. The couple may be married or living together, heterosexual or homosexual, and of any race.

Stepmothers become stepgrandmothers; this woman is playing with her stepdaughter's child.

Nuru and Wang (2014) emphasized the need to include step relationships for those in cohabiting relationships.

Various myths abound regarding stepfamilies, including that new family members will instantly bond emotionally, that children in stepfamilies are damaged and do not recover, that stepmothers are "wicked home-wreckers," that stepfathers are uninvolved with their stepchildren, and that stepfamilies are not "real" families. Stepfamilies are also stigmatized. **Stepism** is the assumption that stepfamilies are inferior to biological families. Like racism, heterosexism, sexism, and ageism, stepism involves prejudice and discrimination.

**Stepism** the assumption that stepfamilies are inferior to biological families.

Stepfamilies differ from nuclear families in a number of ways. These are identified in Table 15.2. These changes impact the parents, their children, and their stepchildren and require adjustment on the part of each member.

The Self-Assessment feature provides a way to assess the degree to which a stepchild accepts and bonds with a stepfather.

## Developmental Tasks for Stepfamilies

A **developmental task** is a skill that, if mastered, allows a family to grow as a cohesive unit. Developmental tasks that are not mastered will move the family closer to the point of disintegration. Some of the more important developmental tasks for stepfamilies are discussed in this section.

**Developmental task** a skill that, if mastered, allows a family to grow as a cohesive unit.

1. Nurture the new marriage relationship
2. Be patient for stepparent/stepchild relationships to develop
3. Have realistic expectations
4. Accept your stepchildren

**TABLE 15.2 Differences between Nuclear Families and Stepfamilies**

| Nuclear Families | Stepfamilies |
|---|---|
| 1. Children are (usually) biologically related to both parents. | 1. Children are biologically related to only one parent. |
| 2. Both biological parents live together with children. | 2. As a result of divorce or death, one biological parent does not live with the children. In the case of joint physical custody, children may live with both parents, alternating between them. |
| 3. Beliefs and values of members tend to be similar. | 3. Beliefs and values of members are more likely to be different because of different backgrounds. |
| 4. The relationship between adults has existed longer than relationship between children and parents. | 4. The relationship between children and parents has existed longer than the relationship between adults. |
| 5. Children have one home they regard as theirs. | 5. Children may have two homes they regard as theirs. |
| 6. The family's economic resources come from within the family unit. | 6. Some economic resources may come from an ex-spouse. |
| 7. All money generated stays in the family. | 7. Some money generated may leave the family in the form of alimony or child support. |
| 8. Relationships are relatively stable. | 8. Relationships are in flux: new adults adjusting to each other; children adjusting to a stepparent; a stepparent adjusting to stepchildren; stepchildren adjusting to each other. |
| 9. No stigma is attached to nuclear family. | 9. Stepfamilies are stigmatized. |
| 10. Spouses had a childfree period. | 10. Spouses had no childfree period. |
| 11. Inheritance rights are automatic. | 11. Stepchildren do not automatically inherit from stepparents. |
| 12. Rights to custody of children are assumed if divorce occurs. | 12. Rights to custody of stepchildren are usually not considered. |
| 13. Extended family networks are smooth and comfortable. | 13. Extended family networks become complex and strained. |
| 14. Nuclear family may not have experienced loss. | 14. Stepfamily has experienced loss. |
| 15. Families experience a range of problems. | 15. Stepchildren tend to be a major problem. |

5. Give parental authority to your spouse/coparent
6. Establish your own family rituals
7. Support the children's relationship with their absent parent
8. Cooperate with the children's biological parent and co-parent
9. Use LAT to reduce the strain and conflict of stepfamily living

*Don't marry someone who you can't see yourself being divorced from.*

KELA PRICE, STEPFAMILY COUNSELOR

Living apart together (LAT) is a structural solution to many of the problems of stepfamily living (see Chapter 5 for a more thorough discussion). By getting two condos (side by side or one on top of the other), a duplex, or two small houses and having the respective biological parents live in each of the respective units with their respective children, everyone wins. The children and biological parent will experience minimal disruption as the spouses transition to the new marriage. This arrangement is particularly useful where both spouses have children ranging in age from 10 to 18. Situations where only one spouse has children from a former relationship or those in which the children are very young will have limited benefit. The new spouses can still spend plenty of time together to nurture their relationship without spending all of their time trying to manage the various issues that come up with the stepchildren.

## SELF-ASSESSMENT | Parental Status Inventory*

The Parental Status Inventory (PSI) is a 14-item inventory that measures the degree to which respondents consider their stepfather to be a parent on an 11-point scale from 0% to 100%. Read each of the following statements and circle the percentage indicating the degree to which you regard the statement as true.

1. I think of my stepfather as my father. (0%, 10, 20, 30, 40, 50%, 60, 70, 80, 90, 100%)
2. I am comfortable when someone else refers to my stepfather as my father or dad. (0%, 10, 20, 30, 40, 50%, 60, 70, 80, 90, 100%)
3. I think of myself as his daughter/son. (0%, 10, 20, 30, 40, 50%, 60, 70, 80, 90, 100%)
4. I refer to him as my father or dad. (0%, 10, 20, 30, 40, 50%, 60, 70, 80, 90, 100%)
5. He introduces me as his son/daughter. (0%, 10, 20, 30, 40, 50%, 60, 70, 80, 90, 100%)
6. I introduce my mother and him as my parents. (0%, 10, 20, 30, 40, 50%, 60, 70, 80, 90, 100%)
7. He and I are just like father and son/daughter. (0%, 10, 20, 30, 40, 50%, 60, 70, 80, 90, 100%)
8. I introduce him as "my father" or "my dad." (0%, 10, 20, 30, 40, 50%, 60, 70, 80, 90, 100%)
9. I would feel comfortable if he and I were to attend a father-daughter/father-son function, such as a banquet, baseball game, or cookout, alone together. (0%, 10, 20, 30, 40, 50%, 60, 70, 80, 90, 100%)
10. I introduce him as "my mother's husband" or "my mother's partner." (0%, 10, 20, 30, 40, 50%, 60, 70, 80, 90, 100%)
11. When I think of my mother's house, I consider him and my mother to be parents to the same degree. (0%, 10, 20, 30, 40, 50%, 60, 70, 80, 90, 100%)
12. I consider him to be a father to me. (0%, 10, 20, 30, 40, 50%, 60, 70, 80, 90, 100%)
13. I address him by his first name. (0%, 10, 20, 30, 40, 50%, 60, 70, 80, 90, 100%)
14. If I were choosing a greeting card for him, the inclusion of the words *father* or *dad* in the inscription would prevent me from choosing the card. (0%, 10, 20, 30, 40, 50%, 60, 70, 80, 90, 100%)

### Scoring

First, reverse the scores for items 10, 13, and 14. For example, if you circled a 90, change the number to 10; if you circled a 60, change the number to 40; and so on. Add the percentages and divide by 14. A 0% reflects that you do not regard your stepdad as your parent at all. A 100% reflects that you totally regard your stepdad as your parent. The percentages between 0% and 100% show the gradations from no regard to total regard of your stepdad as parent.

### Norms

Respondents in two studies (one in Canada and one in America) completed the scale. The numbers of respondents in the studies were 159 and 156, respectively, and the average score in the respective studies was 45.66%. Between 40% and 50% of both Canadians and Americans viewed their stepfather as their parent.

*Developed by Dr. Susan Gamache. 2000. Hycroft Medical Centre, #217, 3195 Granville Street, Vancouver, B.C., Canada, V6H 3K2 gamache@interchange.ubc.ca. Details on construction of the scale including validity and reliability are available from Dr. Gamache. The PSI Scale is used in this text by permission of Dr. Gamache and may not be used otherwise (except as in class student exercises) without written permission.

# The Future of Divorce and Remarriage

Divorce remains stigmatized in our society, as evidenced by the term **divorcism**—the belief that divorce is a disaster. In view of this cultural attitude, a number of attempts will continue to be made to reduce divorce rates. Couple/relationship education (CRE) provides guidelines for instruction in communication, conflict resolution, and parenting skills. Lucier-Greer et al. (2012) found positive outcomes for remarried couples with children who were involved in these programs that were targeted toward stepfamily issues.

**Divorcism** the belief that divorce is a disaster.

Another attempt at divorce prevention is **covenant marriage** (now available in Louisiana, Arizona, and Arkansas), which emphasizes the importance of staying married. Covenant marriage was discussed in more detail in Chapter 6. Although most of a sample of 1,324 adults in a telephone survey in Louisiana, Arizona, and Minnesota were positive about covenant marriage (Hawkins et al. 2002), fewer than 3% of marrying couples elected covenant marriages when given the opportunity to do so (Licata 2002). Although already married couples

**Covenant marriage** emphasizes the importance of staying married.

WHAT IF?

## What If Your Children Will Not Accept Your New Spouse or Partner?

It never occurs to new lovers that their children will not embrace their new spouse or partner. After all, the new spouse or partner represents only joy and happiness for the recently divorced person. The children often see it otherwise. They see the new spouse or partner as the final ending of their previously safe and secure traditional family (mom and dad and the children in one house). Now they must be disrupted every other weekend and, when they are with their nonresident parent (usually the father), the "new woman" is there.

Although most children mature and grow to accept their new stepparent, this usually takes seven years. In the meantime, the biological parent may need to lower expectations and settle for the children just being polite to their stepparent. The following is a script for a divorced father to give children who are barely civil: "I know the divorce is not what you wanted and you would prefer that your mom and I get back together and live in the same house again. That won't happen and we are both moving on. This means a new spouse for me and a stepmother for you. Again, I know you do not want this and did not choose it. I don't require you to love and accept your new stepmother. I do require you to be polite. She will be kind to you and I expect you to be kind and polite to her. That means 'thank you' and 'please.' If you can't at least be polite, it is the end of your cell phone, computer/TV in your room, no friends over and no visiting friends. I will always be polite to your friends and I expect the same from you." The difficulty of being a stepmother is no illusion (Cartwright 2012). Doodson and Davies (2014) compared biological mothers and stepmothers and found higher levels of depression and anxiety among stepmothers.

can convert their standard marriages to covenant marriages, there are no data on how many have done so (Hawkins et al. 2002).

Although most marriages begin with love and hope for a bright future, the reality is that 42% to 50% will end in divorce. Brown (2010) noted the need for stable family structures for children and how researchers might become involved in informing federal policy debates to such stable structures. At the micro level, to protect themselves against a financial disaster should a marriage end in divorce, some individuals purchase divorce insurance. The cost is $16 a month for every $1,250 of coverage. To discourage couples from getting divorced just to collect the insurance money, policy holders must pay premiums for four years before they can collect on a divorce. The idea is the brainchild of John Logan, who watched his "wealth follow his divorce down the drain" (Luscombe 2010b). Persons who are getting remarried have the same hopes and dreams as those getting married for the first time. Since most first marrieds and remarrieds think their love is unique and immune to divorce, they will not be interested in divorce insurance.

The remarriage rate is dropping fast. In 1990, 50 of 1,000 divorced or widowed remarried. By 2011, the number had dropped to 29. Cohabitation is the alternative many once-married individuals seek (Jayson 2013).

# Summary


***What are the factors associated with ending an unsatisfactory relationship and how prevalent is divorce?***

Persons who are reluctant to end a relationship have considerable investment in their current relationship and limited alternatives. Between 40% and 50% of individuals divorce; the percent is lower for college-educated individuals and those with higher incomes. Bartenders and dancer/choreographers are the most likely to divorce.

***What are macro factors contributing to divorce?***

Macro factors contributing to divorce include increased economic independence of women (women can afford to leave), changing family functions (companionship is the only remaining function), liberal divorce laws (it's easier to leave), fewer religious sanctions (churches embrace single individuals), more divorce models (Hollywood models abound), and individualism (rather than familism as a cultural goal of happiness).

***What are micro factors contributing to divorce?***

Micro factors include having numerous differences, falling out of love, negative behavior, lack of conflict resolution skills or satiation, value changes, and extramarital relationships. Those who are separated and going through a divorce say that they wished they had worked harder on their marriage.

***What are the consequences of divorce for spouses/parents?***

Women tend to fare better emotionally after separation and divorce than do men. Women are more likely than men not only to have a stronger network of supportive relationships but also to profit from divorce by developing a new sense of self-esteem and confidence, because they are thrust into a more independent role.

Both women and men experience a drop in income following divorce, but women suffer more. Although custodial mothers are often awarded child support, the amount is usually inadequate, infrequent, and not dependable, and women are forced to work (sometimes at more than one job) to take financial care of their children.

***What are the effects of divorce on children?***

A civil, cooperative, co-parenting relationship between ex-spouses is the greatest predictor of a positive outcome for children whose parents divorce. Other factors associated with minimizing the negative effects of divorce for children include healthy parental psychological functioning, parental attention to the children, allowing children to grieve, each parent encouraging their children to see noncustodial parent, assertion of parental authority, regular/consistent child support payments, stability (e.g., housing/neighborhood), and parents not having a new baby with a new partner.

***What is the nature of remarriage in the United States?***

Two-thirds of divorced women and three-fourths of divorced men remarry. When comparing divorced individuals who have remarried with divorced individuals who have not remarried, the remarried report greater personal and relationship happiness. Trust is a major issue for people getting remarried because they have already had at least one marriage fall apart. This is especially true if the partner had an affair or hid or spent money without the partner's knowledge.

Issues to confront/negotiate in a remarriage include boundary maintenance (not getting entangled with ex), emotional remarriage (loving/trusting again), psychic remarriage (give up freedom of singlehood), community remarriage (shift from single friends to new mate), parental remarriage (adjusting to stepfamily), and economic/legal remarriage (responsibilities to previous and current family).

***What is the nature of stepfamilies in the United States?***

Stepfamilies represent the fastest-growing type of family in the United States. Stepfamilies differ from nuclear families (e.g., the children in nuclear families are biologically related to both parents, whereas the children in stepfamilies are biologically related to only one parent). Stepism is the assumption that stepfamilies are inferior to biological families. Like racism, heterosexism, sexism, and ageism, stepism involves prejudice and discrimination.

Developmental tasks for stepfamilies include nurturing the new marriage relationship, allowing time for partners and children to get to know each other, and supporting the children's relationship with both parents and natural grandparents.


# Key Terms

Arbitration
Binuclear family
Blended family
Collaborative practice
Covenant marriage
December marriages
Developmental task
Divorce
Divorce mediation

Divorcism
Family relations doctrine
Latchkey parents
Legal custody
Litigation
Negative commitment
Negotiation
No-fault divorce
Parental alienation
Parental alienation syndrome
Physical custody
Postnuptial agreement
Satiation
Stepfamily
Stepism

## Web Links

Association for Conflict Resolution
http://www.acrnet.org

Center for Divorce Education
http://www.divorce-education.com/

Divorce Laws by State
http://www.totaldivorce.com/state-laws/default.aspx

Divorce Source (a legal resource for divorce, custody, alimony, and support)
http://www.divorcesource.com/

Divorce Support Page
http://www.divorcesupport.com/index.html

Divorce 360 (divorce advice, news, blogs, and community)
http://www.divorce360.com/

DivorceBusting (solve marriage problems)
http://divorcebusting.com/

North Carolina Association of Professional Family Mediators
http://familymediators.org

First Wives World : The New Face of Divorce
http://www.firstwivesworld.com

A Guide to the Parental Alienation Syndrome
http://www.coeffic.demon.co.uk/pas.htm

Positive Parenting Through Divorce Online Course
http://www.positiveparentingthroughdivorce.com/

National Center for Health Statistics (marriage and divorce data)
http://www.cdc.gov/nchs/

# CHAPTER 16 Relationships in the Later Years

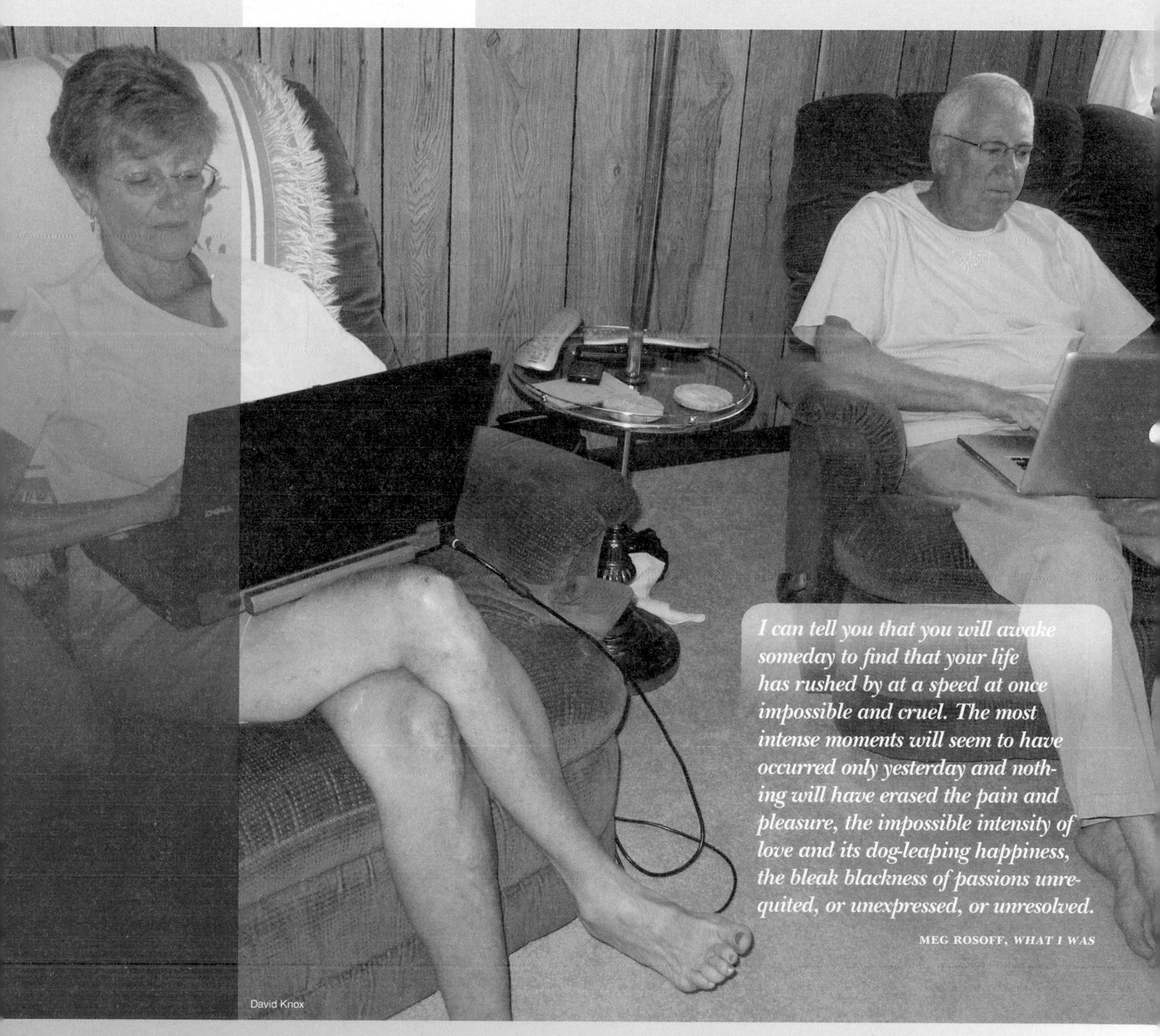

David Knox

*I can tell you that you will awake someday to find that your life has rushed by at a speed at once impossible and cruel. The most intense moments will seem to have occurred only yesterday and nothing will have erased the pain and pleasure, the impossible intensity of love and its dog-leaping happiness, the bleak blackness of passions unrequited, or unexpressed, or unresolved.*

MEG ROSOFF, *WHAT I WAS*

## Learning Objectives

- Specify the meanings of age and ageism.
- Be aware of the theories of aging.
- Discuss the primary issues (income, health, retirement) confronting the elderly.
- Review relationships of the elderly.
- Summarize grandparenting for the elderly.
- Review the end of life issues for the elderly.
- Predict the future for the elderly in the United States.

## TRUE OR FALSE?

1. Good physical health is the single most important determinant of an elderly person's reported happiness.
2. Elderly women are less sexually satisfied than elderly men.
3. Most elderly who attempt suicide are not successful the first time.
4. Dating is more common among elderly men than elderly women.
5. The older the person dating, the greater the concern for the physical health of a potential partner.

*Answers:* 1. T 2. T 3. F 4. T 5. T

Too little emphasis has been given to the advantages of aging. Dr. Ansello (2012) of The Virginia Center for Aging identified some benefits of growing old.

*Things you buy now won't wear out.*
*People no longer view you as a hypochondriac.*
*You can have a party and your neighbors won't realize it.*
*Your sordid secrets are safe with your friends because they won't remember them.*
*Your joints are a more accurate weather predictor than the National Weather Service.*

*An attempt to define old people falls apart as soon as you try.*

GORDON HAWKINS, *AN ERRANT AGE, APPROACHING NINETY*

In 2015, about 14.4% of the 325 million individuals in the United States will be age 65 and older. By 2020, this percentage will grow to 16%. (*Statistical Abstract of the United States, 2012–2013,* Table 9). By 2030, one in five Americans (71 million) will be over the age of 65. Researcher Amy Rauer (2013) noted that we are moving toward a society of more "walkers than strollers."

In this chapter, we focus on the factors that confront individuals and couples as they age and the dilemma of how to care for aging parents. We begin by looking at the concept of age.

*You can't help getting older, but you don't have to get old.*

GEORGE BURNS, COMEDIAN

**Age** term which may be defined chronologically (number of years), physiologically (physical decline), psychologically (self-concept), sociologically (roles for the elderly/retired), and culturally (meaning of age in one's society).

# Age and Ageism

All societies have a way to categorize their members by age. And all societies provide social definitions for particular ages.

### DIVERSITY IN OTHER COUNTRIES

In a global study of concerns about aging, Japan emerged as the country where residents are most likely to report that "growing old is a problem." Nine-in-ten Japanese, eight-in-ten South Koreans, and seven-in-ten Chinese describe aging as a major problem for their country. Americans are among the least concerned, with only one-in-four people expressing this opinion. More social programs are in place for U.S. elderly, which help to account for a lower concern expressed by Americans (Pew Research Center 2014).

## The Concept of Age

A person's **age** may be defined chronologically, physiologically, psychologically, sociologically, and culturally. Chronologically, an "old" person is defined as one who has lived a certain number of years. The concept has obvious practical significance in everyday life. Bureaucratic organizations and social programs identify chronological age as a criterion of certain social rights and responsibilities. One's age determines the right to drive, vote, buy alcohol or cigarettes, and receive Social Security and Medicare benefits.

Age has meaning in reference to the society and culture of the individual. In ancient Greece

| TABLE 16.1 Life Expectancy | | | | |
|---|---|---|---|---|
| Year | White Males | Black Males | White Females | Black Females |
| 2015 | 77.1 | 71.4 | 81.8 | 78.2 |
| 2020 | 77.7 | 72.6 | 82.4 | 79.2 |

Source: *Statistical Abstract of the United States*, 2012–2013. 131th ed. Washington, DC: U.S. Bureau of the Census, Table 104.

and Rome, where the average life expectancy was 20 years, an individual was old at 18; similarly, one was old at 30 in medieval Europe and at 40 in the United States in 1850. In the United States today, however, people are usually not considered old until they reach age 65. However, our society is moving toward new chronological definitions of "old." Three groups of the elderly have emerged—the "young-old," the "middle-old," and the "old-old." The young-old are typically between the ages of 65 and 74; the middle-old, 75 to 84, and the old-old, 85 and beyond. Current life expectancy is shown in Table 16.1.

Gilbert Meilaender's book *Should We Live Forever? The Ethical Ambiguities of Aging* (2013) emphasizes the preoccupation with longevity. Research is being conducted in the Department of Genetics, University of Cambridge that will hopefully add hundreds of years to one's life. And, in case science does not work fast enough, the Cryonics Institute (http://www.cryonics.org/) offers to freeze your body and wake you up (hopefully) in the future when technology is

*First, you believe in Santa Claus. Then you don't believe in Santa Claus. And before you know it, you **are** Santa Claus.*

UNKNOWN

*What can ever equal the memory of being young together?*

MICHAEL STEIN, *IN THE AGE OF LOVE*

David Knox

This woman was still active as a university English professor in her nineties.

**TABLE 16.2 Accomplishments of Older Individuals**

At 80, George Burns won an Academy Award for his performance in *The Sunshine Boys.*

At 81, Benjamin Franklin facilitated the compromise that led to the adoption of the U.S. Constitution.

At 82, Winston Churchill wrote *A History of the English-Speaking Peoples.*

At 85, actress Mae West starred in the film *Sextette.*

At 85, Coco Chanel was the head of a fashion design firm.

At 87, Mary Baker Eddy created the newspaper *Christian Science Monitor.*

At 88, actress Betty White became the oldest person to ever host *Saturday Night Live.*

At 88, B. B. King continues to give performances playing his guitar/singing the blues.

At 89, Doris Haddock, also known as "Granny D," began a 3,200 mile walk from Los Angeles to Washington, D.C., to raise awareness for the issue of campaign finance reform. She walked 10 miles a day for 14 months, skiing 100 miles when snowfall made walking impossible, and completed her cross-country walk at age 90.

At 90, Pablo Picasso was producing drawings and engravings.

At 93, George Bernard Shaw wrote the play *Farfetched Fables.*

At 94, philosopher Bertrand Russell was active in promoting peace in the Middle East.

At 96, fitness guru Jack LaLanne continued to be active and exercised two hours a day.

At 100, Grandma Moses, noted for her rural American landscapes, was painting. She started painting at age 78, but by the time she died at 101 she had created over 1,500 works of art.

---

available to replace human body parts indefinitely and to stop the aging process. The promise of never growing old and living forever is here. This thought certainly gives a new slant on "till death do us part."

Some individuals have been successful in delaying the aging process and are very active in their later years. See Table 16.2 for examples.

Physiologically, people are old when their auditory, visual, respiratory, and cognitive capabilities decline significantly. At age 80 and over, 45% are hearing impaired and 25% have visual impairment (Dillion et al. 2010). Becoming disabled is associated with being old. Sleep changes also occur for the elderly, including going to bed earlier, waking up during the night, and waking up earlier in the morning, as well as such disorders such as snoring and obstructive sleep apnea.

*Attitude is everything. Mae West lived into her eighties believing she was twenty, and it never occurred to her that her math was lousy.*

**SOUNDINGS MAGAZINE**

People who need full-time nursing care for eating, bathing, and taking medication properly and who are placed in nursing homes are thought of as being old. Indeed, successful aging is culturally defined as maintaining one's health, independence, and cognitive ability. It is not death but the slow deterioration from aging that brings the most fear.

People who have certain diseases are also regarded as old. Although younger individuals may suffer from Alzheimer's, arthritis, and heart problems, these ailments are more often associated with aging. As medical science conquers more diseases, the physiological definition of aging changes so that it takes longer for people to be defined as old.

Psychologically, a person's self-concept is important in defining how old that person is. As individuals begin to fulfill the roles associated with the elderly—retiree, grandparent, nursing home resident—they begin to see themselves as aging. Sociologically, once they occupy these roles, others

**DIVERSITY IN OTHER COUNTRIES**

Among Asians, the high status of the elderly in the extended family derives from religion. Confucian philosophy, for example, prescribes that all relationships are of the subordinate-superordinate type—husband-wife, parent-child, and teacher-pupil. For traditional Asians to abandon their elderly rather than include them in larger family units would be unthinkable. However, commitment to the elderly may be changing as a result of the Westernization of Asian countries such as China, Japan, and Korea.

begin to see them as old. Culturally, the society in which an individual lives defines when and if a person becomes old and what being old means. In U.S. society, the period from age 18 through 64 is generally subdivided into young adulthood, adulthood, and middle age. Cultures also differ in terms of how they view and take care of their elderly. Spain is particularly noteworthy in terms of care for the elderly, with most elderly people receiving care from family members and other relatives.

*Age is the only social category identifying subgroups that everyone may eventually join.*

MIKE NORTH AND SUSAN FISKE, RESEARCHERS

## Ageism

Every society has some form of **ageism**—the systematic persecution and degradation of people because they are old. Ageism is reflected in negative stereotypes of the elderly, such as they are slow, they don't change their ways, they are grumpy, they are poor drivers, they can't/don't want to learn new things, they are incompetent, and they are physically and/or cognitively impaired (Nelson 2011). Ageism also occurs when older individuals are treated differently because of their age, such as when they are spoken to loudly in simple language, when it is assumed they cannot understand normal speech, or when they are denied employment due to their age. Another form of ageism—**ageism by invisibility**—occurs when older adults are not included in advertising and educational materials. Ageism is similar to sexism, racism, and heterosexism. The elderly are shunned, discriminated against in employment, and sometimes victims of abuse. Ageism is more likely among those who feel anxious about the negative consequences of aging, and those who lack compassion (Boswell 2012).

*But you do change and the more you accept change and embrace change, the better.*

DIANE KEATON, *LET'S JUST SAY IT WASN'T PRETTY*

**Ageism** the systematic persecution and degradation of people because they are old.

**Ageism by invisibility** when older adults are not included in advertising and educational materials.

Media not only emphasize youth and beauty (Haboush et al. 2012) but also contribute to the negative image of the elderly who are portrayed as difficult, complaining, and burdensome. Some individuals with ageist attitudes attempt to distance themselves from being regarded as old by engaging in risk behaviors (e.g., alcohol, drugs, cigarettes) (Popham et al. 2011).

**Gerontophobia** a shared fear or dread of the elderly.

Negative stereotypes and media images of the elderly engender **gerontophobia**—a shared fear or dread of the elderly. In a study of 1,422 elderly ages 65–70, those who were of low-economic status, living alone, had multiple chronic medical conditions, and were depressed were most likely to have a negative self-perception of aging (Moser et al. 2011). Such a negative view may create a self-fulfilling prophecy. For example, an elderly person forgets something and attributes forgetting to age. A younger person, however, is unlikely to attribute forgetfulness to age, given cultural definitions surrounding the age of the onset of senility.

*To me, fair friend, you never can be old,*

*For as you were when first your eye I eyed,*

*Such seems your beauty still.*

SHAKESPEARE, *SONNET*, 104, 1

The negative meanings associated with aging underlie the obsession of many Americans to conceal their age by altering their appearance. With the hope of holding on to youth a little bit longer, aging Americans spend billions of dollars each year on plastic surgery, exercise equipment, hair products, facial creams, and Botox injections.

**Gerontology** the study of aging.

**Filial piety** love and respect toward parents including bringing no dishonor to parents and taking care of elderly parents.

## Theories of Aging

**Gerontology**, the study of aging, has various theories to help in its understanding (North and Fiske 2012). Table 16.3 identifies several theories, the level (macro or micro) of the theory, the theorists typically associated with the theory, assumptions, and criticisms. As noted, there are diverse ways of conceptualizing the elderly.

### DIVERSITY IN OTHER COUNTRIES

Eastern cultures emphasize **filial piety**, which is love and respect toward their parents. Filial piety involves respecting parents, bringing no dishonor to parents, and taking good care of parents. Western cultures are characterized by filial responsibility emphasizing duty, protection, care, and financial support to one's parents.

**TABLE 16.3 Theories of Aging**

| Name of Theory | Level of Theory | Theorists | Basic Assumptions | Criticisms |
|---|---|---|---|---|
| Disengagement | Macro | Elaine Cumming<br>William Henry | The gradual and mutual withdrawal of the elderly and society from each other is a natural process. It is also necessary and functional for society that the elderly disengage so that new people can be phased in to replace them in an orderly transition. | Not all people want to disengage; some want to stay active and involved. Disengagement does not specify what happens when the elderly stay involved. |
| Activity | Macro | Robert Havighurst | People continue the level of activity they had in middle age into their later years. Though high levels of activity are unrelated to living longer, they are related to reporting high levels of life satisfaction. | Ill health may force people to curtail their level of activity. The older a person, the more likely the person is to curtail activity. |
| Conflict | Macro | Karl Marx<br>Max Weber | The elderly compete with youth for jobs and social resources such as government programs (Medicare). | The elderly are presented as disadvantaged. Their power to organize and mobilize political resources such as the American Association of Retired Persons is underestimated. |
| Age stratification | Macro | M. W. Riley | The elderly represent a powerful cohort of individuals passing through the social system that both affect and are affected by social change. | Too much emphasis is put on age, and little recognition is given to other variables within a cohort such as gender, race, and socioeconomic differences. |
| Modernization | Macro | Donald Cowgill | The status of the elderly is in reference to the evolution of the society toward modernization. The elderly in premodern societies have more status because what they have to offer in the form of cultural wisdom is more valued. The elderly in modern technologically advanced societies have low status because they have little to offer. | Cultural values for the elderly, not level of modernization, dictate the status of the elderly. Japan has high respect for the elderly and yet is highly technological and modernized. |
| Symbolic | Micro | Arlie Hochschild | The elderly socially construct meaning in their interactions with others and society. Developing social bonds with other elderly can ward off being isolated and abandoned. Meaning is in the interpretation, not in the event. | The power of the larger social system and larger social structures to affect the lives of the elderly is minimized. |
| Continuity | Micro | Bernice Neugarten | The earlier habit patterns, values, and attitudes of the individual are carried forward as a person ages. The only personality change that occurs with aging is the tendency to turn one's attention and interest on the self. | Other factors than one's personality affect aging outcomes. The social structure influences the life of the elderly rather than vice versa. |
| Interpersonal | Micro | Palmore Langlois | Negative assumptions based on physical apperance (droopy eyes means sad person) | Some elderly "in shape" |

## Care Giving for the Frail Elderly—the "Sandwich Generation"

About 20 million Americans provide care to their frail parents (Leopold et al. 2014). An elderly parent is defined as **frail** if he or she has difficulty with at least one personal care activity or other activity related to independent living; the severely disabled are unable to complete three or more personal care activities. These personal care activities include bathing, dressing, getting in and out of bed, shopping for groceries, and taking medications. Most (over 90%) frail elderly do not have long-term health care insurance. Most children choose to take care of their elderly parents. The term *children* typically means female adult children taking care of their mothers (fathers often have a spouse or are deceased) (Leopold et al. 2014). These women provide **family care giving** and are known as the **sandwich generation** because they take care of their parents and their children simultaneously.

**Frail** elderly person who has difficulty with at least one personal care activity or other activity related to independent living.

**Family care giving** adult children providing care for their elderly parents.

**Sandwich generation** generation of adults who are "sandwiched" between caring for their elderly parents and their own children.

Care giving for an elderly parent has two meanings. One form of care giving refers to providing personal help with the basics of daily living, such as getting in and out of bed, bathing, toileting, and eating. A second form of care giving refers to performing instrumental activities, such as shopping for groceries, managing money (including paying bills), and driving the parent to the doctor.

The typical caregiver is a middle-aged married woman/mother who works outside the home. High levels of stress and fatigue may accompany caring for one's elders. Lee et al. (2010) studied family members (average age = 46) providing care for elderly parents as well as children and found that the respective responsibilities resulted in not having enough time for both and in an increase in problems associated with stress. When the two roles were compared, taking care of the parents often translated into missing more workdays. The number of individuals in the sandwich generation will increase for the following reasons:

**1.** ***Longevity.*** The over-85 age group, the segment of the population most in need of care, is the fastest-growing segment of our population.

**2.** ***Chronic disease.*** In the past, diseases took the elderly quickly. Today, diseases such as arthritis and Alzheimer's are associated not with an immediate death sentence but with a lifetime of managing the illness and being cared for by others.

**3.** ***Fewer siblings to help.*** The current generation of elderly have fewer children than the elderly in previous generations. Hence, the number of siblings available to help look after parents is more limited. Children without siblings are more likely to feel the weight of caring for elderly parents alone.

**4.** ***Commitment to parental care.*** Contrary to the myth that adult children in the United States abrogate responsibility for taking care of their elderly parents, most children institutionalize their parents only as a last resort. Indeed, most adult children want to take care of their aging parents either in the parents' own home or in the adult child's home. When parents can no longer be left alone and can no longer cook for themselves, full-time nursing care is sought. Asian children, specifically Chinese children, are socialized to expect to take care of their elderly in the home.

**5.** ***Lack of support for the caregiver.*** Caring for a dependent, aging parent requires a great deal of effort, sacrifice, and decision making on the part of more than 20 million adults in the United States who are challenged with this situation. The emotional toll on the caregiver may be heavy. Guilt (over not doing enough), resentment (over feeling burdened), and exhaustion (over the relentless care demands) are common feelings.

Some people reduce the strain of caring for an elderly parent by arranging for home health care. This involves having a nurse go to the home of a parent and provide such services as bathing and giving medication. Other services include taking meals to the elderly (e.g., through Meals on Wheels). The National Family Caregiver Support Program provides support services for

*When she comes to retrieve me [in the nursing home], after the tan-colored pudding with edible oil topping has sat for a while and been removed. . . .*

SARA GRUEN, *WATER FOR ELEPHANTS*

individuals (including grandparents) who provide family care-giving services. Such services include elder-care resource and referral services, caregiver support groups, and classes on how to care for an aging parent. In addition, states increasingly provide family caregivers with a tax credit or deduction.

Offspring who have no help may become overwhelmed and frustrated. Elder abuse, an expression of such frustration, is not unheard of (we discussed elder abuse in Chapter 10). Many wrestle with the decision to put their parents in a nursing home or other long-term care facility. We discuss this issue in the following Personal Choices section.

## PERSONAL CHOICES

### Should I Put My Aging Parents in a Long-Term Care Facility?

Over 1.8 million individuals are in a nursing home. (*Statistical Abstract of the United States, 2012–2013,*Table 73). Factors relevant in deciding whether to care for an elderly parent at home, arrange for nursing home care, or provide another form of long-term care include the following (these factors are often in flux and involve a constant reassessment of the elder care decision; Szinovacz and Davey, 2013):

**1. *Level of care needed.*** As parents age, the level of care that they need increases. An elderly parent who cannot bathe, dress, prepare meals, or be depended on to take medication responsibly needs either full-time in-home care or a skilled nursing facility that provides 24-hour nursing supervision by registered or licensed vocational nurses. Commonly referred to as "nursing homes" or "convalescent hospitals," these facilities provide medical, nursing, dietary, pharmacy, and activity services.

An intermediate-care facility provides eight hours of nursing supervision per day. Intermediate care is less extensive and less expensive and generally serves patients who are ambulatory and who do not need care throughout the night.

A skilled nursing facility for special disabilities provides a protective or security environment to people with mental disabilities. Many of these facilities have locked areas where patients reside for their own protection.

An assisted living facility is for individuals who are no longer able to live independently but who do not need the level of care that a nursing home provides. Although nurses and other health care providers are available, assistance is more typically in the form of meals and housekeeping.

Housing involves a range of options, from apartments where residents live independently to skilled nursing care. These housing alternatives allow older adults to remain in one place and still receive the care they need as they age. In addition, adult children may relocate to be near their aging parents (Leopold et al. 2014; Zhang et al. 2012).

**2. *Temperament of parent.*** Some elderly parents have become paranoid, accusatory, and angry with their caregivers. Family members no longer capable of coping with the abuse may arrange for their parents to be taken care of in a nursing home or other facility. One mother screamed at her son, "You wish I were dead so you would not have to mess with me."

**Filial responsibility** emphasizes duty, protection, care, and financial support for one's parents.

**3. *Philosophy of adult child.*** Most children feel a sense of **filial responsibility**—a sense of personal obligation for the well-being of their aging parents. Theoretical explanations for such responsibility include the norm of reciprocity (adult children reciprocate the care they received from their parents), attachment theory (caring results from positive emotions for one's parents), and a moral imperative (caring for one's elderly parents is the right thing to do).

**4. *Length of time for providing care.*** Time in dependent care can be extensive. Ten or more years is not unusual. The mother of the first author was dependent for 15 years until her death at 90.

**5.** ***Cost.*** Professional care giving in a nursing home is expensive—$75,000 is the average annual cost (Jackson 2011; updated for 2015).

**Medicare**, a federal health insurance program for people 65 and older, was developed for short-term acute hospital care. Medicare generally does not pay for long-term nursing care. In practice, adult children who arrange for their aging parent to be cared for in a nursing home end up paying for it out of the elder's own funds. After all of these economic resources are depleted, **Medicaid**, a state welfare program for low-income individuals, will pay for the cost of care. A federal law prohibits offspring from shifting the assets of an elderly parent so as to become eligible for Medicaid.

**Medicare** federal health insurance program for people 65 and older.

**Medicaid** state welfare program for low-income individuals.

A crisis in care for the elderly is looming. In the past, women have taken care of their elderly parents. However, these were women who lived in traditional families where one paycheck took care of a family's economic needs. Women today work out of economic necessity, and quitting work to take care of an elderly parent is becoming less of an option. As more women enter the labor force, less free labor is available to take care of the elderly. Government programs are not in place to take care of the legions of elderly Americans. Who will care for them when both spouses are working full time?

**6.** ***Other issues.*** Aside from the living situation of the frail elderly parent, the elderly person or those with power of attorney should complete a document called an **advance directive** (also known as a living will), detailing the conditions under which life support measures should be used (do they want to be sustained on a respirator?). These decisions, made ahead of time, spare the adult children the responsibility of making them in crisis contexts and give clear directives to the medical staff in charge of the elderly person. For example, elderly people can direct that a feeding tube should not be used if they become unable to feed themselves. Hence, by making this decision, the children are spared the decision regarding a feeding tube.

**Advance directive (living will)** details for medical care personnel the conditions under which life support measures should be used for one's partner.

A **durable power of attorney**, which gives adult children complete authority to act on behalf of the elderly, is also advised. These documents also help to save countless legal hours, time, and money for those responsible for the elderly. (Appendixes D and E of this text present examples of the living will and durable power of attorney.) Other end-of-life issues include a last will and testament, funeral or burial preplanning (cremation?), and organ and tissue donation (Kelly et al. 2013).

**Durable power of attorney** gives adult children complete authority to act on behalf of the elderly.

Finally, adult children may consider buying long-term care insurance (LTCI) to cover what Medicare does not. Many private health care plans do not cover "non-medical" day-to-day care such as bathing or meals for an Alzheimer's parent, as well as nursing home costs. Costs of LTCI begin at about $1,000 a year but vary a great deal (including over $1,000 a month) depending on the age and health of the insured individual.

## Sources


Jackson, M. 2011. Nursing home placement and caregiver burden. Poster, National Council on Family Relations Annual Meeting, Orlando, November.

Kelly, C. M., J. L. Master, and S. DeViney. 2013. End of life planning activities: An integrated process. *Death Studies* 37: 529–551.

Leopold, T., M. Raab, and H. Engelhardt. 2014. The transition to parent care: Costs, commitments, and caregiver selection among children. *Journal of Marriage and the Family* 76: 300–318.

*Statistical Abstract of the United States, 2012–2013.* 131th ed. Washington, DC: U.S. Bureau of the Census.

Szinovacz, M. E. and A. Davey. 2013. Changes in adult children's participation in parent care. *Ageing and Society* 33: 667–697.

Zhang, Y., M. Engelman, and E. M. Agree. 2012. Moving considerations: A longitudinal analysis of parent–child residential proximity for older Americans. *Research on Aging* online, Sept. 2012.

## Issues Confronting the Elderly

**Age discrimination** older people are often not hired and younger workers are hired to take their place.

Numerous issues become concerns as people age. In middle age, the issues are early retirement (sometimes forced), job layoffs (recession-related cutbacks), **age discrimination** (older people are often not hired and younger workers are hired to take their place), separation or divorce from a spouse, and adjustment to children leaving home. For some in middle age, grandparenting is an issue if they become the primary caregiver for their grandchildren. As couples move from the middle to the later years, the issues become more focused on income, health, retirement, and sexuality. See the Self-Assessment in regard to the level of worry.

**SELF-ASSESSMENT** | Family Member Well-Being (FMWB)

**Purpose**

To measure the adjustment of family members in terms of concern about health, tension, energy, cheerfulness, fear, anger, sadness, and general concerns.

**Directions**

Read each of the eight statements and note the words at each end of the 0 to 10 scale describe opposite feelings. Select the number along the bar which seems closest to how you have generally felt during the past month.

1. How concerned or worried about your health have you been? (During the past month)

Not CONCERNED at all — Very CONCERNED

1 2 3 4 5 6 7 8 9

2. How relaxed or tense have you been? (During the past month)

Very RELAXED — Very TENSE

1 2 3 4 5 6 7 8 9

3. How much energy, pep, vitality have you felt? (During the past month)

No energy at all. LISTLESS — Very energetic. DYNAMIC

1 2 3 4 5 6 7 8 9

4. How depressed or cheerful have you been? (During the past month)

Very DEPRESSED — Very CHEERFUL

1 2 3 4 5 6 7 8 9

5. How afraid have you been? (During the past month)

Not AFRAID — Very AFRAID

1 2 3 4 5 6 7 8 9

6. How angry have you been? (During the past month)

Not ANGRY at all — Very ANGRY

1 2 3 4 5 6 7 8 9

7. How sad have you been? (During the past month)

Not SAD at all — Very SAD

1 2 3 4 5 6 7 8 9

8. How concerned or worried about the health of another family member have you been?(During the past month)

Not CONCERNED at all — Very CONCERNED

1 2 3 4 5 6 7 8 9

**Scoring**

Reverse score items 1, 2, 5, 6, 7, and 8 so that so that 1 = 9, 9 = 1, etc. Then add each of the eight numbers. Higher scores reflect more positive family well-being. The scale was completed by large samples including 297 investment executives and 234 spouses of investment executives, Midwest farm families involving 411 males and 389 females, 813 rural bank employees plus 448 of their spouses, and 524 male members of military families and 465 female members of military families. The means of the FMWB range from 37.5 (SD = 9.1) for Caucasian female members of military families to 45.5 (SD = 10.5) for spouses of investment executives.

**Source**

McCubbin, H. I. and A. I. Thompson (eds.). 1991. Family Member Well-Being (FMWB). In H. I. McCubbin, A. I. Thompson, and M. A. McCubbin (1996). *Family assessment resiliency, coping and adaptation. Inventories for research and practice.* Madison, WI: University of Wisconsin. 753–782. Call the University of Wisconsin, Family Stress Coping, Coping and Health Project at 608-262-5070 for information.

## Income

According to Clarence Kehoe, executive director of accounting firm Anchin, Block & Anchin, the number one regret of retirees is that they did not save enough money for retirement—that they spent more than they should have during their peak earning years (Brown 2014). Indeed, for most elderly, the end of life is characterized by reduced income. Social Security and pension benefits, when they exist, are rarely equal to the income a retired person formerly earned. Many elderly continue working since they can't afford to quit.

*Anyone who lives within their means suffers from a lack of imagination.*

OSCAR WILDE, PLAYWRIGHT

### National **Data**

The median annual income of men aged 65 and older is $25,877; women, $15,282 (*Statistical Abstract of the United States, 2012–2013,* Table 702).

Elderly women are particularly disadvantaged because their out-of-home employment has often been discontinuous, part time, and low-paying. Social Security and private pension plans favor those with continuous, full-time work histories. Hence, their retirement incomes are considerably lower than the retirement income of males.

## Physical Health

There is a gradual deterioration in physical well-being as one ages (Proulx and Ermer 2013). Park-Lee et al. (2013) reported on national data of 8,875 old-old individuals (85 years or older) in home health care, nursing homes, or hospice. Over two-thirds needed assistance in performing three or more activities of daily living (ADLs) and were bladder incontinent. Hypertension and heart disease were the two most common chronic health conditions. Reported physical health also varies by race/ethnicity. Park et al. (2013) compared elderly (65 and older) white, African Americans, Cuban, and non-Cuban Hispanics and found that racial/ethnic minority older adults rated their health more poorly.

David Knox

This 80-year-old man is 60 feet up cutting down this tree.

## FAMILY POLICY | Physician-Assisted Suicide for Terminally Ill Family Members

Paraschakis et al. (2012) studied the elderly in Greece and noted that they had the highest suicide rate of any age group (elderly suicide victims represented 35% of the total suicides). A similar association of aging and suicide has been found in Japan (Ishii et al. 2013). Predictors of suicide for the elderly include being male, over age 75, psychiatric history, depression, and physical illness. The majority (82%) of the elderly suicide victims were successful in their first suicide attempt. The elderly may also ask spouses or adult children for help in ending their life. Requests for physician-assisted suicide also occurs.

**Euthanasia** is from the Greek words meaning "good death," or dying without suffering. Euthanasia may be passive, where medical treatment is withdrawn and nothing is done to prolong the life of the patient, or active, which involves deliberate actions to end a person's life.

Adult children or spouses are often asked their recommendations about withdrawing life support (food, water, or mechanical ventilation), starting medications to end life (intravenous vasopressors), or withholding certain procedures that would prolong life (cardiopulmonary resuscitation). Researchers have found that there is a preference for using life-prolonging treatments among those in the worst health conditions (Winter and Parks 2012).

**Euthanasia** from the Greek meaning "good death" or dying without suffering; either passively or actively ending the life of a patient.

**quality of life** one's physical functioning, independence, economic resources, social relationships, and spirituality

One's aging parents may experience a significant drop in **quality of life**—defined in terms of physical functioning, independence, economic resources, social relationships, and spirituality (Willson 2007). These parents may also not want to be a burden to their children (Schaffer 2007), and may ask for death. The top reasons patients cited for wanting to end their lives are losing autonomy (84%), decreasing ability to participate in activities they enjoyed (84%), and losing control of bodily functions (47%) (Chan 2003).

Physician-assisted suicide (PAS) is legal in the Netherlands and Switzerland. Georges (2008) found that nearly half of a group of general practitioners wanted to avoid physician-assisted suicide because it was against their own personal values or because it was an emotional burden for them to confront the issue. Douglas (2008) observed a "double effect" of sedatives and analgesics administered at the end of life (e.g., morphine)—not only do these relieve pain but can also hasten death. Some physicians find that slow euthanasia is more psychologically acceptable to doctors than active voluntary euthanasia by injection.

All fifty states now have laws for living wills, permitting individuals to decide (or family members to decide on their behalf) to withhold artificial nutrition (food) and hydration (water) from a patient who is wasting away. In practice, this means not putting in a feeding tube. For elderly, frail patients who may have a stroke or heart attack, do-not-resuscitate (DNR) orders may also be put in place. The Supreme Court has ruled that state law will apply in regard to physician-assisted suicide. In January 2006, the Supreme Court ruled that Oregon has a right to physician-assisted suicide. Its Death with Dignity Act requires that two physicians must

*There's an old joke—um . . . two elderly women are at a Catskill mountain resort, and one of 'em says, "Boy, the food at this place is really terrible." The other one says, "Yeah, I know; and such small portions." Well, that's essentially how I feel about life—full of loneliness, and misery, and suffering, and unhappiness, and it's all over much too quickly.*

WOODY ALLEN, *ANNIE HALL*

Good physical health is the single most important determinant of an elderly person's reported happiness. Weight has an effect on one's health as one ages. Felding et al. (2011) studied 1,600 sedentary elderly individuals and found that inability to walk 400 meters was associated with negative outcomes which could include higher morbidity, mortality, hospitalizations, and a poorer quality of life (depression). Although most (over 90%) of the elderly do not exercise, walking at least 20 minutes a day results in physical and cognitive benefits for them (Aoyagi and Shephard 2011). Johnson et al. (2014) reconfirmed the association between greater activity and higher quality of life.

Good physical health has an effect on marital quality. Iveniuk et al. (2014) noted that wives with husbands in fair or poor physical health were more likely to report higher levels of marital conflict, but the reverse was not true. The authors suggested that men with compromised health have limited leverage (in terms of reduced resources and status) to resist changes expected by their wives, so the wives are asked to give in with fewer rewards.

Body image/physical appearance are also concerns of the elderly. Roy and Payette (2012) reviewed the literature on body image among Western seniors agree that the patient is terminally ill and is expected to die within six months; the patient must ask three times for death both orally and in writing; and the patient must swallow the barbiturates themselves rather than be injected with a drug by the physician. The state of Washington also allows the terminally ill patient to swallow barbiturates.

The official position of the American Medical Association (AMA) is that physicians must respect the patient's decision to forgo life-sustaining treatment but that they should not participate in patient-assisted suicide: "PAS is fundamentally incompatible with the physician's role as healer." Rather, the AMA affirms physicians who support life. Arguments against PAS emphasize that people who want to end the life of those they feel burdened by (or worse, they want to inherit the elder's money) can abuse the practice and that, because physicians make mistakes, what is diagnosed as "terminal" may not in fact be terminal.

A concern about PAS is the potential to misuse the law. However, a Dutch study of 5,000 requests per year of euthanasia and physician-assisted suicide in general practice over 25 years concluded, "Some people feared that the lives of increasing numbers of patients would end through medical intervention, without their consent and before all palliative options were exhausted. Our results, albeit based on requests only, suggest that this fear is not justified" (Marquet et al. 2003, p. 202).

**Your Opinion?**

1. Suppose your father has Alzheimer's disease and is in a nursing home. He is 88 and no longer recognizes you. He has stopped eating. Would you have a feeding tube inserted to keep him alive?
2. To what degree do you agree with the Death with Dignity policy operative in Oregon?
3. What do you think the position of the government should be in regard to physician-assisted suicide?

### Sources


Douglas, C., I. Kerridge, and R. Ankeny. 2008. Managing intentions: End of life administration of analgesics and sedatives, and the possibility of slow euthanasia. *Bioethics* 22: 388–402.

Georges, J. J. 2008. Dealing with requests for euthanasia: A qualitative study investigating the experience of general practitioners. *Journal of Medical Ethics* 34:150–63.

Marquet, R., A. Bartelds, G. J. Visser, P. Spreeuwenberg, and I. Peters. 2003. Twenty-five years of requests for euthanasia and physician-assisted suicide in Dutch practice: Trend analysis. *British Medical Journal* 327: 201–202.

Paraschakis, A., A. Douzenis, I. Michopoulos, C. Christodoulou, K. Vassilopoulou, F. Koutsaftis, and L. Lykouras. 2012. Late onset suicide: Distinction between "young-old" vs. old-old" suicide victims. How different populations are they? *Archives of Gerontology & Geriatrics* 54: 136–139.

Schaffer, M. 2007. Ethical problems in end of life decisions for elderly Norwegians. *Nursing Ethics* 14: 242–257.

Willson, A. E. 2007. The sociology of aging. In *21st century sociology: A reference handbook*, ed. Clifton D. Bryant and Dennis L. Peck, 148–155. Thousand Oaks, California: Sage.

Winter, L. and S. M. Parks. 2012. Elders' preferences for life-prolonging treatment and their proxies's substituted judgment. *Journal of Aging and Health*. 24: 1157–1178.


WHAT IF?

## What If Your Spouse Says "No" to Your Mother Living with You?

As parents age and one spouse dies (usually the father), an older adult married child (usually the daughter) may need to take the remaining parent into her home. When a spouse is adamantly against such a change in living space, some alternatives should be explored: get siblings involved so that the widowed parent spends some time with each adult child, share expenses with siblings for the cost of a nursing home facility, or share the cost with siblings of hiring a full-time person to live in with the widowed parent. If none of these alternatives are acceptable, a marriage therapist may be needed to explore other ways of moving past this impasse.

and found that while most were not as concerned about physical appearance as youth, they did feel captive to their aging bodies (there was a discrepancy in their inner self and outward appearance). Women experienced more dissatisfaction about their physical appearance/bodies than did men.

**DIVERSITY IN OTHER COUNTRIES**

In a cross-cultural study of 21,000 adults, ages 40 to 80, in 21 countries, key ingredients associated with reported happiness were good health, decent standard of living, genes that predispose individuals to being optimistic, having happy parents, and continued involvement in a meaningful activity (e.g., taking care of a grandchild, volunteering, and employment) (Coombes 2007).

## Mental Health

Mental health may worsen for some elderly. Mood disorders, with depression being the most frequent, are more common among the elderly. Scheetz et al. (2012) found higher rates of depression among centenarians. Bereavement over the death of a spouse, loneliness, physical illness, and institutionalization may be the culprits. Foreclosures,

## SELF-ASSESSMENT | Alzheimer's Quiz

| | True | False | Don't Know |
|---|---|---|---|
| 1. Alzheimer's disease can be contagious. | ____ | ____ | ____ |
| 2. People will almost certainly get Alzheimer's disease if they live long enough. | ____ | ____ | ____ |
| 3. Alzheimer's disease is a form of insanity. | ____ | ____ | ____ |
| 4. Alzheimer's disease is a normal part of getting older, like gray hair or wrinkles. | ____ | ____ | ____ |
| 5. There is no cure for Alzheimer's disease at present. | ____ | ____ | ____ |
| 6. A person who has Alzheimer's disease will experience both mental and physical decline. | ____ | ____ | ____ |
| 7. The primary symptom of Alzheimer's disease is memory loss. | ____ | ____ | ____ |
| 8. Among people older than age 75, forgetfulness most likely indicates the beginning of Alzheimer's disease. | ____ | ____ | ____ |
| 9. When the husband or wife of an older person dies, the surviving spouse may suffer from a kind of depression that looks like Alzheimer's disease. | ____ | ____ | ____ |
| 10. Stuttering is an inevitable part of Alzheimer's disease. | ____ | ____ | ____ |
| 11. An older man is more likely to develop Alzheimer's disease than an older woman. | ____ | ____ | ____ |
| 12. Alzheimer's disease is usually fatal. | ____ | ____ | ____ |
| 13. The vast majority of people suffering from Alzheimer's disease live in nursing homes. | ____ | ____ | ____ |
| 14. Aluminum has been identified as a significant cause of Alzheimer's disease. | ____ | ____ | ____ |
| 15. Alzheimer's disease can be diagnosed by a blood test. | ____ | ____ | ____ |
| 16. Nursing-home expenses for Alzheimer's disease patients are covered by Medicare. | ____ | ____ | ____ |
| 17. Medicine taken for high blood pressure can cause symptoms that look like Alzheimer's disease. | ____ | ____ | ____ |

Answers: 1–4, 8, 10, 11, 13–16 are False; remaining items are True.

### Source

Neal E. Cutler, Boettner/Gregg Professor of Financial Gerontology, Widener University. Originally published in 1987, in *Psychology Today*, 20th Anniversary Issue, "Life Flow: A Special Report—The Alzheimer's Quiz," 21(5): 89, 93. Reprinted with permission from *Psychology Today* magazine, © 1987 Sussex Publishers, LLC. The scale was completed by 69 undergraduates at East Carolina University in 1998. Of the respondents, 40% identified less than 50% of the items correctly.

associated with depression, during the recession were particularly difficult for the elderly (Cagney et al. 2014). Insomnia in the elderly may also be a precursor to depression (Nadorff et al. 2013).

**Dementia**, which includes Alzheimer's disease, is the mental disorder most associated with aging. Worldwide, 36 million individuals live with dementia like-disease (Hogsnes et al. 2014). The Self-Assessment feature on page 468 reflects some of the misconceptions about Alzheimer's disease.

There has been considerable cultural visibility in regard to the medical use of marijuana. Ahmend et al. (2014) reported data on older adults in regard to its use with Alzheimer's patients; behavioral disturbances and nighttime agitation were reduced.

Hogsnes et al. (2014) interviewed 11 spouses of persons with dementia before and after relocating them to a nursing home. Feelings of shame and guilt and feelings of isolation preceded relocating the spouse to a nursing home. The event which triggered the move was threats and physical violence (some with a knife) directed toward the caring partner. After relocating the spouse to a nursing home partners described feelings of both guilt and freedom, living with grief, feelings of loneliness in the spousal relationship, and striving for acceptance despite a lack of completion.

*There's no point in comforting words, in telling her she'll be all right. She's no fool. Her hand reaches out and I clutch it like a lifeline. As if it's me who's dying instead of Rue.*

SUZANNE COLLINS, *THE HUNGER GAMES*

*The advantage of a bad memory is that one enjoys several times the same good things for the first time.*

FRIEDRICH NIETZSCHE, GERMAN PHILOSOPHER

**Dementia** the mental disorder most associated with aging.

## Divorce

Middle-aged and older adults don't always live "happily ever after." According to Brown and Lin (2012), there is a Gray Divorce Revolution occurring in that the divorce rate for those age 50 and older doubled from 1990 to 2009 (now one in four divorces or 600,000 annually). Blacks, formerly married, non college, and those married less than ten years were most likely to divorce. Reasons include that older people are more likely to be remarried and this group is more likely to divorce, greater acceptance of divorce (e.g., friends likely to be divorced), and wife more likely to be economically independent. Other reasons for later life divorce may include that children are grown (less concern about the effect of divorce on them), healthy (time left for a second life), prenuptial agreements (limit one's economic liability to spouse being divorced), and dating sites (find a new partner quickly).

Some spouses are literally dumped as they age. According to son and biographer Peter Ford, Cynthia Hayword, the third wife of Glenn Ford (movie star of the fifties) enjoyed being on his arm for 11 years of traveling the world and parties with the rich and famous in Hollywood. When he became ill in his mid-eighties, she checked him into a dependency care facility (Pasadena Las Encinas Hospital) and divorced him (Ford 2011).

*Want to know who is living a long healthy life? Ask one question, "Is the person married?"*

AMY RAUER, RAND CORPORATION

## Retirement

Retirement represents a rite of passage through which most elderly pass. Pond et al. (2010) interviewed 60 individuals from 55 to 70 who revealed their reasons for retirement. These included poor health and the "maximization of life." The latter referred to retiring while they were healthy and could enjoy/fulfill other life goals. Another reason was "health protection"—decisions motivated by health protection and promotion. Being able to retire since they could afford to do so was also a reason.

The retirement age in the United States for those born after 1960 is 67. Individuals can take early retirement at age 62, with reduced benefits. Retirement affects an individual's status, income, privileges, power, and prestige. One retiree noted that he was being waited on by a clerk who looked at his name on his check, thought she recognized him and said, "Didn't you use to be somebody?"

People least likely to retire are unmarried, widowed, single-parent women who need to continue working because they have no pension or even Social Security benefits—if they don't work or continue to work, they will have no income, so retirement is not an option. Some workers experience what is called **blurred retirement**

*The trouble with retirement is that you never get a day off.*

ABE LEMONS, OKLAHOMA BASKETBALL COACH

**Blurred retirement** individual works part time before completely retiring or takes a "bridge job" that provides a transition between a lifelong career and full retirement.

**Phased retirement** an employee agrees to a reduced work load in exchange for reduced income.

rather than a clear-cut one. A blurred retirement means the individual works part time before completely retiring or takes a "bridge job" that provides a transition between a lifelong career and full retirement. Others may plan a **phased retirement** whereby an employee agrees to a reduced work load in exchange for reduced income. This arrangement is beneficial to both the employee and employer.

Wang et al. (2011) identified five variables associated with enjoying retirement—individual attributes (e.g., physical and mental health/financial stability), preretirement job-related variables (e.g., escape from work stress/job demands), family-related variables (e.g., being happily married), retirement-transition-related issues (e.g., planned voluntary retirement), and postretirement activities (e.g., bridge employment, volunteer work). Indeed, individuals who have a positive attitude toward retirement are those who have a pension waiting for them, are married (and thus have social support for the transition), have planned for retirement, are in good health, and have high self-esteem. Price and Nesteruk (2013) emphasized that identifying interests to share in retirement is important for couple satisfaction.

*I had a huge list of things to do in retirement, while my husband thought retirement was an endless vacation. But there is still laundry, grocery shopping, cleaning to be done. After a year, we came to a balance of work and entertainment that was acceptable to both of us.*

WIFE IN THE PRICE AND NESTERUK (2013) RESEARCH

Paul Yelsma is a retired university professor. For a successful retirement, he recommends the following:

- Just like going to high school, the military, college, or your first job, know that the learning curve is very sharp. Don't think retirement is a continuation of what you did a few years earlier. Learn fast or you will be bored. For every 10 phone calls you make expect about 1 back; your status has dwindled and you have less to offer those who want something from you.
- Have at least three to five hobbies that you can do almost any time and any place. Feed your hobbies or they will die.
- Having several things to look forward to is so much more exciting than looking backward. If you looked backward to high school while in college, your life would be boring.
- Develop a new physical exercise program that you want to do. Don't expect others to support your activities. The "pay off" in retirement is vastly different from what colleagues expected of you.
- Find new friends who have time to share (colleagues are often too busy).
- Retirement is a NEW time to live and NEW things have to happen.
- Travel—we have traveled to seven new countries in the past few years.

Some retired individuals volunteer—giving back their time and money to attack poverty, illiteracy, oppression, crime, and so on. Examples of volunteer organizations are the Service Corps of Retired Executives, known as SCORE (www.score.org), Experience Works (www.experienceworks.org), and Generations United (www.gu.org).

*Retirement: It's nice to get out of the rat race, but you have to learn to get along with less cheese.*

GENE PERRET, WRITER/PRODUCER

## Retirement Communities—Niche Aging

The stereotype of the elderly residing in an assisted living facility has been replaced by the concept of "niche aging" whereby retirement communities are set up for those with a particular interest (Barovick 2012). For those into country music, there is a community in Franklin, Tennessee, which offers not only the array of housing alternatives but also recording studios and performance venues; for Asian Americans, there is Aegis Gardens in Freemont, California; for gay and lesbian retirees, there is Fountaingrove Lodge in Santa Rosa, California. There are now 100 such communities including those set up close to universities so individuals can continue to take classes (Oberlin and Dartmouth). Some adults seek these contexts and check in as early as age 40.

*Sex at age 90 is like trying to shoot pool with a rope.*

GEORGE BURNS, COMEDIAN

## Sexuality

There are numerous changes in the sexuality of women and men as they age (Vickers 2010). Frequency of intercourse drops from about once a week for those 40 to 59 to once every six weeks for those 70 and older. Changes in men include a decrease in the size of the penis from an average of 5 to about 4 ½ inches. Elderly men also become more easily aroused by touch rather than visual stimulation, which was arousing when they were younger. Erections take longer to achieve, are less rigid, and it takes longer for the man to recover so that he can have another erection. "It now takes me all night to do what I used to do all night" is the adage aging men become familiar with.

Levitra, Cialis, and Viagra (prescription drugs that help a man obtain and maintain an erection) are helpful for about 50% of men. Others with erectile dysfunction may benefit from a pump that inflates two small banana-shaped tubes that have been surgically implanted into the penis. Still others benefit from devices placed over the penis to trap blood so as to create an erection.

Women also experience changes including menopause, which is associated with a surge of sexual libido, an interest in initiating sex with her partner, and greater orgasmic capacity. Not only are they free from worry about getting pregnant, but also estrogen levels drop and testosterone levels increase. A woman's vaginal walls become thinner and less lubricating; the latter issue can be resolved by lubricants like KY Jelly.

Table 16.4 describes these and other physiological sexual changes that occur as individuals age.

Some spouses are sexually inactive. Karraker and DeLamater (2013) analyzed data on 1,502 men and women ages 57 to 85 and found that 29% reported no sexual activity for the past 12 months or more. The longer the couple had been married, the older the spouse, and the more compromised the health of the spouse, the more likely the individual was to report no sexual activity.

**TABLE 16.4 Physiological Sexual Changes as Individuals Age**

| Physical Changes in Sexuality as Men Age |
|---|
| 1. Delayed and less firm erection. |
| 2. More direct stimulation needed for erection. |
| 3. Extended refractory period (12 to 24 hours before rearousal can occur). |
| 4. Fewer expulsive contractions during orgasm. |
| 5. Less forceful expulsion of seminal fluid and a reduced volume of ejaculate. |
| 6. Rapid loss of erection after ejaculation. |
| 7. Loss of ability to maintain an erection for a long period. |
| 8. Decrease in size and firmness of the testes. |
| **Physical Changes in Sexuality as Women Age** |
| 1. Reduced or increased sexual interest. |
| 2. Possible painful intercourse due to menopausal changes. |
| 3. Decreased volume of vaginal lubrication. |
| 4. Decreased expansive ability of the vagina. |
| 5. Possible pain during orgasm due to less flexibility. |
| 6. Thinning of the vaginal walls |
| 7. Shortening of vaginal width and length. |
| 8. Decreased sex flush, reduced increase in breast volume, and longer post orgasmic nipple erection. |

Source: Adapted from Janell L. Carroll. *Sexuality Now: Embracing Diversity*, 4th ed., p. 273. © 2012. Wadsworth, a part of Cengage Learning, Inc. Reproduced by permission. www.cengage.com/permissions.

*My intention– for the remaining years of my life– is to embrace my aging body, honoring its limitations, challenging its capabilities, and celebrating its accomplishments.*

PAM LEWIS, MENTAL HEALTH CONSULTANT

*The aging process has you firmly in its grasp if you never get the urge to throw a snowball.*

DOUG LARSON, NEWSPAPER COLUMNIST

*To keep a fire burning brightly there's one easy rule: Keep the logs together, near enough to keep warm and far enough apart for breathing room. Good fire, good marriage, same rule.*

MARNIE REED CROWEL, POET

Syme et al. (2013) reported on the sexuality of adults aged 63 to 67. Factors associated with lack of sexual satisfaction included spouse in poor health, history of diabetes, and fatigue. Elderly women were less sexually satisfied than men.

As noted above, the most sexually active are in good health. Diabetes and hypertension are major causes of sexual dysfunction. Incontinence (leaking of urine) is particularly an issue for older women and can be a source of embarrassment. The most frequent sexual problem for men is erectile dysfunction; for women, the most frequent sexual problem is the lack of a partner.

### Successful Aging

There is considerable debate about the meaning of successful aging. Torres and Hammarström (2009) noted that at least three definitions have been used: absence of physical problems/disability (no diseases or cognitive impairment), presence of strategies to cope with aging that results in positive emotions, sense of well-being, and not feeling lonely, and having a positive view of aging.

Researchers who worked on the Landmark Harvard Study of Adult Development (Valliant 2002) followed 824 men and women from their teens into their eighties and identified those factors associated with successful aging. Not smoking was "probably the single most significant factor in terms of health." Smokers who quit before age 50 were as healthy at 70 as those who had never smoked. Other behaviors associated with successful aging included developing a positive view of life and life's crises, avoiding alcohol and substance abuse, and maintaining healthy weight and exercising daily. In regard to exercise, Boyes (2013) identified the value of outdoor adventure for the elderly. In interviews and surveys of 80 elderly individuals, he emphasized the physical (healthier), social (connections with other), and psychological (stronger sense of well-being) benefits of involvement in an outdoor adventure program. Physical exercise also helps to promote cognitive function (Wendell et al. 2014). Having a happy marriage was also important for successful aging. Indeed, those who were identified as "happy and well" were six times more likely to be in a good marriage than those who were identified as "sad and sick."

A positive attitude is also critical to aging successfully. One of Clarke and Bennett's (2013) respondents (75-year-old woman with back problems, bursitis, lupus, and arthritis) said:

> *If you don't have a good attitude, that's it. You know? If you're feeling sorry for yourself and you're sitting there thinking, "Oh this is awful today. My knee hurts. I can't walk. I can't do this." Then sure you're going to sit there and you're going to feel worse.*

There are also gender differences in how men and woman age. Men feel a keen sense of loss of control and their increasing dependence on women. Women feel dismayed at their altered physical appearance and how their illness will impact others (Clarke and Bennett 2013).

## Relationships and the Elderly

*In youth the days are short and the years are long; in old age the years are short and the days long.*

NIKITA IVANOVICH PANIN, RUSSIAN STATESMAN

Relationships in the later years vary. Some elderly men and women are single and date.

### Dating

Brown and Shinohara (2013) studied the dating behavior of 3,005 individuals ages 57–85 and found that 14% of singles were in a dating relationship. Dating was more common among men than women and declined with age. Compared to non-daters, daters were more socially advantaged, college educated, and had more assets, better health, and more social connectedness. Some elderly seek dating partners through Internet sites which cater to older individuals. Ourtime.com is an example.

Alterovitz and Mendelsohn (2013) noted the importance of health in seeking a partner for the elderly. In their study of 450 personal ads, they compared the middle aged (40–45), young-old (60–74), and old-old (75–+) and found that the two younger cohorts focused on a partner for adventure, romance, sex, and a soul mate; the older group was more likely to mention the importance of a healthy partner. We (authors of your text) know of an example where a woman in her sixties met a man in his sixties via Match.com. On their second date she asked him to walk with her, which ended up being 1.5 miles. She later said "I was testing to see how healthy he was . . . if he could not walk a mile and a half it would have been our last date."

### Use of Technology to Maintain Existing Relationships

Youth regularly text friends and romantic partners throughout the day. Madden (2010) noted that older adults and senior citizens are increasing their use of technology to stay connected as well. Over 40% of adults over the age of 50 use e-mail. And almost half (47%) of Internet users 50–64 and 25% of users 65 and older use social networking sites such as Facebook. Indeed these older adults and seniors view themselves among the Facebooking and LinkedIn masses. These figures have doubled since 2009.

*Love moderately; long love doth so*

*Too swift arrives as tardy as too slow.*

**SHAKESPEARE, *ROMEO AND JULIET*, II, IV, 14**

Use of Twitter has not caught on among older adults and senior citizens, with only 11% of online adults ages 50–64 reporting such usage. E-mail and online news are the most frequent Internet behaviors of this age group.

### Relationships between Elderly Spouses

Marriages that survive into late life are characterized by little conflict, considerable companionship, and mutual supportiveness. All but one of the 31 spouses over age 85 in the Johnson and Barer (1997) study reported "high expressive rewards" from their mate. Walker and Luszcz (2009) reviewed the literature on elderly couples and found marital satisfaction related to equality of roles and marital communication. Health may be both improved by positive relationships and decreased by negative relationships. Rauer (2013) reported on 64 older couples (married an average of 42 years) and emphasized that being married in late life was the best of all contexts in terms of social/emotional, economic, and behavioral resources (e.g., taking care of each other). She noted that taking care of a spouse actually benefits the caregiver with feelings of self-esteem (but the outcome for the spouse being cared for may be negative since he or she may feel dependent).

Only a small percentage (8%) of individuals older than 100 are married. Most married centenarians are men in their second or third marriage. Many have outlived some of their children. Marital satisfaction in these elderly marriages is related to a high frequency of expressing love feelings to one's partner. Though it is assumed that spouses who have been married for a long time should know how their partners feel, this is often not the case. Telling each other "I love you" is very important to these elderly spouses.

*Love is wiping someone's ass and changing the sheets when they have wet themselves so you can both have enough dignity to keep going.*

**LANDLORD TO HESTOR IN THE MOVIE *THE DEVIL AND THE DEEP BLUE SEA***

## Grandparenthood

**National Data**

Almost 8% (7.8%) of children grow up in three-generation households. Almost half (45%) of single parents live in three-generation households at some time (Pilkauskas 2012).

*Grandchildren don't stay young forever, which is good because Pop-pops have only so many horsey rides in them.*

**GENE PERRET, WRITER/PRODUCER**

Another significant role for the elderly is that of grandparent. Among adults aged 40 and older who had children, close to 95% are grandparents and most have, on average, five or six grandchildren. College students report that their grandparents express affection to them via concern/interest, gifts, interactions, verbal, nonverbal and support/encouragement (Mansson 2012). Factors influencing the quality of the grandparent/grandchild relationship include distance, age of the child/parent, and relationship the grandparent has with the parents (Dunifon and Bajracharya 2012).

Some grandparents actively take care of their grandchildren full time, provide supplemental help in a multigenerational family, assist on an occasional part-time basis, or occasionally visit their grandchildren. Grandparents see themselves as caretakers, emotional and/or economic resources, teachers, and historical connections on the family tree.

## National **Data**

About 4% of children live with their grandparents or other relatives (Carr and Springer 2010).

### Styles of Grandparenting

Grandparents also have different styles of relating to grandchildren. Whereas some grandparents are formal and rigid, others are informal and playful, and authority lines are irrelevant. Still others are surrogate parents providing considerable care for working mothers and/or single parents.

Bob Sammons

The role of most grandfathers is to play with their grandchildren.

Drew and King (2012) surveyed 289 individuals 18–25 in regard to their relationship with their grandparents and confirmed most were closer to their grandmother but satisfied with both sets of relationships. While some grandparents are close with their grandchildren, others are distant and show up only for special events like birthdays and holidays. E-mail and web cams help grandparents stay connected to their grandchildren.

Finally, the quality of the grandparent-grandchild relationship can be affected by the parents' relationship to their own parents. If a child's parents are estranged from their parents, it is unlikely that the child will have an opportunity to develop a relationship with the grandparents. Divorce often shatters the relationship grandparents have with their grandchildren (Doyle et al. 2010).

### Effect of Divorce on the Grandparent-Child Relationship

Divorce often has a negative effect on grandparents seeing their grandchildren. The situation most likely to produce this outcome is when the children are young, the wife is granted primary custody, and the relationship between the spouses is adversarial.

Some grandparents are not allowed to see their grandchildren. In all 50 states, the role of the grandparent has limited legal and political support. By a vote of six to three, the Supreme Court (Troxel v. Granville) sided with the parents and virtually denied 60 million grandparents the right to see their grandchildren. The court viewed parents as having a fundamental right to make decisions about with whom their children could spend time—including keeping them from grandparents. However, some courts have ruled in favor of grandparents. Step grandparents have no legal rights to their step grandchildren.

### Benefits to Grandchildren

Of a sample of 4,535 university students, 83% agreed with the statement, "I have a loving relationship with my grandmother" (72% reported having had a loving relationship with their grandfather) (Hall and Knox 2013). Grandchildren report enormous benefits from having a close relationship with grandparents, including development of a sense of family ideals, moral beliefs, and a work ethic. Feeling loved is also a major benefit. The poem below reflects the love of a grandmother

*Have children while your parents are still young enough to take care of them.*

RITA RUDNER, COMEDIAN

***A Grandma Kind of Love****

*What kind of love is bigger than the sky above?*
*It's that very, very, very, special kind of Grandma love.*

*What kind of love is sweeter than honey? finer than fine? better than best?*
*You're right if "Grandma's love" is what you guessed!*

*What kind of love is playful and fun?*
*It's Grandma's love that shines like the sun.*

*What kind of love sticks like glue?*
*It's the love that Grandma has for you.*

*What kind of love stays sure and strong?*
*It's Grandma's love that lasts so long.*

*What kind of love is off the chart?*
*It's the love that comes from Grandma's heart.*

*What kind of love is always there?*
*It's the love that you and Grandma share.*

*Will Grandma's love ever end?*
*The answer to that is "never"*
*Because Grandma's special kind of love*
*is a love that lasts forever.*

*Caroline Schacht

## The End of Life

**Thanatology** is the examination of the social dimensions of death, dying, and bereavement (Bryant 2007). The end of one's life sometimes involves the death of one's spouse.

**Thanatology** the examination of the social dimensions of death, dying, and bereavement.

*The facts of life and death remain the same. We live and die, we love and grieve, we breed and disappear. And between these essential gravities, we search for meaning, save our memories, leave a record for those who will remember us.*

THOMAS LYNCH, *BODIES IN MOTION AND AT REST*

### Death of One's Spouse

The death of a spouse is one of the most stressful life events a person experiences. Compared to both the married and the divorced, the widowed are the most lonely and have the lowest life satisfaction (Ben-Zur 2012; Hensley 2012).

This 87-year-old husband is telling his 86-year-old wife of 65 years that he loves her.

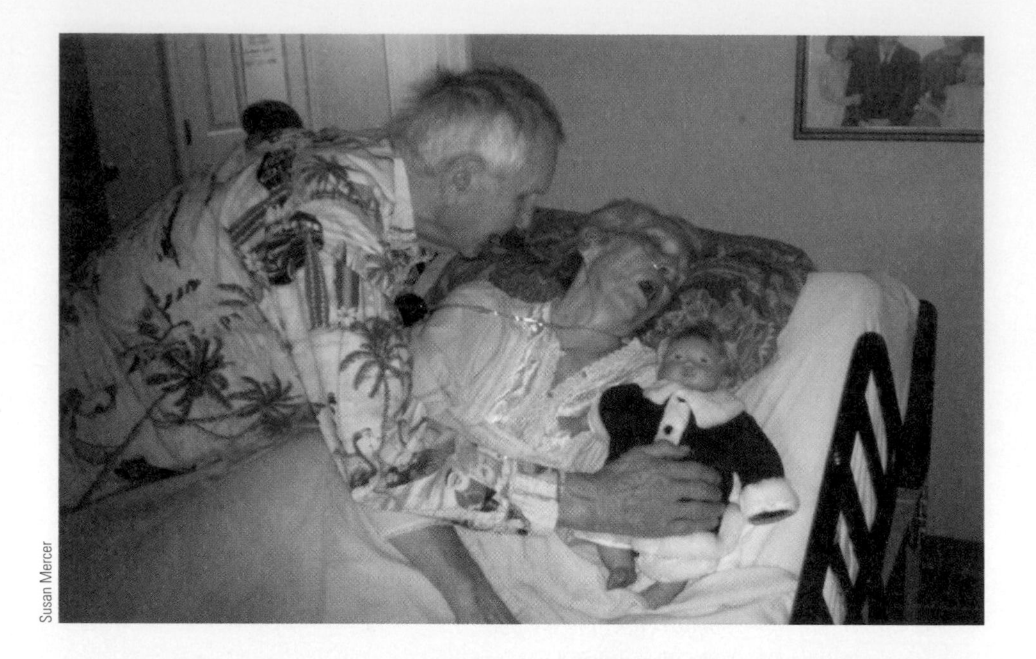

Susan Mercer

*If you get hit by a car, nothing could be better than modern medical technology. But if you are in your eighties suffering from dementia, diabetes, and colon cancer, you are unfortunate to have access to these technologies since they can only prolong your life beyond the point where it has any pleasure or meaning.*

KATY BUTLER, *KNOCKING ON HEAVEN'S DOOR*

*Grief makes one hour ten.*

SHAKESPEARE, ENGLISH PLAYWRIGHT.

*Life does not cease to be funny when people die any more than it ceases to be serious when people laugh.*

GEORGE BERNARD SHAW

Because women tend to live longer than men, they are more likely to experience the widowed role.

Although individual reactions and coping mechanisms for dealing with the death of a loved one vary, several reactions to death are common. These include shock, disbelief and denial, confusion and disorientation, grief and sadness, anger, numbness, physiological symptoms such as insomnia or lack of appetite, withdrawal from activities, immersion in activities, depression, and guilt. Eventually, surviving the death of a loved one involves the recognition that life must go on, the need to make sense out of the loss, and the establishment of a new identity.

Women and men tend to have different ways of reacting to and coping with the death of a loved one. Women are more likely than men to express and share feelings with family and friends and are also more likely to seek and accept help, such as attending grief support groups. Initial responses of men are often cognitive rather than emotional. From early childhood, males are taught to be in control, to be strong and courageous under adversity, and to be able to take charge and fix things. Showing emotions is labeled weak.

Men sometimes respond to the death of their spouse in behavioral rather than emotional ways. Sometimes they immerse themselves in work or become involved in physical action in response to the loss. For example, a widower immersed himself in repairing a beach cottage he and his wife had recently bought. Later, he described this activity as crucial to getting him through those first two months. Another coping mechanism for men is the increased use of alcohol and other drugs.

Women's response to the death of their husbands may necessarily involve practical considerations. Johnson and Barer (1997) identified two major problems of widows—the economic effects of losing a spouse and the practical problems of maintaining a home alone. The latter involves such practical issues as cleaning the gutters, painting the house, and changing the air filters throughout the house.

Whether a spouse dies suddenly or after a prolonged illness has an impact on the reaction of the remaining spouse. The sudden death is associated with being less at peace with death and more angry at life. Death of one's spouse is also associated with one's own death—the widowed are more likely to die sooner than a spouse with a living partner (Simeonova 2013).

The age at which one experiences the death of a spouse is also a factor in one's adjustment. People in their 80s may be so consumed with their own health and disability that they have little emotional energy to grieve. But even after the spouse's death, the emotional relationship with the deceased may continue. Some widows and widowers report a feeling that their spouses are with them for years after the death of their beloved. Some may also dream of their deceased spouses, talk to their photographs, and remain interested in carrying out their wishes. Such continuation of the relationship may be adaptive by providing meaning and purpose for the living or maladaptive in that it may prevent the surviving spouse from establishing new relationships. Research is not consistent on the degree to which individuals are best served by continuing or relinquishing the emotional bonds with the deceased (Stroebe & Schut 2005).

David Knox

Sometimes there is little time to prepare for death. This man went out to cut firewood in the snow and died of a heart attack.

## Use of Technology and Partner's Death

Just as technology has changed the way relationships begin (e.g., text messaging), its use in memorializing the deceased is increasing. Some spouses are so distraught over the death of their partner that they can't weather a traditional funeral. But they can upload a eulogy and videos to a website where others may pay their respects, post their own photos of the deceased, and blog comments of various experiences with the deceased. One such website for the deceased is http://www.respectance.com/.

*It is often thought that the dead see us, and we assume, whether reasonably or not, that if they see us at all they see us more clearly than before.*

C. S. LEWIS, *A GRIEF OBSERVED*

## Preparing for Death

What is it like for those near the end of life to think about death? To what degree do they go about actually preparing for death? Cheng et al. (2013) discussed anticipatory grief therapy where one comes to terms with his or her own death. The applying social research section emphasizes the value of exercise in reducing one's depression and delaying one's own death.

Johnson and Barer (1997) interviewed 48 individuals with an average age of 93 to find out their perspectives on death. Most interviewees were women (77%); of those, 56% lived alone, but 73% had some sort of support from their children or from one or more social support services. The following findings are specific to those who died within a year after the interview.

*We all get sick and we all die.*

ROGER EBERT, FILM CRITIC, DIED IN 2013 OF CANCER

**Thoughts in the Last Year of Life** Most of the respondents had thought about death and saw their life as one that would soon end. Most did so without remorse or anxiety. With their spouses and friends dead and their health failing, they accepted death as the next stage in life. They felt like the last leaf on the tree. Some comments follow:

## APPLYING SOCIAL RESEARCH | Effect of Exercise on Depression and Mortality of the Elderly*

To what degree does exercise in one's late 70s influence the chance of being depressed and of living longer? This was the question a team of researchers attempted to answer.

**Sample and Methods**

Six hundred and twenty-four noninstitutionalized elders (mean age = 77.35) from the Americans' Changing Lives Longitudinal Study were assessed in terms of three exercise levels—gardening, walking, sport—and their reported depression. Six years later the mortality of the respondents was assessed to see if the various levels of exercise had any influence.

**Findings**

Each 1-standard-unit increase on the physical *inactivity* scale significantly predicted adjusted 29%, 30%, and 33% increased risk of depression for gardening, walking, and sport, respectively. Hence not gardening, not walking, and not engaging in sporting activities was associated

*The strange thing about growing old is that the ultimate identification with the here and now is slowly lost. One feels transposed into infinity, more or less, alone.*

ALBERT EINSTEIN

*After all the bills are paid, the shouting and the tumult dies, the band stops playing and all the dancers have gone home, what do we have to show for our lives?*

JOHNNY MERCER, LYRICIST

*You don't stop doing things because you grow old—you grow old because you stop doing things.*

ANONYMOUS

*If I die tomorrow, it would be all right. I've had a beautiful life, but I'm ready to go.*
*My husband is gone, my children are gone, and my friends are gone.*
*That's what is so wonderful about living to be so old. You know death is near and you don't even care.*
*I've just been diagnosed with cancer, but it's no big deal. At my age, I have to die of something. (Johnson and Barer 1997, p. 205)*

The major fear these respondents expressed was not the fear of death but of the dying process. Dying in a nursing home after a long illness is a dreaded fear. Sadly, almost 60% of the respondents died after a long, progressive illness. They had become frail, fatigued, and burdened by living. They identified dying in their sleep as the ideal way to die. Some hastened their death by no longer taking their medications; others wished they could terminate their own life. Competent adults have the legal right to refuse or discontinue medical interventions. For incompetent individuals, decisions are made by a surrogate—typically a spouse or child (McGowan 2011).

**Behaviors in the Last Year of Life** Aware that they are going to die, most simplify their life, disengage from social relationships, and leave final instructions. In simplifying their life, they sell their home and belongings and move to smaller quarters. One 81-year-old woman sold her home, gave her car away to a friend, and moved into a nursing home. The extent of her belongings became a chair, a lamp, and a TV.

Some divorce just before they die. Upon learning he had terminal cancer, actor Dennis Hopper filed for divorce from his fifth wife. One explanation for this behavior is that divorce removes a spouse from automatically getting part of a deceased spouse's estate and allows control of dispensing one's assets (often to one's own children) while one is alive.

Disengaging from social relationships is selective. Some maintain close relationships with children and friends but others "let go." They may no longer send out Christmas cards and stop sending letters. Phone calls become the source of social connections. Some leave final instructions in the form of a will or handwritten note expressing wishes of where to be buried, handling costs associated with disposal of the body (e.g., cremation), and directives about pets. One of with an increased risk of depression. Also, not gardening or walking were associated with an increased risk of death. No significant effects were found for not engaging in sport activities.

**Implications**

It is clear that being sedentary has negative consequences for both mental health and physical well-being (delaying death).

**Source**

Lee, P., W. Lan, and C. C. Lee. 2012. Physical activity related to depression and predicted mortality risk: Results from the Americans' changing lives study. *Educational Gerontology* 38: 678–690.

Johnson and Barer's (1997) respondents left $30,000 to specific caregivers to take care of several pets (p. 204).

Elderly who may have counted on children to take care of them may find that their children have scattered because of job changes, divorce, or both, and may be unavailable for support. Some children simply walk away from their parents or leave their care to their siblings. The result is that the elderly may have to fend for themselves.

*. . . all stories, if continued far enough, end in death, and he is no true story-teller who would keep that from you.*

ERNEST HEMINGWAY, *DEATH IN THE AFTERNOON*

One of the last legal acts of the elderly is to have a will drawn up. Stone (2008) emphasized how wills may stir up sibling rivalry (one sibling may be left more than another), be used as a weapon against a second spouse (by leaving all of one's possessions to one's children), or reveal a toxic secret (name a mistress and several children as heirs).

Some elderly dictate how they wish to be buried. A new option is the "green burial" where the individual is returned to the earth without being embalmed, no casket, etc. It is a way for the soon to be deceased to "return to nature" (Kelly 2012).

Some live full lives up until the very end. Ted Kennedy died of brain cancer at age 77. Up until the last week of his life, he was active in politics (he had sponsored over 300 bills in his Senate career). His words resonate . . . "For all those whose cares have been our concern, the work goes on, the cause endures, the hope still lives, and the dream must never die."

# The Future of the Elderly in the United States

The elderly will increase in number and political clout. By 2030, 30% of the U.S. population will be over the age of 55 (the percentage is now 21). The challenges of old age will be the same—coping with dwindling income, declining health, and the death of loved ones. On the positive side, greater attention will be paid to the health needs of the elderly. Medicare will help pay for some of the medical needs. However, it alone will be inadequate and other private sources will be needed.

*In life, and in sports, we all know that nothing lasts forever.*

PEYTON MANNING, NFL QUARTERBACK

## Summary

***What is meant by the terms age and ageism?***

Age is defined chronologically (by time), physiologically (by capacity to see, hear, and so on), psychologically (by self-concept), sociologically (by social roles), and culturally (by the value placed on elderly). Ageism is the denigration of the elderly, and gerontophobia is the dreaded fear of being elderly. Theories of aging range from disengagement (individuals and societies mutually disengage from each other) to continuity (the habit patterns of youth are continued in old age).

***What is the "sandwich generation"?***

Eldercare combined with child care is becoming common among the sandwich generation—adult children responsible for the needs of both their parents and their children. Two levels of eldercare include help with personal needs such as feeding, bathing, and toileting as well as instrumental care such as going to the grocery store, managing bank records, and so on. Members of the sandwich generation report feelings of exhaustion over the relentless demands, guilt over not doing enough, and resentment over feeling burdened.

Deciding whether to arrange for an elderly parent's care in a nursing home requires attention to a number of factors, including the level of care the parent needs, the philosophy and time availability of the adult child, and the resources of the adult children. Full-time professional nursing care (not including medication) costs an average of $75,000 annually.

Elderly parents who are dying from terminal illnesses incur enormous medical bills. Some want to die and ask for help. Our society continues to wrestle with physician-assisted suicide and euthanasia. Only Oregon and Washington state currently allow for physician-assisted suicide.

***What issues confront the elderly?***

Issues of concern to the elderly include income, mental health, physical health, divorce, retirement, retirement communities, and sexuality. One's health becomes a primary focus for the elderly. Good health is the single most important factor associated with an elderly person's perceived life satisfaction. Mental problems may also occur with mood disorders; depression is the most common.

Though the elderly are thought to be wealthy and living in luxury, most are not. The median household income of people over the age of 65 is less than half of what they earned in the prime of their lives. The most impoverished elderly are those who have lived the longest, who are widowed, and who live alone. Women are particularly disadvantaged because their work history may have been discontinuous, part-time, and low-paying. Social Security and private pension plans favor those with continuous, full-time work histories.

For most elderly women and men, sexuality involves lower reported interest, activity, and capacity. Inability to have an erection and the absence of a sexual partner are the primary sexual problems of elderly men and women, respectively.

Not smoking is "probably the single most significant factor in terms of health." Exercise is one of the most beneficial activities the elderly can engage in to help them maintain good health. Other factors associated with successful aging include a positive view of life and life's crises, avoiding alcohol and substance abuse, continuing to educate oneself, and having a happy marriage.

***What are relationships between elderly spouses like?***

Elderly spouses emphasize that being married in late life is the best of all contexts in terms of emotional, social, and economic well being. Taking care of a spouse may actually benefit the caregiver with feelings of self-esteem.

***What is grandparenthood like?***

There is considerable variation in role definition and involvement. Whereas some delight in seeing their lineage carried forward in their grandchildren and provide emotional and economic support, others are focused on their own lives or on their own children and relate formally and at a distance to their grandchildren. When grandparents are involved in their lives, grandchildren benefit in terms of positive psychological and economic benefits.

***What are some end-of-life issues?***

The death of a spouse is one of the most stressful life events a person ever experiences. Compared to both the married and divorced, the widowed are the most lonely and have the lowest life satisfaction. Because women tend to live longer than men, they are more likely to experience the widowed role.

## Key Terms

Advance directive (living will)
Age
Age discrimination
Ageism
Ageism by invisibility
Blurred retirement
Dementia
Durable power of attorney
Euthanasia
Family care giving
Filial piety
Filial responsibility
Frail
Gerontology
Gerontophobia
Medicaid
Medicare
Phased retirement
Quality of life
Sandwich generation
Thanatology

## Web Links

AARP (American Association of Retired Persons)
http://www.aarp.org

Calculate Your Life Expectancy
http://www.livingto100.com

Channing House Retirement
http://www.channinghouse.org/

ElderSpirit Community
http://www.elderspirit.net

ElderWeb
http://www.elderweb.com

Foundation for Grandparenting
http://www.grandparenting.org/
Generations United http://www.gu.org

National Family Caregiver Support Program
http://www.agingcarefl.org/caregiver/NationalSupport

Senior Corps
http://www.seniorcorps.gov/

Senior Sex
http://www.holisticwisdom.com/senior-sex.htm

Silver Sage Village
http://www.silversagevillage.com

# Careers in Marriage and the Family

David Knox

*Pleasure in the job puts perfection in the work.*

ARISTOTLE, PHILOSOPHER

Students who take courses in marriage and the family sometimes express an interest in working with people and ask what careers are available if they major in marriage and family studies. In this Special Topics section, we review some of these career alternatives, including family life education, marriage and family therapy, child and family services, and family mediation. These careers often overlap, so you might engage in more than one of these at the same time. For example, you may work in family services but participate in family life education as part of your job responsibilities.

For all the careers discussed in this section, having a bachelor's degree in a family-related field such as family science, sociology, or social work is helpful. Family science programs are the only academic programs that focus specifically on families and approach working with people from a family systems perspective. These programs have many different names, including child and family studies, human development and family studies, child development and family relations, and family and consumer sciences. Marriage and family programs are offered through sociology departments; family service programs are typically offered through departments of social work as well as through family science departments. Whereas some jobs are available at the bachelor's level, others require a master's or a Ph.D. degree. More details on the various careers available to you in marriage and the family follow.

## Family Life Education

Family life education (FLE) is an educational process that focuses on prevention and on strengthening and enriching individuals and families. Family life educators empower family members by providing them with information that will help prevent problems and enrich their family well-being. This education may be offered to families in different ways: a newsletter, one on one, online, or through a class or workshop. Examples of family life education programs include parent education for parents of toddlers through a child-care center, a brown-bag lunch series on balancing work and family in a local business, a premarital or marriage enrichment program at the local church, a class on sexuality education in a high school classroom, or a workshop on family finance and budgeting at a local community center. The role of family life educator involves making presentations in a variety of settings, including schools, churches, and even prisons. Family life educators may also work with military families on military bases, within the business world with human resources or employee assistance programs, and within social service agencies or cooperative extension programs. Some family life educators develop their own business providing family life education workshops and presentations.

To become a family life educator, you need a minimum of a bachelor's degree in a family-related field such as family science, sociology, or social work. You can become a certified family life educator (CFLE) through the National Council on Family Relations (NCFR). The CFLE credential offers you credibility in the field and shows that you have competence in conducting programs in all areas of family life education. These areas are individuals and families in societal contexts, internal dynamics of the family, human growth and development, interpersonal relationships, human sexuality, parent education and guidance, family resource management, family law and public policy, and ethics. In addition, you must show competence in planning, developing, and implementing family life education programs.

Your academic program at your college or university may be approved for provisional certification. If you follow a specified program of study at your school, you may be eligible for a provisional CFLE certification. Once you gain work experience, you can then apply for full certification.

## Marriage and Family Therapy

Whereas family life educators help prevent the development of problems, marriage and family therapists help spouses, parents, and family members resolve existing interpersonal conflicts and problems. They treat a range of problems including communication, emotional and physical abuse, substance abuse, and sexual dysfunctions. Marriage and family therapists work in a variety of contexts, including mental health clinics, social service agencies, schools, and private practice. There are about 50,000 marriage and family therapists in the United States and Canada (about 40 percent are members of the American Association for Marriage and Family Therapy).

Currently, all 50 states and the District of Columbia license or certify marriage and family therapists. Although an undergraduate degree in sociology, family studies, or social work is a good basis for becoming a marriage and family therapist, a master's degree and two years of post-graduate supervised clinical work is required to become a licensed marriage and family therapist. Some universities offer accredited master's degree programs specific to marriage and family therapy; these involve courses in marriage and family relationships, family systems, and human sexuality, as well as numerous hours of clinical contact with couples and families under supervision. A list of graduate programs in marriage and family therapy is available at http://www.aamft.org/cgi-shl/twserver.exe?run:COALIST.

Full certification involves clinical experience, with 1,000 hours of direct client, couple, and family contact; 200 of these hours must be under the direction of a supervisor approved by the American Association of Marriage and Family Therapists (AAMFT). In addition, most states require a licensure examination. The AAMFT is the organization that certifies marriage and family therapists. A marriage and family therapist can be found at http://family-marriage-counseling.com/therapists-counselors.htm.

## Child and Family Services

In addition to family life educators and marriage and family therapists, careers are available in agencies and organizations that work with families, often referred to as social service agencies. The job titles within these agencies include family interventionist, family specialist, and family services coordinator. Your job responsibilities in these roles might involve helping your clients over the telephone, coordinating services for families, conducting intake evaluations, performing home visits, facilitating a support group, or participating in grant-writing activities. In addition, family life education is often a large component of child and family services. You may develop a monthly newsletter, conduct workshops or seminars on particular topics, or facilitate regular educational groups.

Some agencies or organizations focus on helping a particular group of people. If you are interested in working with children, youth, or adolescents, you might find a position with Head Start, youth development programs such as the Boys and Girls Club, after-school programs (for example, pregnant or parenting teens), child-care resource or referral agencies, or early intervention services. Child-care resource and referral agencies assist parents in finding child care, provide training for child-care workers, and serve as a general resource for parents and for child-care providers. Early intervention services focus on children with special needs. If you work in this area, you might work directly with children or you might work with the families and help to coordinate services for them.

Other agencies focus more on specific issues that confront adults or families as a whole. Domestic violence shelters, family crisis centers, and employee-assistance programs are examples of employment opportunities. In many of these positions, you will function in multiple roles. For example, at a family crisis center, you might take calls on a crisis hotline, work one-on-one with clients to help them find resources and services, and offer classes on sexual assault or dating violence to high school students.

Another focus area in which jobs are available is gerontology. Opportunities include those within residential facilities such as assisted-living facilities or nursing homes, senior centers, organizations such as the Alzheimer's Association, or agencies such as National Association of Area Agencies on Aging. There is also a need for eldercare resource and referral, as more and more families find that they have caregiving responsibilities for an aging family member. These families have a need for resources, support, and assistance in finding residential facilities or other services for their aging family member. Many of the available positions with these types of agencies are open to individuals with a bachelor's degree. However, if you get your master's degree in a program emphasizing the elderly, you might have increased opportunity and will be in a position to compete for various administrative positions.

## Family Mediation

In Chapter 15 on Divorce and Remarriage, we emphasized the value of divorce mediation. This is also known as family mediation and involves a neutral third party who negotiates with divorcing spouses on the issues of child custody, child support, spousal support, and division of property. The purpose of mediation is not to reconcile the partners but to help the couple make decisions about children, money, and property as amicably as possible. A mediator does not make decisions for the couple but supervises communication between the partners, offering possible solutions.

Although some family and divorce mediators are attorneys, family life professionals are becoming more common. Specific training is required that may include numerous workshops or a master's degree, offered at some universities (for example, University of Maryland). Most practitioners conduct mediation in conjunction with their role as a family life educator, marriage and family therapist, or other professional. In effect, you would be in business for yourself as a family or divorce mediator.

Students interested in any of these career paths can profit from obtaining initial experience in working with people through volunteer or internship agencies. Most communities have crisis centers, mediation centers, and domestic abuse centers that permit students to work for them and gain experience. Not only can you provide a service, but you can also assess your suitability for the "helping professions," as well as discover new interests. Talking with people already in the profession is also a good way to gain new insights. Your instructor may already be in the marriage and family profession you would like to pursue or be able to refer you to someone who is.

*Note:* Appreciation is expressed to Sharon Ballard, Ph.D., CFLE, for the development of this Special Topic section. Dr. Ballard is Chair of Child Development and Family Relations at East Carolina University. She is also a certified family life educator through the National Council on Family Relations.

For more information:

Bourgeois, M. 2013. How to become a certified family life educator. National Council on Family Relations Annual Meeting, San Antonio.

# SPECIAL TOPIC 2 Contraception*

Chelsea Curry

* Appreciation is expressed to Beth Credle Burt for her review/editing/contribution to the content of this special topic.

Once individuals have decided on whether and when they want children, they need to make a choice about contraception. All contraceptive practices have one of two common purposes: to prevent the male sperm from fertilizing the female egg or to keep the fertilized egg from implanting itself in the uterus. About five to seven days after fertilization, pregnancy begins. Although the fertilized egg will not develop into a human unless it implants on the uterine wall, pro-life supporters believe that conception has already occurred.

In selecting a method of contraception, the important issues to consider are pregnancy prevention, STI prevention, opinion of the partner, ease of use, and cost. Often sexual partners use no contraception and live in denial, thinking that "this one time won't end in a pregnancy." In a study of 4,540 undergraduates, 63% reported that the last time they had sexual intercourse, they used a form of birth control (other than withdrawal); thus, almost 40% did not use birth control. Thirty-nine percent used a condom to prevent contracting an STI (almost 60% did not) (Hall and Knox 2013).

With the exception of the male condom, contraceptives have been designed for women, though male non-hormonal contraceptions to impair the effectiveness of sperm are being developed (Nya-Ngatchou and Amory 2013). However, due to the concern by manufacturers that men will be hesitant to buy anything that affects their sperm, these male contraceptives will not reach the marketplace in this decade (Sitruk-Ware et al. 2013).

## Hormonal Methods

Hormonal methods of contraception currently available to women include the "pill," Implanon® or Nexplanon/Implanon NXT®, Depo-Provera® or Depo-subQ Provera 104®, NuvaRing®, and Ortho Evra®.

### Oral Contraceptive Agents (Birth Control Pill)

The birth control pill is the most commonly used method of all the nonsurgical forms of contraception. Although some women who take the pill still become pregnant (about 10%), the pill remains a desirable birth control option.

Oral contraceptives are available in basically two types: the combination pill, which contains varying levels of estrogen and progestin, and the minipill, which is progestin only. Combination pills work by raising the natural level of hormones in a woman's body, inhibiting ovulation, and creating an environment where sperm cannot easily reach the egg.

The second type of birth control pill, the minipill, contains the same progesterone-like hormone found in the combination pill but does not contain estrogen. Progestin-only pills are taken every day with no hormone-free interval. The progestin in the minipill provides a hostile environment for sperm and does not allow implantation of a fertilized egg in the uterus; unlike the combination pill, however, the minipill does not always inhibit ovulation. For this reason, the minipill is somewhat less effective than other types of birth control pills. The minipill has also been associated with a higher incidence of irregular bleeding.

Neither the combination pill nor the minipill should be taken unless prescribed by a healthcare provider who has detailed information about the woman's previous medical history. Contraindications—reasons for not prescribing birth control pills—include hypertension, impaired liver function, known or suspected tumors that are estrogen dependent, undiagnosed abnormal genital bleeding, pregnancy at the time of the examination, and a history of poor blood circulation or blood clotting. The major complications associated with taking oral contraceptives are blood clots and high blood pressure. Also, the risk of heart attack is increased for those who smoke or have other risk factors for heart disease. If they smoke, women over age 35 should generally use other forms of contraception. The progestin in some birth control pills that contain drospirenone, which can raise potassium levels in the blood, may be linked to a higher risk for blood clots and heart problems than other birth control pills. It is important for women to consult/ inform their doctor if they have ever had disease of the kidneys, liver, or adrenal glands and to make the health care provider aware of all medications she they are taking (Planned Parenthood Federation of America, Inc. 2013).

Although the long-term negative consequences of taking birth control pills are still the subject of research, 25% of all women who use them experience short-term negative effects. These side effects include increased susceptibility to vaginal infections, nausea, slight weight gain, vaginal bleeding between periods, breast tenderness, headaches, and mood changes (some women become depressed and experience a loss of sexual desire). Women should also be aware of situations in which the pill is not effective, such as the first month of use, when combined with certain prescription medications, and when pills are missed. The pill is also not as effective if it is not taken at the same time every day. On the positive side, pill use reduces the incidence of ectopic pregnancy and offers noncontraceptive benefits: regular menses, less dysmenorrheal, and reduced incidence of ovarian and endometrial cancers.

Finally, women should be aware that pill use is associated with an increased incidence of chlamydia and gonorrhea. One reason for the association of pill use and a higher incidence of STIs is that sexually active women who use the pill sometimes erroneously feel that because they are protected from becoming pregnant, they are also protected from contracting STIs. The pill provides no protection against STIs; the only methods that provide some protection against STIs are the male and female barrier methods, especially condoms.

Hannaford (2013) summarized decades of research on OC (oral contraceptives):

> *The accumulated evidence does not suggest that oral contraception causes a major public health problem, indeed there may be important long-term benefits. . . . For most women using the contraceptive pill for birth control, the chances of experiencing adverse effects are greatly outweighed by the very high protection against pregnancy, as well as other short- and possibly long-term health benefits (p. 3).*

## Nexplanon/Implanon NXT or Implanon

There have been various subdermal implants (e.g., Norplant and Jadelle), but only Implanon and its newer replacement Nexplanon/Implanon NXT are currently on the market in the United States. Nexplanon/Implanon NXT is a single, flexible plastic rod implant the size of a matchstick that releases a progestin hormone called etonogestrel. A healthcare provider inserts the implant just under the skin of the inner side of a woman's upper arm and it provides pregnancy protection for up to three years, although it may be removed at any time. The main difference between Implanon and Nexplanon/Implanon NXT, other than an improved applicator for placement, is the small amount of barium sulfate that is added to the newer implant, which allows the implant to be seen by X-ray, computed tomography (CT scan), ultrasound scan, or magnetic resonance imaging (MRI). This helps doctors confirm the implant is correctly in place and also aids them in locating it for removal. The implant contains no estrogen, which makes it a viable contraceptive solution for some women who have contraindications to estrogen use. Irregular bleeding is the most common side effect, especially in the first 6–12 months of use (Michielsen and Merck FDA Approved Patent Labeling 2012).

## Depo-Provera® or Depo-subQ Provera 104®

Depo-Provera, also known as "Depo" and "the shot," is a long-acting, reversible, hormonal contraceptive birth control drug that is injected every three months. Side effects of Depo-Provera include menstrual spotting, irregular bleeding, and some heavy bleeding the first few months of use, although 8 out of 10 women using Depo-Provera will eventually experience amenorrhea, or the absence of a menstrual period. Mood changes, headaches, dizziness, and fatigue have also been observed. Some women report a weight gain of 5–10 pounds. Also, after the injections are stopped, it takes an average of 18 months before the woman will become pregnant at the same rate as women who have not used Depo-Provera.

Depo-Provera has been associated with significant loss of bone density, which may not be completely reversible after discontinuing use. The FDA recommends that Depo-Provera be used for longer than two years only if other methods are inadequate (hence a "black box" warning).

Similarly, Depo-subQ Provera 104® is an injection method that offers a 30% lower hormone dosage than Depo-Provera. While Depo-Provera is injected deep into the muscle, Depo-subQ Provera 104 is injected just beneath the skin. Depo-subQ Provera 104 has similar benefits, risks, and side effects—and the same black box warning for risk of significant loss of bone density, as described for Depo-Provera. However, the lower amount of hormone may mean slightly less progestin-related side effects. Less long-term information is available about its effectiveness although short-term studies show similar results to Depo-Provera. Depo-subQ Provera 104 injection is also FDA approved for the treatment of endometriosis-related pain (Mayo Foundation for Medical Education and Research 1998–2013).

## Vaginal Rings

**NuvaRing®**, which is a soft, flexible, and transparent ring approximately two inches in diameter that is worn inside the vagina, provides month-long pregnancy protection. NuvaRing has two major advantages. One, because the hormones are delivered locally rather than systemically, very low levels are administered (the lowest dose of any of the hormonal contraceptives). Second, unlike oral contraceptives, in which the hormone levels rise and fall depending on when the pill is taken, the hormone level from the NuvaRing remains constant. The NuvaRing is a highly effective contraceptive when used according to the labeling. Out of 100 women using NuvaRing for a year, one or two will become pregnant. This method is self-administered. NuvaRing is inserted into the vagina and is designed to release hormones that are absorbed by the woman's body for three weeks. The ring is then removed for a week, at which time the menstrual cycle will occur; afterward the ring is replaced by a new ring.

## Transdermal Applications

**Ortho Evra®** is a contraceptive transdermal patch that delivers hormones to a woman's body through skin absorption. The contraceptive patch is worn for three weeks (anywhere on the body except the breasts) and is changed on a weekly basis. The fourth week is patch free and during this time the menstrual cycle will occur. Ortho Evra is the only contraceptive patch available in the United States.

An alternative to Ortho Evra is being developed: the **NEA-TDS** (norethindrone acetate transdermal system) is a contraceptive patch worn continuously for 7 days and then replaced with a new patch (rotating sites on the abdomen, buttocks, or hips). This patch delivers 0.4 mg each day. Through daily paper diaries kept by 689 women ages 18–47, Simon et al. (2013) evaluated its use. While application site reactions (5%) and menstrual disturbances (6%) were the most common adverse reactions, the researchers found that there are no "major safety concerns" evident following treatment for up to one year of use and that the NEA patch offers a once-weekly, user-controlled, readily reversible, estrogen-free option for contraception.

### Male Hormonal Methods

There is still no male pill available, anywhere. Research to date suggests the success rate of a male hormonal contraception using injectable testosterone alone in clinical trials is actually high and comparable to methods for women. Current studies are attempting to optimize the method of delivery of the hormones and the progestin to use in combination with testosterone (Wang and Swerdloff 2010).

While efforts to develop hormonal methods for males continue, Campo-Englestein (2013) emphasized that women may not trust men to use these new male contraceptives. Three reasons are cited:

> *First, there is a cultural belief that men have an uncontrollable sex drive, which interferes with their ability to contracept. Second, there is a commonly held idea that men are incompetent in domestic tasks, which impairs their ability to correctly use contraception. Third, there is a social perception that men are not committed to pregnancy prevention, or at least not to the degree that women are (p. 283).*

## Barrier Methods

Some women reject the use of hormonal contraceptives on the basis of not wanting to introduce chemicals into their body monthly. Barrier methods provide an alternative: the male and female condom, spermicides, diaphragm, and cervical cap.

### Male Condom

The condom is a thin sheath made of latex, polyurethane, or natural membranes. Latex condoms, which can be used only with water-based lubricants, have been more popular historically. However, the polyurethane condom, which is thinner but just as durable as latex, is growing in popularity. Polyurethane condoms can be used with oil-based lubricants, are an option for some people who have latex-sensitive allergies, provide some protection against the HIV virus and other sexually transmitted infections, and allow for greater sensitivity during intercourse. Condoms made of natural membranes (sheep intestinal lining) are not recommended because they are not effective in preventing transmission of HIV or other STIs. Individuals are more likely to use condoms with casual than with stable partners. Ma et al. (2013) found that condom use appears to increase the "good" bacteria in the vagina that may protect against bacterial vaginosis as well as HIV.

The condom works by being rolled over and down the shaft of the erect penis before intercourse. When the man ejaculates, sperm are caught inside the condom. When used in combination with a spermicidal lubricant that is placed inside the reservoir tip of the condom as well as a spermicidal or sperm-killing agent that the woman inserts inside her vagina, the condom is a highly effective contraceptive. Care should be taken *not* to use nonoxynol-9 as a contraceptive lubricant because it has been shown to provide no protection against STIs or HIV. In addition, nonoxynol-9 products, such as condoms that have N-9 as a lubricant, should not be used rectally because doing so could *increase* one's risk of getting HIV or other STIs.

Like any contraceptive, the condom is effective only when used properly. It should be placed on the penis early enough to avoid any seminal leakage into the vagina. In addition, polyurethane or latex condoms with a reservoir tip are preferable because they are less likely to break. (Even if the condom has a reservoir tip, air should be squeezed out of the tip as it is being placed on the penis to reduce the chance of breaking during ejaculation.) Such breakage does occur, however. (Tip: If the condom breaks, immediately insert a spermicide into the vagina to reduce the risk of pregnancy.) Finally, the penis should be withdrawn from the vagina immediately after ejaculation, before the man's penis returns to its flaccid state. If the penis is not withdrawn and the erection subsides, semen may leak from the base of the condom into the labia. Alternatively, when the erection subsides, the condom will come off when the man withdraws his penis if he does not hold onto the condom. Either way, the sperm will begin to travel up the vagina to the reproductive tract, and fertilization may occur.

To what degree are both the condom AND hormonal methods used? Goldstein et al. (2013) studied a sample of 1,194 women ages 15–24 who attended public family planning clinics. When the female respondents first came to the clinic, 36% reported condom use and 5% reported using a dual method (hormonal contraception plus condom). Once the women began to use hormonal contraception, there was a reduction in condom use (to 27%), and condom use stayed at this lower level throughout the 12-month study. Women who believed their main partner thought condoms were "very important," regardless of perceived sexually transmitted infection risk or participant's own views of condoms, had higher odds of dual method use.

Whenever a condom is used, a spermicide may also be used. In addition to furnishing extra protection, spermicides also provide lubrication, which permits easy entrance of the condom-covered penis into the vagina. If no spermicide is used and the condom is not of the prelubricated variety, a sterile lubricant (such as K-Y Jelly) may be needed. Vaseline or other kinds of petroleum jelly should not be used with condoms because vaginal infections and/or condom breakage may result. Even though condoms should also be checked for visible damage and for the date of expiration, this is rarely done.

## Female Condom

The female condom resembles the male condom except that it fits in the woman's vagina to protect her from pregnancy, HIV infection, and other STIs. The **female condom** is a lubricated, polyurethane adaptation of the male version. It is about 6½ inches long and has flexible rings at both ends. It is inserted like a diaphragm, with the inner ring fitting behind the pubic bone against the cervix; the outer ring remains outside the body and encircles the labial area. Like the male version, the female condom is not reusable. Female condoms have been approved by the FDA and are being marketed under the brand names Femidom and Reality. The one-size-fits-all device is available without a prescription.

**Female condom** a lubricated, polyurethane adaptation of the male version; it is about 6½ inches long and has flexible rings at both ends.

Female condoms are durable and may not tear as easily as latex male condoms. Some women may encounter some difficulty with first attempts to insert the female condom. A major advantage of the female condom is that, like the male counterpart, it helps protect against transmission of the HIV virus and other STIs, giving women an option for protection if their partner refuses to wear a condom. Placement may occur up to eight hours before use, allowing greater spontaneity.

There are problems with the female condom. It can slip out of the woman's vagina, and it is more expensive when compared to the male condom.

## Vaginal Spermicides

A spermicide is a chemical that kills sperm. Vaginal spermicides come in several forms, including foam, cream, jelly, film, and suppository. Previously, in the

United States, the active agent in most spermicides was nonoxynol-9, which had been recommended for added STI protection. However, research has shown that women who used nonoxynol-9 became infected with HIV at approximately a 50% higher rate than women who used a placebo gel. Hence, nonoxynol-9 as a spermicide actually *increases* HIV and offers no protection against gonorrhea/chlamydia infections. Current data suggest that spermicidal creams or gels should be used with a diaphragm. Spermicidal foams, creams, gels, suppositories, and films may be used alone or with a condom. Nonoxynol-9 should *not* be used.

Spermicides must be applied before the penis enters the vagina, no more than 20 minutes before intercourse. Appropriate applicators are included when the product is purchased. Foam is effective immediately. However, suppositories, creams, or jellies require a few minutes to allow the product to melt and spread inside the vagina (package instructions describe the exact time required). Each time intercourse is repeated more spermicide must be applied. Spermicide must be left in place for at least six–eight hours after intercourse; the vagina should not be douched or rinsed during this period.

One advantage of using spermicides is that they are available without a prescription or medical examination. They also do not manipulate the woman's hormonal system and have few side effects. It was believed that a major noncontraceptive benefit of some spermicides is that they offer some protection against sexually transmitted diseases. However, guidelines for prevention and treatment of STIs from the Centers for Disease Control (CDC) suggest that spermicides are not recommended for STI/HIV protection. Furthermore, the CDC emphasizes that condoms lubricated with spermicides offer no more protection from STIs than other lubricated condoms. Furthermore, spermicidally lubricated condoms may have a shorter shelflife, cost more, and be associated with increased urinary tract infections in women.

## Contraceptive Sponge

The Today contraceptive sponge is a disk-shaped polyurethane device containing spermicide. This small device is dampened with water to activate the spermicide and then inserted into the vagina before intercourse begins. The sponge protects for repeated acts of intercourse for 24 hours without the need for supplying additional spermicide. It cannot be removed for at least 6 hours after intercourse, but it should not be left in place for more than 30 hours. Possible side effects that may occur with use include irritation, allergic reactions, or difficulty with removal; the risk of toxic shock syndrome, a rare but serious infection, is greater when the device is kept in place longer than recommended. The sponge provides no protection from sexually transmitted infections. The Today sponge was taken off the market for 11 years due to manufacturing concerns. These problems were eliminated, and national distribution resumed.

## Intrauterine Device (IUD)

Although not technically a barrier method, the IUD is a structural device that prevents implantation and may interfere with sperm and egg transport. The **intrauterine device**, or IUD, is an object inserted into the uterus by a physician to prevent the fertilized egg from implanting on the uterine wall or to dislodge the fertilized egg if it has already implanted. The IUD is only recommended for women who have had at least one child, are in a mutually monogamous relationship, and have no risk or history of ectopic pregnancy or pelvic inflammatory disease. Two common IUDs sold in the United States are ParaGard (copper IUD) and Mirena (hormonal IUD). The copper IUD is partly wrapped in copper and can remain in the uterus for 10 years. The hormonal IUD contains a supply of progestin, which it continuously releases into the uterus in small amounts; after 5 years,

**Intrauterine device (IUD)** an object inserted into the uterus by a physician to prevent the fertilized egg from implanting on the uterine wall or to dislodge the fertilized egg if it has already implanted.

a new IUD must be inserted (American College of Obstetricians and Gynecologists 2012).

As a result of infertility and miscarriage associated with the Dalkon Shield IUD and subsequent lawsuits against its manufacturer by persons who were damaged by the device, use of all IUDs in the United States declined in the 1980s. However, other IUDs do not share the rates of pelvic inflammatory disease (PID) or resultant infertility associated with the Dalkon Shield. Nevertheless, other manufacturers voluntarily withdrew their IUDs from the U.S. market. The IUD was reintroduced in the United States in 2001. The IUD is often an excellent choice for women who do not anticipate future pregnancies, yet do not wish to be sterilized, or for women who are unable to use hormonal contraceptives. However, it is not recommended for women who have multiple sex partners or whose partner has multiple partners (the IUD does not protect against STIs or HIV).

## Diaphragm

Another barrier method of contraception is the **diaphragm**—a shallow rubber dome attached to a flexible, circular steel spring. Varying in diameter from two–four inches, the diaphragm covers the cervix and prevents sperm from moving beyond the vagina into the uterus. This device should always be used with a spermicidal jelly or cream.

**Diaphragm** a shallow rubber dome attached to a flexible, circular steel spring; it covers the cervix and prevents sperm from moving beyond the vagina into the uterus.

To obtain a diaphragm, a woman must have an internal pelvic examination by a health-care provider, who will select the appropriate size and instruct the woman on how to insert the diaphragm. The woman will be told to apply spermicidal cream or jelly on the inside of the diaphragm and around the rim before inserting it into the vagina (no more than two hours before intercourse). The diaphragm must also be left in place for six–eight hours after intercourse to permit any lingering sperm to be killed by the spermicidal agent.

After the birth of a child, a miscarriage, abdominal surgery, or the gain or loss of 10 pounds, a woman who uses a diaphragm should consult her physician or health-care practitioner to ensure a continued good fit. In any case, the diaphragm should be checked every two years for fit.

A major advantage of the diaphragm is that it does not interfere with the woman's hormonal system and has few, if any, side effects. Also, for those couples who feel that menstruation diminishes their capacity to enjoy intercourse, the diaphragm may be used to catch the menstrual flow for a brief time.

On the negative side, some women feel that use of the diaphragm with the spermicidal gel is a messy nuisance, and it is possible that use of a spermicide may produce an allergic reaction. Furthermore, some partners feel that spermicides make oral genital contact less enjoyable. Finally, if the diaphragm does not fit properly or is left in place too long (more than 24 hours), pregnancy or toxic shock syndrome can result.

## Cervical Cap

The **cervical cap** is a thimble-shaped contraceptive device made of rubber or polyethylene that fits tightly over the cervix and is held in place by suction. FemCap is the only brand of cervical cap available in the United States today. Like the diaphragm, the cervical cap, which is used in conjunction with spermicidal cream or jelly, prevents sperm from entering the uterus. Cervical caps have been widely available in Europe for some time and were approved for marketing in the United States in 1988. The cervical cap cannot be used during menstruation because the suction cannot be maintained. The effectiveness, problems, risks, and advantages are similar to those of the diaphragm. (Planned Parenthood Federation of America Inc. 2013).

**Cervical cap** a thimble-shaped contraceptive device made of rubber or polyethylene that fits tightly over the cervix and is held in place by suction.

# Natural Family Planning Methods

**Natural family planning** refraining from sexual intercourse during the one to two weeks each month when the woman is thought to be fertile.

Also referred to as *periodic abstinence*, *rhythm method*, and *fertility awareness*, **natural family planning** involves refraining from sexual intercourse during the one to two weeks each month when the woman is thought to be fertile. Women who use periodic abstinence must know their time of ovulation and avoid intercourse just before, during, and immediately after that time. Calculating the fertile period involves three assumptions: (1) ovulation occurs on day 14 (plus or minus two days) before the onset of the next menstrual period; (2) sperm typically remain viable for two–three days; and (3) the ovum survives for 24 hours.

The time period during which the woman is fertile can be predicted in four ways: the calendar method, the basal body temperature method, the cervical mucus method, and the hormone-in-urine method. These methods may be used not only to avoid pregnancy, but also to facilitate conception if the woman wants to become pregnant. We provide only basic instructions here for using periodic abstinence as a method of contraception. Individuals considering this method should consult with a trained health-care practitioner for more detailed instruction.

## Calendar Method

The calendar method is the oldest and most widely practiced method of avoiding pregnancy through periodic abstinence. The calendar method allows women to calculate the onset and duration of their fertile period. When using the calendar method to predict when the egg is ready to be fertilized, the woman keeps a record of the length of her menstrual cycles for eight months. The menstrual cycle is counted from day one of the menstrual period through the last day before the onset of the next period. She then calculates her fertile period by subtracting 18 days from the length of her shortest cycle and 11 days from the length of her longest cycle. The resulting figures indicate the range of her fertility period. It is during this time that the woman must abstain from intercourse if pregnancy is not desired.

For example, suppose that during an eight-month period, a woman had cycle lengths of 26, 32, 27, 30, 28, 27, 28, and 29 days. Subtracting 18 from her shortest cycle (26) and 11 from her longest cycle (32), she knows the days that the egg is likely to be in the Fallopian tubes. To avoid getting pregnant, she must avoid intercourse on days 8 through 21 of her cycle.

The calendar method of predicting the "safe" period may be unreliable for two reasons. First, the next month the woman may ovulate at a different time from any of the previous eight months. Second, sperm life varies; they may live up to five days, long enough to meet the next egg in the Fallopian tubes.

## Basal Body Temperature (BBT) Method

The BBT method is based on determining the time of ovulation by measuring temperature changes that occur in the woman's body shortly after ovulation. The basal body temperature is the temperature of the body, at rest, on waking in the morning. To establish her BBT, the woman must take her temperature before she gets out of bed each morning for three months. Shortly before, during, or right after ovulation, the woman's BBT usually rises about 0.4–0.8 degrees Fahrenheit. Some women notice a temperature drop about 12 to 24 hours before it begins to rise after ovulation. Intercourse must be avoided from the time the woman's temperature drops until her temperature has remained elevated for three consecutive days. Intercourse may be resumed on the night of the third day after the BBT has risen and remained elevated for three consecutive days. Advantages include being "natural" and avoiding chemicals. Disadvantages include the higher pregnancy rate for persons using this method.

## Cervical Mucus Method

The cervical mucus method, also known as the *Billings method* of natural family planning, is based on observations of changes in the cervical mucus during the woman's monthly cycle. The woman may observe her cervical mucus by wiping herself with toilet paper. **Spinbarkeit** refers to the slippery, elastic, raw eggwhite consistency of the cervical mucus that becomes evident at the time of ovulation. When the cervical mucus becomes this consistency, it is likely that the woman has ovulated.

**Spinbarkeit** the slippery, elastic, raw eggwhite consistency of the cervical musuc that becomes evident at the time of ovulation.

The woman should abstain from intercourse during her menstrual period because the mucus is obscured by menstrual blood and cannot be observed, and ovulation can occur during menstruation. After menstruation ceases, intercourse is permitted on days when no mucus is present or thick mucus is present in small amounts. Intercourse should be avoided just prior to, during, and immediately after ovulation if pregnancy is not desired. Before ovulation, mucus is cloudy, yellow or white, and sticky. During ovulation, cervical mucus is thin, clear, slippery, and stretchy and resembles raw egg white. This phase is known as the *peak symptom*. During ovulation, some women experience ovulatory pain referred to as **Mittelschmerz**. Such pain may include feelings of heaviness, abdominal swelling, rectal pain or discomfort, and lower abdominal pain or discomfort on either side. Mittelschmerz is useful for identifying ovulation but not for predicting it. Intercourse may resume four days after the disappearance of the peak symptom and continue until the next menses. During this time, cervical mucus may be either clear and watery, or cloudy and sticky, or there may be no mucus noticed at all.

**Mittelschmerz** feelings of heaviness, abdominal swelling, rectal pain or discomfort, and lower abdominal pain or discomfort.

Advantages of the cervical mucus method include that it requires the woman to become familiar with her reproductive system, and it can give early warning about some STIs (which can affect cervical mucus). However, the cervical mucus method requires the woman to distinguish between mucus and semen, spermicidal agents, lubrication, and infectious discharges. Also, the woman must not douche because she will wash away what she is trying to observe.

## Hormone-in-Urine Method

A hormone (LH, luteinizing hormone) is released in increasing amounts in the ovulating woman 12 to 24 hours prior to ovulation. Women can purchase over-the-counter ovulation tests, such as First Response and Ovutime. These are designed to ascertain the surge of LH into the urine (signaling ovulation), so the couple will know to avoid intercourse to maximize the chance of preventing pregnancy. Some test kits involve the woman exposing a test stick during urination, whereas others involve collecting the urine in a small cup and placing the test stick in the cup. In practice, the woman conducts a number of urine tests over a period of days because each test kit comes supplied with five or six tests. Some women experience the LH hormone surge within a 10-hour span, so the woman may need to test herself more than once a day. Of course, this method can also be used to predict the best time to have intercourse to maximize chances of becoming pregnant.

# Nonmethods: Withdrawal and Douching

Because withdrawal and douching are not effective in preventing pregnancy, we call them *nonmethods* of birth control (some may argue that *natural family planning methods* are also nonmethods because a high rate of pregnancies results). Also known as **coitus interruptus**, **withdrawal** is the practice whereby the man withdraws his penis from the vagina before he ejaculates. The advantages of

**Coitus interruptus (withdrawal)** the practice whereby the man withdraws his penis from the vagina before he ejaculates.

coitus interruptus are that it requires no devices or chemicals, and it is always available. The disadvantages of withdrawal are that it does not provide protection from STIs, it may interrupt the sexual response cycle and diminish the pleasure for the couple, and it is very ineffective in preventing pregnancy.

Withdrawal is not a reliable form of contraception for two reasons. First, a man can unknowingly emit a small amount of pre-ejaculatory fluid, which may contain sperm. One drop can contain millions of sperm. In addition, the man may lack the self-control to withdraw his penis before ejaculation, or he may delay his withdrawal too long and inadvertently ejaculate some semen near the vaginal opening of his partner. Sperm deposited there can live in the moist labia and make their way up the vagina.

Although some women believe that douching is an effective form of contraception, it is not. Douching refers to rinsing or cleansing the vaginal canal. After intercourse, the woman fills a syringe with water, any of a variety of solutions or a spermicidal agent that can be purchased over the counter, and then flushes (so she assumes) the sperm from her vagina. But in some cases, the fluid will actually force sperm up through the cervix. In other cases, a large number of sperm may already have passed through the cervix to the uterus, so the douche may do little good.

Sperm may be found in the cervical mucus within 90 seconds after ejaculation. In effect, douching does little to deter conception and may even encourage it. In addition, douching is associated with an increased risk for pelvic inflammatory disease and ectopic pregnancy.

# Emergency Contraception

## National **Data**

Approximately half (49%) of the 6.7 million pregnancies in the United States each year are unintended. Just over half (53%) of these unintended pregnancies were the result of a contraceptive failure (Bayer et al. 2013).

**Postcoital contraception (emergency contraception)** various types of combined estrogen-progesterone morning-after pills or post-coital IUD insertion.

Also called **postcoital contraception**, **emergency contraception** refers to various types of combined estrogen-progesterone morning-after pills or post-coital IUD insertion used primarily in three circumstances: when a woman has unprotected intercourse, when a contraceptive method fails (such as condom breakage or slippage), and when a woman is raped. Emergency contraception methods should be used in emergencies for those times when unprotected intercourse has occurred and medication can be taken within five days (120 hours of exposure).

## Combined Estrogen-Progesterone

The most common morning-after pills are the combined estrogen-progesterone oral contraceptives routinely taken to prevent pregnancy. Common names include Plan B One Step, Next Choice One Dose, My Way, and ella. Known as the *Yuzpe method* after the physician who proposed it, this method involves ingesting a certain number of tablets of combined estrogen-progesterone. In higher doses, they serve to prevent ovulation, fertilization of the egg, or transportation of the egg to the uterus. They may also make the uterine lining inhospitable to implantation. *Emergency contraception must be taken within 120 hours of unprotected intercourse to be effective.* Common side effects of combined estrogen-progesterone emergency contraception pills include nausea, vomiting, headaches, and breast tenderness, although some women also experience abdominal pain, headache, and dizziness. Side effects subside within one–two days after treatment is completed. Table ST 2.1 suggests a much lower rate of ending the pregnancy (25%), but this percent includes a wide range of when the pills are taken. In 2013, the FDA approved the availability of the morning-after pill, Plan B One Step, over the counter for purchase without a prescription regardless of age. Some other two-pill brands are available behind the pharmacy counter with age restrictions for purchase for age 17 or older, and available to younger women by prescription only. Ella is restricted "by prescription only" (Planned Parenthood Federation of America 2013; Office of Population Research & Association of Reproductive Health Professionals 2013).

## Postcoital IUD Insertion

Insertion of a copper IUD within five days after ovulation in a cycle when unprotected intercourse has occurred is very effective for preventing pregnancy. This option, however, is used much less frequently than hormonal treatment because women who need emergency contraception often are not appropriate IUD candidates.

**TABLE ST 2.1 Methods of Contraception and Sexually Transmitted Infection Protection**

| Method | Typical Use[1] Effectiveness Rates | STI Protection | Benefits | Disadvantages | Cost[2] |
|---|---|---|---|---|---|
| Oral contraceptive (The Pill) | 92% | No | High effectiveness rate, 24-hour protection, and menstrual regulation | Daily administration, side effects possible, medication interactions | $10–42 per month |
| Nexplanon/Implanon NXT® or Implanon® (3-year implant) | 99.95% | No | High effectiveness rate, long-term protection | Side effects possible, menstrual changes | $400–600 insertion |
| Depo-Provera® (3-month injection) or Depo-subQ Provera 104® | 97% | No | High effectiveness rate, long-term protection | Decreases body calcium, not recommended for use longer than 2 years for most users, side effects likely | $45–75 per injection |
| Ortho Evra® (transdermal patch) | 92% | No | Same as oral contraceptives except use is weekly, not daily | Patch changed weekly, side effects possible | $15–32 per month |
| NuvaRing® (vaginal ring) | 92% | No | Same as oral contraceptives except use is monthly, not daily | Must be comfortable with body for insertion | $15–48 per month |
| Male condom | 85% | Yes | Few or no side effects, easy to purchase and use | Can interrupt spontaneity | $2–10 a box |
| Female condom | 79% | Yes | Few or no side effects, easy to purchase | Decreased sensation and insertion takes practice | $4–10 a box |
| Spermicide | 71% | No | Many forms to choose, easy to purchase and use | Can cause irritation and be messy | $8–18 per box/ tube/ can |
| Today® Sponge[3] | 68–84% | No | Few side effects, effective for 24 hours after insertion | Spermicide irritation possible | $3–5 per sponge |

*(continued)*

**TABLE ST 2.1** *(continued)*

| Method | Typical Use[1] Effectiveness Rates | STI Protection | Benefits | Disadvantages | Cost[2] |
|---|---|---|---|---|---|
| Diaphragm & Cervical cap[3] | 68–84% | No | Few side effects, can be inserted within 2 hours before intercourse | Can be messy, increased risk of vaginal/UTI infections | $50–200 plus spermicide |
| Intrauterine device (IUD): *Paraguard or Mirena* | 98.2–99% | No | Little maintenance, longer term protection | Risk of PID increased, chance of expulsion | $150–300 |
| Withdrawal | 73% | No | Requires little planning, always available | Pre-ejaculatory fluid can contain sperm | $0 |
| Periodic abstinence | 75% | No | No side effects, accepted in all religions/cultures | Requires a lot of planning, need ability to interpret fertility signs | $0 |
| Emergency contraception | 75% | No | Provides an option after intercourse has occurred | Must be taken within 72 hours, side effects likely | $10–32 |
| Abstinence | 100% | Yes | No risk of pregnancy or STIs | Partners both have to agree to abstain | $0 |

[1]Effectiveness rates are listed as percentages of women not experiencing an unintended pregnancy during the first year of typical use. Typical use refers to use under real-life conditions. Perfect use effectiveness rates are higher.

[2]Costs may vary. The Affordable Care Act, health care legislation passed by Congress and signed into law by President Obama on March 23, 2010, requires health insurance plans to cover preventive services and eliminate cost sharing for some services, including "All Food and Drug Administration approved contraceptive methods, sterilization procedures, and patient education and counseling for all women with reproductive capacity." (U.S. Department of Health and Human Services Health Resources and Services Administration, Women's Preventive Services Guidelines, http://www.hrsa.gov/womensguidelines/ August 2013.)

[3]Lower percentages apply to parous women (women who have given birth). Higher rates apply to nulliparous women (women who have never given birth).

*Source: Beth Credle Burt, MAEd, CHES, a health education specialist. Education Services Project Manager, Siemens Healthcare.*

# References


American College of Obstetricians and Gynecologists, FAQ014 (2012). Taken from http://www.acog.org/~/media/For%20Patients/faq014.pdf?dmc=1&ts=20130804T1749212078, January 2012.

Bayer, L. L., A. B. Edelman, A. B. Caughey, and M. L. Rodriguez. 2013. The price of emergency contraception in the United States: What is the cost effectiveness of ulipristal acetate versus single-dose levonorgestrel. *Contraception* 87: 385–390.

Campo-Englestein, L. 2013. Raging hormones, domestic incompetence, and contraceptive indifference: Narratives contributing to the perception that women do not trust men to use contraception. *Culture, Health & Sexuality* 283–295.

Goldstein, R. L., U. D. Upadhyay, and T. R. Raine. 2013. With pills, patches, rings and shots: Who still uses condoms? A longitudinal cohort study. *Journal of Adolescent Health* 52: 77–82.

Hall, S. and D. Knox. 2013. Relationship and sexual behaviors of a sample of 4,567 university students. Unpublished data collected for this text. Department of Family and Consumer Sciences, Ball State University and Department of Sociology, East Carolina University.

Hannaford, P. C. 2013. Mortality among oral contraceptive users: An evolving story. *The European Journal of Contraception and Reproductive Health Care* 18: 1–4.

Ma, L. Z. Lv, J. Su, J. Wang, D. Yan et al. 2013. Consistent condom use increases the colonization of lactobacillus crispatus in the vagina. *Plos ONE 8* (7). e 70716

Mayo Foundation for Medical Education and Research. 1998–2013. https://www.mayoclinic.com/health/depo-provera/MY00995; 1998–2013; Pfizer Patient Information About Depo-subQ Provera 104, https://www.pfizer.com/files/products/ppi_depo_subq_provera.pdf , Pfizer, Inc. NY, NY.

Michielsen, D. and R. Merck. 2012. FDA-Approved Patient Labeling, Taken from http://www.merck.com/product/usa/pi_circulars/n/nexplanon/nexplanon_ppi.pdf, Copyright © 2011 MSD Oss B.V., a subsidiary of Merck & Co., Inc. Revised: 05/2012 2010. State-of-the art of non-hormonal methods of contraception: VI. Male sterilisation. *European Journal of Contraception & Reproductive Health Care, 15:* 136–149.

Moreau, C., J. Trussell, and N. Bajos. 2013. Religiosity, religious affiliation, and patterns of sexual activity and contraceptive use in France. *The European Journal of Contraception and Reproductive Health Care* 18: 168–180.

Nya-Ngatchou, J. and J. K. Amory. 2013. New approaches to male non-hormonal contraception. *Conraception* 87: 296–299.

Office of Population Research & Association of Reproductive Health Professionals. 2013. Taken from http://cc.princeton.edu/questions/qa-otc-access.html, August 2013.
Planned Parenthood Federation of America Inc. 2013. Retrieved August 5, 2013. https://www.plannedparenthood.org/health-topics/birth-control/cervical-cap-20487.htm
Simon, J., K. Caramelli, H. Thomas, and K. Reape. 2013. Efficacy and safety of a weekly progestin-only transdermal system for prevention of pregnancy. *Contraception* 88: 316.
Sitruk-Ware, R., A. Nath, and D. R. Mishell, Jr. 2013. Contraception technology: Past, present and future. *Contraception* 87: 319–330.
Wang, C. and R. S. Swerdloff. 2010. Hormonal approaches to male contraception. Curr Opin Urol. 2010 Nov. 20(6):520-4. doi: 10.1097/MOU.0b013e32833f1b4a.

# Appendix A

## Individual Autobiography Outline

Your instructor may ask you to write a paper that reflects the individual choices you have made that have contributed to your becoming who you are. Check with your instructor to determine what credit (if any) is available for completing this autobiography. Use the following outline to develop your paper. Some topics may be too personal, and you may choose to avoid writing about them. Your emotional comfort is important, so answer only those questions you feel comfortable responding to.

**I. Choices: Free Will Versus Determinism**

Specify the degree to which you feel that you are free to make your own interpersonal choices versus the degree to which social constraints influence and determine your choices. Give an example of social influences being primarily responsible for a relationship choice and an example of you making a relationship choice where you acted contrary to the pressure you were getting from parents, siblings, and peers.

**II. Relationship Beginnings**

**A.** ***Interpersonal context into which you were born.*** If your parents are married, how long had they been married before you were born? How many other children had been born into your family? How many followed your birth? Describe how these and other parental choices affected you before you were born and the choices you will make in regard to family planning that will affect the lives of your children.

**B.** ***Early relationships.*** What was your relationship with your mother, father, and siblings when you were growing up? What is your relationship with each of them today? Who took care of you as a baby? If this person was other than your parents or siblings (for example, a grandparent), what is your relationship with that person today? How often did your mother or father tell you that they loved you? How often did they embrace or hug you? How has this closeness or distance influenced your pattern with others today? Give other examples of how your experiences in the family in which you were reared have influenced who you are and how you behave today. How easy or difficult is it for you to make decisions and how have your parents influenced this capacity?

**C.** ***Early self-concept.*** How did you feel about yourself as a child, an adolescent, and a young adult? What significant experiences helped to shape your self-concept? How do you feel about yourself today? What choices have you made that have resulted in your feeling good about yourself? What choices have you made that have resulted in your feeling negatively about yourself?

**III. Subsequent Relationships**

**A.** ***First love.*** When was your first love relationship with someone outside your family? What kind of love was it? Who initiated the relationship? How long did it last? How did it end? How did it affect you and your subsequent relationships? What choices did you make in this first love relationship that you are glad you made? What choices did you make that you now feel were a mistake?

**B.** ***Subsequent love relationships.*** What other significant love relationships (if any) have you had? How long did they last and how did they end? What choices did you make in these relationships that you are glad you made? What choices did you make that you now feel were a mistake? On a ten-point scale (0 = very distant and 10 = very close), how emotionally close do you want to be to a romantic partner?

**C.** ***Subsequent relationship choices.*** What has been the best relationship choice you have made? What relationship choices have you regretted?

**D.** ***Lifestyle preferences.*** What are your preferences for remaining single, being married, or living with someone? How would you feel about living in a commune? What do you believe is the ideal lifestyle? Why?

**IV. Communication Issues**

**A.** ***Parental models.*** Describe your parents' relationship and their manner of communicating with each other. How are your interpersonal communication patterns similar to and different from theirs?

**B.** ***Relationship communication.*** How comfortable do you feel talking about relationship issues with your partner? How comfortable do you feel telling your partner what you like and don't like about his or her behavior? To what degree have you told your partner your feelings for him or her? To what degree have you told your partner your desires for the future?

**C.** ***Sexual communication.*** How comfortable do you feel giving your partner feedback about how to please you sexually? How comfortable are you discussing the need to use a condom with a potential sex partner? How would you approach this topic?

**D.** ***Sexual past.*** How much have you disclosed to a partner about your previous relationships? How honest were you? Do you think you made the right decision to disclose or withhold? Why?

**V. Sexual Choices**

**A.** ***Sex education.*** What did you learn about sex from your parents, peers, and teachers (both academic and religious)? What choices did your parents, peers, and teachers make about your sex education that had positive consequences? What decisions did they make that had negative consequences for you?

**B.** ***Sexual experiences.*** What choices have you made about your sexual experiences that had positive outcomes? What choices have you made that resulted in sexual regret?

**C.** ***Sexual values.*** To what degree are your sexual values absolutist, legalist, or relativistic? How have your sexual values changed since you were an adolescent?

**D.** ***Safer sex.*** What is the riskiest choice you have made with regard to your sexual behavior? What is the safest choice you have made with regard to your sexual behavior? What is your policy about asking your partner about previous sex history and requiring that both of you be tested for HIV and STIs before having sex? How comfortable are you buying and using condoms?

**VI. Violence and Abuse Issues**

**A.** ***Violent or abusive relationship.*** Have you been involved in a relationship in which your partner was violent or abusive toward you? Give examples of the violence or abuse (verbal and nonverbal) if you have been involved in such a relationship. How many times did you leave and return to the relationship before you left permanently?

What was the event that triggered your decision to leave the first and last time? Describe the context of your actually leaving (for example, did you leave when the partner was at work?). To what degree have you been violent or abusive toward a partner in a romantic relationship?

**B.** ***Family or sibling abuse.*** Have you been involved in a relationship where your parents or siblings were violent or abusive toward you? Give examples of any violence or abuse (verbal or nonverbal) if such experiences were part of your growing up. How have these experiences affected the relationship you have with the abuser today?

**C.** ***Forced sex.*** Have you been pressured or forced to participate in sexual activity against your will by a parent, sibling, partner, or stranger? How did you react at the time and how do you feel today? Have you pressured or forced others to participate in sexual experiences against their will?

**VII. Reproductive Choices**

**A.** ***Contraception.*** What is your choice for type of contraception? How comfortable do you feel discussing the need for contraception with a potential partner? In what percentage of your first-time intercourse experiences with a new partner did you use a condom? (If you have not had intercourse, this question will not apply.)

**B.** ***Children.*** How many children (if any) do you want and at what intervals? How important is it to you that your partner wants the same number of children as you do?

How do you feel about artificial insemination, sterilization, abortion, and adoption? How important is it to you that your partner feels the same way?

**VIII. Childrearing Choices**

**A.** ***Discipline.*** What are your preferences for your use of "time-out" or "spanking" as a way of disciplining your children? How important is it to you that your partner feels the same as you do on this issue?

**B.** ***Day care.*** What are your preferences for whether your children grow up in day care or whether one parent will stay home with and rear the children? How important is it to you that your partner feels the same as do you on this issue?

**C.** ***Education.*** What are your preferences for whether your children attend public or private school or are homeschooled? What are your preferences for whether your child attends a religious school? To what degree do you feel it is your responsibility as parents to pay for the college education of your children? How important is it to you that your partner feels the same as do you on these issues?

**IX. Education or Career Choices**

**A.** ***Own educational or career choices.*** What is your major in school? How important is it to you that you finish undergraduate school? How important is it to you that you earn a master's degree, Ph.D., M.D., or law degree? In what? To what extent do you want to be a stay-at-home mom or stay-at-home dad? How important is it to you that your partner be completely supportive of your educational, career, and family aspirations?

**B.** ***Expectations of partner.*** How important is it to you that your partner has the same level of education that you have? To what degree are you willing to be supportive of your partner's educational and career aspirations? What are the career goals of you and your partner?

# Appendix B

# Family Autobiography Outline

Your instructor may want you to write a paper that reflects the influence of your family on your development. Check with your instructor to determine what credit (if any) is available for completing this family autobiography. Use the following outline to develop your paper. Some topics may be too personal, and you may choose to avoid writing about them. Your emotional comfort is important, so answer only those you feel comfortable responding to.

**I. Family Background**

**A.** Describe yourself, including age, gender, place of birth, and additional information that helps to identify you. On a scale of 0 to 10 (10 is highest), how happy are you? Explain this number in reference to the satisfaction you experience in the various roles you currently occupy (for example, offspring, sibling, partner in a relationship, employee, student, parent, roommate, friend).

**B.** Identify your birth position; give the names and ages of children younger and older than you. How did you feel about your "place" in the family? How do you feel now?

**C.** What was your relationship with and how did you feel about each parent and sibling when you were growing up?

**D.** What is your relationship with and how do you feel today about each of these family members?

**E.** Which parental figure or sibling are you most like? How? Why?

**F.** Who else lived in your family (for example, grandparent, spouse of sibling), and how did they impact family living?

**G.** Discuss the choice you made before attending college that you regard as the wisest choice you made during this time period. Discuss the choice you made prior to college that you regret.

**H.** Discuss the one choice you made since you began college that you regard as the wisest choice you made during this time period. Discuss the one choice you made since you began college that you regret.

**II. Religion and Values**

**A.** In what religion were you socialized as a child? Discuss the impact of religion on yourself as a child and as an adult. To what degree will you choose to teach your own children similar religious values? Why?

**B.** Explain what you were taught in your family in regard to each of the following values: intercourse outside of marriage, the need for economic independence, manners, honesty, importance of being married, qualities of a desirable spouse, importance of having children, alcohol and drugs, safety, elderly family members, people of other races and religions, people with disabilities, people of alternative sexual orientation, people with less or more education, people with and without "wealth," and occupational role (in regard to the latter, what occupational role were you encouraged to pursue?). To what degree will you choose to teach your own children similar values? Why?

**C.** What was the role relationship between your parents in terms of dominance, division of labor, communication, affection, and so on? How has your observation of the parent of the same sex and opposite sex influenced the role you display in your current relationships with intimate partners? To what degree will you choose to have a relationship with your own partner similar to that your parents had with each other?

**D.** How close were your parents emotionally? How emotionally close were/are you with your parents? To what degree did you and your parents discuss feelings? On a scale of 0 to 10, how well do your parents know how you think and feel? To what degree will you choose to have a similar level of closeness or distance with your own partner and children?

**E.** Did your parents have a pet name for you? What was this name and how did you feel about it?

**F.** How did your parents resolve conflict between themselves? How do you resolve conflict with partners in your own relationships?

**III. Economics and Social Class**

**A.** Identify your social class (lower, middle, upper); the education, jobs, or careers of your respective parents; and the economic resources of your family. How did your social class and economic well-being affect you as a child? How has the economic situation in which you were reared influence your own choices of what you want for yourself? To what degree are you economically self-sufficient?

**B.** How have the career choices of your parents influenced your own?

**IV. Parental Plusses and Minuses**

**A.** What is the single most important thing your mother and father, respectively, said or did that has affected your life in a positive way?

**B.** What is the single biggest mistake your mother and father, respectively, made in rearing you? Discuss how this impacted you negatively. To what degree does this choice still affect you?

**V. Personal Crisis Events**

**A.** Everyone experiences one or more crisis events that have a dramatic impact on life. Identify and discuss each event you have experienced and your reaction and adjustment to them.

**B.** How did your parents react to this crisis event you were experiencing? To what degree did their reaction help or hinder your adjustment? What different choice or choices (if any) could they have made to assist you in ways you would have regarded as more beneficial?

**VI. Family Crisis Events**

Identify and discuss each crisis event your family has experienced. How did each member of your family react and adjust to each event? An example of a family crisis event would be unemployment of a primary breadwinner, prolonged illness of a family member, aging parent coming to live with the family, death of a sibling, or alcoholism.

**VII. Family Secrets**

Most families have secrets. What secrets are you aware of in your family and kinship system? To what degree has it been difficult for you to be aware of this family secret?

**VIII. Future**

Describe yourself two, five, and ten years from now. What are your educational, occupational, marital, and family goals? How has the family in which you were reared influenced each of these goals? What choices might your parents make to assist you in achieving your goals?

# Appendix C

# Prenuptial Agreement of a Remarried Couple

Pam and Mark are of sound mind and body and have a clear understanding of the terms of this contract and of the binding nature of the agreements contained herein; they freely and in good faith choose to enter into the PRENUPTIAL AGREEMENT and MARRIAGE CONTRACT and fully intend it to be binding upon themselves.

Now, therefore, in consideration of their love and esteem for each other and in consideration of the mutual promises herein expressed, the sufficiency of which is hereby acknowledged, Pam and Mark agree as follows:

## Names

Pam and Mark affirm their individuality and equality in this relationship. The parties believe in and accept the convention of the wife's accepting the husband's name, while rejecting any implied ownership.

Therefore, the parties agree that they will be known as husband and wife and will henceforth employ the titles of address Mr. and Mrs. Mark Stafford and will use the full names of Pam Hayes Stafford and Mark Robert Stafford.

## Relationships with Others

Pam and Mark believe that their commitment to each other is strong enough that no restrictions are necessary with regard to relationships with others.

Therefore, the parties agree to allow each other freedom to choose and define their relationships outside this contract, and the parties further agree to maintain sexual fidelity each to the other.

## Religion

Pam and Mark reaffirm their belief in God and recognize He is the source of their love. Each of the parties has his/her own religious beliefs.

Therefore, the parties agree to respect their individual preferences with respect to religion and to make no demands on each other to change such preferences.

## Children

Pam and Mark both have children. Although no minor children will be involved, there are two (2) children still at home and in school and in need of financial and emotional support.

Therefore, the parties agree that they will maintain a home for and support these children as long as is needed and reasonable. They further agree that all children of both parties will be treated as one family unit, and each will be given emotional and financial support to the extent feasible and necessary as determined mutually by both parties.

## Careers and Domicile

Pam and Mark value the importance and integrity of their respective careers and acknowledge the demands that their jobs place on them as individuals and on their partnership.

Both parties are well established in their respective careers and do not foresee any change or move in the future.

The parties agree, however, that if the need or desire for a move should arise, the decision to move shall be mutual and based on the following factors:

**1.** The overall advantage gained by one of the parties in pursuing a new opportunity shall be weighed against the disadvantages, economic and otherwise, incurred by the other.

**2.** The amount of income or other incentive derived from the move shall not be controlling.

**3.** Short-term separations as a result of such moves may be necessary.

Mark hereby waives whatever right he might have to solely determine the legal domicile of the parties.

## Care and Use of Living Spaces

Pam and Mark recognize the need for autonomy and equality within the home in terms of the use of available space and allocation of household tasks. The parties reject the concept that the responsibility for housework rests with the woman in a marriage relationship whereas the duties of home maintenance and repair rest with the man.

Therefore, the parties agree to share equally in the performance of all household tasks, taking into consideration individual schedules, preferences, and abilities.

The parties agree that decisions about the use of living space in the home shall be mutually made, regardless of the parties' relative financial interests in the ownership or rental of the home, and the parties further agree to honor all requests for privacy from the other party.

## Property, Debts, Living Expenses

Pam and Mark intend that the individual autonomy sought in the partnership shall be reflected in the ownership of existing and future-acquired property, in the characterization and control of income, and in the responsibility for living expenses. Pam and Mark also recognize the right of patrimony of children of their previous marriages.

Therefore, the parties agree that all things of value now held singly and/or acquired singly in the future shall be the property of the party making such acquisition. In the event that one party to this agreement shall predecease the other, property and/or other valuables shall be disposed of in accordance with an existing will or other instrument of disposal that reflects the intent of the deceased party.

Property or valuables acquired jointly shall be the property of the partnership and shall be divided, if necessary, according to the contribution of each party. If one party shall predecease the other, jointly owned property or valuables shall become the property of the surviving spouse.

Pam and Mark feel that each of the parties to this agreement should have access to monies that are not accountable to the partnership.

Therefore, the parties agree that each shall retain a mutually agreeable portion of their total income and the remainder shall be deposited in a mutually agreeable banking institution and shall be used to satisfy all jointly acquired expenses and debts.

The parties agree that beneficiaries of life insurance policies they now own shall remain as named on each policy. Future changes in beneficiaries shall be mutually agreed on after the dependency of the children of each party has been terminated. Any other benefits of any retirement plan or insurance benefits that accrue to a spouse only shall not be affected by the foregoing.

The parties recognize that in the absence of income by one of the parties, resulting from any reason, living expenses may become the sole responsibility of the employed party, and in such a situation, the employed party shall assume responsibility for the personal expenses of the other.

Both Pam and Mark intend their marriage to last as long as both shall live.

Therefore, the parties agree that, should it become necessary, due to the death of either party, the surviving spouse shall assume any last expenses in the event that no insurance exists for that purpose.

Pam hereby waives whatever right she might have to rely on Mark to provide the sole economic support for the family unit.

## Evaluation of the Partnership

Pam and Mark recognize the importance of change in their relationship and intend that this CONTRACT shall be a living document and a focus for periodic evaluations of the partnership.

The parties agree that either party can initiate a review of any article of the CONTRACT at any time for amendment to reflect changes in the relationship. The parties agree to honor such requests for review with negotiations and discussions at a mutually convenient time.

The parties agree that, in any event, there shall be an annual reaffirmation of the CONTRACT on or about the anniversary date of the CONTRACT.

The parties agree that, in the case of unresolved conflicts between them over any provisions of the CONTRACT, they will seek mediation, professional or otherwise, by a third party.

## Termination of the Contract

Pam and Mark believe in the sanctity of marriage; however, in the unlikely event of a decision to terminate this CONTRACT, the parties agree that neither shall contest the application for a divorce decree or the entry of such decree in the county in which the parties are both residing at the time of such application.

In the event of termination of the CONTRACT and divorce of the parties, the provisions of this and the section on "Property, Debts, Living Expenses" of the CONTRACT as amended shall serve as the final property settlement agreement between the parties. In such event, this CONTRACT is intended to effect a complete settlement of any and all claims that either party may have against the other, and a complete settlement of their respective rights as to property rights, homestead rights, inheritance rights, and all other rights of property otherwise arising out of their partnership. The parties further agree that, in the event of termination of this CONTRACT and divorce of the parties, neither party shall require the other to pay maintenance costs or alimony.

## Decision Making

Pam and Mark share a commitment to a process of negotiations and compromise that will strengthen their equality in the partnership. Decisions will be made with respect for individual needs. The parties hope to maintain such mutual decision making so that the daily decisions affecting their lives will not become a struggle between the parties for power, authority, and dominance. The parties agree that such a process, although sometimes time-consuming and fatiguing, is a good investment in the future of their relationship and their continued esteem for each other.

Now, therefore, Pam and Mark make the following declarations:

**1.** They are responsible adults.

**2.** They freely adopt the spirit and the material terms of this prenuptial and marriage contract.

**3.** The marriage contract, entered into in conjunction with a marriage license of the State of Illinois, County of Wayne, on this 12th day of January 2015, hereby manifests their intent to define the rights and obligations of their marriage relationship as distinct from those rights and obligations defined by the laws of the State of Illinois, and affirms their right to do so.

**4.** They intend to be bound by this prenuptial and marriage contract and to uphold its provisions before any Court of Law in the Land.

Therefore, comes now, Pam Hayes Stafford, who applauds her development that allows her to enter into this partnership of trust, and she agrees to go forward with this marriage in the spirit of the foregoing PRENUPTIAL and MARRIAGE CONTRACT.

Therefore, comes now, Mark Robert Stafford, who celebrates his growth and independence with the signing of this contract, and he agrees to accept the responsibilities of this marriage as set forth in the foregoing PRENUPTIAL and MARRIAGE CONTRACT.

This CONTRACT AND COVENANT has been received and reviewed by the Reverend Ray Brannon, officiating.

Finally, come Vicki Whitfield and Rodney Whitfield, who certify that Pam and Mark did freely read and sign this MARRIAGE CONTRACT in their presence, on the occasion of their entry into a marriage relationship by the signing of a marriage license in the State of Illinois, County of Wayne, at which they acted as official witnesses. Further, they declare that the marriage license of the parties bears the date of the signing of this PRENUPTIAL and MARRIAGE CONTRACT. (Although this document is real, the names are fictitious to protect the real parties.)

# Appendix D

## Living Will

I, [Declarant], ("Declarant" herein), being of sound mind, and after careful consideration and thought, freely and intentionally make this revocable declaration to state that, if I should become unable to make and communicate my own decisions on life—sustaining or life-support procedures, then my dying shall not be delayed, prolonged, or extended artificially by medical science or life-sustaining medical procedures, all according to the choices and decisions I have made and which are stated here in my Living Will.

It is my intent, hope, and request that my instructions be honored and carried out by my physicians, family, and friends, as my legal right.

If I am unable to make and communicate my own decisions regarding the use of medical life-sustaining or life-support systems and/or procedures, and if I have a sickness, illness, disease, injury, or condition that has been diagnosed by two (2) licensed medical doctors or physicians who have personally examined me (or more than two (2) if required by applicable law), one of whom shall be my attending physician, as being either (1) terminal or incurable certified to be terminal, or (2) a condition from which there is no reasonable hope of my recovery to a meaningful quality of life, which may reasonably be referred to as hopeless, although not necessarily "terminal" in the medical sense, or (3) has rendered me in a persistent vegetative state, or (4) a condition of extreme mental deterioration, or (5) permanently unconscious, then in the absence of my revoking this Living Will, all medical life-sustaining or life-support systems and procedures shall be withdrawn, unless I state otherwise in the following provisions.

Unless otherwise provided in this Living Will, nothing herein shall prohibit the administering of pain-relieving drugs to me, or any other types of care purely for my comfort, even though such drugs or treatment may shorten my life, or be habit forming, or have other adverse side effects.

I am also stating the following additional instructions so that my Living Will is as clear as possible: (1) resuscitation (CPR)—I do not want to be resuscitated; (2) intravenous and tube feeding—I do not want to be kept alive via intravenous means, and I do not want a feeding tube installed if I am unable to consume food/liquids naturally; (3) Life-Sustaining Surgery—I do not want to be kept alive by any new or old life-sustaining surgery; (4) new medical developments—I do not want to become a participant in any new medical developments that would prolong my life; (5) home or hospital—I want to die wherever my family chooses, including a hospital or nursing home.

In the event that any terms or provisions of my Living Will are not enforceable or are not valid under the laws of the state of my residence, or the laws of the state where I may be located at the time, then all other provisions which are enforceable or valid shall remain in full force and effect, and all terms and provisions herein are severable.

IN WITNESS WHEREOF, I have read and understand this Living Will, and I am freely and voluntarily signing it on this the day of (month), (year) in the presence of witnesses.

Signed: [Declarant]

Street Address:

County:

City and State:

Witness:

We, the undersigned witnesses, certify by our signatures below, that we are adult (at least 18 years old), mentally competent persons; that we are not related to the Declarant by blood, marriage, or adoption; that we do not stand to inherit anything from the Declarant by any means, including will, trust, operation of law or the laws of intestate succession, or by beneficiary designation, nor do we stand to benefit in any way from the death of the Declarant; that we are not directly responsible for the health or medical care, or general welfare of the Declarant; that neither of us signed the Declarant's signature on this document; and that the Declarant is known to us.

We hereby further certify that the Declarant is over the age of 18; that the Declarant signed this document freely and voluntarily, not under any duress or coercion; and that we were both present together, and in the presence of the Declarant to witness the signing of this Living Will on this the day of (month), (year).

Witness signature:
Residing at:
Witness signature:
Residing at:
Notary Acknowledgment:
This instrument was acknowledged before me on this the day of (month), (year) by [Declarant], the Declarant herein, on oath stating that the Declarant is over the age of 18, has fully read and understands the above and foregoing Living Will, and that the Declarant's signing and execution of same is voluntary, without coercion, and is intentional.
Notary Public
My commission or appointment expires:

# Glossary

## A

**abortion rate**—the number of abortions per thousand women aged 15 to 44

**abortion ratio**—refers to the number of abortions per 1,000 live births. Abortion is affected by the need for parental consent and parental notification

**absolutism**—a sexual value system which is based on unconditional allegiance to tradition or religion (e.g., waiting until marriage to have sexual intercourse)

**acquaintance rape**—nonconsensual sex between adults (of same or other sex) who know each other

**advance directive (living will)**—details for medical care personnel the conditions under which life support measures should be used for one's partner

**agape love style**—also known as compassionate love, characterized by a focus on the well-being of the love object, with little regard for reciprocation

**age**—term which may be defined chronologically (number of years), physiologically (physical decline), psychologically (self-concept), sociologically (roles for the elderly/retired), and culturally (meaning of age in one's society)

**age discrimination**—a situation where older people are often not hired and younger workers are hired to take their place

**ageism**—the systematic persecution and degradation of people because they are old

**ageism by invisibility**—when older adults are not included in advertising and educational materials

**Al-Anon**—organization that provides support for family members and friends of alcohol abusers

**alexithymia**—a personality trait which describes a person with little affect

**alienation of affection**—law which gives a spouse the right to sue a third party for taking the affections of a spouse away

**amae**—expecting a close other's indulgence when one behaves inappropriately

**androgyny**—a blend of traits that are stereotypically associated with masculinity and femininity

**anodyspareunia**—frequent, severe pain during receptive anal sex

**antinatalism**—opposition to children

**anxious jealousy**—obsessive ruminations about the partner's alleged infidelity that can make one's life a miserable emotional torment

**arbitration**—third party listens to both spouses and makes a decision about custody, division of property, child support, and alimony

**arranged marriage**—mate selection pattern whereby parents select the spouse of their offspring. A matchmaker may be used but the selection is someone of whom the parents approve

**artifact**—concrete symbol that reflects the existence of a cultural belief or activity

**asceticism**—sexual belief system which emphasizes that giving in to carnal lust is unnecessary and one should attempt to rise above the pursuit of sensual pleasure into a life of self-discipline and self-denial

**asexual**—the absence of sexual behavior with a partner as well as oneself (masturbation)

**attachment theory of mate selection**—developed early in reference to one's parents, the drive toward an intimate, social/emotional connection

**authentic**—being who one is and saying what one feels

**autistic love**—love conceptualized as being "all about me"

## B

**baby blues**—transitory symptoms of depression in a mother 24 to 48 hours after her baby is born

**battered woman syndrome**—legal term used in court that the person accused of murder was suffering from to justify their behavior. Therapists define battering as physical aggression that results in injury and accompanied by fear and terror

**behavioral couple therapy (BCT)**—therapeutic focus on behaviors the respective spouses want increased or decreased, initiated or terminated

**benevolent sexism**—the belief that women are innocent creatures who should be protected and supported.

**binuclear family**—family that lives in two households as when parents live in separate households following a divorce

**biphobia**—parallel set of negative attitudes toward bisexuality and those identified as bisexual.

**birth control sabotage**—partner interference with contraception

**bisexuality**—cognitive, emotional, and sexual attraction to members of both sexes

**biosocial theory**—also referred to as sociobiology, social behaviors (for example, gender roles) are biologically based and have an evolutionary survival function

**bisexuality**—cognitive, emotional, and sexual attraction to members of both sexes

**blended family**—a family created when two individuals marry and at least one of them brings a child or children from a previous relationship or marriage. Also referred to as a stepfamily.

**blurred retirement**—an individual working part-time before completely retiring or taking a "bridge job" that provides a transition between a lifelong career and full retirement

**boomerang generation**—adult children who return to live with their parents

**brainstorming**—suggesting as many alternatives as possible without evaluating them

**branching**—in communication, going out on different limbs of an issue rather than staying focused on the issue

**bride wealth**—also known as bride price or bride payment, the amount of money or goods paid by the groom or his family to the bride's family for giving her up

## C

**catfishing**—refers to a person on the Internet who makes up an online identify and an entire social facade to trick a person into becoming involved in an emotional relationship

**child abuse**—any behavior or lack of behavior by parents or caregivers that results in deliberate harm to a child's physical or psychological well-being

**child marriage**—marriages in which females as young as 8 to 12 are required by their parents to marry an older man

**childlessness concerns**—the idea that holidays and family gatherings may be difficult because of not having children or feeling left out or sad that others have children

**chronic sorrow**—grief-related feelings that occur periodically throughout the lives of those left behind

**circadian preference**—refers to an individual's preference for morningness-eveningness in regard to intellectual and physical activities

**civil union**—a pair-bonded relationship given legal significance in terms of rights and privileges

**classical conditioning**—an unconditioned stimulus (food) can become a conditioned stimulus (parent's face) in terms of reducing anxiety of child

**closed-ended question**—question that allows for a one-word answer and does not elicit much information

**cohabitation**—two adults, unrelated by blood or by law, involved in an emotional and sexual relationship who sleep in the same residence at least four nights a week for three months

**cohabitation effect**—those who have multiple cohabitation experiences prior to marriage are more likely to end up in marriages characterized by violence, lower levels of happiness, lower levels of positive communication, and depression

**coitus**—the sexual union of a man and woman by insertion of the penis into the vagina

**collaborative practice**—process involving a team of professionals (lawyer, psychologist, mediator, social worker, financial counselor) helping a couple separate and divorce in a humane and cost-effective way

**collectivism**—pattern in which one regards group values and goals as more important than one's own values and goals

**coming out**—being open about one's sexual orientation and identity

**commitment**—the intent to maintain a relationship

**common-law marriage**—a heterosexual cohabiting couple presenting themselves as married

**communication**—the process of exchanging information and feelings between two or more people

**compersion**—sometimes thought of as the opposite of jealousy, the approval of a partner's emotional and sexual involvement with another person

**competitive birthing**—having the same number (or more) of children in reference to one's peers

**complementary-needs theory of mate selection**—states that we tend to select mates whose needs are opposite and complementary to our own needs

**comprehensive sex education program**—learning experience which recommends abstinence but also discusses contraception and other means of pregnancy protection

**conception**—refers to the fusion of the egg and sperm. Also known as fertilization.

**concurrent sexual partnership**—relationship in which the partners have sex with several individuals concurrently

**condom assertiveness**—the unambiguous messaging that sex without a condom is unacceptable

**conflict**—the context in which the perceptions or behavior of one person are in contrast to or interfere with the other

**conflict framework**—the view that individuals in relationships compete for valuable resources

**congruent messages**—message in which the verbal and nonverbal behaviors match

**conjugal love**—the love between married people characterized by companionship, calmness, comfort, and security

**connection rituals**—habits which occur daily in which the couple share time and attention

**consumerism**—economic societal value that one must buy everything and have everything now

**contact hypothesis**—in reference to sexual orientation, heterosexuals have more favorable attitudes toward gay men and lesbians if they have had prior contact with or know someone who is gay or lesbian

**conversion therapy**—also called reparative therapy, focused on changing the sexual orientation of homosexuals

**Coolidge effect**—the effect of novelty and variety on increasing sexual arousal

**corporal punishment**—the use of physical force with the intention of causing a child to experience pain, but not injury, for the purpose of correction or control of the child's behavior

**correct rejections**—both partners are aware that one of them chose not to engage in a negative comment

**cougar**—a woman, usually in her thirties or forties, who is financially stable and mentally independent and looking for a younger man with whom to have fun

**covenant marriage**—type of marriage whereby the spouses agree to have marriage counseling before getting married, to have a cooling off period of two years after children are born before contemplating divorce, and to divorce only for serious faults (such as abuse, adultery, or imprisonment for a felony)

**crisis**—a crucial situation that requires change in one's normal pattern of behavior

**cross-dresser**—individuals who dress or present themselves in the gender of the other sex

**cross-sectional research**—studying the whole population at one time—e.g., finding out from all persons now living together at your university about their experience

**cryopreservation**—the freezing of fertilized eggs for implantation at a later stage

**cybercontrol**—use of communication technology, such as cell phones, e-mail, and social networking sites, to monitor or control partners in intimate relationships

**cybersex**—any consensual, computer-mediated, participatory sexual experience involving two or more individuals

**cybervictimization**—harassing behavior which includes being sent threatening e-mail, unsolicited obscene e-mail, computer viruses, or junk mail (spamming); can also include flaming (online verbal abuse) and leaving improper messages on message boards

**cyclical cohabitation**—the couple live together off and on

## D

**dark triad personality**—term for intercorrelated traits of narcissism (sense of entitlement and grandiose self-view), Machiavellianism (deceptive and insincere), and psychopathy (callous and no empathy)

**date rape**—one type of acquaintance rape which refers to nonconsensual sex between people who are dating or are on a date

**December marriages**—both spouses are elderly

**defense mechanism**—techniques that function without awareness to protect individuals from emotional hurt

**Defense of Marriage Act**—legislation which says that marriage is a "legal union between one man and one woman" and denies federal recognition of same-sex marriage

**demandingness**—the manner in which parents place demands on children in regard to expectations and discipline

**developmental task**—a skill that, if mastered, allows a family to grow as a cohesive unit

**disenchantment**—the transition from a state of newness and high expectation to a state of mundaneness tempered by reality

**displacement**—shifting one's feelings, thoughts, or behaviors from the person who evokes them onto someone else

**divorce**—the legal ending of a valid marriage contract

**divorce mediation**—meeting with a neutral professional who negotiates child custody, division of property, child support, and alimony directly with the divorcing spouses

**divorcism**—the belief that divorce is a disaster

**domestic partnership**—two adults who have chosen to share each other's lives in an intimate and committed relationship of mutual caring. These relationships are given some kind of official recognition by a city or corporation so as to receive partner benefits (for example, health insurance)

**double Dutch**—strategy of using both the pill and condom by sexually active youth in the Netherlands

**down low**—non-gay-identifying African-American men who have sex with men and women

**dual-career marriage**—a marriage in which both spouses pursue careers

**durable power of attorney**—gives adult children complete authority to act on behalf of the elderly

## E

**emergency contraception**—also known as **postcoital contraception**, refers to various types of morning-after pills

**emotional abuse**—nonphysical behavior designed to denigrate the partner, reduce the partner's status, and make the partner feel vulnerable to being controlled by the partner

**emotional competence**—experience emotion, express emotion, and regulate emotion

**empathetic love**—love conceptualized as being "all about you"

**endogamy**—cultural expectation to select a marriage partner within one's own social group

**engagement**—time in which the romantic partners are sexually monogamous, committed to marry, and focused on wedding preparations

**entrapped**—stuck in an abusive relationship and unable to extricate oneself from the abusive partner

**eros love style**—also known as romantic love, the love of passion and sexual desire

**escapism**—the simultaneous denial of and avoidance of dealing with a problem

**euthanasia**—from the Greek meaning "good death" or dying without suffering; either passively or actively ending the life of a patient

**exchange theory**—theory that emphasizes that relations are formed and maintained based on who offers the greatest rewards at the lowest costs

**exogamy**—the cultural pressure to marry outside the family group

**extended family**—the nuclear family or parts of it plus other relatives such as grandparents, aunts, uncles, and cousins

**extradyadic involvement**—refers to sexual involvement of a pair-bonded individual with someone other than the partner; also called extrarelational involvement

**extrafamilial**—child sexual abuse in which the perpetrator is someone outside the family who is not related to the children

**extramarital affair**—refers to a spouse's sexual involvement with someone outside the marriage

## F

**Fafafini**—in Samoan society, some effeminate boys are socialized/reared as females. Fafafini represent a third gender, neither female nor male, and are unique and valued, not stigmatized

**familism**—value that decisions are made in reference to what is best for the family

**family**—a group of two or more people related by blood, marriage, or adoption

**family care giving**—adult children providing care for their elderly parent

**family life course development**—the stages and process of how families change over time

**family life cycle**—stages that identify the various developmental tasks family members face across time

**family of orientation**—also known as the family of origin, the family into which a person is born

**family of procreation**—the family a person begins typically by getting married and having children

**family relations doctrine**—belief that even nonbiological parents may be awarded custody or visitation rights if they have been economically and emotionally involved in the life of the child

**family resiliency**—the successful coping of family members under adversity that enables them to flourish with warmth, support, and cohesion

**family systems framework**—views each member of the family as part of a system and the family as a unit that develops norms of interaction

**female genital alteration**—cutting off the clitoris or excising (partially or totally) the labia minor

**female rape myths**—beliefs that deny victim injury or cast blame on the woman for her own rape

**feminist framework**—views marriage and family as contexts of inequality and oppression for women

**feminization of poverty**—the idea that women (particularly those who live alone or with their children) disproportionately experience poverty

**feral children**—"wild, not domesticated" children thought to have been reared by animals

**filial piety**—love and respect toward parents including bringing no dishonor to parents and taking care of elderly parents

**filial responsibility**—emphasizes duty, protection, care, and financial support for one's parents

**filicide**—murder of an offspring by a parent

**five love languages**—concept made popular by Gary Chapman, these languages are gifts, quality time, words of affirmation, acts of service, and physical touch

**flirting**—to show interest without serious intent

**foster parent**—neither a biological nor an adoptive parent but a person who takes care of and fosters a child taken into custody

**frail**—term used to define elderly people if they have difficulty with at least one personal care activity (feeding, bathing, toileting)

**friends with benefits**—involves platonic friends (i.e., those not involved in a romantic relationship) who engage in some degree of sexual intimacy on multiple occasions. This sexual activity could range from kissing to sexual intercourse and is a repeated part of your friendship such that it is not just a one-night stand

## G

**gatekeeper role**—term used to refer to the influence of the mother on the father's involvement with his children

**gay**—term which refers to women or men who prefer same-sex partners

**gender**—social construct which refers to the social and psychological characteristics associated with being female or male

**gender identity**—the psychological state of viewing oneself as a girl or a boy, and later as a woman or a man

**genderqueer**—the person does not identify as male or female

**gender role ideology**—the proper role relationships between women and men in a society

**gender role transcendence**—abandoning gender frameworks and looking at phenomena independent of traditional gender categories

**gender roles**—social norms which specify the socially appropriate behavior for females and males in a society

**Generation Y**—children of the baby boomers (typically born between 1979 and 1984)

**gerontology**—the study of aging

**gerontophobia**—fear or dread of the elderly, which may create a self-fulfilling prophecy

**GLBT**—general term which refers to gay, lesbian, bisexual, and transgender individuals

**granny dumping**—refers to adult children or grandchildren, burdened with the care of their elderly parent or grandparent, leaving the elder at the entrance of a hospital with no identification

## H

**Hanai adoption**—concept in traditional Hawaiian culture whereby a child may be given to another family or individual to love, rear, and educate

**hanging out**—refers to going out in groups where the agenda is to meet others and have fun

**hate crime**—violence against GLBT individuals to include verbal harassment, vandalism, sexual assault/rape, physical assault, and murder

**hedonism**—the belief that the ultimate value and motivation for human actions lie in the pursuit of pleasure and the avoidance of pain

**HER/his career marriage**—dual career marriage in which wife's career takes precedence

**heterosexism**—the institutional and societal reinforcement of heterosexuality as the privileged and powerful norm. Assumes that homosexuality is "bad"

**heterosexuality**—emotional and sexual attraction to individuals of the other sex

**HIS/her career marriage**—dual career marriage in which husband's career takes precedence

**HIS/HER career marriage**—dual career marriage in which both careers are viewed as equal

**homogamy**—the tendency for an individual to seek a mate who has similar characteristics

**homonegativity**—attaching negative connotations to homosexuality

**homophobia**—negative attitudes and emotions toward homosexuality and those who engage in homosexual behavior

**homosexuality**—refers to the predominance of cognitive, emotional, and sexual attraction to individuals of the same sex

**honeymoon**—the time following the wedding whereby the couple becomes isolated to recover from the wedding and to solidify their new status change from lovers to spouses

**honor crime/honor killing**—refers to unmarried women who are killed because they bring shame on their parents and siblings; occurs in Middle Eastern countries such as Jordan

**hooking up**—a sexual encounter that occurs between individuals who have no relationship commitment

**hypothesis**—a suggested explanation for a phenomenon

**hysterectomy**—form of female sterilization whereby the woman's uterus is removed

## I

**individualism**—making decisions that serve the individual's interests rather than the family's

**induced abortion**—the deliberate termination of a pregnancy through chemical or surgical means

**infatuation**—intense emotional feelings based on little actual exposure to the love object

**infertility**—the inability to achieve a pregnancy after at least one year of regular sexual relations without birth control, or the inability to carry a pregnancy to a live birth

**institution**—established and enduring patterns of social relationships (e.g., the family)

**integral love**—love conceptualized as being a combination of autistic and empathetic love

**integrative behavioral couple therapy (IBCT)**—therapy which focuses on the cognitions or assumptions of the spouses, which impact the way spouses feel and interpret each other's behavior

**internalized homophobia**—a sense of personal failure and self-hatred among lesbians and gay men resulting from social rejection and stigmatization of being gay

**intersexed individuals**—those with mixed or ambiguous genitals

**intimate partner homicide**—murder of a spouse

**intimate partner violence (IPV)**—an all-inclusive term that refers to crimes committed against current or former spouses, boyfriends, or girlfriends

**intimate terrorism (IT)**—behavior designed to control the partner

**intrafamilial**—child sexual abuse referring to exploitive sexual contact or attempted sexual contact between relatives before the victim is 18

**"I" statements**—statements which focus on the feelings and thoughts of the communicator without making a judgment on others

## J

**jealousy**—an emotional response to a perceived or real threat to an important or valued relationship

## L

**laparoscopy**—a form of tubal ligation that involves a small incision through the woman's abdominal wall just below the navel

**latchkey parents**—divorcing parents who spend every other week with the children in the family home so the children do not have to alternate between residences

**legal custody**—judicial decision regarding whether one or both parents will have decisional authority on major issues affecting the children

**leisure**—the use of time to engage in freely chosen activities perceived as enjoyable and satisfying

**lesbian**—a woman who prefers same-sex partners

**litigation**—a judge hears arguments from lawyers representing the respective spouses and decides issues of custody, child support, division of property, etc.

**living apart together (LAT)**—committed couple who does not live in the same home (and others such as children or elderly parents may live in the respective homes of the partners)

**long-distance relationship**—separated from a romantic partner by 500 or more miles, which precludes regular weekly face-to-face contact

**longitudinal research**—studying the same subjects across time—e.g., collecting data from the same couple during each of their four years of living together during college

**lose-lose solution**—a solution to a conflict in which neither partner benefits

**ludic love style**—views love as a game where the player has no intention of getting involved

**lust**—sexual desire

## M

**male rape myths**—beliefs that deny victim injury or make assumptions about his sexual orientation

**mania love style**—the out-of-control love whereby the person "must have" the love object. Obsessive jealousy and controlling behavior are symptoms of manic love

**marital rape**—forcible rape by one's spouse—a crime in all states

**marital success**—refers to the quality of the marriage relationship measured in terms of marital stability and marital happiness

**marriage**—a legal relationship that binds a couple together for the reproduction, physical care, and socialization of children

**marriage benefit**—when compared to being single, married persons are healthier, happier, live longer, have less drug use, etc.

**marriage rituals**—deliberate repeated social interactions that reflect emotional meaning to the couple

**marriage squeeze**—the imbalance of the ratio of marriageable-aged men to marriageable-aged women.

**masturbation**—stimulating one's own body with the goal of experiencing pleasurable sexual sensations

**mating gradient**—norm which gives social approval to men who seek out younger, less educated, less financially secure women and vice versa

**May-December marriage**—age dissimilar marriage (ADM) in which the woman is typically in the spring of her life (May) and her husband is in the later years (December)

**Medicaid**—a state welfare program for low-income individuals

**Medicare**—a federal health insurance program for people 65 and older

**Megan's Law**—law requiring that communities be notified of a neighbor's previous sex convictions

**military contract marriage**—a military person will marry a civilian to get more money and benefits from the government

**millennials**—workers born between 1980 and 1995

**modern family**—the dual-earner family, in which both spouses work outside the home

**mommy track**—stopping paid employment to spend time with young children

**momprenuer**—a woman who has a successful at-home business

## N

**negative commitment**—spouses who continue to be emotionally attached to and have difficulty breaking away from ex-spouses

**negotiation**—spouses discuss and resolve the issues of custody, child support, and division of property themselves

**neonaticide**—killing a baby the first day of life

**no-fault divorce**—neither party is identified as the guilty party or the cause of the divorce

**nomophobia**—the individual is dependent on virtual environments to the point of having a social phobia

**nonverbal communication**—the "message about the message," using gestures, eye contact, body posture, tone, volume, and rapidity of speech

**nuclear family**—consists of you, your parents, and your siblings or you, your spouse, and your children

## O

**obsessive relational intrusion (ORI)**—the relentless pursuit of intimacy with someone who does not want it

**occupational sex segregation**—the concentration of women and men in different occupations

**oophorectomy**—form of female sterilization whereby the woman's ovaries are removed

**open-ended question**—question which elicits a lot of information

**open minded**—an openness to understanding alternative points of view, values, and behaviors

**open relationship**—relationship in which the partners agree that each may have emotional and sexual relationships with those outside the dyad

**operant conditioning**—a method of learning that occurs through reward and punishments for behavior

**oppositional defiant disorder**—disorder in which children fail to comply with requests of authority figures

**overindulgence**—defined as more than just giving children too much, includes overnurturing and providing too little structure

**ovum transfer**—a fertilized egg is implanted in the uterine wall

**oxytocin**—a hormone released from the pituitary gland during the expulsive stage of labor that has been associated with the onset of maternal behavior in lower animals

## P

**palimony**—refers to the amount of money one "pal" who lives with another "pal" may have to pay if the partners end their relationship

**palliative care**—health care for the individual who has a life-threatening illness which focuses on relief of pain/suffering and support for the individual

**pantagamy**—a group marriage in which each member of the group is "married" to the others

**parental alienation**—estrangement of a child from a parent due to one parent turning the child against the other

**parental alienation syndrome**—an alleged disturbance in which children are obsessively preoccupied with deprecation and/or criticism of a parent; denigration that is unjustified and/or exaggerated

**parental consent**—a woman needs permission from a parent to get an abortion if under a certain age, usually 18

**parental investment**—any investment by a parent that increases the offspring's chance of surviving and thus increases reproductive success

**parental notification**—a woman has to tell a parent she is getting an abortion if she is under a certain age, usually 18, but she doesn't need parental permission

**parenting**—defined in terms of roles including caregiver, emotional resource, teacher, and economic resource

**parenting self-efficiency**—feeling competent as a parent

**parricide**—murder of a parent by an offspring

**phased retirement**—an employee agreeing to a reduced work load in exchange for reduced income

**physical custody**—the distribution of parenting time between divorced spouses

**polyamory**—a lifestyle in which two lovers embrace the idea of having multiple lovers. By agreement, each partner may have numerous emotional and sexual relationships

**polyandry**—type of marriage in which one wife has two or more husbands

**polygamy**—a generic term for marriage involving more than two spouses

**polygyny**—type of marriage involving one husband and two or more wives

**pool of eligibles**—the population from which a person selects a mate

**positive androgyny**—a view of androgyny that is devoid of the negative traits associated with masculinity (e.g., aggression) and femininity (e.g., being passive)

**possessive jealousy**—involves attacking the partner or the alleged person to whom the partner is showing attention

**postmodern family**—lesbian or gay couples and mothers who are single by choice, which emphasizes that a healthy family need not be the traditional heterosexual, two-parent family

**postpartum depression**—a more severe reaction following the birth of a baby which occurs in reference to a complicated delivery as well as numerous physiological and psychological changes occurring during pregnancy, labor, and delivery; usually in the first month after birth but can be experienced after a couple of years have passed

**postpartum psychosis**—a reaction in which a woman wants to harm her baby

**power**—the ability to impose one's will on the partner and to avoid being influenced by the partner

**pragma love style**—love style that is logical and rational. The love partner is evaluated in terms of pluses and minuses and is regarded as a good or bad "deal"

**pregnancy**—when the fertilized egg is implanted (typically in the uterine wall)

**pregnancy coercion**—coercion by a male partner for the woman to become pregnant

**prenuptial agreement**—a contract between intended spouses specifying which assets will belong to whom and who will be responsible for paying for what in the event of a divorce or when the marriage ends by the death of one spouse

**primary groups**—small numbers of individuals among whom interaction is intimate and informal

**principle of least interest**—the person who has the least interest in the relationship controls the relationship

**procreative liberty**—the freedom to decide to have children or not

**projection**—attributing one's own thoughts, feelings, and desires to someone else while avoiding recognition that these are one's own thoughts, feelings, and desires

**pronatalism**—cultural attitude which encourages having children

## Q

**quality of life**—one's physical functioning, independence, economic resources, social relationships, and spirituality

**queer**—inclusive term used by individuals desiring to avoid labels. Someone who labels him/herself as "queer" could be gay, lesbian, bisexual, pansexual, trans, intersexed, or non-conforming heterosexual

## R

**rationalization**—the cognitive justification for one's own behavior that unconsciously conceals one's true motives

**reactive attachment disorder**—common among children who were taught as infants that no one cared about them; these children have no capacity to bond emotionally with others since they have no learning history of the experience and do not trust adults, caretakers, or parents

**reactive jealousy**—jealous feelings that are a reaction to something the partner is doing

**red zone**—the first month of the first year of college when women are most likely to be victims of sexual abuse

**reflective listening**—paraphrasing or restating what the person has said to you while being sensitive to what the partner is feeling

**relativism**—value system emphasizing that sexual decisions should be made in the context of a particular relationship

**resiliency**—a family's strength and ability to respond to a crisis in a positive way

**responsiveness**—refers to the extent to which parents respond to and meet the needs of their children

**rite of passage**—an event that marks the transition from one social status to another

**role compartmentalization**—strategy used to separate the roles of work and home so that an individual does not think about or dwell on the problems of one when they are at the physical place of the other

**role overload**—not having the time or energy to meet the demands or responsibilities in the roles of wife, parent, and worker

**role strain**—the anxiety that results from being able to fulfill only a limited number of role obligations

**role theory of mate selection**—theory which focuses on the social learning of roles. A son or daughter models after the parent of the same sex by selecting a partner similar to the one the parent selected

**romantic love**—an intense love whereby the lover believes in love at first sight, only one true love, and that love conquers all

**rophypnol**—causes profound, prolonged sedation and short-term memory loss; also known as the date rape drug, roofies, Mexican Valium, or the "forget (me) pill"

## S

**salpingectomy**—type of female sterilization whereby the fallopian tubes are cut and the ends are tied

**sanctification**—viewing the marriage as having divine character or significance

**sandwich generation**—generation of adults who are "sandwiched" between caring for their elderly parents and their own children

**satiation**—repeated exposure to a stimulus that results in the loss of its ability to reinforce

**second-parent adoption**—legal procedure that allows individuals to adopt their partner's biological or adoptive child without terminating the first parent's legal status as parent

**second shift**—the housework and child care that employed women engage when they return home from their jobs

**secondary groups**—groups in which the interaction is impersonal and formal

**secondary virginity**—a sexually initiated person's deliberate decision to refrain from intimate encounters for a set period of time and to refer to that decision as a kind of virginity (rather than "mere" abstinence)

**self-compassion**—refers to being kind, caring and understanding toward oneself when feelings of suffering are present

**sex**—the biological distinction between females and males

**sex roles**—roles defined by biological constraints and enacted by members of one biological sex only—for example, wet nurse, sperm donor, child-bearer

**sexism**—an attitude, action, or institutional structure that subordinates or discriminates against individuals or groups because of their biological sex

**sexting**—sending erotic text and photo images via a cell phone

**sextortion**—online sexual extortion

**sexual compliance**—an individual willingly agrees to participate in sexual behavior without having the desire to do so

**sexual double standard**—the view that encourages and accepts the sexual expression of men more than women

**sexual identity**—term used synonymously with sexual orientation

**sexual orientation**—classification of individuals as heterosexual, bisexual, or homosexual, based on their emotional, cognitive, sexual attractions, and self-identity

**sexual readiness**—factors such as autonomy of decision (not influenced by alcohol or peers), consensuality (both partners equally willing), and absence of regret (the right time for me) are more important than age as meaningful criteria for determining when a person is ready for first intercourse

**sexual values**—moral guidelines for making sexual choices in nonmarital, marital, heterosexual, and homosexual relationships

**shared parenting dysfunction**—behaviors on the part of both partners focused on hurting the other parent that are counterproductive for the well-being of the children

**siblicide**—murder of a sibling

**sibling relationship aggression**—behavior of one sibling toward another sibling that is intended to induce social harm or psychic pain in the sibling

**single-parent family**—family in which there is only one parent and the other parent is completely out of the child's life through death, sperm donation, or abandonment and no contact is made with the other parent

**single-parent household**—one parent has primary custody of the child/children with the other parent living outside of the house but still being a part of the child's family; also called binuclear family

**situational couple violence (SCV)**—conflict escalates over an issue and one or both partners lose control

**snooping**—investigating (without the partner's knowledge or permission) a romantic partner's private communication (such as text messages, e-mail, and cell phone use) motivated by concern that the partner may be hiding something

**social allergy**—being annoyed and disgusted by a repeated behavior on the part of the partner

**social exchange framework**—views interaction and choices in terms of cost and profit

**socialization**—the process through which we learn attitudes, values, beliefs, and behaviors appropriate to the social positions we occupy

**sociobiology**—theory which emphasizes the biological basis for all social behavior, including mate selection

**sociological imagination**—the influence of social structure and culture on interpersonal decisions

**spatial homogamy**—romantic partners tend to have grown up within six kilometers of each other

**spectatoring**—involves mentally observing your sexual performance and that of your partner

**spillover thesis**—one's work role impacts one's family life in terms of working overtime, being on call on weekends, etc. Rarely does the family role dictate what will happen in the work role

**spontaneous abortion (miscarriage)**—the unintended termination of a pregnancy

**stalking**—unwanted following or harassment of a person that induces fear in the victim

**stepfamily**—family in which spouses in a new marriage bring children from previous relationships into the new home

**stepism**—the assumption that stepfamilies are inferior to biological families

**sterilization**—a permanent surgical procedure that prevents reproduction

**STI (sexually transmitted infection)**—refers to the general category of sexually transmitted infections such as chlamydia, genital herpes, gonorrhea, and syphilis

**storge love style**—also known as companionate love, a calm, soothing, nonsexual love devoid of intense passion

**stress**—reaction of the body to substantial or unusual demands (physical, environmental, or interpersonal)

**structure-function framework**—emphasizes how marriage and family contribute to society

**superwoman/supermom**—a cultural label that allows a woman to regard herself as very efficient, bright, and confident; usually a cultural cover-up for an overworked and frustrated woman

**swinging**—married or pair-bonded couple agree that each will have recreational sex with others while maintaining an emotional allegiance to each other

**symbolic interaction framework**—views marriages and families as symbolic worlds in which the various members give meaning to each other's behavior

## T

**telerelationship therapy**—therapy sessions conducted online, often through Skype, where both therapist and couple can see and hear each other

**texting**—short typewritten messages sent via a cell phone that are used to "commence, advance, and maintain" interpersonal relationships

**thanatology**—the examination of the social dimensions of death, dying, and bereavement

**THEIR career marriage**—dual-career marriage in which spouses share a career or work together

**theoretical frameworks**—a set of interrelated principles designed to explain a particular phenomenon

**therapeutic abortion**—abortions performed to protect the life or health of the woman

**third shift**—the expenditure of emotional energy by a spouse or parent in dealing with various emotional issues in family living

**three-way marriage**—typically one male married to two women (who may be bisexual). Examples have existed in both Brazil and the Netherlands—the arrangements are not full legal marriages

**time-out**—a noncorporal form of punishment that involves removing the child from a context of reinforcement to a place of isolation

**traditional family**—the two-parent nuclear family, with the husband as breadwinner and the wife as homemaker

**transgender**—a generic term for a person of one biological sex who displays characteristics of the other sex

**transition to parenthood**—period from the beginning of pregnancy through the first few months after the birth of a baby during which the mother and father undergo changes

**transphobia**—a set of negative attitudes toward transsexuality or those who self-identify as transsexual

**transracial adoption**—adopting children of a race different from that of the parents

**transsexual**—an individual with the biological and anatomical sex of one gender (for example, male) but the self-concept of the opposite sex (female)

## U

**unrequited love**—love that is not returned

**utilitarianism**—individuals rationally weigh the rewards and costs associated with behavioral choices

**uxoricide**—the killing of one's wife

## V

**vasectomy**—form of male sterilization whereby the vas deferens is cut so that sperm cannot continue to travel outside the body via the penis

**video-mediated communication (VMC)**—individuals are able to see and hear others they are separated from to simulate their presence and enjoy "being with" their beloved

**violence**—physical aggression with the purpose to control, intimidate, and subjugate another human being

## W

**win-lose solution**—outcome of a conflict in which one partner wins and the other loses

**win-win relationship**—relationship in which conflict is resolved so that each partner derives benefits from the resolution

## Y

**"you" statement**—statement that blames or criticizes the listener and often results in increasing negative feelings and behavior in the relationship

# References


## A

Abboud, P. 2013. The third gender: Samoa's Fa'afafine people. http://www.pedestrian.tv/features/arts-and-culture/meet-the-third-gender-samoas-faafafine-people/70b3c7c8-66fc-4453-9f06-1de911d249ee.htm Retrieved Sept 29.

Abowitz, D., D. Knox, and K. Berner. 2011. Traditional and non-traditional husband preference among college women. Annual Meeting Eastern Sociological Society, Philadelphia, PA. March.

Abrahamson, I., H. Rafat, K. Adeel, and M. J. Schofield. 2012. What helps couples rebuild their relationship after infidelity. *Journal of Family Issues* 33: 1494–1519.

Adams, H. L. and L. R. Williams. 2014. "It's not just you two": A grounded theory of peer-influenced jealousy as a pathway to dating violence among acculturating Mexican American adolescents. *Psychology of Violence.* January. Online.

Adolfsen, A., J. Iedema, and S. Keuzenkamp. 2010. Multiple dimensions of attitudes about homosexuality: Development of a multifaceted scale measuring attitudes toward homosexuality. *Journal of Homosexuality* 57: 1237–1257.

Aducci, A. J., J. A. Baptist, J. George, P. M. Barros, and B. S. Nelson Goff. 2012. Military wives' experience during OIF/OEF deployment. Paper, National Council on Family Relations, November, Orlando, FL.

Agate, J. R., R. Zabriskie, S. T. Agate, and R. Poff. 2009. Family leisure satisfaction and satisfaction with family life. *Journal of Leisure Research* 41: 205–223.

Ahmed, A. A., G. A. H. Van den Elsen, M. A. Van der Marck, and M. G. M Olde Rickkert. 2014. Medicinal use of cannabis and cannabinoids in older adults: Where is the evidence? *American Journal of the American Geriatrics Society* 62: 410–411.

Ahnert, L., and M. E. Lamb. 2003. Shared care: Establishing a balance between home and child care settings. *Child Development* 74: 1044–1049.

Alam, R., M. Barrera, N. D'Agostino, D. B. Nicholas, and G. Schneiderman. 2012. Bereavement experiences of mothers and fathers over time after the death of a child due to cancer. *Death Studies* 36: 1–22.

Alexander, P. C., A. Tracy, M. Radek, and C. Koverola 2009. Predicting stages of change in battered women. *Journal of Interpersonal Violence* 24: 1652–1672.

Alford, J. J., P. K. Hatemi, J. R. Hibbing, N. G. Martin, and L. J. Eaves. 2011. The politics of mate choice. *The Journal of Politics* 73: 362–379.

Ali, L. 2011. Sexual self-disclosure predicting sexual satisfaction. Poster, National Council on Family Relations, Orlando, November.

Ali, M. M. and D. S. Dwyer. 2010. Social network effects in alcohol consumption among adolescents. *Addictive Behaviors* 35: 337–342.

Allen, D. W. 2010. A better method for assessing the value of "housewife" services. *American Journal of Family Law* 23: 219–223.

Allen, E. S. and D. C. Atkins. 2012. The association of divorce and extramarital sex in a representative U.S. sample. *Journal of Family Issues* 33: 1477–1493.

Allen, E. S., G. K. Rhoades, S. M. Stanley, H. J. Markman et al. 2008. Premarital precursors of marital infidelity. *Family Process* 47: 243–260.

Allen, K. N. and D. Wozniak. 2011. The language of healing: Women's voices in healing and recovering from domestic violence. *Social Work in Mental Health* 9: 37–55.

Allen, K. R., E. K. Husser, D. J. Stone, and C. E. Jordal. 2008. Agency and error in young adults' stories of sexual decision making. *Family Relations* 57: 517–529.

Allen, R. E. S. and J. L. Wiles. 2013. How older people position their late-life childesssness: A qualitative study. *Journal of Marriage and the Family* 75: 206–220.

Allendorf, K. 2013. Schemas of marital change: From arranged marriages to eloping for love. *Journal of Marriage and the Family* 75: 453–469.

Allendorf, K. and D. J. Ghimire. 2013. Determinants of marital quality in an arranged marriage society. *Social Science Research* 42: 59–70.

Allgood, S. M. and J. Gordon. 2010. Premarital advice: Do engaged couples listen? Poster, National Council on Family Relations annual meeting, November 3–5. Minneapolis, MN.

Alterovitz, S. R. and G. A. Mendelsohn. 2013. Relationship goals of middle-aged, young-old, and old-old Internet daters: An analysis of online person ads. *Journal of Aging Studies* 27: 159–165.

Amato, P. R. 2010. Research on divorce: Continuing trends and new developments. *Journal of Marriage and Family* 72: 650–666.

Amato, P. R. 2004. Tension between institutional and individual views of marriage. *Journal of Marriage and Family* 66: 959–65.

Amato, P. R., A. Booth, D. R. Johnson, and S. F. Rogers. 2007. *Alone together: How marriage in America is changing.* Cambridge, MA: Harvard University Press.

Amato, P. R. and J. B. Kane. 2011. Life-course pathways and the psychosocial adjustment of young adult women. *Journal of Marriage & Family* 73: 279–295.

Amato, P. R., J. B. Kane, and S. James. 2011. Reconsidering the 'Good Divorce' *Family Relations* 60: 511–524.

American Psychological Association. 2012. Lesbian and gay parenting. http://www.apa.org/pi/lgbt/resources/parenting.aspx

Ames, C. M. and W. V. 2012. Preventing repeat abortion in Canada: Is the immediate insertion of intrauterine devices postabortion a cost-effective option associated with fewer repeat abortions? *Contraception* 85: 51–55.

Andersen, C. 2011. *William and Kate: A royal love story.* New York: Gallery.

Anderson, J. R., M. J. Van Ryzin, and W. J. Doherty. 2010. Developmental trajectories of marital happiness in continuously married individuals: A group-based modeling approach. *Journal of Family Psychology* 24: 587–596.

Anderson, K. L. 2010. Conflict, power, and violence in families. *Journal of Marriage and Family* 72: 726–742.

Anderson, S. 2010. The polygamists: An exclusive inside look inside the FLDS. *National Geographic* February 38–51.

Ansello, E. 2012. Humor with aging: From ways to wisdom. Southern Gerontological Society, Nashville, TN, April . Dr. Ansello, Virginia Center on Aging.

Antunes-Alves, S. and J. De Stefano. 2014. Intimate partner violence: Making the case for joint couple treatment. *The Family Journal* 22: 62–68.

Aoyagi, Y. and R. Shephard. 2011. Habitual physical activity and health in the elderly: The Nakanojo study. *Geriatrics & Gerontology International* 10: 236–243.

Aponte, R. and R. Pessagno. 2010. The communications revolution and its impact on the family: Significant, growing, but skewed and limited in scope. *Marriage & Family Review* 45: 576–586.

Armstrong, A. B. and E. F. Clinton. 2012. Altering positive/negative interaction ratios of mothers and young children. *Child Behavior and Family Therapy* 34: 231–242.

Arnold, A. L., C. W. O'Neal, C. Bryant, K. A. S. Wickrama, and C. Cutrona. 2011. Influences of intra and interpersonal factors in marital success. Presentation, National Council on Family Relations, Orlando, November.

Ashwin and Isupova. 2014. "Behind every great man. . . .": The male marriage wage premium examined qualitatively. *Journal of Marriage and Family* 76: 37–55.

Aspinwall, L. G., J. M. Taber, S. L. Leaf, W. Kohlmann, and S. A. Leachman. 2013. Genetic testing for hereditary melanoma and pancreatic cancer: A longitudinal study of psychological outcome. *PsychoOncology* 22: 276–289.

Aubrey, J. S. and S. E. Smith. 2013. Development and validation of the endorsement of the Hookup Culture Index. *Journal of Sex Research* 50: 435–448.

Aulette, J. R. 2010. *Changing American families.* Boston: Allyn and Bacon.

Aumann, K., E. Galinsky, K. Sakai, M. Brown, and J. T. Bond. 2010. The Elder Care Study: Everyday realities and wishes for change. Families and Work Institute. Available at http://www.familiesandwork.org

Averett, S. L., L. M. Argys, and J. Sorkin. 2013. In sickness and in health: An examination of relationship status and health using data from the Canadian National Public Health Survey. *Review of Economics of the Household* 2: 599–633.

Azziz-Baumgartner, E., L. McKeown, P. Melvin, D. Quynh, and J. Reed. 2011. Rates of femicide in women of different races, ethnicities, and places of birth: Massachusetts, 1993–2007. *Journal of Interpersonal Violence* 26: 1077–1090.

## B

Backhans, M. C. and T. Hemmingsson. 2012. Unemployment and mental health – Who is not affected? *European Journal of Public Health* 22: 429–433.

Baer, R. A., J. Carmody, and M. Hunsinger. 2012. Weekly change in mindfulness and perceived stress reduction program. *Journal of Clinical Psychology* 68: 755–765.

Bagarozzi, D. A. 2008. Understanding and treating marital infidelity: A multidimensional model. *The American Journal of Family Therapy* 36: 1–17.

Baker, A. J. L. and D. Darnall. 2007. A construct study of the eight symptoms of severe parental alienation syndrome: A survey of parental experiences. *Journal of Divorce & Remarriage* 47: 55–62.

Baker, J., J. McHale, A. Strozier, and D. Cecil. 2010. Mother–grandmother coparenting relationships in families with incarcerated mothers: A pilot investigation. *Family Process* 49: 165–184.

Balderrama-Durbin, C. M., E. S. Allen, and G. K. Rhoades. 2012. Demand and withdraw behaviors in couples with a history of infidelity. *Journal of Family Psychology* 26: 11–17.

Baldur-Felskov, B., S. K. Kjaer, V. Albieri, M. Steding-Jessen, T. Kjaer, C. Johansen, S. O. Dalton, and A. Jensen. 2013. Psychiatric disorders in women with fertility problems: Results from a large Danish register-based cohort study. *Human Reproduction* 28: 683–690.

Ballard, S. M., C. Sugita, and K. H. Gross. 2011. A qualitative investigation of the sexual socialization of emerging adults. Paper, 73rd Conference of the National Council on Family Relations, Orlando, Florida; November 17.

Ballard, S. M. and A. C. Taylor. 2011. *Family life education with diverse populations.* Thousand Oaks, CA: Sage.

Baltazar, A. M., B. Johnson, D. McBride, G. Hopkins, and C. Vanderwaal. 2011. Parental influence on substance and inhalant use. Poster, National Council on Family Relations, Orlando, November 18.

Baly, A. R. 2010. Leaving abusive relationships: Constructions of self and situation by abused women. *Journal of Interpersonal Violence* 25: 2297–2315.

Barber, L. L. and M. L. Cooper. 2014. Rebound sex: Sexual motives and behaviors following a relationship breakup. *Archives of Sexual Behavior* 43: 251–265.

Barclay, L. 2013. Liberal daddy quotas: Why men should take care of the children, and how liberals can get them to do it. *Hypatia* 28: 163–178.

Barelds, D. P. H., and P. Barelds-Dijkstra. 2007. Love at first sight or friends first? Ties among partner personality trait similarity, relationship onset, relationship quality, and love. *Journal of Social and Personal Relationships* 24: 479–496.

Barelds-Dijkstra, D. P. H., and P. Barelds. 2007. Relations between different types of jealousy and self and partner perceptions of relationship quality. *Clinical Psychology & Psychotherapy* 4: 176–188.

Barnes, H., D. Knox, & J. Brinkley. 2012, March 23. CHEATING: Gender differences in reactions to discovery of a partner's cheating. Poster, Southern Sociological Society, New Orleans.

Barnett, M. A., L. V. Scaramella, T. K. Neppl, L. L. Ontai, and R. D. Conger. 2010. Grandmother involvement as a protective factor for early childhood social adjustment. *Journal of Family Psychology* 24: 635–645.

Barnett, A. E. 2013. Pathways of adult children providing care to older parents. *Journal of Marriage and Family* 75: 178–190.

Barnett, R. C., K. C. Gareis, L. Sabattini, and N. M. Carter. 2010. Parental concerns about after-school time: Antecedents and correlates among dual-earner parents. *Journal of Family Issues* 31: 606–625.

Barovick, H. 2012. Niche aging. *Time,* March 22, pp. 84–87.

Barriger, M. and C. J. Velez-Blasini. 2013. Descriptive and injunctive social norm overestimation in hooking up and their role as predictors of hook-up activity in a college student sample. *Journal of Sex Research* 50: 84–94.

Bartholomew, K., K. V. Regan, M. A. White, and D. Oram. 2008. Patterns of abuse in male same-sex relationships. *Violence and Victims* 23, 617–637.

Bartle-Haring, S. 2010. Using Bowen theory to examine progress in couple therapy. *The Family Journal* 18: 106–115.

Barton, A. L., & M. S. Kirtley. 2012. Gender differences in the relationships among parenting styles and college student mental health. *Journal of American College Health* 60, 21–26.

Barzoki, M. H., N. Seyedroghani, and T. Azadarmaki. 2012. Sexual dissatisfaction in a sample of married Iranian women. *Sexuality and Culture.* 18: 106–115.

Bauer, K. W., M. O. Hearst, K. Escoto, J. M. Berge, and D. Neumark-Sztainer. 2012. Parental employment and work–family stress: Associations with family food environments. *Social Science & Medicine* 75: 496–504.

Bauerlein, M. 2010. Literary learning in the hyperdigital age. *Futurist* 44: 24–25.

Baumle, A. K. 2013. The cost of parenthood: Unraveling the effects of sexual orientation and gender on income. *Social Science Quarterly* 90: 983–1002.

Baumrind, D. 1966. Effects of authoritative parental control on child behavior. *Child Development* 37: 887–907.

Bauserman, R. 2012. A meta-analysis of parental satisfaction, adjustment, and conflict in joint custody and sole custody following divorce. *Journal of Divorce and Remarriage* 53: 464–488.

Becerra, R. M. 2012. The Mexican American family. In *Ethnic families in America 5th ed.*, R. Wright, Jr., C. H. Mindel, T V. Tran, and R. W. Habenstein, eds. 100–111. Upper Saddle River, NJ: Pearson.

Becker, A. B. and M. E. Todd. 2013. A new American family? Public opinion toward family studies and perceptions of the challenges faced by children of same-sex parents. *Journal of GLBT Family Studies* 9: 425–448.

Becker, K. 2010. Risk behavior in adolescents: What we need to know. *European Psychiatry* 25: 85–95.

Becker, K. D., J. Stuewig, and L. A. McCloskey. 2010. Traumatic stress symptoms of women exposed to different forms of childhood victimization and intimate partner violence. *Journal of Interpersonal Violence* 25: 1699–1715.

Beekman, D. 1977. *The mechanical baby: A popular history of the theory and practice of child rearing.* Westport, Ct: Lawrence Hill & Company.

Beerthuizen, R. 2010. State-of-the-art of non-hormonal methods of contraception: V. Female sterilization. European *Journal of Contraception & Reproductive Health Care* 15: 124–135.

Benjamin, Le, N. L. Dove, C. R. Agnew, M. S. Korn, and A.A. Musso. 2010. Predicting nonmarital romantic relationship dissolution: A meta-analytic synthesis. *Personal Relationships* 17: 377–390.

Bennett, S., V. L. Banyard, and L. Garnhart. 2014. To act or not to act, that is the question? Barriers and facilitators of bystander intervention. *Journal of Interpersonal Violence* 29: 3, 476–496.

Bennetts, L. 2007. *The feminine mistake: Are we giving up too much?* New York: Voice/Hyperion.

Ben-Zur, H. 2012. Loneliness, optimism, and well-being among married, divorced, and widowed individuals. *Journal of Psychology* 146: 23–36.

Bergdall, A. R., J. M. Kraft, K. Andes, M. Carter, K. Hatfield-Timajchy, and L. Hock-long. 2012. Love and hooking up in the new millennium: Communication technology and relationships among urban African American and Puerto Rican young adults. *Journal of Sex Research* 49: 570–582.

Berge, J. M., K. W. Bauer, R. MacLehose, M. E. Eisenberg, and D. Neumark-Sztainer. 2014. Associations between relationship status and day-to-day health behaviors and weight among diverse young adults. *Families, Systems, and Health.* (In press.)

Bergman, K., R. J. Rubio, R. J. Green, and E. Padron. 2010. Gay men who become fathers via surrogacy: The transition to parenthood. *Journal of GLBT Family Studies.* 6: 111–141.

Bergstrand, C. and J. B. Williams. 2000. Today's alternative marriage styles: The case for swingers. *Electronic Journal of Human Sexuality* 3: Oct 10. www.ejhs.org

Bermant, G. 1976. Sexual behavior: Hard times with the Coolidge Effect. In *Psychological research: The inside story,* ed. M. H. Siegel and H. P. Zeigler. New York: Harper and Row.

Bersamin, M. M. et al. 2014. Risky business: Is there an association between casual sex and mental health among emerging adults? *The Journal of Sex Research* 51: 43–51.

Bersamin, M. M., M. J. Paschall, R. F. Saltz, and B. L. Zamboanga. 2012. Young adults and casual sex: The relevance of college drinking settings. *Journal of Sex Research* 49: 274–281.

Berscheid, E. 2010. Love in the fourth dimension. *Annual Review of Psychology* 61: 1–25.

Biblarz, T. J. and E. Savci. 2010. Lesbian, gay, bisexual, and transgender families. *Journal of Marriage and Family* 72: 480–497.

Biblarz, T. J. and J. Stacey. 2010. How does the gender of parents matter? *Journal of Marriage and Family* 72: 3–22.

Biehle, S. N. and K. D. Michelson. 2012. First-time parent's expectations about the division of childcare and play. *Journal of Family Psychology* 26: 36–45.

Bisakha, S. 2010. The relationship between frequency of family dinner and adolescent problem behaviors after adjusting for other family characteristics. *Journal of Adolescence* 33: 187–196.

Blackstrom, L., E. A. Armstrong, and J. Puentes. 2012. Women's negotiation of cunnilingus in college hookups and relationships. *Journal of Sex Research* 49: 1–12.

Blackwell, C. W. and S. F. Dziegielewski. 2012. Using the Internet to meet sexual partners: Research and practice implications. *Journal of Social Service Research* 38: 46–55.

Blair, S. L. 2010. The influence of risk-taking behaviors on the transition into marriage: An examination of the long-term consequences of adolescent behavior. *Marriage & Family Review* 46: 126–146.

Bleske-Rechek, A., E. Somers, C. Micke, L. Erickson, L. Matteson, C. Stocco, B. Schumacher, and L. Ritchie. 2012. Benefit or burden? Attraction in cross-sex friendship. *Journal of Social and Personal Relationships* 29: 569–596.

Blomquist, B. A. and T. A. Giuliano. 2012. Do you love me, too? Perceptions of responses to "I love you." *North American Journal of Psychology* 14: 407–418.

Blumer, H. G. 1969. The methodological position of symbolic interaction. In *Symbolic interactionism: Perspective and method.* Englewood Cliffs, NJ: Prentice-Hall.

Bodenmann, G., D. C. Atkins, M. Schär, and V. Poffet. 2010. The association between daily stress and sexual activity. *Journal of Family Psychology* 24: 271–279.

Boehnke, M. 2011. Gender role attitudes around the globe: Egalitarian vs. traditional views. *Asian Journal of Social Science* 39: 57–74.

Bogt, T., R. Engels, S. Boger and M. Kloosterman. 2010. "Shake It Baby, Shake It": Media preferences, sexual attitudes and gender stereotypes among adolescents. *Sex Roles* 63: 844–859.

Boislard P. and F. Poulin. 2011. Individual, familial, friends-related and contextual predictors of early sexual intercourse. *Journal of Adolescence* 34: 289–300.

Booth, C. L., K. A. Clarke-Stewart, D. L. Vandell, K. McCartney, and M. T. Owen. 2002. Child-care usage and mother-infant "quality time." *Journal of Marriage and the Family* 64: 16–26.

Bordoloi, S. D., N. O'Brien, L. Edwards, and R. Preli. 2013. Creating an inclusive and thriving profession: Why the American Association of Marriage and Family Therapy (AAMFT) needs to advocate for same-sex marriage. *Journal of Feminist Family Therapy* 25: 41–55.

Bos, H. and T. G. M. Sandfort. 2010. Children's gender identity in lesbian and heterosexual two-parent families. *Journal Sex Roles* 62: 114–126.

Boss, P. 2013. Myth of closure: What is normal grief after loss, clear or ambiguous? National Council on Family Relations annual meeting, San Antonio.

Boswell, S. S. 2012. Predicting trainee ageism using knowledge, anxiety, compassion, and contact with older adults. *Educational Gerontology* 38: 733–741.

Boyes, M. 2013. Outdoor adventure and successful ageing. *Ageing and Society* 33: 644–665.

Brackett, A., J. Fish, and D. Knox. 2013. Saying goodbye in romantic relationships: Strategies and outcomes. Paper, Southern Sociological Society, Atlanta, April.

Bradford, K., W. Stewart, B. Higginbotham, L. Skogrand, and C. Broadbent. 2012. Validating the Relationship Knowledge Questionnaire. Poster, Annual Meeting of the National Council on Family Relations. Phoenix, Nov 1.

Bradshaw, C., A. S. Kahn, and B. K. Saville. 2010. To hook up or date: Which gender benefits? *Journal Sex Roles* 62: 661–669.

Braithwaithe, S. R., R. Delevi, and F. D. Fincham. 2010. Romantic relationships and the physical and mental health of college students. *Personal Relationships* 17: 1–12.

Brandes, M., C. Hamilton, J. van der Steen, J. de Bruin, R. Bots, W. Nelen, and J. Kremer. 2011. Unexplained infertility: Overall ongoing pregnancy rate and mode of conception. *Human Reproduction* 26: 360–368.

Bratter, J. L. and R. B. King. 2008. "But Will It Last?": Marital instability among interracial and same-race couples. *Family Relations* 57: 160–71.

Braun, M., K. Mura, M. Peter-Wright, R. Hornhung, and U. Scholz. 2010. Toward a better understanding of psychological well-being in dementia caregivers: The link between marital communication and depression. *Family Process* 49: 185–203.

Bredow, C. A., T. L. Huston, and G. Noval. 2011. Market value, quality of the pool of potential mates, and singles' confidence about marrying. *Personal Relationships* 18: 39–57.

Brimhall, A. S. and M. L. Engblom-Deglmann. 2011. Starting over: A tentative theory exploring the effects of past relationships on postbereavement remarried couples. *Family Process* 50: 47–62.

Brimhall, A., K. Wampler, and T. Kimball. 2008. A warning from the past, altering the future: A tentative theory of the effect of past relationships on couples who remarry. *Family Process* 47: 373–387.

Brooks, R. 2014. Learn from retirees learn from biggest regrets. *USA Today*, March 13. 3B.

Brown, A. 2010. How to accurately interpret a peer's social class: Symbols of class status and presentation of self in college students. Paper, Southern Sociological Society, Atlanta, April.

Brown, P. J. and J. Sweeney. 2009. The anthropology of overweight, obesity and the body. *AnthroNotes* Volume 30 No. 1.

Brown, S. and K. Guthrie. 2010. Why don't teenagers use contraception? A qualitative interview study. *European Journal of Contraception & Reproductive Health Care* 15: 197–204.

Brown, S. L. 2010. Marriage and child well-being: Research and policy perspectives. *Journal of Marriage and Family* 72: 1059–1077.

Brown, S. L. and I. F. Lin. 2012. The Gray Divorce Revolution: Rising divorce among middle-aged and older adults, 1990–2009. National Center for Marriage and Family Research. Bowling Green State University. Working Paper Series. WP- 12– 04 March.

Brown, S. L. and S. K. Shinohara. 2013. Dating relationships in older adulthood: A national portrait. *Journal of Marriage and Family* 75: 1194–1202.

Brown, S. M. and J. Porter. 2013. The effects of religion on remarriage among American women: Evidence from the National Survey of Family Growth. *Journal of Divorce and Remarriage* 54: 142–162.

Brownridge, D. A. 2010. Does the situational couple intimate violence terrorism typology explain cohabitors' high risk of intimate partner violence? *Journal of Interpersonal Violence* 25: 1264–1283.

Brucker, H., and P. Bearman. 2005. After the promise: The STD consequences of adolescent virginity pledges. *Journal of Adolescent Health* 36: 271–278.

Bryant, C. D. 2007. The sociology of death and dying. In *21st century sociology: A reference handbook,* ed. Clifton D. Bryant and Dennis L. Peck, 156–66. Thousand Oaks, California: Sage.

Bulanda, J. R. 2011. Gender, marital power, and marital quality in later life. *Journal of Women and Aging* 23, 3–22.

Bulanda, J. R., and S. L. Brown. 2007. Race-ethnic differences in marital quality and divorce. *Social Science Research* 36: 945–959.

Bunger, R. 2014. Department of Anthropology, East Carolina University, Personal communication.

Burke, M. L., G. G. Eakes, and M. A. Hainsworth. 1999. Milestones of chronic sorrow: Perspectives of chronically ill and bereaved persons and family caregivers. *Journal of Family Nursing* 5: 387–394.

Burke, S., M. Wallen, K. Vail-Smith, and D. Knox. 2011. Using technology to control intimate partners: An exploratory study of college undergraduates. *Computers in Human Behavior* 27: 1162–1167.

Burke-Winkelman, S., K. Vail-Smith, J. Brinkley, and D. Knox. 2014. Sexting on the college campus. *Electronic Journal of Human Sexuality.* Vol 17. Feb 3.

Burr, B. K, J. Viera, B. Dial, H. Fields, K. Davis, and D. Hubler. 2011. Influences of personality on relationship satisfaction through stress. Poster, Annual Meeting of National Council on Family Relations, Orlando, November 18.

Burr, W. R. and S. R. Klein.1994. *Reexamining family stress: New theory and research.* Thousand Oaks, CA: Sage.

Burrows, K. 2013. Age preferences in dating advertisements by homosexuals and heterosexuals: From sociobiological to sociological explanations. *Archives of Sexual Behavior* 42: 203–211.

Burton, L. M., E. Bonilla-Silva, V. Ray, R. Buckelew, and E. H. Freeman. 2010. Critical race theories, colorism, and the decade's research on families of color. *Journal of Marriage and Family* 72: 440–459.

Butterworth, P. and B. Rodgers. 2008. Mental health problems and marital disruption: Is it the combination of husbands and wives' mental health problems that predicts later divorce? *Social Psychiatry and Psychiatric Epidemiology* 43: 758–764.

Buunk, A. P., J. Goor, and A. C. Solano. 2010. Intrasexual competition at work: Sex differences in the jealousy-evoking effect of rival characteristics in work settings. *Journal of Social and Personal Relationships* 27: 671–684.

Byrd-Craven, J., B. J. Auer, D. A. Granger, and A. R. Massey. 2012. The father-daughter dance: The relationship between father-daughter relationship quality and daughters' stress response. *Journal of Family Psychology* 26: 87–94.


## C


Cadden, M., and D. Merrill 2007. What married people miss most (based on *Reader's Digest* study of 1,001 married adults). *USA Today,* D1.

Cade, R. 2010. Covenant marriage. *Family Journal* 18: 230–233.

Cagney, K. A., C. R. Browning, J. Iveniuk, and N. English. 2014. The onset of depression during the great recession: Foreclosure and older adult mental health. *American Journal of Public Health* 104: 498–505.

Campbell, K., J. C. Kaufman, T. D. Ogden, T.T. Pumaccahua, and H. Hammond. 2011. Wedding rituals: How cost and elaborateness relate to marital outcomes. Poster, National Council on Family Relations, Orlando, November.

Campbell, K., L C. Silva, and D. W. Wright. 2011 Rituals in unmarried couple relationships: An exploratory study. *Family and Consumer Sciences Research Journal* 40: 45–57.

Camperio Ciani, A. S. and L. Fontanesi. 2012. Mothers who kill their offspring: Testing evolutionary hypothesis in a 110-case Italian sample. *Child Abuse & Neglect* 36: 519–527.

Canario, C., B. Figueiredo, and M. Ricou. 2013. Women and men's psychological adjustment after abortion: A six-months' perospective pilot study. *Journal of Reproductive and Infant Psychology* 29: 262–275.

CareerBuilder.com 2014. Office romance survey. Retrieved April 30 at http://www.careerbuilder.com/share/aboutus/pressreleasesdetail.aspx?sd=2%2F13%2F2014&id=pr803&ed=12%2F31%2F2014

Carey, A. R. and K. Gellers. 2010. Do you believe in the idea of soul mates? *USA Today,* August 31. 1A.

Carey, A. R. and V. Salazar. 2011. Women talk and text more. *USA Today* Feb 1, 2011, p. 1.

Carey, A. and P. Trapp. 2010. Wired vacations. *USA Today,* June 3, A1.

Caron, S. L. and S. P. Hinman. 2012. "I took his V-card": An exploratory analysis of college student stories involving male virginity loss. *Sexuality and Culture.* Online September.

Carpenter, L. M. 2010. Like a Virgin . . . again?: Secondary virginity as an ongoing gendered social construction. *Sexuality and Culture* 14: 253–270.

Carr, K. and T. R. Wang. 2012. "Forgiveness isn't a simple process: It's a vast undertaking": Negotiating and communicating forgiveness in nonvoluntary family relationships. *Journal of Family Communication* 12: 40–56.

Carr, D. and K. W. Springer. 2010. Advances in families and health research in the 21st Century. *Journal of Marriage and Family* 72: 743–761.

Carroll, J. S., L. R. Dean, L. L. Call, and D. M. Busby. 2011. Materialism and marriage: Couple profiles of congruent and incongruent spouses. *Journal of Couple & Relationship Therapy* 10: 287–308.

Carter, C. S. and S. W. Porges. 2013. The biochemistry of love: An oxyticin hypothesis. *Science & Society* 14: 12–16.

Carter, G. L., A. C. Campbell, and S. Muncer. 2014. The dark triad personality: Attractiveness to women. *Personality and Individual Differences* 56: 57–61.

Cartwright, C. 2012. The challenges of being a mother in a stepfamily. *Journal of Divorce and Remarriage* 53: 503–513.

Casey, P., V. Jadva, L. Blake, and S. Golombok. 2013. Families created by donor insemination: Father–child relationships at age 7. *Journal of Marriage and Family* 75: 858–870.

Cassidy, M. L., and G. Lee. 1989. The study of polyandry: A critique and synthesis. *Journal of Comparative Family Studies* 20: 1–11.

Castrucci, B. C., J. F. Culhane, E. K. Chung, I. Bennett, and K. F. McCollum. 2006. Smoking in pregnancy: Patient and provider risk reduction behavior. *Journal of Public Health Management & Practice* 12: 68–76.

Cavaglion, G. and E. Rashty. 2010. Narratives of suffering among Italian female partners of cybersex and cyber-porn. *Sexual Addiction & Compulsivity* 17: 270–287.

Cavazos-Rehg, P. A. et al. 2010. Predictors of sexual debut at age 16 or younger. *Journal Archives of Sexual Behavior* 39: 664–673.

Centers for Disease Control and Prevention. 2012. HIV incidence. Retrieved March 21, 2012, from http://www.cdc.gov/hiv/topics/surveillance/incidence.htm

Chaney, C. and C. N. Fairfax. 2013. A change has come: The Obamas and the culture of Black marriage in America. *Ethnicities* 13: 20–48.

Chaney, C. and K. Marsh. 2009. Factors that facilitate relationship entry among married and cohabiting African Americans. *Marriage & Family Review* 45: 26–51.

Chao, J. K., Y. Lin, M. Ma, C. Lai, Y. Ku, W. Kuo, and I. Chao. 2011. Relationship among sexual desire, sexual satisfaction, and quality of life in middle-aged and older adults. *Journal of Sex & Marital Therapy* 37: 386–403.

Chapman, G. 2010. *The five love languages: The secret to love that lasts.* Chicago: Northfield Publishing.

Chase, L. M. 2011. Wives' tales: The experience of trans partners. *Journal of Gay & Lesbian Social Services* 23: 429–451.

Chatizow, P. M. 2011. Love attitudes and acceptance of intimate partner violence myth among young adults in Poland. Unpublished master's thesis. University of Warsaw.

Chavis, A., J. Hudnut-Beumler, M. W. Webb, J. A. Neely, L. Bickman, M. S. Dietrich, S. J. Scholer. 2013. A brief intervention affects parents' attitudes toward using less physical punishment. *Child Abuse & Neglect* 37: 1192–1201.

Cheng, J. O., R. S. K. Lo, and J. Woo. 2013. Anticipatory grief therapy for older persons nearing the end life. *Aging Health.* 9.1: 103 et passim.

Cherlin, A. J. 2010. Demographic trends in the United States: A review of research in the 2000s. *Journal of Marriage and Family* 72: 403–419.

_______. 2009. *The Marriage-go-round: The State of marriage and the family in America today.* New York: Alfred A. Knopf.

Chih,-Chien, W. and C. Ya-Ting. 2010. Cyber relationship motives: Scale development and motivation.. *Social Behavior & Personality: An International Journal* 38: 289–300.

Childs, G. R. and S. F. Duncan. 2012. Marriage preparation education programs: An assessment of their components. *Marriage & Family Review* 48: 59–81.

Chiu, H. and D. Busby. 2010. Parental influence in adult children's marital relationship. Poster, National Council on Family Relations annual meeting, November 3–5. Minneapolis, MN.

Choi, H. and N. F. Marks. 2013. Marital quality, socioeconomic status, and physical health. *Journal of Marriage and Family* 75: 903–919.

Chonody, J. M., K. S. Smith, and M. A. Litle. 2012. Legislating unequal treatment: An exploration of public policy on same-sex marriage. *Journal of GLBT Family Studies* 8: 270–286.

Claffey, S. T. and K. D. Mickelson 2010. Division of household labor and distress: The role of perceived fairness for employed mothers. *Sex Roles* 60: 819–831.

Clark, S. and C. Kenney. 2010. Is the United States experiencing a "matrilineal tilt"?: Gender, family structures and financial transfers to adult children. *Social Forces* 88: 1753–1776.

Clarke, J. I. 2004. The overindulgence research literature: Implications for family life educators. Poster at the National Council on Family Relations, Annual Meeting, November. Orlando, Florida.

Clarke, L. H. and E. Bennett. 2013. "You learn to live with all the things that are wrong with you": gender and the experience of multiple chronic conditions in later life. *Ageing and Society* 33: 342–360.

Clarke, S. C. and B. F. Wilson. 1994. The relative stability of remarriages: A cohort approach using vital statistics. *Family Relations* 43: 305–310.

Clements, C. M. and R. L. Ogle. 2009. Does acknowledgment as an assault victim impact post assault

psychological symptoms and coping? *Journal of Interpersonal Violence* 24: 1595–1614.
Clunis, D. M. and G. Dorsey Green. 2003. *The lesbian parenting book,* 2nd ed. Emeryville, CA: Seal Press.
Coelho, C. and N. Garoupa. 2014.Do divorce law reforms matter for divorce rates? Evidence from Portugal. *Journal of Empirical Legal Studies* 3: 525–542.
Cohen, S. and D. Janicki-Deverts. 2012. Who's stressed? Distributions of psychological stress in the United States in probability samples from 1983, 2006, and 2009. *Journal of Applied Social Psychology* 42: 1320–1334.
Cohn, A., H. M. Zinzow, H. S. Resnick, and D. G. Kilpatrick. 2013. Correlates of reasons for not reporting rape to police: Results from a national telephone household probability sample of women with forcible or drug-or-alcohol facilitated/incapacitated rape. *Journal of Interpersonal Violence* 28: 455–473.
Cohn, D., G. Livingston, and W. Wang. 2014. After decades of decline, a rise in stay-at home mothers. April 8. Pew Research Center. http://www.pewresearch.org/fact-tank/2014/04/08/7-key-findings-about-stay-at-home-moms/
Cohn, D. 2013. Love and marriage. Pew Research Center. http://www.pewsocialtrends.org/2013/02/13/love-and-marriage/
Cokes, C. and W. Kornblum. 2010. Experiences of mental distress by individuals during an economic downturn: The story of an urban city. *The Western Journal of Black Studies* 34: 24–36.
Confer, J. C. and M. D. Cloud. 2011. Sex differences in response to imagining a partner's heterosexual or homosexual affair. *Personality & Individual Differences* 50: 129–134.
Cook, K., D. Dranove, and A. Sfekas. 2010. Does major illness cause financial catastrophe? *Health Services Research* 45: 418–436.
Cooley, C. H. 1964. *Human nature and the social order.* New York: Schocken.
Coombes, A. 2007. What are the key ingredients to a happier old age? It's easier than you think. http://www.marketwatch.com/news/story/here-key-ingredients-staying-happy/story.aspx?guid=%7BFD3F3A18–7DF3–4BF5-A3A3-B6438317E693%7D (retrieved August 24).
Coontz, S. 2000. Marriage: Then and now. *Phi Kappa Phi Journal* 80: 16–20.
Copeland, L. 2012. Tech keeps tabs on teen drivers. *USA Today.* Oct 23, A3.
_____. 2010. Parent-teen driving contracts. *USA Today.* Oct 19, A1.
Copen, C. E., K. Daniels, W. D. Mosher. 2013. First premarital cohabitation in the United States: 2006–2010. National Survey of Family Growth. April 4. 64. http://www.cdc.gov/nchs/data/nhsr/nhsr064.pdf
Cottle, N. R., R. Hammond, K. Yorgason, K. Stookey, and B. Mallet. 2013. Marital quality among current and former college students. Poster, National Council on Family Relations Annual Meeting. San Antonio, November 5–9.
Covin, T. 2013. Personal communication. Southeastern Council on Family Relations Annual Meeting, Feb 21–23. Birmingham, AL.
Cox, R. B., J. S. Ketner, and A. J. Blow. 2013. Working with couples and substance abuse: Recommendations for clinical practice. *The American Journal of Family Therapy* 41: 160–172
Coyne, S. M., L. M. Padilla-Walker, L. Stockdale, and A. Fraser. 2011. Media and the family: Associations between family media use and family connections. Poster, Annual Meeting of National Council on Family Relations, Orlando, November 18.
Coyne, S. M., L. Stockdale, D. Busby, B. Iverson, and D. M. Grant. 2011, "I luv u :)!": A descriptive study of the media use of individuals in romantic relationships. *Family Relations* 60: 150–162.
Craig, L. and K. Mullan. 2010. Parenthood, gender and work-family time in the United States, Australia, Italy, France, and Denmark. *Journal of Marriage and Family* 72: 1344–1361.
Cramer, R. E., R. E. Lipinski, J. D. Meteer, and J. A. Houska. 2008. Sex differences in subjective distress to unfaithfulness: Testing competing evolutionary and violation of infidelity expectations hypotheses. *The Journal of Social Psychology* 148: 389–406.
Crawford, J. 2012. Changes among parents attending a divorce education program 6 and 12 month follow up. Paper, National Council on Family Relations, Phoenix, November.
Crisp, B. and D. Knox. 2009. *Behavioral family therapy: An evidence-based approach.* Chapel Hill, NC: Carolina Academic Press. http://www.cap-press.com/books/1870.
Crosnoe, R. and S. E. Cavanagh. 2010 Families with children and adolescents: A review, critique, and future agenda. *Journal of Marriage and Family* 72: 594–611.
Crowl, A., S. Ahn, and J. Baker. 2008. A meta-analysis of developmental outcomes for children of same-sex and heterosexual parents. *Journal of GLBT Family Studies* 4: 385–407.
Crutzen, R., E. Nijhuis, and S. Mujakovic. 2012. Negative associations between primary school children's perception of being allowed to drink at home and alcohol use. *Mental Health and Substance Use* 5, 64–69.
Cui, M., F. D. Fincham, and J. A. Durtschi. 2011. The effect of parental divorce on young adults' romantic relationship dissolution: What makes a difference? *Personal Relationships* 18: 410–426.
Cui, M. and F. D. Fincham. 2010. The differential effects of parental divorce and marital conflict on young adult romantic relationships. *Personal Relationships* 17: 331–343.
Curran, M. A., E. A. Utley, and J. A. Muraco. 2010. An exploratory study of the meaning of marriage for African Americans. *Marriage & Family Review* 46: 346–365.
Curtin, S. C., J. C. Abma, S. J. Ventura. 2013. Pregnancy rates for U.S. women continue to drop. U.S. Department of Health and Human Services. NCHS Data # 136. December.

## D

Daigle, L. E., B. S. Fisher, and F. T. Cullen. 2008. The violent and sexual victimization of college women: Is repeat victimization a problem? *Journal of Interpersonal Violence* 23: 1296–1313.
Dailey, R. M., A. A. McCracken, B. Jin, K. R. Rossetto, and E. W Green. 2013. Negotiating breakups and renewals: Types of on-again/off-again dating relationships. *Western Journal of Communication* 77: 382–410.
Dailey, R. M., N. Brody, L. LeFebvre, and B. Crook. 2013. Charting changes in commitment: Trajectories of on-again/off-again relationships. *Journal of Social and Personal Relationships* 30: 1020–1044.
D'Amico, E. and D. Julien. 2012. Disclosure of sexual orientation and gay, lesbian, and bisexual youth's adjustment: Associations with past and current parental acceptance and rejection. *Journal of GLBT Family Studies* 8: 215–242.
Danner-Vlaardingerbroek, G., E. S. Kluwer. E F. van Steenbergen, and T. van der Lippe. 2013. The psychological availability of dual-earner parents for their children after work. *Family Relations* 62: 741–754.
Darlow, S. and M. Lobel. 2010. Who is beholding my beauty? Thinness ideals, weight, and women's responses to appearance evaluation. *Sex Roles* 63: 833–843.
Darnton, K. 2012. Deception at Duke. *Sixty Minutes*/CBS Television Feb 12.
Dattilio, F. M., F. P. Piercy, and S. D. Davis. 2014. The divide between "evidence-based" approaches and practitioners of traditional theories of family therapy. *Journal of Marital and Family Therapy* 40: 5–16.

Davis, S. N., S. K. Jacobsen, and J. Anderson. 2012. From the great recession to greater gender equality? Family mobility and the intersection of race, class, and gender. *Marriage & Family Review* 48: 601–620.

Dearden, K., B. Crookston, H. Madanat, J. West, M. Penny, and S. Cueto. 2013. What difference can fathers make? Early paternal absence compromises Peruvian children's growth. *Maternal & Child Nutrition* 9: 143–154.

De Castro, S. and J. T. Guterman. 2008. Solution-focused therapy for families with suicide. *Journal of Marital and Family Therapy* 34: 93–107.

DeLamater, J., and M. Hasday. 2007. The sociology of sexuality. In *21st century sociology: A reference handbook*, ed. Clifton D. Bryant and Dennis L. Peck, 254–264. Thousand Oaks, California: Sage.

DeLiema, M., Z. D. Gassoumis, D. C. Homeier, and K. H. Wilber. 2012. Determining prevalence and correlates of elder abuse using promotores: Low-income immigrant Latinos report high rates of abuse and neglect. *Journal of American Geriatrics Society*. 60: 1333–1339.

Dema-Moreno, S. and C. Díaz-Martínez. 2010. Gender inequalities and the role of money in Spanish dual-income couples. *European Societies* 12: 65–84.

DeMaris, A., L. A. Sanchez, and K. Knivickas. 2012. Developmental patterns in marital satisfaction: Another look at covenant marriage. *Journal of Marriage and the Family* 74: 989–1004.

Demir, M. 2008. Sweetheart, you really make me happy: Romantic relationship quality and personality as predictors of happiness among emerging adults. *Journal of Happiness Studies* 9: 2, 257–277.

Denney , J. T. 2010. Family and household formations and suicide in the United States. *Journal of Marriage and the Family* 72: 202–213.

Department of Commerce. 2011. *Women in America: Indicators of social and economic well-being.* Retrieved March 15. http://www.whitehouse.gov/sites/default/files/rss_viewer/Women_in_America.pdf

Derby, K., D. Knox, and B. Easterling. 2012. Snooping in romantic relationships. *College Student Journal* 46: 333–343.

De Schipper, J. C., L. W. C. Tavecchio, and M. H Van IJzendoorn. 2008. Children's attachment relationships with day care caregivers: Associations with positive caregiving and the child's temperament. *Social Development* 17: 454–465.

Deutsch, F. M., A. P. Kokot, and K. S. Binder. 2007. College women's plans for different types of egalitarian marriages. *Journal of Marriage and Family* 69: 916–29.

Dew, J. 2011. Financial issues and relationship outcomes among cohabiting individuals. *Family Relations* 60: 178–190.

_____. 2008. Debt change and marital satisfaction change in recently married couples. *Family Relations* 57: 60–71.

Dew, J. and J. Yorgason. 2010. Economic pressure and marital conflict in retirement-aged couples. *Journal of Family Issues* 31: 164–188.

Diamond, A., J. Bowes, and G. Robertson. 2006. Mothers' safety intervention strategies with toddlers and their relationship to child characteristics. *Early Child Development and Care* 176: 271–84.

Diamond, L. M. 2003. What does sexual orientation orient? A biobehavioral model distinguishing romantic love and sexual desire. *Psychological Review* 110: 173–192.

Diem, C. and J. M. Pizarro 2010. Social structure and family homicides. *Journal of Family Violence* 25: 521–532.

Dijkstra, P., D. P. H. Barelds, H. A. K. Groothof, S. Ronner, and A. P. Nautal. 2012. Preferences of the intellectually gifted, *Marriage & Family Review* 48:1, 96–108.

Dillion C. F., Q. Gu, H. Hoffman, and C.W. Ko. 2010. Vision, hearing, balance, and sensory impairment in Americans aged 70 years and over: United States, 1999–2006. NCHS data brief, no 31. Hyattsville, MD: National Center for Health Statistics.

Dir, A. L., A. Coskunpinar, J. L. Steiner, and M. A. Cyders. 2013. Understanding differences in sexting behaviors across gender, relationship status, and sexual identity, and the role of expectancies in sexting. *Cyberpsychology, Behavior, and Social Networking* 16: 568–574.

Dixon, L. J., K. C. Gordon, N. N. Frousakis, and J. A. Schumm. 2012. Expectations and the marital quality of participants of a marital enrichment seminar. *Family Relations* 61: 75–89.

Djamba, Y. K., L. C. Mullins, K. P. Brackett, and N. J. McKenzie. 2012. Household size as a correlate of divorce rate: A county-level analysis. *Sociological Spectrum* 32: 436–448.

Doaring, C. 2014. Justice department extends benefits. *USA Today*. Feb 10. A1.

Dollahite, D. C. and L. D. Marks. 2012. The Mormon American family. In *Ethnic families in America: Patterns and variations, 5th ed.* Edited by Wright, R., C. H. Mindel, T. Van Tran, and R. W. Habenstein. Boston, MA.: Pearson, pp. 461–486.

Dominguez, M. M., J. High, E. Smith, B.Cafferky, P. Dharnidharka, and S. Smith. 2013. The intergenerational transmission of family violence: A meta-analytic review. Poster, National Council on Family Relations Annual Meeting. San Antonio, November 5–9.

Doodson, L. J. and A. P. C. Davies. 2014. Comparison of psychological well-being across stepmothers and biological mothers and across four categories of stepmothers. *Journal of Divorce and Remarriage* 55: 49–63.

Doria, M. V., H. Kennedy, C. Strathie, and S. Strathie. 2014. Explanations for the success of video interaction guidance (VIG): An emerging method in family psychotherapy. *The Family Journal* 22: 78–87.

Dotson-Blake, K., D. Knox, and A. R. Holman. 2010. Reaching out: College student perceptions of counseling. *Professional Issues in Counseling* (Fall). Retrieved from http://www.shsu.edu/~piic/CollegeStudentPerceptions.htm

_____. 2008. College student attitudes toward marriage, family, and sex therapy. Unpublished data from 288 undergraduate/graduate students. East Carolina University, Greenville, NC.

Dotson-Blake, K. P., D. Knox, and M. Zusman. 2012. Exploring social sexual scripts related to oral sex: A profile of college student perceptions. *The Professional Counselor*. 2: 1–11. Online journal http://tpcjournal.nbcc.org/?p=357

Dougall, K. M., Y. Beyene, and R. D. Nachtigall. 2013. Age shock: Misperceptions of the impact of age on fertility before and after IVF in women who conceived after age 40. *Human Reproduction* 28: 350–356.

Dowd, D. A., M. J. Means, J. F. Pope, and J. H. Humphries. 2005. Attributions and marital satisfaction: The mediated effects of self-disclosure. *Journal of Family and Consumer Sciences* 97: 22–27.

Doyle, M., C. O'Dywer, and V. Timonen. 2010. "How can you just cut off a whole side of the family and say move on?" The reshaping of paternal grandparent-grandchild relationships following divorce or separation in the middle generation. *Family Relations* 59: 587.

Drapeau, A. M. M-H Gagne, R. Saint-Jacques, R. Lepine, and H. Ivers. 2009. Post-separation conflict trajectories: A longitudinal study. *Marriage & Family Review* 5: 353–373.

Drefahl, S. 2010. How does the age gap between partners affect their survival? *Demography* 47: 313–326.

Drew, E. N. and J. E. King. 2012. Young adult's perspectives on the difference between grandmother and grandfather involvement. Poster, Research and Creative Achievement Week, East Carolina University, March 26–30.

Drucker, D. J. 2010. Male sexuality and Alfred Kinsey's 0–6 Scale: Toward "A sound understanding of the realities of sex." *Journal of Homosexuality* 57: 1105–1123.

Druckerman, P. 2007. *Lust in translation.* New York: Penguin Group.

Duba, J. D., A. W. Hughey, T. Lara, and M. G. Burke. 2012. Areas of marital dissatisfaction among long-term couples. *Adultspan: Theory Research & Practice* 11: 39–54.

Dubbs, S. L. and A. P. Buunk. 2010. Sex differences in parental preferences over a child's mate choice: A daughter's perspective. *Journal of Social and Personal Relationships* 27: 1051–1059.

Ducanto, J. N. 2013. Why do marriages fail? *American Journal of Family Law* 28: 237–239.

Ducharme, J. K. and M. M. Kollar 2012. Does the "marriage benefit" extend to same-sex union?: Evidence from a sample of married lesbian couples in Massachusetts. *Journal of Homosexuality* 59: 580–591.

Dumka, L. E., N. A. Gonzales, L. A. Wheeler, R. E. Millsap. 2010. Parenting self-efficacy and parenting practices over time in Mexican American families. *Journal of Family Psychology* 24: 522–531.

Dunifon, R. and A. Bajracharya. 2012. The role of grandparents in the lives of youth. *Journal of Family Issues* 9: 1168–1194.

Duntley, J. D. and D. M. Buss. 2012. The evolution of stalking. *Sex Roles* 66: 311–327.

Durtschi, J. A. and K. L. Soloski. 2012. The dyadic effects of coparenting and parental stress on relationship quality. Presentation, National Council on Family Relations, Phoenix, AZ. November.

## E

Eagan, K., J. B. Lozano, S. Hurtado, and M. H. Case. 2013. *The American freshman: National norms fall 2013.* Los Angeles: Higher Education Research.

East, L., D. Jackson, L. O'Brien, and K. Peters. 2011. Condom negotiation: Experiences of sexually active young women. *Journal of Advanced Nursing* 67: 77–85.

Easterling, B. A. 2005. *The Invisible Side of Military Careers: An Examination of Employment and Well-Being Among Military Spouses.* Master's Thesis, University of North Florida.

Easterling, B. A. and D. Knox. 2010. Left behind: How military wives experience the deployment of their husbands. *Journal of Family Life.* July 20, 2010 online. http://www.journaloffamilylife.org/militarywives

Easterling, B., D. Knox, and A. Brackett. 2012. Secrets in romantic relationships: Does sexual orientation matter? *Journal of GLBT Family Studies* 8: 198–210.

Eaton, L., S. Kalichman, D. Skinner, M. Watt, D. Pieterse, and E. Pitpitan. 2012. Pregnancy, alcohol intake, and intimate partner violence among men and women attending drinking establishments in a Cape Town, South Africa, Township. *Journal of Community Health* 37: 208–216.

Ebert, R. 2011. *Life itself.* New York: Grand Central Publishing.

Eck, B. A. 2013. Identity twists and turns: How never-married men make sense of an unanticipated identity. *Journal of Contemporary Ethnography* 42: 31–33.

Edin, K. and R. J. Kissane. 2010. Poverty and the American family: A decade in review *Journal of Marriage and Family* 72: 460–479.

Edwards, T. M. 2000. Flying solo. *Time.* August 28, 47–53.

Eke, A., N. Hilton, G. Harris, M. Rice, and R. Houghton. 2011. Intimate partner homicide: Risk assessment and prospects for prediction. *Journal of Family Violence* 26: 211–216.

Eliason, M. 2012. Lost jobs, broken marriages. *Journal of Population Economics* 25: 1365–1397.

Elliott, L., B. Easterling and D. Knox. 2012. Taking chances in romantic relationships. Poster, Southern Sociological Society Annual Meeting, New Orleans, March.

Ellison, C. G., A. M. Burdette, and W. B. Wilcox. 2010. The couple that prays together: Race and ethnicity, religion, and relationship quality among working-age adults. *Journal of Marriage and Family* 72: 963–975.

Ellison, C. G., A. K. Henderson, N. D. and K. E. Harkrider. 2011. Sanctification, stress, and marital quality. *Family Relations* 60: 404–420.

Ellison, N. B. and J. T. Hancock. 2013. Profile as promise: Honest and deceptive signals in online dating. *Security & Privacy, IEEE* 11: 85–88.

England, P. 2010. The Gender Revolution: Uneven and stalled. *Gender & Society* 24: 149–166.

Enright, E. 2004. A house divided. *AARP The Magazine,* July/August, 60.

Erarslan, A. B. and B. Rankin. 2013. Gender role attitudes of female students in single-sex and coeducational high schools in Istanbul. *Sex Roles* 69: 455–469.

Esmaila, A. 2010. Negotiating fairness: A study on how lesbian family members evaluate, construct, and maintain "fairness" with the division of household labor. *Journal of Homosexuality* 57: 591–609.

Etzioni, A. 2011. The new normal. *Sociological Forum* 26: 779–789.

Eubanks Fleming, C. J. and J. V. Córdova. 2012. Predicting relationship help seeking prior to a marriage checkup. *Family Relations* 61: 90–100.

Euser, E. M., M. H. van Ijzendoorn, P. Prinzie, M. J. Bakermans-Kranenburg. 2010. Prevalence of child maltreatment in the Netherlands. *Child Maltreatment* 15: 5–17.

## F

Fabian, N. 2007. Rethinking retirement—And a footnote on diversity. *Journal of Environmental Health* 69: 85–86.

Faircloth, M., D. Knox, and J. Brinkley. 2012. The good, the bad and technology mediated communication in romantic relationships. Paper, Southern Sociological Society, New Orleans, March 21–24.

Falcon, M., F. Valero, M. Pellegrini, M. Rotolo, G. Scaravelli, J. Joya, and O. Vall. et al. 2010. Exposure to psychoactive substances in women who request voluntary termination of pregnancy assessed by serum and hair testing. *Forensic Science International* 196: 22–26.

Family Caregiver Alliance. http://caregiver.org/caregiver/jsp/content_node.jsp?nodeid=439 (retrieved April 10, 2006).

Farr, D. 2012. Family vacation: Where we've been, where we are and where we need to go. Poster, Annual Meeting of the National Council on Family Relations. Phoenix, Nov 1.

Farris, C., R. J. Viken, and T. Treat. 2010. Alcohol alters men's perceptual and decisional processing of women' s sexual interest. *Journal of Abnormal Psychology* 119: 427–432.

Felding, R. A., W. J. Rejeski, S. Blair, T. Church, M. A. Espeland et al. 2011. The Lifestyle Interventions and Independence for Elders Study: Design and Methods. *Journals of Gerontology Series A: Biological Sciences & Medical Sciences.* 66A: 1226–1237.

Feng, W., Y. Cai, and B. Gu. 2013. Population, policy, and politics: How will history judge China's one-child policy? *Population and Development Review* 38: 115–129.

Fennell, J. 2014. Polyamory. Presentation, Courtship and Marriage, Department of Sociology, East Carolina University, March 18.

Fergusson, D. M., L. J. Horwood, and J. M. Boden. 2013. Does abortion reduce the mental health risks of unwanted or unintended pregnancy? A reappraisal of the evidence. *Australian and New Zealand Journal of Psychiatry* 47: 819–827.

Fernandez-Montalvo, J., J. J. Lopez-Goni, and A. Arteaga. 2012. Violent behaviors in drug addiction: Differential profiles of drug-addicted patients with and without violence problems. *Journal of Interpersonal Violence* 27: 142–157.

Few, A. L. and K. H. Rosen. 2005. Victims of chronic dating violence: How women's vulnerabilities link to their decisions to stay. *Family Relations* 54: 265–79.

Field, C. J., S. R. Kimuna, and M. A. Straus. 2013. Attitudes toward interracial relationships among college students: Race, class, gender and perceptions of parental views. *Journal of Black Studies* 44: 741–776.

Field, T., M. Diego, M. Pelaez, O. Deeds, and J. Delgado. 2012. Depression and related problems of university students. *College Student Journal* 46: 193–202.

Fielder, R. L., J. L. Walsh, K. B. Carey, and M. P. Carey. 2014. Sexual hookups and adverse health outcomes: A longitudinal study of first-year college women. *The Journal of Sex Research* 51: 131–144.

Fieldera, R. L. and M. P. Careya. 2010. Prevalence and characteristics of sexual hookups among first-semester female college students. *Journal of Sex & Marital Therapy* 36: 346–359.

Filisko, G. M. 2010. When global families fail. *ABA Journal* 07470088, July 2010, Vol. 96.

Finer, L. B., L. F. Frohwirth, L. A. Dauphinne, S. Singh, and A. M. Moore. 2005. Reasons U.S. women have abortions: quantitative and qualitative reasons. *Perspectives on Sexual and Reproductive Health* 37: 110–18.

Finer, L. B. and M. R. Zolna. 2014. Shifts in intended and unintended pregnancies in the United States, 2001-2008. *American Journal of Public Health* 104: 43–48.

Finkenauer, C. 2010. Although it helps, love is not all you need: How Caryl Rusbult made me discover what relationships are all about. *Personal Relationships* 17: 161–163.

Finneran, C. and R. Stephenson. 2014. Intimate partner violence, minority stress, and sexual risk- taking among U.S. men who have sex with men. *Journal of Homosexuality* 61: 288–306.

Fiona, C. S., A. W. Chau, and K. Y. Chan. 2012. Financial knowledge and aptitudes: Impacts on college students' financial well-being. *College Student Journal* 46: 114–132.

Fischer H. et al. 2006. Romantic love: A mammalian brain system for mate choice, *Philosophical Transactions of the Royal Society B* 361: 2173–86.

Fish, J., T. Pavkov, J. Wetchler, and J. Bercik. 2011. The role of adult attachment and differentiation in extradyadic experiences. Poster, National Council on Family Relations annual meeting, Orlando, November 18.

Fisher, A. D., E. Bandini, G. Rastrelli, G. Corana, M. Monami, E. Mannucci and M. Maggi. 2012A. Sexual and cardiovascular correlates of male unfaithfulness. *Journal of Sexual Medicine* 9: 1508–1518.

Fisher, A. D. et al. 2012. Stable extramarital affairs are breaking the heart. *International Journal of Andrology* 35: 11–17.

Fisher, H. 2010. The new monogamy: Forward to the past: An author and anthropologist looks at the future of love. *The Futurist* 44: 26–29.

Fisher, H. E., L. L. Brown, A. Aron, G. Strong, and D. Mashek. 2010. Reward, addiction, and emotion regulation systems associated with rejection in love. *Journal of Neurophysiology* 104: 51–60.

Fisher, M. and A. Cox. 2011. Four strategies used during intrasexual competition for mates. *Personal Relationships* 18: 20–38.

Fisher, M. L. and C. Salmon. 2013. Mom, dad, meet my mate: An evolutionary perspective on the introduction of parents and mates. *Journal of Family Studies* 19: 99–107.

Flood, M. 2009. The harms of pornography exposure among children and young people. *Child Abuse Review* 18: 384–400.

Follingstad, D. R. and M. Edmundson. 2010. Is psychological abuse reciprocal in intimate relationships? Data from a national sample of American adults. *Journal of Family Violence* 25: 495–508.

Foran, H. M., K. M. Wright, and M. D. Wood. 2013. Do combat exposure and post-deployment mental health influence intent to divorce? *Journal of Social and Clinical Psychology* 32: 917–938.

Forbes, E. E. and Dahl, R. E. (2012), Research Review: Altered reward function in adolescent depression: what, when and how? *Journal of Child Psychology and Psychiatry* 53: 3–15.

Ford, P. 2011. *Glenn Ford, A Life.* Madison,Wisconsin: University of Wisconsin Press.

Foster, D. G., H. Gould, J. Taylor, and T. A. Weitz. 2012. Attitudes and decision making among women seeking abortions in one U.S. clinic. *Perspectives on Sexual & Reproductive Health* 44: 117–124.

Foster, D. G., K. Kimport, H. Gould, S. C. M. Roberts, and T. A. Weitz. 2013. Effect of abortion protesters on women's emotional response to abortion. *Contraception* 87: 81–87.

Foster, J. 2010. How love and sex can influence recognition of faces and words: A processing model account. *European Journal of Social Psychology* 40: 524–535.

Foster, J. D. 2008. Incorporating personality into the investment model: Probing commitment processes across individual differences in narcissism. *Journal of Social and Personal Relationships* 25: 211–223.

Freeman, D. 1999. *The fateful hoaxing of Margaret Mead: A historical analysis of her Samoan research.* New York: Westview Press.

Freeman, D. and J. Temple. 2010. Social factors associated with history of sexual assault among ethnically diverse adolescents. *Journal of Family Violence* 25: 349–356.

Freud, S. 1905/1938. Three contributions to the theory of sex. In *The basic writings of Sigmund Freud,* ed. A. A. Brill. New York: Random House.

Frias-Navarro, D. and H. Monterde-i-Bort. 2012. A scale on beliefs about children's adjustment in same-sex families: Reliability and validity. *Journal of Homosexuality* 59: 1273–1288.

Frimmel, W., M. Halla, and R. Winter-Ebmer. 2013. Assortative mating and divorce: Evidence from Austrian register data. *Journal of the Royal Statistical Society Series A-Statistics in Society* 176: 907–920.

Frisco, M. L. and M. Weden. 2013. Early adult obesity and U.S. women's lifetime childbearing experiences. *Journal of Marriage and Family* 75: 920–932.

Frye, N. E. 2011. Responding to problems: The roles of severity and barriers. *Personal Relationships* 18: 471–486.

Frye-Cox, N. 2012. Alexithymia and marital quality: The mediating role of loneliness. Paper, National Council on Family Relations, Phoenix. November.

Furukawa, R. and M. Driessnack. 2013. Video-mediated communication to support distant family connectedness. *Clinical Nursing Research* 22: 82–94.

## G

Gaines, S. O., Jr., and J. Leaver. 2002. Interracial relationships. In *Inappropriate relationships: The unconventional, the disapproved, and the forbidden,* ed. R. Goodwin and D. Cramer, 65–78. Mahwah, NJ: Lawrence Erlbaum.

Galambos, N. L., E. T. Barker, and D. M. Almeida. 2003. Parents do matter: Trajectories of change in externalizing and internalizing problems in early adolescent. *Child Development* 74: 578–95.

Galinsky, A. M. and F. L. Sonenstein. 2013. Relationship commitment, perceived equity, and sexual enjoyment among young adults in the United States. *Archives of Sexual Behavior* 42: 93–104.

Galinsky, E. 2010 *Mind in the making.* New York: Harper Collins.

Gallmeier, C. P., M. E. Zusman, D. Knox, and L. Gibson. 1997. Can we talk? Gender differences in disclosure patterns and expectations. *Free Inquiry in Creative Sociology* 25: 129–225.

Ganong, L. H. and M. Coleman. 1999. *Changing families, changing responsibilities: Family obligations following divorce and remarriage.* New York: Lawrence Erlbaum Assoc. Inc.

Ganong, L. H., M. Coleman, R. Feistman, T. Jamison, and M. S. Markham. 2011. Communication technology and post-divorce co-parenting. . Poster, Annual Meeting of National Council on Family Relations, Orlando, November 18.

Gardner, J. and A. J. Oswald. 2006. Do divorcing couples become happier by breaking up? *Journal of the Royal Statistical Society: Series A (Statistics and Society)* 169: 319–36.

Gardner, R. A. 1998. *The parental alienation syndrome.* 2d ed. Cresskill, NJ: Creative Therapeutics.

Garfield, R. 2010. Male emotional intimacy: How therapeutic men's groups can enhance couples therapy. *Family Process* 49: 109–122.

Garrett, T. M., H. W. Baillie, and R. M. Garrett. 2001. *Health care ethics,* 4th ed. Upper Saddle River, NJ: Prentice Hall.

Garrido, A. A., J. C. C. Olmos, A. A. Galende and M. J. G. Moreno. 2012. Same-sex marriages in Spain: The case of international unions. *Anthropological Notebooks* 18: 23–40.

Gartrell, N. , H. M. W. Bos, H. Peyser, A. Deck, and C. Rodas. 2012 Adolescents with lesbian mothers describe their own lives. *Journal of Homosexuality* 59: 1211–1229.

Gates, G. J. 2011. How many people are lesbian, gay, bisexual and transgender? The Williams Institute, UCLA School of Law. http://williamsinstitute.law.ucla.edu/wp-content/uploads/Gates-How-Many-People-LGBT-Apr-2011.pdf

Gatzeva, M. and A. Paik. 2011. Emotional and physical satisfaction in noncohabiting, cohabiting, and marital relationships: The importance of jealous conflict. *Journal of Sex Research* 48: 29–42.

Gault-Sherman, M. and S. Draper. 2012. What will the neighbors think? The effect of moral communities on cohabitation. *Review of Religious Research* 54: 45–67.

Gelabert, E., S. Subirà, L. García-Esteve, P. Navarro, A. Plaza, E. Cuyàs et al. 2012. Perfectionism dimensions in major postpartum depression. *Journal of Affective Disorders* 136: 17–25.

Geller, P., C. Psaros, and S. L. Kornfield. 2010. Satisfaction with pregnancy loss aftercare: Are women getting what they want? *Archives of Women's Mental Health* 13: 111–124.

Gershon, I. 2010. *The breakup 2.0.* New York: Cornell University Press.

Gibbs, N. 2012. Your life is fully mobile: Time Mobility Survey. *Time,* August 27, p. 32 and following.

Gibson, V. 2002. *Cougar: A guide for older women dating younger men.* Boston, MA: Firefly Books.

Gilla, D. L., R. G. Morrow, K. E. Collins, A. B. Lucey, and A. M. Schultze. 2010. Perceived climate in physical activity settings. *Journal of Homosexuality* 57: 895–913.

Gillath, O., M. Mikulincer, G. E. Birnbaum, and P. R. Shaver. 2008. When sex primes love: Subliminal sexual priming motivates relationship goal pursuit. *Personality and Social Psychology Bulletin* 34: 1057–1073.

Girgis, S., R. P. George, and R. Anderson. 2011. What is marriage? *Harvard Journal of Law & Public Policy* 34: 245–287.

Glass, V. Q. 2014. "We are with family": Black lesbian couples negotiate rituals with extended families. *Journal of GLBT Family Studies* 10: 79–100.

Godbout, E. and C. Parent. 2012. The life paths and lived experiences of adults who have experienced parental alienation: A retrospective study. *Journal of Divorce and Remarriage* 53: 34–54.

Goetting, A. 1982. The six stations of remarriage: The developmental tasks of remarriage after divorce. *The Family Coordinator* 31:213–22.

Goldberg, A. E. and K. R. Allen. 2013. Same-sex relationship dissolution and LGB stepfamily formation: Perspectives of young adults with LGB parents. *Family Relations* 62: 529–544.

Goldberg, A. E., J. B. Downing, and A. M. Moyer. 2012. Why Parenthood, and Why Now? Gay Men's Motivations for Pursuing Parenthood. *Family Relations,* 61: 157–174.

Goldberg, A. E., L. A. Kinkler, H. B. Richardson, and J. B. Downing 2011. Lesbian, gay, and heterosexual couples in open adoption arrangements: A qualitative study. *Journal of Marriage and Family* 73: 502–518.

Goldberg, A. E., J. Z. Smith and D. A. Kashy. 2010. Preadoptive factors predicting lesbian, gay, and heterosexual couples' relationship quality across the transition to adoptive parenthood. *Journal of Family Psychology* 24: 221–232.

Gonzaga, G. C., S. Carter and J. Galen Buckwalter. 2010. Assortative mating, convergence, and satisfaction in married couples. *Personal Relationships* 17: 634–644.

Gonzales, G. 2014. Same-sex marriage- A prescription for better health. *New England Journal of Medicine.* 370: 1373–1376.

Gonzalez, K. A., S. S. Rostosky, R. D. Odom, and E. D. B. Riggle. 2013. The positive aspects of being the parent of an LGBTQ child. *Family Process* 52: 325–337.

Goode, E. 1999. New study finds middle age is prime of life. *New York Times,* July 17, D6.

Goodman, M. A., D. C. Dollahite, L. D. Marks, and E. Layton. 2013. Religious faith and transformational processes in marriage. *Family Relations* 62: 808–823.

Gordon, T. 2000. *Parent effectiveness training: The parents' program for raising responsible children.* New York: Random House.

Gotta, G., R. J. Green, E. Rothblum, S. Solomon, K. Balsam, and P. Schwartz. 2011. Heterosexual, lesbian, and gay male relationships: A comparison of couples in 1975 and 2000. *Family Process* 50, 353–376.

Gottman, John. 1994. *Why marriages succeed or fail.* New York: Simon & Schuster.

Gottman, Julie. 2013. The conversation—A discussion of living apart together relationships. June 27, NPR radio. 2000. *Family Process* 50, 353–376.

Gottman, J., and S. Carrere. 2000. Welcome to the love lab. *Psychology Today,* September/October, 42.

Gottman, J. M., R. W. Levenson, C. Swanson, K. Swanson, R. Tyson, and D. Yoshimoto. 2003. Observing gay, lesbian, and heterosexual couples' relationships: Mathematical modeling of conflict interaction. *Journal of Homosexuality* 45: 65–91.
Gottman, J., and S. Carrere. 2000. Welcome to the love lab. *Psychology Today,* September/October, 42.
Graham, A. M., H. K. Kim, and P. A. Fisher. 2012. Partner aggression in high-risk families from birth to age 3 years: Associations with harsh parenting and child maladjustment. *Journal of Family Psychology* 26: 105–114.

Graham, J. M. 2011. Measuring love in romantic relationships: A meta-analysis. *Journal of Social and Personal Relationships* 28: 748–771.

Grases, G., C. Sanchez-Curto, E. Rigo, and D. Androver-Roi. 2012. Relationship between positive humor and state and trait anxiety. *Ansiedad y Estrés* 18: 79–93.

Green, M. and M. Elliott. 2010. Religion, health, and psychological well-being. *Journal of Religion & Health* 49: 149–163.

Gregory, J. D. 2010. Pet custody: Distorting language and the law. *Family Law Quarterly* 44: 35–64.

Greif, G. L. and K. H. Deal. 2012. The impact of divorce on friendships with couples and individuals. *Journal of Divorce and Remarriage* 53: 421–435.

Greitemeyer, T. 2010. Effects of reciprocity on attraction: The role of a partner's physical attractiveness. *Personal Relationships* 17: 317–330.

Grekin, E. R., K. J. Sher, and J. L Krull. 2007. College spring break and alcohol use: Effects of spring break activity. *Journal of Studies on Alcohol and Drugs* 68: 681–693.

Grief, G. L. 2006. Male friendships: Implications from research for family therapy. *Family Therapy* 33: 1–15.

Grogan, S. 2010. Promoting positive body image in males and females: Contemporary issues and future directions. *Sex Roles* 63: 757–765.

Gromoske, A. N. and K. Maguire-Jack. 2012. Transactional and cascading relations between early spanking and children's social-emotional development. *Journal of Marriage and Family* 74: 1054–1068.

Gunnar, M. R., E. Kryzer, M. J. Van Ryzin, and D. A. Phillips. 2010. The rise in cortisol in family day care: Associations with aspects of care quality, child behavior, and child sex. *Child Development* 81: 851–886.

Gupta, K. and T. Cacchioni. 2013. Sexual improvement as if your health depends on it: An analysis of contemporary sex manuals. *Feminism & Psychology* 23: 442–458.

Guttmacher Institute 2012. Abortion facts. Retrieved Jan 13 http://www.guttmacher.org/media/presskits/abortion-US/statsandfacts.html

Guzzo, K. B. and S. R. Hayford. 2014. Fertility and the stability of cohabiting unions: Variation by intendedness. *Journal of Family Issues* 35: 547–576.

## H

Haag. P. 2011. *Marriage confidential.* New York: Harper Collins.

Haandrikman, K. 2011. Spatial homogamy: The geographical dimensions of partner choice. *Journal of Economic and Social Geography* 102: 100–110.

Haas, A. P., et al. 2011. Suicide and suicide risk in lesbian, gay, bisexual, and transgender populations: Review and recommendations. *Journal of Homosexuality* 58: 10–51.

Haboush, A., C. S. Warren, and L. Benuto. 2012. Beauty, ethnicity, and age: Does internalization of mainstream media ideals influence attitudes towards older adults? *Sex Roles* 66: 3–20.

Haley, W. E. 2011. Family care giving for older adults in a changing economic world. Presentation, Annual Meeting of National Council on Family Relations, Orlando, November 18.

Hall, J. A., N. Park, M. J. Cody, and H. Song. 2010. Strategic misrepresentation in online dating: The effects of gender, self-monitoring, and personality traits. *Journal of Social and Personal Relationships* 27: 117–135.

Hall, J. H., W. Fals-Stewart, and F. D. Fincham. 2008. Risky sexual behavior among married alcoholic men. *Journal of Family Psychology* 22: 287–299.

Hall, S. 2010. Implicit theories of the marital institution: Origins and implications. Poster, National Council on Family Relations annual meeting, November 3–5. Minneapolis, MN.

Hall, S. and D. Knox. 2013. Relationship and sexual behaviors of a sample of 4,567 university students. Unpublished data collected for this text. Department of Family and Consumer Sciences, Ball State University and Department of Sociology, East Carolina University.

Hall, S. and D. Knox. 2013. Relationship and sexual behaviors of a sample of 4,590 university students. Unpublished data collected for this text. Department of Family and Consumer Sciences, Ball State University and Department of Sociology, East Carolina University.

Hall, S. and D. Knox. 2012. Double victims: Sexual coercion by a dating partner and a stranger. *Journal of Aggression, Maltreatment & Trauma* 22: 145–158.

Hall, S. and D. Knox 2013. Relationship and sexual behaviors of a sample of 4,540 university students. Unpublished data collected for this text. Department of Family and Consumer Sciences, Ball State University and Department of Sociology, East Carolina University.

Hall, S. S. and R. A. Adams. 2011. Cognitive coping strategies of newlyweds adjusting to marriage. *Marriage and Family Review* 47: 311–325.

Hall, S. S. and R. Adams. 2011. Newlyweds' unexpected adjustments to marriage. *Family and Consumer Sciences Research Journal* 39: 375–387.

Halligan, C., D. Knox, F. J. Freysteinsdóttir, and S. Skulason. 2014. U.S. and Icelandic college student attitudes toward relationships/sexuality. Poster, Southern Sociological Society, Charlotte, NC. April 4.

Halligan, C., D. Knox, F. J. Freysteinsdóttir, and S. Skulason. 2014. U.S. and Icelandic college student attitudes toward relationships /sexuality. *College Student Journal in press.*

Halligan, C., J. Chang, and D. Knox. 2014. Positive effects of parental divorce on undergraduates. Paper, submitted for publication.

Halpern-Meekin, S., W. D. Manning, P. C. Giordano, and M. A. Longmore. 2013. Relationship churning, physical violence, and verbal abuse in young adult relationships. *Journal of Marriage & Family* 75: 2–12.

Hannon, P. A., C. Rusbult, E. Finkel, and M. Kamashiro. 2010. In the wake of betrayal: Amends, forgiveness, and the resolution of betrayal. *Personal Relationships* 17: 253–278.

Hans, J. D. and C. Kimberly. 2011. Abstinence, sex, and virginity: Do they mean what we think they mean? *American Journal of Sexuality Education* 6, 329–342.

Hardie, J. H. and A. Lucas. 2010. Economic factors and relationship quality among young couples: Comparing cohabitation and marriage. *Journal of Marriage and the Family* 72: 1141–1154.

Hartenstein, J. L. and M. S. Markham. 2013. Custody arrangement decisions among divorcing or separating parents. Poster, Annual Meeting National Council on Family Relations, November 6–9. San Antonio.

Hatfield, E., L. Bensman, and R. L. Rapson. 2012. A brief history of social scientists' attempts to measure passionate love. *Journal of Social and Personal Relationships* 29: 143–164.

Hawes, Z. C., K. Wellings, and J. Stephenson. 2010. First heterosexual intercourse in the United Kingdom: A review of the literature. *Journal of Sex Research* 47: 137–152.

Hawkins, A. J., S. L. Nock, J. C. Wilson, L. Sanchez, and J. D. Wright. 2002. Attitudes about covenant marriage and divorce: Policy implications from a three-state comparison. *Family Relations* 51: 166–75.

Hawkins, A. J., B. J. Willoughby, and W. J. Doherty. 2012. Reasons for divorce and openness to marital reconciliation. *Journal of Divorce and Remarriage* 53: 453–463.

Hawkins, D. N., and A. Booth. 2005. Unhappily ever after: Effects of long-term, low-quality marriages on well-being. *Social Forces* 84: 445–65.

Hayes, J., A. T. Chakraborty, S. McManus, P. Bebbington, T. Brugha, S. Nicholson, and M. King. 2012. Prevalence of same-sex behavior and orientation in England: Results from a national survey. *Archives of Sexual Behavior* 41: 631–639.

Healy, M. 2013. The web has transformed adoption, for good and bad. *USA Today.* December 17, 3D.

Healy, M. and P. Trap 2012. Desired changes in one's life. *USA Today,* May 7. 1D.

Healy, M. and V. Salazar. 2010. Dealbreakers. *USA Today.* April 27. D 1.

Heath, N. M., S. M. Lynch, A. M. Fritch, and M. M Wong. 2013. Rape myth acceptance impacts the reporting of rape to the police: A study of incarcerated women *Violence Against Women* 19: 1065–1078.

Hegi, K. E. and R. M. Bergner. 2010. What is love? An empirically based essentialist account. *Journal of Social and Personal Relationships* 27: 620–636.

Heino, R. D., N. B. Ellison, and J. L. Gibbs. 2010. Relation shopping: Investigating the market metaphor in online dating. *Journal of Social and Personal Relationships* 27: 427–447.

Heller, N. 2008. Will the transgender dad be a father? What goes on the birth certificate? http://www.slate.com/id/2193475/ (accessed June 13, 2008).

Helms, H. M., J. K. Walls, A. C. Crouter, and S. M. McHale. 2010. Provider role attitudes, marital satisfaction, role overload, and housework: A dyadic approach. *Journal of Family Psychology* 24: 568–577.

Hendrickx, L. G. and P. Enzlin. 2014. Associated distress in heterosexual men and women: Results from an internet survey in Flanders. *The Journal of Sex Research* 51: 1–12.

Henning, K. and J. Connor-Smith. 2011. Why doesn't he leave? Relationship continuity and satisfaction among male domestic violence offenders. *Journal of Interpersonal Violence* 26: 1366–1387.

Hensley, B., P. Martin, J. A. Margrett, M. MacDonald, I. C. Siegler, and L. W. Poon. 2012. Life events and personality predicting loneliness among centenarians: Findings from the Georgia Centenarian Study. *Journal of Psychology* 146: 173–188.

Herbenick, D. et al. 2011. Beliefs about women's vibrator use: Results from a nationally representative probability survey in the United States. *Journal of Sex & Marital Therapy* 37: 329–345.

Herbenick, D., M. Reece, S. A. Sanders, B. Dodge, A. Ghassemi, and D. Fortenberry. 2010. Women's vibrator use in sexual partnerships: Results from a nationally representative survey in the United States. *Journal of Sex & Marital Therapy* 36: 49–65.

Herman-Kinney, N. J. and D. A. Kinney. 2013. Sober as deviant: The stigma of sobriety and how some college students "stay dry" on a "wet" campus. *Journal of Contemporary Ethnography* 42: 64–103.

Hernandez, R. 2012. The Dominican American family. In *Ethnic families in America, 5th ed.*, ed. R. Wright, Jr., C. H. Mindel, T V. Tran, and R. W. Habenstein. 148–173. Upper Saddle River, NJ: Pearson.

Hersch, J. 2013. Opting out among women with elite education. *Review of Economics of the Household* 2: 469–506.

Hertenstein, M. J., M. J. Hertenstein, J. M. Verkamp, A. M. Kerestes, and R. M. Holmes. 2007. The communicative functions of touch in humans, nonhuman primates, and rats: A review and synthesis of the empirical research. *Genetic Social and General Psychology Monographs* 132: 5–94.

Hertlein, K. M. and F. P. Piercy. 2012. Essential elements of Internet infidelity treatment. *Journal of Marital & Family Therapy* 38: 257–270.

Hertzog, J., R. Rowley, and T. Harpel. 2013. Is it bullying, teen dating violence, or both? Student, school staff, & parent perspectives. Paper, National Council on Family Relations annual meeting, San Antonio.

Hess, J. 2012. Living apart together. *Today*. NBC, March 23. Appreciation is expressed to Dr. Hess for the development of this section.

Hesse, C. and K. Floyd. 2011. The impact of alexithymia on initial interactions. *Personal Relationships* 18: 453–470.

Hetherington, E. M. 2003. Intimate pathways: Changing patterns in close personal relationships across time. *Family Relations* 52: 318–31.

Higginbotham, B. J. and L. Skogrand, L. 2010. Relationship education with both married and unmarried step couples: An exploratory study. *Journal of Couple & Relationship Therapy* 9: 133–148.

Higgins, C. A., L. E. Duxbury, and S. T. Lyons. 2010. Coping with overload and stress: Men and women in dual-earner families. *Journal of Marriage and Family* 72: 847–859.

Higgins, J. A., J. Trussell, N. B. Moore, and J. K. Davison 2010. Virginity lost, satisfaction gained? Physiological and psychological sexual satisfaction at heterosexual debut. *Journal of Sex Research* 47: 384–394.

Hill, E. W. 2010. Discovering forgiveness through empathy: Implications for couple and family therapy *Journal of Family Therapy* 32: 169–185.

Hill, M. R., and V. Thomas. 2000. Strategies for racial identity development: Narratives of black and white women in interracial partner relationships. *Family Relations* 49: 193–200.

Hines, J. 2012. A gender comparison of perceptions of offline and online sexual cheating in middle-aged adults. Dissertation, Department of Psychology, Walden University.

Hirano, Y., N. et al. 2011. Home care nurses' provision of support to families of the elderly at the end of life. *Qualitative Health Research* 21: 199–213.

Hochschild, A. R. 1997. *The time bind.* New York: Metropolitan Books.

Hoffnung, M. and M. Williams. 2013. Balancing act: Career and family during college-educated women's 30s. *Sex Roles* 68: 321–334.

Hogeboom, D. L., R. J. McDermott, K. Perrin, H. Osman, and B. Bell-Ellison. 2010. Internet use and social networking among middle aged and older adults. *Educational Gerontology* 36: 93–111.

Högns, R. S. and M. J. Carlson. 2010. Intergenerational relationships and union stability in fragile families. *Journal of Marriage and Family* 72: 1220–1233.

Hogsnes, L., C. Melin-Johansson, K. Gustaf Norbergh, and E. Danielson. 2014. The existential life situations of spouses of persons with dementia before and after relocating to a nursing home. *Aging & Mental Health* 18: 152–160.

Hogue, M., C. L. Z. Dubois, and L. Fox-Cardamone. 2010. Gender differences in pay expectations: The roles of job intention and self-view. *Psychology of Women Quarterly* 34: 215–227.

Hohmann-Marriott, B. E. and P. Amato. 2008. Relationship quality in interethnic marriages and cohabitation. *Social Forces* 87: 825–855.

Hollander, D. 2010. Body image predicts some risky sexual behaviors among teenage women. *Perspectives on Sexual & Reproductive Health* 42: 67–87.

Hollander, D. 2006. Many teenagers who say they have taken a virginity pledge retract that statement after having intercourse. *Perspectives on Sexual and Reproductive Health* 38:168–173.

Holmes, E. K., T. Sasaki, and N. L. Hazen. 2013. Smooth versus rocky transitions to parenthood: Family systems in developmental context. *Family Relations* 62: 824–837.

Holt-Lunstad J., T. B. Smith, and J. B. Layton. 2010. Social relationships and mortality risk: A meta-analytic review. *PLoS Medicine* 7: 316–333.

Hong, R. and A. Welch. 2013. The lived experiences of single Taiwanese mothers being resilient after divorce. *Journal of Transcultural Nursing* 24: 51–59.

Horowitz, A. D. and L. Spicer. 2013. Definitions among heterosexual and lesbian emerging adults in the U.K. *Journal of Sex Research* 50: 139–150.

Hoser, N. 2012. Making it a dual-career family in Germany: Exploring what couples think and do in everyday life. *Marriage and Family Review* 48: 643–666.

Hou, F. and J. Myles. 2013. Interracial marriage and status-caste exchange in Canada and the United States. *Ethnic & Racial Studies* 36: 75–96.

Hsueh, A. C., K. R. Morrison, and B. D. Doss. 2009. Qualitative reports of problems in cohabiting relationships: Comparisons to married and dating relationships. *Journal of Family Psychology* 23: 236–246.

Huang, H. and L. Leung. 2010. Instant messaging addiction among teenagers in China: Shyness, alienation and academic performance decrement. *Cyber Psychology & Behavior* 12: 675–679.

Hudson, C. G. 2012. Declines in mental illness over the adult years: An enduring finding or methodological artifact? *Aging & Mental Health* 16: 735–752.

Hughes, M. 2011. Hey, we love animals! *Industrial Engineer* 43: p 6.

Human Rights Campaign. 2013. *Resource Guide to Coming Out for African Americans.*

Humble, A. M. 2013. Same-sex weddings in Canada: An ecological analysis of support. Poster, National Council on Family Relations Annual Meeting. San Antonio, November 5–9.

Humphreys, T. P. 2013. Cognitive frameworks of virginity and first intercourse. *Journal of Sex Research* 50: 664–675.

Hunt, A. 2010. Enrollment is up at Grandparent's University. *USA Today,* June 30. 2D.

Huston, T. L., J. P. Caughlin, R. M. Houts, S. E. Smith, and L. J. George. 2001. The connubial crucible: Newlywed years as predictors of marital delight, distress, and divorce. *Journal of Personality and Social Psychology* 80: 237–252.

Huyck, M. H. and D. L. Gutmann. 1992. Thirty something years of marriage: Understanding experiences of women and men in enduring family relationships. *Family Perspective* 26: 249–265.

Hyde, A., J. Drennan, M. Butler, E. Howlett, M. Carney, and M. Lohan. 2013. Parents' constructions of communication with their children about safer sex. *Journal of Clinical Nursing* 22: 3438–3446.

## I

Iantiffi, A. and W. O. Bockting. 2011. View from both sides of the bridge? Gender, sexual legitimacy and transgender people's experiences of relationships. *Culture, Health & Sexuality* 13: 355–370.

Ikegami, N. 1998. Growing old in Japan. *Age and Ageing* 27: 277–278.

Impett, E. A., J. B. Breines, and A. Strachman. 2010. Keeping it real: Young adult women's authenticity in relationships and daily condom use. *Personal Relationships,* 17: 573–584.

Isaacson, W. 2011. *Steve Jobs.* New York: Simon & Schuster.

Isaacson, W. 2007. *Einstein.* New York: Simon & Schuster.

Ishii, N., T. Terao, Y. Araki, K. Kohno, Y. Miokami, M. Arasaki, and N. Iwata. 2013. Risk factors for suicide in Japan: A model of predicting suicide in 2008 by risk factors of 2007. *Journal of Affective Disorders* 147: 352–354.

Israel, T., and J. J. Mohr. 2004. Attitudes toward bisexual women and men: Current research, future directions. In *Current research on bisexuality,* ed. R. C. Fox, 117–34. New York: Harrington Park Press.

## J

Jacinto, E. and J. Ahrend. 2012. Living apart together. Unpublished data provided by Jacinto and Ahrend.

Jackson, J. 2011. Premarital Counseling: An evidence-informed treatment protocol Presentation, Annual Meeting f the National Council on Family Relations, Orlando, November.

Jackson, J. B., R. B. Miller, M. Oka, and R. G. Henry. 2014. Gender differences in marital satisfaction. *Journal of Marriage and Family* 76: 105–112.

Jacobsen, N. and J. Gottman. 2007. *When men batter women: New insights into ending abusive relationships.* New York: Simon & Schuster.

Jalovaara, M. 2003. The joint effects of marriage partners' socioeconomic positions on the risk of divorce. *Demography* 40: 67–81.

James, E. L. 2011. *Fifty shades of grey.* New York: Vintage Books.

James, S. D. 2008. Wild child speechless after tortured life. *ABC News,* May 7.

Jaudesa, P. K., and L. Mackey-Bilaverb. 2008. Do chronic conditions increase young children's risk of being maltreated? *Child Abuse and Neglect* 32: 671–681.

Jayson, S. 2014. Could a date get any more confusing? *USA Today,* January 21, 4B. Stress can test newlyweds. *USA Today,* March 5, 1A and 3D.

_____. 2013. Remarriage in an age of cohabitation. *USA Today,* Sept 13 2A.

_____. 2013. The end of online dating. *USA Today.* Feb 14. P 1A et passim.

_____. 2012. Couples of all kinds are cohabiting. *USA Today,* October 18, A1.

_____. 2011. Is dating dead? *USA Today,* March 30, A1.

Jeanfreau, M. M., A. P. Jurich, and M. D. Mong. 2013. An examination of potential attractions of women's marital infidelity. *American Journal of Family Therapy* 42: 14–28.

Johnson, A., S. Wood, A. C. Taylor, E. Baugh, K. Webster, A. Phoenix, A. Davis, and J. Daves. 2013. The transition from a "talking relationship" to a committed dating couple. Poster, National Council on Family Relations Annual Meeting. San Antonio, November 5–9.

Johnson, C. L. and B. M. Barer. 1997. *Life beyond 85 years: The aura of survivorship.* New York: Springer Publishing.

Johnson, C. W., A. A. Singh and M. Gonzalez. 2014. "It's complicated": Collective memories of transgender, queer, and questioning youth in high school. *Journal of Homosexuality* 61: 419–434.

Johnson, D. W., C. Zlotnick, and S. Perez. 2008. The relative contribution of abuse severity and PTSD severity on the psychiatric and social morbidity of battered women in shelters. *Behavior Therapy* 39: 232–247.

Johnson, J. D., C. J. Whitlatch, and H. L. Menne. 2014. Activity and well-being of older adults: Does cognitive impairment play a role? *Research on Aging* 36: 147–160.

Johnson, M. D. and J. R. Anderson. 2011. The longitudinal association of marital confidence, time spent together and marital satisfaction. Presentation, National Council on Family Relationship, Orlando, November.

Johnson, R. W. and J. M. Wiener. 2006. A profile of frail older Americans and their caregivers. Urban Institute Report. http://www.urban.org/url.cfm?ID–311284 (posted March 1).

Johnson, S., J. Li, G. Kendall, L. Strazdin, and P. Jacoby. 2013. Mothers' and fathers' work hours, child gender, and behavior in middle childhood. *Journal of Marriage & Family* 75: 56–74.

Jonason, P. K. and P. Kavanagh. 2010. The dark side of love: Love styles and the Dark Triad. *Personality & Individual Differences* 49: 606–610.

Jonathan, N. and C. Knudson-Martin. 2012. Building connection: Attunement and gender equality in heterosexual relationships. *Journal of Couple and Relationship Therapy* 11: 95–111.

Jones, K. E. and A. E. Tuttle. 2012. Clinical and ethical considerations for the treatment of cybersex addiction for marriage and family therapists. *Journal of Couple & Relationship Therapy* 11: 274–290.

Jones, R. K. and K. Kooistra. 2011. Abortion incidence and access to services in the United States, 2008. *Perspectives on Sexual & Reproductive Health* 43: 41–50.

Jones, R. K., A. M. Moore, and L. F. Frohwirth. 2011. Perceptions of male knowledge and support among U.S. women obtaining abortions. *Women's Health Issues* 21: 117–123.

Jones, S. L. and M. A. Yarhouse. 2011. A longitudinal study of attempted religiously mediated sexual orientation change. *Journal of Sex & Marital Therapy* 404–427.

Jorgensen, B. L., J. Yorgason, R. Day, and J. A. Mancini. 2011. The influence of religious beliefs and practices on marital commitment. Poster, Annual Meeting of National Council on Family Relations, Orlando, November 18.

Jorm, A. F., H. Christensen, A. S. Henderson, P. A. Jacomb, A. E. Korten, and A. Mackinnon. 1998. Factors associated with successful ageing. *Australian Journal of Ageing* 17:33–37.

Juffer, F., M. van Ijzendoorn, and J. Palacios. 2011. Children's recovery after adoption. *Infancia y Aprendizaje* 34: 3–18.

## K

Kahn, J. R., J. Garcia-Manglano, and S. M. Bianchi. 2014. The motherhood penalty at midlife: Long-term effects of children on women's careers. *Journal of Marriage and Family* 76: 56–72.

Kahneman, D. and A. Deaton. 2010. High income improves evaluation of life but not emotional well-being. Proceedings of the National Academy of Sciences, early edition, September 6, 2010.

Kaiser Family Foundation. 2010. Media use among teens. Retrieved Feb 7, 2010. http://www.kff.org/entmedia/entmedia012010nr.cfm

Kalish, R. and M. Kimmel. 2011. Hooking up. *Australian Feminist Studies* 26: 137–151.

Kamen, C., M. Burns, and S. R. H. Beach. 2011. Minority stress in same-sex male relationships: When does it impact relationship satisfaction? *Journal of Homosexuality* 58: 1372–1390.

Kamiya, Y., M. Doyle, J. C. Henretta et al. 2013. Depressive symptoms among older adults: The impact and later life circumstances and marital status. *Aging and Mental Health* 17: 349–357.

Kamp Dush, C. M. 2013. Marital and cohabitation dissolution and parental depressive symptoms in fragile families. *Journal of Marriage and the Family* 75: 91–109.

Kamp Dush, C. M. 2011. Relationship-specific investments, family chaos, and cohabitation dissolution following a nonmarital birth. *Family Relations* 60: 586–601.

Kamp Dush, C. M. and M. G. Taylor. 2012. Trajectories of marital conflict across the life course: Predictors and interactions with marital happiness trajectories. *Journal of Family Issues* 33: 341–368.

Kaplan, M. S. and R. B. Krueger. 2010. Diagnosis, assessment, and treatment of hypersexuality. *Journal of Sex Research* 47: 181–198.

Karch, D. and K. C. Nunn. 2011. Characteristics of elderly and other vulnerable adult victims of homicide by a caregiver: National violent death reporting system—17 U.S. States, 2003–2007. *Journal of Interpersonal Violence* 26: 137–157.

Karraker, A. and J. DeLamater. 2013. Past year inactivity among older married persons and their partners. *Journal of Marriage and Family* 75: 142–163.

Karlsen, M. and B. Traeen. 2013. Identifying "friends with benefits" scripts among young adults in the Norwegian cultural context. *Sexuality & Culture* 2013: 83–99.

Karten, E. Y. and J. C. Wade. 2010 Sexual orientation change efforts in men: A client perspective. *Journal of Men's Studies* 18: 84–102.

Kasearu, K. 2010. Intending to marry. . . . students' behavioral intention towards family forming. *TRAMES: A Journal of the Humanities & Social Sciences* 14: 3–20.

Katz, P., J. et al. 2011. Costs of infertility treatment: Results from an 18-month prospective cohort study. *Fertility & Sterility* 95: 915–921.

Katz, J. and J. Moore. 2013. Bystander education training for campus sexual assault prevention: An initial meta-analysis. *Violence and Victims* 28: 1054–1067.

Katz, J., and L. Myhr. 2008. Perceived conflict patterns and relationship quality associated with verbal sexual coercion by male dating partners. *Journal of Interpersonal Violence* 23: 798 –804.

Kefalas, M., F. F. Furstenberg, P. J. Carr, and L. Napolitano. 2011. Marriage is more than being together: The meaning of marriage for young adults. *Journal of Family Issues* 32: 845–875.

Keim, B. and A. L. Jacobson. 2011. *Wisdom for Parents: Key ideas from parent educators.* Toronto: de Sitter Publications of Canada.

Keller, G. 2008a. French businesses loath to end 35-hour work week. *Associated Press.* http://www.wtop.com/?nid=105&sid=1471646.

Keller, E. G. 2008b. *The comeback: Seven stories of women who went from career to family and back again.* New York: Bloomsbury.

Kelley, K. 2010. *Oprah: A biography.* New York: Crown Publishers.

Kelly, S. 2012. Dead bodies that matter: Toward a new ecology of human death in American culture. *The Journal of American Culture* 35: 37–51.

Kem, J. 2010. Fatal lovesickness in Marguerite de Navarre's Heptaméron. *Sixteenth Century Journal* 41: 355–370.

Kennedy, D. P., J. S. Tucker, M. S. Pollard, M. Go, and H. D. Green. 2011. Adolescent romantic relationships and change in smoking status. *Addictive Behaviors* 36: 320–326.

Kennedy, G. E. 1997. Grandchildren's memories: A window into relationship meaning. Paper presented at the Annual Conference of the National Council on Family Relations, Crystal City, Virginia.

Kennedy, M. 2013. Gender role observations of East Africa. Written exclusively for this text.

Kennedy, R. 2003. *Interracial intimacies.* New York: Pantheon.

Kerley, K. R., X Xu, B. Sirisunyaluck, and J. M. Alley. 2010. Exposure to family violence in childhood and intimate partner perpetration or victimization in adulthood: Exploring intergenerational transmission in urban Thailand. *Journal of Family Violence* 25: 337–347.

Kesselring, R. G., and D. Bremmer. 2006. Female income and the divorce decision: Evidence from micro data. *Applied Economics* 38: 1605–17.

Kiernan, K. 2000. European perspectives on union formation. In *The ties that bind,* ed. L. J. Waite, 40–58. New York: Aldine de Gruyter.

Killewald, A. 2013. A reconsideration of the fatherhood premium: Marriage, coresidence, biology, and fathers' wages. *American Sociological Review* 78: 96–116.

Kilmann, P. R., H. Finch, M. M. Parnell, and J. T. Downer. 2013. Partner attachment and interpersonal characteristics. *Journal of Sex and Marital Therapy* 39: 144–159.

Kim, H. 2011. Exploratory study on the factors affecting marital satisfaction among remarried Korean couples. *Families in Society* 91: 193–200.

Kimberly, C. and J. Hans. 2012. Sexual self-disclosure and communication among swinger couples. Poster, National Council on Family Relations, Phoenix, AZ, November.

King, A. L. S., A. M. Valenca, A. C. O. Silva, T. Baczynski, M. R. Carvalho, and A. E. Nardi. 2013. Nomophobia: dependency on virtual environments or social phobia? *Computers in Human Behavior.* 29: 140–144.

Kinsey, A. C., W. B. Pomeroy, and C. E. Martin. 1948. *Sexual behavior in the human male.* Philadelphia: Saunders.

Kinsey, A. C., W. B. Pomeroy, C. E. Martin, and P. H. Gebhard. 1953. *Sexual behavior in the human female.* Philadelphia: Saunders.

Klinenberg, E. 2012. *Going solo: The extraordinary rise and surprising appeal of living alone.* New York: Penguin.

Klipfel, K. M., S. E. Claxton, and M. H. M. Van Dulmen. 2014. Interpersonal aggression victimization within casual sexual relationships and experiences. *Journal of Interpersonal Violence* 29: 557–569.

Knog, G., D. Camenga, and S. 2012. Parental influence on adolescent smoking cessation: Is there a gender difference? *Addictive Behaviors* 37: 211–216.

Klos, L. A. and J. Sobal. Weight and weddings. Engaged men's body weight ideals and wedding weight management behaviors. *Appetite* 60: 133–139.

Knox, D., C. Schacht, and M. Whatley. 2014. *Choices in Sexuality, 4th ed.*

Reddington, CA.: Best Value Publishers.
Knox, D., C. Schacht, J. Turner, and P. Norris. 1995. College students' preference for win-win relationships. *College Student Journal* 29: 44–46.
Knox, D., M. E. Zusman, M. Kaluzny, and C. Cooper. 2000. College student recovery from a broken heart. *College Student Journal* 34: 322–324.
Knox, D. and M. Zusman. 2007. Traditional wife? Characteristics of college men who want one. *Journal of Indiana Academy of Social Sciences* 11: 27–32.
Knox, D., M. E. Zusman, L. Mabon, and L. Shivar. 1999. Jealousy in college student relationships. *College Student Journal* 33: 328–329.
Kohlberg, L. 1966. A cognitive-developmental analysis of children's sex-role concepts and attitudes. In *The development of sex differences,* ed. E. E. Macoby. Stanford, CA: Stanford University Press.
———. 1969. State and sequence: The cognitive developmental approach to socialization. In *Handbook of socialization theory and research,* ed. D. A. Goslin, 347–480. Chicago: Rand McNally.
Kolko, D. J., L. D. Dorn, O. Bukstein, and J. D. Burke. 2008. Clinically referred ODD children with or without CD and healthy controls: Comparisons across contextual domains. *Journal of Child and Family Studies* 17: 714–734.
Kondapalli, L. A. and A. Perales-Puchalt. 2013. Low birth weight: Is it related to assisted reproductive technology or underlying infertility? *Fertility and Sterility.* 99: 303–310.
Kornrich, S., J. Brines, and K. Leupp. 2013. Egalitarianism, housework, and sexual frequency in marriage. *American Sociological Review* 78: 26–50.
Kothari, C. L. et al. 2012. Protection orders protect against assault and injury: A longitudinal study of police-involved women victims of intimate partner violence. *Journal of Interpersonal Violence* 27: 2845–2868.
Kotila, L. E., S. J. Schoppe-Sullivan, and C. M. Kamp Dush. 2013. Time parenting activities in dual-earner families at the transition to parenthood. *Family Relations* 62: 795–807.
Kotlyar, I. and D. Ariely. 2013. The effect of nonverbal cues on relationship formation. *Computers in Human Behavior* 29: 544–551.
Kreager, D. A., S. E. Cavanagh, J. Yen, and M. Yu. 2014. "Where Have All the Good Men Gone?" Gendered Interactions in Online Dating. Journal of Marriage and Family. 76: 387–410.
Kulik, L. 2007. Contemporary midlife grandparenthood. In *Women over fifty: Psychological perspectives,* ed. Varda Muhlbauer and Joan C. Chrisler, 131–146. New York: Springer Science and Business Media.
Kulik, L. and E. Heine-Cohen. 2011. Coping resources, perceived stress and adjustment to divorce among Israeli women: Assessing effects. *Journal of Social Psychology* 151: 5–30.
Kulkarni, M., S. Graham-Bermann, S. Rauch, and J. Seng. 2011. Witnessing versus experiencing direct violence in childhood as correlates of adulthood PTSD. *Journal of Interpersonal Violence* 26: 1264–1281.
Kuper, L. E., R. Nussbaum, and B. Mustanski. 2012. Exploring the diversity of gender and sexual orientation identities in an online sample of transgender individuals. *Journal of Sex Research* 49: 244–254.
Kurdek, L. A. 2008. Change in relationship quality for partners from lesbian, gay male, and heterosexual couples. *Journal of Family Psychology* 22: 701–11.
______. 2005. What do we know about gay and lesbian couples? Current Directions in *Psychological Science* 14: 251–254.
______. 2004. Gay men and lesbians: The family context. In *Handbook of contemporary families: Considering the past, contemplating the future,* ed. M. Coleman and L. H. Ganong, 96–115. Thousand Oaks, CA: Sage Publications.
———. 1995. Predicting change in marital satisfaction from husbands' and wives' conflict resolution styles. *Journal of Marriage and the Family* 57: 153–164.
______. 1994. Areas of conflict for gay, lesbian, and heterosexual couples: What couples argue about influences relationship satisfaction. *Journal of Marriage and the Family* 56: 923–934.
______. 1994. Conflict resolution styles in gay, lesbian, heterosexual nonparent, and heterosexual parent couples. *Journal of Marriage and the Family* 56: 705–722.

## L

LaBrie, J. W., J. F. Hummer, T. M. Gaidarov, A. Lac, and S. R. Kenney. 2014. Hooking up in the college context: The event-level effects of alcohol use and partner familiarity on hookup behaviors and contentment. *Journal of Sex Research* 51: 62–73.
Lacayo, R. 2010. Appreciation: J. D. Salinger. *Time,* February 15, p. 66.
Lachance-Grzela, L. and G. Bouchard 2010. Why do women do the lion's share of housework? A decade of research. *Sex Roles* 63: 767–780.
Lacks, M. H., A. L. Lamson, A. Ivanescu, M. B. White, and C. Russoniello. 2013. An exploration of marital status and stress among military couples. Research and Creative Week. East Carolina University, April 8–12.
Lalicha, J. and K. McLarena. 2010. Inside and outcast: Multifaceted stigma and redemption in the lives of gay and lesbian Jehovah's Witnesses. *Journal of Homosexuality* 57: 1303–1333.
Lam, C. B., S. M. McHale, and A. C. Crouter. 2012. The division of household labor: Longitudinal changes and within-couple variation. *Journal of Marriage and Family* 74: 944–952.
Lambert, N. M., S. Negash, T. F. Stillman, S. B. Olmstead, and F. D. Fincham. 2012. A love that doesn't last: Pornography consumption and weakened commitment to one's romantic partner. *Journal of Social & Clinical Psychology* 31: 410–438.
Landor, A. and L. G. Simons. 2010. The impact of virginity pledges on sexual attitudes and behaviors among college students. Paper, Annual Meeting of the National Council on Family Relations, November 3–6. Minneapolis, MN.
Lane, C. D., P. S. Meszaros, and T. Savla. 2012. Developing the family resilience measure Paper, Annual Meeting of National Council on Family Relations, Phoenix, November.
Langeslag, S. J. E., P. Muris, and I. H. A. Fraken. 2013. Measuring romantic love: Psychometric properties of the infatuation and attachment scales. *Journal of Sex Research* 50: 739–774.
LaPierre, T. A. 2010. The legal context of grand families: Meaning and role conflict. Poster, National Council on Family Relations annual meeting, November 3–5. Minneapolis, MN.
Larsen, C. D., J. G. Sandberg, Harper, and R. Bean, R. 2011. The effects of childhood abuse on relationship quality: Gender differences and clinical implications. *Family Relations* 60: 435–445.
LaSala, M. C. and D. T. Frierson. 2012. African American gay youth and their families: Redefining masculinity, coping with racism and homophobia. *Journal of GLBT Family Studies* 8: 428–445.
Laukkanen, J., U. Ojansuu, A. Tolvanen, S. Alatupa, and K. Aunola. 2014. Child's difficult temperament and mothers' parenting styles. *Journal of Child and Family Studies* 23: 312–323.
Lavee, Y., and A. Ben-Ari 2007. Relationship of dyadic closeness with work-related stress: A daily diary

study. *Journal of Marriage and Family* 69: 1021–1035.

Lavner, J. A. and T. N. Bradbury. 2012. Why do even satisfied newlyweds eventually go on to divorce? *Journal of Family Psychology* 26: 1–10.

Lavner, J. A. and T. N. Bradbury. 2010. Patterns of change in marital satisfaction over the newlywed years. *Journal of Marriage and Family* 72: 1171–1187.

Lawrence, Q. and M. Penaloza. 2013. Sexual violence victims say military justice system is broken. NPR, March 21.

Lawyer, S., H. Resnick, V. Bakanic, T. Burkett, and D. Kilpatrick. 2010. Forcible, drug-facilitated, and incapacitated rape and sexual assault among undergraduate women. *Journal of American College Health* 58: 453–460.

Lax, E. 1991. *Woody Allen: A biography.* New York: Knopf.

Lease, S. H., A. B. Hampton, K. M. Fleming, L. R. Baggett, S. H. Montes, and R. J. Sawyer. Masculinity and interpersonal competencies: Contrasting White and African American men. *Psychology of Men & Masculinity* 11: 195–207.

LeCouteur, A. and M. Oxlad. 2011. Managing accountability for domestic violence: Identities, membership categories and morality in perpetrators' talk. *Feminism & Psychology* 21: 5–28.

Leddy, A., N. Gartrell, and H. Bos. 2012. Growing up in a lesbian family: The life experience of adult daughters and sons of lesbian mothers. *Journal of GLBT Family Studies* 8: 243–257.

Lee, D. 2006 Device brings hope for fertility clinics. http://www.indystar.com/apps/pbcs.dll/article?AID=/20060221/BUSINESS/602210365/1003 (retrieved February 22, 2006).

Lee, J. A. 1973. *The colors of love: An exploration of the ways of loving.* Don Mills, Ontario: New Press.

———. 1988. Love-styles. In *The psychology of love,* ed. R. Sternberg and M. Barnes, 38–67. New Haven, CN: Yale University Press.

Lee, J. A., P. Foos, and C. Clow. 2010. Caring for one's elders and family-to-work conflict. *Psychologist-Manager Journal* 13: 15–39.

Lee, J. T., C. L. Lin, G. H. Wan, and C. C. Liang. 2010. Sexual positions and sexual satisfaction of pregnant women. *Journal of Sex & Marital Therapy* 36: 408–420.

Lee, K. S. and H. Ono. 2012. Marriage, cohabitation, and happiness: A cross-national analysis of 27 countries. *Journal of Marriage and Family* 74: 953–972.

Lee, T. and G. R. Hicks. 2011. An analysis of factors affecting attitudes toward same-sex marriage: Do the media matter? *Journal of Homosexuality* 58: 1391–1408.

Lehman, A. D. 2010. Inappropriate injury: The case for barring consideration of a parent's homosexuality in custody actions. *Family Law Quarterly* 44: 115–131.

Lehmiller, J. J., L. E. VanderDrift, and J. R. Kelly. 2014. Sexual communication, satisfaction, and condom use behavior in friends with benefits and romantic partners. *Journal of Sex Research* 51: 74–85.

Leno, J. 1996. *Leading with my chin.* New York: Harper Collins.

Leopold, T., M. Raab, and H. Engelhardt. 2014. The transition to parent care: Costs, commitments, and caregiver selection among children. *Journal of Marriage and the Family* 76: 300–318.

Levchenko, P. and C. Solheim. 2013. International marriages between Eastern European-born women and U.S. born men. *Family Relations* 62: 30–41.

Lever, J. 1994. The 1994 Advocate survey of sexuality and relationships: The men. *The Advocate,* August 23, 16–24.

Levine, S. B. 2010. What is sexual addiction? *Journal of Sex & Marital Therapy* 36: 261–275.

Lewis, K. (2008) Personal communication. Dr. Lewis is also the author of *Five Stages of Child Custody.* Glenside, PA: CCES Press.

Lewis, M. A., D. C. Atkins, J. A. Blayney, D. V. Dent, and D. L. Kaysen. 2013. What is hooking up? Examining definitions of hooking up in relation to behavior and normative perceptions. *Journal of Sex Research* 50: 757–766.

Li, N. P., K. A. Valentine, and L. Patel. 2011. Mate preferences in the U.S. and Singapore: A cross-cultural test of the mate preference priority model. *Personality & Individual Differences* 50: 291–294.

Licata, N. 2002. Should premarital counseling be mandatory as a requisite to obtaining a marriage license? *Family Court Review* 40: 518–532.

Light, A. and T. Ahn. 2010. Divorce as risky behavior. *Demography* 47: 895–921.

Lincoln, A. E. 2010. The shifting supply of men and women to occupations: Feminization in veterinary education. *Social Forces* 88: 1969–1998.

Lindblad-Goldberg, M. and W. Northey. 2013. Ecosylstemic structural family therapy. Theoretical and clinical foundations. *Contemporary Family Therapy: An International Journal:* 35: 147–160.

Lindau, S. T., L. P. Schumm, E. O. Laumann, W. Levinson, C. A. O'Muircheartaigh, and L. J. Waite. 2007. A study of sexuality and health among older adults in the United States. *The New England Journal of Medicine* 357: 762–774.

Lipman, E. L., K. Georgiades, and M. Boyle. 2011. Young adult outcomes of children born to teen mothers: Effects of being born during their teen or later years. *Journal of the American Academy of Child & Adolescent Psychiatry* 50: 232–241.

Lipscomb, R. 2009. Person-first practice: Treating patients with disabilities. *Journal of the American Dietetic Association* 109: 21–25.

Littleton, H., A. Grills-Taquechel, and D. Axsom. 2009. Impaired and incapacitated rape victims: Assault characteristics and post-assault experiences. *Violence & Victims* 24: 439–457.

Liu, S., M. M. Dore, and I. Amrani-Cohen. 2013. Treating the effects of interpersonal violence: A comparison of two group models. *Social Work with Groups* 36: 59–72.

Lo, S. K, A. Y. Hsieh, and Y. P. Chiu. 2013. Contradictory deceptive behavior in online dating. *Computers in Human Behavior* 29: 1755–1762.

Looi, C., P. Seow, B. Zhang, H. So, W. Chen, and L. Wong. 2010. Leveraging mobile technology for sustainable seamless learning: a research agenda. *British Journal of Educational Technology* 41: 154–169.

Lopoo, L. M, and B. Western. 2005. Incarceration and the formation and stability of marital unions. *Journal of Marriage and the Family* 67: 721–35.

Lorber, J. 1998. *Gender inequality: Feminist theories and politics.* Los Angeles, CA: Roxbury.

Lotspeich-Younkin, F. and S. Bartle-Haring. 2012. Differentiation and relationship satisfaction: Mediating effects of emotional intimacy and alcohol/substance use. Paper, National Council on Family Relations, Phoenix, November.

Lucier-Greer, M. and F. Adler-Baeder. 2010. Gender role attitudes during divorce & remarriage: Plastic or plaster? Poster, National Council on Family Relations annual meeting, November 3–5.

Lundquist, J. H. 2007. A comparison of civilian and enlisted divorce rates during the early all-volunteer force era. *Journal of Political and Military Sociology* 35(2), 199–217.

Luscombe, B. 2010a. Divorcing while dying. *Time,* February 15, p. 49.

_____. 2010b. Making divorce pay. *Time,* Sept. 13.

Lyndon, A. E. H. C.Sinclair, J. MacArthur, B Fay, E. Ratajack, and K. E. Collier. 2012. An introduction to issues of gender in stalking research. *Sex Roles* 66: 299–310.

Lyons, M., A. Lynch, G. Brewer, and D. Bruno. 2014. Detection of sexual orientation ("gaydar") by homosexual and heterosexual women. *Archives of Sexual Behavior* 43: 345–352.

Lyons, A., M. Pitts, and J. Grierson. 2013. Growing old as a gay man: Psychosocial well-being of a sexual minority. *Research on Aging* 35: 275–295.

Lyssens-Danneboom, V., S. Eggermont, and D. Mortelmans. 2013. Living apart together (LAT) and law: Exploring legal expectations among LAT individuals in Belgium. *Social & Legal Studies* 22: 357–376.

## M

Ma, J., Y. Han, A. Grogan-Kaylor, J. Delva, and M. Castillo. 2012. Corporal punishment and youth externalizing behavior in Santiago, Chile. *Child Abuse & Neglect* 36: 481–490.

Maatta, K. and S. Uusiautti. 2013. Silence is not golden: Review of studies of couple interaction. *Communication Studies* 64: 33–48.

_____. 2012. Seven rules on having a happy marriage along with work. *The Family Journal* 20: 267–271.

Macleod, C. 2010. China smitten with TV dating. *USA Today,* May 18, p. 8A.

MacLeod, C. 2012. Forced abortion ignites online outrage. *USA Today.* 5A.

MacMillan, H. L., C. N. Wathen, and C. M. Varcoe. 2013. Intimate partner violence in the family: Considerations for children's safety. *Child Abuse & Neglect* 37: 1186–1191.

Madden, M. 2010. Older adults and social media. Pew Internet & American Life Project Posted and retrieved August 27 http://pewresearch.org/pubs/1711/older-adults-social-networking-facebook-twitter.

Maddox, A. M., G. K. Rhoades, and H. J. Markman. 2011. Viewing sexually explicit materials alone or together: Associations with relationship quality. *Archives of Sexual Behavior* 40: 441–448.

MaddoxShaw, A. M., G. K. Rhoades, E. S. Allen, S. M. Stanley, and H. J. Markman. 2013. Predictors of extradyadic sexual involvement in unmarried opposite-sex relationships. *Journal of Sex Research* 50: 598–610.

Magaziner, J. 2010. The new technologies of change. *Psychology Networker.* September/October. 42–47.

Maher, D. and C. Mercer (eds.). 2009. Introduction. *Introduction to religion and the implications of radical life extension.* New York: Palgrave Macmillan.

Mahr, K. 2013. India's shame. *Time,* January 14, p. 12.

Maier, C. and C. R. McGeorge. 2013. Positive perceptions of single mothers and fathers: Implications for therapy. National Council on Family Relations annual meeting, San Antonio.

Maier, T. 2009. *Masters of Sex.* New York: Perseus Books.

Major, B., M. Appelbaum, and C. West. 2008. Report of the APA task force on mental health and abortion. August 13.

Malacad, B. and G. Hess. 2010. Oral sex: Behaviours and feelings of Canadian young women and implications for sex education. *European Journal of Contraception & Reproductive Health Care* 5: 177–185.

Malott, K. M. and C. D. Schmidt. 2012. Counseling families formed by transracial adoption: Bridging the gap in the multicultural counseling competencies. *The Family Journal* 20: 384–391.

Maltby, L E., M. E. L. Hall, T. L. Anderson, and K. Edwards. 2010. Religion and sexism: The moderating role of participant gender. *Sex Roles* 62: 615–622.

Mandara, J., F. Varner and S. Richman. 2010. Do African American mothers really "love" their sons and "raise" their daughters? *Journal of Family Psychology* 24: 41–50.

Maneta, E., S. Cohen, M. Schulz, and R. J. Waldinger. 2012. Links between childhood physical abuse and intimate partner aggression: The mediating role of anger expression. *Violence and Victims* 27: 315–328.

Manning, W. D., J. A. Cohen, and P. J. Smock. 2011. The role of romantic partners, family, and peer networks in dating couples' views about cohabitation. *Journal of Adolescent Research* 26: 115–149.

Manning, W. D., D. Trella, H. Lyons, and N. C. DuToit. 2010. Marriageable women: A focus on participants in a community healthy marriage program. *Family Relations* 59: 87–102.

Manning, W. D. and P. J. Smock. 2000. "Swapping" families: Serial parenting and economic support for children. *Journal of Marriage and the Family* 62: 111–22.

Mansson, D. H. 2012. A qualitative analysis of grandparents' expressions of affection for their young adult grandchildren. *North American Journal of Psychology.* 14: 207–219.

Maple, M., H., E. Edwards, V. Minichiello, and D. Plummer. 2013. Still part of the family: The importance of physical, emotional and spiritual memorial places and spaces for parents bereaved through the suicide death of their son or daughter. *Mortality* 18: 54–71.

Marano, H. E. 1992. The reinvention of marriage. *Psychology Today.* January/February, 49.

Marcus, R. E. 2012. Patterns of intimate partner violence in young adult couples: Nonviolent, unilaterally violent, and mutually violent couples. *Violence & Victims* 27: 299–314.

Markham, M. S. and M. Coleman. 2011. "Part-time parent": Divorced mothers' experiences with sharing physical custody. Poster, National Council on Family Relations annual meeting, Orlando, November 18.

_____. 2010. The good, the bad, and the ugly: Divorced mothers' experiences with coparenting. Poster, National Council on Family Relations annual meeting, November 3–5. Minneapolis, MN.

Markman, H. J., G. K. Rhoades, S. M. Stanley, E. P. Ragan, and S. W. Whitton. 2010. The premarital communication roots of marital distress and divorce: The first five years of marriage. *Journal of Family Psychology* 24: 289–298.

Marquardt, E., D. Blankenhorn, R. I. Lerman, L. Malone-Colón, and W. B. Wilcox. 2012. "The president's marriage agenda for the forgotten sixty percent," The State of Our Unions (Charlottesville, VA: National Marriage Project and Institute for American Values, 2012).

Marshall, A., J. M. Bell, and N. J. Moules. 2010. Beliefs, suffering, and healing: A clinical practice model for families experiencing mental illness. *Perspectives in Psychiatric Care* 46: 197–208.

Marshall, J. 2013. Can we understand that which we cannot define? How marriage and family therapists define the family. National Council on Family Relations Annual Meeting. San Antonio, November 5–9.

Marshall, T. C., K. Chuong, and A. Aikawa. 2011. Day-to-day experiences of amae in Japanese romantic relationships. *Asian Journal of Social Psychology* 14: 26–35.

Martin, A., A. Brazil, and J. Brooks-Gunn. 2013. The socioemotional outcomes of young children of teenage mothers by paternal coresidence. *Journal of Family* Issues 34: 1217–1237.

Martin, B. A. and C. Dula. 2010. More than skin deep: Perceptions of, and stigma against, tattoos. *College Student Journal* 44: 200–206.

Martin-Uzzi, M. and D. Duval-Tsioles. 2013. The experience of remarried couples in blended families. *Journal of Divorce and Remarriage* 54: 43–57.

Masarik, A. S., M. J. Martin, E. Ferrer, and R. D. Conger. 2012. Economic pressure and romantic relationship functioning: The moderating role of effective problem solving. Poster, Annual Meeting of the National Council on Family Relations. Phoenix, Nov. 1.

Masheter, C. 1999. Examples of commitment in post divorce relationships between spouses. In *Handbook of interpersonal commitment and relationship stability,* edited by J. M. Adams and W. H. Jones. New York: Academic /Plenum Publishers: 293–306

Mashoa, S. W., D. Chapmana, and M. Ashbya. 2010. The impact of paternity and marital status on low birth weight and preterm births. *Marriage & Family Review* 46: 243–256.

Mason, A. E., R.W. Law, A. E. Bryan, R. M. Portley, and D. A. Sbarra. 2012. Facing a breakup: Electromyographic responses moderate self-concept recovery following a romantic separation. *Personal Relationships:* 19: 551–568.

Masters, W. H., and V. E. Johnson. 1970. *Human sexual inadequacy.* Boston: Little, Brown.

Mata, J., C. L. Hogan, J. Joormann, C. E. Waugh, and I. H. Gotlib. 2013. Acute exercise attenuates negative affect following repeated sad mood inductions in persons who have recovered from depression. *Journal of Abnormal Psychology* 122: 45–50.

Mathes, E. W. 2013. Why is there a strong positive correlation between perpetration and being a victim of sexual coercion? An exploratory study. *Journal of Family Violence* 28: 783–796.

Maulsby, C., F. Sifakis, D. German, C. P. Flynn, and D. Holtgrave. 2013. HIV risk among men who have sex with men only (MSMO) and men who have sex with men and women (MSMW) in Baltimore *Journal of Homosexuality* 60: 51–68.

McBride, K. and J. D. Fortenberry. 2010. Heterosexual anal sexuality and anal sex behaviors: A review. *Journal of Sex Research* 47: 123–136.

McCabe, M. P. & D. L. Goldhammer. 2012. Demographic and psychological factors related to sexual desire among heterosexual women in a relationship. *Journal of Sex Research* 49*:* 78–87.

McCarry, M. 2013. Becoming a "proper man": Young people's attitudes about interpersonal violence and perceptions of gender. *Gender & Education* 22: 17–30.

McClain, L. R. 2011. Better parents, more stable partners: Union transitions among cohabiting parents. *Journal of Marriage and Family* 73*:* 889–901.

McClure, M. J. and J. E. Lydon. 2014. Anxiety doesn't become you: How attachment anxiety compromises relational opportunities. *Journal of Personality and Social Psychology* 106: 89–111.

McCoy, Shawn P. and M. G. Aamodt. 2009. A comparison of law enforcement divorce rates with those of other occupations. *Journal of Police and Criminal Psychology.* Online journal published Oct. 20, 2009.

McDaniel, B. T., L. E. Philbrook, and D. M. Teti. 2012. Becoming parents: Coparenting quality and salivary cortisol. Paper, National Council on Family Relations, Phoenix, November.

McElvaney, R., S. Greene, and D. Hogan. 2014. To tell or not to tell? Factors influencing young people's informal disclosures of child sexual abuse. *Journal of Interpersonal Violence* 29: 928–947.

McKee, K. S. and A. J. Blow. 2010. Proposed model of decision making for genetic testing in families. Poster, National Council on Family Relations annual meeting, November 3–5. Minneapolis, MN.

McGeorge, C. R., T. S. Carlson, and R. B. Toomey. 2012. Establishing the validity of the Feminist Couple Therapy Scale. Paper, National Council on Family Relations annual meeting, November 3–5. Phoenix, AZ.

McGowan, C. M. 2011. Legal aspects of end of life care. *Critical Care Nurse* 31: 64–69.

McHale, S. M., Updegraff, K. A., and Whiteman, S. D. 2012. Sibling relationships and influences in childhood and adolescence. *Journal of Marriage and Family* 74: 913–930.

McIntosh, J. E., Y. D. Wells, B. M. Smyth, and C. M. Long. 2008. Child-focused and child-inclusive divorce mediation: Comparative outcomes from a prospective study of post separation adjust. *Family Court Review* 46: 105–115.

McIntosh, W. D., L. Locker, K. Briley, R. Ryan, and A. Scott. 2011. What do older adults seek in their potential romantic partners? Evidence from online personal ads. *International Journal of Aging & Human Development* 72: 67–82.

McIntyre, S. L., E. A. Antonucci, and S. C. Haden. 2014. Being white helps: Intersections of self-concealment, stigmatization, identity formation, and psychological distress in racial and sexual minority women. *Journal of Lesbian Studies* 18: 158–173.

McKinney, C., and K. Renk. 2008. Differential parenting between mothers and fathers: Implications for late adolescents. *Journal of Family Issues* 29: 806–827.

McLanahan, S. S. 1991. The long-term effects of family dissolution. In *When families fail: The social costs,* ed. Brice J. Christensen, pp. 5–26. New York: University Press of America for the Rockford Institute.

McLanahan, S. S., and K. Booth. 1989. Mother-only families: Problems, prospects, and politics. *Journal of Marriage and the Family* 51: 557–580.

McLaren, S., P. M. Gibbs, and E. Watts. 2013. The interrelations between age, sense of belonging, and depressive symptoms among Australian gay men and lesbians. *Journal of Homosexuality* 60: 1–15.

McLean, K. 2004. Negotiating (non) monogamy: Bisexuality and intimate relationships. In *Current research on bisexuality,* ed. R. C. Fox, 82–97. New York: Harrington Park Press.

McLennon, S. M., B. Habermann, and L L. Davis. 2010. Deciding to institutionalize: Why do family members cease care giving at home? *Journal of Neuroscience Nursing* 42: 95–104.

McMahon, S. 2010. Rape myth beliefs and bystander attitudes among incoming college students. *Journal of American College Health* 59: 3–11.

McNeil, S., T. Pavkov, A. Tracey. 2012. Child maltreatment, exposure to violence, and adolescent delinquency. Paper, National Council on Family Relations, Phoenix, November.

McQuillan, J., A. L. Greil, K. M. Shreffler, P. A. Wonch-Hill, K. C. Gentzler, and J. D. Hathcoat. 2012. Does the reason matter? Variations in childlessness concerns among U.S. women. *Journal of Marriage and Family* 74: 1166–1181.

Mead, G. H. 1934. *Mind, self, and society.* Chicago: University of Chicago Press.

Mead, M. 1935. *Sex and temperament in three primitive societies.* New York: William Morrow.

Meier, J. S. 2009. A historical perspective on parental alienation syndrome and parental alienation. *Journal of Child Custody* 6: 232–257.

Meilaender, G. 2013. *Should we live forever? The ethical ambiguities of aging.* Grand Rapids, MI: Eerdmans Publishing Co.

Meinhold, J. L., A. Acock, and A. Walker. 2006. The influence of life transition statuses on sibling intimacy and contact in early adulthood. Presented at the National Council on Family Relations Annual meeting in Orlando in 2005.

Meisenbach, M. J. 2010. The female breadwinner: Phenomenological experience and gendered identity in work/family spaces. *Journal of Sex Roles* 62: 2–19.

Melhem, N. M., D. A. Brent, M. Ziegler, S. Iyengar et al. 2007. Familial pathways to early-onset suicidal behavior: Familial and individual antecedents of suicidal behavior. *American Journal of Psychiatry* 164: 1364–1371.

Mellor, D., L. Ricciardelli, M. McCabe, J. Yeow, N. Mamat, and Mohd Hapidzal, 2010. Psychosocial correlates of body image and body change behaviors among Malaysian adolescent boys and girls. *Sex Roles* 63: 386–398.

Meltzer, A. L. and J. K. McNulty. 2014. "Tell me I'm sexy . . . and otherwise valuable": Body valuation and relationship satisfaction. *Personal Relationships* 21: 68–87.

Mena, J. A. and A. Vaccaro. 2013. Tell me you love me no matter what: Relationships and self-esteem among GLBQ young adults. *Journal of GLBT Family Studies* 9: 3–23.

Mendenhall, R., P. Bowman, and L. Zhang. 2013. Single black mothers' role strain and adaptation across the life course. *Journal of African American Studies* 17: 74–89.

Mercadante, C., M. F. Taylor and J. A. Pooley. 2014. "I wouldn't wish it on my worst enemy": Western Australian fathers' perspectives on their marital separation experiences. *Marriage & Family Review.* 50: 318–341.

Merolla, A. J. and S. Zhang 2011. In the wake of transgressions: Examining forgiveness communication in personal relationships. *Personal Relationships* 18: 79–95.

Merrill, J. and D. Knox. 2010. *Finding love from 9 to 5: Secrets of office romance.* Santa Barbara, CA: Praeger.

Meteyer, K. and Maureen Perry-Jenkins. 2012. Father involvement among working-class, dual earner couples. *Fathering: A Journal of Theory, Research, & Practice about Men as Fathers* 8: 379–403.

Meyer, A. S., L. M. McWey, W.McKendrick, and T. L. Henderson. 2010. Substance-using parents, foster care, and termination of parental rights: The importance of risk factors for legal outcomes. *Children & Youth Services Review 32: 639–649.*

Michael, R. T., J. H. Gagnon, E. O. Laumann, and G. Kolata. 1994. *Sex in America: A definitive survey.* Boston: Little, Brown.

Michielsen, D. and R. Beerthuizen. 2010. State-of-the art of non-hormonal methods of contraception: VI. Male sterilization. European *Journal of Contraception & Reproductive Health Care* 15: 136–149.

Miers, D., D. Abbott, and P. R. Springer. 2012. A phenomenological study of family needs following the suicide of a teenager. *Death Studies* 36: 118–133.

Miller, A. J., S. Sassler, S. and D. Kusi-Appouh. 2011. The specter of divorce: Views from working- and middle-class cohabiters. *Family Relations* 60: 602–616.

Miller, E., et al. 2010. Pregnancy coercion, intimate partner violence and unintended pregnancy. *Contraception* 81: 316–322.

Miller, J. D., A. Zeichner, and L. F. Wilson. 2012. Personality correlates of aggression: Evidence from measures of the five-factor model, UPPS model of impulsivity, and BIS/BAS. *Journal of Interpersonal Violence* 27: 2903–2919.

Miller, M. 2012. Love hurts (in more ways than one): Specificity of psychological symptoms as predictors and consequences of romantic activity among early adolescent girls. *Journal of Clinical Psychology* 68: 373–381.

Miller, S., A. Taylor, and D. Rappleyea. 2011. The influence of religion on young adults' attitudes of dating events. Poster, Fifth Annual Research & Creative Achievement Week, East Carolina University, April 4–8.

Minnotte, K. L., M. C. Minnotte, and D. E. Pedersen. (2013). Marital satisfaction among dual- earner couples: Gender ideologies and family-to-work conflict. *Family Relations* 62: 686–698.

Minnotte, K. L., D. E. Pedersena, S. E. Mannonb, and G. Kiger. 2010. Tending to the emotions of children: Predicting parental performance of emotion work with children. *Marriage & Family Review* 46: 224–241.

Mirecki, R. M., J. L. Chou, M. Elliott, and C. M. Schneider. 2013. What factors influence marital satisfaction? Differences between first and second marriages. *Journal of Divorce & Remarriage* 54: 78–93.

Mitchell, B. A. 2010. Midlife marital happiness and ethnic culture: A life course perspective. *Journal of Comparative Family Studies* 41: 167–183.

Mock, S. E. and R. P. Eibach. 2012. Stability and change in sexual orientation identity over a Ten-year period in adulthood. *Archives of Sexual Behavior* 41: 641–648.

Mohr, J., R. Cook-Lyon, and M. R. Kolchakian. 2010. Love imagined: Working models of future romantic attachment in emerging adults. *Personal Relationships* 17: 457–473.

Mongeau, P. A., K. Knight, J. Williams, J. Eden, and C. Shaw. 2013. Identifying and explicating variation among friends with benefits relationships. *Journal of Sex Research* 50: 37–47.

Monro, S. 2000. Theorizing transgender diversity: Towards a social model of health. *Sexual and Relationship Therapy* 15: 33–42.

Montesi, J., B. Conner, E. Gordon, R. Fauber, K. Kim, and R. Heimberg. 2013. On the relationship among social anxiety, intimacy, sexual communication, and sexual satisfaction in young couples. *Archives of Sexual Behavior* 42: 81–91.

Mooney, L., C. Reiser, and K. Wilson. 2010 Pet nation: Demographic correlates of the human- companion animal bond. Poster, Southern Sociological Society, Atlanta, April 23.

Moore, D., S. Wigby, S. English, S. Wong, T. Sekely, and F. Harrison. 2013. Selflessness is sexy: Reported helping behavior increases desirability of men and women as long-term sexual partners. *BMC Evolutionary Biology* 13: 130–182.

Moore, M. M. 2010. Human nonverbal courtship behavior—A Brief Historical Review. *Journal of Sex Research* 47: 171–180.

Morell, V. 1998. A new look at monogamy. *Science* 281: 1982.

Morgan, E. S. 1944. *The Puritan family.* Boston: Public Library.

Morin, R. and D. Cohn. 2008. Women call the shots at home; Public mixed on gender roles in jobs, gender and power. Pew Research Center, September 25.

Morrill, M. 2014. Sibling sexual abuse: An exploratory study of long-term consequences for self esteem and counseling considerations. *Journal of Family Violence* 29: 205–213.

Morrison, T. G., D. Beaulieu, M. Brockman, and C. O. Beaglaoich. 2013. A comparison of polyamorous and monoamorous persons: are there differences in indices of relationship well-being and sociosexuality? *Psychology and Sexuality* 4: 75–91.

Mortensen, O., T. Torsheim, O. Melkevik, and F. Thuen. 2012. Adding a baby to the equation: Married and cohabiting women's relationship satisfaction in the transition to parenthood. *Family Process* 51: 122–139.

Moser, C., J. Spagnoli, and B. Santos-Eggimann. 2011. Self-perception of aging and vulnerability to adverse outcomes at the age of 65–70 years. *Journals of Gerontology* Series B: Psychological Sciences & Social Sciences. 66B: 675–680.

Moskowitz, D. A., G. Rieger, and M. E. Roloff. 2010. Heterosexual attitudes toward same-sex marriage. *Journal of Homosexuality.* 57: 325–336.

Moss, A. R. 2012. Alternative families, alternative lives: Married women doing bisexuality. *Journal of GLBT Family Studies* 8: 405–427.

Mosuo, 2010. http://en.wikipedia.org/wiki/Mosuo#General_Practice

Mouilso, E. R. and K. S. Calhoun. 2013. The role of rape myth acceptance and psychopathy in sexual assault perpetration. *Journal of Aggression, Maltreatment & Trauma* 22: 159–174.

Mullen, P. E., M. Pathe, R. Purcell, and G. W. Stuart. 1999. A study of stalkers. *American Journal of Psychiatry* 56: 1244–1249.

Muraco, J. A. and M. A. Curran. 2012. Marriage for young adults in romantic relationships. *Marriage and Family Review* 48: 227–247.

Muraco, J. A., S. T. Russell, M. A. Curran, and E. A. Butler. 2012. Sexual orientation and romantic relationship quality as moderated by gender, age,

and romantic relationship history. Paper, National Council on Family Relations, Phoenix, AZ, November.
Murdock, G. P. 1949. *Social structure.* New York: Free Press.
Murphy, M. J., L. Deets, and M. Peterson. 2010. The effect of emotional labor and emotional abuse on relationship satisfaction. Poster, National Council on Family Relations annual meeting, November 3–5. Minneapolis, MN.
Murphy-Graham, E. 2010. And when she comes home? Education and women's empowerment in intimate relationships. *International Journal of Educational Development* 30: 320–331.
Murray, C. 2012. *Coming apart: The state of white America, 1960–2010.* New York: Crown Forum.
Musick, K. and L. Bumpass. 2012. Reexamining the case for marriage: Union formation and changes in well-being. *Journal of Marriage and Family* 74: 1–18.
Mutch, K. 2010. In sickness and in health: experience of caring for a spouse with MS. *British Journal of Nursing* 19: 214–219.
Mutran, E. J., D. Reitzes, and M. E. Fernandez. 1997. Factors that influence attitudes toward retirement. *Research on Aging* 19: 251–273.

## N

Nadorff, M. R., A. Fiske, J. A. Sperry, R. Petts, and J. J. Gregg. 2013. Insomnia symptoms, nightmares, and suicidal ideation in older adults. *The Journals of Gerontology: Series B* 68: 145–152.
Nagao, K. et al. 2014. *Journal of Sex & Marital Therapy* 40: 33–42.
Nash, C. L., S. A. Hayes-Skelton, and D. DiLillo. 2012. Reliability and factor structure of the psychological maltreatment and neglect scales of the computer assisted maltreatment inventory (CAMI) *Journal of Aggression, Maltreatment & Trauma* 21: 583–607.
Natale, A. P. and J. E. Miller-Cribbs. 2012. Same-sex marriage policy: Advancing social, political, and economic justice. *Journal of GLBT Family Studies* 8: 55–172.
National Coalition of Anti-Violence Programs. 2012. *2011 National hate crimes report: Anti-lesbian, gay, bisexual and transgender violence in 2011.* New York: National Coalition of Anti-Violence Programs. http://www.avp.org/documents/NCAVPHVReport2011Final6_8.pdf
National Council on Family Relations in Orlando in November 2005.
National Marriage Project, 2012. *The State of Our Unions: Marriage in 2012.* The University of Virginia Center for Marriage and Families at the Institute for American Values.
National Runaway Switchboard (2010). Annual survey. http://www.nrscrisisline.org/news_events/call_stats.html
National Science Foundation, National Center for Science and Engineering Statistics. 2012. Doctorate recipients from U.S. universities. 2010. Special Report NSF 12–305. Arlington, VA.
Nazarinia Roy, R. R. and S. Britt. 2011. The gendered nature of household tasks. Presentation, National Council on Family Relations, Orlando, November.
Neff, K. D. and S. N. Beretvas. 2013. The role of self-compassion in romantic relationships. *Self and Identity* 12: 78–98.
Negash, S., M. Cui, F. D. Fincham and K. Pasley. 2014. Extradyadic involvement and relationship dissolution in heterosexual women university students. *Archives of Sexual Behavior.* 43: 531–539.
Nelms, B. J., D. Knox, and B. Easterling. 2012. The relationship talk: Assessing partner Commitment. *College Student Journal* 46: 178–182.
Nelson, Eve-Lynn and Thao Bui. 2010. Rural telepsychology services for children and adolescents. *Journal of Clinical Psychology: In Session* 66: 490–501.
Nelson, M. K. 2010. *Parenting out of control.* New York: New York University Press.
Nelson, Todd D. 2011. Ageism: the strange case of prejudice against the older you. In *Disability and Aging Discrimination*, R. L. Wiener and S. L. Willborn, eds. (p. 37). New York: Springer Science + Business Media.
Nepomnyaschy, L.and J. Teitler. 2013. Cyclical cohabitation among unmarried parents in fragile families. *Journal of Marriage and the Family* 75: 1248–1265. Commitment. *College Student Journal* 46: 178–182.
Neppl, T. 2012. The effects of economic pressure on family and child outcomes. Poster, Annual Meeting of the National Council on Family Relations. Phoenix, Nov 1.
Nesteruk, O. and A. Gramescu. 2012. Dating and mate selection among young adults from immigrant families. *Marriage & Family Review* 48: 40–58.
Neto, F. 2012. Perceptions of love and sex across the adult life span. *Journal of Social and Personal Relationships,* September 29: 760–775.
Neuman, M. G. 2008. *The truth about cheating: Why men stray and what you can do to prevent it.* New York: John Wiley & Sons.
Newcomb, M. E., M. Birkett, H. L. Corliss, and B. Mustanski. 2014. Sexual orientation, gender, and racial differences in illicit drug use in a sample of US high school students. *American Journal of Public Health* 104: 304–310.
Newton, N. 2002. *Savage girls and wild boys: A history of feral children.* New York: Thomas Dunne Books/St. Martin's Press.
Nielsen, L. 2011. Divorced fathers and their daughters: A review of recent research. *Journal of Divorce & Remarriage*: 52: 77–93.
_____. 2004. *Embracing your father: How to build the relationship you always wanted with your dad.* New York: McGraw-Hill.
Niven, K. J. 2010. *The circle of life: The process of sexual recovery workbook.* (Available at Amazon.com)
Nock, S. L., L. A. Sanchez, and J. D. Wright. 2008. *Covenant Marriage: The Movement to Reclaim Tradition in America.* New Brunswick, NJ: Rutgers University Press.
Nomaguchi, K., S. L. Brown and T. Leyman. 2012. Father involvement and maternal parenting stress: The role of relationship status. Paper, National Council on Family Relations. Phoenix, AZ. November.
North, M. S. and S. T. Fiske. 2012. An inconvenienced youth: Ageism and its potential Intergenerational roots. *Psychological Bulletin* 138: 982–997.
Norton, A. M, and J. Baptist. 2012. Couple boundaries for social networking: Impact of Trust and Satisfaction. Paper, National Council on Family Relations, Phoenix, November.
Nunley, J. M. and Alan Seals. 2010. The effects of household income volatility on divorce. The *American Journal of Economics and Sociology* 69: 983–1011.
Nuru, A. K. and T. R. Wang. 2014. "She was stomping on everything that we used to think of as family": Communication and turning points in cohabiting (step) families. *Journal of Divorce and Remarriage* 55: 145–163.

## O

Oberbeek, G., S. A. Nelemans, J. Karremans, R. C. M. E. Engels. 2013. The malleability of mate selection in speed-dating events. *Archives of Sexual Behavior* 42: 1163–1171.
Ocobock, A. 2013. The power and limits of marriage: Married gay men's family relationships *Journal of Marriage and Family* 75: 191–205.
Offer, S. 2013. Family time activities and adolescents' emotional well-being. *Journal of Marriage & Family* 75: 26–41.
O'Flaherty, K. M. and L. W. Eells. 1988. Courtship behavior of the remarried. *Journal of Marriage and the Family* 50: 499–506.
O'Leary, D. K., H. Foran, and S. Cohen. 2013. Validation of fear of partner scale. *Journal of Marital and Family Therapy* 39: 502–514.
Olmstead, S. B., P. N. E. Roberson, F. D. Fincham, and K. Pasley. 2012. Hooking up and risky sex behaviors among first semester college men:

What is the role of pre-college experience? Presentation, National Council on Family Relations, Phoenix, AZ. November.

Olsen, K. M. and S. Dahl. 2010. Working time: Implications for sickness absence and the work– family balance. *International Journal of Social Welfare* 19: 45–53.

Omarzu, J., A. N. Miller, C. Shultz, and A. Timmerman. 2012. Motivations and emotional consequences related to engaging in extramarital relationships. *International Journal of Sexual Health* 24: online June.

Opree, S. J. and M. Kalmijn. 2012. Exploring causal effects of combining work and intergenerational support on depressive symptoms among middle-aged women. *Ageing & Society* 32: 130–146.

O'Reilly, E. M. 1997. *Decoding the cultural stereotypes about aging: New perspectives on aging talk and aging issues.* New York: Garland.

Ortyl, T. A. 2013. Long-term heterosexual cohabiters and attitudes toward marriage. *Sociological Quarterly* 54: 584–609.

Oswald, D. L. and B. L. Russell. 2006. Perceptions of sexual coercion in heterosexual dating relationships: the role of aggressor gender and tactics. *The Journal of Sex Research* 43: 87–98.

Oswalt, S. B. and T. J. Wyatt. 2013. Sexual health behaviors and sexual orientation in the U.S. National sample of college students. *Archives of Sexual Behavior* 42: 1561–1572.

_____. 2011. Sexual orientation and differences in mental health, stress, and academic performance in a national sample of U.S. college students. *Journal of Homosexuality* 58: 1255–1280.

Overby, L. Marvin. 2014. Etiology and attitudes: Beliefs about the origins of homosexuality and their implications for public policy. *Journal of Homosexuality* 61: 568–587.

Owen, J. J., G. K. Rhoades, S. M. Stanley, and F. D. Fincham. 2010. "Hooking Up" among college students: Demographic and psychosocial correlates. *Journal Archives of Sexual Behavior* 39: 653–663.

Ozay, B., D. Knox, and B. Easterling. 2012. You're Dating Who? Parental attitudes toward interracial dating. Poster, Southern Sociology annual meeting. New Orleans, March.

## P

Padilla, Y. C., C. Crisp, and D. L.Rew. 2010. Parental acceptance and illegal drug use among gay, lesbian, and bisexual adolescents: Results from a national survey. *Social Work* 55: 265–276.

Page, S. 2011. Gay candidates gain acceptance. *USA Today*, July 20, p. 1.

Paik, A. 2010. The contexts of sexual involvement and concurrent sexual partnerships *Perspectives on Sexual and Reproductive Health* 42: 33–43.

Palomer, A., S. A. Corkery, S. V. Gelder, K. Badt, S. Contreras, and M. A. Curran. 2012. Knowing when you are "in a relationship": A qualitative examination. Poster, National Council on Family Relations. Phoenix, AZ. November.

Paluscia, V. J., S. J. Wirtzb, and T. M. Covington. 2010. Using capture-recapture methods to better ascertain the incidence of fatal child maltreatment. *Child Abuse & Neglect* 34: 396–402.

Panetta, S. M., C. L. Somers, A. R. Ceresnie, S. B. Hillman, and R. T. Partridge. Maternal and paternal parenting style patterns and adolescent emotional and behavioral outcomes. *Marriage & Family Review* 50: 342–359.

Paraschakis, A., A. Douzenis, I. Michopoulos, C. Christodoulou, K. Vassilopoulou, F. Koutsaftis and L. Lykouras. 2012. Late onset suicide: Distinction between "young-old" vs. old-old" suicide victims. How different populations are they? *Archives of Gerontology & Geriatrics* 54: 136–139.

Parelli, S. 2007. Why ex-gay therapy doesn't work. *The Gay & Lesbian Review Worldwide* 14: 29–32.

Park, H. and J. M. Raymo. 2013. Divorce in Korea: Trends and educational differentials. Journal of *Marriage and the Family* 75: 110–126.

Park, N. S., Y. Jang, B. S. Lee, and D. A. Chiriboga. 2013. Racial/ethnic differences in predictors of self-rated health: Findings from the survey of older Floridians. *Research on Aging* 35: 201–219.

Park-Lee, E., M. Sengupta, A. Bercovitz, and C. Caffrey. 2013. Oldest old long-term care recipients: Findings from the national center for health statistics' long-term care surveys. *Research on Aging* 35: 296–232.

Parker, K. The boomerang generation. Pew Research Center. 2012. Retrieved on March 31, http://www.pewsocialtrends.org/2012/03/15/the-boomerang-generation/2/#who-are-the-boomerang-kids

Parker, K. and W. Wang. 2013. Modern parenthood: Roles of moms and dads converge as they balance work and family. *Pew Research: Social & Demographic Trends.* March 14. http://www.pewsocialtrends.org/2013/03/14/modern-parenthood-roles-of-moms-and-dads-converge-as-they-balance-work-and-family/

Parker, M., D. Knox, and B. Easterling. 2011. SEXTING: Sexual content /images in romantic relationships. Poster, Eastern Sociological Society, Philadelphia, Feb 24–26.

Parker-Pope, T. 2010. *For better: The science of a good marriage.* New York: E. P. Dutton.

Parry, D. C. and T. P. Light. 2014. Fifty shades of complexity: Exploring technologically mediated leisure and women's sexuality. *Journal of Leisure Research* 46: 38–57.

Parsons, J., T. Starks, S. DuBois, C. Grov, and S. Golub. 2013. Alternatives to monogamy among gay male couples in community survey: Implications for mental health and sexual risk. *Archives of Sexual Behavior* 42: 303–312.

Passel, J. S., W. Wang, and P. Taylor. 2010, June 4. Marrying out: One-in-seven new U.S. marriages is interracial or interethnic. Pew Research Center. Retrieved Nov 22, 2013. http://pewresearch.org/pubs/1616/american-marriage-interracial-interethnic.

Patel, P. and B. Sen. 2012. Teen motherhood and long-term health consequences. *Maternal & Health Journal* 16: 1063–1071.

Patrick, M. E. and C. M. Lee. 2010. Sexual motivations and engagement in sexual behavior during the transition to college. *Journal Archives of Sexual Behavior* 39: 674–681.

Patrick, S., J. N. Sells, F. G. Giordano, and T. Tollerud. 2010. Intimacy, differentiation, and personality variables as predictors of marital satisfaction. *The Family Journal: Counseling and Therapy for Couples and Families* 20 (2). (In press.)

Payne, C. and V. Bravo. 2013. Do you consider your pet to be part of the family? *USA Today* Nov 11. A1.

Payne, K. and P. Trapp. 2013. What makes parenting tougher today? *USA Today* (survey by Survey.com). April 30. D 1.

Pearcey, M. 2004. Gay and bisexual married men's attitudes and experiences: Homophobia, reasons for marriage, and self-identity. *Journal of GLBT Family Studies* 1: 21–42.

Peasant, C., G. R. Parra, and T. M. Okwumabua. 2014. Condom negotiation: Findings and future directions. *The Journal of Sex Research.* Published online March 26.

Pelton, S. L. and K. M. Hertlein. 2011. A proposed life cycle for voluntary childfree couples. *Journal of Feminist Family Therapy* 23: 39–53.

Pendry, P., F. Henderson, J. Antles, and E. Conlin. 2011. Parents' use of everyday conflict tactics in the presence of children: Predictors and implications for child behavior. Poster, Annual Meeting of National

Council on Family Relations, Orlando, November 18.

Penhollow, T. M., A. Marx and M. Young. 2010. Impact of recreational sex on sexual satisfaction and leisure satisfaction. *Electronic Journal of Human Sexuality* 13: March 31.

Perilloux, C., D. S. Fleischman, and D. M. Buss. 2011. Meet the parents: Parent-offspring convergence and divergence in mate preferences. Personality & Individual Differences 50: 253–258.

Perry, M. S. and R. J. Werner-Wilson. 2011. Couples and computer-mediated communication: A closer look at the affordances and use of the channel. *Family & Consumer Sciences Research Journal* 40: 120–134.

Perry, S. L. 2013. Racial composition of social settings, interracial friendship, and whites' attitudes toward interracial marriage. *Social Science Journal* 50: 13–22.

Petrecca, L. 2013. Always working. *USA Today*, March 7. A1 et passim.

_____. 2012. Pregnant CEO tests glass ceiling. *USA Today*, July 19. A.

Pettigrew, J. 2009. Text messaging and connectedness within close interpersonal relationships. *Marriage & Family Review* 45: 697–716.

Pew Research Center. 2010. Social & demographic Trends: The decline of marriage and rise of new families. Published and retrieved on November 18. http://pewresearch.org/pubs/1802/decline-marriage-rise-new-families

Pew Research Center. 2014. Attitudes about aging: A global perspective. http://www.pewglobal.org/2014/01/30/attitudes-about-aging-a-global-perspective/ January 30, 2014.

_____. 2013. In gay marriage debate, both supporters and opponents see legal recognition as "inevitable." June 6. http://www.people-press.org/2013/06/06/in-gay-marriage-debate-both-supporters-and-opponents-see-legal-recognition-as-inevitable/

_____. 2012. A gender reversal on career trends. Young women now top young men in valuing a high-paying career. Published and retrieved on April 19. http://pewresearch.org/pubs/2248/gender-jobs-women-men-career-family-educationalattainment-labor-force-participation?src=prc-newsletterPew Research Center. 2011. The burden of student debt. Retrieved from http://pewresearch.org/databank/dailynumber/?NumberID=1257

_____. 2012a. Half say view of Obama not affected by gay marriage decision: Independents mostly unmoved. May 14, 2012.

_____. 2012b. More support for gay marriage than in 2008 or 2004. April 25.

_____. 2010. Social & demographic Trends: The decline of marriage and rise of new families. Published and retrieved on November 18. http://pewresearch.org/pubs/1802/decline-marriage-rise-new-families

_____. 2010. Religion among the Millennials. Pew Forum on Religion and Public Life. Retrieved from http://pewforum.org/Age/Religion-Among-the-Millennials.aspx

_____. 2010. Social & demographic Trends: The decline of marriage and rise of new families. Published and retrieved on November 18. http://pewresearch.org/pubs/1802/decline-marriage-rise-new-families.

_____. 2008. Pew Forum on Religion and Public Life. The U.S. religious landscape survey. http://pewresearch.org/pubs/743/united-states-religion.

Pfeffer, C. A. 2010. "Women's work"? Women partners of transgender men doing housework and emotion work. *Journal of Marriage & Family* 72: 165–183.

Phillips, L. J., J. Edwards, N. McMurray, and S. Francey. 2012. Comparison of experiences of stress and coping between young people at risk of psychosis and a non-clinical cohort. *Behavioural & Cognitive Psychotherapy* 40: 69–88.

Pietrzak, R. H., C. A. Morgan, and S. M. Southwick. 2010. Sleep quality in treatment-seeking veterans of Operations Enduring Freedom and Iraqi Freedom: The role of cognitive coping strategies and unit cohesion. *Journal of Psychosomatic Research* 69: 441–448.

Pilkauskas, N. V. 2012. Three-generation family households: Differences by family structure at birth. *Journal of Marriage and Family* 74: 931–943.

Pinello, D. R. 2008. Gay marriage: For better or for worse? What we've learned from the evidence. *Law & Society Review* 42: 227–230.

Pines, A. M. 1992. *Romantic jealousy: Understanding and conquering the shadow of love.* New York: St. Martin's Press.

Pizer, J. C., B. Sears, C. Mallory, and N. D. Hunter. 2012. Evidence of persistent and pervasive workplace discrimination against GLBT people: The need for Federal legislation prohibiting discrimination and providing for equal employment benefits. *Loyola of Los Angeles Law Review* 45: 715–779.

Plagnol, A. C., and R. A. Easterlin. 2008. Aspirations, attainments, and satisfaction: Life cycle differences between American women and men. *Journal of Happiness Studies.* Published online July 2008.

Plant, E. A., J. Kunstman, and J. K. Maner. 2010. You do not only hurt the one you love: Self-protective responses to attractive relationship alternatives. *Journal of Experimental Social Psychology* 46: 474–477.

Podgurski, M. J. 2013. *Inside Out: Your body is amazing inside and out and belongs only to you.* Available from www.healthyteens.com

Pompper, D. 2010. Masculinities, the metrosexual, and media images: Across dimensions of age and ethnicity . *Sex Roles* 63: 682–696.

Pope, A. L. and C. S. Cashwell. 2013. Moral commitment in intimate committed relationships: A conceptualization from cohabiting same-sex and opposite sex partners. *Family Journal* 21: 5–14.

Popham, L. E., S. M. Kennison, and K. I. Bradley. 2011. Ageism and risk-taking in young adults: Evidence for a link between death and anxiety and ageism. *Death Studies* 35: 751–763.

Pond, R. L., C. Stephens, and F. Alpass. 2010. How health affects retirement decisions : Three pathways taken by middle-older aged New Zealanders. *Ageing & Society* 30: 527–545.

Poortman, A. and T. Van der Lippe. 2009. Attitudes toward housework and child care and the gendered division of labor. *Journal of Marriage and the Family* 71: 526–541.

Porter, J. & L. M. Williams. 2011. Intimate violence among underrepresented groups on a college campus. *Journal of Interpersonal Violence 26*, 3210–3224.

Porter, N. 2014. The effects of the sexual double standard on the perceptions of college students towards themselves and their peers. Paper, Southern Sociological Society Annual Meeting, April 4.

Potok, M. 2010. Anti-gay hate crimes: Doing the math. *Intelligence Report* 140, Winter, p. 29.

Poverty line, 2014. Retrieved Feb. 23, 2014. http://aspe.hhs.gov/poverty/14poverty.cfm

Power, J. J., A. Perlesz, R. Brown, M. J. Schofield, M. K. Pitts, R. McNair, and A. Bickerdike. 2013. Bisexual parents and family diversity: Findings from the work, love, play study. *Journal of Bisexuality* 12: 519–538.

Powers, R. A. and S. S. Simpson. 2012. Self-protective behaviors and injury in domestic violence situations: Does it hurt to fight back? *Journal of Interpersonal Violence* 27: 3345–3365.

Presser, H. B. 2000. Nonstandard work schedules and marital instability. *Journal of Marriage and the Family* 62: 93–110.

Price, C. A. and O. Nesteruk. 2013. Being married in retirement: Can it be a double-edged sword? National

Council on Family Relations Annual Meeting. San Antonio, November
Priebe, G. and C. G. Svedin. 2013. Operationalization of three dimensions of sexual orientation in a national survey of late adolescents. *Journal of Sex Research* 50: 727–738.
Pritchard, K. M. and L. Kort-Butler. 2012. Multiple motherhoods: The interactive effects of importance of life satisfaction. Poster, National Council on Family Relations, November.
Proulx, C. M. and A. E. Ermer. 2013. Marital quality and health: Longitudinal associations across two generations. National Council on Family Relations Annual Meeting. San Antonio, November 5–9.
Proulx, C. M. and L. A. Snyder. 2009. Families and health: An empirical resource guide for researchers and practitioners. *Family Relations* 58: 489–504.
Purkett, T. 2013. Sexually transmitted infections. Presented to Courtship and Marriage class, Department of Sociology, Fall.
Puri, S. and R. D. Nachtigall. 2010. The ethics of sex selection: A comparison of the attitudes and experiences of primary care physicians and physician providers of clinical sex selection services. *Fertility & Sterility* 93: 2107–2114.

## Q

Qian, Z.and D. T. Lichter. 2011. Changing patterns of interracial marriage in a multiracial society. *Journal of Marriage and Family* 73: 1065–1084.
Quing, M., Z. Li-xia, and S. Xiao-yin. 2011. A comparison of postnatal depression and related factors between Chinese new mothers and fathers. *Journal of Clinical Nursing* 20: 645–652.

## R

Randall, B. 2008. *Songman: The story of an Aboriginal elder of Uluru.* Sydney, Australia: ABC Books.
Randler, C. and S. Kretz 2011. Assortative mating in morningness-eveningness. *International Journal of Psychology* 46: 91–96.
Rapoport, B., and C. Le Bourdais. 2008. Parental time and working schedules. *Journal of Population Economics* 21: 903–33.
Rappaport, A. 2009. Phased retirement—An important part of the evolving retirement scene. *Benefits Quarterly* 25: 38–50.
Rappleyea, D. L., A. C. Taylor, and X Fang. 2014. Gender differences and communication technology use among emerging adults in the initiation of dating relationships. *Marriage & Family Review* 50: 269–284.
Rauer, A. 2013. From golden bands to the golden years: The critical role of marriage in older adulthood. Annual meeting, Southeastern Council on Family Relations, Feb. 22, Birmingham, AL.
Rayne, Karen. 2013. Based of workshop presentations. E-mail Dr. Rayne at karen@karenrayne.com for information about workshops/further use of this material.
Read, S. and E. Grundy. 2011. Mental health among older married couples: The role of gender and family life. *Social Psychiatry and Psychiatric Epidemiology* 46, 331–341.
Ready, B., L. Asare, and E. Long. 2011. Factors that impact a father's philosophy on parenting. Presentation, National Council on Family Relations, Orlando, November.
Reck, K. and B. Higginbotham. 2012. No longer newlyweds: Difficulties experienced by remarried couples over time. Poster, Annual Meeting National Council on Family Relations, November 1.
Reczek, C., H. Liu and D. Umberson 2010. Just the two of us? How parents influence adult children's marital quality. *Journal of Marriage and Family* 72: 1205–1219.
Reece, M., D. Herbenick, B. Dodge, S. A. Sanders, A. Ghassemi, and D. Fortenberry. 2010. Vibrator use among heterosexual men varies by partnership status: Results from a nationally representative study in the United States. *Journal of Sex & Marital Therapy* 36: 389–407.
Rees, C. & G. Pogarsky. 2011. One bad apple may not spoil the whole bunch: Best friends and adolescent delinquency. *Journal of Quantitative Criminology 27,* 197–223.
Reich, S. M., K. Subrahmanyam, and G. Espionoza. 2012. Friending, IMing, and hanging out face-to-face: Overlap in adolescents' online and offline social networks. *Developmental Psychology* 48: 356–368.
Reid, R. C., D. S. Li, R. Gilliland, J. A. Stein, and T. Fong. 2011. Reliability, validity, and psychometric development of Pornography Consumption Inventory in a sample of hypersexual men. *Journal of Sex & Marital Therapy* 37: 359–385.
Reinhardt, R. U. 2011. Bisexual women in heterosexual relationships: A study of psychological and sociological patterns: A reflective paper. *Journal of Bisexuality* 11: 439–447.
Reiss, I. L. 1960. Toward a sociology of the heterosexual love relationship. *Journal of Marriage and Family Living* 22:139–45.
Reissing, E. D., H. L. Andruff, and J. J. Wentland. 2012. Looking back: The experience of first sexual intercourse and current sexual adjustment in young heterosexual adults. *Journal of Sex Research* 49, 27–35.
Reynaud, M., L. Blecha, and A. Benyamina. 2011. Is love passion an addictive disorder? *American Journal of Drug & Alcohol Abuse* 36: 261–267.
Reyome, N. D. 2010. Childhood emotional maltreatment and later intimate relationships: Themes from the empirical literature. *Journal of Aggression, Maltreatment & Trauma* 19: 224–242.
Rhatigan, D. L. and A. M. Nathanson. 2010. The role of female behavior and attributions in predicting behavioral responses to hypothetical male aggression. *Violence Against Women* 16: 621–637.
Rhoades, G. K., S. M. Stanley, and H. J. Markman. 2012. A longitudinal investigation of commitment dynamics in cohabiting relationships. *Journal of Family Issues* 33: 369–390.
_____. 2010. Should I stay or should I go? Predicting dating relationship stability from four aspects of commitment. *Journal of Family Psychology* 24: 543–550.
Rhodes, K V., C. Cerulli, M. E. Dichter, C. L. Kothari, and F. K. Barg. 2010. "I Didn't Want to Put Them Through That": The influence of children on victim decision-making in intimate partner violence cases. *Journal of Family Violence* 25: 485–493.
Richeimer, S. 2011. Love hurts. KABC, Los Angeles http://abclocal.go.com/kabc/story?section=news/health/your_health&id=8039618
Ricks, J. L. 2012. Lesbians and alcohol abuse: Identifying factors for future research. *Journal of Social Service Research 38,* 37–45.
Riela, S., G. Rodriguez, A. Aron, X. Xu, and B. P. Acevedo. 2010. Experiences of falling in love: Investigating culture, ethnicity, gender, and speed. *Journal of Social and Personal Relationships* 27: 473–493.
Rinelli, L. and S. L. Brown. 2010. Race differences in union transitions among cohabiters: The role of relationship features. *Marriage & Family Review* 46: 22–40.
Roberts, A. L., M. M. Glymour, and K. C. Koenen. 2013. Does maltreatment in childhood affect sexual orientation in adulthood? *Archives of Sexual Behavior* 42: 161–171.
Robnett, R. D. and C. Leaper. 2013. "Girls don't propose!: A mixed-methods examination of marriage tradition preferences and benevolent sexism in emerging adults. *Journal of Adolescent Research* 28: 96–121.
Robnett, R. D. and J. E. Susskind. 2010. Who cares about being gentle? The impact of social identity and the gender of one's friends on children's

display of same-gender favoritism. *Sex Roles* 63: 820–832.

Roby, J. and H. White. 2010. Adoption activities on the Internet: A call for regulation. *Social Work* 55: 203–212.

Rodriguez, Y., J. Su, and H. Helms. 2010. Sociocultural context, relationship quality and gender: "His" and "her" experiences in married and cohabiting couples of Mexican origin in the childbearing years. Paper, Annual Meeting of the National Council on Family Relations, November 3–6. Minneapolis, MN.

Rose, H. A. and A. M. Humble. 2013. Family policy from a lifespan perspective: Introducing an analytical tool. National Council on Family Relations 75th Annual Meeting San Antonio.

Rose, K., and K. J. Elicker. 2008. Parental Decision Making about Child Care. *Journal of Family Issues* 29:1161–1179.

Rosenberger, J., D. Herbenick, D. Novak, and M. Reece. 2014. What's love got to do with it? Examinations of emotional perception and sexual behaviors among gay and bisexual men in the United States. *Archives of Sexual Behavior* 43: 119–128.

Rosenfeld, M. J. and R. J. Thomas. 2012. Searching for a mate: The rise of the Internet as a social intermediary. *American Sociological Review* 77: 523–547.

Rosin, H. and R. Morin. 1999. In one area, Americans still draw a line on acceptability. *Washington Post National Weekly Edition* 16(January 11): 8.

Ross, C. B. 2006. An exploration of eight dimensions of self-disclosure on relationship. Paper for Southern Sociological Society, New Orleans, LA. March 24.

Ross, M. W., K. Daneback, and S. Mansson. 2013. Fluid versus fixed: A new perspective on bisexuality as a fluid sexual orientation beyond gender. *Journal of Bisexuality* 12: 449–460.

Rothman, E. F., D. Exner, and A. L. Baughman. 2011. The prevalence of sexual assault against people who identify as gay, lesbian, or bisexual in the United States: A systematic review. *Violence & Abuse* 12: 55–66.

Rothman, E. F., M. Sullivan, S. Keyes, and U. Boehmer. 2012. Parents supportive reactions to sexual orientation disclosure associated with better health: Results from a population-based survey of LGB adults in Massachusetts. *Journal of Homosexuality* 59: 186–200.

Routt, G. G. and L. Anderson. Adolescent violence towards parents. 2011. *Journal of Aggression, Maltreatment & Trauma* 20: 1–18.

Roxburgh, S. 2012. Parental time pressures and depression among married dual-earner parents *Journal of Family Issues* 33: 1027–1053.

Roy, M. and H. Payette. 2012. The body image construct among Western seniors: A systematic review of the literature. *Archives of Gerontology and Geriatrics* 55: 505–521.

Rubinstein, G. 2010. Narcissism and self-esteem among homosexual and heterosexual male students. *Journal of Sex & Marital Therapy* 36: 24–34.

Russell, S. T. 2013. LGBTQ youth well-being: The role of parents and policy. National Council on Family Relations annual meeting, San Antonio.

Russell, S. T., C. Ryan, R. B. Toomey, R. M. Diaz, and J. Sanchez. 2011. Lesbian, gay, bisexual and transgender adolescent school victimization: Implications for young adult health and adjustment. *Journal of School Health* 81: 223–230.

Russell, V. M, L. R. Baker, and J. K. McNulty. 2013. Attachment insecurity and infidelity in marriage: Do studies of dating relationships really inform us about marriage? *Journal of Family Psychology* 27: 241–251.

Russella, D., K. W. Springerb, and E. A. Greenfield. 2010. Witnessing domestic abuse in childhood as an independent risk factor for depressive symptoms in young adulthood. *Child Abuse & Neglect* 34: 448–453.

Rust, Paula. 1993. Neutralizing the political threat of the marginal woman: lesbians' beliefs about bisexual women. *Journal of Sex Research* 30(3): 214–218.

Rutledge, S. E., D. C. Siebert, and J. Chonody.2012. Attitudes toward gays and lesbians: A latent class analysis of university students. *Journal of Social Service Research 38*, 18–28.

Ryan, C. and C. Jetha. 2010. *Sex at dawn.* New York: Harper Perennial.

## S

Sack, K. 2008. Health benefits inspire rush to marry, or divorce. *The New York Times*, August 12.

Saltes, N. 2013. Disability, identity and disclosure in the online dating environment. *Disability & Society* 28: 96–109.

Sammons, R. A., Jr. 2014. First law of therapy. Personal communication. Greenville, NC.

Sanchez, D. T., J. C. Fetterolf, and L. A. Rudman. 2012. Eroticizing inequality in the United States: The consequences and determinants of traditional gender role adherence in intimate relationships. *Journal of Sex Research* 49: 168–183.

Sanday, P. R. 1995. Pulling train. In *Race, class, and gender in the United States*, 3rd ed., ed. P. S. Rothenberg, pp. 396–402. New York: St. Martin's Press.

Sandberg, J. G., J. M. Harper, E. J. Hill, R. B. Miller, J. B. Yorgason, and R. D. Day. 2013. "What happens at home does not necessarily stay at home": The relationship of observed negative couple interaction with physical health, mental health, and work satisfaction. *Journal of Marriage and the Family* 75: 808–821.

Sandberg, S. 2013. *Lean in: Women, work, and the will to lead.* New York: Alfred A. Knopf.

Sandberg-Thoma, S. E. and C. M. Kamp Dush. 2014. Casual sexual relationships and mental health in adolescence and emerging adulthood. *Journal of Sex Research* 51: 121–130.

Sandler, L. 2010. One and done. *Time*, July 19, pp. 34–41.

Sandlow, J. I. 2013. Size does matter: Higher body mass index may mean lower pregnancy rates for microscopic testicular sperm extraction. *Fertility and Sterility* 99: 347.

Sanford, K. and A. J. Grace. 2011. Emotion and underlying concerns during couples' conflict: An investigation of within-person change. *Personal Relationships* 18: 96–109.

Sankar, S. 2012. Adoption of older children: Factors influencing decision to adopt. Poster, Annual Meeting of the National Council on Family Relations. Phoenix, Nov 1.

Sarkar, N. N. 2008. The impact of intimate partner violence on women's reproductive health and pregnancy outcome. *Journal of Obstetrics and Gynecology* 28: 266–278.

Sasaki, T., N. L. Hazen, W. B. Swann Jr. 2010. The supermom trap: Do involved dads erode moms' self-competence? *Personal Relationships* 17: 71–79.

Sassler, S. 2010. Partnering across the life course: Sex, relationships, and mate selection. *Journal of Marriage and Family* 72: 557–575.

Sassler, S. and A. J. Miller. 2011. Class differences in cohabitation processes. *Family Relations* 60: 163–177.

Sayare, S. and M. Ia De La Baume. 2010. In France, Civil Unions gain favor over marriage. *The New York Times.* Dec. 15.

Sayer, L. S. and L. Fine. 2011. Racial-ethnic differences in U.S. married women's and men's housework. *Social Indicators Research* 101: 259–265.

Scarf, M. 2013. *The remarriage blueprint: How remarried couples and their families succeed or fail.* New York: Scribner.

Schacht, T. E. 2000. Protection strategies to protect professionals and families involved in high-conflict divorce. *UALR Law Review* 22(3): 565–592.

Schade, L. C., J. Sandberg, R. Bean, D. Busby, and S. Coyne. 2013. Using technology to connect in romantic relationships: Effects on attachment,

relationship satisfaction, and stability *Educational Interventions* 12: 314–338.
Scheetz, L. T., P. Martin, and L. W. Poon. 2012. Do centenarians have higher levels of depression? Findings from the Georgia Centenarian Study. *Journal of the American Geriatrics Society* 60: 238–242.
Schellekens, J. and D. Gliksberg. 2013. Inflation and marriage in Israel. *Journal of Family History* 38: 78–93.
Schmeer, K. K. 2011. The child health disadvantage of parental cohabitation. *Journal of Marriage & Family* 73: 181–193.
Schmidt, L., T. Sobotka, J. G. Bentzen, and A. Nyboe Andersen. 2012. Demographic and medical consequences of the postponement of parenthood. *Human Reproduction Update* 18: 29–43.
Schmitt, J. and J. Jones. 2012. America's "New Class": A profile of the long-term unemployed. *New Labor Forum*. 21: 57–65.
Schoen, R., N. M. Astone, K. Rothert, N. J. Standish, and Y. J. Kim. 2002. Women's employment, marital happiness, and divorce. *Social Forces* 81: 643–662.
Schoen, R., S. J. Rogers, and P. R. Amato. 2006. Wives' employment and spouses' marital happiness: Assessing the direction of influence using longitudinal couple data. *Journal of Family Issues* 27: 506–528.
Schoppe-Sullivan, S. J., G. L. Brown, E. A. Cannon, S. C. Mangelsdorf, and M. S. Sokolowski. 2008. Maternal gatekeeping, coparenting quality, and fathering behavior in families with infants. *Journal of Family Psychology* 22: 389–397.
Scimeca, G. et al. 2013. Alexithymia, negative emotions, and sexual behavior in heterosexual university students from Italy. *Archives of Sexual Behavior* 42: 117–127.
Schrimshaw, E. W., M. J. Downing, Jr., and K. Siegel. 2013. Sexual venue selection and strategies for concealment of same-sex behavior among non-disclosing men who have sex with men and women. *Journal of Homosexuality* 60: 120–145.
Scott, L. S. 2009. *Two is enough*. Berkeley, CA: Seal Press.
Sears, W. and M. Sears. 1993. *The baby book*. Boston: Little, Brown.
Sefton, B. W. 1998. The market value of the stay-at-home mother. *Mothering* 86: 26–29.
Serovich, J. M., S. M. Craft, P. Toviessi, R. Gangamma, et al. 2008. A systematic review of the research base on sexual reorientation therapies. *Journal of Marital and Family Therapy* 34: 227–39.
Shafer, K. 2010. *Yours, mine, and hours: Relationship skills for blended families. Journal of Comparative Family Studies* 41: 185–186.
Shaley, O., N. Baum and H. Itzhaki. 2013. "There's a man in my bed": The first sex experience among modern-orthodox newlyweds in Israel. *Journal of Sex & Marital Therapy* 39: 40–55.
Shamai, M. and E. Buchbinder. 2010. Control of the self: Partner-violent men's experience of therapy. *Journal of Interpersonal Violence* 25: 1338–1362.
Shapiro, C. H. 2012. Decade of change: New interdisciplinary needs of people with infertility. Paper, National Council on Family Relations, Phoenix, November.
Sharpe, A. 2012. Transgender marriage and the legal obligations to disclose gender history. *Modern Law Review* 75: 33–53.
Shelton, J. N., T. E. Trail, T. V. West, and H. B. Bergsieker. 2010. From strangers to friends: The interpersonal process model of intimacy in developing interracial friendships. *Journal of Social and Personal Relationships* 27: 71–90.
Shin, K., H. Shin, J. A. Yang, and C. Edwards. 2010. Gender role identity among Korean and American college students: Links to gender and academic achievement. *Social Behavior & Personality: An International Journal* 38: 267–272.
Shoemann, A. M., O. Gillath, and A. K. Sesko. 2012. Regrets, I've had a few. Effects of dispositional and manipulated attachment on regret. *Journal of Social and Personal Relationships* 29: 795–819.
Shorey, R. C., J. Febres, H. Brasfield, and G. L. Stuart. 2012. The prevalence of mental health problems in men arrested for domestic violence. *Journal of Family Violence*. Online August 12.
Sidebotham, P. 2013. Rethinking filicide. *Child Abuse Review* 22: 305–310.
SIECUS (Sexuality Information and Education Council of the United States). 2014. http://www.siecus.org/index.cfm?fuseaction=page.viewpage&pageid=605&grandparentID=477&parentID=591 Retrieved April 16.
Siegler, I., and P. Costa. 2000. Divorce in midlife. Paper presented at the Annual Meeting of the American Psychological Association, Boston.
Silveira, H., H. MOraesa, N. Oliveira, E. Freire Coutinho, and A. Jerson. 2013. Physical exercise and clinically depressed patients: A systematic review and metaanalysis. *Neuropsychobiology* 67: 61–68.
Sillito, C. L. 2012. Physical health effects of intimate partner abuse. *Journal of Family Issues* 33: 1520–1539.
Silverstein, L. B., and C. F. Auerbach. 2005. (Post) modern families. In *Families in global perspective*, ed. Jaipaul L. Roopnarine and U. P. Gielen, 33–48. Boston, MA. Pearson.
Simeonova, E. 2013. Marriage, bereavement and mortality: The role of health care utilization. *Journal of Health Economics* 32: 33–50.
Simms, D. C. and E. S. Byers. 2013. Heterosexual dater's sexual initiation behaviors: Use of the theory of planned behavior. *Archives of Sexual Behavior* 42: 105–116.
Simon, R. W. and K. Lively. 2010. Sex, anger, and depression. *Social Forces* 88: 1543–1568.
Simons, L. G., R. L. Simons, M. Lei, and T. E. Sutton. 2012. Exposure to harsh parenting and pornography as explanations for male's sexual coercion and female's sexual victimization. *Violence and Victims* 27: 378–395.
Singer, E. 2010. The "W.I.S.E. Up!" tool: Empowering adopted children to cope with questions and comments about adoption. *Pediatric Nursing* 36: 209–212.
Singh, A. A., D. G. Hays, and L. S. Watson. 2011. Strength in the face of adversity: Resilience strategies of transgender individuals. *Journal of Counseling & Development* 89: 20–27.
Skrbis, Z. M. Western, B. Tranter, D. Hogan, R. Coates, J. Smith, B. Hewitt and M. Mayall. 2012. Expecting the unexpected: Young people's expectations about marriage and family. *Journal of Sociology* 48: 63–83.
Sleeth, D. B. 2013. Three pillars of recovery: The role of integral love in clinical practice. *Journal of Humanistic Psychology* 53: 5–25.
Slinger, M. R. and D. J. Bredehoft. 2010. Relationships between childhood overindulgence and adult attitudes and behavior. Poster, National Council on Family Relations annual meeting, November 3–5. Minneapolis, MN.
Smith, A., A. Lyons, J. Ferris, J. Richter, M. Pitts, and J. Shelley. 2010. Are sexual problems more common in women who have had a tubal ligation? A population-based study of Australian women. *An International Journal of Obstetrics & Gynecology* 117: 463–468.
Smith, A. and M. Duggan. 2013. Dating, social networking, mobile online dating & relationships. *Pew Research Center*. Oct 21. http://pewinternet.org/Reports/2013/Online-Dating.aspx
Smith, A. and J. G. Grzywacz. 2013. Health and well-being in midlife parents of children with special needs. National Council on Family Relations Annual Meeting, San Antonio, TX.
Smith, P. H., C. E. Murray, and A. L. Coker. 2010. The Coping Window: A contextual understanding of the methods women use to cope with battering. *Violence and Victims* 25: 18–28.

Smock, P. J. and F. R. Greenland. 2010. Diversity in pathways to parenthood: Patterns, implications, and emerging research directions. *Journal of Marriage and Family* 72: 576–593.

Snyder, K. A. 2007. A vocabulary of motives: Understanding how parents define quality time. *Journal of Marriage and Family* 69: 320–340.

Sobal, J. and K. L. Hanson. 2011. Marital status, marital history, body weight, and obesity. *Marriage & Family Review* 47: 474–504.

Solomon, A. (2012) *Far from the tree.* New York: Scribner.

Solomon, Z., S. Debby-Aharon, G. Zerach, and D. Horesh. 2011. Marital adjustment, parental functioning, and emotional sharing in war veterans. *Journal of Family Issues.* 32: 127–147.

Soltz, V. and R. Dreikurs. 1991. *Children: The challenge.* New York: Penguin.

Sommers, P., S. Whiteside, and K. Abbott. 2013. Cohabitation: Perspectives of undergraduate students. Poster, National Council on Family Relations Annual Meeting. San Antonio, November 5–9.

Soria, K. M. and S. Linder. 2014. Parental divorce and first-year college students' persistence and academic achievement. *Journal of Divorce and Remarriage* 55: 103–116.

South, A. L. 2013. Perceptions of romantic relationships in adult children of divorce. *Journal of Divorce & Remarriage* 54: 126–141.

South, S. C., B. D. Doss, and A. Christensen. 2010. Through the eyes of the beholder: The mediating role of relationship acceptance in the impact of partner behavior. *Family Relations* 59: 611–622.

Spark, C. 2011. Gender trouble in town: Educated women eluding male domination, gender violence and marriage in PNG. *Asia Pacific Journal of Anthropology* 12: 164–179.

Spencer, J. and P. Amato. 2011. Marital quality across the life course: Evidence from latent growth curves. Presentation, Annual Meeting of the National Council on Family Relations, Orlando, November.

Spitzberg, B. H. and W. R. Cupach. 2007. Cyberstalking as (mis)matchmaking. In *Online Matchmaking,* ed. M. T. Whitty, A J. Baker, and J. A. Inman, 127–46. New York: Palgrave Macmillan.

Sprecher, S. 2014. Evidence of change in men's versus women's emotional reactions to first sexual intercourse: A 23-year study in a human sexuality course at a midwestern university. *The Journal of Sex Research.* Published online, March 10. ations, Orlando, November.

_____. 2002. Sexual satisfaction in premarital relationships: Associations with satisfaction, love, commitment, and stability. *Journal of Sex Research* 39: 190–96.

Stanfield, J. B. 1998. Couples coping with dual careers: A description of flexible and rigid coping styles. *Social Science Journal* 35: 53–62.

Stanik, C. E., S. M. McHale, and A. C. Couter. 2013. Gender dynamics predict changes in marital love among African-American couples. *Journal of Marriage & Family* 75: 795–798.

Stanley, S. M., and L. A. Einhorn. 2007. Hitting pay dirt: Comment on "Money: A therapeutic tool for couples therapy." *Family Process* 46: 293–299.

Stanley, S. M., E. P. Ragan, G. K. Rhoades, and H. J. Markman. 2012. Examining changes in relationship adjustment and life satisfaction in marriage. *Journal of Family Psychology* 26: 165–170.

Stanley, S. M., and H. J. Markman. 1992. Assessing commitment in personal relationships. *Journal of Marriage and the Family* 54: 595–608.

Starks, T. and J. Parsons. 2014. Adult attachment among partnered gay men: Patterns and associations with sexual relationship quality. *Archives of Sexual Behavior* 43: 107–117.

Starr, L.R., J. Davila, C. B. Stroud, P. C. Clara Li, A. Yoneda, R. Hershenberg, and M. Ramsay *Statistical Abstract of the United States, 2012–2013,* 131th ed. Washington, DC: U.S. Bureau of the Census.

*Statistical Abstract of the United States, 2012 –2013.* 131th ed. Washington, DC: U.S. Bureau of the Census.

Stefansson, J., P. Nordström, and J. Jokinen. 2012. Suicide intent scale in the prediction of suicide. *Journal of Affective Disorders* 136: 167–171.

Steinberg, J. R. 2013. Another response to: "Does abortion reduce the mental health risks of unwanted or unintended pregnancy," Fergusson et. al., 2013. *Australian and New Zealand Journal of Psychiatry* 47: 1204–1205.

Sternberg, R. J. 1986. A triangular theory of love. *Psychological Review* 93: 119–135.

Stephenson, K. R. and C. M. Meston. 2012. The young and the restless? Age as a moderator of the association between sexual desire and sexual distress in women. *Journal of Sex & Marital Therapy* 38: 445–457.

Stephenson, K. R., A. H. Rellini, and C. M. Meston. 2013. Relationship satisfaction as a predictor of treatment response during cognitive behavioral sex therapy. *Archives of Sexual Behavior* 42: 143–152.

Stevens, T. H., T. A. More, and M. Markowski-Lindsay. 2014. Declining national park visitation: An economic analysis. *Journal of Leisure Research* 46: 153–164.

Stinehart, M. A., D. A. Scott, and H. G. Barfield. 2012. Reactive attachment disorder in adopted and foster care children: Implications for mental health professionals. *The Family Journal: Counseling and Therapy for Couples and Families* 20: 355–360.

Stone, E. 2008. The last will and testament in literature: Rupture, rivalry, and sometimes rapprochement from Middlemarch to Lemony Sniket. *Family Process* 47:425–439.

Stone, P. 2007. *Opting out?* Berkley: University of California Press.

Strassberg, D. S., R. K. McKinnon, M. A. Sustaita, and J. Rullo. 2013. Sexting by high school students: An exploratory and descriptive study. *Archives of Sexual Behavior* 42: 15–21.

Strauss, J. R. 2013. Care giving for parents and in-laws: Commonalities and differences. *Journal of Gerontological Social Work* 56: 49–66.

Strazdins, L., M. S. Clements, R. J. Korda, D. H. Broom, and M. Rennie. 2006. Unsociable work? Nonstandard work schedules, family relationships, and children's well-being. *Journal of Marriage and Family* 68: 394–410.

Strickler, B. L. and J. D. Hans. 2010. Defining infidelity and identifying cheaters: An inductive approach with a factorial design. Poster, National Council on Family Relations annual meeting, November 3–5. Minneapolis, MN.

Strochschein, L. 2012. Parental divorce and child mental health: Accounting for predisruption differences. *Journal of Divorce and Remarriage* 53: 489–502.

Stroebe, M. and H. Schut. 2005. To continue or relinquish bonds: A review of consequences for the bereaved. *Death Studies* 29: 477–495.

Stulhofer, A. and D. Ajdukovic. 2011. Should we take anodyspareunia seriously? A descriptive analysis of pain during receptive anal intercourse in young heterosexual women. *Journal of Sex and Marital Therapy* 37: 346–358.

Stulhofer, A., B. Traeen, and A. Carvalheira. 2013. Job-related strain and sexual health difficulties among heterosexual men from three European countries: The role of culture and emotional support. *Journal of Sexual Medicine* 10: 747–756.

Su, J. H. 2012. Pregnancy intentions and parents' psychological well-being. *Journal of Marriage and Family* 74: 1182–1196.

Suarez, Z. E. and R. M. Perez. 2012. The Cuban American family. In *Ethnic families in America 5th ed.,* ed. R. Wright, Jr., C. H. Mindel, T. V. Tran, and R. W. Habenstein. 100–128. Upper Saddle River, New Jersey: Pearson.

Sugiura-Ogasawara, M., S. Suzuki, Y. Ozaki, K. Katano, N. Suzumori, and T. Kitaori. 2013. Frequency of recurrent spontaneous abortion and its influence on further marital relationship and illness: The Okazaki Cohort study in Japan. *Journal of Obstetrics and Gynaecology Research* 39: 126–131.

Sullivan, A. 1997. The conservative case. In *Same-sex marriage: Pro and con*, ed. A. Sullivan, 146–154. New York: Vintage Books.

Surjadi, F. F., F. O. Lorenz, R. D. Conger, and K. A. S. Wickrama. 2013. Harsh, inconsistent parental discipline and romantic relationships: Mediating processes of behavioral problems and ambivalence. *Journal of Family Psychology* 27: 762–772.

Sutphina, S. T. 2010. Social exchange theory and the division of household labor in same-sex couples. *Marriage and Family Review* 46: 191–206.

Svab, A. and R. Kuhar. 2014. The transparent and family closets: Gay men and lesbians and their families of origin. *Journal of GLBT Family Studies* 10: 15–35.

Swan, S. C., L. J. Gambone, J. E. Caldwell, T. P. Sullivan, and D. L Snow. 2008. A review of research on women's use of violence with male intimate partners. *Violence and Victims* 23: 301–315.

Sweeney, M. M. 2010. Remarriage and stepfamilies: Strategic sites for family scholarship in the 21st century. *Journal of Marriage and the Family* 72: 667–684.

Sylvestrea, A. and C. Mérettec. 2010. Language delay in severely neglected children: A cumulative or specific effect of risk factors? *Child Abuse & Neglect* 34: 414–428.

Syme, M. L., E. A. Klonoff, C. A. Macera, and S. K. Brodine. 2013. Predicting sexual decline and dissatisfaction among older adults: The role of partnered and individual physical and mental health factors. *The Journals of Gerontology*: Series B 68: 323–332.

## T

Tabi, M. M., C. Doster, and T. Cheney. 2010 A qualitative study of women in polygynous marriages. *International Nursing Review* 57: 121–127.

Takata, Y., T. Ansai, I. Soh, S. Awano, Y. Yoshitake, Y. Kimura, et al. 2010. Quality of life and physical fitness in an 85-year-old population. *Archives of Gerontology & Geriatrics* 50: 272–327.

Tal, G., J. Lafortune, and C. Low. 2014. What happens the morning after? The costs and benefits of expanding access to emergency contraception. *Journal of Policy Analysis and Management* 33: 70–93.

Tan, R., N. C. Overall, and J. K. Taylor. 2012. Let's talk about us: Attachment, relationship-focused disclosure, and relationship quality. *Personal Relationships* 19: 521–534.

Tannen, D. 2006. *You're wearing that? Understanding mothers and daughters in conversation.* New York: Random House.

_____. 1998. *The argument culture.* New York: Random House.

_____. 1990. *You just don't understand: Women and men in conversation.* London: Virago.

Taubman-Ben-Ari, O. and A. Noy. 2011. Does the transition to parenthood influence driving? *Accident Analysis & Prevention* 43: 1022–1035.

Tavares, L P. and A. Aassve. 2013. Psychological distress of marital and cohabitation breakups. *Social Science Research* 42: 1599 1611.

Taylor, A. C., D. L. Rappelea, X. Fang, and D. Cannon. 2013. Emerging adults' perceptions of acceptable behaviors prior to forming a committed, dating relationship. *Journal of Adult Development* 20: 173 184.

Teich, M. 2007. A divided house. *Psychology Today* 40: 96–102.

Temple, J. R., J. A. Paul, P. van den Berg, V. D. Le, Am McElhany, and B. W. Temple. 2012. Teen sexting and its association with sexual behaviors. *Archives of Pediatrics and Adolescent Medicine* 166: 828–833.

Teskereci, G. and S. Oncel. 2013. Effect of lifestyle on quality of life of couples receiving infertility treatment. *Journal of Sex & Marital Therapy* 39: 476–492.

Teten, A. L., B. Ball, L. A. Valle, R. Noonan, and B. Rosenbluth. 2009. Considerations for the definition, measurement, consequences, and prevention of dating violence victimization among adolescent girls. *Journal of Women's Health* 18: 923–927.

Teten, A. L., J. A. Schumacher, C. T. Taft, M. A. Stanley, T. A. Kent, S. D. Bailey, N. Jo Dunn, and D. L. White. 2010. Intimate partner aggression perpetrated and sustained by male Afghanistan, Iraq, and Vietnam veterans with and without posttraumatic stress disorder. *Journal of Interpersonal Violence* 25: 1612–1630.

Thomas, S. G. 2007. *Buy, buy baby: How consumer culture manipulates parents and harms young minds.* Boston: Houghton Mifflin Publisher.

Thomsen, D. and I. J. Chang. 2000. Predictors of satisfaction with first intercourse: A new perspective for sexuality education. Poster at the 62nd Annual Conference of the National Council on Family Relations, Minneapolis, November.

Thomson, E, and E. Bernhardt. 2010. Education, values, and cohabitation in Sweden. *Marriage & Family Review* 46: 1–21.

Todd, M., L. H. Rogers, and C. R. Boyer. 2013. Sexual vs. spiritual? Poster, National Council on Family Relations Annual Meeting. San Antonio, November 5–9.

Toews, M. L., & A. Yazedjian. (2011, November 18). College students' knowledge, attitudes, and behaviors regarding sexuality. Poster session presented at the annual meeting of the National Council on Family Relations, Orlando, FL.

Tong, S. T. 2013. Facebook use during relationship termination: Uncertainty reduction and surveillance. *Cyberpsychology, Behavior, and Social Networking* 16: 788–793.

Toren, P., B. L. Bregman, E. Zohar-Reich, G. Ben-Amitay, L. Wolmer, and N. Laor. 2013. Sixteen-session group treatment for children and adolescents with parental alienation and their parents. *American Journal of Family Therapy* 41: 187–197.

Torres, S. and G. Hammarström. 2009. Successful aging as an oxymoron: older people with and without home-help care: talk about what aging well means to them. *International Journal of Ageing & Later Life* 4: 23–54.

Totenhagen, C., M. Curran, and V. Young. 2012. Relational sacrifices and commitment: A daily diary study. Poster, Annual Meeting of the National Council on Family Relations. Phoenix, Nov. 1.

Totenhagen, C., M. Katz, E. Butler, and M. Curran. 2011. Daily variability in relational experiences. Presentation, National Council on Family Relations, Orlando, November.

Toufexis, A. 1993. The right chemistry. *Time,* February 15, 49–51.

Treas, J., T. Van der Lippe, and T. C. Tai. 2011. The happy homemaker? Married women's well-being in cross-national perspective. *Social Forces* 90: 111–132.

Treat, A. et al. 2013. Variation in health outcomes among co-sleeping mothers and infants. Poster, National Council on Family Relations Annual Meeting. San Antonio, November 5–9.

Trejos, N. 2013. Breaking up is easy to do—in Las Vegas. *USA Today,* December 15, p. 14.

Trucco, E. M., C. R. Colder, and W. F. Wieczorek. 2011. Vulnerability to peer influence: A moderated mediation study of early adolescent alcohol use initiation. *Addictive Behaviors 36,* 729–736.

True Love Waits. 2014. http://www.lifeway.com/ArticleView?storeId=10054&catalogId=10001&langId=-1&article=true-love-waits Accessed January 1, 2014.

Tsunokai, G. T. and A. R. McGrath. 2011. Baby boomers and beyond: Crossing racial boundaries in search for love. *Journal of Aging Studies* 25: 285–294.

Tucker, C. J., G. Cox, E. H. Sharp, K. T. Van Gundy, C. Rebellon and N. F. Stracuzzi. 2013. Sibling proactive and reactive aggression in adolescence. *Journal of Family Violence* 28: 299–310.

Tucker, P., A. Dahlgren, T. Akerstedt, and J. Waterhouse. 2008. The impact of free-time activities on sleep, recovery and well-being. *Applied Ergonomics* 39: 653–661.

Tucker, P. 2005. Stay-at-home dads. *The Futurist* 39: 12–15.

Tumulty, K. 2010. America, the doctor will see you now. *Time*, April 8. 24–32.

Tyagart, C. E. 2002. Legal rights to homosexuals in areas of domestic partnerships and marriages: Public support and genetic causation attribution. *Educational Research* Quarterly 25: 20–29.

Tyler, C. P., M. K. Whitman, J. M. Kraft, et al. 2014. Dual use of condoms with other contraceptive methods among adolescent and young women in the United States. *Journal of Adolescent Health* 54: 169–175.

Tzeng, O. C. S., K. Wooldridge, and K. Campbell. 2003. Faith love: A psychological construct in intimate relations. *Journal of the Indiana Academy of the Social Sciences* 7:11–20.marriages. *International Nursing Review* 57: 121–127.

## U

Uecker, J. and M. Regnerus. 2011. *Premarital sex in America: How young Americans meet, mate, and think about marrying.* United Kingdom: Oxford University Press.

Uji, M., A. Sakamoto, K. Adachi, and T.l Kitamura. 2014. The impact of authoritative, authoritarian and permissive parenting styles on children's later mental health in Japan: Focusing on parent and child gender. *Journal of Child & Family Studies* 23: 293–302.

United Nations. 2011. World marriage patterns. *Population Facts.* December. Retrieved Oct 2, 2013. http://www.un.org/en/development/desa/population/publications/pdf/popfacts/PopFacts_2011-1.pdf

United States Census Bureau, 2014. http://quickfacts.census.gov/qfd/states/00000.html

United States Department of Agriculture, 2013. Parents projected to spend $241,080 to raise a child born in 2012. USDA report. Aug 1, 2013.

Urquia, M., L., P. J. O'Campo, and J. G. Ray. 2013. Marital status, duration of cohabitation, and psychosocial well-being among childbearing women: A Canadian Nationwide Survey. *American Journal of Public Health* 103: 8–15.

U.S. Department of Health and Human Services Releases Report on LGBT Health. 2012.

Uzun, A. K., F. S. Orhon, S. Baskan, and B. Ulukol. 2013. A comparison between adolescent mothers and adult mothers in terms of maternal and infant outcomes at follow-ups. *Journal of Maternal-Fetal & Neonatal Medicine* 26: 454–458.

## V

Vail-Smith, K., D. Knox, and L. M. Whetstone. 2010. The illusion of safety in "monogamous" relationships. *American Journal of Health Behavior 34:* 12–24.

Valle, G. and K. H. Tillman. 2014. Childhood family structure and romantic relationships during the transition to adulthood. *Journal of Family Issues* 35: 97–124.

Valliant, G. E. 2002. *Aging well: Surprising guideposts to a happier life from the landmark Harvard study on adult development.* New York: Little, Brown.

Van Bergen, D. D., H. M. W. Bos, J. V. Lisdonk, S. Keuzenkamp, and T. G. M. Sandfort. 2013. Victimization and suicidality among Dutch lesbian, gay, and bisexual youths. *American Journal of Public Health* 103: 70–72.

Vandell, D. L., J. Belsky, M. Burchinal, L. Steinberg, N. Vandergrift, and NICHD Early Child Care Research Network. Do effects of early child care extend to age 15 years? Results from the NICHD study of early child care and youth development, *Child Development* 81: 737–756.

Van den Brink, F., M. A. M. Smeets, D. J. Hessen, J. G. Talens, and L. Woertman. 2013. Body satisfaction and sexual health in Dutch female university students. *Journal of Sex Research* 50: 786–794.

Vanderkam, L. 2010. *168 hours: You have more time than you think to achieve your dreams.* Portfolio Press (online).

Van Eeden-Moorefield, B., C. R. Martell, M. Williams, and M. Preston, M. 2011. Same-sex relationships and dissolution: The connection between heteronormativity and homonormativity. *Family Relations* 60: 562–571.

Van Geloven, N., F. Van der Veen, P. M. M. Bossuyt, P. G. Hompes, A. H. Zwinderman, and B. W. Mol. 2013. Can we distinguish between infertility and subfertility when predicting natural conception in couples with an unfulfilled child wish? *Human Reproduction* 28: 658–665.

Vannier, S. A. and L. F. O'Sullivan. 2010. Sex without desire: Characteristics of occasions of sexual compliance in young adults' committed relationships *Journal of Sex Research* 47: 429–439.

Vasilenko, S. A., E. S. Lefkowitz and J. L. Maggs. 2012. Short-term positive and negative consequences of sex based on daily reports among college students. *Journal of Sex Research* 49: 558–569.

Vault Office Romance Survey 2014. Retrieved Feb 22 at http://www.vault.com/blog/workplace-issues/love-is-in-the-air-vaults-2014-office-romance-survey/

Vazonyi, A. I. and D. D. Jenkins. 2010. Religiosity, self control, and virginity status in college students from the "Bible belt": A research note. *Journal for the Scientific Study of Sex* 49: 561–568.

Veenhoven, R. 2007. Quality-of-life-research. In *21st century sociology: A reference handbook,* ed. Clifton D. Bryant and Dennis L. Peck, 54–62. Thousand Oaks, CA: Sage Publications.

Vennum, A. 2011. Understanding young adult cyclical relationships. Dissertation, Florida State University. College of Home Economics.

Verbakel, E. and T. A. Diprete. 2008. The value of non-work time in cross-national quality of life comparisons: The case of the United States versus the Netherlands. *Social Forces* 87 679–712.

Vespa, J. 2014. Historical trends in the marital intentions of one-time and serial cohabitors. *Journal of Marriage & Family* 76: 207–217.

_____. 2013. Relationship transitions among older cohabiters: The role of health, wealth, and family ties. *Journal of Marriage and Family* 75: 933–949.

Vickers, R. 2010. Sexuality and the Elderly. Presentation, Sociology of Human Sexuality class, Department of Sociology, East Carolina University, March 22.

Vina, J., F. Sanchis-Gomar, V. Martinez-Bello, and M. C. Gomezx-Cabrera. 2012. Exercise acts as a drug; the pharmacological benefits of exercise. *British Journal of Pharmacology* 167: 1–12.

Vinick, B. 1978. Remarriage in old age. *The Family Coordinator* 27: 359–363.

Von Drehle, D. 2013. How gay marriage won. *Time.* April 8, pp. 17–24.

Vrangalova, Z., and R. C. Savin-Williams. 2010. Correlates of same-sex sexuality in heterosexually identified young adults. *Journal of Sex Research* 47: 92–102.

## W

Wagner, C. G. 2006. Homosexual relationships. *Futurist* 40: 6.

Wagner, S., B. et al. 2013. Perceived stress and Canadian early childcare educators. *Child and Youth Care Forum* 42: 53–70.

Walby, S. 2013. Violence and society: Introduction to an emerging field of sociology. *Current Sociology* 61: 95–111.

Walcheski, M. J. and D. J. Bredehoft. 2010. Exploring the relationship between overindulgence and parenting styles. Poster, National Council on Family Relations annual meeting, November 3–5. Minneapolis, MN.

Walker, J., S. A. Golub, D. S. Bimbi, and J. T. Parsons. 2012. Butch bottom-femme top? An exploration of lesbian stereotypes. *Journal of Lesbian Studies* 6: 90–107.

Walker, R. B. and M. A. Luszcz. 2009. The health and relationship dynamics of late-life couples: A systematic review of the literature. *Ageing and Society* 29: 455–481.

Walker, S. K. 2000. Making home work: Family factors related to stress in family child care providers. Poster at the Annual Conference of the National Council on Family Relations, November. Minneapolis.

Wallerstein, J. and S. Blakeslee. 1995. *The good marriage.* Boston: Houghton-Mifflin.

Wallis, C. 2011. Performing gender: A content analysis of gender display in music videos. *Sex Roles 64,* 160–172.

Walsh, J. L., R. L. Fielder, K. B. Carey, and M. P. Carey. 2013. Changes in women's condom use over the first year of college. *Journal of Sex Research* 50: 128–138.

Walsh, J. L., and L. M. Ward. 2010. Magazine reading and involvement and young adults' sexual health knowledge, efficacy, and behaviors. *Journal of Sex Research* 47: 285–300.

Walsh, S. 2013. Match's 2012 singles in America survey. http://www.hookingupsmart.com/2013/02/07/hookinguprealities/matchs-2012-singles-in-america-survey/

Walters, A. S. and B. D. Burger. 2013. "I love you, and I cheat": Investigating disclosures of infidelity to primary romantic partners. *Sexuality & Culture* 17: 20–49.

Wang, M., K. Henkens, and H. Solinge. 2011. Retirement adjustment: A review of theoretical and empirical advancements. *American Psychologist* 66: 204–213.

Wang, Q., D. Wang, C. Li, and R. B. Miller. 2014. Marital satisfaction and depressive symptoms among Chinese older couples. *Aging & Mental Health* 18: 11–18.

Wang, W., K. Parker and P. Taylor. 2013. Breadwinner moms: Mothers as the sole or primary provider in four-in-ten households with children. *Pew Social & Demographic Trends.* May 29. http://www.pewsocialtrends.org/2013/05/29/breadwinner-moms/

Wang, W. and P. Taylor. 2011. For millennials, parenthood trumps marriage. Pew Research Center Publications. March 9. (See PewSocialTrends.org for full report.)

Warash, B. G., C. A. Markstrom, and B. Lucci. 2005. The early childhood environment rating scale-revised as a tool to improve child care centers. *Education* 126: 240–250.

Ward, J. T., M. D. Krohn, and C. L. Gibson. 2014. The effects of police on trajectories of violence: A group-based, propensity score matching analysis. *Journal of Interpersonal Violence* 29: 440–475.

Warren, J. T., S. M. Harvey, and C. R. Agnew. 2012. One Love: Explicit monogamy agreements among heterosexual young couples at increased risk of sexually transmitted infections. *Journal of Sex Research* 49: 282–289.

Warrener, C., J. M. Koivunen, and J. L. Postmus. Economic self-sufficiency among divorced women: Impact of depression, abuse, and efficacy. *Journal of Divorce & Remarriage* 54: 163–175.

Waterman, J., E. Nadeem, E. Paczkowski, J. Foster, J. Lange, T. Belin and J. Miranda. 2011. Behavior problems & parenting stress for children adopted from foster care over the first five years of placement. National Council on Family Relations, Orlando, November.

Way, N. 2013. Boys' friendships during adolescence: Intimacy, desire, and loss. *Journal of Research on Adolescence* 23: 201–213.

Webley, K. 2012. Cheating Harvard. *Time,* Sept. 17, p. 22.

Weigel, D. J. 2010. Mutuality of commitment in romantic relationships: Exploring a dyadic model. *Personal Relationships* 17: 495–513.

Weisfeld, G. E., N. T. Nowak, T. Lucas, C. C. Weisfeld, E. O. Imamoglu, M. Butovskaya, et al. 2011. Do women seek humorousness in men because it signals intelligence? A cross- cultural test. *International Journal of Humor Research 24,* 435–462.

Weisgram, E. S., R. S. Bigler, and L. S. Liben. 2010. Gender, values, and occupational interests among children, adolescents, and adults. *Child Development* 81: 778–796.

Weiss, K. G. 2010. Male sexual victimization: Examining men's experiences of rape and sexual Assault. *Men and Masculinities* 12: 275–298.

Weisskirch, R. S. and R. Delevi. 2011. "Sexting" and adult romantic attachment. *Computers in Human Behavior* 27: 1697–1701.

Weitz, T. A., D. Taylor, S. Desai, U. D. Upadhyay, J. Waldman, M. F. Battistelli and E. A. Drey. 2013. Safety of aspiration abortion performed by nurse practitioners, certified nurse midwives, and physician assistants under a California legal waiver. *American Journal of Public Health* 103: 454–461.

Wells, R. S., T. A. Seifert, R. D. Padgett, S. Park, and P. Umbach. 2011. Why do more women than men want to earn a four-year degree? Exploring the effects of gender, social origin, and social capital on educational expectations. *Journal of Higher Education 82:* 1–32.

Wendell, C. R., J. Gunstad, S. R. Waldstein, J. G. Wright, L. Ferrucci, and A. B. Zonderman. 2014. Cardiorespiratory fitness and accelerated cognitive decline with aging. *Journal of Gerontology: A Biological Medicine* 69: 455–462.

Wentworth, A. 2012. Jay Leno interview. NBC. *The Jay Leno Show,* Feb 23.

West, P. C. and L. C. Merriam, Jr. 2009. Outdoor recreation and family cohesiveness: a research approach. *Journal of Leisure Research* 41: 351–360.

Westbrook, C. R. A. 2011. Masturbation and sex positive/shameful sexual attitudes. Poster, Easter Sociological Society, Philadelphia, Feb 25.

Wester, K. A. and A. E. Phoenix. 2013. Are there really rules and expectations in talking relationships? Gender differences in relationship formation among young adults. Research and Creative Week. East Carolina University, April 8–12.

Wetherill, R. R., D. J. Neal, and K. Fromme. 2010. Parents, peers, and sexual values influence sexual behavior during the transition to college. *Journal Archives of Sexual Behavior* 39: 682–694.

Weusthoff, S., B. R. Baucom, and K. Hahlweg. 2013. Fundamental frequency during couple conflict: An analysis of physiological, behavioral, and sex-linked information encoded in vocal expression. *Journal of Family Psychology* 27: 212–220.

Wheeler, B., S. Bertagnolli, and J. Yorgason. 2012. Marriage: Exploring predictors of marital quality in husband-older, wife-older, and same-age marriage. Poster, National Council on Family Relations, November, Phoenix.

Wheeler, B. and J. Kerpelman. 2013. Change in frequency of disagreements about money: A "gateway" to poorer marital outcomes among newlywed couples over the first five years of marriage. Poster, National Council on Family Relations Annual Meeting. San Antonio, November 5–9.

Wheeler, L. A., K. A. Updegraff, and S. M. Thayer. 2010. Conflict resolution in Mexican-origin couples: Culture, gender, and marital quality. *Journal of Marriage and the Family* 72: 991–1005.

Whisman, M. A., L. A. Uebelacker, and T. D. Settles. 2010. Marital distress and the metabolic syndrome: Linking social functioning with physical health *Journal of Family Psychology* 24: 367–370.

Whitaker, M. P. 2014. Motivational attributions about intimate partner violence among male and female perpetrators. *Journal of Interpersonal Violence* 29: 517–535.

White, J. M., and D. M. Klein. 2002. *Family theories*, 2d ed. Thousand Oaks, CA: Sage Publications.

White, S. S., N. El-Bassel, L. Gilbert, E. Wu, and M. Chang. 2010. Lack of awareness of partner STD risk among heterosexual couples. *Perspectives on Sexual & Reproductive Health* 42: 49–55.

Whitton, S. W., S. M. Stanley, H. J. Markman, and C. A. Johnson. 2013. Attitudes toward divorce, commitment, and divorce proneness in first marriages and remarriages. *Journal of Marriage and the Family* 75: 276–287.

Whitton, S. W. and B. A. Buzzella. 2012. Using relationship education programs with same-sex couples: A preliminary evaluation of program utility and needed modifications. *Marriage and Family Review* 48: 667–688.

Wick, S. and B. S. Nelson Goff. 2014. A qualitative analysis of military couples with high and low trauma symptoms and relationship distress levels. *Relationship Therapy: Innovations in Clinical and Educational Interventions*. 13: 63–88.

Wickel, K. 2012. Partner support and postpartum depression: A review of the literature. Poster, National Council on Family Relations, Phoenix, AZ, November.

Wickrama, K. A. S., C. M. Bryant, and T. K. Wickrama. 2010. Perceived community disorder, hostile marital interactions, and self-reported health of African American couples: An interdyadic process. *Personal Relationships*, 17: 515–531.

Wienke, C. and G. J. Hill. 2009. Does the "marriage benefit" extend to partners in gay and lesbian relationships? Evidence from a random sample of sexually active adults. *Journal of Family Issues* 30: 259–273.

Wiersma, J. D., J. L. Fischer, B. C. Bray, and J. P. Clifton. 2011. Young adult drinking partnerships: Where are couples 6 years later? Poster, Annual Meeting of National Council on Family Relations, Orlando, November 18.

Williamson, H. C., B. R. Karney, and T. N. Bradbury. 2013. Financial strain and stressful events predict newlyweds' negative communication independent of relationship satisfaction. *Journal of Family Psychology* 27: 65–75.

Willoughby, B. J., J. S. and D. M. Busby. 2012. The different effects of "living together": Determining and comparing types of cohabiting couples. *Journal of Social and Personal Relationships* 29: 397–419.

Willson, A. E. 2007. The sociology of aging. In *21st century sociology: A reference handbook*, ed. Clifton D. Bryant and Dennis L. Peck, 148–55. Thousand Oaks, California: Sage Publications.

Wilson, K. R., S. S. Havighurst, and A. E. Harley. 2012. Tuning in to kids: An effectiveness trial of a parenting program targeting emotion socialization of preschoolers. *Journal of Family Psychology* 26: 56–65.

Wilson, A. C. and T. L Huston. 2011. Shared reality in courtship: Does it matter for marital success? Paper, Annual Meeting of National Council on Family Relations, Orlando, November.

Wilson, C. and R. Rodrigous. 2010. Polyamory. Presentation, Sociology of Human Sexuality Department of Sociology, East Carolina University, Spring.

Wiseman, R. 2013. *Masterminds and wingmen*. New York: Harmony Books.

Witt, M. G. and W. Wood. 2010. Self-regulation of gendered behavior in everyday life. *Journal Sex Roles* 62: 9–10.

Witte, T. H. and R. Kendra. 2010. Risk recognition and intimate partner violence. *Journal of Interpersonal Violence* 25: 2199–2216.

Woertman, L. and F. Van den Brink. 2012. Body image and female sexual functioning and behavior: A review. *Journal of Sex Research* 49: 184–211.

Woldarsky M. and L. S. Greenberg. 2014. Interpersonal forgiveness in emotion-focused couples' therapy: Relating process to outcome. *Journal of Marital and Family Therapy* 40: 49–67.

Wolf, R. and B. Heath. 2013. Rainbow rulings: In two 5–4 rulings, Supreme Court backs recognition of same-sex marriage. *USA Today*, June 24. A-1 et passim.

Woo, J. T., L. A. Brotto, and B. B. Gorzalka. 2012. The relationship between sex guild and sexual desire in a community sample of Chinese and Euro-Canadian woman. *The Journal of Sex Research*. 49: 290–298.

Woszidlo, A. and C. Segrin. 2013. Negative affectivity and educational attainment as predictors of newlyweds' problem solving and marital quality. *Journal of Psychology* 147: 49–73.

Wright, P. J. 2013. U.S. males and pornography, 1973–2010: Consumption, predictors, correlates. *Journal of Sex Research* 50: 60–71.

Wright, P. J. and S. Bae. 2013. Pornography consumption and attitudes toward homosexuality: A national longitudinal study. *Human Communication Research* 39: 492–513.

Wright, P. J., A. K. Randall and J. G. Hayes. 2012. Predicting the condom assertiveness of collegiate females in the United States from the expanded health belief model. *International Journal of Sexual Health* 24: 137–153.

Wright, R. G., A. J. LeBlanc, and L. Badgett. 2013. Same-sex legal marriage and psychological well-being: Findings from the California health interview survey. *American Journal of Public Health* 103: 339–346.

Wright, R., C. H. Mindel, T. Van Tran, and R. W. Habenstein. 2012. *Ethnic families in America: Patterns and variations, 5th ed.* Boston, MA.: Pearson.

Wu, P. and W. Chou. 2009. More options lead to more searching and worse choices in finding partners for romantic relationships online: An experimental study. *CyberPsychology & Behavior* 12: 315–318.

## X

Xu, K. 2013. Theorizing difference in intercultural communication: A critical dialogic perspective. *Communication Monographs* 80: 379–397.

## Y

Yang, J. and A. Gonzalez. 2013. Virtual versus office work. *USA Today*, Feb. 25. Sect. B.

Yang, J. and S. Ward. 2010. Working wives/Ernst & Young Survey. *USA Today*. December 7.

Yanikerem, E., S. Ay, and N. Piro. 2013. Planned and unplanned pregnancy: Effects on health practice and depression during pregnancy. 2013. *Journal of Obstetrics and Gynaecology Research* 39: 180–187.

Yodanis, C., S. Lauer, and R. Ota. 2012. Interethnic romantic relationships: Enacting affiliative ethnic identities. *Journal of Marriage and Family* 74: 1021–1037.

Yoo, H., S. Bartle-Haring, R. Gangamma, F. Lotspeich, R. Patton, and R. Day. 2011. Mediating influences of emotional and physical intimacy on communication and relationship satisfaction. Poster, Annual Meeting of National Council on Family Relations, Orlando,

Yoo, H. C., M. F. Steger, and R. M. Lee. 2010. Validation of the subtle and blatant racism scale for Asian American college students (SABR-A). *Cultural Diversity and Ethnic Minority Psychology*. 16: 323–334.

Yost, M. and G. Thomas. 2012. Gender and binegativity: Men's and women's attitudes toward male and female bisexuals. *Archives of Sexual Behavior* 41: 691–702.

Yu, L. and D. Zie. 2010. Multidimensional gender identity and psychological adjustment in middle childhood: A study in China. *Journal of Sex Roles* 62: 100–113.

## Z

Zaman, M. 2014. Exchange marriages in a community of Pakistan: Adequate social exchange. *The Family Journal: Counseling and Therapy for Couples and Families* 22: 69–77.

Zamora, R., C. Winterowd, J. Koch, and S. Roring. 2013. The relationship between love styles and romantic attachment styles in gay men. *Journal of LGBT Issues in Counseling* 7: 200–217.

Zeitzen, M. K. 2008. *Polygamy: A cross-cultural analysis.* Oxford: Berg.

Zengel, B., J. E. Edlund, and B. J. Sagarin. 2013. Sex differences in jealousy in response to infidelity: Evaluation of demographic moderators in a national random sample. *Personality and Individual Differences* 54: 47–51.

Zhang, Y. 2020. A mixed-methods analysis of extramarital sex in contemporary China. *Marriage & Family Review* 46: 170–190.

Zimmerman, K. J. 2012. Clients in sexually open relationships: Considerations for therapists. *Journal of Feminist Family Therapy* 24: 272–289.

Zinzow, H. M., H. S. Resnick, A. B. Amstadter, J. L. McCauley, K.J. Ruggiero, and D. G. Kilpatrick. 2010. Drug- or alcohol-facilitated, incapacitated, and forcible rape in relationship to mental health among a national sample of women. *Journal of Interpersonal Violence* 25: 2217–2236.

Zusman, M. 2014. Interview, January 12. On social control of choices. Dr. Zusman is retired from Indiana University Northwest. See also Zusman, M. E., D. Knox, and T. Gardner. 2009. *The social context view of sociology.* Durham, NC: Carolina Academic Press.

# Name Index

## E



## F



## G



## H



## I



## J

## W



## X



## Y



## Z

# Subject Index

Note: Page number followed with t refers to tables.

## D

## E



## F

## Q



## R

## T



## U

- HERGÉ -

LES AVENTURES DE TINTIN

# LE TEMPLE DU SOLEIL

CASTERMAN

**Les Aventures de TINTIN et MILOU**
**ont paru dans les langues suivantes:**

| | | |
|---|---|---|
| *afrikaans:* | HUMAN & ROUSSEAU | Le Cap |
| *allemand:* | CARLSEN | Hamburg |
| *alsacien:* | CASTERMAN | Paris/Tournai |
| *anglais:* | METHUEN | Londres |
| | LITTLE BROWN | Boston |
| *arabe:* | DAR AL-MAAREF | Le Caire |
| *asturien:* | JUVENTUD | Barcelone |
| *basque:* | ELKAR | San Sebastian |
| *bengali:* | ANANDA | Calcutta |
| *bernois:* | EMMENTALER DRUCK | Langnau |
| *breton:* | AN HERE | Quimper |
| *bulgare:* | RENAISSANCE | Sofia |
| *catalan:* | JUVENTUD | Barcelone |
| *chinois:* | EPOCH PUBLICITY AGENCY | Taipei |
| *coréen:* | COSMOS | Séoul |
| *corse:* | CASTERMAN | Paris/Tournai |
| *danois:* | CARLSEN | Copenhague |
| *espagnol:* | JUVENTUD | Barcelone |
| *espéranto:* | ESPERANTIX | Paris |
| | CASTERMAN | Paris/Tournai |
| *féroïen:* | DROPIN | Thorshavn |
| *finlandais:* | OTAVA | Helsinki |
| *français:* | CASTERMAN | Paris/Tournai |
| *frison:* | AFUK | Ljouwert |
| *galicien:* | JUVENTUD | Barcelone |
| *gallo:* | RUE DES SCRIBES | Rennes |
| *gallois:* | GWASG Y DREF WEN | Cardiff |
| *grec:* | MAMOUTH | Athènes |
| *hébreu:* | MIZRAHI | Tel Aviv |
| *hongrois:* | EGMONT | Budapest |
| *indonésien:* | INDIRA | Djakarta |
| *iranien:* | UNIVERSAL EDITIONS | Téhéran |
| *islandais:* | FJÖLVI | Reykjavik |
| *italien:* | COMIC ART | Rome |
| *japonais:* | FUKUINKAN | Tokyo |
| *latin:* | ELI/CASTERMAN | Recanati/Paris-Tournai |
| *luxembourgeois:* | IMPRIMERIE SAINT-PAUL | Luxembourg |
| *malais:* | SHARIKAT UNITED | Pulau Pinang |
| *néerlandais:* | CASTERMAN | Dronten/Tournai |
| *norvégien:* | SEMIC | Oslo |
| *occitan:* | CASTERMAN | Paris/Tournai |
| *picard tournaisien:* | CASTERMAN | Paris/Tournai |
| *polonais:* | EGMONT | Varsovie |
| *portugais:* | VERBO | Lisbonne |
| *romanche:* | LIGIA ROMONTSCHA | Cuira |
| *russe:* | CASTERMAN | Paris/Tournai |
| *serbo-croate:* | DECJE NOVINE | Gornji Milanovac |
| *slovaque:* | EGMONT | Bratislava |
| *suédois:* | CARLSEN | Stockholm |
| *tchèque:* | EGMONT | Prague |
| *thaï:* | DUANG-KAMOL | Bangkok |
| *turc:* | YAPI KREDI YAYINLARI | Beyoglu-Istambul |
| *tibétain:* | CASTERMAN | Paris/Tournai |

**ISSN 0750-1110**



ISBN 2 203 00113 5

# LE TEMPLE DU SOLEIL

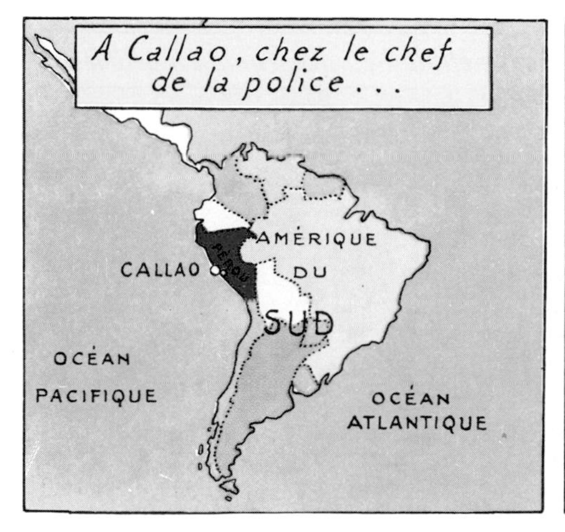

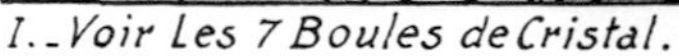
I.-Voir Les 7 Boules de Cristal.

Et quelques minutes plus tard...


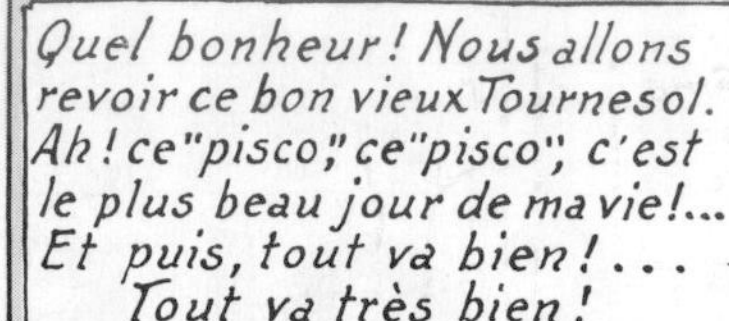

Quel bonheur! Nous allons revoir ce bon vieux Tournesol. Ah! ce "pisco", ce "pisco", c'est le plus beau jour de ma vie!... Et puis, tout va bien!... Tout va très bien!

Allons, ne faites pas cette tête-là, mon ami. Nous allons bientôt revoir Tournesol: tout va donc très bien...
Oui, tout va très bien... N'empêche que, vous l'avez vu, nous sommes sur- veillés...

Ah! la, la! Aucune importance! Regardez plutôt autour de vous. Voyez ce pittoresque, ces Indiens, ces couleurs, ces costumes, ces lamas...

Kilikili! Ah! le beau petit lama!... Il est gentil le petit lama...
Non mais, quel air ça se donne...

Toi faire attention, señor...
Eh bien, quoi?... Je ne vais tout de même pas te le manger, ton lama, non?

Hein, dis, mon brave petit lama, tu n'as pas peur du bon vieux capitaine Haddock...

Quand lama fâché, señor, lui toujours faire ainsi...
En voilà des manières!

Sale vilaine bête de tonnerre de Brest! Qui est-ce qui m'a fabriqué des animaux pareils!

Allons, capitaine, ne faites pas cette tête-là ! Vous le disiez vous-même tout à l'heure : tout va très bien, puisque nous allons retrouver monsieur Tournesol.

HOTEL CRISTOBAL COLON

Hôtel Cristobal Colon. Bueno...

Le lendemain matin.
DRRRING

Allo... Oui, c'est moi... Ah! bonjour, señor inspector superior... Comment ?... Le "Pachacamac" est en vue?... Très bien... Oui, au bassin n° 24... Nous arrivons... A tout à l'heure...

Et quelques minutes plus tard...
Voilà le señor inspector superior et ses hommes, là, au bord du quai...

Mais... Mais je n'ai pas la berlue... Voyez là-bas...
Dupont et Dupond!... Que viennent-ils faire ici, ces olibrius ?...

Messieurs, voilà les amis dont vous me parliez tout à l'heure...

Par quel hasard extraordinaire...
Ce n'est pas le hasard qui nous a envoyé ces messieurs. C'est la police de St Nazaire qui les a délégués ici pour nous aider à retrouver votre ami...

Et alors, ce "Pachacamac", où est-il ?
Là-bas, à gauche de ce petit remorqueur à cheminée rouge...

Ah! oui, je le vois... C'est ça... C'est bien ça... "Pachacamac"... Dire que ce bon Tournesol est à bord !...

Tonnerre de Brest!...
?

Mille sabords! le "Pachacamac" arbore le pavillon jaune et le triangle jaune et bleu : maladie contagieuse à bord !

Sapristi ! et nous qui devons aller à bord pour fouiller le navire . . .
Pas question avant que le service sanitaire n'ait accordé l'autorisation.

Voilà d'ailleurs justement la vedette du service de santé qui se dirige vers le "Pachacamac" . . .

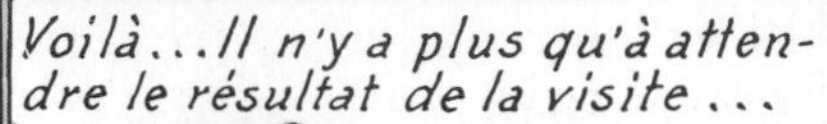
Voilà . . . Il n'y a plus qu'à attendre le résultat de la visite . . .

GUANO

?
GUANO

Au fait, capitaine, qu'est-ce, au juste, que le guano ? . . .
Le guano ? . . . Euh . . . Comment dirais-je ? . . .

PLOC

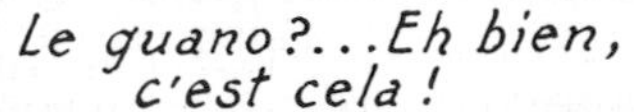
Le guano ? . . . Eh bien, c'est cela !

Tu trouves ça drôle, toi, hein ? . . . Un chapeau presque neuf ! . . . Ça te fait rire, toi !

PLOC

Capitaine ! . . . Capitaine ! . . . Le "Pachacamac" hisse d'autres pavillons !
!

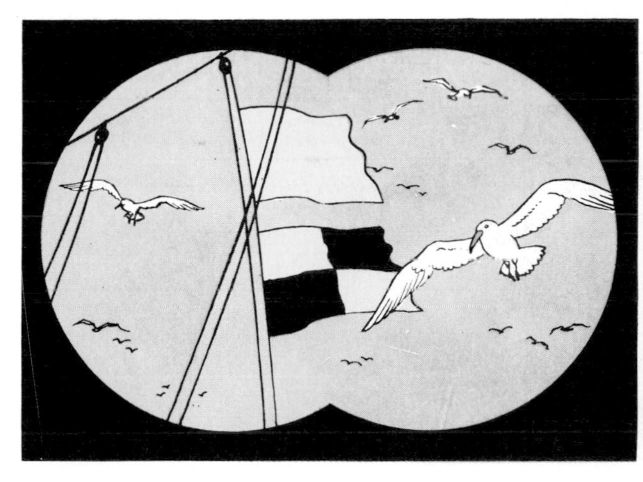

Mille millions de mille milliards de mille sabords ! le signal de la quarantaine !...

C'est pour fêter l'anniversaire du commandant ?
Mettre un navire en quarantaine, marin d'eau douce, signifie le tenir à l'écart pendant un certain temps, afin d'éviter la contagion !

Voilà la vedette qui revient...

Eh bien, docteur ?
Deux cas de peste bubonique à bord !... J'ai ordonné trois semaines de quarantaine !...

Vous avez entendu... J'en suis désolé pour vous... Il va falloir vous armer de patience...
Oui... Evidemment... Dites-moi, ce docteur, n'est-ce pas un Indien ?

Un Quichua, en effet... Mais pourquoi ?
Oh ! Pour rien... Une simple question...

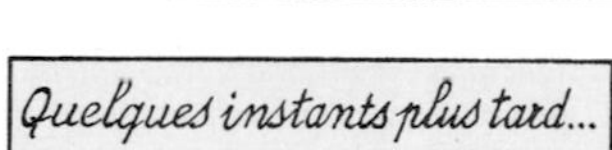
Quelques instants plus tard...

Trois semaines, mille sabords !... Trois semaines avant même de savoir si, oui ou non, Tournesol est à bord de ce cargo de ton-nerre de Brest !...

Pas question d'attendre trois semaines !... Cette nuit-même, nous serons fixés !
Comment, cette nuit-même ?

Parce que, cette nuit, j'irai à bord du "Pachacamac" !
Cette nuit ?... Vous ?... Et la peste, malheureux ?... La peste, hein, vous l'oubliez ?

Capitaine, je parie ce que vous voulez que tout le monde, à bord du "Pachacamac", se porte comme vous et moi !

Mais, saperlipopette ! le docteur a pourtant dit que...
Le docteur est un Indien, capitaine !... Un Indien quichua !... Ça ne vous dit rien, ça ?...

Et la nuit venue...

Stop!... N'allons pas plus loin... On pourrait nous voir...
Bon, ça va... Alors vous êtes toujours décidé?... Je vous ai dit qu'il y avait des requins par ici...

Bah! des requins... A cette heure-ci ils doivent dormir tranquillement, comme tout le monde...
Comme vous voudrez!...

Voilà... Alors, c'est entendu: si, dans deux heures, je ne suis pas revenu, vous avertissez la police... Au revoir, capitaine... Et toi, Milou, sois bien sage...
Bonne chance, fiston...

Tonnerre de Brest!... Quel gaillard, tout de même!

Et maintenant, le plus difficile reste à faire...

¿Qué pasa, ahí abajo?...

¿ Quien es?...

Zut... Encore quelqu'un!

Pas à hésiter!... Vite, entrons dans cette cabine...

Tout va bien!...
Il ne m'a pas vu...
Il passe...

¿ Qué ha pasado, Chiquito?...
No es nada, debe de ser el gato...

Chic! ils croient que c'est un chat!...

Il rentre dans sa cabine...
La porte se referme...
Ouf! sauvé!

RRRON
RRRRON

Sapristi! il y a quelqu'un dans cette couchette!... Vite, filons!...

Pardon... Heu... Un peu plus à l'ouest!...

Il n'y a qu'un homme au monde pour parler de la sorte... Et cet homme, c'est...

Monsieur Tournesol!

Monsieur Tournesol!... Monsieur Tournesol!... Reveillez-vous!... C'est moi, Tintin!... De grâce, réveillez-vous!

Rien à faire!...
Il a certainement été drogué!

Tiens! qu'est-ce que c'est que ça?... Que porte-t-il là, au poignet?

Le bracelet de la momie!!!

Eh! oui, le bracelet de Rascar Capac!...
?

Mais . Mais c'est Chiquito!
Eh! oui, Chiquito...

Que voulez-vous faire de ce malheureux?

Cet homme a commis un sacrilège: il s'est paré du bracelet sacré de l'Inca. Cet homme doit mourir!.. Quant à vous, je n'ai pas encore décidé de votre sort. En attendant vous êtes mon prisonnier...

Alonzo!...

Halte-là, vous!...

Zut! encore un!...

Vite! Vite! par-dessus bord!...

Petite canaille, tu vas me le payer cher!

?
PAN
PAN
?

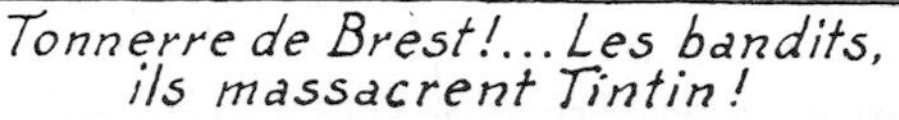

Tonnerre de Brest!... Les bandits, ils massacrent Tintin!
PAN
Iconoclastes!...Pirates!...Encore quelques coups de rames...
... et vous apprendrez à me connaître, bougres de marchands de guano!
PAN

?

Wouah! Wouah!
Mille sabords!...

Wouah! Wouah!
Veux-tu bien te taire, espèce de cornichon!

Ah! voilà Tintin!
PAN
PAN
PAN
Wouah!

Vite, embarquez!... Pas blessé, au moins, fiston?
Rien, pas une égratignure... Mais filons vite...

Tournesol est à bord, capitaine. Je l'ai vu. Ils ont décidé de le mettre à mort parce que, disent-ils, il a commis un sacrilège en passant à son poignet le bracelet de l'Inca!

A terre, maintenant! Et cherchons du renfort, vite!

Vite, courez en ville, alertez la police!... Moi, je reste ici pour observer ce qui se passe...

Et maintenant, ouvrons l'œil...
...Et même les deux!

Rien ne bouge... Auraient-ils l'intention, après ce qui s'est passé... Oh! ils ont mis une embarcation à la mer! Pourvu que le capitaine revienne vite avec du renfort!...

Enfin, une cabine téléphonique!

Allo?... Oui... Police!... Quoi?... Vous voulez parler au señor inspector superior?... A cette heure-ci?... Vous n'êtes pas malade, non?... Il dort, le señor inspector superior...

Je le sais bien, tonnerre de Brest! qu'il dort!... Ce que je vous demande, c'est de le réveiller!... Dites-lui que c'est très, très urgent!

Urgent ou pas urgent, ça m'est égal... On ne réveille pas le señor inspector superior à quatre heures du matin!

Mais puisque je vous dis que... Allo!... Allo!... Allo! Allo!... Ah! le bougre de sauvage de tonnerre de Brest! il a raccroché!

Pendant ce temps-là...
Le canot se rapproche... En avant, Milou... Fais attention, ne te montre pas... Nous allons les observer d'un peu plus près...

Une idée... Je vais téléphoner aux Dupond - Dupont... Quatre... Zéro... Huit... Voilà!...

DRRRING
DRRRING
DRRRING
Dis donc, c'est le téléphone...
Oui, en effet, je crois que c'est le téléphone...

Sapristi!... Mais c'est Monsieur Tournesol qu'on débarque là!

DRRRING
Alors, tu y vas, toi?
Ah! non, jamais de la vie!... Moi, je dors!...

Eh bien! ils y mettent le temps!...

DRRRING
Comment, tu dors?... Tu ne dors pas, puisque tu parles!...
Tu sais bien que je parle toujours en dormant!...

Mille sabords de tonnerre de Brest! est-ce pour aujourd'hui ou pour demain?

C'est bon, j'y vais, mais la prochaine fois, tu iras toi-même...
DRRRING
DRRRING

Allo?... Allo, Dupont?... Enfin, ce n'est pas trop tôt... Ici, le capitaine Haddock...

Comment?... Qui?... Ah! oui, le capitaine Haddock... Je... Comment?... Tournesol?... C'est... Oui... Bien... Bon, nous arrivons immédiatement... Où cela?... Bien...

Et une demi-heure plus tard...
Voilà près de deux heures que je l'ai quitté... Pourvu qu'il ne lui soit rien arrivé!

Voilà notre barque... C'est ici que j'ai laissé Tintin... Mais lui, où est-il?...

Hello, Tintin!...
Tintin!...
Tintin!...

Inutile de nous égosiller davantage. Tintin a disparu. Examinons le sol: nous devons retrouver ses traces au plus vite...

Autant chercher une aiguille dans une botte de foin!...
Je dirais même plus: autant chercher une aiguille dans une fotte de boin!...

Ici, des traces de pas!...

Et ici, d'autres traces... Regardez, il y avait plusieurs hommes... Et des chevaux... Non, des lamas... Voyez ces empreintes, là, dans le sable...

Continuons... Nous sommes sur la bonne piste...

Les traces s'arrêtent à cette route, évidemment... Qu'importe, on devine qu'ils ont continué dans la même direction...

Pardon, minute!... Et si c'était une ruse?... S'ils étaient partis dans la direction opposée?...
Très juste!... Aussi, je propose que la moitié d'entre nous aille d'un côté, l'autre moitié, de l'autre...

Excellente idée, en effet!... Nous sommes trois. La moitié de trois, ça fait un et demi...
Saperlipopette! il y a du vrai dans ce que vous dites!... Comment faire?...

Allez de votre côté, vous deux. Moi, j'irai seul... Et nous verrons qui de nous retrouvera Tintin... Au revoir!... Et ouvrez l'œil!...

Soyez tranquille, nous l'ouvrirons!...
Je dirais même plus, nous...

VIRAGE DANGEREUX

Peu après votre départ, ils ont débarqué Tournesol. Des complices, qui les attendaient sur la plage, ont hissé notre ami sur un lama et l'ont emmené. Moi, je les ai suivis de loin, afin de ne pas me faire voir...

Sans doute avait-il été drogué, car il les suivait docilement, comme un somnambule... Puis, le train est parti... sans moi, hélas! car je n'avais plus assez d'argent pour y prendre place... Alors, je suis revenu sur mes pas, afin de vous rencontrer...

Heureusement que nous sommes arrivés bien à temps: le train va être bondé...

Mais voyons, ce que vous me demandez là est impossible... Je ne puis...
Obéis!... Tu sais ce qu'il en coûte de désobéir aux ordres de... qui tu sais...

Une demi-heure plus tard...
TUUUUT

Nous voilà partis... Curieux, il y avait tant de monde, et personne n'est monté dans notre compartiment...
VOITURE
RESE...ÉE

Bon voyage, señores...

Le train roule depuis plusieurs heures...

Excusez-moi: je reviens dans un instant...

C'est drôle... Figurez-vous que nous sommes absolument seuls dans notre wagon...

Curieux, en effet... Tiens, pendant votre absence, j'ai jeté un coup d'œil sur cette brochure... Savez-vous que cette ligne de chemin de fer atteint une altitude de 15.865 pieds sur 108 milles de trajet et qu'elle est la plus haute du monde?...
Ça ne m'étonne pas: nous grimpons sans arrêt...

Tiens! nous ralentissons... Sans doute arrivons-nous à une gare...

Vite, capitaine, vite, sautons!... Notre wagon s'est détaché et va redescendre la pente...

Sautez, vite!...

A mon tour, maintenant!...

Mon Dieu! j'allais l'oublier!...

Eh bien! pourquoi ne saute-il pas, mille millions de mille sabords?...

Zut! un tunnel!... Milou!... Milou!

Aïe!...

Milou!... Milou!...

Ça, par exemple! il dort!

Allons, vite!
?

Trop tard, maintenant!... Je me tuerais!

Le frein de secours!... Je n'y avais pas pensé!

Allons-y: c'est notre dernière chance!

Du sabotage!... Je comprends tout, à présent!

Que faire?... Que faire?...

Un viaduc... Un cours d'eau... Milou, mon ami, il n'y a pas à hésiter...

Attention!... C'est le moment!

Tintin!...
Où est Tintin?

?

BOUM CRAC
Oh! là-bas, le wagon qui dégringole... Il était temps que nous sautions...
!

Eh bien, mon vieux Milou, nous pouvons nous vanter de l'avoir échappé belle!
A qui le dis-tu!...

D'abord, nous sécher... Ensuite, nous essayerons de retrouver le capitaine...

Allons, Milou, encore un petit effort: nous y sommes...

En route, maintenant!... Allons à la rencontre du capitaine...

Toujours rien!... Se serait-il blessé en sautant?

Que serait-il devenu?...

Le voilà!...
Le voilà!...

Sain et sauf!... Quel bonheur!...

TUUUUT
!

Ohé!...Halte! ..Stop!... Arrêtez!

Vous étiez dans le wagon qui s'est détaché?... Vous avez pu sauter à temps?...Quelle chance!...

Je suis le chef de gare de la station suivante... Lorsque le train est arrivé, on a constaté qu'il manquait un wagon... Je suis désolé: c'est la première fois qu'un accident se produit sur cette ligne...
Un accident?... Vous voulez dire: un attentat...

Un attentat?... C'est impossible, voyons!
C'est cependant ainsi!... Mais ne perdons pas de temps: voulez-vous nous conduire à Jauga, où nous devions nous rendre...

Quelques heures après, à Jauga...
Un homme de petite taille, avec une barbiche noire et des lunettes?...Oui, il me semble...Attendez...Il était accompagné par des Indiens, n'est-ce pas?
C'est à dire qu'il était prisonnier de ces Indiens. Notre ami a été enlevé...

Enlevé par des Indiens?... Je... Hem... Alors, ce n'est pas l'homme que vous cherchez... Celui dont je vous parle avait l'air de suivre ces Indiens de son plein gré...
Evidemment, il avait été drogué!...

Vous croyez?...Non, c'est peu probable... Et puis, maintenant que j'y réfléchis, je me souviens...C'est cela, oui, l'homme qu'on a vu était grand, blond... et il avait le visage rasé...
Mais vous m'avez dit vous-même, il y a un instant...

Je m'étais trompé, voilà tout!... Je regrette de ne pouvoir vous être utile... Messieurs, l'entretien est terminé!...
!
!

Bizarre!... Que signifie la volte-face du commissaire?...On dirait qu'il craint d'être mêlé à cette histoire...Aurait-il peur des Indiens?...
POLICE

Il n'y a plus qu'une solution: interroger, chacun de notre côté, quelques indigènes...
D'accord!...Rendez-vous devant la gare, dans une heure...

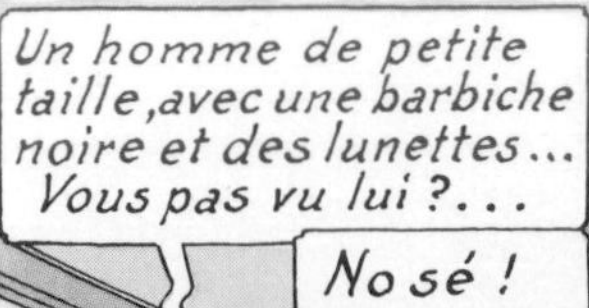

Un homme de petite taille, avec une barbiche noire et des lunettes... Vous pas vu lui ?...

No sé !

Petite taille... barbiche... lunettes... Vous pas vu lui ?...
No sé !

Vous pas vu lui ?...
No sé !

No sé ! No sé !... Ils n'ont que ces mots-là à la bouche, ces bougres de Canaques de tonnerre de Brest !...

?
La charité, mon bon señor...

No sé !
?

Pendant ce temps-là...
No sé ! No sé !... Pas moyen d'en tirer davantage... Ils doivent cependant savoir quelque chose... On dirait qu'ils ont peur...

Tiens, je vais encore interroger ce petit marchand d'oranges...
...Qui te dira, lui aussi, "no sé"...

Tiens, voilà Zorrino... Ne bouge pas: on va rire !...

Ha!ha! ha!ha!
Ha!ha! ha!ha!

Ha!ha! ha! ha!
Ha!ha! ha!ha!

Tu cherches quelque chose, petit ?...
Aaaaaah!

Brute !...

Vous n'avez pas honte?...
Martyriser ainsi ce garçon?...

?

De quoi te mêles-tu, toi ?
Laisse, Pedro, laisse!... Je vais m'expliquer avec Monsieur...

Tiens! attrape ça!...

Ah! canaille! tu crois que... Tiens!...

Sale petite vermine, encaisse ceci!...
Attention! le mur!...

Cette fois, je ne te raterai plus!

AOUW!
Je suis désolé... Je...

!
AAAÏÏE!

AOUW!
Wouah! Wouah!

Milou!... Ici, Milou!... Stop!... Ça suffit.

Tout ça ne nous apprend toujours pas où est passé monsieur Tournesol.

Pstt!... Señor!... Toi t'arrêter, señor... Toi m'écouter...
?
?

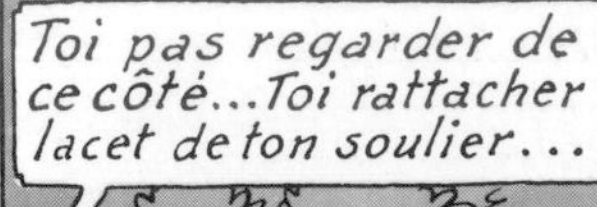

Toi pas regarder de ce côté...Toi rattacher lacet de ton soulier...

Moi savoir où être homme que toi chercher...Toi acheter armes et venir demain, lever du soleil, au pont de l'Inca...Moi te conduire...Toi compris?...Pont de l'Inca, lever du soleil...Toi partir maintenant, vite!...

C'est inouï! voilà un guide qui nous tombe du ciel!

Et si c'était un piège?

Toi écouter moi, señor...
?
?

Moi t'avoir vu prendre défense petit Indien...Toi très bon...Toi très courageux...
Euh...Je...Qui êtes-vous?

Moi te donner bon conseil... Toi pas partir à la recherche de ton ami, sinon toi courir beaucoup dangers...
Comment le savez-vous?

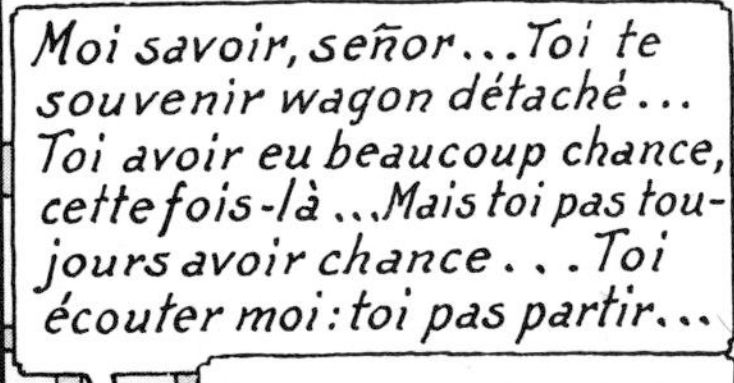

Moi savoir, señor...Toi te souvenir wagon détaché... Toi avoir eu beaucoup chance, cette fois-là...Mais toi pas toujours avoir chance...Toi écouter moi: toi pas partir...
Moi remercier toi, mais moi partir tout de même!

Très dommage pour toi... Mais puisque toi vouloir partir quand même, toi prendre ceci...Très bon, écarter danger...

Une petite médaille...Et puis alors, qu'est-ce que...

!

Le lendemain, à l'aube...
Eh bien, mille sabords! où reste-t-il, celui qui devait nous conduire?

Pssst...Psssst!...
!
!

Vite, señores!...Vous venir vite!
Attention! soyons sur nos gardes...

C'est lui... C'est le petit marchand d'oranges dont je vous ai parlé hier...

C'est donc toi qui...
Oui, c'est moi t'avoir parlé hier, derrière le mur... Si Indiens voir moi te parler, moi mourir tout de suite... Toi venir, maintenant...

Toi m'attendre de l'autre côté du pont... Moi revenir tout de suite...

Eh bien, où court-il?...
Je ne sais pas. Il m'a demandé de l'attendre quelques instants...

Tonnerre de Brest! des lamas!...
Pour porter provisions, señores... Voyage très long!

Tu ne te figures tout de même pas que je vais voyager en compagnie de ces bougres de jets d'eau ambulants, non!...
Lamas très doux, señor... Toi pas avoir peur...

Peur?...Moi?...Peur de ces espèces d'imitations de chameaux?...Il me suffira de les regarder une seule fois dans le blanc des yeux pour qu'ils soient matés à tout jamais...

Comme ça...Voilà!...

OULALAAAH!...

Misérable iconoclaste!...
Toi pas frapper señor!...

Quand lama fâché...
Oui, je sais, mille sabords!... Quand lama fâché, lui toujours faire ainsi!...

Allons, assez perdu de temps!... Nous y sommes?... Au fait, quel est ton nom?...
Zorrino, señor...

Alors, Zorrino, tu sais, toi, où se trouve l'ami que nous cherchons?... Pourtant, pas un seul des Indiens que j'ai interrogés n'avait l'air de le savoir...
Eux savoir comme moi, señor... Mais eux rien dire à señor étranger... Eux peur...

Peur de qui?...
Peur Inca, señor... Vengeance Inca toujours terrible quand Indien dire à Blanc quoi Blanc peut pas savoir...

L'Inca... L'Inca... Il y aurait donc encore un Inca?... A notre époque?... C'est incroyable...
Blancs ignorer señor. Toi seul savoir, maintenant...

Grâce à toi, bien sûr!... Mais dis-moi, Zorrino: tu n'as donc pas peur, toi, de l'Inca?...
Moi seul, alors moi peur: avec toi, moi pas peur!...

Et le soir...
Ça "chulpa", señor, vieux tombeau inca. Nous passer nuit là, et repartir demain matin...

Je prendrai la première garde... Puis, vers minuit, je vous réveillerai, et vous prendrez ma place...
D'accord.

Bonne garde, capitaine... Et n'oubliez pas de me réveiller...
Sois tranquille, fiston! ... Et dormez bien, tous les deux...

Bonne nuit, Zorrino!
Bonne nuit, señor Tintin!

Pas d'erreur, ce sont bien des fleurs d'Inca...

Excusez-moi, monsieur l'Inca, mais avez-vous votre permis de chasse?

Mon permis de chasse?... Misérable sacrilège! que le feu du ciel s'abatte sur ta tête!...

Mon Dieu, quel cauchemar!... Et c'est ce rayon du soleil qui... Mais, au fait...
!

Ça, par exemple! On m'a laissé dormir!... Capitaine!... Eh bien, Capitaine?...

Capitaine!... Capitaine!... Zorrino!...

...Orrino!
...Orrino!
Seul l'écho répond... Que sont-ils devenus?...
Tiens, un écho à deux coups!

Mais soyons prudent: d'abord, ma carabine!

Ah! ça, par exemple! elle a disparu!

Le bonnet de Zorrino: c'est tout ce que je trouve encore...

WOUAH! WOUAH! WOUAH!
?

Qu'a-t-il découvert là?
WOUAH! WOUAH!

!

Me direz-vous enfin...

Eh bien, voilà. Il devait être près de minuit. Je marchais de long en large afin de me réchauffer, lorsque tout à coup une ombre a surgi devant moi. Je n'ai pas eu le temps de faire un geste... pan !... j'ai reçu un coup violent sur la tête... Lorsque je suis revenu à moi, j'étais là, comme vous m'avez trouvé : ligoté, bâillonné, avec ce lézard dans le cou... Et Zorrino où est-il ?

Qu'allons-nous faire, tonnerre de tonnerre de Brest !

Avant tout, essayer de retrouver Zorrino. Ensuite, l'arracher des mains de ses ravisseurs.

Milou!...
Ici, Milou!

Milou, c'est à toi, maintenant, de nous tirer d'affaire... Voici le bonnet de Zorrino... Vas-y!... Cherche!...

En avant!... Suivons-le!
WOUAH! WOUAH!

Eh bien, comme macadam, c'est réussi!

Et deux heures plus tard...
Halte!... Les voilà!...

Le chemin redescend par ici... Ils vont passer juste en dessous de l'endroit où nous sommes...

En coupant par les rochers, nous pouvons les surprendre au passage... Milou, toi, tu resteras ici... Allons, capitaine!
Nous allons nous rompre les os, c'est certain!

Prenez d'un autre côté, capitaine: ce passage-ci est trop difficile.

Il était temps!... Les voilà!... Attention!... Pas de bruit...

AÏE
?

!
?

Aïe! il est tombé!... Pourvu que... Non, il ne s'est rien cassé... Il se relève... Il est pris!
Voilà le dernier qui arrive... Les autres sont hors de vue... N'hésitons plus...

Que se passe-t-il là-bas?...

Toi nous dire si ton ami avec toi?... Où Tintin?...
No sé!...

Toi savoir... Toi nous dire où il est, sinon, toi mourir...
Et moi je vous dis zut!... Et zut!... Et encore zut!... Et...

Et puis... et puis... Et après tout, puisque vous y tenez tellement, le voilà, mon ami, là, derrière vous!
?
?

Les mains en l'air, d'abord!... Parfait!

Vous, capitaine, désarmez d'abord cet Indien... Là, très bien... Maintenant, détachez Zorrino... Moi, je les tiens à l'oeil!...

Content de te revoir, mon petit...

Ça y est?...

All right!... Tout va bien!... En route!...

Attention!

Hourrah!

Allons, vite, capitaine, passez derrière moi . . .
Et vous autres, là-bas, du cal- me! . . .

Et maintenant, vous allez continuer votre route . . . et vite! . . . Le premier qui fait mine de s'arrêter ou de se retourner, je l'abats comme un lapin... Compris?... Allons, en route... et emmenez votre camarade qui reprend connaissance . . .

J'ai dit: vite! . . .
Nous pas pressés . . .

PAN
CLAC

Je crois que, cette fois, ils ont compris... Je puis rejoindre mes compagnons . . .

Tu vois, Zorrino, que nous ne t'avons pas abandonné . . .
Moi savoir, toi me délivrer . . . Et Milou? . . .

Milou, nous l'avons laissé un peu plus haut: il n'aurait pas pu nous suivre... Tiens, le voilà! . . .
Bonjour, Milou! . . .

Wouah!
Wouah!

?
Wouah!
Wouah!

Moi, j'aime voir les choses de haut!

Oooh! un condor . . .
!

Wouaaah!

Tonnerre de Brest!
Que faire, mon Dieu! que faire?... Je ne puis pourtant pas tirer...

WOUAH!
Milou! Mon pauvre, pauvre Milou!...

Là... Voyez... Il se pose sur un rocher... C'est le moment ou jamais... Au nom du ciel, visez juste!

PAN

Hourrah!

Et maintenant, vite! des cordes, un foulard... Il s'agit d'aller au secours de Milou...
Malheureux! Vous n'allez pas faire ça!

Vous ne pensez pas, capitaine, que je vais laisser Milou là-haut, blessé, mourant peut-être...
Tintin, je vous l'assure, vous allez vous tuer!

Milou!... Milou!... Pas de réponse...

Milou!... Milou!...

Toujours rien!...

Ah! c'est toi?... Tu sais, ils ont un garde-manger magnifique, ces oiseaux-là!
!?

Ouf! je respire!... Le voilà sauvé!... Momentanément du moins, car le malheureux va devoir redescendre...

Sauvé!

Pirate!... Doryphore!... Moule à gaufres!... Attends que je te déplume, espèce de chouette mal empaillée !

Quelques minutes plus tard...

Pays de sauvages, mille sabords!...Des montagnes, toujours des montagnes, et des tas de sales animaux!...

C'est encore loin, Zorrino?
Loin, señor, très loin!... Encore longtemps marcher, très longtemps, beaucoup de jours!... Encore franchir hautes montagnes de neige...

Les jours passent...

Et un matin...

Nous arriver col, señores... Là beaucoup danger... Vous pas faire bruit, vous pas parler, sinon avalanches...
Compris, fiston, on y veillera.

Brrr! quel froid de canard... Pas de doute, je vais m'enrhumer... Qu'est-ce que je disais... Là, ça y est... Aaaah!... Aaaah!...

AAAAAAH...

TCHOUM

BRRROUM
BRRROUM
Une avalanche!...
?

Vite !... Derrière ce rocher...

Ouf ! je respire... Eh bien, nous avons eu de la chance !... Maintenant, vite ! dégageons Zorrino !

Où lamas ?... Où capitaine ?...
Je ne sais pas, Zorrino... Ensevelis quelque part sous la neige... Pourvu qu'on les retrouve...

Capitaine !... Capitaine !...
Attention !... Toi pas crier !...

Sapristi, c'est vrai !... Pourvu que... Non, plus rien ne bouge !

Wouah ! Wouah !

Le capitaine !... Il a senti le capitaine...

Allons-y... Au travail... Cherchons...

Le voilà !...

Mon Dieu ! il ne donne plus signe de vie... Vite, dégageons-le...

Le malheureux ! il est gelé !
!

Il faudrait tout de suite le frictionner à l'alcool, mais où en trouver?... Oh! il doit certainement en avoir un flacon dans sa poche-revolver...

Voilà!... J'en étais sûr.

Voyons ce que c'est...

?
?!

Du whisky... Ca ira.

!!!

?

Attention, capitaine, pas trop vite!... Et ne buvez pas tout!...

Là-haut, señores... Lamas pas morts...

Bon!... Hic... Ça va!... Je... Je... Je... vais les chercher!...
Non, non, capitaine! J'irai moi-même...

Silence! mille sabords! ou j'éternue... C'est m-m-moi qui s-s-suis... hic... la c-c-cause de tout ce qui est arrivé... C'est moi qui... qui... hic... qui irai les chercher!
Mais...

Ici, espèces de mérinos mal peignés!... Ici!...

Bougres de phénomènes de tonnerre de Brest! Ils s'enfuient dès que je m'approche d'eux!...

Ici, bougres de zouaves!... Ici, mille sabords!... Ici!...
Quelle nouvelle catastrophe va-t-il encore déclencher?

Les voilà!... Ils ont sans doute été surpris par une avalanche: ils ne sont plus que deux.
Tant mieux! nous n'en règlerons que plus facilement leur compte!

Ah! ça... Je... Je n'ai pas la berlue!... Quels sont ces individus?... Mille sabords! ce sont les Indiens qui avaient enlevé Zorrino!

Au large, canailles!... Au large, flibustiers!...

Au large, bande de Zapotèques!...
A qui peut-il adresser toutes ces injures?... Allons voir!

Patagons!... Bachi-Bouzouks!... Marchands de tapis!... Tchouk-tchouk nougats!...
Vas-y!... Tire!...
J'attends qu'il soit plus près...

Mon Dieu!... Encore ces Indiens!... Mais regarde-les détaler, Zorrino!... Et ce pauvre capitaine, où va-t-il s'arrêter?

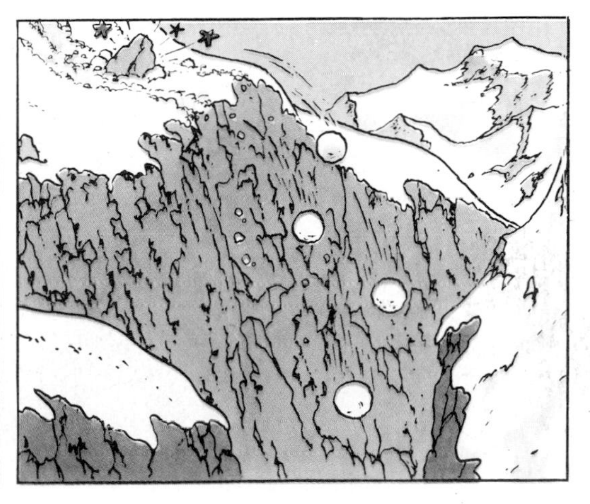

Eh bien! Zorrino, il y a un dieu pour les amateurs de whisky...

Rien de cassé, capitaine?... Non?... Ah! tant mieux!... Sans le faire exprès, je crois que vous nous avez définitivement débarrassés de ces bandits... A présent, retournons là-haut... Ça ira?
Oui, oui...

Mais j'y songe, où est Milou?... Voilà un bon moment déjà que je ne le vois plus auprès de nous... Milou?... Milou?...

Milou!... Milou!... Où a-t-il bien pu passer?

Brave Milou! qui est parvenu à dénicher la casquette du capitaine...

Votre casquette est retrouvée, c'est parfait. Mais, malheureusement, nous avons perdu les lamas, et, par conséquent, plus de vivres, plus de munitions...
Comment, plus de munitions?

De ce côté-là, soyez tranquille. Regardez: j'ai deux boîtes de cartouches dans ma poche.
Quelle chance! Ainsi nous pourrons, s'il le faut, vivre du produit de notre chasse... Et gardez soigneusement le papier: il pourra nous servir pour allumer du feu.

Après de longues heures de marche...

Là, vous regarder. Demain, nous entrer forêt vierge...

C'est dans cette forêt que se trouve le Temple du Soleil?
Non, señor, encore plus loin. Nous traverser forêt, puis encore hautes montagnes...

Mille sabords! ça n'en finira donc jamais?... Je commence à en avoir assez, moi, de cette petite excursion!

Stop!... Regardez, là, une grotte!... Si on s'arrêtait là pour passer la nuit?
C'est une idée, mais...

Soyez tranquille. Je vais d'abord y jeter un coup d'œil...

Ça va!...

Ça va... Vous pouvez venir... L'endroit est très confortable...

Eh bien, quoi?... Qu'est-ce que vous avez à gesticuler ainsi?...
Comment?... Quoi?... Qu'est-ce que vous dites?... Criez plus fort, je ne comprends pas!...

Quoi?... Mais criez donc plus fort, tonnerre de Brest!

Un ours!... Là, derrière vous!...

!!!!

Le lendemain...

Ça va, capitaine?...

Non, ça ne va pas!... On dirait vraiment qu'il n'y a que des moustiques dans ce pays de tonnerre de Brest!

Mille millions de sabords! Attrape, sale bête!

HA-HA-HA-HA-HA-HA
HA-HA-HA-HA-HA-HA-HA-HA
!

HA-HA-
HA-HA
HA-HA-HA-
HA-HA-HA

Ça, par exemple!...Des singes hurleurs!... C'est à croire qu'ils se payent ma tête, ces bougres de zouaves d'anthropopithèques!

Vous avez de la chance que nos cartouches sont comptées, sinon...

PLOUF
!

Mille millions de mille sabords de tonnerre de Brest!...Tout ça à cause de ces espèces de macaques!...Que le diable les emporte!...

Non, non, ce n'est rien...Un simple faux pas dans une petite mare...
Ah! bon! Tant mieux!

AU SECOURS
?
Ça, c'est Zorrino!

Vite!

PAN

Eh bien, Zorrino, mon petit, il était moins cinq...
Toi encore sauvé moi, señor Tintin...

CRAC
CRAC
CRAC
?
?

¿

CRAC
CRAC
CRAC

Dites-moi toute la vérité ! Ne me cachez rien! C'est un autobus, n'est-ce pas, qui m'a renversé ?
Mais non, capitaine, c'est un tapir...
Quand tapir pressé, señor, lui aller droit devant lui, lui prendre garde aucun obstacle. Mais tapir pas méchant, lui pouvoir facilement être apprivoisé.
Ah! oui?... Eh bien, le prochain que je rencontrerai, c'est à coups de carabine que je l'apprivoiserai, moi!

En tout cas, une chose est certaine : mes prochaines vacances, c'est ici que je viendrai les passer. C'est vraiment un endroit de tout repos...

Ah! ces sales moustiques, mille sabords!

Ici, clairière. Bon endroit pour passer la nuit...
Excellente idée...

Et la nuit venue...

Le lendemain, à l'aube...
RRRON

RRRON RRRON
RRRON

!

Mmmmh... Milou... Allons, M-M-Milou... Laisse-moi tranquille...
?

?
!

AU SECOURS!
!

Au large! espèce de Cyrano à quatre pattes!...
!

Rassurez-vous: ce n'est qu'un brave fourmilier tamanoir qui est venu vous dire bonjour...
Toi être couvert fourmis... Tamanoirs manger fourmis...

Et les jours passent...

Là-bas, bientôt, grande rivière... Nous devoir traverser...
Et comment?... A la nage?...
Sales bêtes!...

Vous attendre ici, señor... Zorrino bientôt revenir...
Ça va...

Comme c'est curieux, tous ces troncs d'arbres qui flottent sur la rivière...
Des troncs d'arbres?... Ne vous y fiez pas, fiston!... Ce sont des alligators!

Des alligators!... Ça, par exemple!... J'aurais juré que...
Oui, oui, évidemment... Mais moi, je m'y connais!

A MOI, TINTIN!

PAN

Hem!... Je... Euh... Merci, fiston... Je... Vous voyez bien que...
En effet, capitaine... Mais, à présent, le voilà aussi inoffensif qu'un tronc d'arbre véritable...

CRAC

Ce n'est rien... C'est Zorrino qui a fait craquer une branche morte.
Vous venir, señores. Moi trouvé pirogue.

Voilà...

Mes enfants, attention! Je crois que ça va chauffer... Les voilà!... Ils nous ont repérés.

PAN

PAN
PAN

PAN

PAN

!

PAN

PAN

Sales bêtes!...Je m'en vais les exterminer...
Non, non! ne gaspillez pas nos munitions...

Sale pays de tonnerre de Brest!... Quand donc en sortirons-nous?...
Demain, señor capitaine, nous quitter forêt.

Et le lendemain soir...
Nous camper ici, cette nuit... Là-haut, derrière montagnes, Temple du Soleil...

Le lendemain matin...
En route!... Ça, par exemple! Où as-tu déniché ces cordes?
Sans doute nous avoir besoin de cordes... Moi les fabriquer ce matin avec lianes...

Diable! le Temple du Soleil est bien défendu... Ce torrent est infranchissable... Plus loin, peut-être, trouverons-nous un passage...

Et deux jours plus tard...

Rien à faire, capitaine... Nous allons essayer de passer ici... Il y a là un piton rocheux autour duquel nous pourrons lancer une corde...
Ça va!

Allons-y!

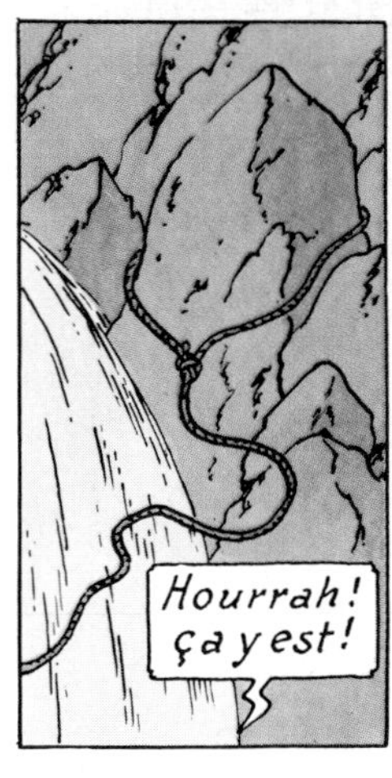
Hourrah! ça y est!

Voilà l'autre bout fixé à un arbre... Et maintenant, qui passe le premier?

Zorrino, avec fusil Tintin, pour prouver corde très solide!

Il a du cran, ce petit!
Sois prudent, Zorrino!

Et voilà!
Bien... A mon tour maintenant...

Tonnerre de Brest! Il s'agit d'avoir le pied marin!

Mille sabords! Ma casquette!

Au nom du ciel, lâchez cette casquette, capitaine !... Vous allez tomber !...

Jamais de la vie !... J'y tiens, moi, à ma casquette !

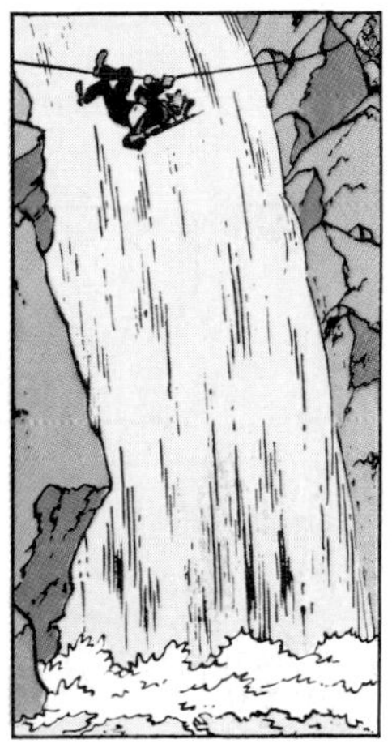

Ouf ! ça y est !
A moi, maintenant.
Aïe ! aïe ! aïe ! encore des acrobaties !...

Allons, Milou, tiens-toi tranquille... Nous y sommes...
Wouaaah !
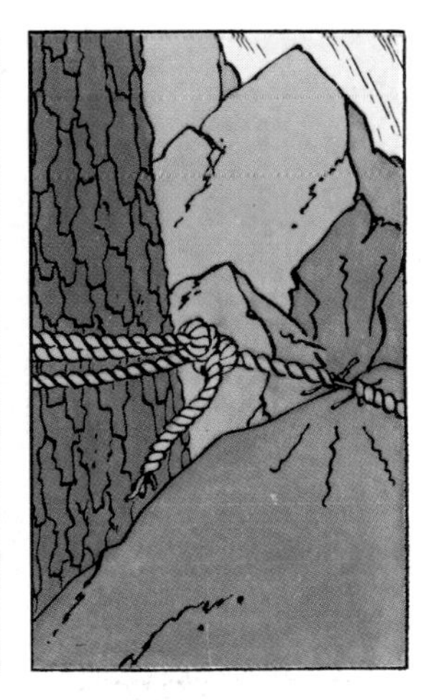

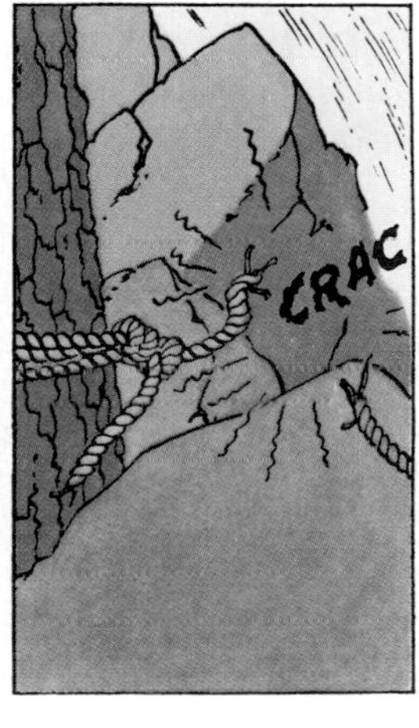
CRAC

?

Horreur !...

Tintin!
Tintin!

Non...Rien...Je ne le vois pas... Voyons, ce n'est pas possible... Il est excellent nageur... Il va reparaître à la surface...

Rien...Rien...C'est fini...Il s'est noyé...Mon Dieu, c'est épouvantable!
!

Noyé?...Noyé?...Tintin?...Tintin pas mort, n'est-ce pas, capitaine?
Hélas! Zorrino!...

Hélas! mon pauvre Zorrino, nous ne le verrons plus... C'est fini... Fini!

Ohé!
?
?

Voyons... Cette voix!... Est-ce que je rêve?... Ce n'est pas lui...
Si, si!... Ça voix de Tintin!...
Capitaine!... Zorrino!...

Tintin!...Tintin!...Est-ce bien vous?... Où donc êtes-vous?
Wouah! Wouah!
Ici, derrière la chute d'eau...

Derrière la chute d'eau?... Comment, derrière la chute d'eau?
Descendez!... Vous allez voir...

?
Descendez!...Descendez encore...

Approchez!...Bon...Regardez maintenant le bas de la chute... Je vais lancer une pierre pour vous indiquer où je suis...

Voilà!
!
!

Vous avez vu?...Bien...Je crois que j'ai découvert quelque chose de très intéressant!... Remontez chercher la corde... Attachez-y une grosse pierre et lancez-la moi...
Ça va.

Voilà qui est fait...
Je vous la lance...

Très bien!

Attachez solidement le bout de la corde à un rocher. Moi, je fais la même chose de ce côté-ci.
O.K.

Voilà, ça y est!

Parfait. Et maintenant, venez me rejoindre.
?

Com... Com... Comment?... Vous rejoindre?... N'est-ce pas vous, au contraire, qui...
Non, non! Tenez-vous fermement à la corde et passez à travers la chute... Ce n'est qu'un mince rideau d'eau, vous verrez...

Mais... Mais... Vous êtes sûr que...
Oui, oui! Allez-y!

A Dieu vat, puisqu'il le faut!

Et voilà!
!

Ça mille sabords!... Mais où sommes-nous donc?
Un instant... J'appelle Zorrino...

C'est extraordinaire... C'est inouï... C'est incroyable...
A ton tour, Zorrino!...

Et hop! ça y est!
!

Content de te retrouver, Zorrino !
Tintin!... Ah Tintin!... Zorrino a eu beaucoup peur.... Toi pas blessé?


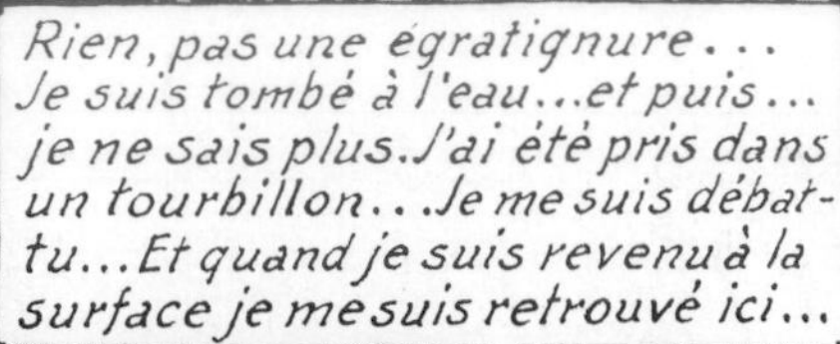

Rien, pas une égratignure... Je suis tombé à l'eau...et puis... je ne sais plus. J'ai été pris dans un tourbillon... Je me suis débattu... Et quand je suis revenu à la surface je me suis retrouvé ici...

Et je crois que, par un hasard vraiment providentiel, j'ai découvert une ancienne entrée du Temple du Soleil... Une entrée probablement oubliée des Incas eux-mêmes... D'ailleurs, nous verrons bien.

Mais, bon sang! il doit faire noir comme à l'intérieur d'un cachalot, là-dedans!
Je le croyais comme vous. Mais je suis allé voir. Les rochers sont couverts d'une matière phosphorescente qui diffuse une certaine lumière... Nous y allons?

Surtout, pas de bruit!... Soyons prudents!... J'ai l'impression que nous ne sommes plus loin de monsieur Tournesol.

Allons-y... Continuons...

Où allons nous aboutir ?...

Continuons toujours... Nous verrons bien...

Oh! Oh! voilà qui est plus grave!... Le couloir est obstrué!... Plus moyen de continuer!...

C'est probablement à la suite d'un de ces tremblements de terre, si fréquents dans ce pays, que l'éboulement a dû se produire... Nous voilà donc bloqués... A moins que...
Wouah! Wouah!

Wouah! Wouah! Par ici la sortie!

Milou semble nous indiquer quelque chose... Là, en effet, on dirait qu'il y a un passage... Prends ceci, Zorrino, je vais essayer de m'y glisser...

Ça ira ?
Je l'espère...

Ça va ?
Jusqu'à présent, oui...

?

Je viens de déboucher dans une sorte de grotte... Je vais voir s'il y a moyen de... OH!...
Mon Dieu qu'y a-t-il ?

!

Je... Hem!... Euh... Beau temps, n'est-ce pas ?

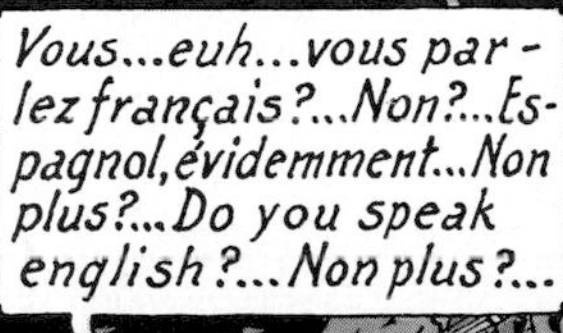
Vous... euh... vous parlez français ?... Non ?... Espagnol, évidemment... Non plus ?... Do you speak english ?... Non plus ?...

Mais, ma parole ! ce visage immobile... Ce regard fixe... Je me demande si...

?

Sapristi ! C'est tout le contenu d'une tombe qui vient de dégringoler là !

L'hypothèse du tremblement de terre se confirme... Allons voir ce qu'il y a de l'autre côté...

?

Des momies incas... Nous sommes bel et bien dans un tombeau...

Si l'on parvenait à faire basculer cette dalle... Mais, seul, je n'y arriverai pas... Je vais appeler les autres...
Il a mauvaise mine, celui-ci...

Allo!... Capitaine! ...Zorrino!... Venez, j'ai besoin de votre aide...
Ça va, nous arrivons...

Vas-y le premier, Zorrino... Je te passerai ensuite les carabines et les ponchos...

Toi donner moi les fusils, señor capitaine...
Les voici...

Ici fusils, Tintin...
Merci, Zorrino...

Oh! chambres des morts, ici!
Eh! oui, Zorrino, nous n'avons pas le choix...

Ah! à mon tour, maintenant...

!
?
TUUUT

Mais, sapristi! c'est Milou qui a fait ce bruit-là!... Qu'est-ce que cela signifie?...
Ça, par exemple! ils ont des os à musique, dans ce pays...

Flûte des morts, Tintin... Incas fabriquer flûtes avec os des morts...
Une flûte sculptée dans un tibia!... Et c'est Milou qui a soufflé dedans par mégarde...

Et bien, capitaine, où restez-vous?

Ah! ça, mille sabords! mais c'est une tombe, ici!... Eh bien, c'est gai!...
Que voulez-vous, capitaine, il n'y a pas moyen de faire autrement.

Dites donc, c'est pour me présenter à cette paire de joyeux trompe-la-mort que vous m'avez fait venir ici?

Non, non, capitaine c'est pour autre chose... Je suis persuadé que nous touchons au but... Voyez-vous cette dalle?... Nous allons essayer de la faire basculer... Et qui sait? Peut-être que derrière...
Peut-être, en effet...

Eh bien, allons-y... Une... Deux... Trois... Hop!...

Ça va!... Elle a bougé!... Encore un effort... Une... Deux... Hop!...

!

Qu'on se saisisse de ces sacrilèges!

Arrière, poussières!... Au large, espèces d'Incas de carnaval!...

Va-nu-pieds!...Bandes de zapotèques!... Moules à gaufres!...Anthropopithèques!... Lâchez-moi, bande de sauvages!...

Bien...Qu'on les enferme avant de les faire comparaître devant l'Inca.

Bien. Quoi qu'il en soit, notre loi ne prévoit qu'un châtiment pour ceux qui se risquent à pénétrer dans le temple sacré où nous perpétuons le culte du Soleil, et ce châtiment, étrangers, c'est la mort!

La mort, la mort, la mort!... Et vous croyez que nous allons nous laisser massacrer ainsi, espèce de cannibale emplumé!
De grâce, capitaine, taisez-vous!

Noble Fils du Soleil, permettez-moi de vous expliquer ce qui s'est passé. Notre but n'a jamais été de commettre un sacrilège. Simplement, nous étions à la recherche de notre ami, le professeur Monsieur Tournesol qui...

Votre ami a osé se parer du bracelet sacré de Rascar Capac. Votre ami sera, lui aussi, mis à mort!...

Vous n'avez pas le droit de tuer cet homme, mille sabords! Pas plus que de nous tuer, nous, mille tonnerres de Brest! C'est de l'assassinat pur et simple!

Aussi n'est-ce pas nous qui vous mettrons à mort. C'est le Soleil lui-même qui, de ses rayons, mettra le feu au bûcher qui vous est destiné.

Quant à ce jeune Indien qui a guidé ces étrangers et qui, de cette façon, a trahi sa race, il subira le châtiment réservé aux traîtres!... Qu'il soit immédiatement égorgé sur l'autel du Soleil!

Mille millions de mille sabords! le premier qui ose toucher à un seul cheveu de la tête de ce garçon, celui-là est un homme mort!
Grrrr!...

Mon Dieu, j'y pense!... Ta médaille, Zorrino!... Montre la médaille que je t'ai donnée...

?

Où as-tu volé cette médaille, misérable petite vipère!...

Moi pas volé, noble Fils du Soleil, moi pas volé!... Lui donné moi cette médaille!... Moi pas volé!

Et toi, chien d'étranger, où l'as-tu prise?... Sans doute, comme tes pareils en ont l'habitude, en violant la sépulture d'un de nos ancêtres!

Noble Fils du Soleil, je demande la parole...
!

C'est moi, ô noble Fils du Soleil, qui ai donné à ce jeune étranger la médaille sacrée.

Comment, toi, Huascar, un grand prêtre du Soleil, tu as commis le sacrilège de donner ce talisman à un ennemi de notre race?...

Ce n'est pas un ennemi de notre race, Seigneur... Je l'ai vu, de mes yeux vu, prendre tout seul la défense de cet enfant, que brutalisaient deux de ces infâmes étrangers que nous haïssons. C'est pour cela, sachant qu'il allait au devant de graves dangers, que je lui ai donné cette médaille. Ai-je mal fait, ô noble Fils du Soleil?

Non, Huascar, tu as agi noblement. Mais ton geste n'aura servi qu'à sauver la vie de ce jeune Indien, puisque le voilà protégé par ce talisman...

... et non celle du jeune étranger qui, par sa générosité, s'est privé de sa seule chance de salut. Nos lois sont formelles: il sera mis à mort, ainsi que son compagnon!...
Cependant, je désire leur accorder une grâce...
Allons! Il n'est pas si méchant qu'il en a l'air!

Et cette grâce, la voici... Dans les trente jours à venir, ils pourront choisir eux-mêmes le jour et l'heure où les rayons de l'astre sacré enflammeront leur bûcher. Je leur donne jusqu'à demain pour réfléchir et me porter leur réponse.

Quant à ce jeune Indien, il sera séparé de ses compagnons et il aura donc la vie sauve. Mais il restera jusqu'à sa mort dans ce temple afin que notre secret ne soit point divulgué au dehors.

A présent, qu'on emmène ces étrangers et qu'on les mette au secret jusqu'à demain... Tel est la volonté du Fils du Soleil!...
Eh bien, nous voilà dans de beaux draps!
Oui, c'est vrai... Mais il est heureux déjà que Zorrino soit sauvé, lui...

Tas de sauvages!... Je m'en vais fumer une pipe... cela me calmera les nerfs... Où est-elle? Ah! la voici... Et ça, qu'est-ce que c'est?...

Ah! oui, je me souviens... Le journal qui a servi à emballer nos cartouches...

Fini, maintenant... Nous n'en aurons plus besoin... Ce n'est plus nous qui devrons allumer du feu, à présent...
On l'allumera pour nous, tonnerre de Brest!

Que faire?... Comment sortir d'ici?...

Ces barreaux peut-être?... Non, hélas! ils sont solidement fixés...
?

Et puis, même si nous arrivions à les desceller, cette fenêtre donne sur un précipice.

Mille sabords! j'ai perdu mes allumettes!

Donnez-moi votre pipe, capitaine. J'ai une petite loupe...
Une loupe?...Ah! oui...

Mille sabords! Ça prend!
Oui, voilà, ça y est...

Ça y est!... C'est merveilleux!... C'est magnifique!...

Magnifique, oui... Et c'est certainement de la même manière que les Incas bouteront le feu au bûcher sur lequel nous serons grillés...

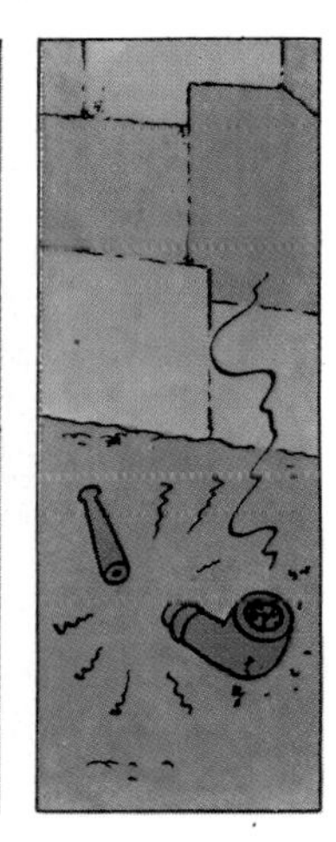

...A moins qu'ils n'utilisent des miroirs paraboliques, comme le fit Archimède pour incendier les vaisseaux romains qui assiégeaient Syracuse...
Ma pipe!

Ma pipe!... Ma pauvre pipe!... Elle est cassée, mille sabords!

Eh bien, Milou, que fais-tu là?... Où as-tu déniché ce papier?

Pendant ce temps, en Europe...
Chef, nous avons fouillé de fond en comble toute l'Amérique du Sud, sans aucun résultat. Tintin, le capitaine et le professeur Tournesol sont restés introuvables.
Je dirai même plus: introuvables.

Aussi avons-nous décidé d'entreprendre de nouvelles recherches, sur des bases toutes neuves, et avec des méthodes entièrement inédites.
Je dirai même plus: c'est ce que nous avons décidé.

Ah?... Et en quoi consistent ces méthodes?
Permettez-nous, chef, de garder à ce sujet un silence aussi compact que discret... Vous le savez, botus et mouche cousue, telle est notre devise.

La radiesthésie, mon cher, comme Monsieur Tournesol, voilà qui va nous mettre sur leur piste...

Qu'est-ce que c'est que ce morceau de journal?
Eh bien, c'est celui que vous m'aviez demandé de conserver et qui devait nous servir à allumer du feu.
Tu vas me le rendre, dis?...

Tiens! tiens! intéressant, cet article... Dommage que...
Wouah! Wouah!

!?

Milou!... Ici, Milou!... Lâche ce papier!

Milou, veux-tu m'obéir...

Milou!... Ici!

Milou! Milou!... Rends-moi ce journal!

Milou!... Saperlipopette!

Milou!... Veux-tu, oui ou non, cesser ce petit jeu?... Allons, ici!
Mais, c'est qu'il à l'air d'être réellement fâché...

Ah! je crois que je vais parvenir à la raccommoder...

Ça, par exemple, c'est vraiment curieux!... Quelle coïncidence!
Alors, quoi? On ne peut même plus jouer...

???!!!

Ça y est: elle est réparée!... Tintin, est-ce que vous ne pourriez pas...

YOUPIIIIE!

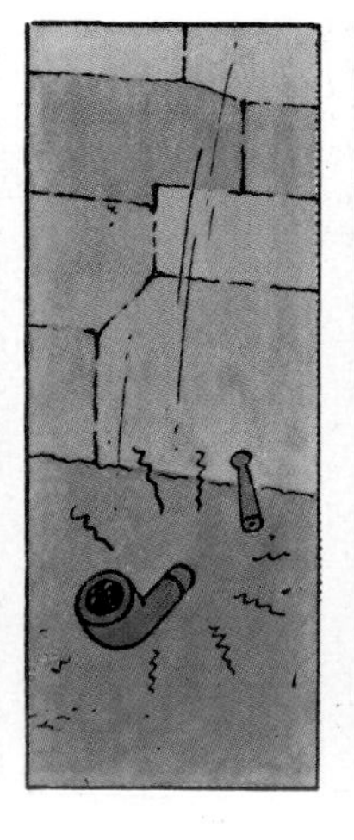

Au nom du Ciel, Tintin, que se passe-t-il?
Tra la la-itou!

Le lendemain matin...

Et maintenant, allez-vous m'expliquer ce que tout cela signifie ?
Pas encore, capitaine, pas encore !... Je ne puis vous dire qu'une seule chose : tout va bien !

Tout va bien !... Tout va très bien, Madame la Marquise !... Nous allons être grillés vifs dans dix-huit jours... Mais à part ça, tout va très bien !... Je dirai même plus, comme nos amis Dupont et Dupond : tout va bien !

Les jours ont passé...
Plus que six jours, demain matin, tonnerre de Brest !... Ah ! misère de misère !...

Et le lendemain...
Comment en sortir ?... Qui pourrait nous aider ?... Zorrino, peut-être...

Le jour suivant...

C'est formidable !... Nous n'avons plus que cinq jours à vivre, mille sabords ! et vous faites de la culture physique !... Le moment est vraiment bien choisi !

Que voulez-vous, capitaine, il faut bien conserver sa souplesse...

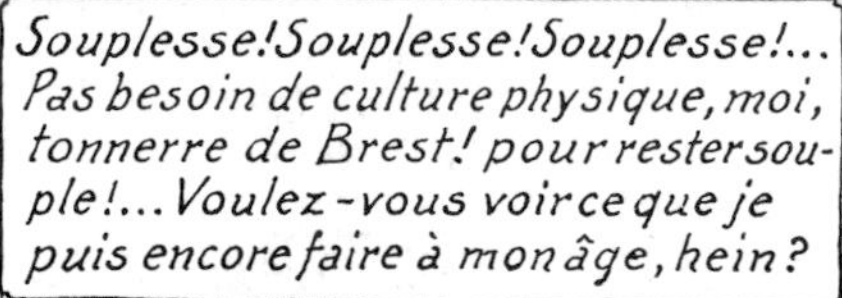
Souplesse ! Souplesse ! Souplesse !... Pas besoin de culture physique, moi, tonnerre de Brest ! pour rester souple !... Voulez-vous voir ce que je puis encore faire à mon âge, hein ?

Voilà : à pieds joints, sans élan, au-dessus de cette table...

HOP !
Chic !...

Vous trouvez ça drôle, vous ?

Plus que quatre jours...
Il ne sera pas dit que je me laisserai rôtir ainsi comme un vulgaire poulet... Il faut absolument s'évader.
Vous savez bien que c'est impossible.

Plus que trois...
Que faire, bon sang de bon sang?...
Il va finir par me donner le vertige...

Plus que deux...
Et vous restez là à vous prélasser, mille millions de sabords!... Il faut faire quelque chose!...
Ayez confiance, capitaine. Dans deux jours, nous serons sauvés...

Plus qu'un jour...
C'est fini!... Plus rien à espérer!... Jamais je n'ai touché à ce point le fond du désespoir!

Au même moment...
D'après le pendule, ils doivent être bien bas...

Et le lendemain...
Plus que quelques heures à vivre, mille sabords!... Et tout ce que vous trouvez à faire, c'est de relire, pour la centième fois, ce morceau de journal!

"...L'expédition suisse est en route pour la Cordillère des Andes. Elle sera..." Et le reste est déchiré...

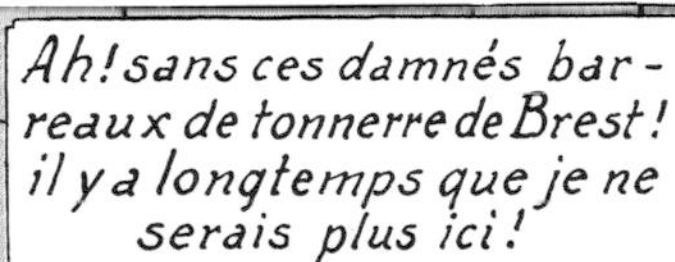
Ah! sans ces damnés barreaux de tonnerre de Brest! il y a longtemps que je ne serais plus ici!

CRAC
BING
BOUM
?

Libres!... Nous sommes libres!... Vite, Tintin, vite!... Filons!...
Ne faites pas ça, capitaine, vous allez vous rompre les os!

Ah! ah! nous arrivons à temps!
Trop tard! Tonnerre de Brest...
!
!

L'heure est venue. Veuillez passer la robe du sacrifice.
Moi passer cette espèce de chemise de nuit ?... Jamais !...

C'est la loi. Il faut obéir...
Je vous dis: jamais !... Et quand je dis: jamais, c'est jamais !
Voyons, capitaine...

Qu'on lui passe la robe du sacrifice.

CLAC
Jamais !
CRAC
BOUM

Le bûcher ! Voilà notre bûcher ! Et vous croyez que je vais monter là-dessus, moi ?... Jamais !

Il faut, coûte que coûte, échapper à cette bande d'enragés !

!

Rien de cassé, au moins, capitaine ?
J'ai l'impression qu'il se passe quelque chose de pas naturel, ici...

BOUM
BOUM
Quelle est donc cette musique?
Vous appelez ça de la musique, vous?

BOUM BOUM BOUM BOUM

Pacharurac - Pachacamac Viracocha

Cayhinapac Churasunqui Camasunqui

Voilà Monsieur Tournesol, capitaine!... Tournesol, que nous avons cherché pendant si longtemps!... Le voilà!... On va l'attacher auprès de nous...

Vous, capitaine?... Ah! quelle bonne surprise!... Comment allez-vous?
Très bien, merci, comme vous voyez!...

Et vous aussi, mon cher Tintin, vous voilà!... Je suis si heureux de vous retrouver!... Mais, dites-moi, que signifie cette sorte de mascarade?... Où sommes-nous, ici?...
Chez les Incas...

Ah! du cinéma!... Bon, je comprends!... C'est une reconstitution historique, sans doute... Ces gens-là sont costumés, comme... comme des Aztèques, dirait-on... Ou plutôt, non, comme des Incas!
Précisément, des Incas, vous l'avez deviné...

Ah! oui, admirablement grimés... Et voyez cette danse: quel naturel, quelle conviction chez le moindre de ces figurants...
Pourvu que tout se passe comme je l'espère...

Noble Fils du Soleil, voici venue l'heure du sacrifice...

Et pendant ce temps là...
Cependant, d'après le pendule, ils doivent être quelque part où il y a beaucoup de soleil...

Que le sacrifice commence!... Que le Grand Prêtre du Soleil s'approche du bûcher!...

Qu'est-ce que c'est que cet instrument-là ?
Ça, c'est la loupe qui doit mettre le feu à notre bûcher.
Non?...

Laissez-moi!... Je ne veux pas qu'on les tue...

O Pachacamac, puissant astre du jour, toi qui as fait le monde, toi le dieu qui l'anime, frappe ce bûcher de tes rayons vengeurs!...

Arrête, ô Huascar!... Tes invocations ne seront pas entendues par le dieu souverain.
?
?
Grrrr

Et toi, ô puissant Soleil, montre à tous, par un signe tangible, que tu ne désires pas notre mort.

Silence! chien d'étranger!... De quel droit oses-tu t'adresser au Soleil?

O sublime Pachacamac! je t'adjure de manifester ta toute-puissance!... Si tu ne veux pas de ce sacrifice, voile ici, devant tous, ta face étincelante...
Pauvre petit! il a perdu la raison!
Mais non, mais non: votre chapeau est très beau, lui aussi.

Merci, ô astre souverain!... Merci, ô Soleil!... Tu as entendu ma prière... Voici que tes rayons déclinent...

Mais... Mais, ma parole, il a raison... Que se passe-t-il?... Est-ce que je deviens fou, moi aussi?... C'est de la sorcellerie!...

!

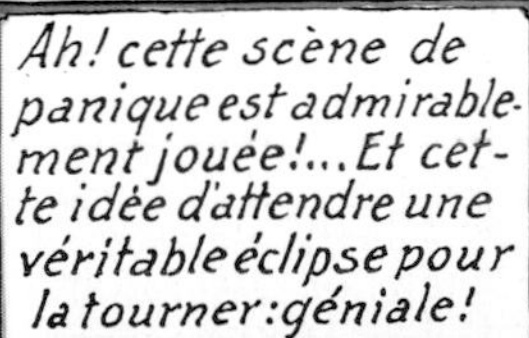
Ah! cette scène de panique est admirablement jouée!... Et cette idée d'attendre une véritable éclipse pour la tourner: géniale!

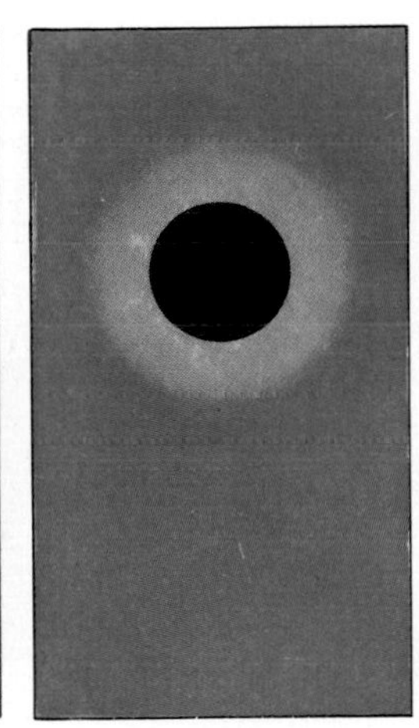

Une éclipse!... Une éclipse!!... Une éclipse!!!
Inutile de s'alarmer: c'est tout simplement une éclipse...
Wou--ou-ouh!...

Grâce, Etranger, je t'en supplie!... Fais que le soleil luise à nouveau... Et je t'accorderai tout ce que tu me demanderas!

Soit, noble Inca, j'ai confiance en ta parole... Sois donc rassuré: je vais ordonner au soleil de réapparaître...
Wou-ou-ou-ouh!

O Soleil, puissant astre du jour, je t'en conjure, sois clément!... Aie pitié de tes fils et que ta lumière réapparaisse!
Wou-ou-ou-ouh!

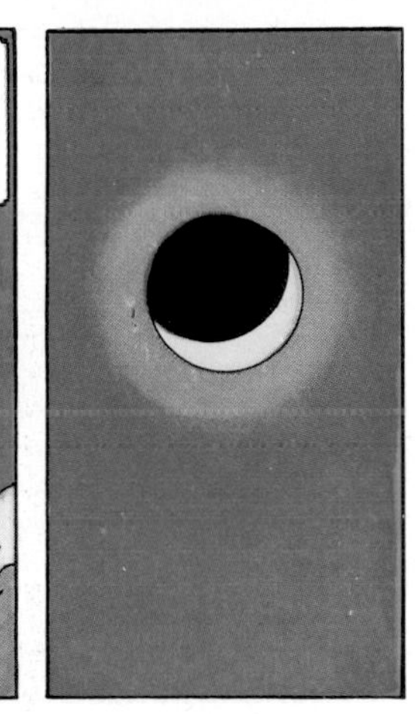

Par Pachacamac! le soleil lui obéit... Vite! vite! qu'on les délivre à l'instant!

Eh bien, capitaine?... Le journal... Comprenez-vous, à présent?...
C'est... c'est magnifique!...

Merci, ô puissant astre du jour!... Merci d'avoir daigné répondre à l'appel du jeune étranger!

"Le Soleil a rendez-vous avec la lune" ♫♫
Un peu de majesté, capitaine, comme il sied à des gens qui commandent au soleil.

!!!!

?
!!!

Pendant ce temps-là...
Personne... Et pourtant, ils sont quelque part où ils ont été fort secoués...

Le lendemain...
Je n'ai qu'une parole, ô nobles étrangers: vous êtes libres... Et je vais vous faire reconduire jusqu'au pied des montagnes...
Je te remercie, noble Fils du Soleil, mais j'ai encore une demande à t'adresser...

Il y a dans mon pays, sept savants qui, je le suppose, continuent à endurer, à cause de toi, de terribles souffrances. Quelle que soit la manière dont tu les tiens en ton pouvoir, je demande que tu mettes fin à leurs tourments.

Ces hommes sont venus ici comme des hyènes, pour violer les tombeaux et piller nos richesses sacrées. Ces hommes méritent le châtiment que je leur ai réservé.

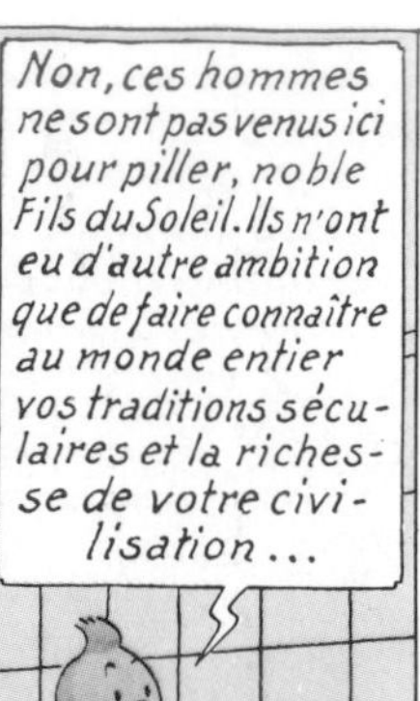
Non, ces hommes ne sont pas venus ici pour piller, noble Fils du Soleil. Ils n'ont eu d'autre ambition que de faire connaître au monde entier vos traditions séculaires et la richesse de votre civilisation...

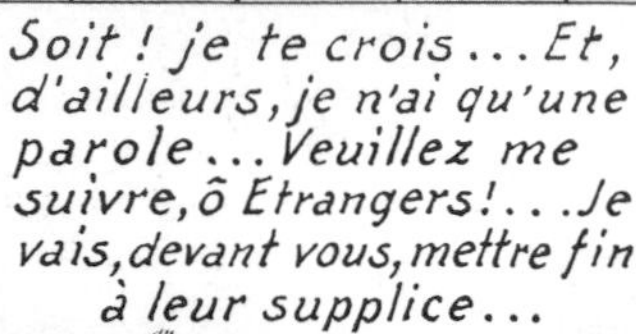

Soit! je te crois... Et, d'ailleurs, je n'ai qu'une parole... Veuillez me suivre, ô Etrangers!... Je vais, devant vous, mettre fin à leur supplice...

Voilà sept statuettes de cire. Chacune d'elles représente un des hommes dont vous m'avez demandé la grâce. C'est d'ici, de ce sanctuaire, que, par magie, nous les faisons souffrir. C'est d'ici que nous allons mettre fin à leur supplice...
De l'envoûtement!... Je m'en doutais!... Mais les boules de cristal, à quoi servaient-elles?

Ces boules contenaient un liquide sacré, tiré de la coca, qui plongeait les victimes dans un profond sommeil pendant lequel le Grand Prêtre les tenait en son pouvoir jusqu'au moment où l'envoûtement commençait...

Je comprends tout maintenant!... Les boules de cristal, la léthargie, les souffrances endurées par les explorateurs au moment où, ici même, le Grand Prêtre torturait les statuettes qui représentaient chacun d'eux...
Détruis ces statuettes, Huaco...

Au même moment, en Europe...
Qu'est-ce que je fais ici?...

Que s'est-il passé?... Comment se fait-il que je me trouve dans cette clinique?...

Que fais-tu ici, Charlet?...
Je me le demande, Sanders... Et toi?...

Vous ici, Laubépin?...
Clairmont!... Comment se fait-il que...
Hein! que m'est-il arrivé?...

Avant de vous quitter, nobles étrangers, j'ai, moi aussi, une grâce à vous demander...
Je sais laquelle, noble Fils du Soleil, et je tiens à vous rassurer tout de suite à ce sujet...

Je jure que jamais je ne révélerai à quiconque l'emplacement du Temple du Soleil!...
Moi aussi, vieux frère, je le jure!... Que le grand Cric me croque et me fasse avaler ma barbe si je lâche un mot à ce sujet!

**FIN**

Imprimé en Belgique par Casterman, s.a., Tournai.
Dépôt légal: 2e trimestre 1955; D. 1966/0053/55.